AF841330

Turbulent Transport in Magnetized Plasmas

Second Edition

Turbulent Transport in Magnetized Plasmas

Second Edition

Wendell Horton
University of Texas at Austin, USA

World Scientific

NEW JERSEY • LONDON • SINGAPORE • BEIJING • SHANGHAI • HONG KONG • TAIPEI • CHENNAI • TOKYO

Published by

World Scientific Publishing Co. Pte. Ltd.
5 Toh Tuck Link, Singapore 596224
USA office: 27 Warren Street, Suite 401-402, Hackensack, NJ 07601
UK office: 57 Shelton Street, Covent Garden, London WC2H 9HE

British Library Cataloguing-in-Publication Data
A catalogue record for this book is available from the British Library.

TURBULENT TRANSPORT IN MAGNETIZED PLASMAS
Second Edition

Copyright © 2018 by World Scientific Publishing Co. Pte. Ltd.

All rights reserved. This book, or parts thereof, may not be reproduced in any form or by any means, electronic or mechanical, including photocopying, recording or any information storage and retrieval system now known or to be invented, without written permission from the publisher.

For photocopying of material in this volume, please pay a copying fee through the Copyright Clearance Center, Inc., 222 Rosewood Drive, Danvers, MA 01923, USA. In this case permission to photocopy is not required from the publisher.

ISBN 978-981-3225-88-6

Desk Editor: Christopher Teo

Printed in Singapore

Foreword to the Second Edition

Research in the subject of plasma physics has developed rapidly since the first edition in 2012. Larger and technically more advanced machines have begun operation in the intervening period. Larger and more advanced simulation models and computers are widely used. Plasma physics combines the areas of mechanics, electromagnetic field theory, and statistical mechanics. The subject is a rich and rewarding course of study. Plasma science shows in a natural and beautiful setting how the laws of mechanics, electricity and magnetism, and statistical theory work together to predict the behavior of high temperature gases in the presence of externally applied magnetic fields combined with the magnetic fields produced by the electric currents arising from the differing trajectories of the ions and electrons in the plasma.

To keep the new volume a manageable size the topics of (i) ion acoustic turbulence, (ii) the compact high-field ignition machines and (iii) some topics special to the ITER program have been omitted in this edition. The 2015 book ITER Physics, ISBN:978-981-4678-66-7, covers many topics of importance to ITER such as resonant magnetic perturbation coils (RMPs) for control of magnetic islands.

The author would like to express his gratitude to his colleagues and students who have contributed to the knowledge and work presented in this book. Owing to the forty-year history of the research presented here, however, the author must confine the list of acknowledgments to his closest colleagues. The author wishes to thank and acknowledge the contributions from A. Aydemir, H. Berk, D. W. Ross, K. Gentle, P. J. Morrison, T. Tajima, D.-I. Choi, J. W. Van Dam, S. M. Mahajan, F. L. Waelbroeck, B. N. Breizman, H. V. Wong, D. Brower, R. D. Bengtson, W. Rowan and L. Leonard. He also benefited from including work developed in collaboration with A. Beklemishev, S. Hamaguchi, H. Sugama, I. Holod, A. Ishizawa, Y. Kishimoto, H. Miura, and P. Zhu. Many useful suggestions were contributed from The Directors of the Institute for Fusion Studies, M. N. Rosenbluth, D. Baldwin, R. D. Hazeltine, J. W. Van Dam, and F. L. Waelbroeck.

As with all modern science and engineering fields, fusion plasma physics has many specialized terms and acronyms. A glossary that defines most of the terms used in this book can be found at http://www.iter.org/glossary. A widely-used handbook for the formulas and constants used in plasma physics can be found at http://www.ppd.nrl.navy/nrlformulary. Chapter 1 has an abbreviated list of the most frequently used physical constants in this book. The plasma literature has a mixture of cgs and mks units.

The typing and LaTeX formatting of the book was performed by the long-time contributor to The University of Texas Fusion Research group, Suzy Mitchell.

Contents

Chapter 1

Basic Concepts and Historical Background

Plasmas are created by heating gases through the absorption of ionizing radiation in the visible through the ultraviolet part of the electromagnetic spectrum and through collisional ionization of molecules with energetic electrons. The absorption and heating processes are typically localized in space so that large spatial gradients occur in the plasma density, temperature and other characterizing parameters. Due to the high electrical conductivity of ionized gases, large electrical currents typically flow producing large, self-generated magnetic fields in the plasma. In other plasmas strong magnetic fields exist, produced by external currents such as those flowing in the cores of the planets or those in the conducting coils designed to produce plasma confinement in laboratory machines. Thus, the focus of this book is on the transport processes in these hot magnetized plasmas.

Due to the 50 years of intense international pursuit of the laboratory confinement of high temperature ($T > 10^7$ K) hydrogenic plasmas for the purpose of obtaining controlled nuclear fusion, there is an enormous resource to be found in the scientific literature on the experiments, the theory, and the numerical simulations of the plasma transport processes. While there are some books, most notably those by Mikhailovskii (1992), Wakatani (1998), Weiland (1999), Wesson (2004), and Friedberg (2007) developing the subject, and numerous review articles on the transport problem, there is need for a book with a global and integrated presentation of the extensive transport research. In this book we attempt to provide a general presentation starting from elementary considerations, while including the most recent results that one may view as firmly established in the understanding of the nonlinear transport processes in ionized gases.

The collisional transport produced by the binary Coulomb interactions between the charges in a plasma is a well-developed subject treated in the works of Balescu (1988) and Helander and Sigmar (2002). When the plasma temperature is high and density is low, the distance between these nearby encounters is long with complex magnetized orbits of the particles between the collisions. The mean distance between the Coulomb collisions is called the mean-free-path and the complex orbits have names like banana and potato particle orbits [Helander and Sigmar (2002)]. In general, the turbulent transport greatly exceeds the collisional transport in magnetic fusion plasmas. The few exceptions are discussed in Chapter 14 of Helander and Sigmar (2002).

A clear and reliable way to understand transport in fluids and plasmas is to follow the evolution of test particles, or a passive field, typically called a *dye* or *trace impurity* field, in the system. Test particle transport theory, simulations, and experiments are emphasized here, perhaps more than in the existing plasma literature. The limitations of the test particle method are serious, however, which accounts for their being less well developed in earlier works on plasma transport. The limitations arise because the motion of the charged particles modifies the electric charge and current distributions in the plasma. These changes in the electric charge density $\rho_q(\mathbf{x}, t)$ and current density $\mathbf{j}(\mathbf{x}, t)$ in turn modify the electromagnetic fields $\mathbf{E}(\mathbf{x}, t)$ and $\mathbf{B}(\mathbf{x}, t)$ in the plasma. Thus, the orbits of the particles are themselves changed again due to the collective effect of the accumulated charges and currents from the fields, due to their motions in these same fields. This problem is known as the self-consistent field problem. The self-consistent field problem forms the essence of the theoretical and simulation modeling of plasma dynamics. It is this circular sequence of interactions of the motion of the charged particles in the self-consistent electromagnetic fields that contains both the externally-applied fields and those fields generated by the particle motions that determine the actual transport of the plasma. Thus, the large body of works on plasma transport theory and modeling have their complexity rooted in the problem of determining, as clearly as possible, the solutions of the orbital equations with the self-consistent plasma electric and magnetic fields. Many different theoretical approaches can be and have been made to extract key transport information from this problem. Often, the approach is set by *a priori* assumptions about the nature of the fields; coherent waves or vortices, weak turbulent fields or states of strong turbulence, for example. Observational and simulation data gives evidence for long flights of vortex structures in Chapters 9 and 10.

While the most practical information desired may be contained in turbulent transport coefficients such as the particle diffusivity D and the electron and ion thermal diffusivities χ_e and χ_i, respectively, there are important processes where no such simple diffusivities exist. In such regimes there can be bursty large transport events, the dominance of coherent structures such as streamers, zonal flows, the existence of long flights of the particles through small gaps in transport barriers, and other exotic transport processes.

1.1 Space and Astrophysics

Early plasma physics research began in about 1900 when scientists puzzled over the high altitude lights in the northern sky and the systematic deflections of the magnetic compass needle recorded in careful measurements taken across the northern latitudes of the Scandinavian countries. Birkeland, whose name is attached to the large plasma currents known to flow into and out of the northern ionosphere from the magnetosphere, pioneered this research by building models of the magnetosphere in the laboratory, in an effort to reproduce the auroral observations.

The power driving these large plasma currents arises from the solar wind plasma interacting with the planetary magnetic fields. The properties of the Birkeland plasma currents are now measured with spacecraft that are equipped with high-resolution particle and field instruments [Kivelson and Russell (1995)]. Birkeland set up a laboratory magnetic confinement experiment and reproduced the effect of the auroral rings, but uncovered many more problems of complicated physics than could be answered with the knowledge and equipment of that era.

Alfvén and Falthammer, along with Stoermer, pursued these complex issues and realized that these were generic problems for the solar system and what they called the cosmos in their subsequent books that give many of the basic plasma physics concepts used today. Two of these concepts developed by Alfvén that are used extensively in the drift wave transport are (i) the drifts of the guiding centers of the cyclotron or gyro orbits of the trajectories of the charged particles in the confining magnetic fields and (ii) the bending oscillations of the magnetic field. In these plasmas the low-frequency electromagnetic Alfvén plasma waves and the guiding-center drift of the centers are used to describe the trajectories of charged particles in the confining magnetic fields in plasmas. For these and other discoveries, Hannes Alfvén received the Nobel Prize in 1970 in plasma physics.

The plasma physics of the solar corona [Aschwanden (2006)] has similarities to that of the toroidally confined plasmas in tokamaks. The flux ropes observed with high resolution solar observatories such as SOHO, TRACE, and SDO show both smaller ($L < 10\,\mathrm{Mm}$) structures in the chromosphere to lower corona and large ($L < 100\,\mathrm{Mm}$) structures that extend into the upper corona that last for many hours. These large flux ropes are called prominences and appear as EUV radiation lines, particularly from ionized iron (Fe) atoms. A gallery of the missions from ACE to WIND and MMS (magnetosphere multiscale mission, `mms.gsfc.nasa.gov`) and XMM Newton spacecraft is at `www.nasa.gov` with beautiful images of plasma dynamics. The images from solar-observing spacecraft, including TRACE, SOHO and STEREO show magnetic flux loops with plasma currents that are embedded in the photosphere. In these large coherent structures one can see strands of more and less dense plasma trapped in smaller ($< 10\,\mathrm{Mm}$) helical flux tubes that show the winding number of the helical magnetic field. In terms of the tokamak magnetic field the twist is specified by the inverse twist number q. These solar magnetic fields appear as structures with $q < 0.5$. The corresponding plasma currents are then of order 10^{11} Amps. Laboratory tokamaks keep $q \gtrsim 1.0$ for more stable magnetic structure and define q as the "safety factor." The solar corona has an electron density n_e and temperature T_e such that the pressure $p_e = n_e T_e$ is less than the magnetic field pressure $P_M = B^2/2\mu_0$. The solar coronal plasma extending many solar radii beyond the solar radius $R_\odot$ has an electron density $n_e \simeq 10^9/\mathrm{cm}^3$ and temperature $T_e = 100\,eV$ with plasma $\beta \simeq 0.1$ [Aschwanden (2005)]. These coronal plasma currents are driven by the injection of magnetic flux from vertical convection of the inner solar plasma up through the photosphere surface from the subsurface solar convection zone. The injected magnetic flux is of order 10^8 Wb/s producing the accelerating toroidal electric field. The foot-points of the flux tubes are brighter in EUV and soft X-ray emission than the surrounding plasma. Thus, we see that a tokamak-like toroidal structure occurs in nature as a self-organized natural state of the plasma. This observation may be consistent with the fact that across the spread of laboratory magnetic confinement geometries the simple ohmic heated tokamak has the highest energy confinement time τ_E and the largest Lawson product $n_e\tau_E$ of any of the man-made confinement devices.

In the laboratory fusion experiments the auxiliary heating power, $P^{\rm aux}$, is applied with values far above the natural ohmic heating power, $P_{\rm oh}$, from the toroidal plasma current and the confinement product of $n_e\tau_E$ decreases with approximately the inverse square root of the ratio of $P^{\rm aux}/P_{\rm oh}$. The physics behind this behavior is that the drift wave turbulence, and the turbulent transport increases as the temperature profiles become steeper with increased auxiliary power $P^{\rm aux}$. The core plasma temperature must reach temperatures in the 10-20 KeV range needed to produce the nuclear reactivity sufficient for a net power gain in the driven laboratory confinement vessel. The fusion reactivities for the D-T and D-D collisions as a function of temperature are shown in Fig. 1.1.

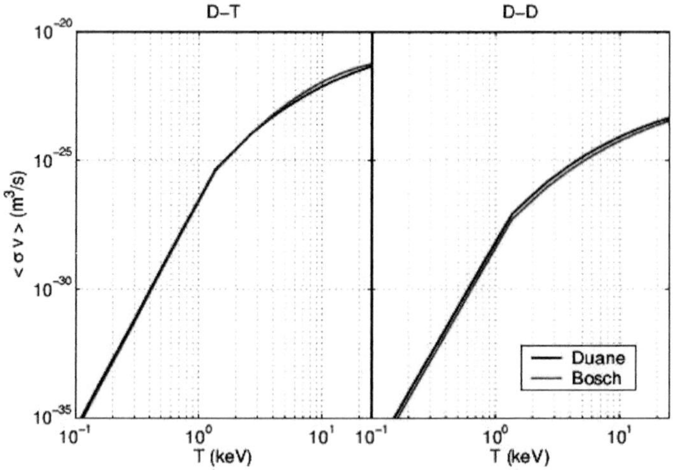

Fig. 1.1 The nuclear fusion reactivity$\langle\sigma v\rangle$ of deuterium D with tritium T in frame (a) and deuterium alone in frame (b) as the temperature of the Maxwellian distribution functions increases from 100 eV to 20 KeV [Duane (1972), Bosch and Hale (1992)].

The average fusion reactivity $\langle\sigma v\rangle$ shown in Fig. 1.1 is for a Maxwell-Boltzmann distribution function at the temperature given on the horizontal axis in KeV. A database from nuclear physics is used for the fusion-reaction cross-sections $\sigma(v)$, which are in the range of a few barns, where one barn is defined as $10^{-24}\,\text{cm}^2$. In contrast, the typical cross-sections for atomic processes including ionization of the neutral atoms are of order $10^{-16}\,\text{cm}^2$.

The principal difficulty in achieving fusion in the laboratory arises from the small value of the nuclear cross-section for fusion reactions compared with much larger cross-sections for atomic and Coulomb scattering of the ions. Thus, it becomes essential to use tritium as the target and fuel material for the experiments and thermonuclear reactors, owing to the much larger fusion cross-section, as seen by comparing the two reactivities shown in Fig. 1.1. Tritium is not naturally occurring, owing to its 12.6 year half-life for decay into deuterium. Thus, the tritium is made from lithium in other nuclear reactors, principally as a by-product from fission reactors. For fusion power reactors the tritium consumption for a 1 GW steady-state reactor would be a few kilograms of tritium per year. The tritium fuel would be produced by the fusion system itself by surrounding the reacting plasma with a blanket of lithium, which is naturally occurring, and "breeding" the tritium through the interaction of the escaping fusion neutrons with the lithium. The reaction of the fusion neutron with the naturally occurring $\text{Li}(A = 7, Z = 3)$ nucleus yields a helium nucleus $\text{He}(A = 4, Z = 2)$ plus a tritium $\text{T}(A = 3, Z = 1)$ nucleus. The cross-section for this reaction is large, so the fusion power system will produce its own fuel from lithium. Alternatively, one may say that lithium is the primary fuel for the D-T fusion reactor, following Maisonnier (1989). The issues of the breeding of tritium, the neutronics of fusion reactors and the radiological aspects of the problem, are described in Benkadda (2010).

1.2 World War II, Teller 1952

The path followed in this book comes from a second major line of plasma research developed directly as a natural follow-up step from the development of the enormous release of fusion energy in producing the hydrogen bomb with the nuclear fusion reactions from deuterium and tritium after World War II. The project led by Edward Teller was pushed aggressively by a cadre of his co-workers and there are many interesting books that describe the developments of nuclear fusion research. Herken (2002) describes in his book, *The Brotherhood of the Bomb*, the political and scientific developments of this era. After this military science project was concluded, the developers turned their effort to producing a controlled thermonuclear fusion reaction in the laboratory. The work took place under the project name Sherwood supported by the Atomic Energy Commission in the United States. Similar projects were developed in the Soviet Union (by the Kurchatov group) and in Great Britain. In 1958 the projects were declassified and a general science meeting was held in Geneva where all three countries showed their results. The results from the three countries were similar and showed that the plasma was typically unstable to strong electromagnetic-particle instabilities, the description of which is a large part of this book.

The fastest and largest length scale of these plasma instabilities is also common to space and astrophysical plasmas. Due to their large scale, the instabilities can often be described by a truncated description of the mean particle motion called the magnetohydrodynamic or fluid description of the plasma. There are different levels of the approximations used in the reduction of the particle orbits to the fluid models. The most extreme level of reduction, which gives the most tractable model and can be likened to the Bohr model – which is not a quantum mechanics theory – in that the single fluid with infinite conductivity replacing the many-body problem of interacting charged electrons and ions. Rather, the MHD model is an extension of the ideal gas behavior and the inviscid (no viscosity) limit of the fluid equations with infinite electrical conductivity of the fluid. The ideal MHD model, like the Bohr atom, is never correct, but always useful. There are well-known books by Bateman (19798), Friedberg (1987), and Hazeltine and Meiss (1992) on the MHD models and their generalizations to take into account the real, or non-ideal, aspects of the plasma. We assume the reader is familiar with these descriptions.

1.3 Controlled Nuclear Fusion

The search for controlled thermonuclear fusion is dominated by the approach that uses strong magnetic fields to confine high-temperature laboratory plasmas. From the 1970s to the 1990s the prevailing thought was that the low-amplitude drift wave turbulence would not seriously impede reaching the critical Lawson condition in MHD stable machines when heated with sufficient auxiliary (external) plasma heating power, P_{aux}. Exceeding the Lawson condition, $n_e\tau_E \sim 2\times 10^{20}\,\text{m}^{-3}\text{s}$ at 15-25 KeV, is required for D-T ignition. In the Lawson condition on $n_e\tau_E$, the product of the plasma density n_e and $\tau_E = W_p/(P_{\text{in}} - dW_p/dt)$ is the energy confinement time where W_p is the stored plasma energy, P_{in} is the total power input and dW_p/dt is the time variation of the stored plasma energy (dW_p/dt). For a steady-state plasma dW_p/dt vanishes.

The plasma fusion community now realizes that merely increasing the auxiliary heating is not sufficient to reach ignition. As external heating is added to the plasma, the turbulence level increases dramatically and more coherent structures appear so that the confinement time is seriously degraded in comparison to that at lower levels of auxiliary heating. A driven steady-state fusion reaction with the power amplification ratio $Q \sim$ 5-10 can be achieved more readily than ignition.

1.4 Magnetic Confinement Conditions for Nuclear Fusion

A less stringent condition is derived from the fusion power P_f amplification factor defined by $Q_f = P_f/P_{\text{loss}}$. The fusion Q_f is maximized at a lower temperature between $10 - 20\,\text{KeV}$ where the nuclear reaction rate is $\langle\sigma v\rangle_{\text{DT}} \propto T_i^2$ so that $Q_f \sim n_e\tau_E T \gtrsim 10^{21}\,\text{m}^{-3}s\,\text{KeV}$. Here, to reduce the amplification Q_f to the fusion triple product $n_e\tau_E T$, we use $P_f = (1/4)E_f n_e^2\langle\sigma v\rangle_{\text{DT}}$ where $E_f = 17.6\,\text{MeV}$ is the fusion energy release by a D-T reaction and $P_{\text{loss}} = nT/\tau_E$. The reactivity function $\langle\sigma v\rangle_{DT}$ is shown in Fig. 1.1, where at lower temperatures the increase is proportional to the square of the ion temperature T_i^2.

For a mixture of deuterium of density n_D and tritium of density n_T, the reactivity is the product of $n_D n_T$ and $\langle\sigma v\rangle$. For charge balance between the electrons and ions we have $n_e = n_D + n_T$ and, thus, the maximum reactivity occurs when $n_D = n_T = n_e/2$. The inevitable presence of impurity ions from the plasma-wall interactions reduces the reactivity from this ideal rate described in Chapters 8 and 15.

Reference values for the drift wave turbulence parameters in the International Thermonuclear Experimental Reactor (ITER) device are given in Table 1.4 currently under construction. There is a broad level of recognition in the fusion community that an understanding of the chaos and coherent structures created by plasma drift wave instabilities is required to reach practical fusion energy power-producing regimes. The monograph, *Chaos and Structures in Nonlinear Plasmas* [Horton and Ichikawa (1996)] examines in detail how the basic drift wave mechanism creates plasma structures. Chapters 11 to 15 give the relevant background science needed to understand the design and operation of ITER and other magnetic confinement fusion machines.

The condition for the onset of fusion power is set by the values of the cross-sections for atomic scattering and nuclear fusion reactions. Atomic scattering cross-sections are of order $10^{-16}\,\text{cm}^2$ and the nuclear cross-sections are of order $10^{-24}\,\text{cm}^2$. The nearly elastic scattering of the hot electrons from the ions results is an intrinsic loss of electron energy loss mechanism by radiation from high frequency electromagnetic waves in the process called Bremsstrahlung radiation [Jackson (1999)]. Both processes (fusion reactions and radiation losses) increase with the square of the plasma density so that the ignition temperature is set by a temperature around 7 KeV for the fusion power to dominate in the absence of turbulent transport. Chapter 15 contains a derivation of the condition for ignition including turbulence.

1.5 Nature of Plasma Turbulence

Plasmas with electrically interacting ions and electrons naturally have a variety of collective modes of oscillations occurring at low frequencies and long wavelengths compared with the more familiar oscillations of electron plasma waves. In the spatially nonuniform plasma in a magnetic field the electric fields associated with the low frequency collective modes produce drifts of the guiding centers of the charges. A characterization is given of the spatial scale of the collective modes by the gyroradius ρ_s of the ions $s = i$ and electrons $s = e$ and by the scale length L_p of the inhomogeneity of the plasma. Typically, but not always, the gyroradius scale is small compared to the inhomogeneity scale length L_n for the density n and L_T for the temperature T fields. Mathematically this is expressed by introducing epsilon as a small parameter where $\epsilon = \rho_i/L_n \ll 1$. The magnetized plasma will $\mathbf{E}\times\mathbf{B}$

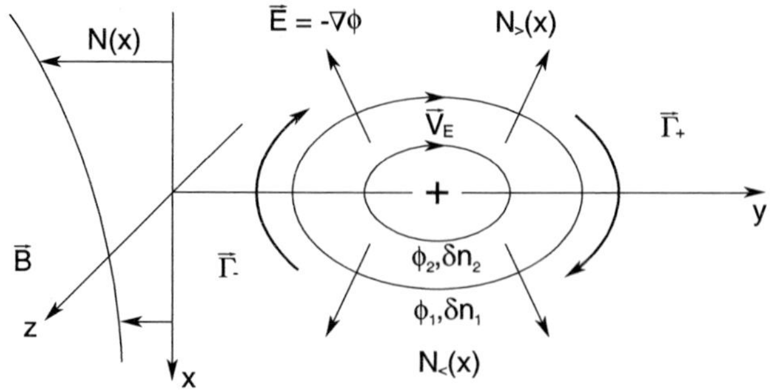

Fig. 1.2 Drift-Wave mechanism showing $\mathbf{E} \times \mathbf{B}$ convection in a nonuniform, magnetized plasma. This vortex structure commonly occurs in magnetized plasmas and is called the drift wave vortex described in Chapter 10.

convect as shown schematically in Fig. 1.2 around the maximum and minimum of the electric potential variations in the plane perpendicular to the magnetic field.

The parallel (along z) motion of the particles allows the density variations parallel to the magnetic field to align with the potential variations. The plasma ion and electron densities adjust to the electric potential $\varphi(\mathbf{x}, t)$ to be close to the local thermodynamic equilibrium Boltzmann relation, where the density, for example, becomes close to a function of the potential with $n_s(\mathbf{x}, t) = n_0 \exp\left[-q_s\varphi(x,t)/k_BT_s\right]$. This situation is defined as the "adiabatic regime" in the plasma physics literature and the electrons in particular will move to form this situation as shown in Fig. 1.2. This adiabatic density response, while not exact, is often a useful model especially when forces like the temperature gradients are driving the plasma turbulence. When the density response is adiabatic, we see from Fig. 1.2 that there is no net density convection integrated across a given radial surface. The density flux Γ_+ brought in on the right-hand side of the structure is balanced by the plasma density flux Γ_-, returned on the left side of the structure, in both the ion temperature gradient (ITG) gradient turbulence and the electron temperature gradient (ETG) driven turbulence described in Chapters 12 and 13.

In more general situations, however, the electron density is not able to stay in phase with the electric potential fluctuation as in the case with the strong temperature gradient-driven instabilities. This general case is shown in Fig. 1.3 where there is a phase shift that develops between the density and potential minima and maxima. Owing to the phase shift $\Psi_{\delta n,\phi}$ shown in the figure there is a net transport of the density down the density gradient from the convection of higher density region that is shifted from positive potential region. The phase shift of the density from the potential is a general property analyzed in Chapters 3 and 4. The phase shifts determine the extent to which electric potentials produce a net cross-field transport, as opposed to vortex or wave motion without a net plasma transport. When the temperature gradient is driving the fluctuations, the phase shift may be positive, as shown, or negative and give rise to outward or inward convection of the plasma density.

Calculating the local particle flux, $\delta n_e v_r$, gives the particle flux Γ and averaging over y or time gives $\Gamma = \sum_{\mathbf{k}} [k_y \phi_k \delta n_k / B] \sin \Psi_k$. The electron response to the potential determines the phase shift Ψ_k and amplitude ratio $\delta n_e / \phi_k$. The special case where the phase shift is zero gives $\delta n_e / n_e = e\phi_k / T_e$ and is called the adiabatic response, owing to its being the linearization of the equilibrium Boltzmann relation.

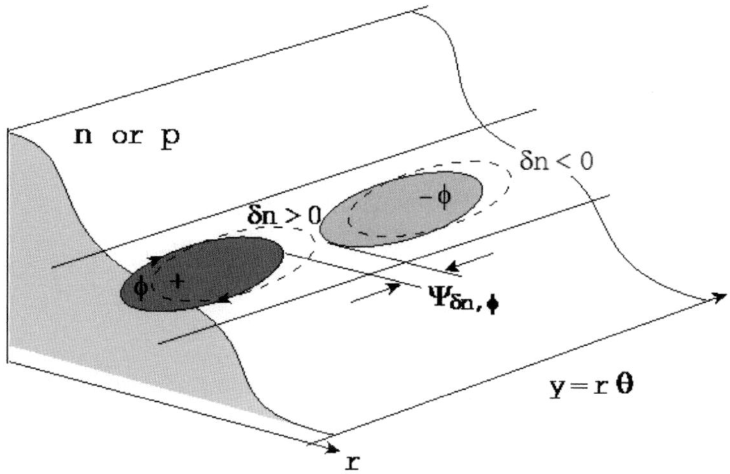

Fig. 1.3 The general case where the drift wave density fluctuations are shifted in phase from the potential fluctuations by Ψ. This results in a net flux of density down the density gradient when the particle flux is integrated over a fixed radial surface $r =$ const outside of the annulus of the fluctuation zone. This is the general situation that is measured in the drift-wave experiments analyzed in Chapters 3 and 4.

This adiabatic response is used in many of the classical drift wave models, e.g., when the temperature gradient drives the drift waves unstable rather than the density gradient ($n \to T$) as in Fig. 1.2. In turbulent plasmas this convective flux occurs at each $\mathbf{k}$ with the total transported flux being a sum over all scales of k.

The typical size of the turbulent cells is $\lambda_\perp = 2\pi / k_\perp \sim 6\rho_s$ and the frequency is approximately $\omega_k = k_y v_{de}$, where $v_{de} = -(T_e/eBn_e)(dn_e/dr) = T_e/eBL_n = c_s\rho_s/L_n$. The scale length, $\rho_s = c_s/\omega_{ci}$, is called the ion inertial scale length and determines the scale of the waves and vortex structures. In the quasi-geostrophic flow in planetary atmospheres the equivalent quantity is the Rossby radius [Horton and Ichikawa (1996)].

For reference tokamak parameters, we consider the superconducting Tore Supra [Hoang, *et al.* (2003)] that is now upgraded to WEST which has $R/a = 2.4\,\mathrm{m}/0.72\,\mathrm{m}$, $B = 2\,\mathrm{T}$, $n_e = 3 \times 10^{19}\,\mathrm{m}^{-3}$ and $T_e = 1\,\mathrm{KeV}$. The machine parameters are given in Table 1.1 in more detail. For these plasma parameters $\rho_s/L_n = 0.1\,\mathrm{cm}/80\,\mathrm{cm}$, the sound speed $c_s = 3 \times 10^7\,\mathrm{cm/s}$ and thus $v_{de} = 3 \times 10^4\,\mathrm{cm/s}$ and the drift wave frequency for $k_y = 20\,\mathrm{cm}^{-1}$ is about 6×10^5 rad/s or 100 kHz. Reference values for the ITER machine and plasma are given in Table 1.4. Drift wave fluctuations have been studied in detail in a wide variety of laboratory machines.

Table 1.1 WEST superconducting RF-driven Tokamak in Cadarache, France. Tore Supra (1988-2014) was remodeled from 2014-2016 to test steady operation with configured magnetic divertor with tungsten (W) plasma-facing wall.

major radius R	2.37 m
minor radius a	0.8 m
toroidal magnetic field	2 T
plasma current	1 MA
current duration	1000 s
solenoid flux swing	21 Wb
real-time feedback control of plasma	
niobium-titanium windings	1.8 K
LHCD ICRH	6 MW 5 MW
Heat Load	10 MW/m^2

Tore Supra, constructed during 1981-1988, reported successful long pulse (7 seconds) Ohmic discharges at the 1988 International Conference on Plasma Physics and Controlled Fusion in Nice, France [Equipe Tore Supra (1989)]. The machine's chief architect, R. Aymar (1989), reported for the Equipe Tore Supra, the results for plasma parameters up to $T_e = 1.5\,\mathrm{KeV}$ and $n_e = 4 \times 10^{13}/\mathrm{cm}^3$ for Ohmic discharges with loop voltage $V_{\mathrm{loop}} = 1\,\mathrm{V}$ Volt and plasma currents $I_p = 600\,\mathrm{kAmp}$ maintained for four seconds. The machine parameters are given in Table 1.1. A twenty-year perspective on the research from Tore Supra is given in Aymar (2001). The machine is now operating as WEST with a tungsten (W) divertor chamber.

The Tore Supra machine was rebuilt from 2014 through 2016 to create a model of the ITER machine. The redesigned machine WEST uses the single null divertor geometry with tungsten [W] coated chamber walls to test the plasma-wall interactions expected in ITER. A similar new Experimental Super Advanced Steady-State Torus (EAST) is operating in Heifei, China. The EAST tokamak parameters are in given in Table 1.2 [Zhang, *et al.* (2014)].

The first deuterium-tritium (D-T) controlled fusion power experiments were reported at the 1988 IAEA in Nice from the Tokamak Fusion Test Reactor (TFTR) [Bell, *et al.* (1989), Hawryluk (1998)], with 30 MW of balanced deuterium neutral beam heating, achieved an ion temperature of 32 KeV and the triple product of $n_e(0)\tau_E T_i(0) = 4 \times 10^{20}\,\mathrm{m}^3\,\mathrm{s\,KeV}$. Since the electron temperature remained at about 8-10 KeV, this type of hot ion mode of operation is called the *hot ion supershot regime*, and is one of the promising approaches to fusion power. The Bell, *et al.* (1989) TFTR article describes how power from deuterium-beams into deuterium plasmas with the measured neutron flux of 3×10^{16}/s were converted

Table 1.2 EAST (Experimental Super Advanced Steady-State Torus) parameters operating from 2006 in IPP, Heifei, China.

major radius R	1.7-1.9 m
minor radius a	0.4-0.45 m
triangularity δ	0.4-0.7
toroidal magnetic field	3.5 T
plasma current	1 MA
current duration LHCD	
solenoidal flux swing	30 s
LHCD 2.45 GHz	4 MW
ICRF 25-70 MHz	6 MW
NBI	8 MW
lithium wall coating	
real-time feedback control	

into what the corresponding nuclear power would be with deuterium-tritium fusion power experiment. The conversion predicted that the tritium experiments should yield a fusion of $Q_{\text{fus}} \cong 0.5$. In 1998 the tritium experiments in TFTR were conducted. The highest fusion power amplification achieved was $Q_{\text{fus}} = 0.5 \pm 0.13$.

Owing to the complexity of using tritium fuels many fusion experiments use D-D rather than D-T plasmas. The increase in the neutron yield is approximately a factor of 200 for typical toroidal fusion power experiments.

Tokamaks reported encouraging results at the Nice 1988 IAEA meeting including Doublet III, PDX, ASDEX, Alcator C-Mod, TEXTOR, ISX-B, PBX-M and then the new, large Japanese tokamak JT-60. The encouraging results from these experiments led to formation of international groups to study a design for the NEXT-STEP machine that eventually became ITER. An important step towards demonstrating steady-state plasma confinement was obtained by the Tore Supra tokamak with parameters given in Table 1.1. In Tore Supra superconductivity coils allow confinement for 10^3 s with a gigajoule of plasma energy passing through the system at KeV plasma temperatures for periods of minutes as described in Chapter 13. The EAST tokamak is described in Table 1.2 which has record long-time quasi steady state RF driven plasmas. The new KSTAR tokamak is described in Table 1.3 with record plasma confinement states.

Currently, ITER is several years under construction at the Cadarache Laboratory in the Luberon Mountains outside of Aix-en Provence. The parameters for ITER are given in Table 1.4. A detailed description of the ITER project and its plasma dynamics is described in *ITER Physics* [Horton and Benkadda (2015)].

Coherent structures are now understood to produce intermittent plasma turbulence. Figure 1.4 shows a perspective view of the ITER machine from `https://www.iter.org/` and is described in some detail in ITER Physics (Horton and Benkadda, 2015). Since the publication of ITER Physics 2015 the ITER Director has changed to the head of the Commissariat of Atomic Energia (CAE)

Table 1.3 KSTAR (Korea Superconducting Tokamak Advanced Research) operating since 2008 in Daejeon, Korea

major radius R	1.8 m	
minor radius a	0.45 m	
triangularity δ	0.8	
toroidal magnetic field	3.5 T	
plasma current	I_p	2 MA
pulse length	20 s (present)	300 (target)
heating power	(present)	(target)
NBI	5.5 MW	12 MW
ECH	1 MW	6 MW
ICRH	1 MW	
LHCH/HCD	0.8 MW	8 MW

Table 1.4 ITER parameters with typical drift wave frequencies and wavenumbers.

Installed auxiliary heating	73 MW
Fusion power	500 MW
Plasma major radius (R)	6.2 m
Plasma minor radius (a)	2.0 m
Vertical elongation (κ)	1.5
Plasma volume	830 m^3
Plasma current(I_p)	15 MA
Central solenoid flux	200 Wb
Toroidal field at 6.2 m radius (B)	5.3 T
Density gradient length (L_n)	1 m
Electron temperature (T_e)	10 KeV
Electron Diamagnetic Drift Velocity (v_{de})	1.9 km/s
Ion inertial radius (ρ_s)	0.3 cm
Gyro-Bohr diffusivity ($D_{gB} = \rho_s v_{de}$)	5.7 m^2/s
Drift wave frequency ($\omega = k_y v_{de}$)	190 KHz at $k_y \rho_s = 0.3$

of France and the cost and time line for the first D-T experiments has been extended. To keep up with the ITER project subscribe to the ITER Newsletter at `https://www.iter.org/`.

A comprehensive review of the ITER machine and technology is given in the special issue of *Fusion Science and Technology* [Fusion Science Technology (2011)].

A new generation of tokamaks commencing operation since 2010 are superconducting for steady-state operation with results just now arriving. These new machines are the KSTAR tokamak (Korean Superconducting Tokamak Advanced Research) (`http://www.nfri.re.kr`) in South Korea and EAST (Experimental Advanced Superconducting Tokamak) (`http://www.chinatechgadget.com/east-ht-7u-super-conducting-tokamak-fusion-reactor.html`) in China. Both machines are producing new results with neutral beam injection and radio frequency of driving the plasma current as described in Chapter 13. A superconducting upgrade of the JT-60U machine is under construction at the time of writing.

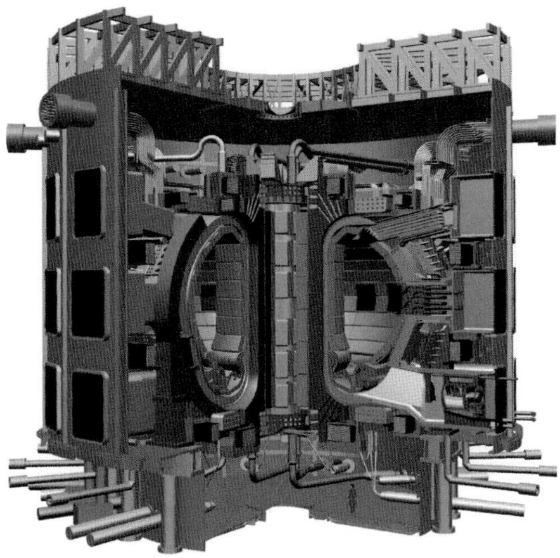

Fig. 1.4 The $800\,m^3$ ITER machine under construction in Provence, France neighboring the Cadarache Nuclear Laboratory on the Durance river. The project is a seven-country partnership with the European Union leading project with Japan, India, China, Russia, South Korea and the United States, under a 2007 treaty agreement. By 2020 the approximately US $15 billion machine will create plasmas that will set records for controlled nuclear fusion confinement. The D-T plasmas testing may start in 2027 with the projection of creating 500 MW of fusion power from 5 MW of injected power from RF and neutral beam heating (`www.iter.org`) and ITER Physics [Horton and Benkadda (2015)].

This new tokamak is called JT-60SuperAdvanced and will test the integrity of the first walls to the intense neutron fluxes from the fusion reactions. The JT-60SA is a joint Japan-Euro project called the Broader Approach which is in reference to a step beyond the ITER science machine. The JT-60SA tokamak machine would test reactor design issues including the lifetime of the walls of the vacuum vessel. The WEST [Tungsten (W) WEST Tungsten (W) environment in steady-state tokamak] machine with monoblocks of tungsten W in the divertor chamber is now a test-bed facility for the plasma-wall interactions described in Bucalossi (2011) research for simulating the ITER plasma-wall design.

There are other new toroidal and mirror fusion plasma machines that do not use the central solenoid or core transformer to confine the plasma as in tokamaks. These machines are stellarators, mirrors, helicity driven plasma chambers.

We open this introductory chapter with Fig. 1.5 of the Wendelstein 7-X machine that started operation in 2016 with ECRH heating of 4 MW in a helium plasma. The April 2016 Newsletter reported a plasma with core temperatures $T_e = 8\,\mathrm{KeV}$ and $T_i = 2\,\mathrm{KeV}$ and central density of $n_e = 3 \times 10^{19}\mathrm{m}^{-3}$. In the next campaign (2017) with a new plasma divertor unit and neutral beam injection (NBI) of 8 MW for 10 seconds, the W7-X stellarator will set new fusion plasma confinement records.

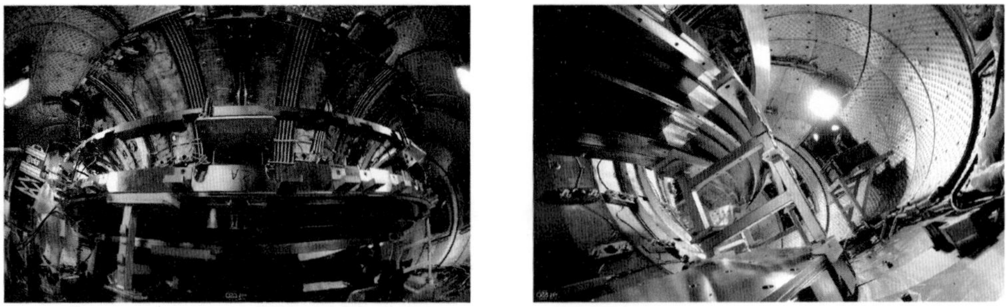

Fig. 1.5 Two views of the complex super coil structures and plasma chamber structure of the Wendelstein 7-X machine in Max-Plank Institute für Plasmaphysik in Greifswald, Germany. The first plasma (helium) was produced 10 Dec 2015. Six gyrotrons provide 8 MW of electron cyclotron heating power (`http://www.ipp.mpg.de`). April 2016 the plasmas reached 8 KeV electron temperatures from 4 MW of ECRH. The divertor TDU is being installed for higher heating power for 2017.

The dominant energy loss mechanism in the toroidal plasma confinement is from drift waves. Drift waves are sufficiently slow so that the electrons adjust to the electric potential fluctuations according to

$$E_{\|} + \frac{1}{en}\nabla_{\|}p_e = 0 \tag{1.1}$$

as verified in basic laboratory experiments.

1.6 Dominant Plasma Energy Loss Mechanism

Equation (1.1) states that the forces acting on the electron fluid parallel to the magnetic field line are in a state of balance. Out of balance, the electrons will adjust rapidly by motion parallel, or along the magnetic field lines. Thus, the waves have the opposite polarization of the MHD waves, which have $E_{\|} = 0$. The parallel electric field, measured in detail in the experiments, is a key part of the drift waves and the related trapped electron mode instabilities. The fluctuating electron current, $\delta j_{\|}$, driven by the parallel electric field takes the energy $\langle \delta j_{\|} E_{\|} \rangle < 0$ out of the nonuniform electron distributions and produces the wave growth. The effect is as though the plasma had a negative resistivity/conductivity $\sigma_{\|}(\omega, \mathbf{k})$ for the drift wave.

When there is zero or negligible phase shift between the density and the potential fluctuation, the $\mathbf{E} \times \mathbf{B}$ convection produces vortex structures shown in Fig. 1.2. These plasma structures can have large amplitudes forming nonlinear coherent vortices in the plasma. The study of these nonlinear structures and their propagation speeds is a rich area that is explored in Chapter 9. Under the presence of unfavorable field line magnetic curvature these vortices are accelerated out of the plasma down the effective gravity created by the charge polarization from the magnetic curvature drifts of the ion and electron guiding centers. The effect is to produce a bursty, intermittent transport that adds to the background of turbulent wave transport.

Detailed measurements and comparisons with theory suggest that, in fact, one of the most serious limiting factors for the success of ITER and similar machines is the control of drift wave turbulent transport. The seriousness of the turbulence becomes more evident in machines with increasing injected power $P_{\rm aux}$ and clearly evident when the injected core power densities reach $1\,{\rm MW/m}^3$. The plasma temperature rises only as a fractional power law, like $P_{\rm aux}^{1/2}$, owing to the associated increased turbulence levels, and the energy confinement time, τ_E, thus decreases even though the plasma temperature increases. Thus, the triple fusion power product may remains essentially constant with increasing auxiliary heating $P_{\rm aux}$.

As a consequence of this stronger heating power, $P_{\rm aux}$, the turbulence is amplified, which shortens the plasma energy confinement time, $\tau_E = W/P_{\rm aux}$, in agreement with the extensive ITER database [I.P.E. Group (1999), Kaye, *et al.* (1997), L-H Mode Database (1994)]. The drift wave turbulence has explained the empirical database laws for both helical systems [Yamada, *et al.* (2005)] and tokamaks [I.P.E. Group (1999), Kaye, *et al.* (1997)], showing that $\tau_E = \tau_E^0(P_0/P_{\rm aux})^{2/3}$. Thus, doubling the $P_{\rm aux}$ shortens the confinement time, $\tau_E \to \tau_E^0/1.59 = 0.64\tau_E^0$. For example, the reference design for ITER (see Table 1.1) has $P_{\rm ICRF} = P_{\rm ECE} = 20\,{\rm MW}$ and $P_{\rm NBI} = 16\,{\rm MW}$ injected into the plasma volume of $V = 2\pi^2\,{\rm Rab} \simeq 800\,{\rm m}^3$. The average power density is then approximately $0.1\,{\rm MW/m}^3$. To reach the power density of $1\,{\rm MW/m}^3$ at the core, as in current large tokamaks, the high-power radio frequency (RF) sources are applied by antennas on the vessel walls driving both ICRH and ECE electromagnetic plasma waves focused in the central core volume of $40\,{\rm m}^3$, corresponding to the core plasma radius $r_c = 0.6\,{\rm m}$. Antennas launching tens of megawatts of radio frequency (RF) heating power are used to maintain the fusion plasmas in a steady state.

In large toroidal magnetic geometry there must be external control mechanisms for stabilizing the plasma from interchange motions that are analogous to the interchange instability that occurs when a dense liquid or gas is supported above a light, less dense gas or liquid. This instability is called the Rayleigh-Taylor (described in Fig. 1.6 and in Table 1.5) instability in hydrodynamics. In plasmas there is analogous instability called the magnetic Rayleigh-Taylor instability. The effective plasma acceleration $g_{\rm eff}$ shown in Fig. 1.6 arises from electric forces created by the vertical separation of the ion and electron charges.

1.7 Optimization of Plasma Cross-Sectional Shape

The DIII-D machine shown in Fig. 1.7 with simulations of the 3D drift wave turbulence embedded in the machine vessel give a realization of the importance of the drift wave turbulent transport.

The detailed description of the plasma interchange modes is complicated and derived in Chapter 4. The direction of the effective gravity is horizontal and

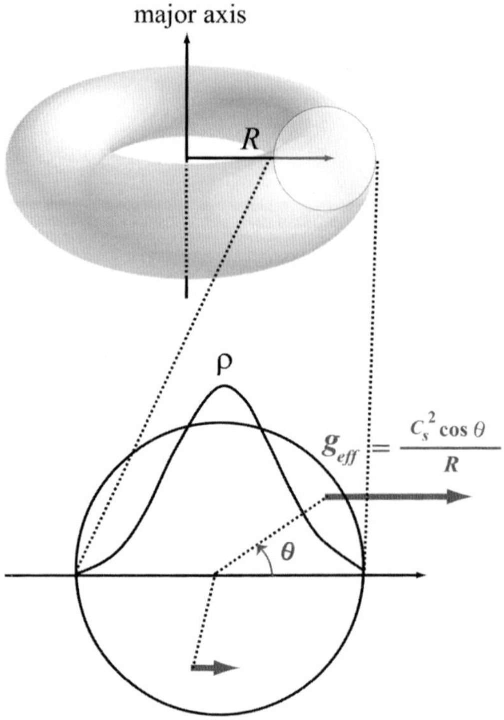

Fig. 1.6 Schematic showing the effective gravity that arises from separation of electric charge in a plasma in a toroidal confinement magnetic field. The effective acceleration represented by g_{eff} in the lower figure arises from the motion of the particles in the curved magnetic field lines of the torus. The sound speed is $c_s = \sqrt{[(T_e + T_i)/m_i]}$ and R is the distance from the axis of symmetry to the observation point. The profile ρ is the mass density of the plasma.

directed outward from the axis of symmetry. Thus, for the peaked plasma mass density profile ρ shown in the lower diagram, there is a strong acceleration determined by the sound speed c_s squared divided by the major radius R of the torus. This configuration is unstable to the onset of an interchange of plasma from the higher density core to the lower-density outer region. The methods of maintaining the stability are described in Chapter 5 in the context of the MHD model. There it is shown that the tension in the magnetic field attached to the plasma provides the restoring force. This balance between the magnetic tension force and the acceleration from g_{eff} give the fundamental limit on the maximum plasma pressure that a torus with a specified magnetic field B_ϕ and major radius R can contain. This pressure limit is the called the beta limit and is given by an expression called the Troyon formula [Troyon, *et al.* (1984)]. The limit is thoroughly validated by the extensive tokamak database accumulated from the many well-documented tokamak experiments performed over the past forty years. Chapter 6 shows that the plasma loss is by the drift waves in the MHD regimes found in the laboratory in Chapter 7. Figure 1.6 gives a few of the key equations that will be derived in Chapter 5. The

Table 1.5 Table with key formulas showing the relationship of the pressure gradient driven interchange instability in a magnetized plasma with the classical neutral fluid gravitation instability in an inverted mass density profile. Key equations for the magnetized plasma instability are on the left and the neutral fluid analog on the right. The role of gravity is played by the centrifugal force the charged particles experience moving along the curved magnetic field lines, and the role of the mass density gradient is played by the plasma pressure gradient. The outer half of a torus has an unstable stratification of the pressure gradient, and the instability operates in the drift waves on the micro-scales of the ion and electron gyroradius as explored in Chapters 13 and 14.

$p' \equiv \frac{p}{L_p}$ and $\mathbf{E}_\perp = -\boldsymbol{v} \times \mathbf{B}$	Gravity g and density gradient
$\gamma^2_{\mathrm{MHD}} = -\frac{p'}{\rho}\frac{2}{R}$	$\gamma^2_{\mathrm{RT}} = -\frac{g}{\rho}\frac{d\rho}{dr} = -\frac{\boldsymbol{g} \cdot \nabla\rho}{\rho}$
Blobs accelerate out	
$g_{\mathrm{eff}} = \frac{2c_s^2}{R} \sim 2(3 \times 10^5\mathrm{m/s})/(3\mathrm{m}) \sim 6 \times 10^{10}\mathrm{m/s}^2$	
Over outside of torus	
Sets limits on three scales:	
MHD p', ρ_i − ITG and ρ_e − ETG	

physical constants used in the calculations for the plasma parameters such as the frequencies, growth rates and diffusivities are given in Table 1.6.

The results on the effect of optimizing the plasma cross-sectional shape did not come in time to influence the design of TFTR. These shaping results started with the Doublet II and Doublet III tokamaks at General Atomics in San Diego. Followed by the Princeton divertor experiment (PDX) at Princeton, higher confinement and larger plasma currents were achieved for given major radius and toroidal magnetic field by using elliptically-elongated plasma cross-sections.

Fig. 1.7 A simulation from the continuum gyrokinetic codes GYRO/CGYRO for the drift wave turbulence in the DIII-D superimposed on a photo of the interior of that tokamak at General Atomics. The machine has $R/a = 1.7\,\mathrm{m}/0.7\,\mathrm{m}$ with $B = 3\,\mathrm{T}$, [http://fusion.gat.com].

Table 1.6 Physical constants in mks units.

Physical Quantity	Symbol	Value	mks Units
Boltzmann constant	k_B	1.3807×10^{-23}	JK^{-1}
Elementary charge	e	1.6022×10^{-19}	C
Electron mass	m_e	9.1094×10^{-31}	kg
Proton mass	m_p	1.6726×10^{-27}	kg
Gravitational constant	G	6.6726×10^{-11}	$m^3s^{-2}kg^{-1}$
Planck constant	h	6.6261×10^{-34}	Js
	$h = h/2\pi$	1.0546×10^{-34}	Js
Speed of light in vacuum	c	2.9979×10^{8}	ms^{-1}
Permittivity of free space	ϵ_0	8.8542×10^{-12}	Fm^{-1}
Permeability of free space	μ_0	$4\pi \times 10^{-7}$	Hm^{-1}
Proton/electron mass ratio	m_p/m_e	1.8362×10^{3}	
Electron charge/mass ratio	e/m_e	1.7588×10^{11}	Ckg^{-1}
Bohr radius	$a_0 = \epsilon_0 h^2/\pi m e^2$	5.2918×10^{-11}	m
Atomic cross-section	πa_0^2	8.7974×10^{-21}	m^2
Classical electron radius	$r_e = e^2/4\pi\epsilon_0 m c^2$	2.8179×10^{-15}	m
Thomson cross-section	$(8\pi/3)r_e^2$	6.6525×10^{-29}	m^2
Compton wavelength	$h/m_e c$	2.4263×10^{-12}	m
of electron	$\hbar/m_e c$	3.8616×10^{-13}	m
Fine-structure constant	$\alpha = e^2/2\epsilon_0 hc$	7.2974×10^{-3}	
	α^{-1}	137.04	

The loss of high-energy ions from Alfvén eigenmodes is analyzed by Van Zeeland, *et al.* (2012) in reversed magnetic shear plasmas. The possible significance for ITER from the DIII-D data is discussed.

The tokamak configuration variable (TCV) in Ecole Polytechnique Lausanne extensively documented the effect of plasma shaping in detail, as described further in Chapter 12. In the 1980s the new machines JT-60U, JET and now ITER were designed with elliptical and D-shaped cross-sections to take advantage of the higher plasma current limits. The final design parameters of ITER were settled in a "Final Design Report," [ITER-EDA (1998)]. The world record for the Q-value was achieved for the ion temperature and the fusion triple product in JET as described in Sections 1.8 and 1.9.

The TCV machines in Ecole Polytechnique Lausanne documented the effect of plasma shaping [Goodman, *et al.* (2003)]. The TCV results are described and analyzed in Chapter 13. The list of new plasma-shaping machines includes EAST and KSTAR (`https://fusion.gat.com/global/KSTAR`), the new superconducting tokamaks in China and South Korea.

1.8 Reversed Magnetic Shear Control of Transport

The JT-60U tokamak demonstrated the use of a minimum in the q-profile to create an internal transport barrier. As shown in Chapter 6 the formation of a minimum in the $q(r)$-profile separates the density of rational surfaces producing a transport barrier at $q_{\min}$. The JT-60U tokamak did not operate with tritium (T). When the neutron yield from the D-D collisions was measured and converted to what the yield would have been if the working gas were 50% D and 50% T. The record value of the fusion multiplication would have been achieved with the equivalent $Q_{DT} = 1.25$ according to the JAERI publications [Fujita, *et al.* (1999)]. These record discharges had neutron emission rates of 3.6×10^{16} n/s at the end of a 4 s current ramp-up of I_p, starting at 0.5 MA and rising to 2.5 MA.

Feedback control using the measured neutron production to control the neutral beam power injection rate was used to get the plasma through the unstable region where the minimum value of the safety factor q reached the value 3 in a reversed shear current profile. The profiles of density, ion, and electron temperature all showed a "box" shape with a steep gradient occurring at the magnetic surface, where the safety factor $q(r)$ reached its minimum value, which was just above 2 at the end of the discharge. The record discharge was shot number E31872, which at the final time, reached the values of $n_e(0) = 8.5 \times 10^{19}/\mathrm{m}^3$, $T_i(0) = 16.8\,\mathrm{KeV}$ and $T_e(0) = 7.2\,\mathrm{KeV}$. The energy confinement time was reported as $\tau_E = 1.07\,\mathrm{s}$, giving a significant multiple over the emperical tokamak scaling law for energy confinement, called the ITER89P formula for τ_E of $H_{\mathrm{ITER89P}} = 3.2$. The actual fusion power produced from the D-D nuclear reactions driven by the 10 MW of neutral beam power was $Q_{\mathrm{DD}} = 5.6 \times 10^{-3}$ which when multiplied by 220 for the ratio of the energy released in the two different nuclear reactions (DT/DD) gives the published value of the equivalent performance of $Q_{DT} = 1.25$. There are issues as to how reliable this conversion between two different working gases, along with higher energy production from the D-T fusion reactions, can be carried out. More energetic nuclear decay products produce a higher influx of impurities from the walls of the vessel to the plasma volume.

The large shaped cross-section tokamak called JT-60U for the Japanese Tokamak with approximate volume of $60\,\mathrm{m}^3$ was operated at the JAEA facility near Mito with parameters given in Table 1.7 from 1990 until 2007. In 1997, outstanding results from the JT-60 Team were reported in Koide, *et al.* (1997). All JT-60U experiments were performed in deuterium plasmas so an extrapolation of order 200 is required to convert the neutron yield from the D-D plasma to the D-T plasma assuming the only difference is with replacing the nuclear cross-sections.

Under the assumption on the equivalence of the D-D experiments to an equivalent D-T plasma, Koide and the JT-60U team [Koide and JT-60 Team (1997)] published that they achieved the equivalent reactivity of a break-even experiment with $Q_{\mathrm{fusion}} = 1$. In this measure of performance, the Q_{fusion} is the maximum nuclear power P_{fus} from D-D nuclear fusion reactions to the injected heating power P_{aux} after using formulas to replace the D-D plasma with a D-T plasma.

The high-performance of the JT-60U plasmas was achieved by having reversed magnetic shear (RMS) plasmas which forms an internal transport barrier (ITB) in both the electron and ion thermal flux channels as described in Chapters 4 and 5.

Table 1.7 Parameters of the large JT-60U tokamak at the Japan Atomic Energy Research Institute, Naka, Japan.

A. Device Concept and Components	
Vacuum vessel: Plasma facing material[1]:	carbon
Divertor/limiter facing material	carbon-fiber composites
Working gas	deuterium
Boundary configuration:	divertor
Toroidal magnetic field $B(T)$	4.2
B. Plasma Geometry	
Major radius, $R(m)$:	~ 3.4
Minor radius, $a(m)$:	~ 1.0
Plasma volume (m^3):	< 100
C. Plasma Parameters (Maximum Values)	
Central ion temperature[2], T_i (KeV):	45
Central electron temperature, T_e (KeV):	26
Line averaged plasma density, $\langle n \rangle$ (10^{20} m^{-3}):	-
Central electron density, n (10^{20} m^{-3}):	2.8
Plasma current, I (MA):	5
Discharge pulse duration (s):	15
Plasma beta[3], $\langle \beta \rangle (\%)$:	2.7
Stored energy[4], (MJ):	10.9
Energy confinement time[5], $\tau_e(s)$:	1.1
D. Plasma Heating and Current Drive	
NBI heating[6], (MW):	40
RF heating:[7] -IC: (MW):	7
RF heating: -EC:[8] (MW):	3
RF heating: -LH[9]: (MW):	8
colrule E. Reactor relevant plasma performance parameters	
Triple Fusion Product, (10^{20} KeV sm^{-3}):	15.3

[1]N. Hosogane, *et al.* "Development and operational experiences on the JT-60U tokamak and power supplies," Fusion Sci. and Tech., 42, 368 (2002).

[2]S. Ishida, *et al.* "High performance experiments in JT-60U high current divertor discharges," Proc. 16th Inter. Conf. on Fusion Energy, Montreal, Vol. 1, 315 (1996).

[3]Y. Kamada, *et al.* "High triangularity discharges with improved stability and confinement in JT-60U," Proc. 16th Inter. Conf. on Fusion Energy, Montreal, Vol. 1, 247 (1996).

[4]S. Ishida, *et al.* "Achievement of high fusion performance in JT-60U reversed shear discharges," Phys. Rev. Lett., 79, 3917 (1997).

[5]T. Fujita, *et al.* "High performance reversed shear plasmas with a large radius transport barrier in JT-60U," Nucl. Fusion, 38, 207 (1998).

[6]M. Kuriyama, *et al.* "Operation and development on the positive ion-based neutral beam injection system for JT-60 and JT-60U," Fusion Sci. and Tech., 42, 410 (2002).

[7]H. Kimura, *et al.* "ICRF Heating and TAE Modes in Reactor Relevant JT-60U Discharges," Proc. 16th Inter. Conf. on Fusion Energy, Montreal, Vol. 3, 295 (1996).

[8]Y. Ikeda, *et al.* "The 110 GHz electron cyclotron range of frequency system on JT-60U: design and operation," Fusion Sci. and Tech., 42, 435 (2002).

[9]Y. Ikeda, *et al.* "Development and operation of JT-60 LHRF launchers," Plasma Devices and Operations, Vol. 1, 155 (1991).

1.9 Designs for High-Performance H-mode Confinement

The JET tokamak with $R/a = 2.88\,\text{m}/0.9\,\text{m}$ with $B/I_p = 4\,\text{T}/4.5\,\text{MA}$ produced record discharges by 1995-1999 with what has become a standard regime for the ITER design called the H-mode for high confinement. During the 1997 campaign with DT experiments the best results were obtained in the H-mode with an edge transport barrier with an edge pedestal [Cordey, *et al.* (1999), Gibson and the JET Team (1998)].

The H-mode plasma has bursty releases of stored plasma energy across the edge pedestal that are called edge localized modes or ELMs for short. The ELMy H-mode in the D-T plasma had the longest energy confinement time reaching $\tau_E \sim 700\,\text{ms}$ at densities of $\langle n_e \rangle \gtrsim 10^{19}\,\text{m}^{-3}$. By 2000 extensive confinement databases were assembled and elaborate scaling laws with long names were published. The scaling laws will not be presented in this book, but the interested reader is referred to the Nuclear Fusion articles on ITER design. For example, the Cordey, *et al.* (1999) article compares their DT shots from JET with the ITERH-EPS 97(y) and the ITERH-93P scaling laws given in their Eq. (1.1) from the world-wide hydrogen and deuterium data. The first detailed engineering and physics design studies for the original international test reactor are given in Aymar, *et al.* (2001), where the device was called ITER-FEAT with larger major radius ($R = 8\,\text{m}$) than the final machine under construction as specified in Table 1.1 with a major radius of $R = 6.2\,\text{m}$.

1.10 Helical Magnetic Confinement Systems

Large helical device (LHD) was developed in Kyoto University and built in the National Institute for Fusion Science in Toki, Japan. The heating power and the pumping/fueling capability [Motojima, *et al.* (2007)] are designed for steady-state fusion power. The operational regime has higher densities and longer pulse lengths than tokamaks. Quasi-steady-state plasmas with internal transport barriers (ITB) were produced by a combination of fast pumping in the local island divertors and by core fueling with pellet injection that produced dense core plasmas with core densities of $5 \times 10^{20}\text{m}^{-3}$. Discharges of 54 min duration were achieved with volume-averaged plasma beta of $\langle \beta \rangle = 4.5\%$ and with total energy input of 1.6 GJ from 500 kW input power, as described in Chapter 11 on the helical machines.

Fig. 1.8 Final installation and adjustments by machine engineer inside the complex-shaped toroidal Wendelstein 7-X torus before start-up in December 2015. The elliptical shape of the chamber and the complex magnetic field coils are evident.

The new advanced stellarator, Wendelstein 7-X, has reported long-pulse plasmas driven by ECH heating for periods of minutes. In 2016-2017 the NBI and ICRH driven plasmas will be produced. Figure 1.8 shows the last stage of assembly of the Wendelstein 7-X stellarator in 2015. The complex helical superconducting coils produce the magnetic confinement of most ion orbits without the toroidal plasma currents.

Wolf (2003) reviews the various types of internal transport barriers (ITBs) in different plasma confinement modes and geometries. The regimes with higher ion than electron temperature are analyzed separately from those with higher electron than ion temperatures. The types of confinement regimes such as the RS regime with reversed magnetic shear are analyzed. The complex variety of the different types of MHD oscillations are discussed by Wolf (2003). The edge-localized modes (ELMs) that are categorized in the literature as Type-I, Type-II and Type-III are

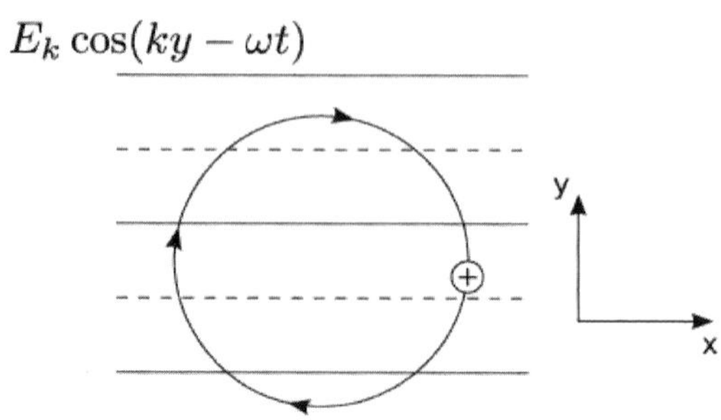

Fig. 1.9 The gyro orbit of a charged particle in a plasma wave giving the amplitude reduced by $J_0(k\rho)$.

described. The confinement modes such as ELMy H-modes and the ELM-free H-mode are characterized in Wolf (2003). The avoidance of large Type I ELMs has become a high priority for the operation of ITER so numerous ideas have been advanced to explain the onset of these large ELMs and methods of controlling the ELMs with external coils has grown to be a major research area.

The sudden appearance of hot plasma filaments or "blobs" by convection from the hot-core plasma into the cool-edge plasma results in a radiation spike in the visible (red) line radiation from the neutral hydrogen or deuterium gas (outside the core plasma) that is a readily-measured diagnostic for the edge plasma. The red line radiation measured with the wavelength 656 nm arises from the $n = 3$ to $n = 2$ electron transition in the neutral hydrogen gas component in the edge plasma.

Worked Problem 1.1: Gyro-Averaged Fields

In the lab or observer's reference frame, the charge q with mass m moves in the constant electromagnetic fields $\mathbf{E}$ and $\mathbf{B}$ with velocity $\boldsymbol{v}$.

(1) Use the Lorentz force to show that in the frame moving with the $\mathbf{E} \times \mathbf{B}$ velocity there is no perpendicular electric field so the particle moves in a helical orbit with gyroradius $\rho = v_\perp/\Omega$ where $\Omega = qB/m$ is the gyrofrequency. In the local frame with $\mathbf{B} = B\widehat{\boldsymbol{e}}_z$ the motion is $x(t) = x_{\rm gc} + \rho\cos(\zeta - \Omega t)$, $y(t) = y_{\rm gc} + \rho\sin(\zeta - \Omega t)$. Check that $\dot{v}_x = \Omega v_y$, $\dot{v}_y = -\Omega v_x$ and sketch the orbit for a 1 KeV proton in a 1 T magnetic field.

(2) For a wave $E = E_k\cos(ky - \omega_k t)$ find the electric force on the charge averaged over one gyro-period $T = 2\pi/\Omega$. Show that the effective electric field $\langle E_k\rangle$ acts on the guiding center $(x_{\rm gc}, y_{\rm gc})$ is reduced by the factor $J_0(kv_\perp/\Omega)$ where J_0 is the Bessel function.

Solution:

(1) The motion perpendicular to the magnetic field has the Lorentz force

$$\boldsymbol{F} = q(\mathbf{E} + \boldsymbol{v} \times \mathbf{B}) \tag{1.2}$$

in the lab frame. Transforming to the $\mathbf{E} \times \mathbf{B}$ frame with

$$\boldsymbol{v} = \frac{\mathbf{E} \times \mathbf{B}}{B^2} + \boldsymbol{v}' \tag{1.3}$$

removes the electric force and the equation of motion is

$$\frac{d\boldsymbol{v}'}{dt} = \Omega \boldsymbol{v}' \times \widehat{\boldsymbol{e}}_z \tag{1.4}$$

with solutions given in the statement of the problem. Note if the Lorentz transformation is made the magnetic field is reduced by $B' = B/\gamma$ where $\gamma = (1 - v_E^2/c^2)^{-1/2}$ and the cyclotron frequency is lowered to $\Omega = eB/m\gamma$.

(2) Figure 1.9 shows the cyclotron orbit in the $\cos(ky - \omega t)$ wave with $\rho = v_\perp/\Omega > 1/k_\perp$ to see the canceling direction of the electric force over a

gyro-period. The average of $\cos[k_y y(t) - \omega t]$ with $y(t) = y_{\rm gc} + \rho\sin(\zeta - \Omega t)$ requires the expression

$$e^{ix\sin\theta} = \sum_{n=-\infty}^{\infty} J_n(x) e^{in\theta} \tag{1.5}$$

from the generating function for Bessel function $J_n(x)$. Now assuming $\omega \ll \Omega$, the average over the cyclotron period of the electric force gives

$$\begin{aligned}\langle F\rangle &= q\int_0^{2\pi/\Omega} dt E_k \cos\left[k\left(y_{\rm gc} + \rho\sin(\zeta - \Omega t)\right) - \omega t\right] \\ &= qE_k J_0(k\rho)\cos\left(ky_{\rm gc} - \omega t\right)\end{aligned} \tag{1.6}$$

after using the real part of Eq. (1.5) and taking the integral over one cyclotron period. Thus, plasma waves give charge separation and plasma currents with the electric fields acting on the large gyroradius ions is reduced by the Bessel function as form factors that reduce the net force with increasing gyroradius.

Chapter 2

Alfvén and Drift Waves in Plasmas

In this chapter we analyze the dispersion relations that describe the linear wave propagation of drift waves in inhomogeneous plasmas. We show how drift waves relate to plasma waves of the infinite, uniform plasma. The objective of this chapter is to develop an overview of the more elementary and universal aspects of drift waves, before studying their realization in particular plasma geometries.

There are now many approaches to achieving magnetic plasma confinement and designs for reaching ignition or a burning fusion plasma. The simplest geometries are linear with magnetic mirrors [Beklemishev, *et al.* (2013)] which maybe either in simple symmetric geometry or complex tandem mirror geometries with sections having quadrupole magnetic fields as in the GAMMA 10 machine [Cho, *et al.* (2005, 2016)]. Magnetic mirror geometry with a large ion current producing a reversal of the magnetic field along the core axis is called a RFC and is used in the new Tri Alpha machine [Binderbauer, *et al.* (2015)]. While the more complex toroidal geometries described in Chapter 1 have the record plasma temperature and confinement times, the linear machines have advantages for lower cost neutron sources.

Miyamoto (1978) describes well the variety of confinement geometries from linear to toroidal with their numerous variations in principles and designs. All the magnetic confinement systems have forms of the Alfvén and drift waves coupled by the finite plasma pressure. Thus, in this chapter we analyze the stability of these two intrinsic low-frequency waves and their eigenmodes in the confinement systems.

For clarity, in this chapter we recall from available references [Krall and Trivelpiece (1973); Mikhailovskii (1967, 1974, 1992)] the required plasma electrical conductivities and charge susceptibilities, rather than interrupting the presentation with their lengthy derivations from the Vlasov-Maxwell system of equations. In Chapter 11 we discuss and develop some aspects of these derivations for particular magnetic geometries. For the stellarator and large helical device the equations are so complicated by the magnetic field structure one needs to use the simulation codes developed for the particular machine to find the eigenmodes and instabilities.

2.1 Low-Frequency Wave Dispersion Relations

The study of drift waves, as with other plasma waves, begins with the study of small amplitude sinusoidal solutions of the self-consistent field equations for a system of ions and electrons interacting through their mutual Coulomb forces. The self-consistent plasma electric fields $\mathbf{E}(\mathbf{k},\omega)$ are determined by the solutions of Maxwell's equations with the sources of charge density $\rho(\boldsymbol{k},\omega)$ and current density $\mathbf{j}(\mathbf{k},\omega)$ calculated from the plasma dynamics due to these same electromagnetic fields. Of course, the time varying electric fields and plasma currents induce time varying magnetic fields $\mathbf{B}(\mathbf{k},\omega)$ which are often of a relatively small amplitude and produce relatively weak transport compared with that from the time varying electric fields.

The plasma dynamics in the presence of the oscillating electric field is described by the linearized kinetic equations for the perturbed distribution functions $\delta f_j(\mathbf{v},\mathbf{k},\omega)$. In certain cases, particularly when collisions dominate the particle motion along the magnetic field, the fluctuating charge and current densities are adequately described by the fluid equations with parallel transport coefficients. This hydrodynamic regime is studied in Chapter 3. In both cases the theoretical results required for the study of small amplitude motion are conveniently expressed as a frequency and wavenumber dependent plasma conductivity tensor $\sigma_{\alpha\beta}(\mathbf{k},\omega)$ and the scalar charge density susceptibility function $\chi(\mathbf{k},\omega)$. The theoretical results for these quantities and their derivations are well known from the widely available reference works on plasma waves in nonuniform media [Braginskii (1965), Mikhailovskii (1967, 1992), Krall 1973)].

We begin here by simply introducing these conductivity $\sigma_j(\mathbf{k},\omega)$ formulas with the appropriate explanatory material rather than repeating their lengthy derivations from the kinetic equation. We shall give the first principles kinetic derivations in later chapters (cf. Chapter 11) where particular confinement geometries are introduced after the relevant estimates from elementary considerations have been established.

Let us begin very simply by considering plasma that is nonuniform only because of a density variation in a direction perpendicular to that of the confining magnetic field. We introduce a coordinate system with $\widehat{\mathbf{e}}_z$ in the direction of the magnetic field $\boldsymbol{B}$, $\widehat{\mathbf{e}}_x$ in the direction of the density gradient, and $\widehat{\mathbf{e}}_y$ in the direction orthogonal to $\widehat{\mathbf{e}}_x$ and $\widehat{\mathbf{e}}_z$. The orthogonal coordinate system is taken so as to have $\widehat{\mathbf{e}}_x$, $\widehat{\mathbf{e}}_y$, $\widehat{\mathbf{e}}_z$ right-handed. The plasma pressure gradient ∇p produced by the density gradients is balanced by the $\mathbf{j}\times\mathbf{B}$ force where the current density is $\mathbf{j} = j_y\widehat{\mathbf{e}}_y$ is the sum of the partial currents from the drift velocities v_{dj} of each charge species

$$j_y = \sum_j e_j n_j v_{dj}. \tag{2.1}$$

Here, the jth charge species has charge e_j, mass m_j, density n_j and temperature T_j. The diamagnetic drift velocity v_{dj} is given by

$$v_{dj} = \frac{T_j}{e_j B} \frac{1}{n_j} \frac{dn_j}{dx} \tag{2.2}$$

when the temperature is uniform. More generally, from force balance $\mathbf{j} \times \mathbf{B} = \nabla \Sigma p_j$ we find that v_{dj} is proportional to the partial pressure gradient. The positive and negative charges drift in opposite directions and combine in Eq. (2.1) to give the net current across the magnetic field. This current j_y is called the diamagnetic current and v_{dj} is called the diamagnetic drift velocity. The direction of the plasma current is such as to produce a magnetic field component ΔB that reduces the given $B\widehat{\mathbf{e}}_z$ which accounts for the meaning of the name diamagnetic. The change in the magnetic field B from Ampére's law with the diamagnetic plasma current then gives $d(p + B^2/8\pi)/dx = 0$ in absence of magnetic curvature. Here the total pressure $p = \sum_j n_j T_j$ as follows from Eqs. (2.1), (2.2), and Ampére's law.

The measurement of ΔB is one of the basic diagnostic signals used to determine the total stored plasma energy $W_p(t)$. The flux from $\Delta B(t)$ is called the diamagnetic signal and is measured as the voltage induced in a coil surrounding the plasma column.

Clearly, the current density in Eq. (2.1) satisfies the force balance equation

$$\mathbf{F} = \left(j_y B - \frac{dp}{dx} \right) \mathbf{e}_x = 0 \tag{2.3}$$

after using Eq. (2.2). The electric force $\rho_q E_x$ does not enter force balance since the equilibrium is charge neutral with $\rho_q = \sum_j e_j n_j(x) = 0$ which requires that $\sum_i Z_i n_i(x) = n_e(x)$. As a consequence the drift velocities v_{dj} are related by

$$\sum_j \frac{n_j e_j^2}{T_j} v_{dj} = 0. \tag{2.4}$$

Equation (2.4) follows by substituting Eq. (2.2) into Eq. (2.1) using $d/dx(\sum_j e_j n_j)$ $= 0$. For a pure electron-ion plasma, Eq. (2.4) reduces to

$$\sum_{j=i,e} \frac{e_j}{T_j} v_{dj} = 0. \tag{2.5}$$

Thus, the higher-temperature species have the larger diamagnetic drift velocities. Often the drift waves propagate in the direction of the higher-temperature species.

Let us consider a single Fourier component $\mathbf{k}\omega$ of an oscillating electric field perturbation of the form

$$\mathbf{E}(x,t) = \widehat{\mathbf{E}}_{\mathbf{k}\omega} e^{i\mathbf{k}\cdot\mathbf{x} - i\omega t} + \widehat{\mathbf{E}}^*_{\mathbf{k}\omega} e^{-i\mathbf{k}\cdot\mathbf{x} + i\omega t} \tag{2.6}$$

with the vector wavenumber

$$\mathbf{k} = k_y \widehat{\mathbf{e}}_y + k_z \widehat{\mathbf{e}}_z. \tag{2.7}$$

Here we allow the angular frequency ω to be complex $\omega = \omega_r + i\gamma$. The linearized Vlasov equation for the perturbation $\delta f_j(\mathbf{k}, \omega, \mathbf{v})$ of the phase space density function f_j is induced by the electric field Eq. (2.6) in the particle distributions. The linear fluctuations can be calculated exactly and used to calculate the complex dispersive plasma conductivities $\sigma_j(\mathbf{k}, \omega, \mathbf{v})$. Maxwell's equations then yield the following dielectric tensor defined by

$$\varepsilon_{\alpha\beta}(\mathbf{k}, \omega) = I_{\alpha\beta} + 4\pi i \frac{\sigma_{\alpha\beta}(\mathbf{k}, \omega)}{\omega} \tag{2.8}$$

where for low-frequency drift-Alfvén waves ($\omega \ll \omega_{ci}$) the values of $\varepsilon_{\alpha\beta}$ are given by

$$\varepsilon_{xx} \cong 1 + \frac{\omega_{pi}^2}{\omega_{ci}^2}\left(1 - \frac{\omega_{*i}}{\omega}\right)\frac{1 - \Gamma_0(b)}{b} \tag{2.9}$$

$$\varepsilon_\perp \equiv \varepsilon_{yy} \cong 1 + \frac{\omega_{pi}^2}{\omega_{ci}^2}\left(1 - \frac{\omega_{*i}}{\omega}\right)\frac{1 - \Gamma_0(b)}{b} \tag{2.10}$$

$$\varepsilon_\parallel \equiv \varepsilon_{zz} \cong 1 - \frac{\omega_{pe}^2}{k_\parallel^2 v_e^2}\left(1 - \frac{\omega_{*e}}{\omega}\right) Z'\left(\frac{\omega}{k_\parallel v_e}\right) - \frac{\omega_{pi}^2}{\omega^2}\left(1 - \frac{\omega_{*i}}{\omega}\right)\Gamma_0(b) \tag{2.11}$$

where the diamagnetic drift frequencies and gyroradius averaging are

$$\omega_{*j} \equiv k_y v_{dj} \qquad \text{and} \qquad \Gamma_0(b) \equiv e^{-b} I_0(b) \tag{2.12}$$

with I_0 being the modified Bessel function of zeroth order. In these equations we use b to denote the square of the wavenumber perpendicular to the magnetic field $k_\perp$ measured with respect to the thermal ion gyroradius; i.e.

$$b = \frac{1}{2}\frac{k_\perp^2 v_i^2}{\omega_{ci}^2} = \frac{k_\perp^2 T_i}{m_i \omega_{ci}^2}. \tag{2.13}$$

The physical origin of the Bessel functions in Eq. (2.9) to Eq. (2.12) is made clear by considering the average over one gyro or cyclotron orbit period of the electric field in Eq. (2.6) using the procedure in Worked Problem 1.1. The average is performed using the vector from the guiding center to the particle $\mathbf{X}(t) = \mathbf{X}_{gc} + \mathbf{b} \times \mathbf{v}/\omega_c$. Integrating around one gyro circle at fixed guiding center position $\mathbf{X}_{gc}$ gives J_0 and then computing the integral over the Maxwellian velocity distribution yields the modified Bessel function $I_0(b)e^{-b}$. The calculation is carried out explicitly in Chapter 10. In Eqs. (2.9)–(2.11) the complex plasma dispersion function $Z(\zeta)$ and its derivative $Z'(\zeta)$ appear from the parallel velocity integrals over the wave-particle propagator $g_{k\omega} = (\omega - k_\parallel v_\parallel + i0^+)^{-1}$. The properties of the dispersion function are given in Fried and Conte (1961) and the derivation is given in Worked Problem 2.2 and Figs. 2.1-2.2.

Note that Eq. (2.8) is written in cgs units. In mks units the 4π is replaced with $\mu_0 = 4\pi \times 10^{-7}\,\mathrm{mT/A}$ where m is meters, T is Tesla, A is amperes.

The dielectric components from the electrical conductivities in Eqs. (2.9)–(2.11) are valid in the frequency-wavelength regime defined by

$$\begin{aligned} k_\parallel v_i \ll \omega \ll \omega_{ci} \ll \omega_{ce} \\ k_\perp v_e \ll \omega_{ce}, \end{aligned} \tag{2.14}$$

which is sufficient for present purposes.

To see that conditions Eq. (2.14) are satisfied for plasmas with weak density gradients over the scale of the ion gyroradius we make the following estimates. It is standard to define the density gradient scale length L_n by

$$\frac{1}{n}\frac{dn}{dx} = \frac{-1}{L_n} \tag{2.15}$$

which gives from Eqs. (2.2) and (2.12)

$$\omega_{*e} = k_y \frac{T_e}{eBL_n} = k_y \rho_s \frac{c_s}{L_n} \tag{2.16}$$

where we define the ion sound-inertial scale length $\rho_s = c_s/\omega_{ci} = c(m_i T_e)^{1/2}/eB$ with the ion sound, or acoustic speed c_s defined by electron temperature and the ion mass by $T_e = m_i c_s^2$. For $T_e \sim T_i$ the spatial scale length ρ_s is similar to the ion thermal gyroradius ρ_i. For $k_y \rho_i \leq 1$ the drift wave frequency is bounded above by

$$\omega_k \leq \max_k(\omega_k) = \frac{c_s}{L_n} \tag{2.17}$$

which gives

$$\frac{\omega_k}{\omega_{ci}} \leq \frac{\rho_s}{L_n}. \tag{2.18}$$

Thus for $\rho_s/L_n \ll 1$, as is typical of magnetized plasmas, the drift wave frequencies are well below the ion cyclotron frequency ω_{ci}.

After eliminating the perturbed magnetic field $\mathbf{B}_{\mathbf{k}\omega}$ in Ampére's law by substituting $\mathbf{B}_{\mathbf{k}\omega} = c\mathbf{k} \times \mathbf{E}_{\mathbf{k}\omega}/\omega$ from Faraday's law, we obtain the system of self-consistent field equations

$$\left(k_\parallel^2 - \frac{\omega^2}{c^2}\,\varepsilon_{yy}\right) E_y - k_y k_\parallel E_z = 0 \tag{2.19}$$

$$-k_y k_\parallel E_y + \left(k_\perp^2 - \frac{\omega^2}{c^2}\,\varepsilon_{zz}\right) E_z = 0 \tag{2.20}$$

$$\left(k_y^2 + k_\parallel^2 - \frac{\omega^2}{c^2}\,\varepsilon_{xx}\right) E_x = 0 \tag{2.21}$$

which are written here in the approximation that $\sigma_{xz} = \sigma_{xy} = \sigma_{yz} = 0$.

The coordinates in Eqs. (2.19), (2.20), and (2.21) are rotated from those in the Stix-matrix and follow Mikhailovski (1992). The Stix matrix uses the $\boldsymbol{k} - \mathbf{B}_0$ plane to define the $\widehat{\mathbf{x}} - \widehat{\mathbf{z}}$-direction. In laboratory problems $\widehat{\mathbf{x}}$ is in the radial direction and the dominant $\boldsymbol{k}$-vector is in the $\widehat{\mathbf{y}} - \widehat{\mathbf{z}}$ plane.

From the field equations (2.19)-(2.21) we find that there are compressional modes that have $B_{z\mathbf{k}\omega} = k_y E_x/\omega$ with the dispersion relation

$$k^2 = \frac{\omega^2}{c^2}\,\varepsilon_{xx} \qquad \text{and} \qquad E_y = E_z = 0 \tag{2.22}$$

separate from the approximately incompressional modes with $B_{z\mathbf{k}\omega} \equiv 0$. These modes with $E_y, E_z \neq 0$ from Eqs. (2.19) and (2.20) are the coupled Alfvén-drift waves given by

$$k_{\|}^2\varepsilon_{\|} + k_y^2\varepsilon_{\perp} = \frac{\omega^2}{c^2}\,\varepsilon_{\|}\varepsilon_{\perp} \qquad \text{and} \qquad E_x = 0 \tag{2.23}$$

as derived from Eqs. (2.19)–(2.21). The compression here refers to the change in $B_z \equiv B_0 + \delta B_z$ which produces a large perturbed magnetic pressure force $\mathbf{F}_B = -\nabla(B^2/2\mu_0)$ along the ambient magnetic field B_0.

Observe that the incompressible modes follow from the parallel component of Ampére's law and Poisson's equation. For the incompressible mode, Faraday's law gives only a perpendicular component of the perturbed $B_{x\mathbf{k}\omega}$ field perpendicular to the ambient magnetic field $B\widehat{\mathbf{e}}_z$. The parallel component of Ampére's law, consequently, reduces to

$$\begin{aligned}(\nabla\times\mathbf{B})_{\|} &= -ik_yB_x = \frac{ick_y}{\omega}\,(k_{\|}E_y - k_yE_{\|})\\ &= \frac{4\pi\sigma_{\|}}{c}\,E_{\|} - \frac{i\omega}{c}\,E_{\|} = -i\,\frac{\omega}{c}\,\varepsilon_{\|}E_{\|}.\end{aligned} \tag{2.24}$$

From Poisson's equation we have $\rho_q = (i/4\pi)(k_{\|}E_{\|} + k_yE_y)$, and thus from charge conservation it follows that

$$\begin{aligned}\frac{\partial\rho_q}{\partial t} + \nabla\cdot\mathbf{j} = &+\frac{\omega}{4\pi}\,(k_yE_y + k_{\|}E_{\|})\\ &+ i(k_y\sigma_{\perp}E_y + k_{\|}\sigma_{\|}E_{\|}) = 0.\end{aligned} \tag{2.25}$$

In the limit where $\mathbf{E}_{k\omega} = -i\mathbf{k}\phi_{k\omega}$ the charge conservation Eq. (2.25) reduces to

$$\left(k_{\|}^2\epsilon_{\|} + k_{\perp}^2\epsilon_{\perp}\right)\phi_k = 0 \tag{2.26}$$

which gives the low-frequency limit of Eq. (2.23).

Worked Problem 2.1: Low-Frequency Energy Conservation Theorem

Problem:

Find the electric current j_p produced by a low-frequency oscillation of an electric field $E(t)$ in a plasma that accelerates the charged particles across the ambient magnetic field $B_z = \text{const}$. From the formula for the plasma current, which is called the polarization current, show that the rate of change dW_E/dt of the kinetic energy W_E in the $\mathbf{E}\times\mathbf{B}$ drift velocities of the particles equals the electric power deposited in the plasma through the polarization current.

Solution:

Low frequency is defined in the (drift) wave literature as frequencies $\omega \ll \omega_c = qB/m$ where the motions are then given by the $\mathbf{E} \times \mathbf{B}$ drift velocity $\mathbf{v}_E$. The acceleration of the charge from the increase of $\mathbf{v}_E$ in time then gives rise to a smaller drift velocity $\mathbf{v}_{p,j}$ of the particle through Newton's law

$$m_j \frac{d\mathbf{v}_E}{dt} = q_j \mathbf{v}_p \times \mathbf{B}. \tag{2.27}$$

We solve for the polarization velocity $\mathbf{v}_{p,j}$ by crossing Eq. (2.27) with $\mathbf{B}$ and working out the double vector cross product to show that

$$\mathbf{v}_{p,j} = \frac{m_j}{q_j B^2} \frac{d\mathbf{E}}{dt}. \tag{2.28}$$

The drift velocity $\mathbf{v}_p$ is the polarization drift velocity which, owing to the factor m_j/q_j, is in the opposite directions for ions and electrons so as to create an electric field polarization of the plasma. This polarization of the plasma gives the dielectric constant $\epsilon_\perp$ used in deriving the Alfvén wave from the electromagnetic waves in dielectric materials from electromagnetic wave theory.

The polarization drift is in opposite directions for positive and negative charges and thus produces the plasma polarization current $\mathbf{j}_p$ given by

$$\mathbf{j}_p = \sum_j \frac{n_j m_j}{B^2} \frac{d\mathbf{E}}{dt}. \tag{2.29}$$

The energy conservation theorem follows by computing the rate of change of particle kinetic energy

$$W_E = \frac{1}{2} \sum_j m_j n_j \, v_E^2 \tag{2.30}$$

in the $\mathbf{E} \times \mathbf{B}$ drift motion of the particles

$$\frac{dW_E}{dt} = \frac{n_j m_j \mathbf{v}_E}{B^2} \cdot \frac{d\mathbf{E}}{dt} \tag{2.31}$$

$$= \mathbf{j}_p \cdot \mathbf{E}(t) \tag{2.32}$$

where to obtain the identity we used the dot product of Eq. (2.29) with $\mathbf{E}$ for the electric power transfer from the electric field to the particle motion.

The form of the energy conservation theorem, Eq. (2.32), is closely related to the Poynting theorem when we introduce the plasma dielectric function $\epsilon_\perp$. As in the electrodynamics of materials the polarization current $\mathbf{j}_p$ and the polarization charge density ρ_p are related by the conservation of charge as in Eq. (2.25). For magnetized plasmas the polarization at low frequencies is high which gives a high index of refraction to electromagnetic waves slowing the velocity of the transverse waves from the speed of light to the Alfvén speed.

We see that we can interpret $\mathbf{j}_p$ in terms of the perpendicular dielectric function $\epsilon_\perp$ given by

$$\epsilon_\perp = \epsilon_0 + \sum_j \frac{n_j m_j}{B^2} \tag{2.33}$$

where ϵ_0 is the dielectric of the vacuum in mks units. The student will want to show that the units of the plasma density contribution are the same as the vacuum contribution which is usually expressed in Farad/meter. At a relatively low-plasma density, the plasma contribution exceeds the vacuum contribution and slows the electromagnetic waves through the standard dispersion relation

$$\epsilon_\perp \omega^2 = c^2 k^2 \tag{2.34}$$

as we will see in greater detail in Chapter 4. The form of the Poynting theorem in a dielectric in MKS system now called SI MKS material requires $\mathbf{E} \cdot \mathbf{D}$ with $\mathbf{D} = \epsilon \mathbf{E}$. Standard International units are abbreviated as SI units.

There are two systems of units used in the plasma physics literature: the cgs units of centimeters, grams and seconds and the mks units of meters, kilograms, seconds. The electrical units corresponding to the cgs system are E(statvolt/cm), B(gauss), j(abamperes) and to the mks units E(volts/m), B(Tesla) and j(amps/m^2). The speed of light enters in the cgs units for Maxwell's equations and the Lorentz force whereas in the mks units enters the fundamental constants $\epsilon_0 = 8.854 \times 10^{-12}\,\mathrm{F/m}$ and $\mu_0 = 4\pi \times 10^{-7}\,\mathrm{T\cdot m/A}$ enter the corresponding mks equations. The speed of light determines μ_0 in the mks system through $c^2 = 1/\epsilon_0\mu_0$. It is important for plasma physicists to know both systems of units since the astrophysical and early fusion literature is written in the cgs units while laboratory and space plasma physics is written in terms of the mks system of units.

The NRL formulary is a valuable tool for going between the units and finding reference values in both systems of units for typical plasma systems, and can be found at `http://www.ppd.nrl.navy.mil/nrlformulary/`. Errors are often made by the novice in evaluating electric and magnetic phenomenon in either cgs or mks units. As an example, we give the reference values for a typical laboratory plasma of the drift speed and electric polarization current for a plasma with a radial electric field E_r of $1\,\mathrm{kV/m}$= ($1/300 \times 100$cm/m) statvolt/cm increasing at the rate of $1\,\mathrm{kHz}$ in a cylindrical plasma with $\mathbf{B} = B\mathbf{e}_z$ of $B = 1T = 10^4\,\mathrm{G}$. The value of the $\mathbf{E} \times \mathbf{B}$ drift velocity is $1000\,\mathrm{m/s} = 10^5\,\mathrm{cm/s}$ and for a density of $n = 10^{20}$ protons/m^3 $= 10^{14}$ protons/cm^3, the plasma polarization current is $j_p = [10^{20}/\mathrm{m}^3\, 1.67 \times 10^{-27}\,\mathrm{kg}/(1\mathrm{T}^2)] \times 10^6\,\mathrm{V/ms} = 0.167\,\mathrm{A/m}^2$. By the energy conservation theorem, the power deposited in the particle kinetic drift velocities is $0.167\,\mathrm{kW/m}^3$, which is a typical Alfvén wave heating power in laboratory plasmas.

Worked Problem 2.2

Problem:

Find the collisionless damping rate γ for (1) electron plasma oscillations $\omega = \omega_{pe} k_\| / k$ and (2) ion-acoustic waves $\omega = k_\| c_s$ in a strongly magnetized plasma with temperature T_e.

Solution:

Strongly magnetized means that the cyclotron frequency $\omega_{ce} = eB/m_e \gg \omega_{pe}$, the plasma frequency, so the electron motion is one-dimensional. We describe the motion in the wave with electron Vlasov equation for the electron phase space density $f(x, v_\|, t)$, which satisfies

$$\frac{\partial f}{\partial t} + v_\| \frac{\partial f}{\partial z} + v_E \frac{\partial f}{\partial x} - \frac{eE_\|}{m_e} \frac{\partial f}{\partial v_\|} = 0. \tag{2.35}$$

(1) Electron plasma oscillations

The electron density $n_e = \int f dv_\|$ and we solve for the part of $f = f_0 + \delta f$ that oscillates with the wave electric field $\mathbf{E} = \mathbf{E}_{k\omega} \exp(i\mathbf{k} \cdot \mathbf{x} - i\omega t)$. The fluctuating δf follows from solving

$$\left(-i\omega + ik_\| v_\|\right) \delta f = \frac{eE_\|}{m_e} \frac{\partial f_0}{\partial v_\|} - \frac{E_y}{B} \frac{\partial f_0}{\partial x} \tag{2.36}$$

and the associated oscillating density in Poisson's equation

$$\nabla \cdot \mathbf{E} = -k^2 \phi_{k\omega} = -\frac{e}{\varepsilon_0} \int_{-\infty}^{+\infty} dv_\| \delta f_{k\omega}(v_\|). \tag{2.37}$$

Substituting δf from Eq. (2.36) into Eq. (2.37) gives the dispersion relation

Kinetic Dispersion Relation

$$k^2 \varepsilon_k(\omega) \phi_{k\omega} = \left[k^2 + \frac{ne^2}{\varepsilon_0 m_e} \int_{-\infty}^{+\infty} dv_\| \frac{\left(k_\| \frac{\partial f_0}{\partial v_\|} + \frac{k_y}{eB} \frac{\partial f_0}{\partial x} \right)}{\omega - k_\| v_\| + i0^+} \right] \phi_{k\omega} = 0 \tag{2.38}$$

where the wave-particle resonance at $\omega - k_\| v_\| = 0$ is resolved by taking the adiabatic growth of the wave from $E_k \exp(-i(\omega + i0^+)t)$ or equivalently from using the Landau initial value problem solution with a Laplace transform. The residue of the complex, analytic density response function in Eq. (2.38) gives the collisionless damping of the wave when $df_0/dx = 0$. The resonance $\omega_k - k_\| v_\| = 0$ provides the drift wave growth rate when df_0/dx is large.

For waves with $\omega_k \gg \gamma_k$ we solve Eq. (2.38) by a Taylor series expansion about the real root $\omega = \omega_k$. For electron plasma waves with $\omega_k \gg k_\| v_e$ we take the high-phase velocity limit of Eq. (2.38) $\varepsilon_k(\omega) = 1 -$

$(\omega_{pe}^2/k^2)(k_\parallel^2/\omega^2) = 0$ to obtain the plasma waves $\omega_k = \pm\omega_{pe}k_\parallel/k$. Now, with this frequency we can evaluate the residue from complex part of the response function to obtain from Eq. (2.38)

$$\frac{2i\gamma\omega_{pe}^2}{\omega^3} - \frac{i\pi\omega_{pe}^2}{k^2}\int dv_\parallel \delta\left(\omega - k_\parallel v_\parallel\right) k_\parallel \frac{\partial f_0}{\partial v_\parallel} = 0.$$

This Landau resonance problem occurs throughout plasma wave theory so it is important to define the associated plasma dispersion function $Z(\zeta)$ and $dZ/d\zeta$ that occur over and over. The Maxwellian velocity distribution function gives

$$\int_{-\infty}^{+\infty} \frac{dv_\parallel f_M(v_\parallel)}{k_\parallel v_\parallel - \omega} = \int_{-\infty}^{+\infty} \frac{dx}{\pi^{1/2}} \frac{\exp\left(-x^2/2\right)}{x - \omega/k_\parallel v_e + i0^+} \equiv Z(\zeta)$$

which defines the analytic complex dispersion function $Z(\zeta)$ with $\zeta = \omega/|k_\parallel|v_e$being the complex variable continued by analytic continuation from the upper half of the complex frequency or ζ plane. Figure 2.1 gives $Z(\zeta)$.

Thus we arrive at the electron plasma wave damping rate γ_k^{el} as

$$\gamma_k^{el} = -\left(\frac{\pi}{8}\right)^{1/2} \frac{\omega_{pe}^2}{k^3 v_e^3}\omega_k^2 \exp\left(\frac{-\omega_k^2}{k_\parallel^2 v_e^2}\right).$$

(2) Ion-acoustic waves

For the second question, the low-frequency ion-acoustic waves have $\zeta = \omega/|k_\parallel|v_e \ll 1$ and the expansion of the plasma response function requires some analysis skipped here. One finds that the result is $Z(\zeta) \cong -2\zeta - i\pi^{1/2} - \zeta^2$. Using this small phase velocity response function in Eq. (2.38) gives

$$\omega^2 = \frac{\omega_{pi}^2}{1 + \frac{\omega_{pe}^2}{k_\parallel^2 v_e^2}\left(1 + i\pi^{1/2}\frac{\omega}{|k_\parallel||v_e}\right)}.$$

Expanding up the small resonant electron contribution gives the damping rate of the ion-acoustic (IA) wave as

$$\gamma_k^{ia} = -\left(\frac{\pi}{8}\right)^{1/2} |\omega_k| \left(\frac{T_e}{T_i}\right)^{3/2} \exp\left(-\frac{\omega_k^2}{2k_\parallel^2 v_e^2}\right).$$

We note that the plasma waves $\omega_k = \pm\omega_{pe}(k_\parallel/k) = \pm\omega_{pe}\cos(\theta)$ have the same kinematic structure as internal gravity waves with $\omega_{pe} \to N-$ the Brunt-Väisäla frequency and the angle θ is between the **k**-vector and the horizontal. The Brunt-Väisäla or buoyancy frequency for a stable stratification of the mass density $\rho(z)$ is given by $N(z) = \sqrt{(-gd\rho/pdz)}$.

The plasma dispersion function $Z(\zeta)$ is shown in Fig. 2.1 for real values of $\zeta = x = \omega/kv_e$. For real $\zeta = x$, the response from the $\mathrm{Re}\, Z'$ gives the continuous change from adiabatic shielding in the low-frequency dynamics with $|x| \ll 1$ to the

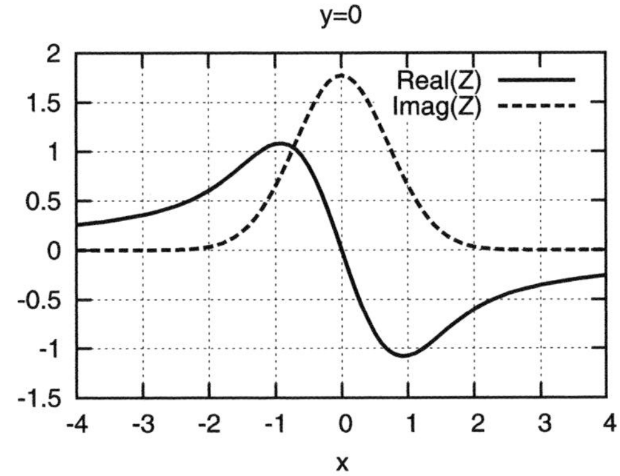

Fig. 2.1 The real and imaginary parts of the plasma dispersion function $Z(x)$ in the range $x = [0, 5]$ for the dimensionless phase velocity $x = \omega/k_{\|}v_e$ for real ω in the Maxwellian plasma velocity distribution $v_e = \sqrt{[2T_e/m_e]}$ with temperature T_e. The real part of $Z(x)$ is anti-symmetric in $x \to -x$ and the imaginary part of $Z(x)$ is symmetric in $x \to -x$.

hydrodynamic reactive oscillatory response for $|x| \gg 1$. Plasma waves or plasma oscillations occur when $|x| \gg 1$ and the (exponentially) small imaginary part of Z' then describes the response of the resonant particles with $\omega = k_{\|}v_e$.

In the adiabatic regime defined by $|x| \ll 1$ the imaginary resonant response is order of $|x|$ and thus there is strong dissipation or growth from these wave-particle interactions. This is the regime of the electrons in the drift wave and thus when $\omega < \omega_*$ there is a strong growth of the drift wave from the resonant electrons.

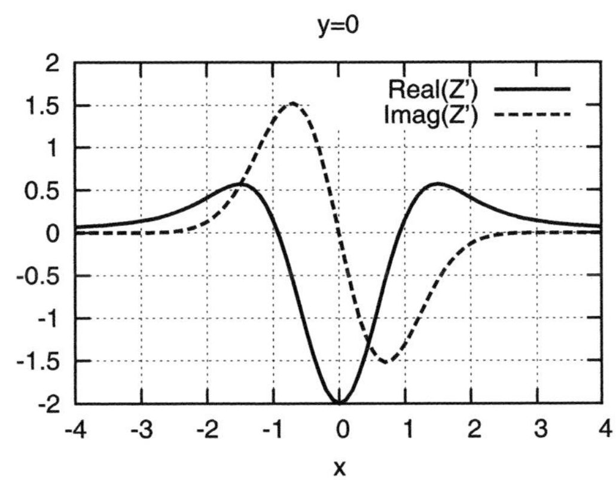

Fig. 2.2 The real and imaginary part of the plasma dispersion function $Z'(x) = dZ/dx$ in the range $x = [0, 5]$ for the dimensionless phase velocity $x = \omega/kv_e$ for real ω in the Maxwellian plasma velocity distribution $v_e = \sqrt{[2T_e/m_e]}$ with temperature T_e. The real part of $Z'(x)$ is symmetric in $x \to -x$ and the imaginary part of $Z(x)$ is anti-symmetric in $x \to -x$. The response at $|x| \ll 1$ is $Z'(x) = -2$ which gives the Debye shielding of charges in the plasma. This regime in the fluid description is called the adiabatic response of the species. The regime $|x| \gg 1$ is $Z'(x) = 1/x^2$ and is called the hydrodynamic or cold plasma response giving electron plasma oscillations.

When the perturbed parallel electron current is calculated and the Poynting theorem is used to calculate the averaged transfer of power from the particles to the waves as $-\langle \mathbf{j} \cdot \mathbf{E} \rangle$ we find that this is positive or negative and of order of $\omega/k_\parallel v_e$ for the electron drift wave. The condition for the negative value of $\langle \mathbf{j} \cdot \mathbf{E} \rangle$ is that $\omega < \omega_*$, which is then the condition for the instability to occur. The proof of this statement is most carefully given by construction of a Nyquist diagram that maps the upper-half of the complex ω-plane with the dispersion function $\epsilon(\omega, k)$ onto the complex ϵ plane. When the mapped image encircles the origin of the ϵ plane, the Cauchy theorem guarantees there is a root of the $\epsilon(\omega, k) = 0$ inside the contour in the ω-plane. This provides one method of checking the stability of the system and deriving the critical conditions for the onset of an instability.

In Fig. 2.2 the derivative $Z' = dZ/d\zeta$ of the Z-function is shown for the dimensionless phase velocity $\zeta = \omega/kv_e$ in the range $[-5, +5]$.

2.2 Reduction of the Kinetic Dispersion Relation

For the low-frequency and long parallel wavelength modes of interest the vacuum term (displacement current) in $\varepsilon_\parallel$ is negligible compared to the large parallel electron currents produced by the fluctuating electric field. Consequently, we make the following reduction:

$$\begin{aligned}\varepsilon_\parallel &= 1 - \frac{\omega_{pe}^2}{k_\parallel^2 v_e^2}\left(1 - \frac{\omega_{*e}}{\omega}\right) Z'\left(\frac{\omega}{k_\parallel v_e}\right) - \frac{\omega_{pi}^2}{k_\parallel^2 v_i^2}\left(1 - \frac{\omega_{*i}}{\omega}\right) \Gamma_0(b) Z'\left(\frac{\omega}{k_\parallel v_i}\right) \\ &\cong -\frac{\omega_{pi}^2}{2k_\parallel^2 c_s^2}\left(1 - \frac{\omega_{*e}}{\omega}\right) Z'\left(\frac{\omega}{k_\parallel v_e}\right) - \frac{\omega_{pi}^2}{k_\parallel^2 v_i^2}\left(1 - \frac{\omega_{*i}}{\omega}\right) \Gamma_0 Z'\left(\frac{\omega}{k_\parallel v_i}\right).\end{aligned} \tag{2.39}$$

Likewise, the perpendicular ion current, and at high densities $\omega_{pe} > \omega_{ce}$ the perpendicular electron current, dominates the vacuum displacement current in most cases of practical interest. That is the particle currents dominate the vacuum displacement current. This dominance of the plasma current allows the approximation

$$\begin{aligned}\varepsilon_\perp &= 1 + \frac{\omega_{pi}^2}{\omega_{ci}^2}\left(1 - \frac{\omega_{*i}}{\omega}\right) \frac{1 - \Gamma_0(b)}{b} + \frac{\omega_{pe}^2}{\omega_{ce}^2}\left(1 - \frac{\omega_{*e}}{\omega}\right) \\ &\cong \frac{\omega_{pi}^2}{\omega_{ci}^2}\left(1 - \frac{\omega_{*i}}{\omega}\right)\left(\frac{1 - \Gamma_0(b)}{b}\right) + \frac{\omega_{pe}^2}{\omega_{ce}^2}\left(1 - \frac{\omega_{*e}}{\omega}\right)\end{aligned} \tag{2.40}$$

to be made. With these reductions in Eqs. (2.39) and (2.40) substituted into the field equations, Eqs. (2.24) and (2.25), we derive a simpler drift wave dispersion relation. The magnetic part of the wave arises from

$$(\nabla \times \mathbf{B})_\parallel = -ik_y B_x = \frac{ick_y}{\omega}(k_\parallel E_y - k_y E_\parallel) = \frac{4\pi\sigma_\parallel}{c} E_\parallel \tag{2.41}$$

and the electric field is constrained by

$$\nabla \cdot \mathbf{j} = ik_y \sigma_\perp E_y + ik_\parallel \sigma_\parallel E_\parallel = 0. \tag{2.42}$$

Substituting Eqs. (2.39) and (2.40) into Eqs. (2.41) and (2.42), and canceling the overall plasma density factor ω_{pi}^2/c_s^2, we obtain the drift-Alfvén wave dispersion relation

Drift-Alfvén Waves

$$\frac{T_e}{T_i}\left[1-\Gamma_0(b)\right]\left(1-\frac{\omega_{*i}}{\omega}\right)+\left[1-\frac{1-\Gamma_0(b)}{b}\frac{\omega(\omega-\omega_{*i})}{k_\parallel^2 v_A^2}\right]$$
$$\times\left[-\frac{1}{2}\left(1-\frac{\omega_{*e}}{\omega}\right)Z'\left(\frac{\omega}{k_\parallel v_e}\right)-\frac{k_\parallel^2 c_s^2}{\omega^2}\left(1-\frac{\omega_{*i}}{\omega}\right)\Gamma_0(b)\right]=0. \tag{2.43}$$

For $b \ll 1$ the first term in Eq. (2.43) is small since $\Gamma_0(b) \simeq 1-b$. Consequently, at sufficiently small b there are two modes that are given by the two-second factors within the square brackets in Eq. (2.43). The ion drift-Alfvén mode is defined by

$$\omega(\omega-\omega_{*i}) = k_\parallel^2 v_A^2 \tag{2.44}$$

arising from the first factor in Eq. (2.43). The electron drift-ion acoustic mode with $\omega < k_\parallel v_e$ defined by

$$\omega(\omega-\omega_{*e}) - k_\parallel^2 c_s^2\left(1-\frac{\omega_{*i}}{\omega}\right)=0 \tag{2.45}$$

arising from the second factor in Eq. (2.43). Equation (2.45) is a cubic equation with the ion-acoustic waves and a thermal mode. The roots of the ion-acoustic wave become unstable when ω_{*i}/ω_{*e} is large and the instability is called the ion temperature gradient mode. Chapter 13 develops the theory and application of this ion temperature gradient (ITG) instability. Clearly, the first term proportional to $1-\Gamma_0(b)$ in Eq. (2.43) couples the two modes given in Eqs. (2.44) and (2.45) giving a fifth-order polynomial in ω for each $\boldsymbol{k}$.

The precise description of the complex $\omega(\mathbf{k})$ roots of Eq. (2.43) is a difficult problem requiring numerical analysis. Rather than pursue such an analysis at this point in the presentation let us simply write down the four modes that are described by Eqs. (2.44) and (2.45) in the small b limit. The motivated reader can independently pursue both the problems (1) of determining the wavelength domains in $k_y - k_\parallel$ where the following approximation solutions are valid and (2) the form of the roots $\omega_k^\alpha(\alpha = 1,2,3,4)$ in regions outside the range of the $k_y - k_\parallel$ domains where the modes have the following simple forms. With these cautions, then, we may write the following formulas for the four modes of propagation that follow from Eq. (2.43) in the $k_\parallel v_e > \omega$ domain where electron Landau damping is the dominant dissipation mechanism, except for the ion-acoustic mode for which ion Landau damping dominates unless T_e/T_i is large. Here we assume that $c_s \ll v_A \ll v_e$.

Shear Alfvén Waves

$$\omega_k^a = \pm k_\parallel v_A - i\left(\frac{\pi}{8}\right)^{1/2} \frac{k_\parallel^2 v_A^2 k_\perp^2 \rho_s^2 (1+T_i/T_e)}{|k_\parallel| v_e}. \tag{2.46}$$

Ion-Acoustic Waves

$$\omega_k^s = \pm k_\parallel c_s - i\left(\frac{\pi}{8}\right)^{1/2} \frac{k_\parallel^2 c_s^2}{|k_\parallel| v_e}\left[1 + \left(\frac{m_i}{m_e}\right)^{1/2}\left(\frac{T_e}{T_i}\right)^{1/2} \exp\left(-\frac{T_e}{2T_i}\right)\right]. \tag{2.47}$$

Drift Waves

$$\omega_{\mathbf{k}}^{\mathrm{dw}} = \omega_{*e} + i\left(\frac{\pi}{2}\right)^{1/2} \frac{\omega_{*e}^2 k_\perp^2 \rho_s^2 (1+T_i/T_e)}{|k_\parallel| v_e}. \tag{2.48}$$

Ion Drift Modes

$$\omega_{\mathbf{k}}^{id} = \omega_{*i} - i\left(\frac{\pi}{2}\right)^{1/2} \frac{\omega_{*i}^2 (1+T_e/T_i)}{|k_\parallel| v_e}. \tag{2.49}$$

The formulas (2.46)-(2.49) apply where $k_\parallel v_e > \omega$. On the other hand in the electron hydromagnetic limit where $k_\parallel v_e < \omega$ the plasma modes of propagation obtained from Eq. (2.43) are as follows:

Inertial Alfvén Waves

$$\omega_{\mathbf{k}}^a = \pm \frac{k_\parallel v_A}{(1 + k_\perp^2 c^2/\omega_{pe}^2)^{1/2}}. \tag{2.50}$$

Ion Drift Modes

$$\omega_{\mathbf{k}}^{id} = \omega_{*i} \tag{2.51}$$

where both the electron and ion damping are now exponentially small and hence not included in Eqs. (2.50) and (2.51). The damping is small since the phase velocity $\omega/k_\parallel$ is faster than the electron thermal velocity. The ion-acoustic waves do not occur in this regime and are replaced by the lower-hybrid waves and electron-acoustic waves. Abbreviated as IAW, wave in Eq. (2.50) is important in the ionosphere.

2.3 Drift Waves

The drift wave formulas in Eqs. (2.47), (2.48), (2.49), (2.50) and (2.51) are the first of a number of drift wave formulas needed to understand the wave-induced transport in magnetized, inhomogeneous plasma. Both hydrodynamic and kinetic descriptions will be used in the following chapters for the drift waves. The shorter wavelength dispersion terms to the frequency in Eq. (2.51) are derived in Section 2.5 and give a downward shift in the frequency from the ion pressure gradient proportional to $k_\perp^2 \rho_i^2 (1+\eta_i)$.

2.4 Kinetic Alfvén Waves

Alfvén waves are ubiquitous in space and laboratory plasmas. There are two kinds of wave dispersion relations that are important for Alfvén waves. The form for the wave dispersion in the cold plasma arises from the electron inertial skin depth $\delta = c/\omega_{pe}$. In hot plasmas the wave dispersion comes from the ion sound-inertial scale length ρ_s and the ion thermal gyroradius ρ_i for the hot ion plasmas.

2.5 Coupling of the Drift Wave, Ion-Acoustic and Shear Alfvén Waves

A precise description of the coupling of the modes defined in Eqs. (2.46)–(2.48) requires that the full dispersion relation be retained. Here we give the fifth order polynomial in ω that describes the coupling of these modes in the $k_\parallel v_e > \omega$ regime where the electron Landau damping is weak. Rationalizing the fractional expression in Eq. (2.43), we obtain the polynomial dispersion

5th Order Dispersion Reduction

$$D_5(\mathbf{k},\omega) = \sum_{n=0}^{5} a_n \omega^n = 0 \tag{2.52}$$

where polynomial coefficients are

$$\begin{aligned}
a_0 &= -\frac{k_\parallel^2 c_s^2}{\tau}\,\omega_*\Gamma_0(b) \\
a_1 &= k_\parallel^2 c_s^2 \Gamma_0 - \frac{c_s^2}{v_A^2}\,\frac{1-\Gamma_0(b)}{b}\,\Gamma_0(b)\,\frac{\omega_*^2}{\tau^2} \\
a_2 &= -\omega_*\Gamma_0\left(1 - \frac{2c_s^2}{v_A^2\tau}\right) \\
a_3 &= -\frac{\omega_*^2}{\tau k_\parallel^2 v_A^2}\left(\frac{1-\Gamma_0}{b}\right) - 1 - \tau(1-\Gamma_0) - \frac{c^2}{v_A^2}\,\Gamma_0
\end{aligned} \tag{2.53}$$

$$a_4 = \frac{\omega_*(\tau-1)}{\tau k_\parallel^2 v_A^2}\cdot\frac{1-\Gamma_0(b)}{b} \tag{2.54}$$

$$a_5 = \frac{1-\Gamma_0(b)}{k_\parallel^2 v_A^2 b}. \tag{2.55}$$

In Fig. 2.3 we show the disposition of the roots of Eq. (2.52) as a function of the $k_\parallel$ for $k_y\rho_s \ll 1$. We notice that the roots are complex along the dashed curve region where the waves resonate with the thermal ions. The fluid expansions of the dispersion relation are not capable of describing the waves accurately in the regions where $\omega \simeq k_\parallel v_j$ where $v_j = (T_j/m_j)^{1/2}$ is the thermal speed. For the limit of a

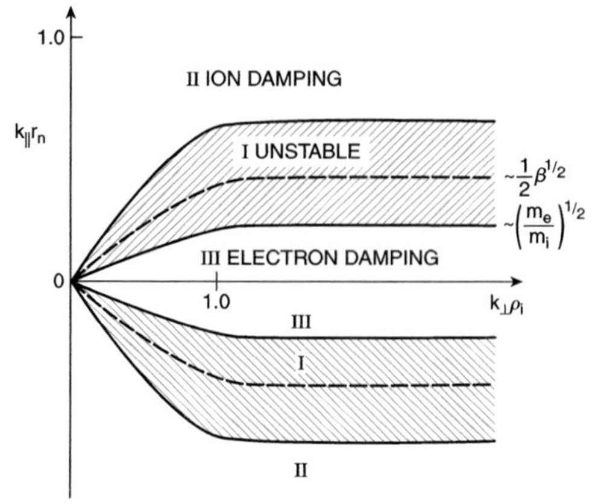

Fig. 2.3 The stable and unstable (shaded) regions of **k**-space for the electron drift wave driven by density gradient specified by the gradient scale length $L_n = r_n$.

uniform plasma $\omega_* = 0$ and the polynomial reduces due to $a_0 = 0$ to a fourth degree with the standard forward and backward ion-acoustic and shear Alfvén waves.

From the analysis presented in this chapter, and particularly from Fig. 2.3, it is evident that for typical laboratory plasmas where $c_s \ll v_A \lesssim v_e$ the decoupling of the ion drift-Alfvén modes given in Eq. (2.44) from the electron drift-acoustic modes given in Eq. (2.45) is a good approximation.

The relative error made in neglecting the Alfvén wave effects on the drift wave-acoustic modes is of order

$$\widehat{\beta} = \frac{\omega_{pi}^2 \omega^2}{(c^2 c_s^2 k_\|^2 k_\perp^2)} \simeq \frac{\omega_*^2}{(k_\|^2 v_A^2 b)} = \frac{\beta}{2k_\|^2 L_n^2} \tag{2.56}$$

where we use $\omega \simeq \omega_*$ and $c_s^2/v_A^2 = \beta/2$. The ratio $\widehat{\beta}$ follows comparing the terms $a_5\omega^5$ and $a_3\omega^3$ in Eq. (2.52). For electrostatic motion $\widehat{\beta}$ is small compared to unity. Neglecting these Alfvén coupling effects altogether reduces the dispersion relation to the electrostatic modes with the polarization $\mathbf{E} = -\nabla\phi = -\Sigma_{\mathbf{k}} i\boldsymbol{k}\phi_{\boldsymbol{k}\omega}$ with $\delta\mathbf{B} = 0$.

2.5.1 *Electrostatic drift waves*

For $\widehat{\beta} = m_i\beta/m_e \ll 1$ the plasma modes are electrostatic (ES) with

$$D_{\text{ES}} = k_y^2\varepsilon_\perp + k_\|^2\varepsilon_\| = 0 \tag{2.57}$$

with $\varepsilon_\|$ and $\varepsilon_\perp$ given by Eqs. (2.39) and (2.40), respectively. In this case the electric field is purely electrostatic with

Electrostatic Polarization

$$E_y = \left(\frac{k_y}{k_\|}\right) E_\|.$$

As is apparent from Eq. (2.48) the drift wave branch is intrinsically unstable. The approximate solution given in Eq. (2.48) is valid for $k_\| c_s < \omega_{*e} < k_\| v_A$. From these bounds on $|\omega_{*e}|$, it follows that there is instability only for $k_\| L_n < k_y \rho_s$ for $k_y \rho_s \leq 1$, and since $\omega_{_k} \leq c_s/L_n$ for larger $k_y \rho$ the requirements that $k_\| c_s < \omega_k$ becomes $k_\| L_n < 1$ at $k_y \rho > 1$.

In simple geometries where the length $L_\|$ of the magnetic field line determines $k_\|$, the stability diagram in Fig. 2.3 gives stable drift waves with $L_\| < 2\pi L_n$ for high $k_y \rho_s \lesssim 1$. Ion drift mode wavelength waves $k_y \rho_s < 1$ require even longer magnetic field lines to be drift wave unstable. The stability condition is approximately $L_\| < 2\pi L_n / k_y \rho_s$.

In Fig. 2.3 we show the stability boundary implied by the electrostatic dispersion relation Eq. (2.57) computed numerically illustrates these conditions on $k_\| r_n$ versus $k_y \rho$ for instability. In Fig. 2.3 the symbol $r_n = L_n$ is used for the density gradient scale length as in the early drift wave literature.

The stability boundary is a function of the parallel electron drift velocity $u = -j_\| / e n_e$. Confined plasmas often contain parallel currents with u/c_s of the order unity or so. The case shown is for $T_e = T_i$ and the plasma is sufficiently collisionless to neglect finite $\lambda_{\rm mfp} = v_e/\nu_e$ effects altogether. We notice that although electron drift $u/c_s \sim 1$ is destabilizing, the effect is rather weak.

In toroidal confinement systems this limiting of the parallel wavelength to smaller than critical values is not a possible solution. In fact in toroidal systems with a nonuniform rotational transform in the magnetic field there always occur small radial regions where the parallel wavelengths $\lambda_\|$ can approach infinity. Thus, we may expect the toroidal system to have many unstable drift waves.

For magnetic fields with a substantial gradient in the rotational transform, a property which is defined as magnetic shear, the regions of long parallel wavelengths is spatially localized. In such systems the locally unstable region is consequently connected to neighboring regions of stable $k_\|(\mathbf{x})$ due to the spatial nonuniformity of the helical magnetic field. The net stability of the modes is determined by the strength of the connection between the stable and unstable regions due to the wave propagation. We now investigate this stability problem of the drift wave in a sheared magnetic field.

2.6 Drift Wave Eigenmodes in a Sheared Magnetic Field

As a simple example of the localizing and stabilizing effect of the magnetic shear on drift waves we consider a local radial region at $r = r_0$ with $x = r - r_0$ forming a slab of plasma in which the magnetic field varies locally according to

$$\mathbf{B}(\mathbf{x}) = B_0 \left(\widehat{\mathbf{e}}_z + \frac{x}{L_s} \widehat{\mathbf{e}}_y \right) \tag{2.58}$$

where L_s is the magnetic shear length. Clearly the magnetic field, which is along the z axis at $x = 0$, rotates by $\pm 45°$ in going a distance $x = \pm L_s$ from the $x = 0$

plane. This magnetic field is created by external currents for $B_0\widehat{\mathbf{e}}_z$ and a constant plasma current $\nabla \times \mathbf{B} = B_0/L_s\,\widehat{\mathbf{e}}_z = (4\pi j_z/c)\widehat{\mathbf{e}}_z$ for $B_y(x)$ in Eq. (2.58). Thus the model applies where dj_z/dx is negligible for the waves of interest. For the modes associated with dj_z/dx see Chapter 15.

Due to the infinite uniform medium property of the system in the y- and z-directions the fields for the plasma modes must vary according to

$$E_\alpha(\mathbf{x},t) = E_{\alpha,k}(x)\exp\left(ik_y y + ik_z z - i\omega t\right) + \text{c.c.} \tag{2.59}$$

with all allowed k_y and k_z values considered. The derivation of the equations for $E_\alpha(x)$ from first principles is lengthy. The results are perhaps apparent from Eq. (2.39) with the observation that b in that equation is now the square of the local perpendicular wavenumber which, due to the x dependence of the magnetic field vector in Eq. (2.39), is $k_\perp^2 = k_y^2 + k_x^2$ and, furthermore, the k_x wavenumber is properly defined from the fields by $\partial E_\alpha/\partial x = ik_x E_\alpha$. Thus, we expect that the first principle calculation should simply yield Eq. (2.39) with

$$b \to \frac{v_i^2}{2\omega_{ci}^2}\left(k_y^2 - \frac{\partial^2}{\partial x^2}\right). \tag{2.60}$$

Making the Ansätz that the relevant modes $E_{\alpha,k}$ have wavelengths in the x-direction that are long compared with the ion gyroradius, the operator $\rho_i^2\partial^2/\partial x^2$ is small compared with $k_y^2\rho_i^2$ and is Taylor series-expanded out of the Bessel functions in Eq. (2.39) to obtain the second-order differential equation governing the field amplitude variation for the nonlocal mode. Performing the small $\rho_i^2\partial^2/\partial x^2$ expansion in Eq. (2.39) yields the wave equation

Drift Wave Eigenmodes

$$\begin{aligned}\rho_s^2\frac{\partial^2 E_y(x)}{\partial x^2} &+ \left\{-k_y^2\rho_s^2 + \left[1 - \frac{\omega(\omega-\omega_{*i})}{k_\parallel^2 v_A^2}\right]\right.\\ &\left.\times\left[\frac{1}{2}\left(1-\frac{\omega_{*e}}{\omega}\right)Z'\left(\frac{\omega}{k_\parallel v_e}\right) - \frac{k_\parallel^2 c_s^2}{\omega^2}\right]\right\}E_y(x) = 0.\end{aligned} \tag{2.61}$$

Here $\rho_s^2 = \tau\rho_i^2$ with $\tau = T_e/T_i$ is the ion inertia scale length squared. In Eq. (2.61) the parallel wavenumber is a function of x due to the rotation of the magnetic field vector with changing position in x. In fact, we have from Eq. (2.58) that $k_\parallel(x) = k_z + k_y(x/L_s)$ so that for given k_y and k_z the different x positions define a set of straight lines on the marginal stability plane in Fig. 2.3 which pass from the unstable region to the stable region as x is varied. From the marginal stability diagram we should expect the unstable mode attempts to localize its amplitude to the small $k_\parallel(x)$ region. To look for such unstable localized modes we shift the origin of the x coordinate to the point where $k_\parallel = 0$ or $x' = x - x_0$ where $x_0 = -k_zL_s/k_y$.

Or equivalently, for the given origin defined by Eq. (2.58), the k_y, k_z modes which are most localized to the unstable region are those modes which have $k_z = 0$.

The physical meaning of this shift $x \to x - x_0$, or the choice of k_y/k_z for a given x_0, is to bring the pitch of the magnetic field twist to equal the pitch of the phase fronts $k_y y + k_z z =$ constant of the wave.

For the modes with $k_\parallel(x) = k_y x/L_s$ we have the wave equation

Sheared Slab Model

$$\rho_s^2 \frac{\partial^2}{\partial x^2} E_y(x) + \left\{ -k_y^2\rho_s^2 + \left[1 - \frac{\omega(\omega-\omega_{*i})L_s^2}{k_y^2 x^2 v_A^2} \right] \right. \\ \left. \times \left[\frac{1}{2}\left(1 - \frac{\omega_{*e}}{\omega}\right) Z'\left(\frac{\omega L_s}{k_y v_e |x|}\right) + \frac{c_s^2 k_y^2 x^2}{\omega^2 L_s^2} \right] \right\} E_y(x) = 0 \tag{2.62}$$

where for sufficiently large x the hydrodynamic approximation for the ions $\omega \gg k_\parallel v_i$ fails and the mode enters a region of strong ion Landau damping which exponentially damps the amplitude of the mode. For large values of x the mode equation Eq. (2.62) reduces to

$$\left(\rho_s^2 \frac{\partial^2}{\partial x^2} + \frac{c_s^2 k_y^2}{\omega^2 L_s^2}\, x^2 \right) E_y(x) \cong 0 \tag{2.63}$$

with the two linearly independent asymptotic solution given by

$$E_y^{\pm}(\mathbf{x}, t) = \frac{C_{\pm}(\omega)}{|x|^{1/2}} \exp\left(i k_y y \pm \frac{i k_y c_s}{2\omega L_s}\, x^2 - i\omega t \right) + \text{c.c.} \tag{2.64}$$

For a wave packet with given $C_{\pm}(\omega)$ the group velocity of the wave in the x-direction is clearly opposite to the phase velocity of the waves. The constant wave fronts follows from Eq. (2.64) are given by

Phase Front Propagation

$$S_{k,\omega}^{\pm}(x, t) = \pm \frac{k_y c_s^2}{2\omega L_s}\, x^2 - \omega t = \text{const.} \tag{2.65}$$

where $S^{\pm}$ is called the eikonal of the wave function.

The stationary phase point of the wave packet follows from the derivative of the phase in Eq. (2.65) with respect to ω giving

Group Velocity Propagation

$$\frac{dS_{k,\omega}^{\pm}}{d\omega} = \mp \frac{k_y c_s^2}{2\omega^2 L_s}\, x^2 - t = 0. \tag{2.66}$$

Equation (2.66) follows by considering $\int d\omega C_{\pm}(\omega)\exp\left(iS^{\pm}_{k,\omega}(x,\omega,t)\right)$ evaluated by the method of stationary phase [Jackson (1999)] in Section 10.12.

2.7 Symmetries of the Drift Wave Eigenmodes

In the derivation of Eq. (2.62) we assume that the plasma flow from $v_E = -E_r/B$ is constant so as to Doppler shift wave frequency and that there is no significant gradient of the drift wave frequencies $\omega_{*i,e}$ over the width of the eigenmodes. In general, this is not the case and these background gradients break the alternating even and odd symmetries of the eigenmodes given by Eq. (2.62). When the background shearing effects are weak we treat them with perturbation theory as in quantum mechanics finding second order shifts of the eigenvalues.

The symmetry breaking due to the background shearing flows has a second more subtle effect that is important at finite plasma pressure. The modes with odd reflection symmetry for the electrostatic potential have an even symmetry of the associated magnetic perturbation of the $\delta B_r(x)$ field arising from the oscillating electron parallel currents in the drift waves. These magnetic perturbations from the odd symmetry potential modes create small magnetic islands that can then grow from nonlinear driving terms and energy associated with the shear in the ambient magnetic field. The calculations for this effect are postponed until Chapter 15 where the instabilities that arise from magnetic reconnection of twisted magnetic fields are analyzed. Simulations showing the transformation at finite plasma pressure $\beta_e \geq 0.005$ over the magnetic pressure of the electrostatic drift waves to the magnetic tearing mode with odd parity are shown in Muraglia, *et al.* (2009, 2011).

2.8 Outgoing Wave Boundary Conditions

In the time interval from t to $t + dt$ the wave phase fronts move from x by $dx = \pm(\omega^2 L_s/xk_y c_s^2)dt$ whereas the wave packet maximum moves from x by $dx = \mp(\omega^2 L_s/xk_y c_s^2)dt$. Consequently, the asymptotic solution that represents a propagation of the wave energy in a wave packet from the unstable region about $k_\parallel(x) \simeq 0$ to the regions of wave absorption at large $|k_\parallel|$ is the negative branch (lower signs) in Eq. (2.64). This negative branch solution has outgoing wave energy and incoming phase fronts [Pearlstein and Berk (1969)].

The explicit form of the wave absorption at large $|k_\parallel|$ is from the ion-wave resonance given by Im $Z'(\omega/k_\parallel v_i)$ in Eq. (2.25). In writing Eq. (2.61) this imaginary part of the ion Z'-function has been dropped. Eigenvalue calculations retaining the function $Z'(\zeta_i)$ where $\zeta_i = \omega/|k_\parallel|v_i$ are given in Gladd and Horton (1973). The component of the wavenumber vector $\mathbf{k}$ parallel to the magnetic field is critical in determining the response of the plasma to the wave.

With these results we are able to determine the physically acceptable boundary conditions (B.C.) on the solutions of Eq. (2.62) for large positive and negative values of x. With these outgoing wave boundary conditions which we may write as

Outgoing Energy B.C.

$$\lim_{|x|\to\infty}\left[\rho_s\frac{dE_y}{dx}+i\,\frac{k_y c_s x}{\omega L_s}\,E_y\right]=0 \tag{2.67}$$

a unique solution of Eq. (2.63) is specified. The unique nontrivial solution only occurs for the eigenvalues of ω. In the approximation where the electron response is $Z'=-2$ and $v_A/c_s\to\infty$ the differential equation reduces to such an extent that we can write down the well-known eigenvalues and eigenfunctions. Retaining the full complex potential in Eq. (2.63) requires lengthy analytical perturbation theory calculation or numerical "shooting method" for the eigenvalue analysis.

For modes sufficiently broad that $\langle |k_\parallel(x)|v_e/\omega\rangle > 1$ we may approximate $Z'(\omega/k_\parallel(x)v_e)\simeq -2$ and reduce Eq. (2.61) to

$$\rho_s^2\,\frac{\partial^2 E_y}{\partial x^2}+\left[-k_y^2\rho_s^2-\left(1-\frac{\omega_{*e}}{\omega}\right)+\frac{k_\parallel^2 c_s^2}{\omega^2}\right]E_y(x)=0. \tag{2.68}$$

The spectrum $n=0,1,2,3,\ldots$ of solutions of the wave equation Eq. (2.68) satisfying the boundary condition given in Eq. (2.67) are

Eigenmodes

$$\varphi_{k_y,n}(x)=H_n\left[\left(\frac{ik_y\rho_s c_s}{\omega_k L_s}\right)^{1/2}x\right]\exp\left(-\frac{ic_s k_y x^2}{2\omega_k\rho_s L_s}\right) \tag{2.69}$$

where the eigenfrequency $\omega_{k,n}$ is shifted from $\omega_k=\omega_{*e}/(1+k_y^2\rho_s^2)$ to the spectrum of waves

Eigenfrequencies

$$\omega_{k,n}=\frac{\omega_{*e}[1-i(L_n/L_s)(2n+1)\mathrm{sg}(k_y)]}{1+k_y^2\rho_s^2}. \tag{2.70}$$

Here $n=0,1,2,\ldots$ is radial mode number and $H_n(\xi)$ are the Hermite polynomials. The physical effect of the shear is manifested by the damping at the rate $|\omega_*|(L_n/L_s)$ produced by the outward propagation of energy from the locally unstable region. The damping in Eq. (2.70) from the propagation of the wave energy from the unstable region to the stable region is the basis of the important principle of magnetic shear stabilization of the drift waves. The boundary condition Eq. (2.67) assumes that all outgoing wave energy is lost. In reality the wave energy is absorbed by the ion Landau damping in the outer regions where $|k_\parallel|v_i > |\omega_k|$. This radial region is in the tails of the eigenmode wave functions far from the mode rational surface where $k_\parallel=0$.

The numerical solutions of the drift wave eigenvalue problem with a magnetic shear are shown in Fig. 2.4. The growth rate decreases with increasing magnetic shear for $k_y\rho_s < 0.5$. In this figure from Gladd and Horton (1973) the symbol $a = \rho_s = c_s/\omega_{ci}$ and $r_n = L_n$ and $T_i = T_i$. For $S = 0.8$ the higher $k_y\rho_s > 0.8$ are unstable.

A recent reconsideration of the eigenvalue problem is given in Landreman, *et al.* (2015). Computer simulations show the universal nature of the density gradient driven drift wave instability. The Landreman, *et al.* (2015) results for the eigenvalue problem use two independent gyrokinetic codes GS2 and GENE. A reduced mass ratio of $m_i/m_e = 25$ is used for the code comparisons.

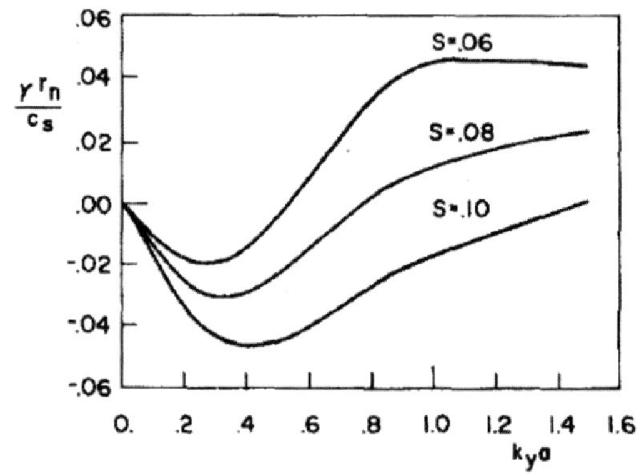

Fig. 2.4 Eigenmode growth rate γ as a function of wavenumber k_y for three values of the magnetic shear $S = L_n/L_s$ parameters. In this figure $a = \rho_s$ and $r_n = L_n$.

We demonstrate that the universal mode driven by the density gradient in a plasma slab can be absolutely unstable even in the presence of reasonable magnetic shear. Previous studies from the 1970s that reached the opposite conclusion used an eigenmode equation limited to $L_x \gg \rho_i$, where L_x is the scale length of the mode in the radial direction, and ρ_i is the ion Larmor radius. The gyrokinetic simulations with GS2 and GENE eliminate this restriction and showed the unstable drift waves. Instability is found for perpendicular wavenumbers k_y in the range $0.7 \lesssim k_y\rho_i \lesssim 100$, and for sufficiently weak magnetic shear, $L_s/L_n \gtrsim 17$, where L_s and L_n are the scale lengths of magnetic shear and density. Thus, the plasma drift wave in a sheared magnetic field is unstable even with no temperature gradients, no trapped particles, and no magnetic curvature as shown in Fig. 2.5 from Landreman, *et al.* (2015).

Let us consider the nature of the eigenfunction produced by the magnetic shear and determine the implications of the drift wave mode structure. In Chapter 10.1.1 we will see that electron dissipation $i\delta_k$ shifts the denominator of Eq. (2.70) to $1 + k_y^2\rho_s^2 - i\delta_k$. The electron dissipation then gives $\gamma = \text{Im}(\omega_{k,n}) > 0$ corresponding

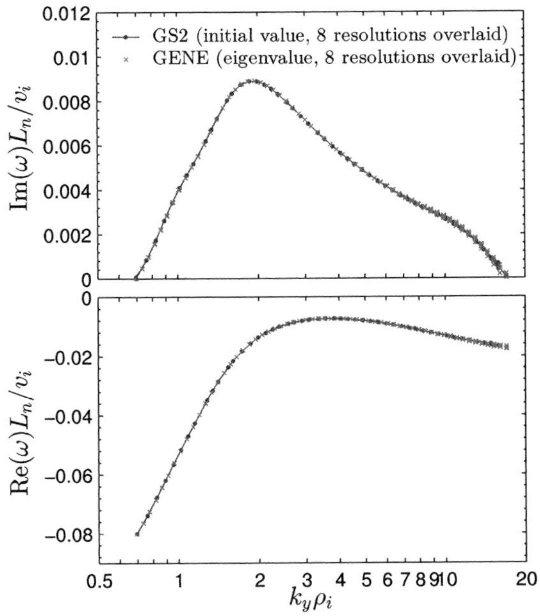

Fig. 2.5 Demonstration that the universal instability is numerically robust. The positive growth rate and real frequency are plotted for the unstable range of k_y using calculations from two independent codes – GS2 run as an initial-value simulation and GENE run as a direct eigenvalue solver – with results that agree to high precision. For each code, results from eight different numerical resolutions are overlaid in the figure. A reduced mass ratio $m_i/m_e = 25$ is used to further minimize the possibility of numerical artifacts [Landreman, *et al.* (2015)].

to exponential growth in time for the wave for the k_y, n eigenmodes with

Unstable Growth

$$\delta_k > \frac{r_n}{L_s}(2n+1)\text{sg}(k_y) \tag{2.71}$$

where $n = 0, 1, 2, \ldots$ are the radial eigenvalue mode numbers in Eq. (2.69).

2.8.1 *Localized ion drift modes*

From the mode equation given in Eq. (2.61) we see that there also occurs the possibility of narrower modes localized to regions such that $\langle |k_\parallel(x)| v_e \rangle \ll \omega$. Such hydrodynamic modes would encompass the FLR ion drift modes given by the dispersion relation

$$\left(k_\perp^2 \rho_i^2\right) \omega(\omega - \omega_{*i}) + k_y^2 \sum_j v_{d_j} v_{D_j} = 0 \tag{2.72}$$

where $v_{D_j} = cT_j/eBR_c$ is the particle guiding center drift velocity induced by curvature of the magnetic field line. The strength of the magnetic curvature is measured by the radius of curvature R_c. These modes given by Eq. (2.72) are extremely important in system with weak or no magnetic shear such as mirror machines and field reversed confinement (FRC) machines. The modes are called "flute modes."

Often the eigenvalue analysis is clearer in dimensionless variables. Thus we introduce the dimensionless variables $\xi = x/\rho_s, \Omega = \omega/\omega_{*e}, K = k_y\rho_s, S = L_n/L_s$, and $\beta = 8\pi p/B^2$.

The ion mode equation in such regions is

$$\frac{\partial^2\varphi}{\partial\xi^2} + \left[-K^2 + \frac{S^2\xi^2}{\Omega^2} - \frac{1}{2}\beta + \left(1 - \frac{1}{\Omega}\right)\left(\frac{m_i}{m_e}\right)\left(\frac{S^2\xi^2}{\Omega^2} - \frac{1}{2}\beta\right)\right]\varphi = 0 \quad (2.73)$$

where $E_y(x) = -ik_y\varphi(x)$, $\xi = x/\rho$, $\Omega = \omega/\omega_{*e}$ and $S = L_n/L_s$. The eigenvalue Ω is a function of four parameters: $K = k_y\rho_s$, $S = r_n/L_s$, m_i/m_e, $\beta = 8\pi p/B_0^2$.

The ground state eigenmode of Eq. (2.73) is

$$\varphi = \exp\left(-\frac{\sigma\xi^2}{2}\right) \quad (2.74)$$

with

$$\sigma = \left(\frac{m_i}{m_e}\right)^{1/2}\frac{|S|}{\Omega}\left(\frac{1}{\Omega} - 1\right)^{1/2}. \quad (2.75)$$

The eigenvalue Ω for the ground state is given by

$$\begin{aligned} &-\frac{S}{\Omega}\left(\frac{1}{\Omega} - 1\right)^{1/2}\left(\frac{m_i}{m_e}\right)^{1/2} - K^2 - \frac{1}{2}\beta \\ &+\frac{1}{2}\left(\frac{m_i\beta}{m_e}\right)\left(\frac{1}{\Omega} - 1\right) = 0. \end{aligned} \quad (2.76)$$

For $0 < \Omega_k < 1$ there is one mode with real Ω: outside this frequency range there are growing and decaying solutions. All roots of Eq. (2.76) must satisfy the localization criterion $|k_\parallel' v_e \Delta x/\omega| < 1$ to satisfy the validity condition for reducing Eq. (2.62) to Eq. (2.72). Typically, only the root close to $\Omega = 1/[1 + K^2 2m_e/m_i\beta]$, which in dimensional form, is

$$\omega_k = \frac{\omega_{*e}}{(1 + k_\perp^2 c^2/\omega_{pe}^2)} \quad (2.77)$$

satisfies the localization requirement. The dispersion of the waves from the electron inertial scale length or skin depth $\delta_s = c/\omega_{pe}$ is important for small-scale drift-wave turbulence. We will see the shielding effect of this electromagnetic wave field in the case of tearing modes and electron temperature gradient driven modes in Chapters 14 and 15. The dispersion relation applies to auroral drift waves.

2.9 Ion-Acoustic Drift Wave Turbulence

Many space and laboratory plasmas have spectra of ion-acoustic waves. When a small density gradient is present across an ambient magnetic field the forward and backward ion-acoustic waves become split in to a drift wave with a phase velocity faster than the ion-acoustic speed and another slower wave. For the density gradient small or vanishing, the solutions of electrostatic dispersion given ion oscillations with adiabatic shielding $\delta n_e = n_0(e\varphi/T_e)$ behavior of the electrons. This ion oscillation is defined as the ion-acoustic waves and the ion fluid is compressed and rarefracted during the oscillations. The electric field $\mathbf{E} = -\nabla\varphi$, however, dominates the ion pressure gradient force giving $\omega^2/k^2 = c_s^2 = T_e/m_i$. It is useful to study the measurements of the ion-acoustic wave spectra as an introduction to measuring the spectrum of the drift waves.

Renormalized kinetic theory for the long range collective oscillations of turbulent plasmas is developed through infinite order summations of the interactions in the Vlasov-Poisson self-consistent field equations. Three levels of renormalized turbulence theory are known and in the lowest level the (simply renormalized) equations are readily solved for the problem of ion-acoustic instability. The physical predictions of renormalized turbulence theory were used to interpret several experiments on ion-acoustic turbulence [Horton and Choi (1979)] and drift waves [Horton and Rowan (1994)].

2.9.1 *Electromagnetic scattering measurements of plasma waves*

The spectral distribution of ion-acoustic fluctuations and drift wave fluctuations are measured by electromagnetic wave scattering from the electron density fluctuations $\delta n_e(\boldsymbol{x}, t)$ in the plasma waves. In laboratory plasmas the predicted wavelengths are sufficiently short that the method of scattering laser light and infrared electromagnetic microwaves lend themselves to the detection of the electron density fluctuations in the plasma. Strong scattering maximum occur at the Bragg angle where the $\Delta\mathbf{k}$ of the scattered electromagnetic wave equals the wavenumber vector of the fluctuation in the electron density $\delta n_e(\Delta\mathbf{k})$. From the scattering experiments an understanding of the wave vector distribution of the density fluctuations in ion-acoustic and drift wave turbulence is obtained. The measured spectral distributions agree with the theoretical distributions.

Non-perturbative measurements of the plasma fluctuations are carried out with Bragg scattering, $\mathbf{k} = \mathbf{k}_{\text{out}} - \mathbf{k}_{\text{in}}$, where an incoming RF or laser wave, $\mathbf{E}_{\text{in}}\exp(i\mathbf{k}_{\text{in}} \cdot \mathbf{x} - i\omega_{\text{in}}t)$, sets the electron into quiver motion with $\delta\mathbf{v}_e = -e\mathbf{E}_{\text{in}}/im_e\omega_{\text{in}}$ that then radiates the out-going wave $\mathbf{E}_{\text{out}}\exp(i\mathbf{k}_{\text{out}} \cdot \mathbf{x} - i\omega_{\text{out}}t)$, with current $\delta\mathbf{j} = -e\delta n_{\mathbf{k}}\delta\mathbf{v}_e(\mathbf{k}_{\text{in}})$. The scattered wave, $\mathbf{E}_{\text{out}}\exp(i\mathbf{k}_{\text{out}} \cdot \mathbf{x} - i\omega_0 t)$ has $\mathbf{k}_{\text{out}} = \mathbf{k} + \mathbf{k}_{\text{in}}$ and $\omega_{\text{out}} = \omega_{\mathbf{k}} + \omega_{\text{in}}$. Since $\omega_{\mathbf{k}} \ll k_{\text{in}}c$, the frequency shift is often too small to be resolved. The scattering cross-section is

$$\frac{d\sigma}{d\Omega} = \left(\frac{e^2}{m_e c^2}\right)^2 \left\langle |\delta n_e(\mathbf{k})|^2 \right\rangle \frac{\Omega_{\text{scat}}}{n_e}$$

where Ω_{scat} is the scattering volume defined by the intersection of the two electromagnetic waves. The cross-section gives a non-perturbative measurement of $\langle|\delta n_e(\mathbf{k})^2\rangle$ fluctuations. Both X- and O-mode polarizations of $\mathbf{E}_{\text{in}}$ are useful. Here the dimensions of the cross-section are given by the square of the classical radius of the electron $r_0 = e^2/m_e c^2 = 2.8 \times 10^{-13}$ cm.

In the scattering experiments, the enhancement of the scattered power due to correlations in the electron density is measured by the dimensionless form factor $S(\mathbf{k})$ where the wavenumber $\mathbf{k}$ of the fluctuation is determined by the scattering geometry. For scattering geometries where there are a large number of wavelengths in the scattering volume V_s, knowledge of the form factor $S(\mathbf{k})$ is equivalent to determining the spectral distribution of $N(\mathbf{k})$ of the electron density correlation function.

For a specified type of small amplitude fluctuations, theory is used to relate the measured density correlation function $N(\mathbf{k})$ to the potential fluctuation spectrum $I(\mathbf{k})$ used in nonlinear theory. Here we review the formulas used in the analysis of the scattering data. The phase of the scattered RF or laser wave $\mathbf{E}_{\text{out}}$ relative to that of the incident wave $\mathbf{E}_{\text{in}}$ is used in the phase contrast imaging (PCI) method described in Chapter 12.2 to yield further information about the turbulence.

The differential cross-section for scattering the incident electromagnetic wave $\widehat{\varepsilon}_0 E(x)\exp(i\mathbf{k}_0 \cdot \mathbf{x} - i\omega_0 t)$ from the scattering volume V_s into the direction $\widehat{\mathbf{n}}$ of observation is

$$\frac{d\sigma}{d\boldsymbol{\Omega}} = r_0^2 F(\mathbf{q})\,|\widehat{\varepsilon}_0 \times \widehat{\mathbf{n}}|^2 \tag{2.78}$$

where $\mathbf{q} = (\omega/c)\widehat{\mathbf{n}} - \mathbf{k}_0$, $r_0 = e^2/m_e c^2$ and $F(\mathbf{q})$ is the scattering form factor defined by

$$F(\mathbf{q}) = \left\langle \left| \int_{V_s} d\mathbf{x}\, n_e(\mathbf{x}) e^{-i\mathbf{q}\cdot\mathbf{x}} \right|^2 \right\rangle. \tag{2.79}$$

For randomly distributed electrons the form factor $F(\mathbf{q})$ clearly reduces to $F = N_s = n_e V_s$, the number of electrons in the scattering volume. The dimensionless scattering form factor $S(\mathbf{q})$, also called the static form factor, is defined relative to this incoherent scattering function of F, that is

$$S(\mathbf{q}) = \frac{F(\mathbf{q})}{n_e} V_s. \tag{2.80}$$

For fluctuations with $q^3 V_s \gg 1$ it is evident from Eq. (2.79) that $S(\mathbf{q})$ is related to the spectral distribution of the electron density fluctuations $N(\mathbf{k})$ spectral density defined by

$$\langle n_e(\mathbf{x}') n_e(\mathbf{x})\rangle = \int_{V_s} d\mathbf{k}\; N(\mathbf{k}) e^{i\mathbf{k}\cdot(\mathbf{x}'-\mathbf{x})}. \tag{2.81}$$

Upon using the relations for the Fourier transforms in the limit $k^2 k_\| V_s \to \infty$ we obtain

$$S(\mathbf{k}) = (2\pi)^3 \frac{N(\mathbf{k})}{n_e} \tag{2.82}$$

from Eqs. (2.79), (2.80), and (2.81). When the electron density fluctuations are due to ion-acoustic or drift wave fluctuations we have the simple relation

$$\frac{N(\mathbf{k})}{n_e^2} = \frac{e^2 I(\mathbf{k})}{T_e^2} \tag{2.83}$$

between the spectral distribution of the electron density fluctuations and the electric potential fluctuations.

The total cross-section for high-frequency electromagnetic wave scattering from a single electron is the Thomson cross-section $\sigma_T = (8\pi/3)r_0^2 \simeq 2/3\times10^{-24}\,\mathrm{cm}^2$ and is valid for photon energies less than the rest mass of the electron.

2.9.2 *Two-point correlation functions in turbulence*

Plasma turbulence is characterized by the two-point correlation functions and their spectral distributions defined in the previous Sec. 2.9.1 on wave scattering.

As an example we review the stationary uniform fully ionized krypton plasma in which ion-acoustic turbulence was driven by a constant return current [Stenzel (1978)]. Two-point probes allowed detailed measurements of the characteristics of the turbulence were used to derive the correlation functions between diagnostic probes. The correlation data verified that the frequency and wavelength relations of the maximum of $I(\mathbf{k},\omega)$ for the measured signals satisfy the real part of the linear dispersion relation for ion-acoustic modes in the presence of the turbulence. The waves were shown to have a finite lifetime due to the presence of the turbulence as predicted by theory through the line width ν_k calculation. In space the waves first grow exponentially and then decay due to the finite lifetime $1/\nu_k$ determined by the turbulence.

A typical current-driven ion-acoustic wave frequency spectrum measured by Stenzel for which this comparison was made is shown in Fig. 2.6. The Fourier

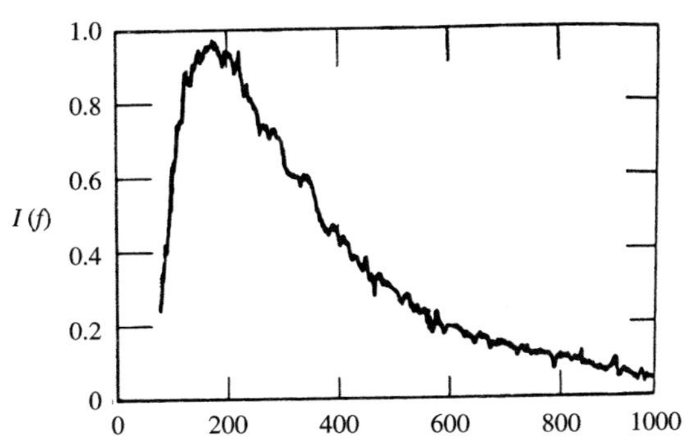

Fig. 2.6 Current driven ion-acoustic turbulence measured in a large volume collisionless Krypton plasma experiment by Stenzel (1978). The turbulent spectrum is produced by a uniform plasma return current with $u \approx 40c_s \approx 0.08v_e$ in a steady state where $T_e/T_i \approx 20(m_{Kr} = 84m_H)$.

transform of the two-space correlation function shows a broad **k** space spectrum peaked at $k\lambda_{De} \approx 0.2$ and with an angular width that approaches 90°. The experiments reported that the observed frequency spectrum is consistent with that obtained from the wavenumber spectrum derived in Horton, *et al.* (1977). The level of the turbulent energy density $W/nT_e \simeq 10^{-2}$ compared to the electron thermal energy density.

The experiment shows that the process of induced nonlinear wave scattering is a relatively strong process. The validation involves the observed wave amplitudes when the beat wave phase velocity is appropriately resonant with either the thermal ions or electrons. The wave scattering is verified by test wave propagation in the experiments. The waves produce an effective or turbulent resistivity $\eta_{\text{eff}} = m_e\nu_{\text{eff}}/n_e e^2$ by angle scattering of the thermal electrons.

2.9.3 *Drift wave spectral distributions*

Since drift waves are closely related in their properties to the ion-acoustic waves, further understanding of the predictions of renormalized turbulence theory is obtained from studying the experiments with stationary spectra of drift wave fluctuations. Drift waves are the fast wave branch of the ion-acoustic waves in a nonuniform magnetized plasma [Eqs. (2.45)–(2.48)] as described further in Chapter 5.

For drift modes, a relatively simple formula [Horton (1976)] for the spectral distribution $I(k_\perp)$ of the turbulence at a given radial position is derived from renormalized plasma turbulence theory applied to drift wave normal modes. In the derivation of the spectrum the linear electron growth rate $\gamma^\ell(k_\perp)$ of the drift waves is balanced by the induced wave scattering arising through the $\mathbf{E}\times\mathbf{B}$ nonlinearities in the ion dynamics. The nonlinear ion response is obtained from renormalized turbulence theory. The ion response is through the renormalized particle propagator is $[\omega - \omega' - (k_\parallel - k'_\parallel)v_\parallel + i\nu_{k-k'}]^{-1}$ where ν_k is determined through the Feynman diagrams in Choi and Horton for the nonlinear $\delta n\,\mathbf{E}\times\mathbf{B}$ convective flux. For this problem one finds that even at the low-fractional turbulence levels observed in the experiments and predicted by the theoretical formula for $I(k_\perp)$ we have $\langle\omega_\mathbf{k}\rangle \gtrsim \langle\nu_k\rangle \gg \langle k_\parallel\rangle v_i$ so that the wave scattering resonance is dominated by the turbulent $\mathbf{E}\times\mathbf{B}$ motions $\nu_k = \langle(\mathbf{k}\cdot\mathbf{v}_E)^2\rangle^{1/2}$. The turbulence balance occurs between the rate of nonlinear transfer of energy from the mode $k_\perp, \omega_{k\perp}$ and the rate of energy transfer into the mode through the electron growth rate $\gamma^\ell(k_\perp)$.

In the ATC tokamak experiment at Princeton the drift wave spectrum was measured with scattering of a 90 GHz ($\lambda_0 = 1/3\,\text{cm}$) microwave beam from the plasma. The fastest-growing linear modes were predicted by theory to have wavelengths across the magnetic field of approximately one centimeter. The experiments of Mazzucato (1976, 1978) in the ATC experiment are shown in Fig. 2.7 with scattered microwave intensities. Six values of $k_\perp$ are measured along with the frequency spectrum of the scattered RF wave signal at each $k_\perp$ determined by angle between the microwave receiver and the antenna. The measured frequency spectrum was

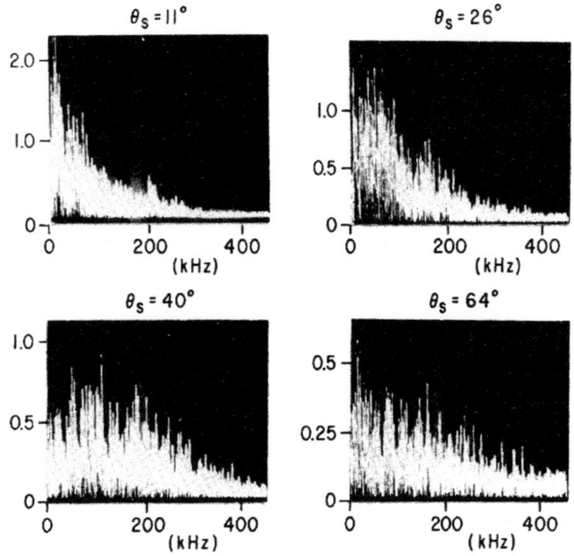

Fig. 2.7 Spectra of microwaves scattered by density fluctuations with wavelengths of $\lambda(11°) = 2.3\,\text{cm}$, $\lambda(26°) = 1.2\,\text{cm}$, $\lambda(40°) = 0.6\,\text{cm}$, $\lambda(64°) = 0.4\,\text{cm}$. θ_s is the scattering angle; the ordinate is in arbitrary units and is proportional to the electric field of the scattered microwaves [Goldston, *et al.* (1977)].

broadly peaked at the $\omega_{k\perp}$ corresponding to drift wave theory, and the width of the spectrum is consistent with the ν_k formula of renormalized theory. Changing the direction of the magnetic field produces the predicted shift of the frequency by $\omega_{k_\perp} \to -\omega_{k_\perp}$ and $\nu_{k_\perp} \to +\nu_{k_\perp}$ and provides a null measurement of $2\omega_{k_\perp}$. The frequency spectrum and the determination of $2\omega_{k_\perp}$ are shown in Goldston, *et al.* (1977).

In these rather collisionless plasmas the drift wave normal modes are predicted by theory to be partially localized to the outside region of the tokamak where the trapped electron density is maximum. In Fig. 2.7 from Goldston, *et al.* (1977) the increase in the scattered signal observed as the scattering volume is moved from the inside to the outside position of the tokamak is shown. The relative increase in the scattered power predicted from theory was observed in the experiment.

The theoretical spectral distribution was calculated in detail for the radial plasma profiles measured in the experiment. With these parameters and profiles the dispersion relation and the nonlinear terms used in the energy transfer rate were evaluated to obtain $I(\mathbf{k}_\perp)$ in the scattering volume. The dimensional scale for the wavenumber is determined by the ion inertia through $\rho_s = c(m_i T_e)^{1/2}/eB$, which is approximately one millimeter in this experiment.

Further reports on microwave scattering from drift waves in tokamaks was given at the Sixth Conference on Plasma Physics and Nuclear Fusion in Berchtesgaden, along with additional theoretical analysis was performed to predict the

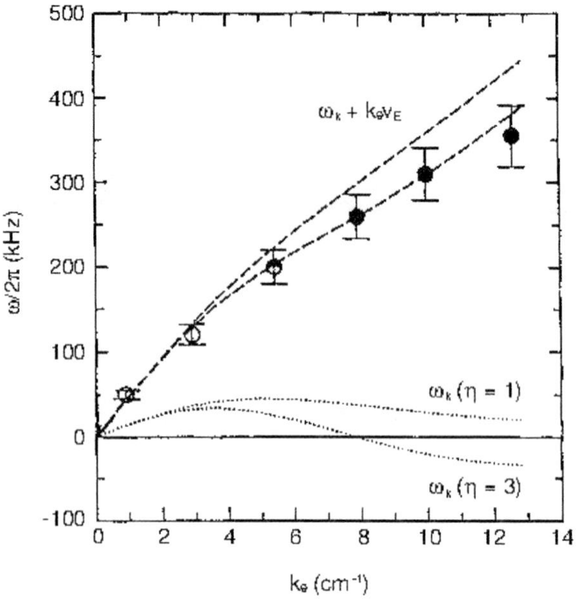

Fig. 2.8 Frequency spectrum from far-infrared laser scattering from the electron density fluctuations associated with drift waves in the TEXT tokamak [Bravenec, *et al.* (1992)]. The dotted line is the frequency in the plasma rest frame and the dashed line is the frequency in the laboratory frame. The ion temperature gradient is specified by η as described in Chapter 3.3.

spectral distribution of the fluctuations in the tokamak at the C.E.A. Fontenay-aux-Roses laboratory called TFR. The theoretical prediction published by Horton (1977) explains the spectrum obtained by microwave scattering measured by the TFR Equipe. The theoretical prediction is shown in Fig. 1 of Horton, *et al.* (1977) and was predicted to be $\widetilde{n}_e/\overline{n}_e \approx 10^{-2}$ with the associated turbulent energy density is $W/nT_e \approx 10^{-4}$. This level was somewhat higher than the level of $\widetilde{n}_e/\overline{n}_e =$(3-5)$\times 10^{-3}$ reported by the TFR Equipe. The fixed scattering volume, however, was located in azimuth at 105° around from the outside midplane of the torus where the mode theoretically has its maximum amplitude.

In the TEXT tokamak Brower, *et al.* (1987) used far-infrared laser scattering experiments with the six wavenumbers $k_\perp = 1, 2, 4.5, 7, 9$, and $12\,\mathrm{cm}^{-1}$ to measure the turbulence producing the electron thermal diffusivity. TEXT was an ohmic tokamak with $R = 1\,\mathrm{m}$, $a = .27\,\mathrm{m}$, $B = 2.8\,\mathrm{T}$ for $I_p = 150\,\mathrm{kA}$ to $350\,\mathrm{kA}$. The density increased from $n = 2 \times 10^{13}\,\mathrm{cm}^{-3}$ to $9 \times 10^{13}\,\mathrm{cm}^{-3}$ with the reference discharge at $I_p = 300\,\mathrm{kA}$ and $\overline{n}_e = 3.5 \times 10^{13}\,\mathrm{cm}^{-3}$. The discharges had sawtooth oscillations except at the lowest currents. The reference discharge had $T_{e0} = 10^3\,\mathrm{eV}$, $T_{i0} = 550\,\mathrm{eV}$, $Z_{\mathrm{eff}} = 2$ and $\tau_E = 12\,\mathrm{ms}$.

Bravenec, *et al.* (1992) used the measured density fluctuation level $\widetilde{n}(r)$ and the centroid of the frequency spectrum $\omega(k)$ are shown in Fig. 2.8. The fluctuation level $\widetilde{n}$ was found to linearly increase with mean density with $\widetilde{n} \simeq 0.02\overline{n}$.

The radial electric field was accurately known in the TEXT tokamak from the heavy ion beam probe (HIBP) diagnostic [Hallock, *et al.* (1994)]. The HIBP probe diagnostic uses the abrupt change in the heavy neutral particle orbit at the point of the charge exchange collision with the plasma ion to determine electrostatic potential of the plasma $\Phi(\boldsymbol{r}, z)$ at this time and the position of the charge exchange collision with the ambient working gas ion. This HIBP diagnostic gave profiles of the $E_r(r,t)$ electric field from which the Doppler shift $k_\theta v_{E_r}$ was determined as shown in Fig. 2.8 from Bravenec, *et al.* (1992). From the inward-pointing radial electric field the plasma $\mathbf{E}_r \times \mathbf{B}$ rotates the plasma in the electron diamagnetic direction increasing the wave frequency measured in the laboratory frame for drift waves propagating in the electron diamagnetic direction. For the ion temperature gradient driven drift waves rotating in the ion diamagnetic direction in the plasma rest frame, the Doppler shift was large enough to change the wave direction to the electron diamagnetic direction in the laboratory frame of the TEXT tokamak. Figure 2.8 shows the Doppler shifted drift wave frequency for the discharge taking into account the ion temperature gradient parameter $\eta_i = 1$ and 3 for the far-infrared laser scattering experiment of Brower, *et al.* (1987). Including the E_r-Doppler shift is clearly important for identifying the drift waves in the presence of the ion temperature gradients. New experiments have demonstrated laser wave scattering in the large helical device (LHD) by Tanaka, *et al.* (2012) with drift waves with different characteristics rotating both in the ion and electron diamagnetic directions described further in Chapter 12.

2.10 RF Scattering from ETG Turbulence

Small electron gyroradius ($k_y\rho_e \sim 1$) and electron inertial scale ($k \sim \omega_{pe}/c$) turbulence was measured using scattering with ultra-high frequency (UHR) RF beams of diameter less than one centimeter at frequencies that resonate with the upper hybrid plasma resonance frequency. In the FT-2 tokamak [Gurchenko, *et al.* (2007)] the upper-hybrid frequency $f_{\rm UH}^2 = f_{ce}^2(R) + f_{pe}^2$ corresponding to $f_{\rm UH} \cong 52 - 69\,\rm GHz$.

The backscattering diagnostics have the antenna on the inside wall launching the X-mode beam toward two receiving horns – one for X-mode $\mathbf{E}_{\rm RF} \perp \mathbf{B}_\phi$) with the electric field perpendicular to the toroidal magnetic field and one for O-mode polarization with $\mathbf{E}_{\rm RF}$ parallel to the magnetic field. The plasma in the FT-2 Frascati Torus with $R/a = 0.55\,\rm m/0.08\,m$ for $B/I_p = 2.2\,\rm T/32\,kA$ at $n_e = 3 \times 10^{19}\rm m^{-3}$ has the electron temperature $T_e(r,t)$ measured with Thomson scattering. Off-axis ETG drift wave turbulence has $\omega_{*e} \sim 2 - 3\,\rm MHz$ and $k_r \sim 150\,\rm cm^{-1}$ giving a backscattered Bragg radial wave vector of $q_r = -2k_r$. The presence and strength of the ETG turbulence was shown to vary directly with the ECRH heating power that controls the electron temperature gradient R/L_{T_e} parameter that drives the ETG turbulence.

In DIII-D the ETG turbulence was measured with high-k backscattered using X-mode polarized waves at 94 GHz for wavenumbers in the range $k = 35 - 40\,\mathrm{cm}^{-1}$ corresponding to ETG turbulence with $k\rho_i \sim$ 4-10. There were two UHR beams viewing the millimeter-wave backscattered probe beam providing high spatial resolution of the turbulence.

The DIII-D plasmas were lower single-null plasmas with $B/I_p = 1.9\,\mathrm{T}/700\,\mathrm{kA}$ at $n_e = 3 \times 10^{19}\mathrm{m}^{-3}$. Electron cyclotron heating at 110 GHz was injected every 300 ms and absorbed near $r/a = 0.5$ where $T_e \sim T_i \sim 500\,\mathrm{ev}$ and $a/L_n \sim a/L_{T_e}$ with $q \sim 2$ and $s(r) = d\ln q/d\ln r \sim 0.8$. The electron temperature profile showed that sharp radial gradients formed during the periods of the ECH heating. The measured strong increases in the electron temperature gradient and the electron thermal diffusivity χ_e are synchronized in time with the ECH pulses during which there was no significant change in the plasma density nor the ion temperature. The low and intermediate-k turbulence measured with FIR diagnostic decreased during the pulses while the high-k turbulence increased both in frequency and amplitude during the ECH pulses. A stability analysis of the experimental profiles shows that the electron temperature gradient R/L_{T_e} exceeded the ETG critical gradient by a wide margin in the region just beyond the ECH resonance absorption layer.

2.11 Internal Ion Transport Barriers from Reversed Rotational Transforms

Chapters 17 and 18 develop theory for the formation of radial regions of suppressed turbulent transport around local minimum of the profile of the safety factor $q(r)$ which is the reciprocal of the rotational transform $\iota(r) = 1/q(r) = RB_\theta/rB_\phi$ in the helical toroidal magnetic field. At a minimum or maximum of $q(r)$ there is a large gap between the mode rational surfaces where $q(r) = m/\ell$. Away from the minimum of $q(r)$ there is a high density of mode rational surfaces so that the drift waves centered at each resonant surface defined by $q(r_{m/l}) = m/\ell$ overlap and interact with the neighboring modes as in a periodic lattice. This interaction of overlapping resonant modes makes the convective or turbulent transport continuous over broad radial regions. Where the magnetic rotational profile has a maximum or minimum, however, the modes do not overlap and there is region of suppressed radial transport. In the fusion literature this has become known as a "an internal transport barrier" or ITB. The region is not a true transport barrier but a region of strongly-reduced transport since the $\mathbf{E} \times \mathbf{B}$ convection vanishes across the layer, as shown in detail in Beklemishev and Horton (1992) and demonstrated numerically in Horton, *et al.* (1998).

The ITB phenomenon was discovered experimentally in JT-60U as described in Ninomiya, *et al.* (1992) and is very pronounced in the radial profiles of both the ion and electron temperatures measured in JT-60U shot E27969 and subsequent discharges with reversed $q(r)$ profiles. A similar effect, but not as clearly identified, was observed in the TFTR machine in Grove and Meade (1985). Subsequent to the

JT-60U experiment, internal transport barriers (ITBs) were discovered in a wide variety of toroidal confinement devices.

The reversed shear experiments carried out in JET with DD and DT plasmas suggest that this regime of weakly reversed magnetic shear is one method for obtaining high ion core temperature plasmas with enhanced fusion neutron power production. The modeling of plasmas with ITBs is described in detail in Chapter 18 for JET, along with projections for ITER [Keilhacker, *et al.* (1999)].

For transient tokamak discharges with evolving plasma currents one may design time histories of the ITB from an evolving rotational transform profile to produce record values of the fusion $Q_{\rm DT}$ parameter. Antennas designed with 48 $\rm TE_{1,0}$ wave guides powered with 16 klystrons launching 5 MW of carefully phased 3.7 GHz waves drive the lower hybrid waves into plasma. The waves or plasmons transfer their moment to the electrons maintaining the toroidal plasma current as described by Ekedahl, *et al.* (2009). The RF lower hybrid waves is the proven method to maintain steady-state tokamaks toroidal current profiles. These klystron driven lower-hybrid antennas give the possibility of maintaining steady states with controlled rotational transform profiles. Proper control of the toroidal current profile significantly reduces the turbulent radial transport by producing targeted $q(r)$-profiles. For example, regimes with relatively flat rotational transform profiles were used earlier in stellarators and in the first ECRH-driven plasmas in Wendelstein 7-X to reduce the radial transport are described in Chapter 12.

2.12 Drift Waves in LHD and the Wendelstein 7-X Stellarator

Stellarators give a complementary perspective on the nature of drift wave turbulent transport and plasma confinement. The parametric variation of the plasma energy and particle confinement times τ_E and τ_p are sometimes difficult to establish in the magnetic confinement experiments. One example [Meyer and Mahn (1981)] in a stellarator confinement system where the inverse correlation particle and energy τ_p and $\tau_E \propto 1/\langle \delta n^2/n^2 \rangle$ with the drift wave fluctuations were verified using far-infrared laser scattering from the electron density fluctuations in a stellarator equipped with both ohmic heating and neutral beam heating. A discharge where the initial ohmic heating was gradually reduced over 100 ms and the neutral beam heating was ramped up while the scattering data was continuously measured showed a clear inverse correlation of the plasma confinement energy W_p with the turbulent fluctuation level at constant total heating power.

The experiments were performed in the Wendelstein VII-A Stellarator at the Max-Planck Institute in Garching. The VII-A stellarator had $B = 3T$, $R = 2\,\rm m$, $a = 0.1\,\rm m$ and negligible plasma current except when ohmic heating was applied. There were three neutral H_2 beam injectors operating at 27 kV giving $P_N = 1\,\rm MW$.

For both D and He working gases the confinement time goes up with increasing density from the low-density limit and the fluctuation $\langle \delta n^2/n^2 \rangle$ level goes down. The energy confinement time showed the variation $\tau_E \propto 1/\langle \delta n^2/n^2 \rangle$ with the measured fluctuation levels.

The time history of the plasma stored energy $W_p(t)$ and the fluctuation energy density $W(t)$ showed this inverse relation when the ohmic power was ramped down and the compensating neutral beam power was ramped up over periods of 100 ms. Comparison of the fluctuation levels with stored plasma energy using two versus three neutral beams showed the correlations of the increasing fluctuation level with the fraction of beam heating to the Ohmic heating for a fixed total heating power.

2.13 Edge-Localized Modes (ELMs) Imaged with ECE Diagnostics

Toroidal plasmas form coherent high-pressure filaments called edge-localized modes (ELMs) from the steep density and temperature gradients at their edge or just inside last closed magnetic flux surfaces.

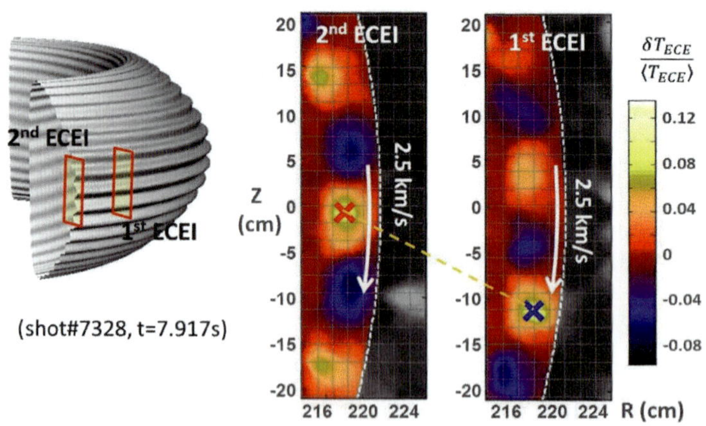

Fig. 2.9 Example of 3D images of an edge-localized mode in KSTAR. The color scale is the normalized fluctuation amplitude of the emission temperature (TECE). The measured pitch (~6°) agrees with that of the magnetic field line estimated by equilibrium reconstruction [Yun, *et al.* (2014)].

The diagnostic system imaging electron cyclotron emission (ECEI) system gives images of these structures as shown in KSTAR in Fig. 2.9 [Yun, *et al.* (2014)].

This diagnostic has 24 diodes measuring both the X mode and O mode electron cyclotron plasma waves at frequencies above the electron plasma frequency.

This gives clear images of the sawteeth, tearing modes, Alfvén waves and the edge localized mode (ELM). Figure 2.9 shows the ELMS which are propagating edge-localized modes that appear as narrow higher temperature filaments emitted from the interior hot plasma [Lee, *et al.* (2014)].

Toroidal plasmas form coherent high-pressure filaments called edge-localized modes (ELMs) from the steep density and temperature gradients at their edge or just inside last closed magnetic flux surfaces. The diagnostic electron cyclotron emission imaging (ECEI) system shows the formation and outward propagation of the edge-localized modes that appear as narrow higher temperature filaments emitted from the interior hot plasma. Lee, *et al.* (2014) developed and reported the ability of a two-independent ECEI diagnostic system to measure and track the emission of the hot plasma filaments from the outer edge plasma in the KSTAR tokamak.

Chapter 3

Mechanisms for Drift Waves

Drift waves occur universally in magnetized plasmas producing the dominant mechanism for transport of particles, energy and momentum across magnetic field lines. A wealth of information is obtained from laboratory experiments for plasma confinement for drift waves driven unstable by density gradients, temperature gradients and trapped particle effects. The modern understanding of the origin of the scaling laws for Bohm and gyro-Bohm transport fluxes measured in toroidal plasmas arises from drift wave turbulence. The role of sheared flows and magnetic shear in controlling the transport fluxes is discussed and illustrated with large scale computer simulations. Plasmas turbulence models are derived with reduced magnetized fluid descriptions. The types of theoretical descriptions include weak turbulence theory and anisotropic Kolmogorov-like spectral indices and mixing lengths. A number of standard turbulent diffusivity formulas are known for the various space-time scales of the drift wave turbulent mixing.

3.1 Drift Wave Turbulence

We start by deriving the self-consistent field equation for the electrostatic potential $\phi_{\mathbf{k}\omega}$ and the electrostatic kinetic dispersion function $\epsilon_{\mathrm{ES}}(\mathbf{k}\omega)$.

Worked Problem 3.1

Problem:

Find a Hamiltonian description of the drift waves with $d = 2$ degrees of freedom for the parallel motion $p_1 = mv_{\|}$, $q^1 = z$ and the cross-field guiding-center motion $dp_2 = Bdx$, $dq^2 = dy$. Use the cross-field polarization dielectric $\epsilon_{\perp}$ derived in Worked Problem 2.1 when writing the Poisson equation.

Solution:

The guiding-center motion derived in plasma textbooks follows from Hamilton's equations of motion $\dot{p}_\alpha = -\partial H/\partial q^\alpha$, $\dot{q}^\alpha = \partial H/\partial p_\alpha$ with the

Hamiltonian

$$H(p_\alpha, q^\alpha) = \frac{1}{2} m v_\parallel^2 + q\phi(x,y,z,t) + \mu B(x,y) \tag{3.1}$$

where μ is the first adiabatic invariant that is constant to the first order in the small parameters $\epsilon = \omega/\Omega_i$ and $k_\perp^2 \rho_i^2$. In the potential energy q is the particle charge, which we take as $q = e$ to avoid the use of the symbol q for both Hamiltonian coordinate and the electric charge.

The particle phase space density $f(x, y, z, v_\parallel, t)$ is conserved by the volume preserving Hamiltonian flow, so that

$$\frac{\partial f_j}{\partial t}(p, q, t) + [H, f_j] = 0 \tag{3.2}$$

for each species j compute the guiding-center motion from

$$\frac{dz}{dt} = v_\parallel \quad \frac{dv_\parallel}{dt} = -\frac{e}{m}\frac{\partial \phi}{\partial z} \tag{3.3}$$

$$\frac{dx}{dt} = \frac{E_y}{B} = -\frac{1}{B}\frac{\partial \phi}{\partial y} \tag{3.4}$$

$$\frac{dy}{dt} = -\frac{E_x}{B} = \frac{1}{B}\frac{\partial \phi}{\partial x} \tag{3.5}$$

and re-expressing the motion in terms of p_α, q^α as defined in the statement of the problem gives

$$[H, f] = v_\parallel \frac{\partial f}{\partial z} - \frac{e}{m}\frac{\partial \phi}{\partial z}\frac{\partial f}{\partial v_\parallel} + \frac{1}{B}\frac{\partial \phi}{\partial x}\frac{\partial f}{\partial y} - \frac{1}{B}\frac{\partial \phi}{\partial y}\frac{\partial f}{\partial x} \tag{3.6}$$

for the directional or Lie derivative of the Hamiltonian flow in the phase space [Jose and Saletan (2002)].

The self-consistent electric potential satisfies Poisson's equation taking into account the polarization charge density $\rho_p = -\nabla \cdot \boldsymbol{j}_p$ from Worked Problem 2.1. The field equation in configuration space is

$$\nabla_\perp \cdot (\epsilon_\perp \nabla_\perp \phi) + \partial_z^2 \phi = -4\pi \sum_j q_j \int f_i d^3 v. \tag{3.7}$$

Now we used Worked Problem 2.1 and Eq. (2.33) where $\epsilon_\perp = 1 + \sum_j \omega_{pi}^2/\Omega_j^2$ is the dielectric function. In the space of the fluctuations where

$$\phi(\boldsymbol{x}, t),\ \delta f_j\left(\boldsymbol{x}, v_\parallel, t\right) \sim \phi_{\boldsymbol{k}}, \delta f_j\left(\boldsymbol{k}, v_\parallel, \omega\right) \times \exp(i\boldsymbol{k}\cdot\boldsymbol{x} - i\omega t) + \text{c.c.}$$

the partial differential equations (3.6) and (3.7) become algebraic with complex dispersion function

$$k^2 \epsilon_{\text{ES}} = k_\perp^2 \epsilon_\perp + k_\parallel^2 + \sum_j \frac{4\pi n_j e_j^2}{m_j} \int \frac{d^3 v \left(\omega \frac{\partial f_j}{\partial \epsilon} + \frac{k_y}{e_j B}\frac{\partial f_j}{\partial x}\right)}{\omega - k_\parallel v_\parallel - k_y v_D + i0^+} \tag{3.8}$$

where $\epsilon = mv^2/2$ is the particle energy, $\mu = mv_\perp^2/2B$ is the particle magnetic moment, and $v_{Dj} = (\mu/e_j)\partial_x \ell n\, B(x)$ is the guiding-center drift velocity.

The dispersion relation in Eq. (3.8) reduces to that of Worked Problem 2.2 when the plasma becomes uniform in space.

3.2 Drift Wave Mechanism

One of the most intellectually interesting problems in plasma physics is the problem of turbulence and the associated transport of the plasma properties including density, temperature and momentum. Plasma turbulence can be rather different from that in neutral fluids due to the numerous waves in plasmas. Plasma turbulence is closer to that of geostrophic flows on rotating planets where the Rossby wave is closely analogous to the plasma drift wave. At finite amplitude the wave dispersion often balances the wave nonlinearity to form solitary structures and coherent vortex structures. This aspect of plasma turbulence is explored in numerous workshop proceedings [Horton and Reichl (1984)], the proceedings of the Marseilles Workshops published as Benkadda, *et al.* (1995) and Horton and Ichikawa (1996). Drift waves are the most widely-studied form of plasma turbulence in magnetized plasmas.

3.3 Energy Bounds for Turbulence Amplitudes

3.3.1 *Density gradients*

The electron drift wave and trapped electron instability (trapped particle mode) [Kadomtsev and Pogutse (1970)] are driven by the square of the plasma density gradient in plasmas with no background temperature gradient. The wave with a temperature gradient is analyzed in Section 3.3.2. In the early literature the electron drift wave was called the universal instability since the instability drive from $(\nabla n)^2$ is always present in confined plasmas.

Growth Rate of Density Gradient Driven Drift Wave

$$\gamma_k = \left(\frac{\pi}{2}\right)^{1/2} \frac{\omega_{*e}^2 k_\perp^2 \rho_s^2}{|k_\parallel| v_e} \propto \left(\frac{dn}{dx}\right)^2$$

The energy released from the density gradient or the temperature gradient is understood from the Carnot engines powered by the gradients, as shown in Fig. 3.1.

Figure 3.1 shows the model for the energy release for the density gradient. The left higher density n_1 (and pressure $p_1 = n_1 T$) box from $x = [0, x_i]$ is allowed to expand to $[0, x_f]$ and compress the gas in the right box until the force balance is reached with

$$P_f = P_1 \left(\frac{x_i}{x_f}\right)^\Gamma = P_2 \left(\frac{a - x_i}{a - x_f}\right)^\Gamma \tag{3.9}$$

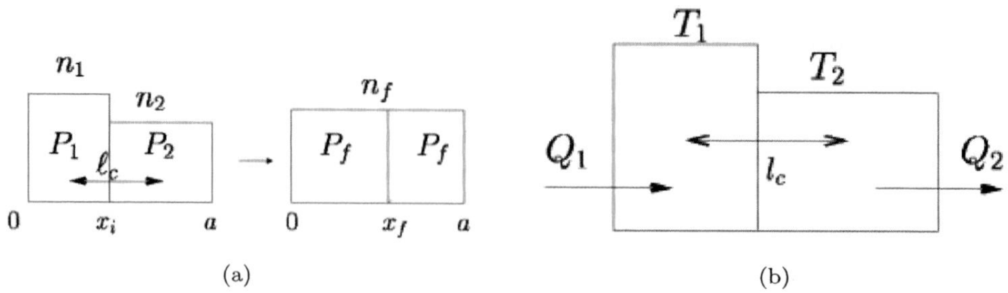

Fig. 3.1 (a) Expansion diagram under density gradient $n_1 > n_2$ for calculating the released plasma energy in the left figure. (b) The right figure is the energy diagram for the temperature gradient driven turbulence. The two regions are one radial correlation length ℓ_c apart. The partition at x_i moves in frame (a) while the plasma cycles between the T_1 and T_2 chambers in frame (b).

using the adiabatic gas constant Γ. Solving for the new position x_f and calculating the work, $W = 1/a \int_{x_i}^{x_f} (P_1 - P_2)dx$ per unit volume done by the gas during the expansion gives the free energy. For $\Delta x/a \ll 1$, the turbulent energy per unit volume is

$$W_{\text{turb}} \leq W_{\text{max}} = \frac{(\Delta P)^2 x_i(a - x_i)}{\Gamma P a^2} \leq \frac{(\Delta P)^2}{4\Gamma P}. \tag{3.10}$$

The limit $\Gamma \to \infty$ corresponds to an incompressible gas for which the partition between the pressures P_1 and P_2 will not move.

If the drift wave energy W_{turb} is almost entirely in the $\mathbf{E} \times \mathbf{B}$ kinetic energy of the ions, then we have the bound for the average turbulent convection velocity,

$$\left\langle v_E^2 \right\rangle^{1/2} \leq \left(\frac{\ell_c}{L_n} \right) c_s. \tag{3.11}$$

When the correlation length ℓ_c is limited by the ion inertial scale length $\ell_c = \rho_s$, the bound for the mean square turbulent plasma velocity is

$$\left\langle v_E^2 \right\rangle \leq \left(\frac{\rho_s}{L_n} \right)^2 c_s^2 \equiv v_{de}^2. \tag{3.12}$$

The limit of $\left\langle v_E^2 \right\rangle \leq v_{de}^2$ follows from another argument called the mixing length limit. This argument states that when the gradient of the fluctuating density, $\nabla \delta n$, reaches the value $dn/dx = -n/L_n$ of the ambient density gradient, the gradient source, on average, is eliminated and growth stops. Then $\delta n/n = 1/k_x L_n$, with $\ell_c = 1/k_x = \rho_s$. The same amplitude limit applies and the scale of the turbulent diffusion gives

$$D_{gB} = \left\langle v_E^2 \right\rangle^{1/2} \ell_c = (\rho_s/L_n)(T_e/eB) \tag{3.13}$$

which is defined as the gyro-Bohm diffusivity. To derive the famous Bohm diffusivity, $D_B = T_e/eB$, one instead assumes ℓ_c increases to be the geometric mean of L_n and ρ_s [Bohm, *et al.* (1949)].

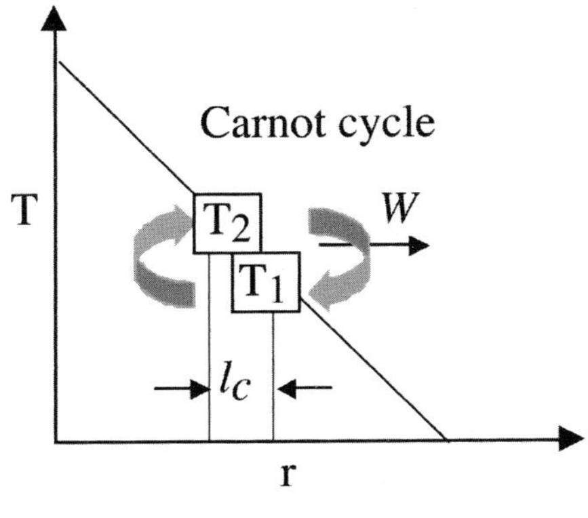

Fig. 3.2 Carnot cycle diagram for temperature-gradient driven drift waves. The maximum energy W released to the plasma turbulence is $W = Q_1 - Q_2 = n\Delta T(\Delta T/T - (\Gamma - 1)\Delta n/n)$.

3.3.2 *Temperature gradients*

For auxiliary heated plasmas that develop a temperature gradient ∇T (either ion or electron) from input power P_{aux}, the relevant thermodynamic argument is the Carnot engine. As shown in Fig. 3.2, there is an input of energy Q_1 in the left box. The vortices and streamers cycle the plasma over the correlation length ℓ_c, connecting the T_1 and T_2 plasmas. Now, for each cycle, the temperature gradient drives the drift wave to release the energy,

$$W_{\text{turb}} \leq \Delta W_{\text{carnot}} = \Delta S \Delta T \tag{3.14}$$

where ΔS is the change in the entropy of the plasma gas. From kinetic theory, the value of the gas constant, Γ, depends on the type of drift wave $\Gamma = 5/3$ when the wave-particle resonance is $\omega_k = \omega_{\nabla B} + \omega_{\text{curv}}$ and $\Gamma = 3$ where $\omega_k = k_\parallel v_\parallel$ and $k_\perp \rho_i \ll 1$. For toroidal ITG and ETG, a simple calculation for modes with $\Gamma = 5/3$ yields

$$\Delta S = n k_B \left(\frac{3}{2}\frac{\Delta T}{T} - \frac{\Delta n}{n} \right) \tag{3.15}$$

which in Eq. (3.14) together with $\Delta T = -\ell_c dT/dr$ give

$$\Delta W_{\text{carnot}} = \frac{3}{2} n T \ell_c^2 \frac{\partial \ln T}{\partial r} \left[\frac{\partial \ln T}{\partial r} - \frac{2}{3}\frac{\partial \ln n}{\partial r} \right] \tag{3.16}$$

where $\epsilon = E/T_i$ and μ denotes system parameters such as η_i and T_e/T_i in the kinetic theory-wave dispersion function. Figure 3.3 shows the roots of the Vlasov-Poisson dispersion relation, $D_{\vec{k}}(\omega, \mu) = 0$, with increasing $\eta \equiv \nabla \ln T / \nabla \ln n$ where for the kinetic dispersion relation,

$$D_{\vec{k}}(\omega, \mu) = \frac{1}{T_e} + \frac{1}{T_i}\left[1 - \left\langle \frac{\omega - \omega_{*i}(\eta, \epsilon)}{\omega - \omega_{Di}\epsilon + i0^+} \right\rangle \right] \tag{3.17}$$

with $\langle f(\omega, k, \epsilon) \rangle$ is defined as the integral over the energy distribution $F(\epsilon)$.

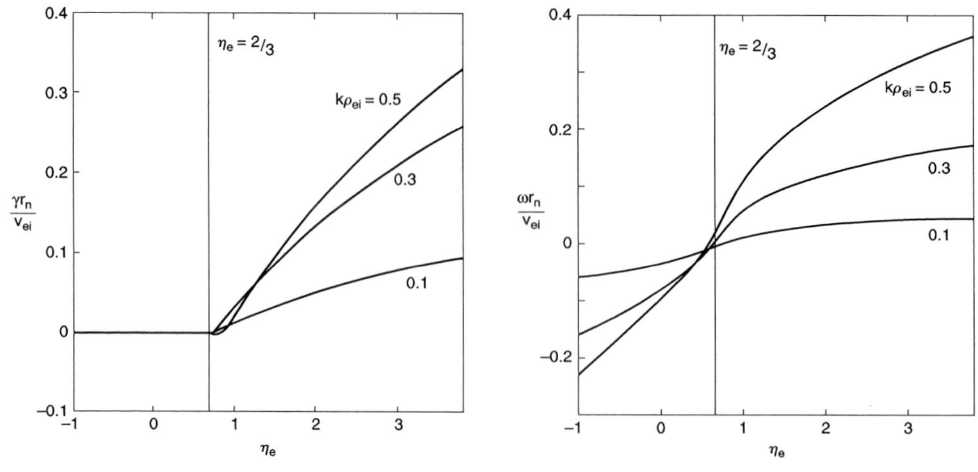

(a) (b)

Fig. 3.3 The Vlasov (GKE) equation growth rate and frequency as a function of the temperature gradient divided by the density gradient [Horton, *et al.* (1988)]. The roots of Eq. (3.17) give (a) the growth rate and (b) the frequency for the three wavenumbers.

The critical temperature gradient from the Nyquist analysis is given by

$$\eta_{\text{crit}} = \frac{\partial \ln T}{\partial \ln n} = \frac{2}{3} \tag{3.18}$$

as shown in Fig. 3.3. The agreement between the thermo-dynamical formula and the stability condition of the kinetic dispersion relations can be understood when one sees from Fig. 3.3 that the wave frequency, $\omega_{\vec{k}} \simeq 0$ at the critical point. Here the relevant gradient of the phase-space density function $f(\epsilon)$ is proportional to $\partial/\partial r[n/T^{d/2}]$ where

$$\frac{\partial f}{\partial r} = \frac{\partial}{\partial r}\left(\frac{n(r)}{T^{d/2}(r)}\right) \quad \text{for} \quad E \ll T \tag{3.19}$$

with $\Gamma = (d+2)/d$ for a plasma with d degrees of freedom. For one-dimensional waves like Langmuir waves, $d = 1$ and $\Gamma = 3$, and for toroidal drift waves $d = 3$ and $\Gamma = 5/3$. The key, universal result given by Eq. (3.16) for W_{carnot} is that there is only a window of η values, $0 < \eta \leq \Gamma - 1$, which are generally stable. Drift waves are unstable for both inverted gradients $\nabla n \cdot \nabla T < 0$, and strong temperature gradients $\nabla T \gg \nabla n$.

Worked Problem 3.2

Problem:

Express the drift wave dispersion relation in Eq. (3.8) for a Maxwellian distribution function in terms of the plasma dispersion function.

Solution:

We need to find the integral over Maxwellian velocity distribution with a singular resonant point. In the case of 1D plasma wave, the dispersion is

$$D(\omega, k) = 1 + \frac{n_0 e^2}{m_e k^2 \epsilon_0} \int_{-\infty}^{\infty} \left(\frac{\frac{\partial f_0}{\partial v_\parallel} + \frac{k_y}{k_\parallel} \frac{1}{eB} \frac{\partial f}{\partial x}}{\omega / k_\parallel - v_\parallel} \right) dv \tag{3.20}$$

where

$$f_0(v) = n_0 \sqrt{\frac{m_e}{2\pi T_e}} \exp\left(-\frac{m_e v^2}{2T_e} \right) \tag{3.21}$$

$$\frac{\partial f_0}{\partial v_\parallel} = -\sqrt{\pi} \frac{2 v_\parallel}{v_{th}^3} \exp\left(-\frac{v^2}{v_{th}^2} \right) \tag{3.22}$$

and $v_{th} = \sqrt{2T_e/m_e}$. The $v_\parallel$-integral is evaluated as in Problem 2.2

$$\int_{-\infty}^{\infty} \frac{\partial f_0/\partial v}{\omega/k - v} dv = \frac{n_0}{\sqrt{\pi}} \frac{2}{v_{th}^3} \int_{-\infty}^{\infty} \frac{v \exp(-v^2/v_{th}^2)}{\omega/k - v} dv \tag{3.23}$$

$$= \frac{2n_0}{\sqrt{\pi}\, v_{th}^3} \int_{-\infty}^{\infty} \left(1 + \frac{\omega/k}{v - \omega/k} \right) e^{-v^2/v_{th}^2} dv \tag{3.24}$$

$$= \frac{2n_0}{\sqrt{\pi}\, v_{th}^3} \sqrt{\pi}\, v_{th} \left(1 + \frac{\omega}{k v_{th} \sqrt{\pi}} \int_{-\infty}^{\infty} \frac{e^{-t^2}}{t - \omega/k v_{th}} dt \right) \tag{3.25}$$

$$= \frac{2n_0}{v_{th}^2} [1 + \zeta Z(\zeta)], \tag{3.26}$$

where $t = v/v_{th}$, $\zeta = \omega/k v_{th}$ and $Z(\zeta)$ is the complex Plasma Dispersion function or Z function

$$Z(\zeta) = \pi^{-1/2} \int_{-\infty}^{\infty} \frac{\exp(-t^2)}{t - \zeta} dt \qquad \text{for } \mathrm{Im}(\zeta) > 0 \tag{3.27}$$

as defined and tabulated in Fried and Conte (1961). For real value k, ω the function requires analytic continuation in the complex ζ-plane. For real ζ the real part of the Z function is the Hilbert transform defined by

$$\mathcal{H}[f(y)] \equiv \frac{1}{\pi} PV \int_{-\infty}^{\infty} \frac{f(x)}{x - y} dx. \tag{3.28}$$

High- and low-phase velocity limits

There are two limits of the Z function useful for analytic analysis:
(i) low-phase velocities $|\zeta| = |\omega|/k v_{th} \ll 1$

$$Z(\zeta) = i\sqrt{\pi}\, e^{-\zeta^2} - 2\zeta \left(1 - 2\zeta^2/3 + 4\zeta^4/15 - 8\zeta^6/105 + \cdots \right) \tag{3.29}$$

$$= i\sqrt{\pi}\, e^{-\zeta^2} - 2\zeta \sum_{n=0}^{\infty} \frac{(-2\zeta^2)^n}{(2n+1)!!} \tag{3.30}$$

(ii) high-phase velocities asymptotic series ($|\zeta| \gg 1$)

$$Z(\zeta) = i\sqrt{\pi}\,\sigma e^{-\zeta^2} - \zeta^{-1}\left(1 + 1/2\zeta^2 + 3/4\zeta^4 + 15/8\zeta^6 + \cdots\right) \quad (3.31)$$

$$= i\sqrt{\pi}\,\sigma e^{-\zeta^2} - \zeta^{-1}\sum_{n=0}^{\infty}\frac{(2n+1)!!}{(2\zeta^2)^n} \quad (3.32)$$

where

$$\sigma = \begin{cases} 0 & y > 0 \\ 1 & y = 0 \\ 2 & y < 0 \end{cases}. \quad (3.33)$$

Note that $Z(\zeta)$ is still continuous when crossing the real axis, i.e. Im $Z(\zeta) \approx i\sqrt{\pi}\,e^{-|\zeta|^2}$ when y approaches 0 from both sides. When $\zeta \to -i\infty$, the Z function blows up.

If ζ is real ($\zeta = x$), we have

$$Z(x) = i\sqrt{\pi}\,e^{-x^2} - 2xY(x), \quad (3.34)$$

where $Y(x) = [e^{-x^2}/x]\int_0^x e^{t^2}\,dt$ and

$$Y(x) = 1 - 2x^2/3 + 4x^4/15 - 8x^6/105 + \cdots \quad \text{when } |x| \ll 1, \quad (3.35)$$

$$Y(x) = (1/2x^2)[1 + 1/2x^2 + 3/4x^4 + \cdots \quad \text{when } |x| \gg 1, \quad (3.36)$$

In Worked Problem 3.3 at the end of the chapter, the methods used to compute the plasma Z-function are given in more detail.

3.3.3 *Drift wave eigenmodes in toroidal geometry*

The first effect of toroidal geometry that changes the stability of the drift waves is the unfavorable magnetic field line curvature vector, $\boldsymbol{\kappa} = -(\widehat{\mathbf{b}}\cdot\nabla)\widehat{\mathbf{b}} \simeq (\widehat{\boldsymbol{r}}\cos\theta + \widehat{\boldsymbol{\theta}}\sin\theta)/R$ which covers the outer half ($\cos\theta > 0$) of the toroidal surfaces. This magnetic curvature creates small scale interchange turbulence with waves that are only partially stabilized by the magnetic shear, $s = rq'/q$, where $q(r) = rB_\phi/RB_\theta$ gives the helical pitch of the field lines. The E_r-shear can also stabilize the modes as discussed in Section 3.3.4. In addition, and closely related, the weakening of the magnetic field with increasing radius (see Fig. 3.4) leads to an outward acceleration given by the force, $F = -\mu\nabla B$ acting on the plasma.

In the toroidal drift wave equations the local gradients of the density and pressure are normalized with respect to the gradient length, R, of the magnetic field. Thus, the key stability parameters are

$$\epsilon_n = \frac{L_n}{R}, \; \epsilon_T = \frac{L_T}{R} \quad \text{and} \quad \epsilon_p = \frac{L_p}{R} \quad (3.37)$$

throughout the toroidal plasma equations.

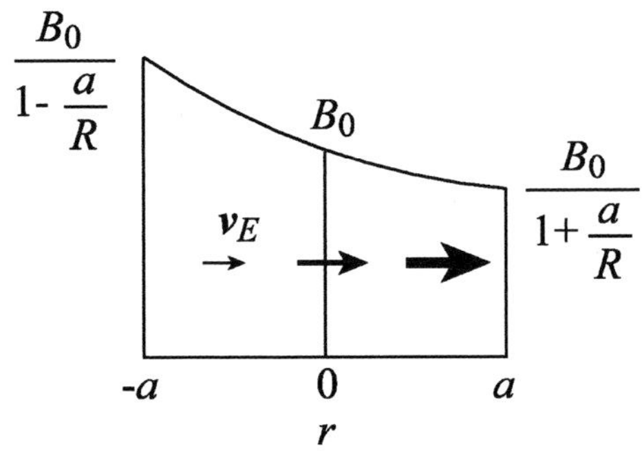

Fig. 3.4 Compression/rarefraction of the plasma by $\boldsymbol{E} \times \boldsymbol{B}$ convection in the toroidal magnetic field.

The faster radial flows shown in Fig. 3.4 give the plasma compression,

$$\nabla \cdot \vec{v}_E = \frac{2v_x \cos\theta}{R} = -\frac{2\cos\theta}{RB}\frac{\partial\phi}{\partial y} \tag{3.38}$$

compared with the plasma convection,

$$\vec{v}_E \cdot \nabla n = -\frac{nv_x}{L_n} = \frac{n}{L_n B}\frac{\partial\Phi}{\partial y} \tag{3.39}$$

which gives rise to the two terms proportional to $(1-2\epsilon_n\cos\theta)\omega_{*e}$ in the drift wave eigenmode equation. For flat density profiles toroidal compression-expansion term in Eq. (3.38) dominates the density gradient term in Eq. (3.39).

The density and potential fluctuations set up along the magnetic field line give rise to ion acoustic waves with $\omega^2 = k_\parallel^2 c_s^2$, where

$$k_\parallel^2 \to -\frac{1}{q^2R^2}\left(\frac{\partial}{\partial\theta} + q(r)\frac{\partial}{\partial\phi}\right)^2 \tag{3.40}$$

is the differential operator $\nabla_\parallel^2$ in the nonuniform plasma. Including both the ion parallel and convection motion gives $\omega^2 = k_\parallel^2 c_s^2 + \omega\omega_{*e}(1-2\epsilon_n\cos\theta)$.

Similarly, the shear-Alfvén wave $\omega^2 = k_\parallel^2 v_A^2$ propagates with the $k_\parallel^2$ operator. The periodicity of the torus means that a physical field is the same at $(\theta+2\pi m, \phi+2\pi n)$ for all integer values of m, n which defines the set of rational magnetic surfaces. Rational surfaces appear when the pitch of the magnetic field is the same as that of the mode $\exp(im\theta - in\phi)$, which occurs when the right-hand side of Eq. (3.40) vanishes. The compressional term in Eq. (3.38) and convection term in Eq. (3.39) combine with the ion acoustic waves to make the toroidal drift wave eigenmode waves of the form

$$\begin{aligned}\Psi(\mathbf{x}, t) &= \sum_m \Phi_{m,n}(r)e^{im\theta - in\phi - i\omega t} + \text{c.c.} \\ &= \Phi(r,\theta)e^{-in\phi - i\omega t} + \text{c.c.}\end{aligned} \tag{3.41}$$

The toroidal mode solution, $\Phi(r,\theta)$, satisfies a Mathieu-type equation for each given r given by

$$\begin{aligned}&\frac{c_s^2}{\omega^2 q^2 R^2}\left(\frac{\partial}{\partial\theta} - inq(r)\right)^2 \Phi \\ &+ \left[1 - i\delta_{\vec{k}} - \rho_s^2\nabla_\perp^2 - \frac{\omega_{*e}}{\omega}(1-2\epsilon_n\cos\theta)\right]\Phi = 0.\end{aligned} \tag{3.42}$$

Here we used the simple phase-shifted, non-adiabatic electron response, $\delta n_e = n_e(e\phi/T_e)(1-i\delta_{\vec{k}})$. Introducing the phase variation $\Phi \to \Phi(\theta)e^{inq(r)\theta}$, required to keep $k_\parallel^2\Phi$ small, we find that $\nabla_\perp^2\Phi = (\partial_r^2 + r^{-2}\partial_\theta^2)\Phi e^{inq\theta} = -(k_r^2+k_\theta^2)\Phi = -k_\theta^2(1+s^2\theta^2)\Phi$, where $k_r = sk_\theta\theta$ and $k_\theta = nq/r$ with $s = rq'/q$.

The drift-eigenmode equation for the one-dimensional ansatz $\Phi(\theta)$ is called the "ballooning mode" equation, which from Eqs. (3.40)–(3.42) is

$$\begin{aligned}&\frac{c_s^2}{\omega^2 q^2 R^2}\frac{\partial^2\Phi}{\partial\theta^2} + \left[1 - i\delta_{\vec{k}} + k_\theta^2\rho_s^2\left(1+s^2\theta^2\right)\right. \\ &\left. - \frac{\omega_{*e}}{\omega}(1-2\epsilon_n\cos\theta)\right]\Phi = 0.\end{aligned} \tag{3.43}$$

The solution of Eq. (3.43) with a shooting code requires as a boundary condition that when $\theta^2 \to \infty$ the solution $\Phi \to \exp(i\sigma\theta^2/2)$ where $\sigma = \omega qR|sk_\theta|\rho_s/c_s$ with $\mathrm{Im}\,\sigma \propto \mathrm{Im}\,\omega > 0$. When the mode is strongly localized to the small θ^2 region, we may use $\cos\theta \simeq 1-\theta^2/2$ to reduce Eq. (3.43) to the analytic harmonic oscillator problem with eigenvalues,

$$\omega_{k_\theta,n} = \frac{\omega_{*e}(1-2\epsilon_n) - ic_s|sk_\theta|(\rho_s/qR)(2n+1)}{1+k_\theta^2\rho_s^2 - i\delta_{\vec{k}}} \tag{3.44}$$

and eigenfunctions, $\Phi_n(\theta) = \exp(i\frac{\sigma}{2}\theta^2)H_n\left((-i\sigma)^{1/2}\theta\right)$ with $n = 0,1,2,\ldots$ The $n=0$ mode is a symmetric ground state eigenmode with flow vortex center on the rational surface. This symmetry is called the interchange. The $n=1$ mode is $\theta\exp(i\sigma\theta^2/2)$ with odd parity and corresponds to the varicose mode in neutral fluids with the flows into and out of the rational surface at $\theta = 0$. For finite $\beta_e = 8\pi\, p_e/B^2$ this odd mode develops a significant reconnection magnetic field component. Small magnetic islands are formed as described in Chapters 6 and 11. Thus, the magnetic shear, $|s|$, introduces a damping of the waves due to the coupling to the ion acoustic waves that the propagate wave energy to region of Landau damping. Marginal stability conditions follow from Eq. (3.44) once the details of the dissipative $\delta_{\vec{k}}(\mu)$ function are specified from wave-particle resonances or collisional dissipation.

The structure of the eigenmode Eq. (3.43) can be transformed to a structure like the sheared slab eigenmode equation as in Chapter 2 by taking the Fourier transform with respect to the variable θ defined on the real line $[-\infty \le \theta \le +\infty]$. The first term becomes the ion acoustic wave term $k_\parallel^2$ of Eq. (2.59) and the term $(k_\theta^2\, s\theta)^2$ becomes the radial derivative term in Eq. (2.59). The toroidal term is absent so that an important destabilization mechanism is absent in the slab drift waves compared with the toroidal drift waves. The two eigenmode wave functions are clearly related when the toroidal mode is localized to the outside of the torus with $|\theta| < \pi$.

The presence of sheared flows from a nonuniform radial electric field E_r in the plasma and from sheared diamagnetic frequencies $\omega_*(r)$ breaks the even and odd symmetries of the eigenmodes and shifts the wave function off the rational magnetic surfaces. At finite plasma electron pressure this results in driving up a magnetic fluctuation δB_r and coupling the tearing and ballooning mode symmetries as described in Chapters 11-15.

Mazzucato (1976) measured the density fluctuations level $\delta n/n$ in TFTR density as a function of increasing neutral beam (NB) heating power $P_{\rm NB}$. The data for $P_{\rm NB} = 0$ for the ohmic discharge gave the low $\delta n/n \simeq 0.001$ fluctuation level. As the beam heating power increased, the fractional fluctuation levels increased to $\delta n/n \simeq 0.003$ at 5 MW to 0.006 at 9 MW and 0.015 at 14 MW. This increase of $\delta n/n$ is consistent with power balance $\langle nT\rangle/\tau_E = (P_{\rm NB} + P_{\rm Oh})/{\rm Vol}$ with $1/\tau_E \propto \chi_{gB}/a^2 \propto T^{3/2}$. Here Vol is the mean plasma volume determined by the mean value theorem,

$$\int \frac{3}{2} nT {\rm d}^3x = \frac{3}{2}\langle nT\rangle \cdot {\rm Vol}. \tag{3.45}$$

In plasmas with higher core temperatures and thus stronger radial temperature gradients, the drift wave changes character from a density gradient instability to a temperature gradient driven instability. For hot ion plasmas the drift waves are destabilized by the ion temperature gradient, ∇T_i. In this regime, called the ion temperature gradient (ITG) instability, the electron dissipation $i\delta_{\vec{k}}$ can be dropped and the electron response taken as adiabatic which is defined by $\delta n_e/n_e = e\phi_{\vec{k}}/T_e$.

Simple basic experiments validating the ITG model and then the ETG model were carried out in the Columbia Linear Machine (CLM) as described in Sen, *et al.* (1991, 2001). These hydrogen plasmas, produced by RF heating, have $n_e \sim 2 \times 10^{14}\,{\rm m}^{-3}$ and $T_i \sim 10\,{\rm eV}$. There is a temperature anisotropy $T_{i\|} \neq T_{i\perp}$ that complicates the linear dispersion but is useful for both NBI or RF heated toroidal experiments and the central cell plasma in the mirror machine. In the GAMMA-10 tandem mirror resonant ion cyclotron heating in the central cell produced $T_{i\perp} \sim 6.5\,{\rm KeV}$ and $T_{i\|} \sim 2.5\,{\rm KeV}$ [Cho, *et al.* (2005, 2006, 2008, 2016)]. The details of the two temperature components of the ITG theory with the simple cylindrical central-cell plasma data can be found in Hojo (1995) and Sen, *et al.* (1991).

In Section 3.3.1 and 3.3.2, we analyzed the free energy for the ITG modes that limits their intensity. Simulations with gyrokinetic Vlasov codes confirm these amplitude limits. The literature concerning the numerical modeling of ITG modes is extensive due to their role as the standard model for explaining the anomalous ion thermal diffusivity χ_i.

The ion thermal diffusivity is strongly influenced by the sheared mass flows in the background plasma [Hamaguchi and Horton (1990)]. Thus we now consider the dynamics of a plasma with a strongly nonuniform $\mathbf{E} \times \mathbf{B}$ sheared flow in the equilibrium. The dimensionless shear-flow parameter Υ_s derived here will be important in Chapter 17 for analyzing the formation of internal transport barriers in tokamaks.

Worked Problem 3.3: KH Instability

Find the eigenvalue equation for electrostatic potential waves in a plasma with $\epsilon_\perp \gg 1$ that has a sheared $\mathbf{E} \times \mathbf{B}$ flow velocity $\boldsymbol{V}_0 = v_0(x)\widehat{\boldsymbol{e}}_y$ across the constant $\mathbf{B} = B\widehat{\boldsymbol{e}}_z$ from a nonuniform $E_r(r)$.

Solution:

From Problem 2.1 we know that the net plasma current is the polarization current

$$\boldsymbol{j}_P(\boldsymbol{x}, t) = \frac{\rho}{B^2}\frac{d\mathbf{E}_\perp}{dt} \tag{3.46}$$

which must close, in the absence of drift wave parallel currents, in loops across $\mathbf{B}\widehat{\boldsymbol{e}}_z$ according to

$$\nabla \cdot \boldsymbol{j}_p = \nabla \cdot \left[\frac{\rho}{B^2}\frac{d(-\nabla_\perp \varphi)}{dt}\right] = 0 \tag{3.47}$$

for electrostatic fluctuations $\mathbf{E} = -\nabla\varphi(\boldsymbol{x}, t)$.
Using the convective derivative with $\mathbf{v}_E \cdot \nabla\varphi$, one finds that

$$\frac{d}{dt} = \frac{\partial}{\partial t} + \frac{1}{B}\left(\frac{\partial \varphi}{\partial x}\frac{\partial}{\partial y} - \frac{\partial \varphi}{\partial y}\frac{\partial}{\partial x}\right) \tag{3.48}$$

from the $\mathbf{E} \times \mathbf{B}$ flow which reduces Eq. (3.47) to the nonlinear scalar equation

$$\frac{\rho}{B^2}\frac{d}{dt}\nabla_\perp^2 \varphi = 0. \tag{3.49}$$

Equation (3.49) is equivalent to the equation of 2D incompressible flow in hydrodynamics whose properties are extensively explored in fluid turbulence. In the uniform plasma the vorticity from the $v_x v_y$-flow is the origin of the vorticity

$$\omega = \partial_x v_y - \partial_y v_x = \frac{\nabla_\perp^2 \varphi(\boldsymbol{x}, t)}{B} \tag{3.50}$$

so that Eq. (3.49) states that the vorticity parallel to $\mathbf{B}$, and all functions of the vorticity $F(\omega)$, are constant in the frame moving with the $\mathbf{E} \times \mathbf{B}$ flow.

Waves $(\boldsymbol{k}, \omega)$ in the one-dimensional sheared flow $v(x)\mathbf{e}_y$ are of the form

$$\varphi(x, y, t) = \varphi_0(x) + \delta\varphi_k(x) e^{ik_y y - i\omega t} + \text{c.c.} \tag{3.51}$$

Linearizing Eq. (3.49) yields the eigenvalue equation

$$(\omega - k_y v(x)) \left[\partial_x^2 \delta\varphi_{k_y} - k_y^2 \delta\varphi_{k_y}\right] + k_y v''(x)\varphi_{k_y}(x) = 0. \tag{3.52}$$

Equation (3.52) is isomorphic with the famous equation for the Kelvin-Helmholtz instability in hydrodynamics.

The simplest analytic solution showing the stable and unstable wave spectrum is given by the model of $v(x)$ relevant to the "vee-shaped" radial electric $E_r(r)$ occurring in H-mode edge plasmas. For the vee-flow profile we use the model

$$v(x) = v_0 \left(1 - \frac{|x|}{a}\right) \text{ for } |x| \leq a \text{ and } v(x) = 0 \text{ for } |x| > a \tag{3.53}$$

and follow Chandrasekhar (1961) to write out the quadratic equation dispersion relation with unstable modes for $|k_y|a < 1$ and stable modes for higher wavenumbers $k_y a > 1$. Note that the Rayleigh theorem requires an inflection point ($v'' = 0$) in the flow profile for linearly unstable eigenmodes. However, the K-H operator is not self-adjoint so that there can be substantial convective instability from the initial value problem with $v(x)$ and $v'(x)$ profiles without inflection points.

The LAPD experiments of a plasma with a "vee-shaped" radial field where the Kelvin-Helmholtz instability was produced and controlled in the laboratory, are analyzed in Horton, *et al.* (2005b) and Carter and Maggs (2009).

The nonlinear dynamics of the Kelvin-Helmholtz instability results in the formation of a chain of tilted vortices, as shown in the laboratory data in Fig. 3.5 and Fig. 3.6 from the simulation of Horton, *et al.* (2005b). There is a phase shift between the two cross-field velocity components so that the average along the chain of vortices produces a net transport of the sheared velocity through the Reynolds stress tensor $\langle v_x v_y \rangle$. The radial divergence of this stress tensor then gives the rate of transport of the positive flow momentum on the right-hand side of the vortex street to the negative momentum on the left-hand side. In other examples in Chapters 13-17 we show how the pressure gradient instabilities can generate Reynolds stress fluxes and thus create the sheared plasma flows. The creation of sheared flows from density and temperature gradient is of key importance, and is measured, in numerous hot confined plasmas.

3.3.4 *Effect of magnetic shear and E_r shear on drift waves*

The amplitude of the ITG turbulence is controlled by the magnetic shear, $s = rq'/q$, and the shear in the azimuthal (or poloidal) flow velocity, $v_E = -E_r/B$. For example, Hamaguchi and Horton (1990) solve the drift wave problem with magnetic shear and flow shear with the FLR fluid model. The variation of the amplitudes and the resulting turbulent ion heat flux, $q_i = \langle n_i T_i v_r \rangle$ with the temperature gradient η_i are extensively investigated with magnetic and E_r shears. Kishimoto, *et al.* (1996) showed similar results with particle-in-cell (PIC) codes. In the toroidal code, the

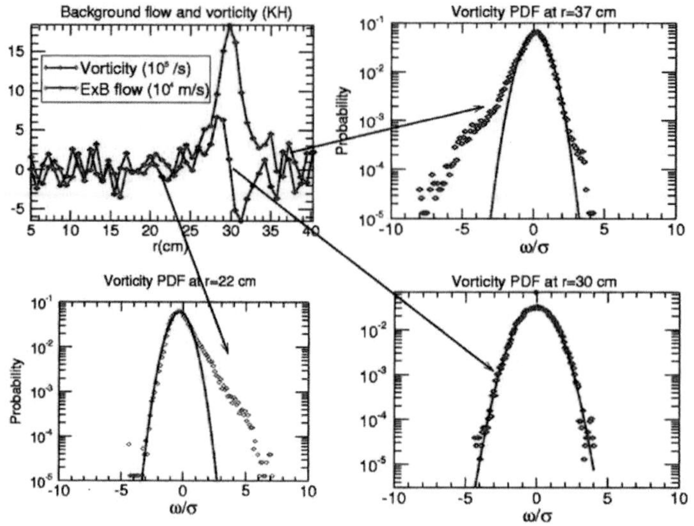

Fig. 3.5 Vorticity probability distribution functions for representative radius in strong the E_r-shear regime [Perez, *et al.* (2006)].

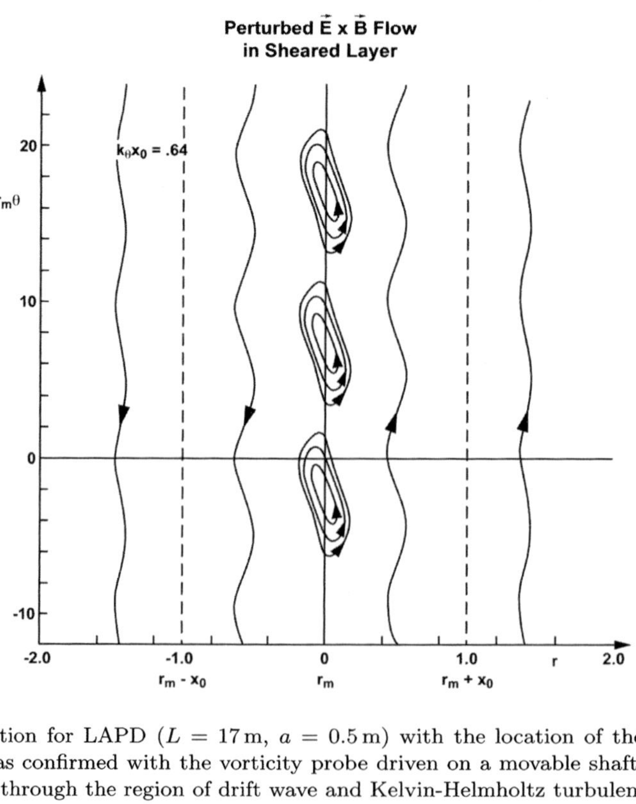

Fig. 3.6 Simulation for LAPD ($L = 17\,\mathrm{m}$, $a = 0.5\,\mathrm{m}$) with the location of the vortices in the shear-flow layer as confirmed with the vorticity probe driven on a movable shaft taking 10 shots per $\Delta r = 0.8\,\mathrm{cm}$ through the region of drift wave and Kelvin-Helmholtz turbulence.

flow shear, v'_E, from E_r stifles the χ_i transport, whereas suppression of χ_i in the local slab model is given by

$$\chi_i(v'_E) = \frac{\chi_i}{1 + (L_s v'_E/c_s)^2}. \tag{3.54}$$

The flow shear transport suppression factor in the denominator of Eq. (3.54) is defined as $\Upsilon_s = L_s v'_E/c_s$ arises from the eigenmode calculation of the driftwaves in the sheared flows [Hamaguchi and Horton (1992)]. Here $L_s = qR/s(r)$ is the magnetic shear length defined by the distance spanned by the magnetic field when the $\vec{B}$-vector rotates by approximately $\pi/4$.

The physical picture of the velocity-shear effect is given in Biglari, *et al.* (1988), where the large eddies or vortices of size $\Delta r = \rho_s(L_s/L_T)$ in the absence of shear flows are reduced in the shearing time, $\tau_{\rm sh} = 1/v'_E$, to small eddies. The details of the shear flow stabilization are complicated [Diamond, *et al.* (2005)]. A kinetic theory calculation is given by Dong, *et al.* (1992) and the role of shear flow in magnetized plasma is reviewed by Itoh and Itoh (1996), Terry (2000), and Diamond, *et al.* (2005).

Experiments for validation of drift wave theory were performed on LAPD. Variable shear-flow regimes were created by varying the radial electric field, E_r, from weak to moderate and then to strong. In the moderate and strong E_r regime, the electron-density transport barrier was measured in numerous experiments including the LAPD, TEXT, and CLM. The main characteristic of the drift wave was identified clearly in the weak E_r-shear regime and as the mixed Kelvin-Helmholtz drift wave regime in the strong E_r-shear regime.

In the strong E_r-shear regime there is a deep well of negative E_r, as occurs in the edge of H-mode tokamak plasmas. The nature of edge H-mode plasma depends on the structure of the last closed flux surfaces as described in Chapters 13-15. The vorticity probe measures $\omega_z = \widehat{\mathbf{b}}\cdot\nabla\times\boldsymbol{v}_E$ as shown in Fig. 3.5 for the long cylindrical LAPD plasma.

As the probe sweeps across the shear layer, the vorticity changes from counterclockwise to clockwise rotating vortices, just as expected from simple considerations and computer simulations. The region between these two oppositely spinning vortices forms the partial transport barrier as described by Biglari, *et al.* (1988) and Terry (2000).

The details of the effects of both magnetic and E_r shear are subtle since, like for the KH instability, the entire shape of flow profile, including its curvature, needs to be analyzed [Balmforth and Morrison (1999)]. In some circumstances shear flow can be destabilizing [Chen and Morrison (1991)]. Also, the actual region of transport suppression, the location of the transport barrier, sometimes occurs where the shear is a minimum, as suggested by Horton, *et al.* (1998) and Marcus, *et al.* (2008). Sheared rotational flows play an important role in the Gas Dynamic Trap, GAMMA-10 mirror machine and the TAE mirror machine as described in Chapter 17. Figure 3.7 shows the growth and death of a vortex in a mirror machine plasma.

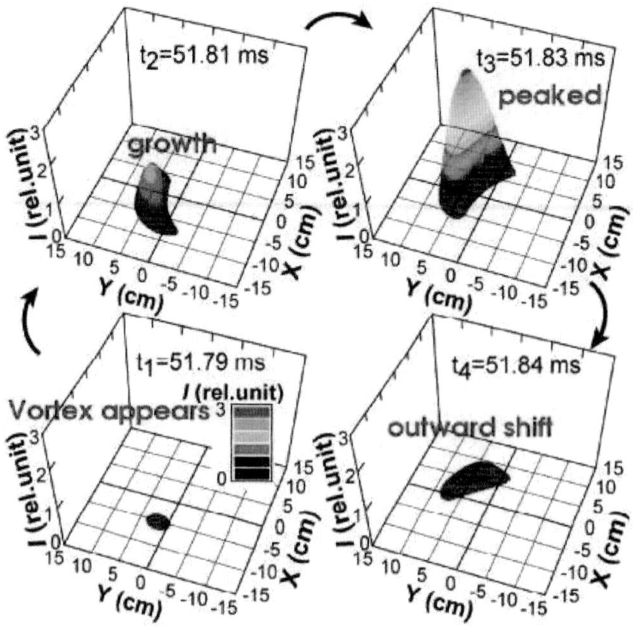

Fig. 3.7 X-ray tomography of the growth and decay of a vortex structure. Intermittent coherent structures have typical lifetimes ranging around a hundred microseconds to milliseconds and follow the rotational motion of the $\mathbf{E} \times \mathbf{B}$ drift [Cho, *et al.* (2005, 2006, 2016)].

Strong shear flow in a plasma creates temporal and spatial intermittency as shown by the probability distribution function with large skewness and excess probability levels (kurtosis) over the Gaussian level as shown in Fig. 3.5. In the central cell of the GAMMA-10 tandem mirror, where the electron energies reached $T_e \sim 800\,\text{eV}$, there was sufficient soft-X-ray emission to image the growth and decay of the plasma vortices using two micro-channel plates and tomographic image reconstruction [Cho, *et al.* (2005, 2006, 2016)] as shown in Fig. 3.7. Electrodes outside the central plasma are used to control the $E_r(r,t)$ field in mirror geometries. The high-ion temperature mirror-confined plasma have explored stability limits using electric biasing of end plates to control the plasmas. This control system is called the cusp-anchored gas-dynamic trap Anikeev, *et al.* (1997).

3.4 Weak Turbulence Theory for Drift Waves

For turbulent states with a broad range of wavenumbers, $\Delta\mathbf{k}$, and frequencies, $\Delta\omega_{\mathbf{k}}$, in the fluctuation spectrum the weak turbulence theory is developed by assuming that the deviation of the fluctuations from Gaussian statistics is weak. The deviations from Gaussian statistics are the skewness and kurtosis with examples in Figures 3.5 and 3.7. Under this condition the chain of coupled correlation functions is broken off at fourth order with what is called the "quasi-normal closure

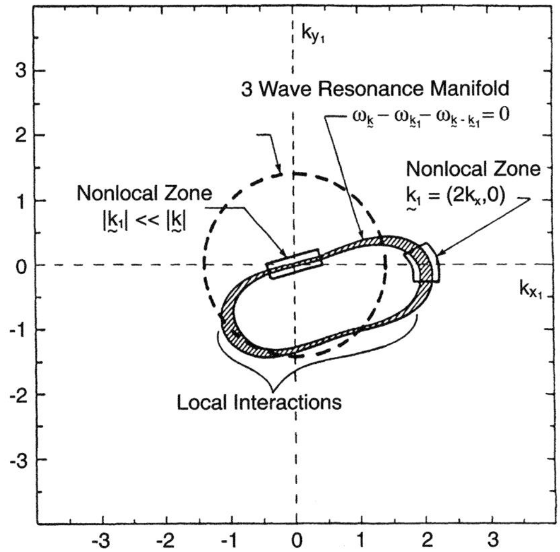

Fig. 3.8 The resonant three-wave manifold for the drift-wave interaction calculations in weak turbulence theory. The shaded ellipse defines the resonant manifold for 3-wave interactions.

approximation." This means that the ensemble average of the product of four $\phi_{\mathbf{k}}$ fields is factorized as

$$\begin{aligned}\langle\phi_{\mathbf{k}_1}\phi_{\mathbf{k}_2}\phi_{\mathbf{k}_3}\phi_{\mathbf{k}_4=-\mathbf{k}_1-\mathbf{k}_2-\mathbf{k}_3}\rangle &= \langle\phi_{\mathbf{k}_1}\phi_{\mathbf{k}_2}\rangle\langle\phi_{\mathbf{k}_3}\phi_{\mathbf{k}_4}\rangle \\ &+ \langle\phi_{\mathbf{k}_1}\phi_{\mathbf{k}_3}\rangle\langle\phi_{\mathbf{k}_2}\phi_{\mathbf{k}_4}\rangle + \langle\phi_{\mathbf{k}_1}\phi_{\mathbf{k}_4}\rangle\langle\phi_{\mathbf{k}_2}\phi_{\mathbf{k}_3}\rangle\end{aligned} \tag{3.55}$$

which closes the equations for the dynamics of $I_{\mathbf{k}}(t) = \langle\phi^*_{\mathbf{k}}\phi_{\mathbf{k}}\rangle$.

For drift waves the turbulent energy spectrum is

$$W_{\mathbf{k}} = \left(1 + k_\perp^2\rho_s^2\right) I_{\mathbf{k}}(t) \tag{3.56}$$

and the resulting weak-turbulence wave kinetic equation is

$$\begin{aligned}\frac{\mathrm{d}W_{\mathbf{k}}}{\mathrm{d}t} = 2\gamma_{\mathbf{k}}W_{\mathbf{k}} + \sum_{\mathbf{k}_1,\mathbf{k}_2=\mathbf{k}-\mathbf{k}_1}\frac{(\mathbf{k}\times\mathbf{k}_1\cdot\widehat{\mathbf{z}})^2}{B^2}(k_2^2 - k_1^2)R_{\mathbf{k},\mathbf{k}_1,\mathbf{k}_2}\cdot \\ \left[\left(k_2^2 - k_1^2\right)W_{\mathbf{k}_1}W_{\mathbf{k}_2} - \left(k^2 - k_1^2\right)W_{\mathbf{k}}W_{\mathbf{k}_1} + \left(k^2 - k_2^2\right)W_{\mathbf{k}}W_{\mathbf{k}_2}\right]\end{aligned} \tag{3.57}$$

with the three-wave resonance function given by

$$\begin{aligned}R_{\mathbf{k},\mathbf{k}_1,\mathbf{k}_2} &= -\operatorname{Im}\left[\frac{1}{\omega_{\mathbf{k}} - \omega_{\mathbf{k}_1} - \omega_{\mathbf{k}_2} + i(\nu_{\mathbf{k}} + \nu_{\mathbf{k}_1} + \nu_{\mathbf{k}_2})}\right] \\ &\simeq \pi\delta(\omega_{\mathbf{k}} - \omega_{\mathbf{k}_1} - \omega_{\mathbf{k}_2})\end{aligned} \tag{3.58}$$

for $\nu_{\mathbf{k}} \ll \max(\omega_{\mathbf{k}})$. In the classical model with adiabatic electrons, the wave-kinetic equation obeys the conservation laws of energy, $d/dt\sum_{\mathbf{k}} W_{\mathbf{k}} = 0$, and enstropy, $d/dt\sum_{\mathbf{k}}(1 + k_\perp^2)W_{\mathbf{k}} = 0$. Detailed derivations (and additional examples) of Eqs. (3.56), (3.57), and (3.58) are given in Horton and Ichikawa (1996) (*Chaos and Structures* in Nonlinear Plasma, ISBN 9810226365, World Scientific, 1996).

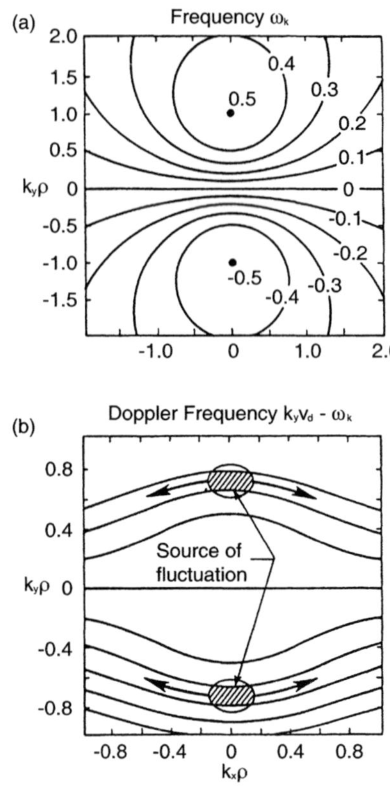

Fig. 3.9 (b) The 2D drift wave frequency spectrum in (a) with both positive negative propagating in the v_d direction. The flux of wave energy from the source regions to the damped modes given by the wave-kinetic equation.

These two invariants, energy and enstropy, lead to the cascade of enstropy, $\Omega = \sum_{\mathbf{k}} k_\perp^2 W_{\mathbf{k}}(t)$ to high $\mathbf{k}_\perp$ and to the inverse cascade of energy W to low $\mathbf{k}_\perp$. The equilibrium solution of the wave-kinetic equation with no growth/damping, i.e. $\gamma_{\mathbf{k}} \equiv 0$, is the equipartition of the energy $W_{\mathbf{k}} = W_{\text{total}}/N_{\text{modes}}$, where N_{modes} is the total number of modes.

In reality, there are limited regions of $\mathbf{k}$-space where $\gamma_{\mathbf{k}} > 0$ and hence where modes can grow from the gradient mechanisms. In all other $\mathbf{k}$-space regions the modes are damped. While some approximate analytic solutions to Eqs. (3.56)–(3.58) are known in terms of anisotropic Kolmogorov-type spectra $1/|k_x|^{m_x}|k_y|^{m_y}$, the general solution to the problem can only be found numerically. For magnetized plasma the turbulent coupling to $k_\| = k_z$ is weaker than the $k_x - k_y$ coupling.

Figure 3.9 shows by the shaded loop in $\mathbf{k}$-space the region of resonant three-wave interaction defined by the subspace $\omega(\mathbf{k}) - \omega(\mathbf{k}_1) - \omega(\mathbf{k}_2) = 0$. There are two special parts of this resonant manifold: one where $\mathbf{k}_1 = (2k_x, 0)$, which is a zonal flow type of state with $u_y(x) = k_x \phi_{\mathbf{k}_1} \sin(2k_x x)$ and $u_x = 0$, and the second region is the quasi-adiabatic part of the resonant manifold where two neighboring waves $(\mathbf{k}, \mathbf{k}_2)$ resonate, i.e. $(\mathbf{k}, \omega_{\mathbf{k}}) - (\mathbf{k}_2, \omega_{\mathbf{k}_2}) = (\mathbf{k}_1, \omega_{\mathbf{k}_1})$ with small values of $\mathbf{k}_1$ and $\omega_{\mathbf{k}_1}$. This second class of interactions gives a scale separation where the symbols $\boldsymbol{q} = \mathbf{k}_1 \ll |\mathbf{k}|$ and $\Omega_{\mathbf{q}} = \omega_{\mathbf{k}_1} \ll |\omega_{\mathbf{k}}|$ are often introduced for the large-scale slowly-varying part of the spectrum. These cross-scale $\mathbf{k} \leftrightarrow \boldsymbol{q}$ interactions give a long-range energy transfer in $\mathbf{k}$-space. These two regions of special interest are in Fig. 3.9.

Figure 3.10 shows an example of the anisotropic Kolmogorov spectrum that applies when there is only an unstable ring for $0.35 < (k_x^2 + k_y^2)^{1/2} \leq 0.4$ with small $\gamma_0 > 0$ and damping with $-\gamma_0$ at both high $\mathbf{k}$ and low $\mathbf{k}$. The steady-state spectra of $W(k_x) = \sum_{k_y} W(k_x, k_y) \simeq 1/|k_x|^{m_x}$ and $W(k_y) = \sum_{k_x} W(k_x, k_y) = 1/|k_y|^{m_y}$ are shown in Fig. 3.10. A majority of the turbulent energy has accumulated in $\mathbf{k}$'s below the growth region $(k_\perp \sim 0.35)$, while the region above the growth region has a power-law decay with $m_x = 3.7 \pm 0.4$ and $m_y = 4.1 \pm 0.4$.

These results are typical but not universal for drift wave turbulence. The adiabatic electron model leads to the nonlinear $\mathbf{E} \times \mathbf{B}$ convection of the density by $\mathbf{v}_E \cdot \nabla n = 0$ vanishing. In general however, there is an appreciable non-adiabatic

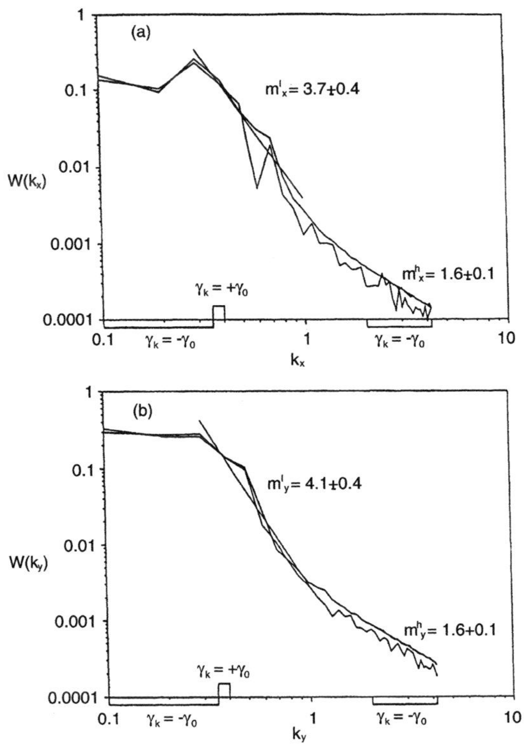

Fig. 3.10 The anisotropic spectral distribution of wave energy for the driven-damped Hasegawa-Mima equation. Spectral indices are computed for comparison with weak turbulence theory.

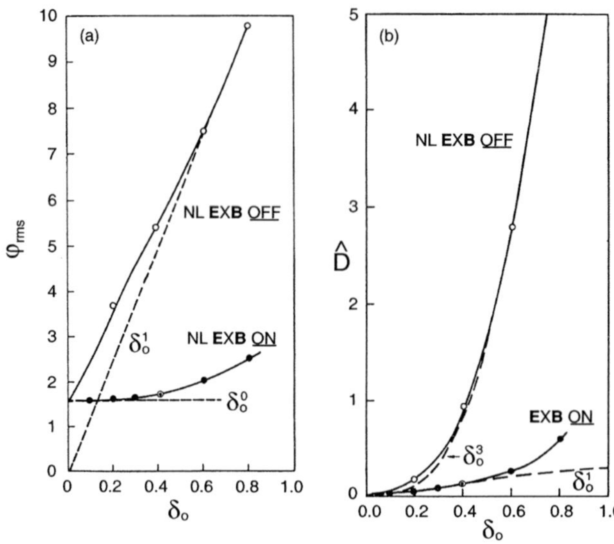

Fig. 3.11 Scaling of the drift-wave turbulence with δ_0. In frame (a) the scaling of the rms potential fluctuation level and (b) the normalized diffusivity D/D_{dw} for the Hasegawa-Mima nonlinearity (open circles) and the self-consistent nonlinear drift-wave Eq. (99) (solid circles) containing the $\mathbf{E}\times\mathbf{E}$ nonlinearity in the convection of the density [Waltz (1983)].

part of the density response to the potential ϕ. The nonadiabatic part, for example, $\delta n = (1+\delta_0\partial/\partial y)\phi \to (1+i\delta_0 k_y)\phi_{\mathbf{k}}$ leads to a nonlinear $\mathbf{E}\times\mathbf{B}$ drift $\mathbf{v}_E\cdot\nabla n_e \neq 0$ and turbulent transport for δn_e. Waltz (1983) has shown that the additional nonlinear term from $\mathbf{v}_E\cdot\nabla n$ leads to a low nonlinear (NL) amplitude as shown in Fig. 3.11.

The limits in Figure 3.11 are consistent with "mixing length rule" used both in neutral fluid and plasma turbulence.

The values of $\phi_{\mathrm{rms}} = \left\langle\phi^2\right\rangle^{1/2}$ and the diffusivity D are shown as a function of the non-adiabatic parameter δ_0. Many dissipative processes can lead to the phase shift represented by the operator $ik_y\delta(k_\perp,\mu)$ in the electron density response [Horton (1976)] where $\{\mu\}$ is the list of system and plasma parameters.

Experiments in the Brazilian tokamak at the University of Sao Paulo Sao Paulo identified the three-wave coupling of drift waves using sophisticated signal processing of the probe data in the edge region of the plasma [Batista, *et al.* (2006)].

3.5 Ion Temperature Gradient Driven Turbulence

There are numerous nonlinear formulations of the ion temperature gradient mode instability in the literature. The models range from one-field models (e.g. Hasegawa-Mima) with one nonlinearity $\mathbf{v}_E\cdot\nabla$, to two-field and three-field models with the fluctuating electrostatic potential ϕ and fluctuating ion temperature. The one-field model, however, uses the adiabatic electron response $\delta n_e/n_e = e\phi/T_e$ for the density fluctuation whereas the two-field model uses either the fluctuating ion pressure, δp_i, or the fluctuating ion temperature, δT_i, dynamical equation. These gyro-fluid models assume a closure that can be either an equation of state with adiabatic gas constant, Γ, for the ion fluid or closure formulas for the parallel ion-thermal-heat flux, $\delta q_{i\,\|}$ [Hammett and Perkins (1990), Mattor and Parker (1997), Mattor (1999), Sugama, *et al.* (2001)].

The value of the adiabatic gas constant Γ should be chosen to be relevant to the specific regime of the plasma and the geometry for the wave-particle interactions. For example, in the collisionless slab the appropriate value for the ITG modes is that $\Gamma = 3$ due to the one-degree-of-freedom ($f_{\text{dof}} = 1$) motion of the ions along the magnetic field lines with the resonance $\omega = k_\| v_\|$. In the torus, however, the resonance of the ions with the waves has $f_{\text{dof}} = 3$ so the appropriate choice of Γ is $\Gamma = 5/3$. In 3D systems the ITG modes become critical in the simplest kinetic dispersion relations when η_i exceeds $\Gamma - 1 = 2/3$. This is the point in the Carnot cycle without dissipation where no net work is done in cycling the plasma from the high to low-temperature regions. Similar considerations apply to the ETG instabilities.

The use of a local or effective Γ is also required for astrophysical gases by Shapiro and Tuekolsky (1983) in their Chapter 3 where the symbol Γ_1 for this locally relevant adiabatic gas constant is introduced. For electrons there is fast thermal transport along the magnetic field lines so the relevant value of Γ_e is often barely exceeding unity. Unity applies when the gas is strictly isothermal. A clear example for the need of a electron gas constant near unity is given by the temperature of the solar wind at the Earth (1 AU) where $T_e \sim 10\,\text{eV}$ for the greatly-expanded flux tube volumes that are connected to the solar corona. The value of 5/3 for the electron gas would have the electron temperature below the recombination value at one astronomical unit rather than 10 eV.

References describing different models are Horton, *et al.* (1980), Hamaguchi and Horton (1990), and Dimits, *et al.* (1991, 2000) for the slab ITG mode and Horton, *et al.* (1981), Waltz, *et al.* (1994), and Beer and Hammett (1999) for the toroidal ITG mode. The toroidal ITG mode is typically stronger than the slab ITG mode and has a mathematical structure similar to a small scale ($\lambda_\perp \gtrsim \rho_i$) Rayleigh-Taylor instability. The Rayleigh-Taylor MHD instability,

$$\omega^2 = -\gamma_{\text{MHD}}^2 = \frac{k_y^2}{k_\perp^2}\left(\frac{1}{\rho_s}\frac{\mathrm{d}p}{\mathrm{d}x}\frac{1}{R_c}\right) \gg k_\|^2 v_A^2 \tag{3.59}$$

changes in drift wave dynamics, to a slower growing mode with $\gamma(k_y = 0) = 0$ and a well-defined maximum growth rate, γ_{max}, at a finite $k_y\rho_s \simeq [(1 - 2\epsilon_n)/(1 + \eta_i)]^{1/2}$. Here $\epsilon_n = L_n/R$ and $\eta_i = \mathrm{dln}T_i/\mathrm{dln}n_i$. The electrostatic shielding by the adiabatic electron fluid introduces a dielectric constant, $\epsilon = 1 + 1/k_\|^2\lambda_{De}^2 \gg 1$. Since $k_\|\lambda_{De} = k_\| v_e/\omega_{pe} \ll 1$. Thus, the ambient background dielectric $\epsilon \gg 1$ reduces the electric fields in the drift wave-interchange instability dynamics.

Quantitative validation studies of flux-tube gyrokinetic Vlasov simulations on ion and electron heat transport are carried out for the JT-60U tokamak experiment. The ion temperature gradient (ITG) and/or trapped electron modes (TEM) driven turbulent transport and zonal flow generations are investigated for an L-mode plasma in the local turbulence limit with a sufficiently small normalized ion thermal gyroradius and weak mean radial electric fields. Nonlinear turbulence simulations by the GKV code (gyro-kinetic-Vlasov) successfully reproduce radial profiles of the ion and electron energy fluxes in the core region. The numerical results show that the TEM-driven zonal flow generation in the outer region is more significant than that in the core region with ITG- and ITG-TEM-dominated turbulence, leading to a moderate transport shortfall of the outward edge region ion energy flux. Error levels in the prediction of the ion and electron temperature gradient profiles in the core region are estimated as less than $\pm 30\%$, based on a multiple flux matching technique, where the simulated ion and electron energy fluxes are simultaneously matched to the experimental values [Nakata, *et al.* (2016)].

The impact of the isotope ion mass on ion-scale and electron-scale microinstabilities such as ion temperature gradient (ITG) mode, trapped electron mode (TEM), and electron temperature gradient (ETG) mode in helical plasmas are investigated by using gyrokinetic Vlasov simulations with a hydrogen isotope and real-mass kinetic electrons. Comprehensive scans for the equilibrium parameters and magnetic configurations clarify the transition from ITG mode to TEM instability, where a significant TEM enhancement is revealed in the case of inward-shifted plasma compared to that in the standard configuration. The GKV simulations show that the ion-to-electron mass ratio dependence on the ratio of the electron one, i.e. $\nu_{ei}/\omega_{ti} \propto (m_i/m_e)^{1/2}$, leads to a stabilization of the TEM for heavier isotope ions. The ITG growth rate indicates a gyro-Bohm-like ion-mass dependence, where the mixing-length estimate of diffusivity yields $\gamma/k_\perp^2 \propto m_i^{1/2}$. On the other hand, there is a weak ion isotope dependence of the ETG growth rate and thermal flux. A collisionality scan showed that the TEM stabilization by the isotope ions becomes more significant for higher collisionality within the banana regime [Nakata, *et al.* (2016b)].

The energy source for ion temperature gradient instability is explained in Section 3.3.2. The typical fastest-growing mode has $\gamma_{\max} = 0.1c_s/(L_{Ti}R_c)^{1/2}$ for a toroidal system with a magnetic radius of curvature, $R_c = R$, on the outside of the torus (of radius R) where the mode balloons in amplitude. The ballooning-mode equation and a simple nonlinear model for the toroidal ITG are given in Horton, *et al.* (1981).

Balescu (2005) gives a four-field extension of the three-fold Hamaguchi-Horton slab ITG formulation. Balescu (2005) describes the ITG mode with the four partial differential equations,

$$\frac{\partial \delta n_i}{\partial t} = -v_{de}\frac{\partial \delta\phi}{\partial y} - \nabla_\parallel \delta v_\parallel - \frac{c_s^2}{\Omega_i}\left[\delta\phi, \delta n_i\right] \tag{3.60}$$

$$\frac{\partial}{\partial t}\left(\delta n_i - \rho_s^2 \nabla_\perp^2 \delta\phi\right) = -v_{de}\frac{\partial}{\partial y}\left(1-\tau(1+\eta_i)\rho_s^2\nabla_\perp^2\right)\delta\phi - \nabla_\parallel \delta v_\parallel$$
$$-\frac{c_s^2}{\Omega^i}\left[\delta\phi, (\delta n_i - \rho_s^2\nabla_\perp^2\delta\phi)\right] \quad (3.61)$$

$$\frac{\partial \delta v_\parallel}{\partial t} = -\frac{1}{m_i n_0}\nabla_\parallel \delta p_i - c_s^2 \nabla_\parallel \delta\phi - \frac{c_s^2}{\Omega_i}\left[\delta\phi, \delta v_\parallel\right] \quad (3.62)$$

$$\frac{\partial \delta p_i}{\partial t} = -v_{de}\tau(1+\eta_i)\frac{\partial \delta\phi}{\partial y} - \Gamma p_{io}\nabla_\parallel \delta v_\parallel - \frac{c_s^2}{\Omega_i}\left[\delta\phi, \delta p_i\right] \quad (3.63)$$

where $\tau = T_i/T_e$, $v_{de} = (\rho_s/L_n)c_s$ and Γ is the ion adiabatic gas constant. Each equation has a convective nonlinearity given by the Poisson brackets.

The earliest example of the slab ITG turbulence is given in Horton, *et al.* (1980) with three partial differential equations. Two versions are given: 1) with full compressional nonlinearity as in MHD for the ion pressure and 2) with reduced nonlinearity of purely $\mathbf{E}\times\mathbf{B}$ convection of nonlinear pressure as in Eqs. (3.62) and (3.63) in the Balescu model. The energy integral for the system takes two different forms for these two versions of the ITG model. The Horton, *et al.* (1980) work solved the nonlinear convection terms with upwind finite difference formulas and found fast relaxation of the pressure profile that was fixed with Dirichlet boundary conditions at the core and the edge of the simulation box. The CRAY-1 computer was used in these early 3D simulations of drift wave turbulence.

There are a number of global and flux-tube codes that simulate the full gyrokinetic equations for the ITG turbulence [Lin, *et al.* (1998), Dimits, *et al.* (2000), Jenko and Dorland (2001), Sugama, *et al.* (2003)]. Here we show the distribution of the electrostatic potential fluctuations in three dimensions in Fig. 3.12 from the Watanabe, *et al.* (2007) code. In this figure where saturation has occurred, the tendency for the fluctuations to establish a zonal-flow pattern is evident from the ringed structure of the potential.

The adiabatic closure in Eq. (3.63) with adiabatic ion gas constant Γ allows the turbulence to be computed with well-developed spectral codes that produce high accuracy in the fluid conservation laws. Dissipation in ITG turbulence occurs through ion-wave resonances and ion-ion collisions ν_i, not shown in Eqs. (3.61)–3.63).

In short-mean-free-path $\lambda_{\rm mfp}$ plasmas, the heat flux is calculated by an expansion in the small parameter $k_\parallel \lambda_{\rm mfp}$. There is a well-accepted and widely-used set of transport formulas given by Braginskii (1965) for the collisional dominated regime with small $k_\parallel \lambda_{\rm mfp}$. Thus, in this book we discuss some problems in terms of the Braginskii fluid equations. In the collisional regime the parallel ion thermal transport is given by

$$q_\parallel = -3.9\frac{n_i T_i}{m_i \nu_i}\nabla_\parallel T_i. \quad (3.64)$$

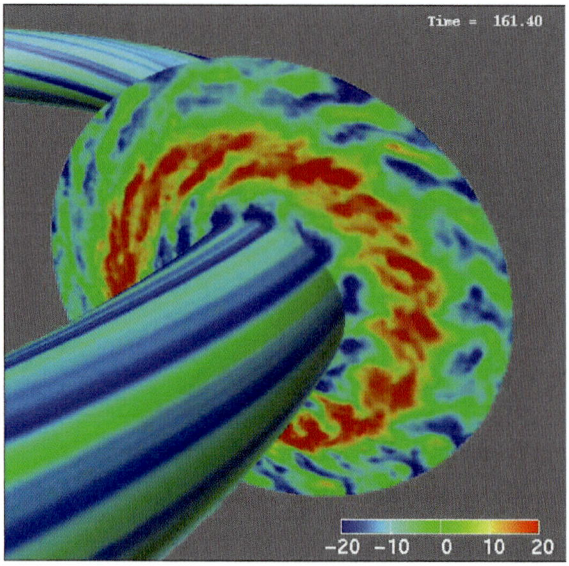

Fig. 3.12 ITG simulation from Watanabe GKV showing the contours of constant electrostatic potential and density ($\delta n/n = e\delta\varphi/T_e$) in the saturated state.

In the problem of nonlinear simulations of ITG turbulence Hamaguchi and Horton (1990) used this formula with a small value of ν_i such that $\max(k_\parallel v_i) = \nu_i$. This cut-off method is known to be an adequate approximation for the nonlocal thermal flux. The alternative is to use Padé approximates for $q_\parallel$.

The critical gradient for the onset of the ITG instability depends on the exact nature of the wave-ion resonance at the marginal stability wave frequency $\omega_{\rm crit}(\boldsymbol{k})$. Nyquist diagrams are used to derive these conditions from a complex dispersion relation $D_k(\omega)$ in both fluid and kinetic theory. In Chapters 13 and 14 the collisionless kinetic dispersion relations are analyzed with Nyquist diagrams. Here we show in Fig. 3.13 the results for the critical ion temperature expressed as R/L_{T_i} for a particular set of plasma parameters representative of the mid-radius of tokamak. The parameters and modeling are described in Dimits, *et al.* (1991) from the team project to compare the results of seven simulation groups listed in the box. The codes used by the seven groups varied from two-component fluid codes to different types of kinetic theory codes. These simulation codes are described further in Chapter 12. Simulation closure models that include the residual zonal flows predicted by Rosenbluth and Hinton (1998), are described by Beer and Hammett (1999) and by Sugama, *et al.* (2007).

3.6 First Experiments for Drift Waves

The first experimental studies of drift waves were carried out in steady-state, low-temperature cylindrical plasmas by Chen (1965a,b) and D'Angelo and Motley

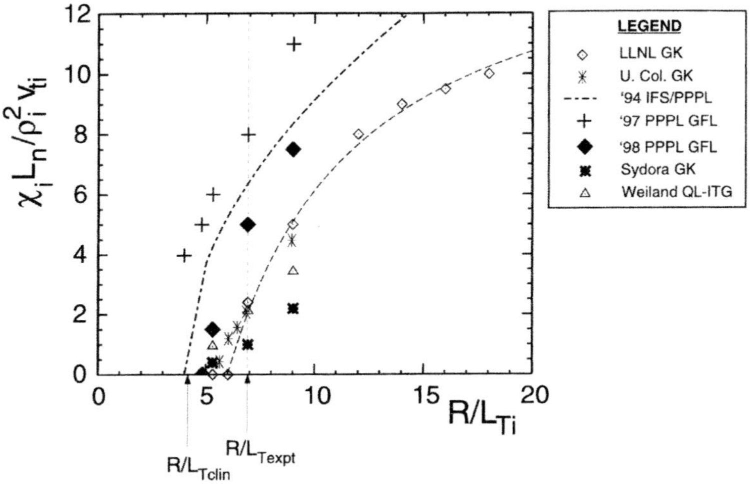

Fig. 3.13 Comparisons of the turbulent ion thermal diffusivity χ_i in units of $\rho_i^2 v_{ti}/L_n$ versus the ion temperature gradient parameter R/L_{T_i} from nonlinear simulation codes with different architectures with the developers listed in the box. The reference plasma parameters such as $a, R, B, q, \nabla T_i, \nabla n, T_e$ were adopted from mid-radius value of a typical tokamak discharge [Dimits, *et al.* (2000)] in a SciDAC project. The electron response was modeled as adiabatic and other simplifications introduced so that the seven codes could simulate the same plasma with the sme reductions. The critical gradient inferred from the experimental data is indicated with subscript expt and that from linear kinetic theory with subscript $T_{\rm clin}$.

(1963). These low-temperature, cylindrical plasmas had a low level of fluctuations in them compared with the toroidal plasmas so they were called Q-machines where Q designated "quiet" plasmas meaning low noise or ambient fluctuation levels. In Q-machine plasma waves could be launched with antennas or produced by instability mechanisms and studied in detail with Langmuir probes owing to the steady, low temperature nature of the plasma. The plasmas were singly ionized gases produced from easily ionized alkali metals of cesium and potassium [D'Angelo and Motley (1963)]. Figure 3.14 shows a schematic of the plasma fields in a typical Q-machine used to study drift waves and ion acoustic waves. There were many of these devices in the US and Europe during the 1960s and 1970s. Figure 3.15 shows in detail the design of one such Q-machine. There is an oven at the right end with a tungsten hot plate at a temperature of 2800 K that ionizes the cesium atoms that strike the plate in what is known as contact ionization. The entire magnetized cylinder then fills with singly ionized cesium ions with a temperature of about $T = 0.25\,\text{eV} = 2700\,\text{K}$. The plasma diffuses by collisions to the region beyond the hot plate radius of $r_p = 12\,\text{cm}$ in this particular device developing a peaked density profile and a uniform temperature profile. This situation spontaneously produces the drift wave instability described in Chapter 3.2. The strength of the instability was controlled by changing the density and the magnetic field in the system. Figure 3.14 shows a typical set of fluctuating density and potential profiles associated

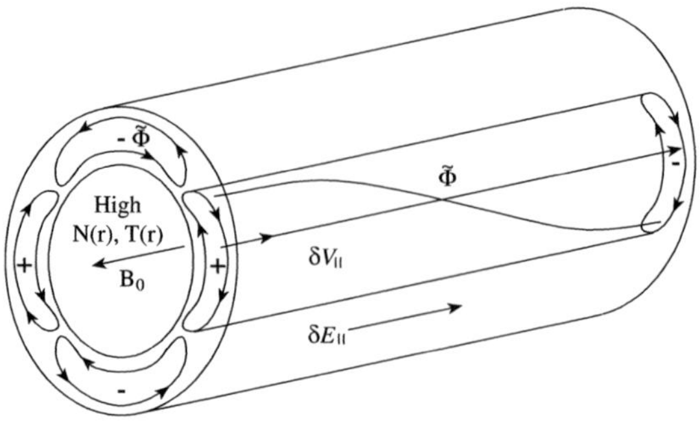

Fig. 3.14 Three-dimensional configuration of the drift-wave fields in a cylinder.

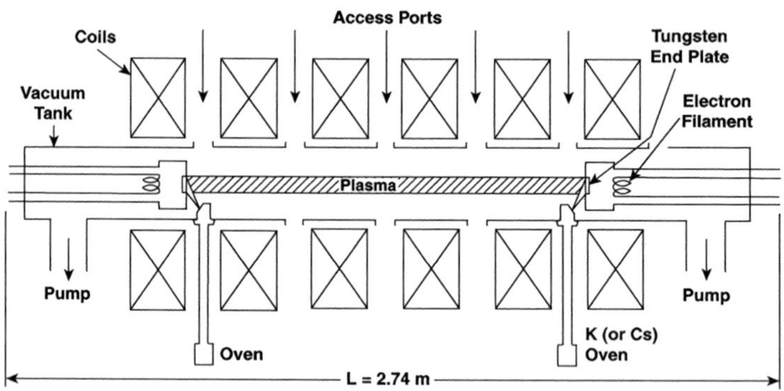

Fig. 3.15 Schematic and key parameters of the Q-machine used in the research of the linear and nonlinear properties of drift waves. Note the spray indicated from the ovens for the potassium (K) or cesium (Cs) atoms on to the tungsten hot plates at both ends of the 1.8 m plasma column [MATT-586 (1968) and Hendel, *et al.* (1968)].

with the drift waves. The measured phase shift with fluctuating density leading the potential is given by the dispersion relation in Chapter 3 and produces the net convection by the $n\mathbf{E}\times\mathbf{B}$ flux from the core radii of higher density to the lower density edge region. The plasma flux was measured in detail and agreed with drift wave theory.

3.7 Drift Waves Paradigms: Hasegawa-Mima and Hasegawa-Wakatani Models

The basic physical mechanisms of drift wave instabilities and the associated plasma transport described in Secs. 3.1-3.6 are complicated so that reduced nonlinear models or paradigms are needed. Reduced models called the Hasegawa-Mima and Hasegawa-Wakatani Models are often used to describe the drift waves. The Hasegawa-Mima model is a one-field first-order 2D partial differential equation with the convection of the vorticity $\omega = \hat{\boldsymbol{z}} \cdot \nabla \times \mathbf{V}_E = \nabla_\perp^2 \varphi / B$ as the single plasma field. The Hasegawa-Wakatani model is a pair of nonlinear differentiation equations for the plasma density and the vorticity then gives a quadratic equation $A_{\mathbf{k}}\omega^2 + B_{\mathbf{k}}\omega + C_{\mathbf{k}} = 0$ in 3D for the linear waves. The model contains the $\mathbf{V}_E$-convection of the plasma density and the vorticity.

Drift waves develop into a mixture of wave turbulence and coherent structures with the details of the mixture depending on the driving forces ∇n_e, ∇T_e and ∇T_i and the amount of magnetic shear and background velocity shear in the system. The mixture of zonal flows, waves and vortices has certain similarities to the Rossby waves in planetary atmospheres and the equatorial-to-polar region temperature gradient driven turbulent heat transport. The principal difference is in the replacement of the gravitational force on planets with a electric force from charge separation in the curved magnetic fields. The Coriolis force has the same mathematical form as the Lorentz force in the momentum balance equations.

To see the reduction of the ITG turbulence to the simple one-field Hasegawa-Mima model, the density fluctuation δn_e is taken as adiabatic so that with quasineutrality $\delta n_i = \delta n_e = n_0(e\delta\varphi/T_e)$ in Eq. (3.61) and the ion acoustic waves are dropped with $\nabla_\| \delta v_\| \to 0$. The vorticity Eq. (3.61) then reduces to a single, simple nonlinear partial differential equation for $d\omega/dt = 0$ with $\omega = \delta\varphi - \rho_s^2 \nabla_\perp^2 \delta\varphi - v_d x$ and the convective derivative given by $\boldsymbol{v}_E \cdot \nabla\omega = [\delta\varphi, \omega]$. In this limit the drift wave equation is isomorphic to the Charney equation of geostrophic flows. This subject is developed in detail in Horton and Hasegawa (1994) and Horton and Ichikawa (1996).

The free-energy sources created by the radial gradients of the plasma density and the temperature are useful to characterize the strength of the drift wave turbulence. Real time diagnostic signals from microwave interferometers and Thompson scattering data are used to compute the thermodynamic formulas for the upper bounds of the turbulence given in Section 3.3 for the density and temperature fluctuations. The derivations of the free energy formulas treat the plasma as an ideal gas with a compressibility gas constant, $\Gamma = (d + 2)/d$, where d is the number of degrees of freedom associated with the type of instability plasma regime under consideration. The Carnot-cycle description of the auxiliary heated plasma is particularly relevant. The method is used in Horton, *et al.* (2005) for the turbulence driven by the electron temperature profiles in Tore Supra and is being used for real-time data analysis in NSTXU and with the CRONOS suite of codes [Artaud, *et al.* (2010)] for ITER.

The structure of the drift wave eigenmodes in the double-periodic toroidal geometry is described after a discussion of the early experiments on drift waves in steady-state machines. Experiments in the Columbia Linear Machine (CLM) [Sen, *et al.* (1991)] explored the ion temperature gradient (ITG) mode without magnetic shear. Six-channel CO_2 laser scattering experiments in the TEXT tokamak [Brower, *et al.* (1987)] clearly showed the role of the hybrid of the ITG mode and

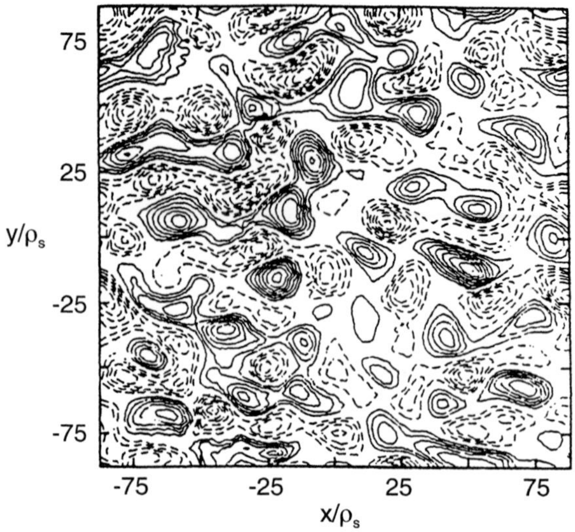

Fig. 3.16 Typical isopotential contours computed from spectral numerical simulations of the driven-damped ion temperature gradient driven drift wave. A saturated state with solid lines for positive and dashed lines for negative values of $\varphi(x,y,z,t)$ shows the complex structure of vortices and elongated radial convection patterns [Horton, *et al.* (1981)].

the trapped electron mode TEM. An example of the isolines of the electrostatic potential contours for the direct numerical simulations is shown in Fig. 3.16. The spectral method of fluid simulations is described in detail in Chapters 6 and 7 of Horton and Ichikawa (1996).

The ITG mode has become the standard model for describing ion transport in basically all toroidal experiments with strong ion temperature gradients. The basic toroidal mechanism and nonlinear models are developed in Horton, *et al.* (1981). Currently, large-scale Tera-flop computers commonly compute the ITG turbulence from direct numerical solutions of the gyrokinetic-Vlasov-Poisson system. Examples of the state of the turbulence are shown in Chapters 13-15 from the large-scale simulations. The standard article on the comparison of direct numerical simulation for the ITG turbulence is Dimits, *et al.* (1991). Review of the state of the numerical simulations of the ETG turbulence is Nevins, *et al.* (2006). The electron turbulent transport is more demanding to compute but shows larger enhancements over the quasilinear fluxes due to the formation of extended radial structures with correlation lengths, $\ell_c \gg \rho_e$. The structures are insensitive to the radial electric-field shear but do depend sensitively on the magnetic shear profile.

The empirical energy confinement scaling laws for both helical system [Yamada, *et al.* (2005)] and tokamaks [Kaye, *et al.* (1997)] follow the drift wave formulas and simulations to within the accuracy of the experimental data. Some of the more detailed comparisons of theoretical models [Kinsey, *et al.* (1996a), Kinsey and Bateman (1996b), Redd, *et al.* (1998), Bateman, *et al.* (1998), Snyder, *et al.* (2012),

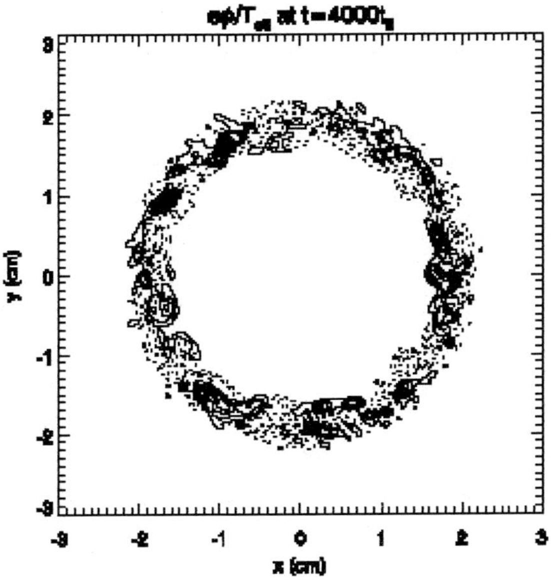

Fig. 3.17 Example of an electron temperature gradient driven drift wave spectrum in the nonlinear state with a spectrum of vortex structures with outward extending plumes that produce an anisotropy in the radial and poloidal drift wave fluctuation spectrum [Fu, *et al.* (2012)].

Sun, *et al.* (2016)] with tokamak databases are described in Chapters 13 to 15. Tokamaks [Kaye, *et al.* (1997)] follow the drift wave formulas and simulations to within the accuracy of the experimental data. Some of the more detailed comparisons of theoretical models [Kinsey, *et al.* (1996a), Kinsey and Bateman (1996b), Redd, *et al.* (1998), Bateman, *et al.* (1998), Snyder, *et al.* (2012), Sun, *et al.* (2016)] with tokamak databases are described in Chapters 18.

The validation of a gyrokinetic code for the electron temperature gradient driven turbulence in a cylindrical plasma with constant B is shown in Fig. 3.17. The simulations with the gyrokinetic code (GTC) agree with the steady-state 10 eV hydrogen ETG experiments carried out in the Columbia Linear Machine (CLM) by Sen, *et al.* (1991). The GTC codes were developed by Lin, *et al.* (1998). The validation of the GTC code with the CLM experiment was a SciDAC project [Fu, *et al.* (2012)]. In the initial linear growth phase of the instability high mode numbers $m > 50$ dominate – and then a rapid coalescence of the smaller vortices. The steady turbulence with the dominant $m = 11, 12, 13$ modes in Fig. 3.17 follows in the final stage which agrees with the measured turbulence in the CLM plasma. The close agreement with the steady-state data validates the ETG turbulence model and the simulation code.

Practical gyrofluid codes named GS-2 and GENE are available for modeling and interpreting tokamak data with the ITG and TEM instabilities. An early success of these gyrofluid simulations is shown in Fig. 3.18, where neutral beam

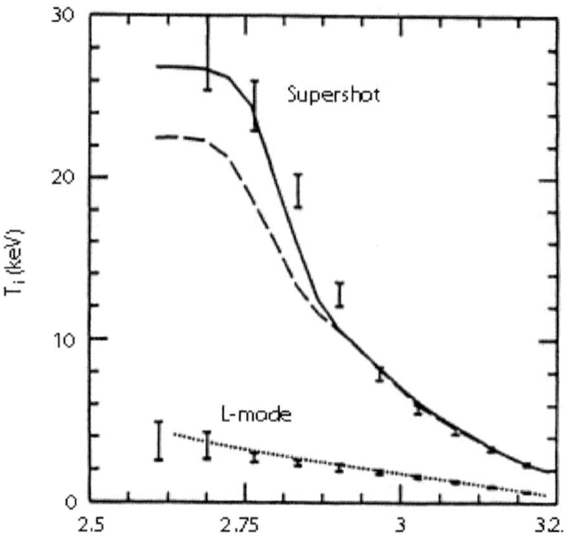

Fig. 3.18 Comparison of the IFS-PPPL model for an L mode and a supershot discharge in TFTR. The complex χ_i formula is able to account for the large increase in the core ion temperature from the experimental values of T_i with the boundary condition from edge data [Kotschenreuther, *et al.* (1995a,b)].

driven discharges in the TFTR tokamak were shown to be interpreted by using nonlinear gyrofluid simulations to construct a complicated, but comprehensive interpolating formula $\chi_i(\{\mu\})$ for the ion thermal conductivity. The χ_i formulas with a large parameter set $\{\mu\}$ taken from the TRANSP transport code were able to reproduce the bifurcation of the plasma to the supershot regime shown in Fig. 3.18. The formula derived from a large number of nonlinear simulations by Kotschenreuther, *et al.* (1995a,b) are useful for modeling and predicting the turbulent ion transport.

The topic of submarginal turbulence and self-organized criticality is reviewed by Krommes (1999, 2002).

The GS-2 [Kotschenreuther, *et al.* (1995b)] is currently fully electromagnetic with three ion species designed for running on massively parallel computers. The electron temperature profile in NSTX is resilient [Gates, *et al.* (2002, 2007)] and consistent with the fast ETG turbulence.

Reduced magnetohydrodynamics is a simplified subset of the full MHD equations. The equations are not in the canonical Hamiltonian form but do have a number of conserved functionals. Kraus, *et al.* (2016) developed a discretization for the reduced MHD equations using a variational principle applied to a Lagrangian. The Lagrangian equations then conserve the plasma energy, magnetic helicity and cross helicity exactly. The integrator is free of numerical resistivity, preventing spurious magnetic reconnection. When the electron inertia term is added, magnetic reconnection is allowed through the Hamiltonian structure.

Worked Problem 3.4

Problem: Find Computational Methods for the Dispersion Function $Z(\zeta)$

(a) Direct Path Integral

According to Landau's prescription, the integral path needs to be deformed properly to make $Z(\zeta)$ analytic when ζ moves across the real axis. This requires that the singular point of the v-integral be always on the same side of the integration path. For $\mathrm{Im}(\zeta) > 0$, there is no problem. However, for $\mathrm{Im}(\zeta) = 0$, the path of integration is a deformed semi-circle under the singular point. Using Cauchy's theorem for analytic functions, the integral over this semi-circle gives $i\pi$ times the residue. Similarly, for $\mathrm{Im}(\zeta) < 0$, the integral over an extra circle around the singular point is needed and gives $2i\pi$ times the residue. We derive

$$\begin{aligned} Z(\zeta) &= \pi^{-1/2} \int_{-\infty}^{\infty} \frac{\exp(-t^2)}{t-\zeta} dt \\ &= \pi^{-1/2} \left(PV \int_{-\infty}^{\infty} \frac{\exp(-t^2)}{t-\zeta} dt + R \right) \end{aligned} \tag{3.65}$$

where R is the residue from the singularity $\omega = kv$

$$R = \begin{cases} 0 & \mathrm{Im}(\zeta) > 0 \\ i\pi \exp(-\zeta^2) & \mathrm{Im}(\zeta) = 0 \\ 2i\pi \exp(-\zeta^2) & \mathrm{Im}(\zeta) < 0 \end{cases} \tag{3.66}$$

given by the Cauchy Theorem and PV designates the principal value integral. This method will encounter difficulty when evaluating the principal-value (PV) integrals for $y = 0$. The key trick to evaluate PV integrals is to use symmetric intervals on both sides of the singular point. GNU GSL (`https://www.nersc.gov/users/software/programming-libraries/math-libraries/gsl/`) library provides a special integrator QAWC for PV integrals.

(b) Continued Fraction Evaluation

The plasma dispersion function can be related to the complex error function through

$$Z(\zeta) = 2ie^{-\zeta^2}[1 + \mathrm{erf}(i\zeta)]. \tag{3.67}$$

The error function $\mathrm{erf}(z)$ can be calculated using Gauss' continued fraction [Jones and Thron (1980)]

$$\frac{\sqrt{\pi}}{2} e^{z^2} \mathrm{erf}(z) = \cfrac{z}{1 - \cfrac{z^2}{\frac{3}{2} + \cfrac{z^2}{\frac{5}{2} - \cfrac{\frac{3}{2}z^2}{\frac{7}{2} + \cfrac{2z^2}{\frac{9}{2} - \cfrac{\frac{5}{2}z^2}{\frac{11}{2} + \cfrac{3z^2}{\frac{13}{2} - \ddots}}}}}}} \tag{3.68}$$

Therefore, the plasma dispersion function can be numerically evaluated by (a) the path integral or (b) the continued fraction.

3.8 Rotational Instabilities and Momentum Transport

Analyzing the sheared flow through the modal approach, commonly used by the neutral fluid turbulence community, reveals that the linear operators describing sheared flow are non-normal and that the non-normality results in transient growth and mode coupling with the intrinsic drift waves and Alfvén waves. In cases where the nonlinear feedback is positive, this linear transient growth can produce self-sustaining vortices that affect the drift wave turbulent transport [Kim, *et al.* (2006)]. The momentum in the rotational flows is transported by drift waves through the turbulent $\mathbf{E} \times \mathbf{B}$ convection as is plasma mass and momentum. Thus the turbulent plasma transport problem becomes generalized to include the turbulent transport of the momentum. In hydrodynamics the turbulent flux of momentum is given the name of Reynolds stress tensor. When the turbulence is driven by temperature gradients then the Reynolds stress tensor can create a net circulation of fluid flows. This is the mechanism that creates dust devil vortices in the atmosphere [Horton, *et al.* (2016)].

In experiments at low Prandtl number $\Pr = \mu/X$ of the Benard convection, Krishnamurti and Howard (1981) showed the onset of a large vortex of sheared flow once the thermally driven transport exceeded a critical level. Low-order models of the plasma physics equivalent of this situation have been developed in a number of magnetic confinement theories including Diamond, *et al.* (1997), Horton, *et al.* (1996), Hu and Horton (1997), Sugama and Horton (1998), Kishimoto, *et al.* (2000), and Uzawa, *et al.* (2009). In magnetic confinement in tokamaks the interest has been driven by the discovery of the transition from the L-mode to the H-mode of plasma confinement when the heating power exceeds a critical level. The first L-H transitions were observed by Wagner, *et al.* (1982) in ASDEX and then subsequently in most major tokamak facilities. These transitions are consistent with a drop in the thermal gradient driven turbulence level and the onset of strongly sheared rotation in the edge region of the plasma. In Chapter 12.4 these types of transitions are called edge transport barriers. The 1st Edition analyzes turbulent transport in two JET and DIII-D fusion-grade plasmas with transport barriers.

The situation in the toroidal plasma is complicated since there are two components of the flow. The toroidal rotation is in a direction of rotational symmetry and is arbitrary until the symmetry breaking external heating and transport processes themselves are included in the transport equations. Sufficiently strong plasma heating produces a secondary parametric instability that creates toroidal plasma rotation. The fastest toroidal rotations are up to a substantial fraction of the sound speed, that is with speeds of order a 100 km/s, which are common in the neutral beam driven plasmas. For beam (NBI) driven plasmas the degree of imbalance between the two rotational directions is often varied which is described as balance or unbalanced injection. Rotation in the direction of the toroidal current is called co-rotation.

Experiments on DIII-D have focused on elucidating the drive mechanisms for intrinsic rotation in tokamak fusion plasmas. In H-mode plasmas, the effective torque at the edge ($\rho > 0.8$) associated with the intrinsic rotation shows a dependence on the pedestal pressure gradient ∇P_{ped}, which is qualitatively consistent with models describing $\mathbf{E} \times \mathbf{B}$ shear as a means of creating 'residual stress' and driving intrinsic rotation. However, direct measurement of the turbulent Reynolds stress using probes suggests that this is not the full picture. Specifically, there is a significant mismatch between the plasma spin-up and the inferred torque from the Reynolds stress at the edge, indicating that additional mechanisms are necessary to understand intrinsic edge rotation generation. A narrow rotation layer is observed near the separatrix, which can qualitatively be explained using a model of thermal ion orbit loss. Parametrically, the torque from such a process is expected to increase with $T_i^{1/2}$. A good predictor of the edge intrinsic torque is obtained by including this dependence, together with the previously observed ∇p_{ped} dependence, in a regression fit of a wide range of H-mode conditions. The intrinsic torque in the core ($\rho < 0.5$) of H-mode plasmas tends to be much smaller than observed at the edge, although some examples have been found where it is large enough to modify the rotation profile. For instance, in certain plasmas with electron cyclotron heating, a significant counter-intrinsic torque has been observed in the inner region of the plasma [Solomon, *et al.* (2011)].

In tokamaks the plasma current defines a preferred direction around the torus. Thus, the rotation is described as co-rotation in the direction the current and counter rotation in the opposite direction. One can show that for up-down symmetric toroidal flux surfaces there is a symmetry in the transport of angular momentum in the co- and counter direction. But, in single null divertor machines like JET, ASDEX and most modern tokamaks, the top-to-bottom symmetry is broken with the consequence there is net radial flux of toroidal angular momentum a torque force at zero rotation speed driving the plasma into rotation spontaneously. One sees the preferred direction from the cross product of the vector to the closer divertor separatrix x-point with vector in the direction of the toroidal plasma current. The two configurations are shown in Fig. 3.19.

The mechanisms of nonlinear interaction of external helical fields [Lazzaro, *et al.* (2002)] with a rotating plasma are investigated analyzing the results of a series of experiments in the Joint European Torus (JET) [Gibson and the JET Team (1998)]. The equilibrium neoclassical transport equations for these rotating plasmas are described in Hirshman and Sigmar (1997), Sugama and Horton (1998), Helander and Sigmar (2005), and Newton and Helander (2006), and the broken symmetry arguments are given in Peeters and Angioni (2005) and Parra and Catto (2010) shown in Fig. 3.19. There is a generalization of the Grad-Shafranov equilibrium

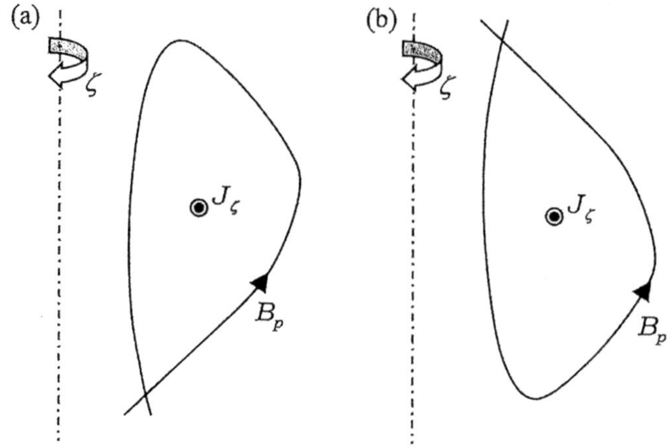

Fig. 3.19 Example of single null divertor configuration with broken top-down symmetry [Parra and Catto (2010)].

equation that gives a large class of equilibrium solutions for rotating ideal MHD plasmas [Goedbloed, *et al.* (2010), Sugama and Horton (1997, 1998)]. Chapter 9 describes the symmetry-breaking effect on the impurity transport.

In the torus the population of mirror-trapped ions in the outboard weak magnetic field regions gives rise to a frictional drag on the plasma rotation in the poloidal direction. This mirror-trapped ion-population friction results in a low-poloidal drift speed with respect to the high-speed toroidal rotation. The equation for the residual poloidal rotation rate is specified by the ion temperature gradient and the ion-ion collisionality regime of the plasma as derived in Hirshman and Sigmar (1997), Balescu (1988), Sugama and Horton (1997, 1998) and Helander and Sigmar (2005).

generalizations of the Grad-Shafranov equilibria are required when there is strongly peaked neutral beam injection [Cooper, *et al.* (1980)], strong LHCD [Hoang, *et al.* (2009)] heating, or a strong runaway electron component to the plasma [Knoepfel and Spong (1979)] as shown in these references.

3.9 Rotation in Cylindrical Plasmas

For a uniform axial magnetic field $\mathbf{B} = B\widehat{e}_z$ and a radial electric field $E_r(r)$ the plasma rotates at the angular frequency $\Omega(r) = -E_r/rB$ with respect to the laboratory frame in which the electric field is specified. In the frame moving with the plasma at this angular rotational velocity $\Omega(r)$ the electric field vanishes and there is a centrifugal force radially outward given by $F_r(r) = n_i m_i v_\theta^2/r$ and there is the Coriolis force $\boldsymbol{F}_{\rm cor} = -n_i m_i 2\boldsymbol{v} \times \boldsymbol{\Omega}$ acting on the plasma. The centrifugal force adds to the effective gravity term $g_{\rm eff}$ introduced in Chapter 4 for describing the effect of the curvature of the magnetic field. So we expect and find instability of the interchange type analyzed in Chapter 4 when the rotational velocity is large and density profile is monotonically decreasing. When the ion density profile is hollow, however, plasmas with large rotation velocities are stabilizing as shown in experiments in mirror machines at the Budker Institute of Nuclear Physics and the University of Maryland [Beklemishev and Horton (1992)].

Now we consider the drift wave equations in the presence of the radial electric field. First we consider electrostatic dynamics in a low-dimensionless pressure, $\beta \ll 1$, plasma. Taken into account are the cross-field currents driven by the electrostatic potential through the divergence of the collisionless ion momentum stress tensor [Rosenbluth, *et al.* (1962), Rosenbluth and Simon (1965)]. The eigenmode equation for the rotating plasma with an effective radially outward acceleration $g(r)$ from the sum of the magnetic curvature and the centrifugal force is

$$\nabla \cdot \left[n \left(1 - \frac{\omega_{*i}(r)}{\omega} \right) \nabla_\perp \varphi \right] + \left[\frac{1}{r} \frac{d}{dr} \left(\frac{n\omega_{*i}(r)}{\omega} \right) + \frac{m^2 g}{r^2 \omega^2} \frac{dn}{dr} \right] \varphi = 0. \tag{3.69}$$

The eigenvalues of the higher wavenumber modes $m \gg 1$ are found by using the local approximation to reduce Eq. (3.69) to the local algebraic dispersion relation

$$k_\perp^2 \omega \left(\omega - \omega_{*i} \right) - k_\theta^2 \frac{g}{n} \frac{dn}{dr} = 0 \tag{3.70}$$

with $k_\theta \gg k_r$.

In the absence of the FLR term ω_{*i} the dispersion relation is the same as that for internal gravity waves in hydrodynamics. Stably-stratified layers in hydrodynamics have the buoyancy frequency $N^2 = -g(d \ln n/dr) = g/L_n$ [Gill (1982)].

From the quadratic Eq. (3.70) one sees that the ion diamagnetic frequency ω_{*i}^2 is stabilizing against the interchange term from $g d \ln n(r)/dr$. Thus, from the condition of a negative discriminant of the quadratic equation for ω, one finds that the instabilities are limited to mode numbers below a critical mode number $m < m_{\text{crit}}$ determined by ω_{*i} relative to the growth rate at $\omega_{*i} = 0$. Stably-stratified layers corresponding to the atmospheric waves would have $n(r)$ increasing with radius.

To determine the stability and transport, one solves the eigenvalue problem. The problem is generally numerical requiring the use of a "shooting code" that repeatedly integrates the second-order differential equation from the $r = 0$ to $r = r_{\text{edge}}$ with appropriate boundary conditions at both end points [Gladd and Horton (1973)]. The exception occurs for a rigidly rotating plasma with $\Omega = \text{const}$. For rigid rotation, there is an analytic solution of Eq. (3.69). Shown in Fig. 13.20 for constant angular velocity Ω in the ion diamagnetic direction as typical for mirror cells.

For the case of a rigidly rotating plasma with $\Omega = \text{const}$, the electric field is given by

$$E_r = -r\Omega B \tag{3.71}$$

with a plasma potential Φ given by

$$\Phi(r) = \frac{1}{2} r^2 \Omega B \tag{3.72}$$

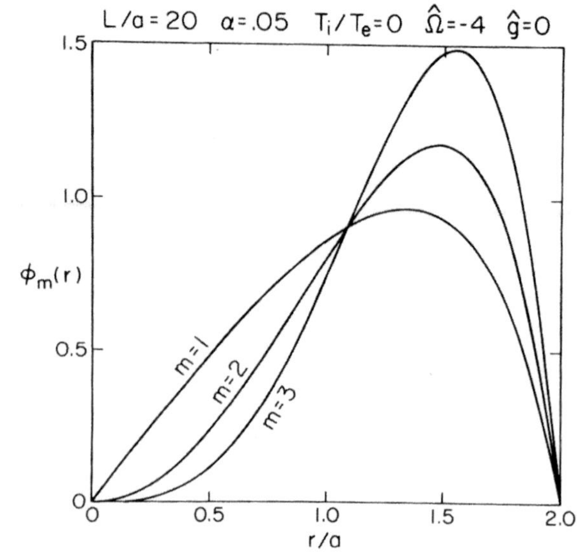

Fig. 3.20 Radial eigenfunctions for the rigidly-rotating plasma with cold ions. The density profile is gaussian so the electron diamagnetic frequency is constant in radius. In this example the $\Omega = -4$ means the rotation is four times the electron diamagnetic frequency and in the ion diamagnetic direction. The choice of positive E_r from a positive plasma potential with $\Phi(r = 0) \sim 4T_e$ is characteristic of open-ended systems. In open systems electrons rapidly leave on the open magnetic field lines until the plasma potential becomes sufficiently positive to slow their loss rate to balance that of ions. This typically requires $\Delta\phi = \phi(0) \sim 4T_e$.

that is a quadratic potential well or hill, depending on the direction of rotation. For example, core ohmic tokamak plasmas have $E_r < 0$ with rotation in the electron diamagnetic direction. Core mirror plasmas have $E_r > 0$ to confine electrons escaping through the mirror loss cone with rotation in the ion direction in Fig. 3.20.

For the rigid rotation potential, the Boltzmann phase space density for the ions is a Gaussian density profile

$$n(r) = n_0 \exp\left(-\frac{r^2}{a^2}\right). \tag{3.73}$$

The electrons rapidly adjust to have charge neutrality $n_e = n_i$ with $a^2 = k_B T_i/eB|\Omega|$. In simple mirror machines (without biasing end plates as used often) the electrons rapidly leave the ends of the machines so that $E_r > 0$ and the rotation is in the ion diamagnetic direction. An example for the TAE machine is described in Chapter 15. In the GDT machine the E_r field is externally controlled in the large-end expansion tanks by electric biasing of a concave screen intersected by the expanding magnetic field [Bagryansky, *et al.* (2015)].

For this equilibrium with solid body rotation the differential Eq. (3.69) reduces to a form that can be solved with confluent hypergeometric function $F(a, c, x)$ or the Whittaker function $W_{\mu,\kappa}(x)$. The details are given in Rognlien (1973) and

Mikhailovskii (1992). The eigenvalues and eigenfunctions for solutions of Eq. (3.69) for rigid body rotation equilibrium are found by a transformation to the Whittaker functions with

$$\varphi = \frac{e^{\frac{r^2}{2a^2}}}{r} y\left(\frac{r^2}{a^2}\right) = \frac{e^{x/2}}{x^{1/2}} y(x) \tag{3.74}$$
$$y = C_1 W_{\frac{(1+\nu)}{2}, \frac{m}{2}}(x) + C_2 W_{\frac{(1+\nu)}{2}, \frac{-m}{2}}(x)$$

following Rognlien (1973). The boundary condition gives the eigenvalues

$$\nu = m + 2n \qquad n = 0, 1, 2, \ldots \sim \frac{1}{\Gamma\left(\frac{1}{2} + \frac{m}{2} - \frac{1+\nu}{2}\right)} \longrightarrow 0.$$

Using these eigenvalues, the dispersion relation becomes

$$\omega^2 - \omega_{*i}\omega\left(1 - \frac{1}{\nu}\right) + \frac{m^2 g_0}{a(\nu)} = 0 \tag{3.75}$$

where we define g_0 by $g = r g_0/a$ where $g_0 = aV_T^2/L_t^2$. The distance between the minimum and maximum of the mirror magnetic field is characterized as L_t and the plasma radius is a.

The roots of Eq. (3.75) are FLR stabilized for

$$\frac{\rho_i}{a}\frac{m-1}{\sqrt{m}} > \frac{a}{L_t}. \tag{3.76}$$

This FLR stabilization condition in Eq. (3.76) is a famous method for stabilizing mirror machines and field-reversed confinement systems. However, the $m = 1$ mode behaves differently from the $m = 2, 3$ and higher-order azimuthal modes. In the experiments the $m = 1$ modes appears in the nonlinear state as a "wobbling" of the plasma column in the central cell of a long mirror machine. This topology of the $m = 1$ mode is a displacement of the core plasma. The $m = 1$ mode appears in the simulations and the laboratory as coherent displacement of the entire plasma rotating at the mean rotation rate Ω of the plasma. As the rotation rate increases above a critical value the plasma accelerates into the wall of the confining vessel.

Here we take the plasma wall as a perfect conductor of radius $r = b$ giving the Dirichlet boundary condition $E_\theta(r = b) = 0$. Fitzpatrick, *et al.* (1993) and others consider resistive walls which are more realistic but give rise to slow resistive wall modes.

Figures 3.20 and 3.21 shows the lowest radial eigenmodes for the azimuthal mode numbers $m = 1$, 2 and 3 in a collisionless plasma with a gaussian density profile and constant electron temperature profile. The profile of the annular rotation frequency $\Omega(r)$ is first taken as constant in Fig. 3.20 and then taken as monotonically decreasing from a core value of $\Omega(0)$ to an edge value of $\Omega(r = b)$ in Fig. 3.21. The rotational shear weakens the interchange dynamics and the growth rates.

In Figs. 3.20 and 3.21 there is a conducting wall at $r = b$ at which $\phi_m(r = b) = 0$ since otherwise the wall would have an tangential electric field

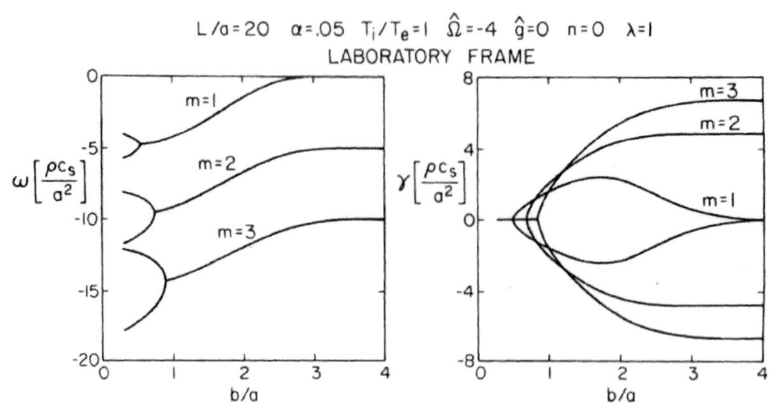

Fig. 3.21 For the case shown in Fig. 3.20 except the ion temperature is taken equal to the electron temperature and the distance b to the conducting wall relative to the plasma radius a of the plasma is taken as the variable parameter. The left panel shows how the two stable frequencies at sufficiently small b/a bifurcate into the complex plane as the wall is moved outward. The first mode to go unstable is the $m = 1$ mode whose frequency subsequently drops to zero as the wall is move far out from the plasma. This is due to a rigid displacement being a symmetry of the equilibrium in the absence of the wall. Next the $m = 2$ and $m = 3$ modes are destabilized and their growth rates shown in the right panel increase monotonically to the "no wall" values. By using coils with the $m = 2$ and $m = 3$ symmetries on the walls a stabilizing feedback control system can be designed to control these instabilities.

$E_\theta = \mathrm{Re}[(-im/b)\varphi(b)]$ contrary to the surface being a perfect conductor. If the wall is thin and the resistivity of the wall is finite, then we must modify this boundary condition as in Fitzpatrick, *et al.* (1993) to be that of a resistive wall. The resistive skin time of the wall compared to the growth/damping rate of the plasma wave then becomes a key parameter in the problem. Here we give the results for the perfectly conducting wall meaning that the skin time is long compared with the growth time.

Figure 3.20 shows the eigenfunctions and Fig. 3.21 shows the frequencies $\omega(m, b/a)$ and growth rates $\gamma(m, b/a)$ for the lowest azimuthal mode numbers in a conducting wall cylinder with a constant temperature gaussian density hot plasma. The $m = 1$ wobble mode has a qualitatively different behavior from the $m = 2$ and 3 modes. The $m = 2$ and 3 modes are unstable with the wall well away from the plasma, while the $m = 1$ mode is a stable oscillation in this case. When the wall is moved closer to the plasma by decreasing the wall radius b divided by the radius a of the plasma profile, the $m = 2$ and 3 modes are stabilized, while the $m = 1$ mode is destabilized, as shown in Fig. 3.21. This behavior is confirmed in the laboratory experiments.

The $m = 1$ mode is a displacement of the plasma towards the conducting wall with zero frequency in the rest frame of the plasma. Thus we know from electrostatics that there is an induced image charge in the conducting wall. There is then an attractive force between the charges in the conductor and those in the plasma that give rise to the motion of the plasma toward the wall. The effect saturates at a certain relatively low amplitude of the displacement and so a new nonlinear state of the rotating plasma is created. This type of nonlinear rotating displaced state was routinely observed in the central cells of the TMX experiment at LLNL [Hooper, *et al.* (1982)] and in other rotating plasma experiments. The dangerous rotational instabilities with respect to magnetic fusion experiments then starts with the $m = 2$ instability. Many small ion gyroradius experiments end in the crash created by the fast-growing $m = 2$ instability. The large ion gyroradius experiments like the Tri Alpha RFC are stable (`http://www.trialphaenergy.com/`) but also see the $m = 2$ component in the plasma rotation.

Figure 3.21 gives an overview of the instabilities for the central cell of the mirror confined plasma with $E_r > 0$ as the conducting wall at b increases with respect to radius a of the plasma. The analysis shows that at higher mode numbers the growth rate comes back to zero. This effect is due to the large ion gyroradius effect producing a stabilizing charge separation from the electrons. The effect is contained in the dispersion relation through the $B^2 > 4AC$ condition for $\omega_\pm$ in the discriminant of the quadratic equation. The B^2-term arises from the frequency ω_{*i} in Eq. (3.75). This key stabilizing effect is called FLR stabilization for the finite-Larmor-radius stabilization arising from the electron-ion charge separation. The effect is contained in Chapter 2 in the Bessel functions I_0 and I_1 in, for example, Eqs. (2.10)–(2.20) that arise from the gyro-orbit averaging described in Worked Problem 1.1 in Chapter 1.

The clearest data for fusion grade plasma comes from the central cells of Tandem Mirror Machines which are hot, MHD-stable mirror-confined plasmas. Examples are from GAMMA 10 at the University of Tsukuba, the TMX at LLNL from 1970-1980 and the field reversed configuration (FRC) experiments at University of Washington and Tri Alpha Energy machine [Binderbauer, *et al.* (2015)].

The axisymmetric tandem mirror in the plasma is stabilized by having sufficient plasma pressure beyond the end-cells to have the total line integral of the pressure gradient times the magnetic curvature strength throughout the length of the machine to have the MHD stability. The condition on how much plasma pressure and injected power is required to satisfy this global stability condition is developed in Berk and Pratt (2011).

3.9.1 *Shear Alfvén wave stabilization of the rotational instabilities*

The eigenvalue equation for mode number m with sheared rotation $\omega_E(r)$ and with Alfvén waves is

$$\begin{aligned}\frac{1}{r}\frac{d}{dr}\left[rn\left(\omega-\omega_E\right)\left(\omega-\omega_E-\omega_{*i}\right)\frac{d\varphi}{dr}\right]-\nabla_\perp\cdot\left(nk_\parallel^2v_A^2\nabla_\perp\varphi\right)\\+\left[\left(1-m^2\right)\left(\omega-\omega_E\right)\left(\omega-\omega_E-\omega_{*i}\right)n+\left(r\Omega^2+g(r)\right)\frac{dn}{dr}\right]\varphi=0.\end{aligned}\tag{3.77}$$

This form of the eigenmode equation makes clear that the $m = 1$ mode behaves differently from the $m = 2$ and 3 modes as shown in Fig. 3.21 where $\widehat{\Omega} = \omega_E/\omega_{*i}$.

Now, the radial plasma displacement ξ, derived within the approximation of $\mathbf{E} + \boldsymbol{v} \times \mathbf{B} = 0$ motion bends the magnetic field lines which launches shear Alfvén waves with $\omega^2 = k_\parallel^2 v_A^2$. The dispersion relation with Alfvén waves for $a \gg c/\omega_{pi}$ becomes

$$\omega\left(\omega - \omega_{*i}\right) - k_\parallel^2 v_A^2 + \frac{g_{\text{eff}}}{r_p} \frac{k_\theta^2}{k_\perp^2} = 0 \tag{3.78}$$

where $g_{\text{eff}} = r\Omega^2 + g(r)$; where the $g(r)$ is from the curvature of the magnetic field lines. In the region close to the axis ($r = 0$) of a cylindrically-symmetric field $\mathbf{B} = B_r(r,z)\boldsymbol{e}_r + B_z(r,z)\boldsymbol{e}_z$, one derives that $g(r) = rC_s^2/L_t^2$ where $1/L_t^2 = -\frac{1}{B_z}\frac{d^2B_z}{dz^2}$ is positive at B_z maxima in the throat of the magnetic mirror. On the symmetry axis $r = 0$, the field lines are straight with $B_r = 0$. Thus, both the magnetic curvature of the field lines and the centrifugal force from the rotation combine to form the magnetic form of the Rayleigh-Taylor instability in the plasma column. Any variation of plasma motion along the field lines gives the Alfvén wave stabilization from $k_\parallel^2 v_A^2$ arising from the tension force.

The instability now requires that the wave function be sufficiently flute-like that

$$\frac{g}{r_p}\frac{k_\theta^2}{k_\perp^2} > k_\parallel^2 v_A^2 + \frac{\pi}{4}\,\omega_*^2 \tag{3.79}$$

is satisfied for the $m \geqq 2$ in stable mirror plasmas. The boundary conditions at the ends of the flux tubes determine when this line bending stabilizes the rotational instabilities. The condition of a vanishing plasma displacement at the ends of the flux tube is called the "line tying" boundary condition and can eliminate the flute modes that are most strongly unstable. The boundary conditions at the ends of actual machines, or flux tubes in the magnetosphere and solar corona, are complicated as shown in Berk, *et al.* (1983) and for the ambipolar trap by Dimov (2005).

The details of the analysis of the eigenvalue analysis with the line-bending stabilization is given by Janssen (1982). The stability diagram of $k_\parallel v_A/\omega_{*i}$ versus Ω_E/ω_{*i} is given for the $m = 2$ mode and lowest-order radial eigenmode. A nonlinear perturbation calculation for small deviations into the unstable domain leads to a nonlinear oscillator equation that shows both shear angular velocity profile and flattened density profile developed in the nonlinear saturated state.

Hooper, *et al.* (1982) report that there were nondisruptive $m = 1$ oscillations at 7 kHz and $m = 0$ oscillations at 13 kHz in the TMX experiments. The $m = 0$ mode is an ion acoustic mode with a phase shift of 180° degrees at the two end cells. The $m = 1$ mode characteristics change with variations in the density profile but remain marginally stable drift mode given in Eqs. (3.74)–(3.76). This $m = 1$ mode is a nonlinear wobble of core plasma. The eigenmode is shown in Fig. 3.21 and the dependence of the frequency on the conducting wall radius relative the density profile radius is shown in Fig. 3.20.

Table 3.1 Summary of stability condition for rotating plasmas.

Summary of Rotational Stability		
Rotation + curvature	$g \to$	$g + v_\theta^2/r$
	over L_t	over length L_c
FLR stabilization	$\frac{1}{4}\omega_{*i}^2 > \frac{g}{r_p}$	
line bending stabilization	$\frac{g}{r_p} < k_\parallel^2 v_A^2$	
parallel wavenumber $k_\parallel L_c = \pi$	$\beta < \pi \frac{r_p R_c}{L_c^2}$	
Beta stability limit	$\frac{1}{\beta} \geq \frac{1}{\beta_{\text{balloon}}} + \frac{1}{\beta_{\text{rot}}}$	
	$\beta_{\text{rot}} = \pi \frac{r_p r}{L_c^2}\frac{a^2}{\rho^2} = \pi\left(\frac{ra}{L_c\rho}\right)^2$	

A summary of the key features of the rotational instabilities is given in Table 3.1.

Magnetic confinement systems with large electron temperature gradients universally show anomalously fast loss of electron thermal energy across the confining magnetic field. The turbulent electron transport has prevented a number of large tokamaks and helical devices from reaching their engineering designed level of fusion power production. There are numerous geometries where the electron temperature gradient (ETG) turbulence has been identified and validated. The simplest example of the ETG turbulence is that produced in the cylindrical geometry of the Columbia Linear Machine [Fu, *et al.* (2012)] analyzed with theory, fluid and gyrokinetic simulations. More recently, the IFS team and the IRFM team [Horton, *et al.* (2013)] evaluated the turbulence driven in the long – up to 6-min steady-state discharges – driven by two types of multi-megawatt lower-hybrid antennas reported in van Houtte, *et al.* (2004). The minute-long RF-driven plasmas have infrared and X-ray diagnostics that show the high temperature erosion of the plasma facing components particularly the RF antenna surfaces.

In the magnetic confinement experiments the electron confinement is anomalous by a large factor across all confinement geometries and confinement regimes, as emphasized by Kadomtsev (1992) and validated in numerous confinement experiments including TFTR, NSTX, DIII-D, EAST and LHD. The temperature gradient driven turbulence is such a mechanism that occurs in slab, cylindrical and toroidal geometry over a wide range of plasma collisionality conditions. In tokamaks, there is the second limit on plasma performance given by the internal kink instability from the $m, n = (1, 1)$ global modes that has a variety of forms producing various types of plasma limitations and the ELMs. The Type I ELMs set the limit on the plasma wall interactions and must be mitigated. Research currently is emphasizing the advantage of a less disruptive confinement regime called the intermediate or improved I-mode confinement regime. Important new impurity transport data and dust particle transport data is arriving from Wendelstein 7-X and the LHD owing to their long-time pulse lengths. The chamber walls emit micron-sized clumps called plasma dust particles reducing the fusion power performance.

Chapter 4

Two-Component Magnetohydrodynamics

The two-component fluid equations are used to describe the low-frequency-large-scale dynamics of plasmas. The two-component adjective means that the ion and electrons are described separately as fluids with their own characteristic properties. This formulation – allowing for the ions and electrons to behave differently – is of key importance making the dynamics richer and more accurate than possible with the one-component MHD model. Many strong assumptions must be made to reduce the two-component fluid theory to the MHD model. Both ion acoustic waves and drift waves are lost when using the MHD model. Generally, lower frequency waves including magnetic reconnection dynamics require the two-component fluid theory.

The two-component fluid equations are valid for disturbances with frequencies ω much less than the ion cyclotron frequency ω_{ci} and wavelengths perpendicular $\lambda_\perp$ and parallel $\lambda_\parallel$ to the magnetic field such that $\lambda_\perp \gg \rho_i$ and $\lambda_\parallel \gg \lambda_{\rm mfp}$ where ρ_i is the average ion gyroradius and $\lambda_{\rm mfp}$ is the mean-free-path determined by the collision frequency $\nu_{\rm coll}$. The plasma equilibrium is determined by the cross-field transport processes of diffusion and convection which lead to a collisional decay at a rate proportional to $\nu_{\rm coll}/B^2$. The electrostatic modes are basically drift waves and sound waves with frequencies $\omega \simeq k_y v_{de}$ and $\omega \simeq k_\parallel c_s$ which are fast compared with the equilibrium time scales and slow compared with the electromagnetic fast modes that establish the equilibrium plasma. The corrections due to the coupling of the electrostatic modes to the faster Alfvén dynamics is of order $\omega^2/k_\parallel^2 v_A^2 \simeq \beta$ where the toroidal geometry determines the scale of the parallel wavelengths. The neglect of the electromagnetic coupling compared to the toroidal effects restricts the plasma beta to $\beta < (r/R)^2$. Here the plasma $\beta = 8\pi p/B^2[\text{cgs}] = 2\mu_0 p/B^2[\text{mks}]$ is the ratio of the plasma pressure to the magnetic pressure.

4.1 Collisional Transport Equations

In this chapter the domains of plasma instability are delineated from the dispersion relation derived from the linearization of the ion and electrons fluid equations given in Eqs. (4.1)–(4.7) in toroidal systems. In the analysis the equilibrium radial gradients are weak over the scale of the gyroradius and the dispersion relations are correct to within small corrections arising from the radial normal mode eigenvalue problem. The different drifts of the ions and electrons from their different temperatures and variation in ion and electron coupling to the electromagnetic fields is responsible for the rich behavior of the plasma.

The plasma fluid is described by the electron and ion number densities n_e, n_i, the velocities $\boldsymbol{v}_e, \boldsymbol{v}_i$, and temperatures T_e, T_i. The transport equations for these moments are

$$\frac{\partial}{\partial t} n_i + \nabla \cdot (n_i \boldsymbol{v}_i) = 0 \tag{4.1}$$

$$\frac{\partial}{\partial t} n_e + \nabla \cdot (n_e \boldsymbol{v}_e) = 0 \tag{4.2}$$

$$m_i n_i \left(\frac{\partial}{\partial t} + \boldsymbol{v}_i \cdot \nabla \right) \boldsymbol{v}_i + \nabla \cdot \boldsymbol{\pi}_i = e n_i \left(\mathbf{E} + \frac{\boldsymbol{v}_i}{c} \times \mathbf{B} \right) - \nabla p_i - \boldsymbol{R} \tag{4.3}$$

$$m_e n_e \left(\frac{\partial}{\partial t} + \boldsymbol{v}_e \cdot \nabla \right) \boldsymbol{v}_e + \nabla \cdot \boldsymbol{\pi}_e = -e n_e \left(\mathbf{E} + \frac{\boldsymbol{v}_e}{c} \times B \right) - \nabla p_e + \boldsymbol{R} \tag{4.4}$$

$$\frac{3}{2} n_i \left(\frac{\partial}{\partial t} + \boldsymbol{v}_i \cdot \nabla \right) T_i + n_i T_i \nabla \cdot \boldsymbol{v}_i = -\nabla \cdot \boldsymbol{q}_i - \boldsymbol{\pi}_i : \frac{\partial \boldsymbol{v}_i}{\partial \boldsymbol{x}} + Q_i \tag{4.5}$$

$$\frac{3}{2} n_e \left(\frac{\partial}{\partial t} + \boldsymbol{v}_e \cdot \nabla \right) T_e + n_e T_e \nabla \cdot \boldsymbol{v}_e = -\nabla \cdot \boldsymbol{q}_e + Q_e \tag{4.6}$$

and the quasineutrality condition

$$n_i = n_e \tag{4.7}$$

for singly charged ions. The case of two ion species is treated in Chapter 9.

The effects of particle collisions are contained in the frictional force $\boldsymbol{R}$, the momentum flux $\boldsymbol{\pi}$, the heat flux $\boldsymbol{q}$ and the heat generation Q.

The momentum equation Eq. (4.4) for the electrons will often be written in the limit that the electron mass $m_e \to 0$. This is a good approximation for low-frequency waves when $k_\perp^2 c^2/\omega_{pe}^2 \ll 1$. There are exceptions where the electron inertia is important in both Alfvén waves and tearing modes, as described in detail in Chapter 15.

As is well known from collisional transport theory [Balescu (1988), Helander and Sigmar (2002)], the form of the collisional transport $\boldsymbol{R}, \boldsymbol{q}, \boldsymbol{\pi}$ fluxes depends strongly on the character of the particle orbits between collisions. In the collisional regime defined by $\omega_b < \nu_{\text{col}} < \omega_{ci}$ where the collision frequency is intermediate to the average particle bounce frequency ω_b and the cyclotron frequency ω_{ci}, the collisional fluxes are taken from the approximation that $\mathbf{B} = \text{const}$. The bounce frequency ω_b arises from the reflection of the parallel motion of the particles by the mirror force $F_\parallel = -\mu \nabla_\parallel B$. In the constant B limit, where the orbits are helical gyro-orbits with constant parallel speeds, the collisional fluxes [Braginskii (1965)] parallel to **B** and the finite gyroradius fluxes (FLR) across **B** are

$$\boldsymbol{R} = en\left(\eta_\parallel \boldsymbol{j}_\parallel + \eta_\perp \boldsymbol{j}_\perp\right) - \beta n \nabla_\parallel T_e \tag{4.8}$$

$$\boldsymbol{\pi} = \boldsymbol{\pi}_{\mathrm{FLR}} - \frac{nT_i}{\nu_i}\nabla_\parallel v_{\parallel i} \tag{4.9}$$

$$\boldsymbol{q}_e = -\beta\frac{T_e}{e}\boldsymbol{j}_\parallel - \alpha\frac{nT_e}{m_e\nu_e}\nabla_\parallel T_e - \frac{5}{2}\frac{cnT_e}{eB}\widehat{\boldsymbol{b}}\times\nabla T_e \tag{4.10}$$

$$\boldsymbol{q}_i = -\gamma_i\frac{nT_i}{m_i\nu_i}\nabla_\parallel T_i + \frac{5}{2}\frac{cnT_i}{eB}\widehat{\boldsymbol{b}}\times\nabla T_i \tag{4.11}$$

$$Q_i = \Delta Q = 5.9\frac{m_e}{m_i}n\nu_e\left(T_e - T_i\right) \tag{4.12}$$

$$Q_e = \frac{\boldsymbol{R}\cdot\boldsymbol{j}}{n_e} - \Delta Q = \eta_\parallel j_\parallel^2 + \eta_\perp j_\perp^2 - \beta\frac{j_\parallel \nabla_\parallel T_e}{e} - \Delta Q \tag{4.13}$$

where α, β, γ are numerical coefficients of order unity. The parallel resistivity is given by $\eta_\parallel = m_e\nu_e/ne^2$ and $\eta_\perp = 2\eta_\parallel$. For reference, the collision frequency ν_e defined here is 0.51 times the collision frequency defined in Braginskii (1965) and $\alpha = 1.6$, $\beta = 0.71$ and $\gamma_i = 3.0$. Note that this transport coefficient $\gamma = 3.0$ defined by Braginskii is distinct from the growth rate symbol conventionally taken as γ and distinct from the adiabatic gas constant γ. Capital Γ is used for the adiabatic gas constant and one seldom needs the constant symbol γ_i in the Braginskii ion heat flux equation occurring in Eq. (4.11).

Before linearizing the equations, some reductions are made in the transport equations appropriate for the low-frequency modes. For the modes considered we have $\omega/\omega_{ci} \sim \varepsilon^2$ and $k_\perp a_i \sim \varepsilon \ll 1$, and the perpendicular particle fluxes are solved for by iteration in ε. The first-order ion and electron particle fluxes are

$$n\boldsymbol{v}_{\perp i,e}^{(1)} = \frac{nc\mathbf{E}\times\mathbf{B}}{B^2} \pm \frac{c\mathbf{B}\times\nabla p_{i,e}}{eB^2}. \tag{4.14}$$

The divergence of the next order ion particle flux is given by

$$\nabla\cdot\left(n\boldsymbol{v}_{\perp i}^{(3)}\right) = \nabla\cdot\left[\frac{nm_ic^2}{eB^2}\left(\frac{\partial}{\partial t} + \boldsymbol{v}_i^{(1)}\cdot\nabla\right)\mathbf{E}_\perp\right] \tag{4.15}$$

where in calculating this term the convective derivative and the finite gyroradius part of the momentum tensor $\boldsymbol{\pi}_{\mathrm{FLR}}$ in Eq. (4.3) are combined as discussed in detail in Smolyakov (1998) and for drift waves in Hinton and Horton (1971). In a constant magnetic field similar reductions occur in the finite gyroradius contributions to the heat balance equations. To the lowest order in the aspect ratio, where $|\mathbf{B}| = \text{const.}$ the equations governing the perturbations can be written as follows:

$$\frac{\partial n_i}{\partial t} + \frac{c\mathbf{E}\times\mathbf{B}}{B^2}\cdot\nabla n_i + \nabla\cdot\left(\frac{n_im_ic^2}{eB^2}\left(\frac{\partial}{\partial t} + \boldsymbol{v}_i^{(1)}\cdot\nabla\right)\mathbf{E}_\perp\right)$$
$$+ \frac{\partial}{\partial x_\parallel}(n_iv_{\parallel i}) = 0 \tag{4.16}$$

$$\frac{\partial n_e}{\partial t} + \frac{c\mathbf{E}\times\mathbf{B}}{B^2}\cdot\nabla n_e + \frac{\partial}{\partial x_\parallel}(n_ev_{\parallel e}) = 0 \tag{4.17}$$

$$m_i n_i \left(\frac{\partial v_{\|i}}{\partial t}\right) + \left(\boldsymbol{v}_E + v_{\|i}\widehat{\boldsymbol{b}}\right) \cdot \nabla v_{\|i} = -\nabla_{\|}\left(p_i + p_e\right) + \frac{\partial}{\partial x_{\|}}\left(\frac{n_i T_i}{\nu_i}\frac{\partial v_{\|i}}{\partial x_{\|}}\right) \tag{4.18}$$

$$j_{\|} = \frac{1}{\eta_{\|}}\left(E_{\|} + \frac{T_e}{e n_e}\frac{\partial n_e}{\partial x_{\|}} + \frac{(1+\beta)}{e}\frac{\partial T_e}{\partial x_{\|}}\right) \tag{4.19}$$

$$\frac{3}{2} n_i \left(\frac{\partial T_i}{\partial t} + (\boldsymbol{v}_E + v_{\|i}\widehat{\boldsymbol{b}}) \cdot \nabla T_i\right) + n_i T_i \frac{\partial v_{\|i}}{\partial x_{\|}} = \frac{\partial}{\partial x_{\|}}\left(\frac{\gamma_i n_i T_i}{m_i \nu_i}\frac{\partial T_i}{\partial x_{\|}}\right) + \Delta Q \tag{4.20}$$

$$\frac{3}{2} n_e \left(\frac{\partial T_e}{\partial t} + (\boldsymbol{v}_E + v_{\|e}\widehat{\boldsymbol{b}}) \cdot \nabla T_e\right) + n_e T_e \frac{\partial v_{\|e}}{\partial x_{\|}} = \frac{\partial}{\partial x_{\|}}\left(\alpha \frac{n_e T_e}{m_e \nu_e}\frac{\partial T_e}{\partial x_{\|}}\right) + \frac{\beta}{e} T_e \frac{\partial j_{\|}}{\partial x_{\|}} + \eta_{\|} j_{\|}^2 - \Delta Q. \tag{4.21}$$

The self-consistent field problem is closed by the condition of quasineutrality:

$$n_i = n_e \tag{4.22}$$

where n_i and n_e are computed from Eqs. (4.16)–(4.21) with $\boldsymbol{v}_E = c\mathbf{E} \times \mathbf{B}/B^2$ and $\widehat{\boldsymbol{b}} = \mathbf{B}/|\mathbf{B}|$.

The linear solution of the self-consistent field equation gives the dispersion relations and the polarizations of the waves $\omega_{k\alpha} = \omega_{\boldsymbol{k}}(\alpha)$ and instabilities of wavenumber $\boldsymbol{k}$ and type α. For electromagnetic disturbances the field closure includes Ampere's law with the plasma electrical currents $j_{\perp}$ and $j_{\|}$ giving rise to the wave magnetic field components.

The derivation of the two-component fluid equations is complicated. These equations are widely known as the Braginskii equations from their publication in Braginskii (1965). A careful description and derivation is given in Chapter 4 of Helander and Sigmar (2002), along with a proof of the Onsager symmetries possessed by the transport matrix. A useful physical picture of the content of the transport equations along with references to other works on this problem, particularly those of Balescu, is given in Helander and Sigmar (2002).

The fundamental view of what these fluid equations mean in terms of the reduction of the many-body Coulomb interaction in statistical mechanics is developed in Elskens and Escande (2003). The relationship to the Hamiltonian formulation of the particle dynamics and the reduction to the Vlasov-Poisson self-consistent field equations is analyzed in detail by Elskens and Escande (2003) (Chapters 3, 4 and 5). Their analysis is concerned with simpler uniform plasmas but shows the nature of the Landau damping to the singular eigenfunctions of the Vlasov-Poisson field equations and the role of Coulomb interactions on the dynamics in a fundamental way. Such a detailed formulation, however, is not practical for use in the study of drift waves in tokamak plasmas. Thus, in this chapter we reduce the particle simulations following the guiding-center orbits to a closed chain of fluid equations. These reduced partial differential equations for the fluid moments are found to be generally accurate for a variety of drift wave instabilities and turbulent transport. The complete analysis of the forces from the divergence of the FLR momentum stress tensor and the dynamics of the corresponding heat flux $\boldsymbol{q}$ is an intricate problem which is best left to the literature. A simple example is found in the drift wave work of Hinton and Horton (1971) for the Q-machine experiments. The general analysis including the FLR deduction from the dynamical equations for both the momentum stress tensor $\boldsymbol{\Pi}$ and the heat flux tensor $\boldsymbol{q}$ is given in Smolyakov (1998).

4.2 Current, Density and Temperature Gradient Driven Drift Modes

The density gradient, the temperature gradients, and the plasma current $j_\parallel$ in the equilibrium are driving forces for instability in the drift modes of the plasma. In addition to the driving forces, the form of instability is determined by the dissipative processes in the plasma. The role of the dissipative processes can be viewed as providing constraints in the dynamics of the plasma. For example, the degree of thermal transport determines the variation in the ion and electron temperatures in the wave. A fast thermal transport rate constrains the wave motion to occur at nearly constant temperature and eliminates some of the terms causing instability.

At high densities and low temperatures both ions and electrons behave adiabatically with $\omega \gg (k_\parallel^2 T_e/m_e\nu_e)$, $(k_\parallel^2 T_i/m_i\nu_i)$. At the higher temperatures of interest in fusion and space physics, however, the electron transport of heat becomes rapid compared with the wave frequency, and electrons behave approximately isothermally. For sound waves, for example, the wave frequency is $\omega \simeq k_\parallel v_i$, and the ions respond adiabatically for $k_\parallel \lambda_{\rm mfp} < 1$, and the electrons respond approximately isothermally for $k_\parallel \lambda_{\rm mfp} > (m_e/m_i)^{1/2}$. In the following analysis this domain of slightly collisional waves defined by $(m_e/m_i)^{1/2} < k_\parallel \lambda_{\rm mfp} < 1$ is examined in detail.

In the limiting case of perfect electron transport and negligible thermal ion transport along the magnetic field, the dispersion relation obtained is

Reactive Limit of Dispersion Relation

$$
\begin{aligned}
D_R(\omega) &= \omega^2 - \frac{k_\parallel^2}{m_i}\left(T_e + \frac{5}{3}T_i\right) - \omega\left[k_\perp v_{de} - (k\rho_s)^2(\omega - k_\perp v_{di})\right] \\
&+ \frac{2}{3}\frac{k_\parallel^2 T_i}{m_i}\frac{k_\perp v_{de}}{\omega}\left(1 - \frac{3}{2}\eta_i\right) - \frac{2}{3}\frac{k_\parallel^2 T_i}{m_i}(k_\perp\rho_s)^2\left(1 - \frac{k_\perp v_{di}}{\omega}\right) = 0. \qquad (4.23)
\end{aligned}
$$

The first two terms determine the form of the compressible sound wave in the magnetized plasma with isothermal electrons and adiabatic ions. The three modes in Eq. (4.23) arise from the time dependence of $\widetilde{n}, \widetilde{v}_{\parallel i}$, and $\widetilde{T}_i$. The fast electron thermal transport makes $\widetilde{T}_e$ negligible for these lower-frequency modes.

4.2.1 *Ion acoustic waves and the thermal mode*

In the absence of a temperature gradient ($\eta_i = d\ell n\, T_i/d\ell n\, n_i = 0$) the three roots of Eq. (4.23) are real. For cold ions the modes are the coupled drift-ion acoustic waves with

$$\omega = \frac{1}{2}\left[k_y v_{de} \pm \left((k_y v_{de})^2 + 4k_\parallel^2 c_s^2\right)^{1/2}\right]. \tag{4.24}$$

For finite ion temperature an additional stable mode arises from $\partial\widetilde{T}_i/\partial t$ and has a frequency $\omega \simeq k_y v_{di}$ for small $k_\parallel$ and a frequency $\omega \simeq \frac{2}{5}\, k_\perp v_{de}/(1 + \frac{3}{5}\, T_e/T_i)$ for large $k_\parallel$.

4.2.2 *Ion temperature gradient instability*

In the presence of an ion temperature gradient such that $\eta_i > 2/3$, an unstable mode appears in Eq. (4.23). For small $k_\parallel$ the mode is purely growing with

$$\gamma \sim |k_\parallel|\, v_i \left(\eta_i - \frac{2}{3}\right)^{1/2}. \tag{4.25}$$

As $k_\parallel$ increases the mode has a growth rate

$$\gamma \sim \left(|k_\parallel|\, v_i\right)^{2/3} \left(k_\perp v_{de}\eta_i\right)^{1/3} \quad \text{with} \quad \gamma_{\max} \lesssim k_\perp v_{di}\eta_i. \tag{4.26}$$

The compressional effect of the ion-sound waves stabilizes the ion temperature gradient mode for

$$k_\parallel c_s > k_\perp v_{di}\eta_i. \tag{4.27}$$

The use of magnetic shear to achieve the necessary $k_\parallel$ to stabilize the mode as given by condition Eq. (4.27) is found [Coppi, *et al.* (1968a,b)], to require an order unity value for the magnetic shear parameter $S = L_n/L_S$. Such a large magnetic shear is outside the range of the perturbation expansion in power of $(k_x\rho_i)^2$. This shear stabilization problem leads to the development of integral eigenvalue formulations of the problem developed in Dong, *et al.* (1992) and Dong and Horton (1993). Thus, the ion temperature gradient mode is one of the fundamental means of driving the low-frequency plasma turbulence universally seen in laboratory plasma confinement devices. Part of the slowing down in the progress between 1970 to the present time towards reaching a fusion reactor can be attributed to the fact that as more and more ion heating is applied to driving up the ion nuclear fusion rate, the ion temperature gradient turbulence becomes stronger and stronger. One solution to this problem seems to be to keep the parameter $\eta_i < 2/3$ by peaking the density profile. Experiments in C-Mod Alcator [Greenwald, *et al.* (2007)] with peaked density profiles from pellet injection fueling seems to confirm this method for achieving improved thermal confinement. Recall that $\eta_i = \frac{d \ln T_i}{d \ln n_i}$ and the nature of these unstable modes is analyzed in further detail in Chapter 13.

4.3 Closure Models for Coupled Chain of Fluid Moments

The use of partial differential equations rather than phase space densities from the Vlasov-Maxwell system of kinetic equations as used in Chapter 2, depends on a truncation of the chain of derived fluid moments from the kinetic equations. This is a well-known problem explored for example by Balescu (1988) and by many other authors. The problem is universal to both laser plasma for inertial confinement, magnetic confinement and space-astrophysical problems. Here we discuss the methods used for the drift wave turbulence. Other methods apply to shocks and inertial confinement problems.

4.4 Pressure Gradient Driven Instabilities

As shown in Fig. 1.5 in Chapter 1 there is a pressure gradient driven instability which has the curved magnetic field producing an effective gravity from the associated charge dependent drifts of the ion and electron guiding centers in magnetized plasmas. This instability is known as the resistive-g instability since the mathematical structure and the physics is closely analogous to the gravitation instabilities known as the Rayleigh-Taylor instability. In the toroidal geometry the effective gravity is given by

$$\boldsymbol{g} = -\left(\left\langle v_\parallel^2 \right\rangle + \frac{1}{2}\left\langle v_\perp^2 \right\rangle\right)\boldsymbol{\kappa} \tag{4.28}$$

where $\boldsymbol{\kappa}$ is the magnetic curvature vector given by

$$\boldsymbol{\kappa} = (\widehat{\boldsymbol{b}} \cdot \nabla)\widehat{\boldsymbol{b}} = -\frac{\widehat{\boldsymbol{n}}}{R_c} \tag{4.29}$$

where $\widehat{\boldsymbol{b}} = \mathbf{B}(\boldsymbol{x})/|B|$ is the unit vector along the magnetic field. This effective g-vector has a large horizontal component from the toroidal magnetic field and a small radially outward component from B_θ^2/r. Both components produce the effect of driving the convection cells that produce the interchange of hot core plasma with outer cooler plasma. The acceleration $\boldsymbol{g}$ is produced by the particle drifts in the nonuniform magnetic fields. As shown in Fig. 1.5 in Chapter 1, the resulting time scale for growth of the convective cells is the geometric mean of the pressure gradient scale length L_p and the radius of curvature R_c of the magnetic field divided by the sound speed in the plasma. This time for growth of a small fluctuation is on the scale of microseconds in unstable plasmas. Thus, for the establishment of a confined plasma state in which drift waves exists the magnetic geometry must have stability to the interchange or magnetic Rayleigh-Taylor instability. In Chapter 5 we show examples where there are regions outside a critical layer which are interchange unstable so the qualitative differences in the confined interior plasma and MHD unstable exterior plasma are clearly manifested.

Next we consider what happens after the system is stable to the interchange instability but the plasma is described by the transport equations in this chapter, namely the resistive-viscous two-component plasma transport equations. The plasma has a remnant of the Rayleigh-Taylor instability but with a growth rate controlled by a hybrid exponent of the Rayleigh-Taylor growth rate and the resistive diffusion rate for the magnetic field to diffuse a localized plasma current density $j_\parallel$ filament. The parallel plasma current controls the rate of change of the plasma vorticity and thus the creation and decay of the vortex structures in the stabilized Rayleigh-Taylor system. Drift wave mode structures spanning the region from the hot core plasma to the cooler edge plasma develop in the plasma.

Essentially all methods of stabilizing MHD instabilities rely on the restoring force from bending the magnetic field exceeding the destabilizing force from the perturbed pressure gradient. This force condition translates directly into an energy condition. The MHD stability is often presented in terms of an energy principle for the dynamics since this gives information without solving the eigenvalue-eigenvector problem for the linear wave equations. For drift waves the energy principle is not useful and for transport calculations one needs the eigenvalues and eigenvectors.

The basic method used to produce stability to interchange instabilities on the fast time scale is to use the principle that for fast motions the plasma carries the magnetic field with it from the frozen in flux condition [Kulsrud (2005)] of

$$\mathbf{E} + \boldsymbol{v} \times \boldsymbol{B} = 0 \tag{4.30}$$

which together with Faraday's law leads to the frozen in magnetic flux

$$\frac{\partial \mathbf{B}}{\partial t} = \nabla \times (\boldsymbol{v} \times \mathbf{B}). \tag{4.31}$$

By introducing the displacement vector $\boldsymbol{\xi}(x,t)$ such that $\boldsymbol{v} = \partial\boldsymbol{\xi}/\partial t$, we can solve for $\delta\mathbf{B}$ from Eq. (4.31) to get

$$\delta B_r = (\mathbf{B}\cdot\nabla)\xi_r = iB\,k_\parallel \xi_r. \tag{4.32}$$

The radial acceleration of the plasma is given by

$$\rho\,\frac{\partial^2 \xi_r}{\partial t^2} = F_r = \frac{(\mathbf{B}\cdot\nabla)^2\xi_r}{\mu_0} - \frac{2\delta p}{R_c} \tag{4.33}$$

giving Alfvén waves in the absence of the pressure fluctuation δp and the field line curvature $2/R_c$. The pressure fluctuation is given by convection $\xi_r\, dp_0/dr$ and compression $p_0 \nabla\cdot\boldsymbol{\xi}$. For incompressible motion with $\nabla\cdot\boldsymbol{\xi} = 0$, the radial force reduces to

$$F_r = \left(-\frac{k_\parallel^2 B^2}{\mu_0} - \frac{2}{R_c}\frac{dp_0}{dr}\right)\xi_r \tag{4.34}$$

where the first term is a restoring force producing Alfvén oscillations at the frequency

$$\omega_A^2 = \frac{k_\|^2 B^2}{\mu_0 \rho} \tag{4.35}$$

and the second force drives the interchange instability with

$$\omega^2 = -\gamma^2 = \frac{2}{R_c}\frac{1}{\rho}\frac{dp_0}{dr} \tag{4.36}$$

where $\xi_r = \xi_r(0)\exp(\gamma t)$.

For radially distributed motions we construct the energies from the plasma motion

$$W_{\text{alf}} = \frac{1}{2\mu_0}\langle \delta B_\perp^2 \rangle = \frac{B^2}{2\mu_0}\left\langle (k_\| \xi_r)^2 \right\rangle \tag{4.37}$$

$$W_{\text{int}} = -\left(\frac{2}{R_c \rho}\frac{dp_0}{dr}\right)\langle \xi_r^2 \rangle \tag{4.38}$$

and kinetic energy

$$W_{KE} = \frac{1}{2}\left\langle \rho_0 \left(\frac{\partial \xi_r}{\partial t}\right)^2 \right\rangle. \tag{4.39}$$

The interchange motion releases the energy W_{int} so the system is stable when the increase in the magnetic energy W_{alf} is greater than the energy released by the convection across the pressure gradient W_{int}. This, is the basic method by which all magnetic confinement systems toroidal and mirror geometries provide MHD stability. The $W_{\text{int}} < W_{\text{alf}}$ is satisfied when the pressure gradient, and thus the maximum pressure itself p, is less than the magnetic energy density W_{mag} times the geometrical factors controlling the minimum parallel wave number $k_\|$ and the pressure gradient scale length L_p and the radius of curvature R_c of the magnetic field lines. The minimum $k_\|$ can be controlled in number of ways. In mirror machines and field-reversed configurations [Binderbauer, *et al.* (2015)] the distance between the $B_{\max}$ mirror regions controls $k_\|$.

In the tokamak the value is set by the major radius and the safety factor $q(r)$ and the magnetic shear $s(r)$ which control the profile of the twisting magnetic field. At each radius the distance along the magnetic field line

$q(r)R =$ length to go once around the torus following the magnetic field line*is*

$$L_c = \frac{rB_\phi}{B_p} \tag{4.40}$$

and the magnetic shear function for radial variation of the twisting is

$$s(r) = \frac{r}{q}\frac{dq}{dr} \tag{4.41}$$

where B_ϕ and B_p are the toroidal and poloidal magnetic fields at minor radius r from the magnetic axis.

The connection length is defined as $L_c = qR/2$ which separates the region of favorable curvature on the inside of the torus and the unstable region on the outside of the torus. This formula is widely used to estimate the parallel wavenumber $k_\parallel = 1/qR$ in both drift wave and MHD calculations. The $k_\parallel$-formula is adequate on average but fails to recognize the role of the special rational magnetic surfaces at $r_{m/n}$ where $k_\parallel = 0$. Of course, this applies only at a point in $r = r_{m/n}$ and thus one needs a measure δ_r for the weight of the effect over the radial zone surrounding the rational surface. The weight is the width of the eigenmode that exists from the dynamical equations for the unstable motion. For the drift waves we found the mode width in Chapters 2.6 and 2.7. The width was proportional to the inertial scale length ρ_s and inversely proportional to the magnetic shear parameter s. Thus, for low magnetic shear the mode width is wide and there is a large radial width to the interchange motion. While for strong magnetic shear the mode is narrow and the interchange motion is restricted to be close to the rational surface. When these considerations and factors are used in the estimates for the released energies in the equation $W_{\rm int} < W_{\rm alf}$, we arrive at the limit on the plasma pressure-to magnetic pressure given by

$$\beta_{\rm tor} = \frac{2\mu_0 p}{B^2} = \frac{L_p R_c}{L_c^2} = \frac{L_p}{q^2 R} \tag{4.42}$$

for the toroidal confinement system. The pressure gradient is $d\ell n\, p/dr = -1/L_p$.

For the mirror confinement geometry the situations changes as the minimum $k_\parallel$ is determined by the length of the field line set either by the distances to the conducting end walls or the distance to the stabilizing mirror cells. The radius of curvature R_c of the field lines also changes becoming infinite on the axis of the mirror cell and decreases with the radius r as one moves off the axis to give

$$\frac{1}{R_c} = \frac{r}{L_m^2} \tag{4.43}$$

as the curvature of the magnetic field lines. Repeating the balance condition $W_{\rm int} < W_{\rm alf}$ with Eqs. (4.37) and (4.38) now yields the plasma pressure limit

$$\beta_{\rm mirror} = \frac{L_p R_c}{L_m^2} = \frac{L_p}{r}. \tag{4.44}$$

This behavior is borne out in the Earth's magnetosphere where the plasma beta exceeds unity in much of the night-side plasma sheet. When the magnetic fields lines become stretched, the curvature is the strongest in the region where the plasma $\beta \sim 1$ becomes unstable [Crabtree, *et al.* (2003), Horton, *et al.* (2005)] with the interchange motions observed by spacecraft plasma instruments.

For helical toroidal systems like the large helical device (LHD), the pitch of the helical magnetic field is controlled by the external coils and fixed. There is a rational surface corresponding to the $q_{\rm ext} = 2$ winding pitch of the magnetic field. In this case the plasma pressure limit given by Eq. (4.42) reduces to the simple result of

$$\beta_{\text{LHD}} = \frac{a}{2R}. \tag{4.45}$$

For Large Helical Device (LHD) this limit is a volume averaged $\beta = 0.05$ which seems to be a firm limit in the plasma pressure achieved in LHD [Yamada, *et al.* (2005)].

Now, we consider that the plasma pressure is below the limits set by Eq. (4.42) and see how the effect of finite resistivity produces the resistive-g instabilities. These resistive-g modes are important in the outer layers where the plasma is weakly collisional. The multi-mode code used in Chapters 16 and 18 for transport includes transport formulas for the resistive-g and these formulas and their generalizations are now used to predict the presence of edge localized modes, called ELMs for short, in toroidal confinement machines.

In Chapters 13 and 14 the stability analysis shows that magnetized plasmas with significant temperature gradients across the magnetic field are the standard situation owing to the thermal insulating properties of the magnetic field. Such temperature gradients occur in the magnetosphere, the solar corona and with particularly high gradients in the laboratory experiments aimed at magnetic confinement for nuclear fusion. Thus, we often have situations where the pressure gradient is below the critical value for the onset of the MHD modes, however, the temperature gradients are above the critical values required for the onset of intense thermal convection from the drift wave scale turbulence. The situations differ according to whether the dominant gradient is the ion temperature gradient or the electron temperature gradient that is driving the turbulent thermal convection. The ion temperature gradient case is known as the ITG mode and is treated in Chapter 13. The case of the electron temperature gradient driving the turbulence is called the ETG mode and is treated in Chapter 14.

The simulations for transport must consider the stable MHD equilibrium with the resistive-g modes in the edge plasma, the ITG and ETG turbulence in the core plasma as intrinsic to determining the performance of a fusion machine like JET, KSTAR, EAST or ITER. For reference Table 4.1 gives for JT-60U some of the key parameters at the confinement limit.

In the large tokamaks experiments reviewed in Chapter 1 we see that the core ion temperatures have reached 30 KeV in the TFTR device and 38 KeV in the JT-60U machine with $R/a = 3.1\,\text{m}/0.7\,\text{m} \simeq 4$. Thus, we know that the magnetic field can provide strong thermal insulation withstanding gradients greater than 30 KeV/m under conditions with closed irrational magnetic surfaces. Table 4.1 gives the parameters for a well-known JT-60U shot that reached a core ion temperature of 38 KeV and core electron temperature of 12 KeV in a deuterium plasma.

The ion (deuterium) quantities T_i and v_ϕ are derived from spectroscopic measurement of the partially ionized carbon ions in the plasma. There are few direct measurements of the fully ionized hydrogen isotopes in fusion plasma other than the neutron yields and the gamma rays from the nuclear reactions.

Table 4.1 JT-60U high-β_p experiment with internal transport barrier: parameters in Phase I of shot 17110.

R/a	3.1 m/0.7 m
B_ϕ	4.4 T
I_p	2 MA
$P_{\rm NBI}$	27 MW
$n_D(0)$	$4.1 \times 10^{19}\,{\rm m}^{-3}$
$T_i(0)/T_e(0)$	38 KeV/12 KeV
$n_D \tau_E T_i$	$1.1 \times 10^{21}\,{\rm m}^{-3} \cdot s \cdot$ KeV
$v_\phi(0)$	-100 km/s

The whole concept of magnetic confinement for thermonuclear fusion of plasmas depends on this thermal insulating principle. From the deuterium-tritium reactivity function shown in Fig. 1.1 one finds that the core ion plasma temperatures of order $10^8\,{\rm K} \sim 10\,{\rm KeV}$ are required for significant fusion nuclear reactivity. This high-temperature plasma must be partially insulated from the vessel walls which must remain below 5000 K to retain their integrity. Thus, steep temperature gradients of order 10 KeV/m must be stably maintained in the thermonuclear fusion device. The associated ion thermal diffusivity χ_i is then of order or less than $1\,{\rm m}^2/{\rm s}$ in devices with minor radius of a few meters.

Such magnetic thermal insulation is consistent with the results from the latest generation of tokamak and spherical tokamak experiments under favorable conditions. The thermal flux across the magnetic surfaces is large $q_i = -n_i \chi_i dT_i/dr$ reaching $0.1\,{\rm MW/m}^2$ requiring large heating rates of many MW to maintain the high-temperature plasma against turbulent losses. For the alpha particle decay products from the fusion reactions to provide the power to maintain the ion temperature the core electron temperature T_e must also be sufficiently high in the 10 KeV range. Chapter 13 is devoted to instabilities driven by the electron temperature gradient and the associated turbulent transport of electron thermal energy across the magnetic flux tubes. This electron thermal transport problem is more difficult to control in the magnetic confinement experiments than the ion thermal turbulent transport as emphasized by Kadomtsev in his monograph [Kadomtsev (1992)].

In both laboratory plasmas and in astrophysical plasmas where a strong local heating takes place there is clearly a potential source of plasma instability. There can be MHD (single component conducting fluid in Chapter 5) instabilities controlled by the constraints of frozen-in magnetic flux and the adiabatic equation of state. These fast instabilities limit the plasma pressure gradients with respect to the confining magnetic pressure. These fast, single fluid instabilities have be eliminated in the magnetic confinement devices developed for nuclear fusion. Here, we consider that the plasma is stable to ideal MHD motions and study the remaining smaller scale drift wave instabilities. The drift waves have dispersive frequencies and finite parallel electric fields which means that the associated waves and vortices have resonant wave-particle interactions.

The drift wave dispersion relation in a collisionless plasma does not assume a local thermal Maxwell-Boltzmann distribution function. In the high temperatures of fusion and geospace plasmas the particle distributions in energy typically have high-energy tails and different effective temperatures across and parallel to the magnetic field. In low beta $\beta = p/(B^2/2\mu_0)$ plasmas the ion temperature gradient instability, known as the ITG mode, has nearly an electrostatic polarization for the coupled drift waves and the ion acoustic waves that gives rise to the collective instability. In higher pressure plasmas with $\beta > m_e/m_i$ the ITG mode couples with the shear Alfvén wave becoming a dispersive oscillation with an electromagnetic polarization. As the plasma pressure increases the inductive electric field from the time changing δB_x begins canceling part of the electrostatic component of the parallel electric field. This cancellation reduces the energy transfer rate $\langle j_\| E_\| \rangle$ and reduces the growth rate of the instability. There are numerous studies of these weakly electromagnetic ITG modes including Rewoldt, *et al.* (1999) and Hong, *et al.* (1989a,b) that detail how the ITG modes change with increasing plasma beta. At certain intermediate values of plasma beta both the ITG and the kinetic ballooning mode of the MHD polarization are present with different mode frequencies and growth rates.

4.4.1 *Scale invariance properties arising from Ohm's law with electron inertia*

The dynamics for electrostatic pressure gradient driven turbulent fluctuations for a large aspect ratio tokamak (in a coordinate system $r, \theta, \phi = \zeta$ where r and θ are cylindrical coordinates about the magnetic axis and ζ is the toroidal angle) can be written as the vorticity equation

$$\frac{\rho}{B_0}\frac{d}{dt}\boldsymbol{\nabla}_\perp^2 \delta\phi - \mathbf{B}\cdot\boldsymbol{\nabla}\delta j_\| = \frac{\widehat{\mathbf{e}}_\zeta \cdot \boldsymbol{\nabla} R^2 \times \boldsymbol{\nabla}\delta p}{R^2} \tag{4.46}$$

with incompressible convection

$$\frac{d}{dt}\delta p = 0 \tag{4.47}$$

of the pressure fluctuation. Here $\widehat{\mathbf{e}}_\zeta = \widehat{\mathbf{e}}_\phi$ is the unit vector in the toroidal direction. In the right-hand side of Eq. (4.46) we use $\nabla \times \mathbf{E} = 0$ to reduce the divergence of the diamagnetic current.

For low-collision frequencies, the electron inertia becomes important so the collisionless Ohm's law

$$\frac{\mathbf{B}| \cdot \boldsymbol{\nabla}\delta\phi}{B_0} = \frac{m_e}{ne^2}\frac{d}{dt}\delta j_\| \tag{4.48}$$

where

$$\frac{d}{dt} = \frac{\partial}{\partial t} + \frac{\boldsymbol{e}_\zeta \times \boldsymbol{\nabla}\delta\phi \cdot \boldsymbol{\nabla}}{B_0} \tag{4.49}$$

$$\mathbf{B} \cdot \boldsymbol{\nabla} = \frac{B_0}{R}\left(\frac{\partial}{\partial \zeta} + \frac{1}{q}\frac{\partial}{\partial \theta}\right) \qquad R = R_0 + r\cos\theta. \tag{4.50}$$

Here $\delta\phi, \delta j_\parallel$ and δp are the fluctuations in electrostatic potential, parallel current and pressure, respectively.

Considering fluctuations with short-perpendicular wavelengths comparable to the collisionless skin depth $\delta = c/\omega_{pe}$, we introduce

$$x = \frac{r - r_0}{\delta} \qquad y = \frac{r_0\theta}{\delta} \tag{4.51}$$

as local coordinates perpendicular to the magnetic field at a minor radius r_0. Defining dimensionless forms for the turbulent fluctuations $\delta\phi, \delta j_\parallel$ and δp by

$$\frac{\delta\phi}{B} = \frac{\delta^2}{Rq} v_A \delta\phi \qquad \delta j = \frac{B_0}{Rq\mu_0} j \qquad \delta p = \delta\left(-\frac{d\delta p_0}{dr}\right)\delta p \tag{4.52}$$

where the dimensionless time variable is $\tau = v_A t/Rq$, these nonlinear resistive-g equations can be simplified to

$$\frac{d}{d\tau}\nabla_\perp^2 \delta\phi = \left(\frac{\partial}{\partial \theta} + sx\frac{\partial}{\partial y}\right)\delta j + \alpha\left(\cos\theta\frac{\partial}{\partial y} + \sin\theta\frac{\partial}{\partial x}\right)\delta p \tag{4.53}$$

$$\frac{d\delta p}{d\tau} + \frac{\partial \delta\phi}{\partial y} = 0 \tag{4.54}$$

$$\left(\frac{\partial}{\partial \theta} + sx\frac{\partial}{\partial y}\right)\delta\phi = \frac{d}{d\tau}\delta j \tag{4.55}$$

where

$$\frac{d}{d\tau} = \frac{\partial}{\partial \tau} + \frac{\partial \delta\phi}{\partial y}\frac{\partial}{\partial x} \qquad \nabla_\perp^2 = \frac{d^2}{dx^2} + \frac{d^2}{dy^2}. \tag{4.56}$$

and the dimensionless pressure gradient is

$$\alpha = -2\mu_0 R_0\left(\frac{dp_0}{dr}\right)\frac{q^2}{B_0^2}. \tag{4.57}$$

Scale invariance techniques [Connor (1993)] can be applied to these equations to derive the scaling of the thermal diffusivity. The invariance scaling gives that the diffusivity scaling χ is given by

$$\chi \sim \frac{\langle (r - r_0)^2 \rangle}{t} \sim \frac{v_A}{Rq}\left(\frac{c}{\omega_{pe}}\right)^2 \frac{\langle x^2 \rangle}{\tau}. \tag{4.58}$$

The solutions of Eqs. (4.53), (4.54), and (4.55) must have the property

$$\frac{\langle x^2 \rangle}{\tau} = F(\alpha, s) \tag{4.59}$$

where F is, as yet, an arbitrary function of dimensionless pressure gradient α and magnetic shear s.

The equations (4.53), (4.54), and (4.55) are invariant under a single scale transformation

$$\begin{aligned}
&\delta\phi \to \lambda^2 \delta\phi \quad \delta j \to \lambda \delta j \quad \delta p \to \delta p \quad x \to \lambda x \\
&y \to \lambda y \quad \tau \to \lambda^{-1}\tau \quad \alpha \to \lambda^2 \alpha
\end{aligned} \tag{4.60}$$

which allows us to show that

$$F = \chi\, \alpha^{3/2} f(s). \tag{4.61}$$

Following Connor (1993) we can use periodic equilibrium variation in θ and the assumption 'flute'-like so that a two-scale analysis along the field line to find the form of the function $f(s)$ can be determined. The analysis shows the $f(s) = \text{const}/s^{1/2}$ and the function $F(\alpha, s)$ in Eq. (4.59) must be

$$F = \frac{C\alpha^{3/2}}{s^{1/2}} \tag{4.62}$$

in Eq. (4.59) with a constant C.

In terms of the g-modes the effect of unfavorable curvature can be modeled by the replacement

$$\alpha \left(\cos\theta \frac{\partial}{\partial y} + \sin\theta \frac{\partial}{\partial x} \right) \to g \frac{\partial}{\partial y} \tag{4.63}$$

where g is the analogue of α in Eq. (4.53). In the resistive-g (rg) limit ($m_e = 0$) scale invariance allows one to derive

$$F_{rg} = C_{rg} \frac{g^{3/2}}{s^2}. \tag{4.64}$$

With electron inertia ($m_e \neq 0$), the turbulent thermal diffusivity is

$$\chi \sim \frac{v_A}{qR} \frac{c^2}{\omega_{pe}^2} \frac{\alpha^{3/2}}{s^{1/2}} \tag{4.65}$$

which agrees with the formula derived by Itoh and Fukuyama (1992).

4.4.2 *Scaling of correlation length and time*

From scale invariance we can also deduce scalings for the perpendicular correlation length ℓ_c and time τ_c of the turbulence from the scaling results Eq. (4.60). The scaling analysis gives

$$\ell_c \sim \left(\frac{\alpha}{s}\right)^{1/2} \frac{c}{\omega_{pe}} \qquad \tau_c \sim \left(\frac{\alpha}{s}\right)^{-1/2} \frac{Rq}{v_A} \tag{4.66}$$

in agreement with Itoh and Fukuyama (1992). This supposes that the fluctuations are not coupled by toroidal effects to generate radially extended structures with global correlation lengths $\ell_c = (\rho_s L_p)^{1/2}$ which increases the transport coefficients.

4.4.3 *Magnetic flutter thermal transport*

For a high density of rational surfaces, the perturbed magnetic fields $\mathbf{B}_0 + \delta\mathbf{B}$ produce stochastic magnetic field lines. The electron thermal transport is due to the stochastic magnetic field turbulence in the strong turbulence limit is

$$\chi_e^{\rm st} \sim v_e \left(\frac{\delta B_r}{B}\right) \ell_c \tag{4.67}$$

where v_e is the electron thermal velocity and ℓ_c is the correlation length along the magnetic field from $\ell_c \Delta k_\parallel = 1$. Here δB_r is the rms value of the magnetic fluctuation. From Ampére's law for δB_r one shows that δj leads to

$$\chi_e^{\rm st} \sim \frac{v_e}{Rq}\left(\frac{c}{\omega_{pe}}\right)^2 \frac{\alpha^2}{s} \tag{4.68}$$

for the electron thermal diffusivity for pressure gradient (α) driven magnetic field stochasticity.

4.4.4 *Electron inertia replacing Ohm's law*

The use of the electron equation of motion, Eq. (4.48), applies for 'cold' electrons, i.e. when $v_e \tau_c / Rq \ll 1$. In the opposite limit [Kadomtsev and Pogutse (1984)] introduced an Ohm's law involving Landau damping. Ignoring diamagnetic effects, valid for turbulence with $\omega_* \tau_c \ll 1$ where ω_* is a diamagnetic frequency, the collisionless Ohm's law is

$$\mathbf{B} \cdot \boldsymbol{\nabla}\phi = \frac{m_e v_e}{ne^2} \mathbf{B} \cdot \boldsymbol{\nabla} j \tag{4.69}$$

or, using the field line following the variables, Eq. (4.55),

$$\left(\frac{\partial}{\partial\theta} + sx\frac{\partial}{\partial y}\right)\left(\phi - \frac{v_e}{v_A} j\right) = 0. \tag{4.70}$$

Scaling arguments can be applied to Eqs. (4.53), (4.54), and (4.70), which replaces Eq. (4.55) as a collisionless Ohm's law. The result for χ is that

$$\chi \sim \left(\frac{c}{\omega_{pe}}\right)^2 \frac{v_e}{Rq} \alpha g(s) \tag{4.71}$$

where $g(s)$ = constant in the two-scale approximation. Connor shows that the result in Eq. (4.71) remains valid when the Ohm's law Eq. (4.69) is replaced by a drift-kinetic description of electron dynamics, provided the same conditions on v_e and ω_* hold.

The correlation length and time are given by

$$\ell_c \sim \left(\frac{v_e}{v_A}\right)^{1/2} \alpha^{1/4} \left(\frac{c}{\omega_{pe}}\right) \sim \alpha^{1/4}\left(\frac{c}{\omega_{pe}}\rho_s\right)^{1/2} \qquad \tau_c \sim \frac{Rq}{v_A}\alpha^{-1/2} \tag{4.72}$$

where $\rho_s = (m_i T_e)^{1/2}/eB$ is the 'ion sound' Larmor radius and α from Eq. (4.57). In Chapter 14 the length ρ_s is invoked to explain the sawtooth reconnection process in a collisionless plasma as given by Coppi and Detragiache (1992).

For typical tokamak conditions,

$$\frac{v_e}{Rq}\tau_c \sim \frac{v_e}{v_A}\alpha^{-1/2} \gg 1 \tag{4.73}$$

so that the result Eq. (4.71) is more appropriate than Eq. (4.65). The associated stochastic magnetic field electron thermal diffusivity is then given by

$$\chi_e^{\rm st} \sim \left(\frac{c}{\omega_{pe}}\right)\rho_s \frac{v_e}{Rq}\frac{\alpha^{3/2}}{s^{1/2}}. \tag{4.74}$$

4.5 Momentum Stress Tensor Stability Analysis

The contribution to the plasma currents from the force arising from the divergence of the off-diagonal momentum stress tensor or the gyro-orbits of the particles gives a key stabilizing term to the magnetic ballooning and interchange modes and to the rotational instabilities. The effect of this term in the dispersion relation is given by

$$\omega^2 \to \omega\,(\omega - \omega_{*i}) \qquad \text{where} \qquad \omega_{*i} = \frac{k_y}{eBn_i}\frac{dp_i}{dr}. \tag{4.75}$$

The effect of this ion diamagnetic pressure gradient term on the interchange-ballooning mode is to stabilize the higher wavenumbers. This effect is called the finite-Larmor radius stabilization of MHD modes in the early literature. The effect is strong and important for geometries with large ion gyroradius as in the Field Reversed Configurations.

Worked Problem 4.1: FLR stabilization of Interchange Modes

Show that the interchange instability dispersion relation is modified by the divergence of the current from the force $\boldsymbol{F}_{\rm gyro} = -\nabla\cdot\boldsymbol{\Pi}_{\rm FLR}$ to become $k_\perp^2\rho^2\,\omega(\omega-\omega_{*i}) + \omega_D\,\omega_{*p} = 0$.
Follow the solutions in Smolyakov (1998), Kim, *et al.* (1991), and Hinton and Horton (1971).

4.6 Kinetic Ballooning Mode Instability

The coupled drift-shear Alfvén mode including the complete Bessel function gyroradius effect and the $\nabla_\perp\mathbf{B}$-curvature guiding center drift resonance of kinetic theory is solved for the toroidal ballooning mode eigenvalues and eigenfunctions.

The basic physical mechanisms of drift wave instabilities and the associated plasma transport are now known and understood in detail. Drift waves develop into a mixture of wave turbulence and coherent structures with details of the mixture depending on the driving forces ∇n_e, ∇T_e and ∇T_i and the amount of magnetic shear and background velocity shear in the system. The mixture of zonal flows, waves and vortices has certain similarities to the waves in planetary atmospheres and the equatorial-to-polar temperature gradient driven turbulent heat. The principal difference is in the replacement of the gravitational force on planets with a electric forces from charge separation in the curved magnetic fields. The role of the Coriolis force on planets for waves with periods longer than the rotation period of the planet is analogous mathematically to the Lorentz force from the ambient magnetic field.

The free energy sources created by the radial gradients of the plasma density and the temperature are used to characterize the strength of the drift wave turbulence. Diagnostic signals from microwave interferometers and Thompson scattering data described in Chapter 2 can be used to compute the thermodynamic formulas for the upper bounds of the turbulence as shown in Chapters 13 and 14 for the density and temperature fluctuation levels. The derivations of the free energy formulas treat the plasma as an ideal gas with a compressible gas constant, $\Gamma = (d+2)/d$, where d is the number of degrees of freedom associated with the type of instability under investigation.

The Carnot-cycle description of the auxiliary heated plasma is particularly relevant for the instabilities driven by the temperature gradients. The electron temperature profiles drive the temperature gradient instabilities, as given in Horton, *et al.* (2005) for Tore Supra. The structure of the drift wave eigenmodes in the double-periodic toroidal geometry is described in detail in Chapters 13 and 14.

Experiments for the ion temperature gradient (ITG) mode were carried out in the Columbia Linear Machine (CLM) [Sen, *et al.* (1991)] and in TEXT [Brower, *et al.* (1987)]. Both experiments show the role of the ITG drift modes in driving turbulent transport.

The ITG mode is the standard model for describing ion thermal transport in basically all toroidal magnetic fusion confinement experiments. The toroidal instability mechanism and nonlinear models were derived in Horton *et al.* (1981). High-ion temperatures profiles can be stable by simultaneously maintaining sufficient density profile peaking.

Large-scale simulations with Tera-flop and Peta-flop computers are used to develop the ITG turbulence from direct numerical solutions of the gyrokinetic-Vlasov-Maxwell system. These simulations codes allow the computation of the drift wave turbulence in complicated geometries, including the nonaxisymmetric systems like the Large Helical Device at the National Institute for Fusion Science (NIFS) and the Wendelstein 7X machine at IPP Greifswald, Germany.

The electron turbulent transport is more demanding to compute but shows large enhancements over the quasilinear fluxes due to the formation of extended radial convective structures with correlation lengths, $\ell_c \gg \rho_e$. The convective structures in the turbulence called streamers are insensitive to the radial electric-field shear but do depend sensitively on the magnetic shear profile. A review of the numerical simulations of the ETG turbulence is Nevins *et al.* (2006).

The empirical energy confinement time scaling laws for both helical system [Yamada, *et al.* (2005)] and tokamaks [Kaye, *et al.* (1997)] follow the drift wave formulas and simulations to within the accuracy of the experimental data. Some of the more detailed comparisons of theoretical models [Kinsey, *et al.* (1996a,b), Redd, *et al.* (1998), Bateman, *et al.* (1998)] with tokamak databases are referenced but a detailed discussion of the correspondence between the data and simulations is beyond the scope of this chapter.

We move to a description of the drift turbulence in simpler basic laboratory experiments in Chapter 5.

Chapter 5

Magnetohydrodynamics and Magnetic Confinement Geometries

The MHD-magnetohydrodynamic model plays a key role in plasma physics in both the laboratory and in space physics. From the perspective of Chapter 2 that gives the kinetic theory and Chapter 4 that develops the two-component (ion and electron) fluid equations, the MHD model is a further step down in the accuracy and detailed description of plasma dynamics. In fact, it is not possible to derive precisely the conditions under which the MHD model is valid. It is not unusual in science, and physics in particular, that the simpler less detailed description, is more useful. In physics two famous examples of this rule come to mind: Ohm's law and the Bohr atom. From the perspective of the more complex models introduced in Chapters 2 and 4, one can say that the ideal MHD model is to plasma physics as the Bohr atom model is to quantum atomic physics. One concludes that the MHD model is rarely precisely correct, but is always useful just as with the Bohr model of the atom. There are phenomena, of course, that are well-described by the idealized Bohr atomic and MHD plasma models. For plasma MHD the key properties that make the model so useful, and often adequate, are the conservation laws the model embodies. The MHD model is a set of seven partial differential equations advancing in time a single perfectly conducting fluid of mass density ρ, flow velocity $\boldsymbol{u}$, and isotropic pressure p that obeys the adiabatic gas law with a user-specified adiabatic gas constant Γ. The MHD model is needed for various purposes. Essentially all laboratory plasma confinement experiments are designed to have large parameter regions that are MHD stable. The drift wave and Kelvin-Helmholtz instabilities are of interest in the MHD stable plasmas.

5.1 The MHD-Magnetohydrodynamic Model

First and foremost, the study of drift waves and anomalous transport assumes that there already exists a stable MHD equilibrium. The point is that, even if the details of the MHD stability description are not precisely known, it is unlikely that drift waves and microinstabilities are of interest, or even well-posed mathematically, if the plasma is already unstable in the MHD description. There are some exceptions to this rule at marginal stability, and we discuss such a case below. But, the exceptions are rare. The very formulation of the equilibrium, or even the evolving ambient plasma state, requires the MHD equations be satisfied locally on the time scale of transport. The reconstruction of the density and temperature profiles used in the analysis of drift wave turbulent transport will use the conditions of local MHD equilibrium in the inversion of the magnetometer, interferometer and Thomson scattering data, for example.

5.1.1 *MHD equations*

To obtain the MHD equations from the kinetic equations or even the two-component FLR-fluid equations requires a long list of approximations. For those interested in such an effort, see the discussions in Krall and Trivelpiece (1968), Miyamoto (1976), White (2014), and Braginskii (1965). Here we write down the ideal equations and then discuss some of the key physical processes missing from the description with respect to the subject of drift waves and tearing modes.

The conservation laws for a single ideal, perfectly conducting fluid with mass density ρ and center of mass flow velocity $\boldsymbol{u}$ are

$$\frac{\partial \rho}{\partial t} + \nabla \cdot (\rho \boldsymbol{u}) = 0 \tag{5.1}$$

$$\rho \left(\frac{\partial \boldsymbol{u}}{\partial t} + \boldsymbol{u} \cdot \boldsymbol{\nabla} \boldsymbol{u} \right) = \boldsymbol{j} \times \mathbf{B} - \nabla p + \rho \boldsymbol{g} \tag{5.2}$$

$$\frac{\partial \mathbf{B}}{\partial t} = \nabla \times (\boldsymbol{u} \times \mathbf{B}) \tag{5.3}$$

$$\frac{\partial p}{\partial t} + \boldsymbol{u} \cdot \boldsymbol{\nabla} p + \Gamma p \nabla \cdot \boldsymbol{u} = 0 \tag{5.4}$$

where the current density follows from $\mu_0 \boldsymbol{j} = \nabla \times \mathbf{B}$ and the electric field is determined by idealization of a plasma gas with infinite electrical conductivity

$$\mathbf{E} + u \times \mathbf{B} = 0. \tag{5.5}$$

Equations (5.1) and (5.4) are combined to show that the specific entropy of the plasma gas $S = p/\rho^\Gamma$ called the specific entropy is constant in the frame $\boldsymbol{u}$ of the moving plasma. The equation of state (EOS) of the ionized gas is $S = \text{const}$ in the rest frame of the gas. Combining this statement with the mass conservation law one defines the entropy density $s(\boldsymbol{x}, t) = n k_B \ln(p/\rho^\Gamma)$ is a conserved density satisfying

$$\frac{ds}{dt} = \frac{\partial s}{\partial t} + \nabla \cdot (\boldsymbol{u} s) = 0. \tag{5.6}$$

Here n is the number density, k_B the Boltzmann constant and Γ is the adiabatic gas constant.

Equation (5.5) is the condition that the electric field $\mathbf{E}'$ vanishes in the local rest frame of the moving plasma. Equation (5.6) is the statement that in a fluid element following the fluid flow, there is no change in the entropy. These two equations are the strong statements that fail in most of the plasma dynamics studied in this book. They become reasonable approximation only for very fast, and thus strongly out of force balance dynamics.

Equation (5.5) then produces the motional EMF in the laboratory frame that is the fixed initial frame in which the velocity $\boldsymbol{u}$ is measured with respect to. For example, in the frame attached to Earth, the solar wind plasma has the velocity $\boldsymbol{u}_{\mathrm{sw}}$ that is directed radially outward from the Sun with speeds that range from 200 km/s to 700 km/s. This plasma contains the interplanetary magnetic field $\mathbf{B}_{\mathrm{IMF}}$ and then in magnetosheath surrounding the Earth's magnetosphere has an additional $\Delta\mathbf{B}$ component. In the Earth's frame of reference this solar wind flowing across the ambient magnetic field $\mathbf{B}(\boldsymbol{x}, t)$ drives the electric field

$$\mathbf{E}_{\mathrm{sw}} = -\boldsymbol{u}_{\mathrm{sw}} \times \mathbf{B} \tag{5.7}$$

that maps down the magnetic field lines, since $\mathbf{B} \cdot \mathbf{E} = 0$, to impose the measured electric field in the ionospheric polar caps. For this discussion, the polar cap is defined as the region where a field line has only one end attached to Earth and the other end goes to the interplanetary magnetic field $\mathbf{B}_{\mathrm{IMF}}$. These solar wind generated potentials range from 40 kV in quiet times to hundreds of kilovolts during magnetic storms. The electric fields are measured by spacecraft surrounding the Earth, by spacecraft in the ionosphere and inferred by ground based radar measurements of ionospheric plasma convection [Kelley (2009), Kivelson and Russell (1983)].

In the laboratory plasma the most important example of Eq. (5.5) is of the radial electric field $E_r(r, t)$ that is produced by the high-speed toroidal rotational velocities (200 to 600 km/s) driven by the neutral beam injection (NBI) into tokamak plasmas. For toroidal plasma flow velocities u_ϕ, these rotation driven electric fields are then radial with

$$E_r^{(0)} = u_\phi B_\theta \tag{5.8}$$

and are known to be very important in influencing the plasma confinement.

The laboratory measurements of E_r are complex and indirect. Thus, the errors in the radial profile of E_r are rather large. The zero superscript indicates the dominant component of E_r for large toroidal plasma flows $\boldsymbol{u} = \boldsymbol{u}_\phi = R\Omega\widehat{\boldsymbol{e}}_\phi$. However, what is known is that, even at the highest velocities, the MHD approximation $E_r = E_r^{(0)}$ given by Eq. (5.8) is incomplete in rotating tokamaks. There are carefully-documented measurements that show that E_r is controlled not only by u_ϕ but by the flows driven by the ion pressure $\nabla p_i / e n_i$ and the impurity pressure gradients as well as the parallel force $F_{\|ab}$ between the working gas ions and the impurity ions. Grierson, *et al.* (2010) develop the measurements of E_r and correlate the profiles of E_r with the predictions of neoclassical theory Hirshman and Sigmar (1997).

5.1.2 *Conservation form of MHD*

The success of the MHD model in describing a multitude of plasma dynamics is due to the conservation of mass, momentum, energy and charge. Numerical implementations are most reliable when those properties are built into the time advancement scheme as emphasized by Barnes and Brackbill (1977) and Boris and Book (1973). For example, shocks are a universally-observed, complex phenomenon that are well described outside the dissipation layer by the conservation laws of MHD. In conservation form the equations are

$$\frac{\partial \rho}{\partial t} + \nabla \cdot (\rho \boldsymbol{u}) = 0 \tag{5.9}$$

$$\frac{\partial}{\partial t}(\rho \boldsymbol{u}) + \nabla \cdot \left[\rho \boldsymbol{u}\boldsymbol{u} + \left(p + \frac{B^2}{2\mu_0}\right)\boldsymbol{I} - \frac{\mathbf{B}\mathbf{B}}{\mu_0}\right] = 0 \tag{5.10}$$

and

$$\frac{\partial}{\partial t}\left(\frac{1}{2}\rho u^2 + \frac{B^2}{2\mu_0} + \frac{p}{\Gamma - 1}\right) + \nabla \cdot \left[\rho \boldsymbol{u}\left(\frac{u^2}{2} + \frac{\Gamma}{\Gamma - 1}\frac{p}{\rho}\right)\right. \tag{5.11}$$

$$\left. + \boldsymbol{u}\frac{B^2}{\mu_0} - \mathbf{B}\,\frac{\boldsymbol{u} \cdot \mathbf{B}}{\mu_0}\right] = 0 \tag{5.12}$$

where $(I)_{ij} = \delta_{ij}$. Care must be taken to maintain $\nabla \cdot \mathbf{B} = 0$ exactly in using Eq. (5.10) since a residual (numerical) error creates a large magnetic monopole force $\mathbf{B}(\nabla \cdot \mathbf{B})$ in Eq. (5.10). Charge is conserved as long as $\nabla \cdot \boldsymbol{j} = \nabla \cdot (\nabla \times \mathbf{B}) \equiv 0$ is maintained. Morrison and Greene (1980) show that the conservation form is naturally Hamiltonian and Kulsrud (1983) gives generalizations of MHD for astrophysics where the gravity field is included, for analogous momentum conservation law and the virial theorem.

In advancing Eqs. (5.9)–(5.12) one uses the ideal conductivity law in Eq. (5.3) which adds two degrees of freedom. Thus, the system has $5 + 2 = 7$ degrees of freedom. Linearization about a uniform state yields a cubic equation in ω^2 and a zero frequency $\omega = 0$ mode.

The great difficulty, as well as beauty, of the system (5.9)-(5.12) is that finite amplitude disturbances steepen as the disturbances propagate leading to finite time singularities. These singularities are the formation of the MHD shocks [Burgess (1995)]. Methods for handling the shock problem are the Godunov method and the flux corrected transport FTC problem. In terms of the physics, however, the MHD description clearly fails for the transition region of shocks. Here, we know that kinetic particle descriptions are required to describe the reflected ions, the turbulent fluctuations and anomalous resistivity and viscosity. The lower-hybrid drift waves are measured with CO_2 laser scattering experiments in numerous laboratory shock experiments. Spacecraft passing through the bow shock measure many types of plasma wave waves and complex particle dynamics. For shock physics the reader is referred to Sagdeev and Galeev (1969), Burgess (1995), and Kivelson and Russell (1983, p. 129).

Worked Problem 5.1: Grad-Shafranov Equilibrium

In cylindrical geometry (r, ϕ, z) with symmetry (axisymmetric equilibria) about the vertical z-axis

(1) Find the magnetic flux function $\psi(r, z)$ that describes the specified magnetic fields $B_r(r, z)$ and $B_z(r, z)$ arising both from external coils and the plasma currents.

(2) Show that the contours of $\psi(r, z) = \text{const}$ describe their "poloidal" magnetic field vectors.

(3) When there is no flow ($u = 0$) and no external force ($g = 0$), express the force balance in terms of magnetic flux ψ. The resulting equation is Grad-Shafranov equation. Note that the pressure p is a function of the flux function $p = p(\psi)$ and the toroidal magnetic field B_ϕ can be expressed in terms of ψ through $I = rB_\phi = I(\psi)$.

Solution:

(1) From the vanishing divergence of the magnetic field vector $\mathbf{B}$ and the axisymmetry, we have

$$\frac{1}{r}\frac{\partial}{\partial r}(rB_r) + \frac{\partial B_z}{\partial z} = 0. \tag{5.13}$$

This constraint is satisfied by a scalar function $\psi(r, z)$ and is a sufficient condition for the existence of the flux function, such that

$$B_r = \frac{-1}{r}\frac{\partial \psi}{\partial z} \qquad B_z = \frac{1}{r}\frac{\partial \psi}{\partial r}. \tag{5.14}$$

(2) Consider the equations for trajectories tangent to the magnetic field vector

$$\frac{dr}{du} = B_r = \frac{-1}{r}\frac{\partial \psi}{\partial z} \tag{5.15}$$

$$\frac{dz}{du} = B_z = \frac{1}{r}\frac{\partial \psi}{\partial r}. \tag{5.16}$$

The trajectories are along the level contours defined by

$$\psi(r, z) = \text{const}. \tag{5.17}$$

In fact, the trajectories are Hamiltonian with $H(p_1 = r^2/2,\ q^1 = z) = \psi(r^2, z)$.

The situation is isomorphic to that of the stream function $\psi(r, z)$ for 2D incompressible $\nabla \cdot \boldsymbol{u} = 0$ fluid flows. Furthermore, the vorticity $\nabla \times \boldsymbol{u}$ in the symmetry direction $\widehat{e}_\phi$ is isomorphic with the plasma current $\mu_0\, \boldsymbol{j} = \nabla \times \mathbf{B}$ in the toroidal direction.

The current now has a poloidal component $\boldsymbol{J}_p$ in addition to the toroidal component J_ϕ. The total current is

$$\mu_0 \boldsymbol{J} = \nabla \times \mathbf{B} = -\left[\frac{\partial}{\partial r}\left(\frac{1}{r}\frac{\partial \psi}{\partial r}\right) + \frac{1}{r}\frac{\partial^2 \psi}{\partial z^2}\right] \widehat{\boldsymbol{e}}_\phi + \nabla\left(rB_\phi\right) \times \nabla \phi. \quad (5.18)$$

The second-order operator acting on the flux function ψ is defined in the literature as $\Delta^* \psi$ and called the Grad-Shafranov operator. The definition of Δ^* is

$$\Delta^* = r\frac{\partial}{\partial r}\frac{1}{r}\frac{\partial}{\partial r} + \frac{\partial^2}{\partial z^2} \quad (5.19)$$

with which the toroidal component of Eq. (5.18) becomes

$$\Delta^* \psi = -\mu_0\, r J_\phi. \quad (5.20)$$

In addition to the toroidal J_ϕ there is the poloidal current from the last term in Eq. (5.18)

$$\mu_0\, \boldsymbol{J}_p = \nabla\left(rB_\phi\right) \times \nabla \phi \equiv \nabla I \times \nabla \phi \quad (5.21)$$

where the literature defines the scalar function $I \equiv rB_\phi(r,z)$.
(3) Now the force resulting from these currents $\boldsymbol{J}_p$ and J_ϕ is

$$\boldsymbol{f} = \boldsymbol{J} \times \mathbf{B}. \quad (5.22)$$

Carrying out the cross products one arrives at the force

$$\boldsymbol{f} = \frac{J_\phi}{r} \nabla \psi - \frac{I}{\mu_0 r^2} \nabla I. \quad (5.23)$$

The force $\boldsymbol{f}$ is perpendicular to the level lines of the flux function when it can be balanced by a pressure gradient. When $\boldsymbol{f}$ is balanced by the plasma pressure, $\mathbf{B} \cdot \nabla p = 0$ gives $p = p(\psi)$. And $\boldsymbol{J} \cdot \nabla p = 0$ gives $I = I(p) = I(\psi)$. Then Eq. (5.23) becomes

$$\frac{dp}{d\psi}\nabla\psi = \frac{J_\phi}{r}\nabla\psi - \frac{I}{\mu_\theta r^2}\frac{dI}{d\psi}\nabla\psi.$$

Thus, for MHD equilibria without flow ($\rho \boldsymbol{u} = 0$) we have two functions that determine the static equilibrium $p(\psi)$ and $I(\psi)$ for the pressure and toroidal magnetic field function. Calculating $\boldsymbol{J} \times \mathbf{B} = \nabla p$ then gives that

$$\Delta^* \psi = -\mu_0 r^2 \frac{dp}{d\psi} - I\frac{dI}{d\psi} \quad (5.24)$$

called the Grad-Shafranov equation. There is extensive computer software to solve the Grad-Shafranov equation for specified profile functions $p(\psi)$ and $I(\psi)$. The turbulent transport is the mechanism by which these profile functions are established from the externally applied heating and fueling sources.

More comprehensive analysis and derivations of the Grad-Shafranov equation are given in Goedbloed, *et al.* (2010) on advanced magnetohydrodynamics with general methods for solving the G-S equation. We return to the G-S equation for transport calculations in Chapter 16 and note that the collisional transport theory in tokamaks is also formulated in terms of the G-S equation in Helander and Sigmar (2002), Chapter 6. There is extensive software for solving the G-S equation both for theory and for real-time for shot analysis of diagnostic data.

There is a continuum of solutions to the G-S equation so that external forces control the realized laboratory plasma equilibrium along with the bifurcations from one equilibrium to a second equilibrium. The particular G-S equilibrium and the bifurcations are controlled by the external drivers such as neutral beam injectors, RF drivers and the central solenoid primary flux in the case of tokamaks.

5.1.3 *MHD stable plasmas and bifurcations*

Plasmas are created by various atomic and molecular ionization methods in the laboratory and in nature. Often the creation process is rather violent, forming plasmas with fast flows, large electric currents and/or with high energy particles. After the passage of some time, typically measured by the number of Alfvén wave oscillations, the plasma will settle into a local, slowly evolving quasi-equilibrium state. In this second stage we define the characteristic time scale $\tau_{\rm eq}$ and space scale $L_{\rm eq}$ for the plasma time and space scale of evolution. When the quasi-equilibrium state contains a significant magnetic field B such that $\beta \equiv$ energy density ratio of thermal particle energy density $p = k_B nT$ to the magnetic energy density $B^2/2\mu_0$ is of order unity, the plasma scale length s becomes $L_\perp \ll L_\|$. This inequality $L_\perp \ll L_\|$ is satisfied in laboratory magnetic confinement experiments.

For each problem some modeling judgment is required to determine when the profiles have entered into this slowly varying quasi-equilibrium phase of evolution. Even if the system is slightly above the critical state according to stability analysis of ideal MHD for the corresponding frozen state it is still important to determine the microinstabilities of the evolving system. This stability analysis can be carried out with sufficient rigor on the evolving background for wave frequencies $\omega(\boldsymbol{k})$ and wavenumbers $\boldsymbol{k}$ for modes that have $\omega\tau_{\rm eq}$ and $kL_{\rm eq}$ are large compared with unity.

For example, the shock fronts structures are often analyzed locally for the small scale lower-hybrid drift wave turbulence. The transport bifurcations in the tokamak system also are sufficiently slow to allow drift wave stability analysis to apply locally in radius and time in the transport barrier region for determining the turbulent transport in the barrier region development and evolution. A key point is that the MHD equilibrium analysis alone will yield a large (infinite) class of equilibrium solutions even when the wall boundary conditions are fixed. It is the turbulent transport coupled to the sources and sinks of particles, energy and momentum that determine the particular MHD equilibrium that the plasma settles into. The large equilibrium forces of $\boldsymbol{j} \times \mathbf{B}$ and ∇p will be equilibrated first through compressional motions and maintain the plasma close to force balance equilibrium throughout the transport evolution. This is a self-consistency loop called the problem of equilibrium, stability and transport. Transport codes such as TRANSP, CRONOS, and BALDUR compute the coupled evolution dynamics. What has become clear with running these codes is that the coupling of the dynamical loops of back-reaction of the transport (turbulent and collisional) on the profiles and thus the turbulence is a fundamental part of dynamical feedback and feed-forward loop. These loops have critical points in which the fractional flows of energy-momentum from the input to the output take different pathways through the system. Many names are associated with the different plasma states such as L-mode, H-mode, ELMy H-mode, ELM-free H-modes, and the I-modes for the different plasma states. The bifurcation or transitions between the plasma states can be viewed as second-order phase transitions.

The importance of tracking the self-consistent interactions in the transport problem has only been fully realized and taken into account in the past few decades. A key laboratory experiment helped start this revolution in transport studies as

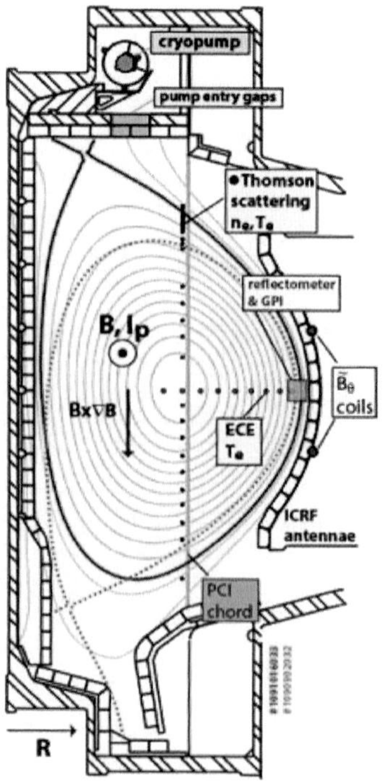

Fig. 5.1 Example of upper single-null equilibrium used for I-mode studies (solid black separatrix) with the ion $B \times \nabla B$ pointed away from the X-point. Also shown is the separatrix (grey dashed line) for the H-mode with the $B \times \nabla B$ pointed towards the X-point with the lower single null at the divertor chamber at the bottom of the vessel. The upper divertor cryo-pump, entry gaps, and key diagnostic locations are shown for the C-Mod tokamak.

well as the maturing of the transport simulation models that track the stability and transport of the different instabilities that occur in the different plasma states was the ASDEX. The ASDEX laboratory experiments made clear the occurrence of bifurcations in the transport was the discovery of the high-confinement mode, called the H-mode. The pre-existing confinement state is then called the L-mode for low confinement. The L-mode state is the base confinement level of standard tokamak operation and is the most thoroughly documented confinement state of high temperature toroidal plasmas.

The tokamak plasma data describing the various confinement states is available at `www.fusion.magnetique.cea.fr` and described in *ITER Physics* (2015), ISBN 978-981-4678-66-7. The original $R = 8\,\text{m}$ ITER machine design was based on L-mode confinement and only when it became clear that the factor of roughly two-fold confinement improvement from the H-mode would produce large costs savings in the design size of the machine was the L-mode scaling law abandoned and the H-mode confinement state adopted as the baseline performance model for the ITER plasma. The projected confinement times for the L- and H-modes in the original and reduced-sized ITER experiment are shown in Table 1.1. To achieve the L-to-H bifurcation, the injected heating power must exceed a critical value defined as $P_{L\to H}$. In this way we see that the change of confinement phase is analogous to a phase transition. The power threshold for the H-mode is given in Snipes, *et al.* (1996) and McDonald, *et al.* (2007).

The L-mode data is studied and parameterized to empirical scaling laws for baseline tokamak operation for more than two decades. Tokamaks can be routinely designed and brought on line with the performance given by the L-mode scaling laws. There are many means known for enhancing the confinement over that given by the L-mode and this enhancement is quantified as the H-factor for the global energy confinement time τ_E. For example, with the use of reversed magnetic shear the JT-60U experiment, and many others subsequently, obtained H-factors of 2 to 2.5 meaning that the global energy confinement time was 2 to 2.5 times longer than the time predicted by the L-mode confinement formula for the same 9 system parameters. These predictions are studied in detail in Chapters 13-15 since they are some of the most thoroughly documented of the laboratory plasma confinement results.

Bifurcations of plasma states are to be expected of course as discussed already by Kadomtsev (1992). The phenomena of bifurcations is studied in detail in neutral fluid experiments for many years. The most important correspondence between the plasma L- to H-mode bifurcation and those in neutral fluids is for the Rayleigh-Bénard convection system. As described qualitatively in Fig. 1.5 and shown mathematically in Chapter 4, there is a close correspondence between the resistive-g plasma model and the Rayleigh-Bénard system. In toroidal plasmas there is an effective gravity from $g = 2c_s^2/R$ where R is the major radius of the torus. The ion temperature gradient ITG turbulence is the collisionless limit of the resistive $-g$ instability, and thus is also related the Rayleigh-Bernard convection system.

In terms of drift waves, the ITG system applies when there is significant coupling of the plasma vorticity equation to the ion thermal balance equation through the magnetic curvature ($\kappa = 1/R_c = 2/R$) and $\nabla_\perp B$ drift particle currents produced by the toroidal magnetic field geometry. In a straight system there is also a weak ITG mode but now the coupling comes through an inverted parallel pressure gradient $\nabla_\| \delta p_i$ that drives the parallel ion mass flow R.

The Rayleigh-Bernard experiments of Krishnamurti and Howard (1981) and Howard and Krishnamurti (1986) in fluids heated against gravity show that as the temperature gradient is turned up or the fluid viscosity is lowered there is a bifurcation of the fluid flows to new states with large scale sheared flows. In higher flow states there is a global circular flow around the entire confinement chamber of the fluid. This global sheared mass flow then interferes with and thus reduces the convection in the pre-existing temperature gradient driven vortices. The vortices become tilted in the presence of the sheared flows. The fluid simulations of the plasmas with ITG show exactly this behavior even in the presence of magnetic shear [Hamaguchi and Horton (1991)]. Furthermore, Howard and Krishnamurti (1986) develop a low order model for the onset of the flow with the tilted vortices interacting with the sheared mass flow. The low-order model is such that below the bifurcation point, called in the subcritical state, the system reduces to the Lorenz system of the three odes that describe the nonlinear cellular convection.

The same bifurcation phenomenon happens in plasmas but in more varied and complex ways. There are simulations of the process in two and three dimensions with fluid models in Carreras, *et al.* (1987), Newman, *et al.* (1998), and Wakatani (1998). There are gyrokinetic simulations of the process. Analytic models of the L to H bifurcations are given in Diamond, *et al.* (2005), Newman, *et al.* (1998), Sugama, *et al.* (2001, 2003, 2007).

The universality of the rotational shear flow coupling to the wave spectrum has now been extended to stellarators and large helical confinement devices.

5.2 Interchange Stability Condition

In this section we derive the energy released from the interchange of two magnetic flux-tubes in a mirror magnetic field geometry. If the potential energy of the final state, upon interchange, is lower, than the energy in the initial states then it is assumed that a suitable plasma motion, $\boldsymbol{v} = d\boldsymbol{\xi}/dt$ will arise to enable the interchange dynamics. In MHD it is easy to justify that such a motion exists because the equations of motion follow from minimizing the potential energy, as in classical mechanics.

The potential energy is the sum of the magnetic field energy $B^2/2\mu_0$ and the internal energy of the plasma $p/(\Gamma - 1)$. The internal plasma energy in a flux-tube is given by

$$E_p = \int \frac{p}{\Gamma - 1} d^3x = \int \frac{p}{\Gamma - 1} ds dA = \int \frac{p}{\Gamma - 1} \Phi \frac{ds}{B} = \frac{p\Phi}{\Gamma - 1} \int \frac{ds}{B} \qquad (5.25)$$

where s is the length along the flux-tube, dA is the cross-sectional area of the flux-tube, and $\Phi = BdA$ is the magnetic flux which is a constant for each flux-tube as is the plasma pressure $p(\boldsymbol{x})$. The most familiar regime is a gas with three degrees of freedom for which $\Gamma = 5/3$ and $1/(\Gamma - 1) = 3/2$. Let $V\Phi$ be volume of the flux tube $\Phi(+\Psi)$.

In any adiabatic motion we have PV^Γ = const so that on interchange of flux-tubes labeled 1 and 2 we have

$$P_1' = P_2 \left(\frac{V_2}{V_1}\right)^\Gamma \quad \text{and} \quad P_2' = P_1 \left(\frac{V_1}{V_2}\right)^\Gamma \tag{5.26}$$

where the prime signifies the final state. Thus, the change in internal plasma energy due to the interchange of two flux-tubes is given by

$$\Delta E_p = \frac{1}{\Gamma - 1}\left[\left\{P_1 \frac{V_1^\Gamma}{V_2^\Gamma} V_2 + P_2 \frac{V_2^\Gamma}{V_1^\Gamma} V_1\right\}_{\text{final}} - \{P_1 V_1 + P_2 V_2\}_{\text{init}}\right]. \tag{5.27}$$

The magnetic energy in a flux-tube is

$$E_M = \int \frac{B^2}{2\mu_0} dV = \frac{\Phi^2}{2\mu_0} \int \frac{ds}{dA} \tag{5.28}$$

and upon interchanging two flux-tubes labeled 1 and 2 the change in the magnetic energy is given by

$$\Delta E_M = \frac{1}{2\mu_0}\left[\left\{\Phi_1^2 \int_1 \frac{ds}{dA} + \Phi_2^2 \int_2 \frac{ds}{dA}\right\}_{\text{final}} - \left\{\Phi_1^2 \int_2 \frac{ds}{dA} + \Phi_2^2 \int_1 \frac{ds}{dA}\right\}_{\text{initial}}\right]. \tag{5.29}$$

Thus, if we exchange two flux-tubes with the same amount of magnetic flux, i.e., $\Phi_1 = \Phi_2$, then there is no change in magnetic energy. Thus, the $\Delta E_p > 0$ stability condition reduces to the condition that is called negative V'' - in the fusion literature - and is given by

$$\frac{dp}{d\psi}\frac{d}{d\psi}\int \frac{ds}{B} + \frac{\Gamma p}{V}\left(\frac{d}{d\psi}\int \frac{ds}{B}\right)^2 \geqq 0 \tag{5.30}$$

where the volume of the flux tube per unit of magnetic flux $d\psi$ is given by $V' = \int(ds/B)$ where we use that $dV = \Delta A ds$ where ΔA is the cross-sectional area of the flux tube such that $\Delta\psi = B\Delta A$ = const along the tube. The $\Gamma p(V')^2$ term is stabilizing and sets the limit on the first driving term from $dp/d\psi$.

5.2.1 *Plasma energy functional of MHD*

The plasma energy functional of MHD is given by

$$\delta W_p^{\text{MHD}} = \frac{1}{2}\int \frac{d\Psi dy ds}{B}\left[\Gamma p(\nabla \cdot \boldsymbol{\xi})^2 + \boldsymbol{Q}_\perp^2 + \boldsymbol{Q}_L^2 - 2\boldsymbol{\xi} \cdot \boldsymbol{\kappa}\boldsymbol{\xi} \cdot \nabla p\right] \tag{5.31}$$

as derived and analyzed in Friedberg (2007). The functional in Eq. (5.31) is the potential energy for the displacement $\boldsymbol{\xi}$ of the system from an MHD equilibrium. The equilibrium is stable if all possible displacements ξ give a positive value of $\delta W_p^{\mathrm{MHD}}$. This method is highly developed in the plasma physics community in both magnetic confinement studies [Troyon (1987)] and in the astrophysics community [Shapiro and Tuekolsky (1983)].

5.2.2 *Limitations of the MHD model*

For the MHD model to give (i) accurate growth rates for an optimized displacement vector ξ and (ii) marginal stability boundaries in agreement with the first principle stability calculations places severe constraints on the system parameters.

Namely, the motions predicted from the MHD model must be sufficiently unstable, and thus fast in the dynamics, for the MHD model to be a valid description of the plasma motions. The statement of marginal stability in the context of MHD means the plasma is stable with respect to fast-large scale motions. This constraint on applicability gives rise to the notion of FAST MHD.

When the MHD model predicts motions that are fast compared with the drift mode frequencies and the transit or bounce time of the particles in their motion along the field lines, then the MHD description is confirmed by the kinetic theory. Often this condition is not satisfied in a relaxed system. The condition is especially not satisfied in the case of confinement geometries that are carefully constructed to be MHD stable. When such a system is operated in the MHD stable window of system parameters, the remaining instabilities are of the kinetic resonant particle variants of the MHD oscillations. These oscillations and instabilities require kinetic theory to describe adequately. These instabilities are entirely missed by the MHD description with a single fluid and a vanishing parallel electric field.

For motion of frequency ω and wavenumber k there are natural collective oscillations at the diamagnetic drift frequencies $\omega_* = k_y v_d$ as described in Section 2.1 associated with each species ω_{*i} and ω_{*e}. The MHD growth must be fast compared to both the ω_*'s and the transit/bounce frequency for the kinetic description to reduce to MHD. In addition, there are the requirements that (1) the parallel electric field be negligible and that (2) the distinction between the ion and electron pressure gradients is not important. These conditions are difficult to meet unless the motion is rather violent.

In the intermediate state where MHD gives a good guide, but not an accurate value for the growth rate and stability condition, the kinetic theory may give a maximum growth rate $\gamma_{\max}$, for example, that is one half the value $\gamma_{\max}^{\mathrm{mhd}}$ of the MHD calculation. More significantly, however, is that the kinetic growth rate $\gamma(k)$ will be non-vanishing and with a significant magnitude for wavenumber k values where the MHD growth rate is stable $\gamma_k^{\mathrm{mhd}} = 0$. There are both low and high wavenumber values for which the first principle calculations give unstable oscillations whereas the MHD oscillations are only unstable in the MHD unstable window. This phenomenon is well known in the branch of the MHD literature that deals with the resistive/viscous single fluid equations.

A typical result, for example, is that below the critical beta of $\beta_{\rm crit} = L_p R_c / L_c^2 \simeq 5 I_p / a B_T$ for the onset of the ballooning MHD mode, there is a resistive interchange mode with growth rate given by $\gamma^{rg} \leq c_s / (L_p R_c)^{1/2}$ [Sugama and Horton (1994), Carreras, *et al.* (1987)].

The fast internal kink instability that limits the maximum toroidal current I_p that a tokamak contains has this same limitation. Below the MHD critical current there are important kinetic-drift wave instabilities associated with the toroidal current density profiles. These slower modes are seen in the precursor oscillations before the strong current disruptions. This dynamics is discussed in Chapter 14.

Laboratory plasma imaging by Park, *et al.* (2004) shows movies of the precursor motions and then the final disruption dynamics in the TEXTOR plasmas. The movies are made from tomographic reconstruction of the plasma electron temperature derived from soft-X ray emission with high resolution spectra. For currents that are below the MHD onset of the internal kink $m = 1/n = 1$ mode there is the resistive tearing mode unstable growth rate. The calculations for this resistive MHD mode are famous and elaborate since they provide one of the intrinsic limits to the high-performance operation of tokamaks.

A detailed calculation of the stability theory of the 1/1-mode known as the internal kink mode is given by Crew and Ramos (1983) and Hegna, *et al.* (1996). They calculate $\delta W(\xi)$ for each toroidal mode number n and the internal kink mode growth rates, $\gamma = n v_{A1}/R_0$, for shifted circular magnetic surfaces by an expansion in low $\epsilon = r/R_0$ for $\beta_p = 2\mu_0 \langle p \rangle / B_\theta^2(a) > 1$ provided $\epsilon \beta_p < 1$. Here v_{A1} is the Alfvén velocity calculated with poloidal magnetic field from the plasma current at the rational surface $q(r_{1/1}) = 1$ where the field line twist matches the magnetic perturbation $\delta B_r = \delta B_{1,1} \cos(\theta - \phi)$ so that a magnetic island is formed at $r_{1/1}$. This special magnetic surface is where the field line is returning to exactly the same point, thus being the shortest such length in the torus, after one revolution in both the toroidal (long way) and poloidal (short way) around the torus. Experimentally, from the beginning of the tokamak experiments, strong oscillations and disruptions of the toroidal plasma current originate from the vicinity of this $r_{1/1}$ magnetic surface. Images of the plasma from X-ray tomography from H. Park are available at `http://orion.ph.utexas.edu/~starpower`. Analysis and simulations [Aydemir (1990)] show there is an unstable window for $\beta_{p1} < \beta_p < \beta_{p2}$ with the internal kink mode structure. The nonlinear effect of this mode is the displacement of the core plasma inside the $q = 1$ rational surface into a helical $\times$-point. At the $\times$-point magnetic reconnection occurs as given by the Kadomtsev theory [Aydemir (1990), White (2014)]. The core plasma temperature drops abruptly in what is called the sawtooth instability in MHD stable domains.

5.3 Double Adiabatic Pressure Tensor for Anisotropically Heated Plasmas

The use of the scalar pressure in the MHD model is a severe restriction for plasmas with strong external heating when the mean-free-path is long compared to the parallel scales $L_\|$ of the background plasmas. Two important effects and their associated instabilities are missed with the isotropic pressure model. Heating methods such as neutral beam injection and ion cyclotron resonance heating drive up a temperature anisotropy with $T_\| > T_\perp$ for NBI heating and with $T_\perp > T_\|$ for the cyclotron resonance heating. To include these anisotropies in the wave dynamics the scalar pressure is replace in the momentum equation with a pressure tensor

$$\boldsymbol{p} = p_\perp(\boldsymbol{I} - \boldsymbol{bb}) + p_\| \boldsymbol{bb} \tag{5.32}$$

where the force from the pressure tensor in the momentum equation becomes

$$\boldsymbol{F}_p = -\nabla \cdot \boldsymbol{p} = -\nabla p_\perp - \left(p_\| - p_\perp\right)(\boldsymbol{b} \cdot \nabla)\boldsymbol{b} - [\mathbf{B} \cdot \nabla\left(p_\| - p_\perp)/B\right))]\widehat{\boldsymbol{b}}. \tag{5.33}$$

The parallel component of the force $\boldsymbol{F}$ is used in the collisional transport theory in Chapter 10. So we give here the form $F_\|$ that is used in that chapter and in general for the bootstrap current calculations. The parallel part of the force can be written in several different forms, one of which is

$$\mathbf{B} \cdot \boldsymbol{F} = -\left(p_\perp - p_\|\right)(\boldsymbol{b} \cdot \boldsymbol{\nabla} B) - B\boldsymbol{b} \cdot \nabla p_\|. \tag{5.34}$$

In the low-collisionality regime where the trapped particles dominate the transport the fluxes are computed from the average over the magnetic surface of the parallel force from the anisotropic pressure tensor. The divergence of the anisotropic pressure tensor produces a parallel force given by

$$F_\| = -\left(p_\perp - p_\|\right) \boldsymbol{b} \cdot \boldsymbol{\nabla} \ln B - \boldsymbol{b} \cdot \nabla p_\| = -\frac{p_\perp}{\mathbf{B}} \boldsymbol{b} \cdot \nabla B - \mathbf{B} \cdot \nabla \left(\frac{p_\|}{B}\right). \tag{5.35}$$

When this force is averaged over the entire flux surface with periodicity in the toroidal coordinates the net parallel force in plasma between ψ and $\psi + \delta\psi$ is given by

$$\langle \mathbf{B} \cdot \boldsymbol{\nabla} \cdot \boldsymbol{\pi} \rangle = \left\langle \left(p_\perp - p_\|\right) \nabla_\| B \right\rangle. \tag{5.36}$$

Thus, the anisotropy gives the average parallel force in Eq. (5.36) that drives the banana-plateau radial particle flux from the parallel variation of magnetic field strength, $\nabla_\| B \neq 0$, in combination with a difference between the parallel and perpendicular pressures. This pressure difference can be written in terms of the Legendre polynomial $p_2(\xi) = (3\xi^2 - 1)/2$

$$p_\| - p_\perp = \int m v^2 p_2(\xi) f d^3 v. \tag{5.37}$$

Here ξ is the cosine of the particle pitch angle with $\xi = v_\|/v$.

Thus we see that the various heating methods as NBI, ICRH and ECH contribute to the pressure anisotropy driving a parallel force that produces transport.

The simplest cases of this are familiar from in the collisional regime where the force due the parallel resistivity and thermoelectric effects determine the Pfirsch-Schlüter diffusion, derived in Chapter 8.7 and the generalizations from thermoelectric effects. This name "diffusion" is a misnomer in that these fluxes are actually from cross field convective flows that are set up by the forces acting parallel to the magnetic fields. The calculations for the $\mathbf{E} \times \mathbf{B}$ flows and the parallel plasma flows are then the first part of the calculation for the parallel forces that drive these net cross-field particle fluxes. When the parallel flow is related to the cross field pressure gradients and perpendicular electric field this calculation is called the calculation for the parallel viscosity coefficients of the plasma. The works of Hirshman and Sigmar (1997) and Helander and Sigmar (2002) derive lengthy formulas for the transport effects in the absence of the fluctuations. The works of Sugama and Horton (1994), Sugama, *et al.* (1996) and Balescu (2005) derive general transport formulas in the presence of the fluctuations.

The value of the form in Eqs. (5.34)–(5.37) of the parallel pressure gradient force is that when the surface average is taken. The average of the second term proportional to $B\boldsymbol{b} \cdot \boldsymbol{\nabla} F$ vanishes for any F, leaving the anisotropy in the pressure as the net driving force. The result is used extensively in both collisional and turbulent transport calculations.

Now the pressure force is further understood by introducing the volume element of the flux magnetic tube $dAds = d\psi ds/B$ where $d\psi$ is the element of the flux tube volume defined by the product of its area and the local magnetic field vector $d\psi = \mathbf{B} \cdot \boldsymbol{dA}$. Here $\boldsymbol{dA}$ is the vector element of area and thus $d\psi$ is the magnetic flux through this surface element. For plasma in the given flux tube the cross-sectional area decreases as the magnetic field strength increases. The plasma becomes compressed in the perpendicular direction and the perpendicular pressure increases strongly. To describe the microscopic kinetic physics correctly one takes into account that the particle's magnetic moments $\mu = mv_\perp^2/2B$ is conserved if the dynamics is slow on the space-time scale of the cyclotron orbit. This means that the fluid motion must be such as to have

$$\frac{dp_\perp}{dt} + 2p_\perp \nabla \cdot \boldsymbol{u} - p_\perp \boldsymbol{b} \cdot (\boldsymbol{b} \cdot \boldsymbol{\nabla})\boldsymbol{u} = 0 \tag{5.38}$$

which is consistent with the conservation of the average magnetic moment expressed in fluid variables as $p_\perp/\rho B =$ constant.

For the evolution of the parallel pressure we must use a weaker adiabatic invariant associated with the parallel motion, namely the second particle adiabatic invariant from the action $J = \int v_\parallel ds$. Averaged over the particles motion this statement is that $p_\parallel B^2/\rho^3 =$ constant. Thus the partial differential equation derived for the evolution of the parallel pressure is

$$\frac{dp_\parallel}{dt} + p_\parallel \nabla \cdot \boldsymbol{u} + 2p_\parallel \left(\boldsymbol{b} \cdot (\boldsymbol{b} \cdot \nabla)\boldsymbol{u}\right) = 0. \tag{5.39}$$

To understand that (5.38) and (5.39) are consistent with $p_\perp$ and $p_\parallel$ equations of state, one uses that the $\boldsymbol{b}$-component of $\partial_t \mathbf{B} = \nabla \times (\boldsymbol{u} \times \boldsymbol{B})$ gives $\boldsymbol{b} \cdot (\boldsymbol{b} \cdot \boldsymbol{\nabla} \boldsymbol{u}) = B \boldsymbol{\nabla} \cdot \boldsymbol{u} + dB/dt$.

These equations are known as the Chew-Goldberger-Low equations and are described in the usual text books and in Volkov (1966). The linearization of these equations gives the quadratic equation in ω^2 with

$$\omega^4 - B\omega^2 - C = 0 \tag{5.40}$$

with $B(\boldsymbol{k})$ and $C(\boldsymbol{k})$ given by

$$B(\boldsymbol{k}) = \left(\frac{2p_\perp}{\rho} + v_A^2\right) k^2 + \frac{(2p_\parallel - p_\perp)}{\rho} k_\parallel^2 \tag{5.41}$$

and

$$C(\boldsymbol{k}) = \left(\frac{p_\perp^2}{\rho^2} - 3\left(\frac{p_\parallel}{\rho_o}\right)\left(\frac{2p_\perp}{\rho} + v_A^2\right)\right) k^2 k_\parallel^2 + \left(\frac{3p_\parallel^2 + p_\perp(3p_\parallel - p_\perp)}{\rho^2}\right) k_\parallel^4. \tag{5.42}$$

For small anisotropies these waves give the kinetically correct dispersion relations without ad hoc choices for the adiabatic gas constant Γ. For larger anisotropies and plasma pressures comparable to the magnetic pressure the discriminant of Eq. (5.40) goes negative for certain directions of propagation. The instabilities are (i) the firehose mode for parallel propagation and (ii) the mirror mode for perpendicular propagation. Figures 5.2 and 5.3 show the unstable k-space domains for the firehose and mirror mode instabilities. Both instabilities are important and common in the magnetospheres and solar wind plasmas. The mirror mode limits the maximum beta $\beta = 2\mu_0 p_\perp / B^2$ for a stable mirror trapped plasma for a specified mirror ratio $R_M = B_{\max}/B_{\min}$.

To understand the nonlinear dynamics of either the firehose or the mirror mode one needs to introduce the kinetic theory dispersion terms. This requirement can be seen from the fact that once dispersion relation Eq. (5.40) gives a growing mode instabilities, all modes with $k \to \infty$ are unstable in the model given in Eq. (5.40). Kinetic theory shows that the firehose mode becomes a right-hand polarized wave that connects to the whistler dispersion relation, while the mirror mode becomes a compressional wave with FLR and ω_* corrections.

The pressure anisotropies change the MHD equilibrium from that given by the Grad-Shafranov equation to a complex pair of partial differential equations. The MHD stability analysis now becomes complicated with numerous special cases with the mirror mode stability limiting the high beta minimum-B mirror equilibrium and the firehose mode limiting the parallel pressure in toroidal systems with parallel NBI heating. The variational formulation of the MHD stability analysis for the anisotropic pressure equilibrium is developed in the Nelson, *et al.* (1978). In that analysis one finds the firehose and mirror mode criterion limit further the maximum pressure that can be stability confined in a magnetic confinement system. A comprehensive extension of the MHD to include anisotropy and subsonic flows is

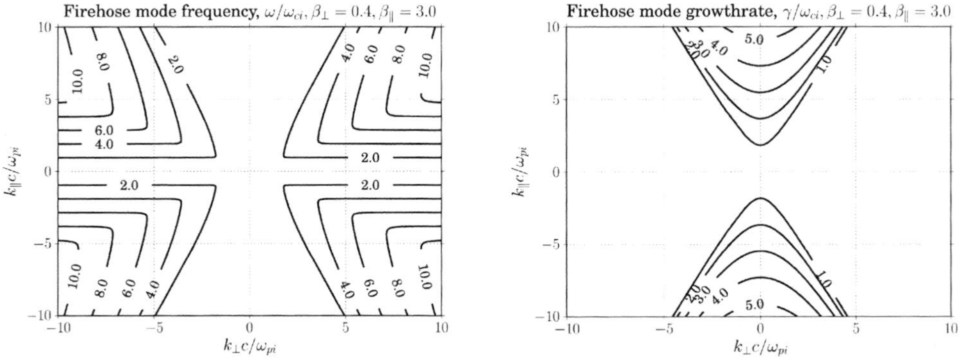

(a) (b)

Fig. 5.2 The left figure above shows the frequency contours when $\beta_\perp = 0.4$, $\beta_\parallel = 3.0$. ($\beta_\perp = 8\pi P_\perp/B^2$, $\beta_\parallel = 8\pi P_\parallel/B^2$). The blank regions, which form two triangle-like shapes, are the firehose instability regions. The criterion for firehose mode instability is $\beta_\parallel - \beta_\perp > 2$. The right figure above shows the growth rate contours for the same parameters. Firehose mode mainly propagates along the magnetic field.

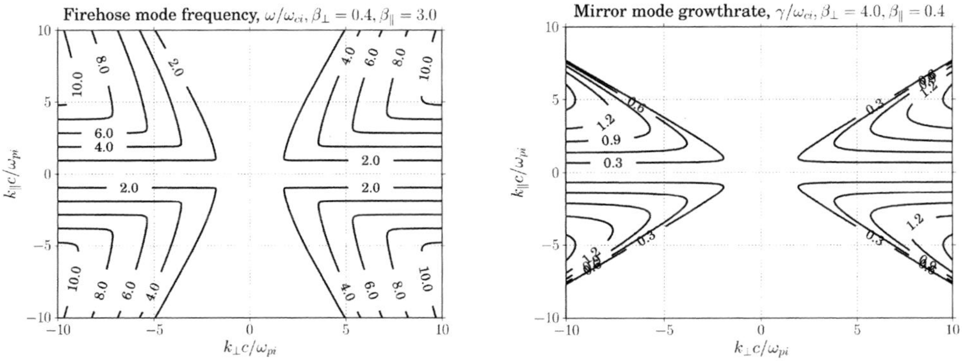

(a) (b)

Fig. 5.3 The left figure above shows the frequency contours when $\beta_\perp = 4, \beta_\parallel = 0.4$. In contrast to firehose instability, we observe another kind of instability which mainly propagates perpendicular to the magnetic field. The double adiabatic criterion for mirror mode instability is $\beta_\perp^2 > 6\beta_\parallel(1+\beta_\perp)$. The right figure shows the growth rate contours. In kinetic theory the factor 6 becomes unity.

developed in Ito, *et al.* (2007, 2011) where the anisotropic MHD is extended by calculating the electron parallel heat flux, whose moment is then closed in such manner as to closely model the corresponding dispersion relation derived from kinetic theory. The authors call the resulting single fluid model anisotropic Hall MHD model. In an equilibrium with poloidal flow V_p, an analysis of the two coupled equilibrium equations shows that the system is elliptic (wave-like) provided the poloidal flow is subsonic by a margin that depends on the degree of anisotropy in the pressures $p_\perp$ and $p_\parallel$.

Kinetic theory gives a lower threshold for the mirror mode and is required for finding the maximum growth rate $\gamma_{\max}$ and corresponding wavenumber $k_{\max}$. The kinetic theory stability parameter is $K = T_\perp/T_\| - 1 - 1/\beta_\perp > 0$ and $\gamma_{\max}/\omega_{ci} \propto \beta_\perp^{1/2} K^{3/2}$ [Pokhotelov, *et al.* (2004)].

5.4 Ballooning Interchange Modes and the Trapped Particle Instability

Once the spatial variation of the magnetic field vector $\mathbf{B}(\boldsymbol{x}) = B(\boldsymbol{x})\boldsymbol{b}(\boldsymbol{x})$ – where $\boldsymbol{b}$ is the unit tangent vector – is taken into account the force accelerating the plasma becomes a parallel mirror force and a perpendicular field line bending or curvature force. The mirror force traps particles with large magnetic moments in the local minimum of the magnetic field strength which separates their dynamics from the high parallel velocity particles that pass over the maxima of the magnetic field. This separation of the trapped and passing particles exists until a Coulomb collision rotates the velocity vector sufficiently to allow the particle to change its class from trapped or passing. This difference means that the trapped particles may behave as though they were in flute modes with $k_\| = 0$ and thus become interchange unstable for magnetic field line curvature concave towards the higher pressure plasma.

The lowest-order eigenmode of the plasma with a changing sign of the curvature of the magnetic field from concave to convex and remains as a flute mode with $k_\| = 0$ and will still be unstable when the condition $(dp/d\psi)(dV/d\psi) < 0$ is satisfied. This condition expresses the appropriate averaging of the magnetic field line curvature κ through the gradient $\frac{dV}{d\psi}$ of the flux tube volume. This is the stability condition for fast growing interchange modes that are relatively constant along the magnetic field line the displacement ξ_r is measured in magnetic flux. That is the contravariant displacement $\xi^\psi = B\xi_r(s)$ is constant s.

Here $V(\psi)$ is the flux tube volume given by

Flux Tube Volume

$$V(\psi) = \int \left[\frac{ds}{B(s,\psi)}\right].$$

This instability condition can be understood from either thermodynamics or kinetic theory. In thermodynamics the condition follows from requiring the energy released in a compressional motion down the pressure gradient exceed the work required to compress the plasma. In kinetic theory the condition follows from the flux tube average of the microscopic particle wave resonance condition of

Line Average Interchange

$$\int \frac{ds}{B(s)} \left[\omega_* \omega_D\right] > 0$$

which is the condition for the guiding center drift velocity to separate charge from over the wave in such a wave to produce an $\mathbf{E} \times \mathbf{B}$ drift of the plasma down the pressure gradient. This interchange condition was understood early by Longmire and Rosenbluth (1956).

Including the stabilizing term from plasma compression with an ideal gas constant of Γ we get the condition for instability is

Gradient of Entropy

$$\frac{dpV^{\Gamma}}{d\psi} < 0.$$

This is a famous stability condition and applies to a wide variety of confinement geometries including magnetospheres and solar magnetic flux tubes. The condition is used to understand how the dipole magnetic fields with unfavorable magnetic curvature and guiding center drifts are stable until the pressure gradient becomes very steep. This is because for a dipole field the volume per unit of flux

$$V(r) = \int \frac{ds}{B(s,\psi)} = \text{const}\,\frac{r^4}{B_{dp}} \tag{5.43}$$

so the pressure gradient must exceed $-d\ln p/dr > 4\Gamma/r$ for the MHD motion to release plasma thermal energy.

5.5 Trapped Particle Dynamics and Instabilities

Dipole magnetic fields have everywhere unfavorable magnetic curvature for interchange stability were not considered for laboratory plasma confinement. The simple extension to a kind of "double dipole" with two concentric loops with parallel currents, formally called the quadrupole, gives rise to alternating regions of favorable and unfavorable magnetic curvature inside the critical flux surface ψ_c. Inside the surface ψ_c the field lines have average favorable magnetic curvature where the stability condition above Eq. (5.43) is satisfied, and now the system is one suitable for magnetic confinement. There are regions of favorable curvature near the field null is clear from geometry of the quadrupole field lines as well as the fact that B^2 increases away from the magnetic $\times$-point. In fact, moving away from the magnetic null or $\times$-point the system has local minimum $B(\boldsymbol{x})$ surfaces not just average minimum-B properties. Thus the separatrix region is one of strong single particle and plasma confinement.

A number of linear and toroidal quadrupole and octupole experiments were built and operated in the late 1960s and early 1970s. These machines showed for the first time that confinement better than the Bohm value by more than an order of magnitude could be achieved. The experiments also clearly demonstrated that drift wave fluctuations are universal in most regions of an MHD stable plasma.

An important example of the quadrupole experiments was the 5-meter long, 0.45 meter diameter cylindrical device. The plasma was produced by inductively coupling 200 kHz of RF power to the helium working gas. The typical after-glow plasma density was decaying on the time scale of a 1 msec with temperature T_e dropping from 10 to 1 eV and $T_i \sim T_e/5$.

The fluctuations in the density and the potential were measured by the ion saturation current and the floating potential measurements with Lagrangian probes. The probe is movable and had a variable bias to determine the plasma fluctuating δn_e and $\delta\varphi$. Common to all the quadrupole and octupole experiments there is an abrupt drop from the high fluctuation level $\delta n/n \sim$ 50-100% in the flute mode unstable region where

MHD Unstable

$$\frac{dV}{d\psi} \cdot \frac{dp}{d\psi} < 0 \tag{5.44}$$

to a small amplitude 10-15% turbulence in the MHD stable region when the sign of Eq. (5.44) reverses. The fluctuations with frequencies $f \sim 60$ to 200 kHz are drift waves. These drift wave oscillations occur in the $V(\psi) = \oint ds/B$-stable regions where $(dV/d\psi)(dp/d\psi) > 0$. The drop in the fluctuation level occurs right at the critical magnetic surface ψ_c where the derivative of the flux tube volume $dV/d\psi$ changes sign.

The $ds/v_{\|}$-averaged sum of the ∇B and curvature drift also changes sign at ψ_c. The average of the guiding center drifts over the flux tube is given by the gradient of the flux tube volume $V(\psi) = \int ds/B$. Thus, the critical surface where the average electron and ion drift changes from stabilizing to destabilizing is where

$$\frac{dV}{d\psi} = \frac{d}{d\psi} \int \frac{ds}{B}\bigg|_{\psi=\psi_c} = 0. \tag{5.45}$$

Inside of ψ_c the average magnetic curvature is favorable, and the fluctuations are of the drift wave type with $k_{\|} \neq 0$. Thus, $E_{\|} = -\boldsymbol{b} \cdot \nabla\phi = \mathrm{Re}[-ik_{\|}\widetilde{\phi}_k \exp(i\boldsymbol{k} \cdot \boldsymbol{x} - i\omega t)]$ is finite for the fluctuations inside ψ_c with $\omega \lesssim \omega_*$.

For single null divertor configuration the broken symmetry creates a significant angular momentum flux that does not vanish at zero rotation speed. This flux creates a spontaneous rotation that develops further with auxiliary heating from the strong density and temperature gradients observed in the pear-shaped single null tokamaks. A kinetic theory analysis of the equilibrium of rotating toroidal plasmas with a temperature anisotropy is given by Cremashini and Tessarotto (2011). The work develops the kinetic theory description of the collisionless rotating tokamak plasmas with anisotropic temperatures. The evaluation of these collisional fluxes in single null divertor configuration is given in Casson, *et al.* (2010, 2015) and Angioni, *et al.* (2007, 2012, 2014, 2015) for JET and ASDEX-U, as described in Chapter 9.

5.6 Discovering the Trapped Particle Instability

Meade and Yoshikawa (1967) were the first to report the details of this transition from large amplitude flute mode with $\delta n/n \approx 0.5$ outside ψ_c with frequencies in the 1-20 kHz, to the ballooning drift waves with $\delta n/n \lesssim 0.15$ with frequencies $f \geq 60\,\text{kHz}$ inside the ψ_c surface. The word "flute" is used to describe the topology of the $k_\parallel = 0$ fluctuations. The topology of the flute plasma motion is analogous to the fluted columns in Greek architecture.

The trapped particle mode structure is found to be an odd parity ballooning mode with its nodes at the points of maximum favorable curvature and anti-nodes at the point of maximum bad curvature. This parallel variation means that there is one parallel wavelength $2\pi/k_\parallel$ along the closed field line. For the He plasma at the condition of $n_e = 2 \times 10^{12}\text{cm}^{-3}$ and $T_e = 5\,\text{eV}$ the mean-free-path $\lambda_{\text{mfp}} = 5\,\text{m}$ and the field line length is 0.7 m. Thus, $k_\parallel \lambda_{\text{mfp}} \simeq 5$ and the drift wave was that of the collisionless trapped electron mode. The local Boltzmann relation $\delta n_e/n \simeq e\delta\phi/T_e$ was satisfied. The transverse (axial) wavelength was $6 - 10\,\text{mm}$ so that $k_\perp \rho_i = 0.5$ to 1.0 with $\rho_i = (m_i T_i)^{1/2}/eB_{\text{min}}$. There are approximately $10\rho_i$ between the separatrix ψ_s surface and the critical surface ψ_c for the baseline operation with 45 kA in the internal conductors that are 20 cm apart. The internal conductors had a stainless steel shield that was grounded and supplied with 60 kA for 200 ms. Subsequent work of Coppi (1968a,b) refer to the Meade and Yoshikawa (1967) experiment and a closely related experiment of Ohkawa and Yoshikawa (1968) in the GA octupole. In the GA octupole the fluctuations in the MHD stable region are described as having $k_\perp \rho_i > 1$ and interpreted as high-frequency drift waves that are a FLR modified pressure gradient ballooning interchange mode [Horton, *et al.* (2002)].

In summary the multipole experiments clearly demonstrated the existence of both (i) the flute-type interchange modes with high amplitude fluctuation levels in the MHD unstable region beyond the critical surface ψ_{crit} defined by

$$\left. \frac{dV}{d\psi} \frac{dp}{d\psi} \right|_{\text{crit}} = 0 \tag{5.46}$$

and (ii) the ballooning drift wave turbulent fluctuations in the MHD stable regions. In terms of the bounce-averaged guiding-center drift velocities $\omega_{Dj}(\lambda)$ the condition Eq. (5.46) is equivalent to

$$\langle \overline{\omega_{D_j}}(\lambda) \omega_{*p_j} \rangle_{\text{MB}} \geq 0 \tag{5.47}$$

where the average is over the Maxwell-Boltzmann (MB) distribution functions.

In the inner region, inside ψ_{crit}, of flute stability boundary where $V'p' > 0$, the experiments showed a wide spectrum of high frequency oscillations that were partially localized to the regions of locally unfavorable curvature. These localized modes were called ballooning modes in the title of the Meade and Yoshikawa (1967) article. Both groups report that the confinement time was up to ten times the value expected from the tradition Bohm transport formula. To beat the Bohm value by an order magnitude was a major breakthrough at the time. The stellarator experiments had for years produced Bohm confinement times. Bohm confinement would preclude building a reasonably-sized nuclear fusion reactor.

5.6.1 *Parallel particle dynamics*

The condition that the growth time, $1/\gamma_{\max}$, be shorter that the time for the particles to travel over the inhomogeneous of the perturbed fields, $1/k_{\|}v_T$, and the mirror time of the single particles is required for the fluid model to be valid. In the case of no mirror force, if this condition is not satisfied then the time scale is comparable to the particle transit time over the parallel wavelength. In this regime there is phase mixing and Landau growth or damping from the interaction of the particles with the k, ω mode. These are effects that require the kinetic equation to describe. In the case that the parallel wavelength is very long, but the equilibrium magnetic field increases from some minimum value $B_{\min}$, then the bouncing motion of the mirroring particles averages the perturbed fields over the bounce motion. The self-consistent field equations are then integral equations over this bounce motion. When the integral field equation yields unstable motions with growth times shorter than the bounce time of the particles, the integral equations reduce to the local differential equations of the fluid models.

The first theory that extended the MHD stability formulation of the perturbed plasma potential energy $\delta W(\boldsymbol{\xi})$ to take these effects in to account is the Kruskal and Oberman (1958) stability theory. To correctly calculate the perturbed plasma energy in this regime, the theory introduces the bounce averaging operator for particles using that their energy and magnetic moment are constant. The magnetic moment, $\mu = mv_\perp^2/2B(\boldsymbol{x})$, is a constant of the motion in both the equilibrium and perturbed electromagnetic fields and that the energy E is constant $E = 1/2\, mv^2$ in the equilibrium (but not the perturbed state). Then the response of the plasma is determined by the bounce averaging of the perturbed field, for example, the perturbed electrostatic potential $\phi_k(s)$ and compressional change in the magnetic field $\delta B_{\|}$ through

Bounced Averaged Perturbations

$$\overline{\phi}_k = \frac{1}{\tau}\int_{s_1}^{s_2} \frac{ds'\ \phi_k(s')}{v_{\|}(E,\mu,s')} \tag{5.48}$$

$$\overline{\delta B_{\|}(k)} = \frac{1}{\tau}\int_{s_1}^{s_2} \frac{ds'\ \delta B_{\|}(k,s')}{v_{\|}(E,\mu,s')} \tag{5.49}$$

where $\tau = \int_{s_1}^{s_2} ds'/v_{\|}(E,\mu,s')$. The parallel velocity is given by $v_{\|} = [2/m(E - \mu B(s) - q\Phi(s))]^{1/2}$. In Eqs. (5.48) and (5.49) where $q_s\Phi$ is often small compared with μB. The turning points s_1, s_2 are then given by the roots of $E = \mu B(s_{1,2})$. Often the pitch angle-like variable $\lambda = \mu B_{\min}$ is used in toroidal geometry for the calculations of $\overline{\phi}$ and $\delta\overline{B}_{\|}$ and the other bounce averaged fields occurring in the low-frequency dynamics.

The mirror bounce frequencies $\omega_{be} \gg \omega_{bi}$ introduce new structure to the frequency spectrum as shown in Fig. 5.4. The upper shaded region is that of the electron bounce frequency and the collisional detrapping ν_{eff} destabilizes the waves.

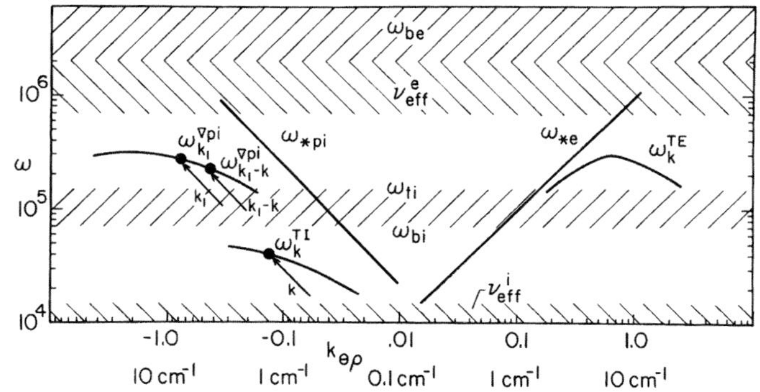

Fig. 5.4 A schematic of the frequency ranges given by shaded zones for the electron and ion bounce frequencies ω_b and the collisional detrapping frequencies $\nu_{\rm eff}$ for the electrons and the ions in a typical high-temperature tokamak. Between these shaded regions are the unstable drift waves propagating in the electron diamagnetic direction on the right and in the ion diamagnetic direction on the left. On the left is a typical three-mode interaction between wavenumbers $\mathbf{k}, \mathbf{k}_1$ and $\mathbf{k}_2 = \mathbf{k} - \mathbf{k}_1$. The three frequencies may be resonant or nonresonant in such three-mode interactions. The transit frequencies have lower frequency given by the bounce frequency ω_b and upper frequency given by the transit frequency ω_t for the small pitch-angle particles.

Clearly, this orbit-averaged response can be very different from the local response depending on the symmetry and number of spatial oscillations of the fields $\phi_k(s)$ and $\delta B_{\parallel k}(s)$ along the magnetic field line. The bounce averaged stability problem is particularly important for determining the stability of the magnetic dipole trapped plasmas in nature and the multipole magnetic plasma confinement systems.

In Eqs. (5.48) and (5.49) the s_1 and s_2 are the mirroring points of the particle where $v_\parallel = 0$. The weight $ds/v_\parallel$ is the time the orbit spends in ds at s and this integral is then divided by $\tau = \int_1^2 ds/v_\parallel$ the total time to move from the turning point s_1 to that at s_2. The particle is decelerated and stopped at s_1 and s_2 by the combination of the parallel decelerating force from $F_\parallel = -\mu\frac{dB}{ds} - qd\Phi/ds$. In the absence of an equilibrium potential $\Phi(s)$ the mirror points are only a function of the particle pitch angle α through $\sin^2\alpha = \lambda = \mu B_{\rm min}/E$.

The bounced-averaged field depends only on the dimensionless pitch angle variable $\lambda = \mu B_{\rm min}/E$. Thus, the bounce averaged fields are

$$\overline{\Phi}(\lambda) = \frac{1}{\tau}\int_1^2 \frac{ds\ \phi(s)}{\sqrt{q(1-\lambda B(s))}} \tag{5.50}$$

$$\overline{\delta B}(\lambda) = \frac{1}{\tau}\int_1^2 \frac{ds\delta B_\parallel(s)}{\sqrt{2(1-\lambda B(s))}}. \tag{5.51}$$

This averaging operator cannot be inverted to give $\phi(s)$ from $\overline{\phi}(\lambda)$ uniquely. To be able to make an inversion one needs to compute the whole sequence of pitch angle averaged fields weighted by $\sin(n\tau(s)/2\tau(\lambda))$ and $\cos(n\tau(s)/2\tau(\lambda))$ where the half-period $\tau(s) = \int_1^s ds'/v_\parallel(E,\mu,s')$ maps s to particle time τ. The presence of a weak rotation of the plasma Doppler shifts the frequency from the plasma rest frame ω to the laboratory frame ω_{lab}. For a rotation with significant shear involved, the situation becomes more complicated.

Multi-scale simulations show that the small-scale electron temperature gradient (ETG) driven turbulence modifies the large-scale ion and thus slower time scale turbulence driven by the ion temperature gradient and trapped electron response to the drift wave turbulence (ITG and TEM). Howard, *et al.* (2015) show the effect in detail using the gyrokinetic simulations of L-mode plasma data from Alcator C-Mod where both the boron ions and the deuterium ions are included in the simulations. The simulations are thus state-of-the-art taking 37 days (approximately 15 M CPU hours) using MPI with 17,000 processors on the NERSC Edison supercomputer. The separation of ion and electron scales is given by $(m_D/m_e)^{(1/2)} = 60$ and simulations in k-space to $k\rho_s = 48$ in boxes that are $44\rho_s \times 44\rho_s$.

These huge simulations make possible the study of the interaction of the small-scale turbulence on the large-scale turbulence. The work shows that the small scale and fast ETG turbulence strongly reduce the effect of zonal flows in the ion gyroradius large-scale turbulence. There has been significant disagreement in the ion thermal flux with that measured in the high power plasmas when only the large-scale turbulence is used in the comparisons. The small-scale turbulence reduces the build-up of the zonal flows in the large-scale turbulence, giving better agreement with the measured thermal fluxes. An example is given with C-Mod data where the ETG streamers reduce the zonal flows and enhance the low k_1 driven transport by about 100%. The work shows that the result is to bring both the thermal flux of electrons and ions up to the measured levels to within the experimental uncertainties. The work expresses these uncertainty estimates as $Q_{i,\text{exp}} = 0.07 \pm 0.021$ and $Q_{e,\text{exp}} = 0.23 \pm 0.069$ within about 30% in both the ion heat and electron thermal fluxes. The simulation with the smallest ion temperature gradient has a very large enhancement when the small-scale ETG turbulence is included in the simulation, with the increase being a factor of three. The conclusion is that keeping the small-scale ETG turbulence is essential to have accurate thermal fluxes in both the ions and electron channels as argued by Horton, *et al.* (1988) in early works on drift wave turbulence and from experience with fluid turbulence inverse cascade problems.

Drift wave turbulence produces a particle flux that has both a diffusive part D_s and a convective part V_s with the particle flux of species as given by

$$\Gamma_s = -D_s \nabla n_s + V_s n_s \tag{5.52}$$

that is perpendicular to the magnetic surface $\psi_p(x,y,t)$. We call this $\nabla\psi_p$ the radial direction, as would be the case for circular magnetic flux surfaces.

The turbulent particle flux has several contributions in a toroidal geometry owing to the magnetic shear and the change of the direction of the magnetic curvature vector with respect to the local flux surface normal vector over the surface. In a single null divertor configuration the particle fluxes in the top and bottom regions are rather different as well as on the inside and outboard side of the flux surface.

The flux Γ_s can be characterized as having two main components: one is due to the pinch effect described earlier that attempts to minimize the entropy production and bring the density distribution to the Einstein state with minimal entropy production. This contribution is called the TEP for "turbulent equipartition" [Bourdelle (2005)]. The second contribution to the particle flux is from the temperature gradient and is often called the "thermo-diffusion particle flux". Bourdelle writes the particle convection velocity with coefficients C_q and C_T to distinguish these two contributions. This division is not unique and does not include other possible sources such as the DC electric fields – either radial or toroidal – to the turbulent particle flux.

Typically, the particle sources are from neutral gas ionization in the outer or edge plasma. In this case the core plasma may relax to a state of vanishing or nearly vanishing particle flux Γ_s in a steady-state plasma. With hot fusion plasma there is often a significant release of wall material giving an influx of particular impurity ions (carbon, beryllium, lithium or tungsten). This complex problem is described in detail in Chapter 8.

In the literature with the coefficients C_q and C_T defined for the convective velocity V_s in the particle flux is written as

$$\frac{V_s}{D_s} = -C_q \frac{\nabla q}{q} + C_T \frac{\nabla T_e}{T_e}. \tag{5.53}$$

The trapped electron turbulence (TEM) drives an inward particle flux. The details are given in Bourdelle (2005) and the discovery of the effect is given in Coppi, *et al.* (1968a,b). The analysis for the formulas for the velocity coefficients C_q and C_T becomes involved as described in the review article of Bourdelle (2005). More recent descriptions for single null divertor configurations with plasma rotation and the radial electric fields taken into account is given in Casson, *et al.* (2015) and Angioni, *et al.* (2015). The calculations are sufficiently complicate that we leave the reader to find the relevant regime dependent formulas from the references. Additional references for the derivations and simulations are given in the publications of Casson, *et al.* (2015), Angioni, *et al.* (2015) and Bourdelle (2005).

Chapter 6

Turbulence in Helical Magnetic Fields

In this chapter we explore how drift waves and Alfvén waves develop in sheared helical magnetic field configurations. We start with a relatively simple toroidal magnetic field configuration that does not require the metric tensors $g_{\alpha\beta}$ used in general toroidal confinement geometries. The simplest toroidal configuration plasma system with magnetic shear $s = rq'/q$ and toroidal curvature $\kappa(r)$ is the Helimak geometry shown in Fig. 6.1. Extensive plasma experiments with inert gases helium He, neon Ne and argon Ar were performed in the Helimak at the Institute for Fusion Studies at The University of Texas at Austin. By using the heavy Ar^{+1} plasmas with atomic mass A $= 40$ the Alfvén velocity v_A is lowered to become comparable to the electron thermal velocity giving plasma parameters similar to those in toroidal fusion plasmas in deuterium plasmas with $v_A \sim v_e$ equivalent to $\beta_e = 2m_e/m_i \simeq 10^{-4}$. The connection length L_c for the parallel wavenumber $k_\parallel^{-1} \simeq qR$ is variable from 50-500 m. The Alfvén waves $\omega^2 = k_\parallel^2 v_A^2$ are a natural mode of the system and the magnetic fluctuations are an important aspect of the plasma dynamics in contrast to the earlier Q-machines described in Chapter 3.

The linear growth rate, $\gamma_{\vec{k}} = \operatorname{Im}\omega$, is produced by the phase shift, $\psi_{\delta n,\phi}$, of the density response from the electron collisions, ν_e, resulting in the density response,

$$\frac{\delta n_e}{n_e} = \frac{\omega_{*e} + ik_\parallel^2 v_e^2/\nu_e}{\omega + ik_\parallel^2 v_e^2/\nu_e}\frac{e\phi_{\vec{k}}}{T_e} \tag{6.1}$$

from the electric potential wave $\phi_{\vec{k}}$. The growth rate, $\gamma_{\vec{k}}$, is controlled in the experiments by varying the ion viscosity μ, where $\mu/\rho_m = (3/10)\nu_i\rho_i^2$, ν_i is the ion collision frequency, ρ_i the ion gyroradius and ρ_m the mass density. Since $\mu/\rho_m \propto 1/B^2$, the experimentalists varied the **B**-field through the critical value, $\mathbf{B}_{\rm crit}$, to measure directly the growth rate as well as the wave frequency ω_k in the laboratory. The nonlinear saturation occurs when the nonlinearity in the ion dynamics balances the growth rate from the electron dynamics in Eq. (6.1). The growth is from ω_{*e}^2, which is the driving term of the instability. The $\boldsymbol{v}\cdot\nabla f$ convective nonlinear saturation is generic to drift waves since the rapid parallel electron motion leaves the electron-wave phase as in the linear response theory.

Formula Eq. (6.1) shows the key feature for density transport. When $\omega = \omega_{*e}$ the complex fraction is unity and the density is the adiabatic Boltzmann response. For $0 < \omega < \omega_{*e}$ the density wave leads the potential wave and there is net $\mathbf{E} \times \mathbf{B}$ transport down the density gradient. The high-frequency regime $\omega > \omega_{*e}$ is characterized by evanescent wave structures that become the vortex structures described in Chapter 9, when the amplitude of the potential fluctuation exceeds a critical level.

6.1 Helimak Geometry and Plasma Dynamics

The Helimak system produces laboratory simulations of space and astrophysical plasma with combined mass flow and sheared magnetic flux tubes as in the magnetosphere and solar corona. The Helimak geometry is shown in Fig. 6.1. The electron temperature is similar to the corona at $T_e \sim 10\,\text{eV}$ and the electron density is $n_e \sim 10^{11}\text{cm}^{-3}$ a factor of 100 higher than the lower solar coronal. The magnetic field is strong so the plasma has a low plasma pressure-to-magnetic pressure ratio. By combined flow and magnetic shears, the Helimak creates a situation in which a sheared flow field and a sheared magnetic field are both present. This type of system is perhaps best illustrated by providing examples of the two canonical cases: the magnetized jet [Dahlburg, *et al.* (1998)] and the plane current vortex sheet [Dahlburg, *et al.* (1997)]. Both of these systems have sheared magnetic field structures. In the case of the magnetized jet, there is a fluid jet superimposed on the magnetic field structure. The magnetized jet is used to model a variety of physical systems, including solar surges [Carbone, *et al.* (1987), Dahlburg and Karpen (1994)] and the fast flows in the Earth's magnetotail [Sato and Walker (1982)]. The magnetized jet can be turned into a magnetized wake by application of a Galilean transform. This configuration is used to model the formation and acceleration of the slow solar wind [Einaudi and Rubini (1986, 1989, 1999)], the heliospheric current sheet [Wang, *et al.* (1988)], and nonthermal galactic jets [Dahlburg, *et al.* (2001, 2002)]. This plasma system with sheared magnetic fields and fast flows is used to model solar surges [Carbone, *et al.* (1987)] and tokamak plasmas with equilibrium flows [Paris and Sy (1988)].

Laboratory plasmas experiments with combined flows and magnetic shear are the Helimak [Perez, *et al.* (2006)] and the LAPD [Horton, *et al.* (2005)]. In general, these two laboratory plasmas share the characteristic that they are bounded, in contrast to astrophysical plasmas in which boundaries are often so far away from the region of interest that they may be ignored. Hence it is of interest to determine how the presence of walls alters the stability of magnetofluids with combined flow and magnetic shears. Are walls a stabilizing influence, or do they serve as regions for destabilization and turbulence production? For example, many of the unstable modes found in earlier studies are ideal, dissipationless instabilities. These modes are generally of large spatial extent in the cross-stream direction. Walls disrupt the structure of these modes and alters their stability properties. Conversely, it is

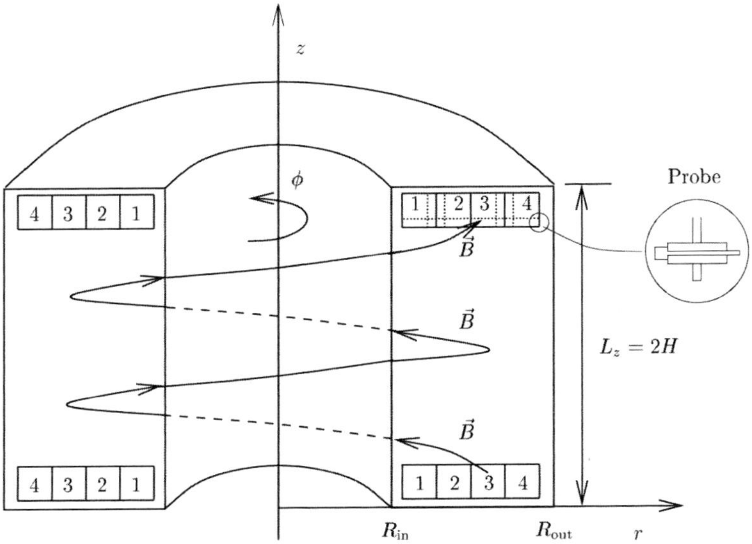

Fig. 6.1 A schematic drawing of the Helimak illustrating the laboratory coordinate system. The mathematical analysis uses a rotated coordinate system that brings $\widehat{e}_z$ parallel to the magnetic field at $r = (R_{\rm in} + R_{\rm out})/2$ and $\widehat{e}_y$ as the binormal direction perpendicular to $\widehat{e}_z$ and $\widehat{e}_r$. The probes labeled 1, 2, 3, 4 are biased to control the radial electric field and thus the plasma flow velocity $W_0(x)$ along $\widehat{e}_z$. The trajectory of the field lines is $z = 2\pi\phi R B_z / B_\phi(R)$.

found in both fluid and fusion confinement systems that walls can provide a source region for instabilities, since large stresses can develop in response to the velocity or magnetic field striving to meet the imposed wall boundary condition.

For example, in magnetized plane Poiseuille flow, large Reynolds and Maxwell stresses develop at the channel walls for the case of a DC magnetic field [Nachtsheim and Reshotko (1965)]. Kent (1966) showed that sheared magnetic fields also strongly influenced the stability of MHD plane Poiseuille flow. Early studies of magnetized channel flows indicate that there can be a significant interaction between sheared flows and magnetic fields in magnetofluids bounded in channels [Stern (1963), Kent (1966)].

The Helimak experiment at The University of Texas at Austin is a toroidal magnetic confinement device designed to realize in the laboratory the sheared magnetic field slab geometry used by theorists in the analysis of localized modes in tokamaks [Perez, *et al.* (2006)]. For thin resistive boundary layers it is not necessary for theory to keep the boundary layer radial profiles to study the plasma instabilities and turbulence. By virtue of the helical magnetic field structures, modes with long parallel wavelengths $\lambda_\parallel > 50 - 500\,\mathrm{m}$ are formed in the experiment. The Alfvén wave dynamics $\omega^2 = k_\parallel^2 v_A^2(r)$ is thus a natural mode of the system and the fluctuations of the magnetic field are an important aspect of the system's dynamics as opposed to experiments with short parallel wavelengths.

The plasma flow velocity is externally induced by the radial electric fields applied by segmented endplate electrodes numbered 1,2,3,4 in Fig. 6.1. Included are the effects of resistivity and viscosity, as well as no-slip walls [Dahlburg, *et al.* (1998)]. Linear theory provides a guide to the parameter space of the Helimak plasma. In particular the stress analysis gives an understanding of the energetics of the plasma flow and the fluctuating magnetic fields.

6.1.1 *Turbulence data in Argon plasmas*

Figure 6.2 shows three examples of the vertical cross-field flow $v_z = w(x)$ configurations created in the Helimak plasma with the data points and error bars from

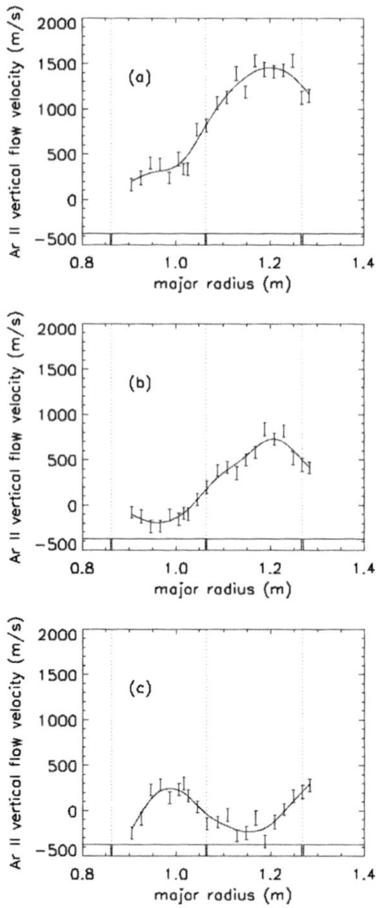

Fig. 6.2 Vertical flow profiles of a Helimak plasma at bias voltage values (a) $V_{\text{bias}} = 0$, (b) $V_{\text{bias}} = -10\,\text{V}$, and (c) $V_{\text{bias}} = -15\,\text{V}$. The error bars correspond to the statistical uncertainty in spectroscopic Doppler shift measurements of spectral atomic emission lines from a working gas of singly ionized argon gas $\text{ArII} = \text{Ar}^{+1}$ ions with atomic mass 40.

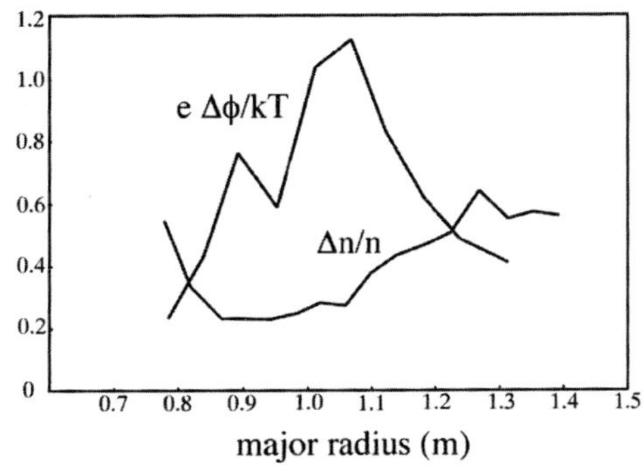

Fig. 6.3 Fluctuation amplitudes of the electric potential and the electron density measured with Langmuir probes in a 10 eV argon Helimak plasma.

spectroscopic Doppler shift measurements of the spectral lines from the working gas of singly ionized Argon. The flow profiles are controlled by the bias voltage $V_{\rm bias}$ given in the figure which changes the cross-field electric field E_r. The plasma responds in a complex manner to give the measured velocity profiles shown in frames a, b and c. The profiles have shear and the associated vorticity dV_z/dr of order 5×10^3 to 10^4/s changes sign at the inflection points in the profile. By virtue of the long lengths L of the magnetic field lines given by $L = B_\phi H/B_z > 50\,\mathrm{m}$ the Alfvén wave frequency is low $k_\| v_A/2\pi \sim 10\,\mathrm{kHz}$.

Figure 6.3 shows the fluctuation amplitudes of the electric potential and the electron density measured with Langmuir probes in the 10 ev Argon plasma. The probes show that $e\phi/T_e \gg \delta n_e/n_e$ over much of the flow profile which means that we may drop the drift wave terms $|\omega_{*e}/\omega|$ to first order. The Ohm's law then simplifies to $E_\| = \eta j_\|$ in the modeling of the experiment. The key fluctuations are the electric potential and the perpendicular component of the magnetic field in these flow dominated plasmas. Owing to the large mass of the Argon ions ($m_{\rm Ar}/m_p = 40$) these low-temperature plasmas have $m_i\beta_e/m_e \sim 1-2$ where β_e is the ratio of the electron thermal energy density to the magnetic energy density.

A range of potential profiles is shown in Gentle and He (2008). The convective derivative $\boldsymbol{v}\cdot\nabla n$ is the dominant saturation mechanism.

6.1.2 *Varicose and sinuous M-modes in the Helimak*

Fluid jets can be transformed into a wake field problem by a Galilean transformation. This transformation is achieved by moving to a frame of reference moving at the maximum jet speed in the original frame of reference. Unbounded magnetized jets and wakes are studied in a variety of problems. In general, researchers have investigated systems which are either magnetically dominated or flow dominated. This is significant because the Alfvén number ($A = v_A/v_0$ where v_A is the characteristic Alfvén velocity and v_0 is the characteristic flow velocity) is important in determining the evolution of the plasma. Note that the inverse of the Alfvén number A, is the Alfvén Mach number, or $M_A = v_0/v_A$. For example, for the low-A, or flow dominated plasmas, researchers have looked at the heliospheric current sheet, the Earth's magnetotail, and at nonthermal galactic filaments. For the high-A, or magnetically dominated case, researchers have looked at solar surges and NBI-driven tokamaks. An important problem that requires investigation of both the high-A and low-A regimes is that of slow solar wind formation and acceleration. In this case the plasma originates in a magnetically dominated region below the solar Alfvén point, and then moves to a flow dominated region above the Alfvén point.

In the linear analysis, three unstable modes are found [Dahlburg, *et al.* (1998)]. For flow-dominated systems, with $0 \leq A \leq 1$, the two unstable modes are the ideal varicose mode and the sinuous mode. A varicose mode is one in which the cross-stream flow is antisymmetric, and a sinuous mode is one in which the cross-stream flow is symmetric. As A is increased these modes are stabilized by the stiffness of the magnetic field. A generalization of the Howard semicircle theorem to the case of the magnetized jet shows that a strong enough magnetic field suppresses the ideal (vanishing resistivity) instabilities. As A increases above one, the system becomes magnetically dominated. In this case a resistive varicose instability occurs. In the fusion literature the varicose mode is called the tearing mode parity. In more graphic terms the varicose mode is a squeezing variation while the sinuous mode is a snake-like motion.

Analysis of the fluctuation energy-stress balance [Dahlburg, *et al.* (1997, 1998)] shows that there are three stress components: (i) the Reynolds stress, (ii) the Maxwell stress, and (iii) the cross-field stress. The values of the stresses and direction of energy flow are sensitive to the Alfvén number. The Reynolds stress moderates the transfer of energy between the basic flow field and the perturbed flow field. The Maxwell stress moderates the transfer of energy between the basic flow field and the perturbed magnetic field. The cross-stress is the dynamo term that moderates the transfer of energy between the perturbed flow and magnetic fields fluctuations.

The resistive varicose mode is characterized by a large positive cross-stress contribution to the perturbed energy in the high-$A = v_A/v$ regime. For low-A the cross-stress contribution is negative. An increase in a positive Maxwell stress contribution is not enough to compensate for increased Ohmic and viscous dissipation for low-A. The ideal varicose mode has a large positive Maxwell stress contribution for all values of A. Again, there is increased dissipation for high-A, resulting in damping.

For the sinuous mode, the Reynolds stress is very large for low-A. There is also a more modest contribution to the perturbed energy due to the Maxwell stress. Both processes are strongly damped as A becomes large leading to increased stability. Extensive analysis for ideal and resistive MHD waves and instabilities is given for fusion, magnetospheric and astrophysical plasmas in *Advanced Magnetohydrodynamics* [Goedbloed, *et al.* (2010)].

Research on the nonlinear incompressible problem focused on the magnetized wake, due to its use as a model for several solar physics, space physics, and astrophysical situations. In two spatial dimensions, it is found that moving plasmoids form and accelerate [Einaudi, *et al.* (1999)]. Of interest is finding how rapidly the wake (or jet) spreads in the cross stream direction. When the span-wise direction is included, secondary instabilities occur, followed by a transition to turbulence [Dahlburg and Karpen (1994)].

Compressible Dynamics

Compressibility introduces a significant increase in the complexity for this problem. The linearized equations are far more extensive and difficult to solve. The three unstable modes still occur. Increasing the Mach number leads to decreased growth rates. In contrast with the current-vortex sheet, no second modes are found at high Mach numbers [Dahlburg and Einaudi (2000)]. An interesting feature is that, for certain situations, the three-dimensional ideal varicose modes grow at a faster rate than their two-dimensional counterparts [Einaudi, *et al.* (2001)]. In the nonlinear compressible problem, massive accelerated plasmoids are observed [Einaudi, *et al.* (2001)]. Rappazzo, *et al.* (2005) analyze the diamagnetic and expansion effects on the formation of plasmoids.

6.1.3 *Unstable magnetized flows in the Helimak*

The Helimak Experiment

The Helimak geometry shown in Fig. 6.1, is similar to the magnetohydrodynamic (MHD) Poiseuille flow geometry [Stuart (1954), Velikhov (1959)] with concentric conducting walls at $R_{\rm in} = a = 0.6\,\rm m$ and $R_{\rm out} = b = 1.6\,\rm m$ and a height of $H = 2\,\rm m$. The dominant toroidal field, $B_\phi \simeq 0.1\,\rm T = 1\,kG$ is produced by a set of 16 toroidal field coils around the vacuum chamber. Three poloidal field coils are used to produce a weaker vertical field, B_z, which may be varied up to 10% of the toroidal field by changing the ratio of the current flowing in the toroidal coils to the current flowing in the poloidal field coils. The magnetic field lines are thus helices spiraling from bottom to top and whose pitch varies with the radius as the toroidal field decreases as $1/r$, where r is measured from the vertical symmetry axis of the device. The field line length $L_\parallel$ may be varied from less than 10 m to more than 1 km by reducing the B_z field.

The plasma jet is a localized flow $v_z = w(x)$, controlled with the biased endplates at $z = \pm H/2$. Plasma flows across the magnetic field from the steady-state radial electric field E_r. Probe measurements of the radial electric field show localized E_r structures, which means that there is a localized high-speed vertical plasma stream $v_z = v_0(r)$ in the vertical direction $\widehat{e}_z$.

The plasma has toroidal $B_\phi = 0.1\,\rm T$ from the sixteen poloidal coils and a variable vertical or poloidal field from $B_z = 0$ to $0.05\,T = 500\,\rm G$ from the three

Table 6.1 Helimak viscoresistive MHD parameters in Argon plasma.

Toroidal Magnetic Field	$B_\phi = 0.1\,\mathrm{T}$
Poloidal Magnetic Field	$B_z = 0.01\,\mathrm{T}$
Half Height	$H = 1.0\,\mathrm{m}$
Half Width	$L = 0.5\,\mathrm{m}$
Aspect ratio	$H/L = 2$
Magnetic Field Line Length	$L_\parallel = 20\,\mathrm{m}$
Argon Density	$n_e = 10^{11}\,\mathrm{cm}^{-3}$
Electron Temperature	$T_e = 10\,\mathrm{eV}$
Chamber Volume	$V = 13.8\,\mathrm{m}^3$
Space-Time Scale	$(L,T) = (L, L/V_A) = (0.5\,\mathrm{m},\ 0.6\mu\,\mathrm{s})$
Alfvén velocity	$v_A = 1 - 3 \times 10^8\,\mathrm{cm/s}$
Sound speed	$c_s = 4 \times 10^5\,\mathrm{cm/s}$
Dimensionless pressure	$\frac{m_i}{m_e}\,\beta_e = 3$
Ion sound gyroradius	$\rho_s = 2\,\mathrm{cm}$
Magnetic Diffusivity	$\eta/\mu_0 = 30\,\mathrm{m}^2/\mathrm{s}$
Magnetic Reynolds Number	$S_M = V_A L/(\eta/\mu) = 10^4$
Ion kinematic viscosity	$\nu = 0.3\rho_i^2\nu_i = 0.3\,\mathrm{m}^2/\mathrm{s}$
Viscous Reynolds Number	$S_v = 10^4$
Bohm Diffusivity	$\chi_e = 1/16(T_e/B) = 6\,\mathrm{m}^2/\mathrm{s}$

toroidal coils. The ionized Argon plasma is produced with ECRH at 2.45 GHz heating with temperatures of up to $T_e = 10\,\mathrm{eV}$ at the peak of the temperature profile. The plasma density is approximately $2 \times 10^{17}/\mathrm{m}^3 = 2 \times 10^{11}/\mathrm{cm}^3$. From these parameters the Alfvén velocity is $v_A(r) = 1 - 3 \times 10^8\,\mathrm{cm/s}$, the resistivity is $\eta = 10^{-5}\Omega\,\mathrm{m}$ and the kinetic viscosity $\nu = 10^4\mathrm{cm}^2/s$. Details on the plasma parameters are given in Table 6.1.

For this plasma we define the magnetic Reynolds $R_M = S$

Magnetic Reynolds Number

$$S = S_M = \frac{V_A a}{(\eta/\mu_0)} \tag{6.2}$$

and the viscous flow Reynolds number $R_\nu = S_\nu$

Prandtl Number

$$S_v = \frac{V_A a}{\nu}. \tag{6.3}$$

For the plasma in Table 6.1 the magnetic Reynolds number as $S = 5000$ and the viscous Reynolds number or Prandtl number as $S_v = 5000$ for the baseline case. The velocity $\boldsymbol{v}(x,t)$ of the singly-ionized argon atoms Ar^{+1} is measured spectroscopically.

Singly ionized argon has emission lines in the blue part of the visible spectrum. The Doppler shift of these lines is measured to give accurate, well-resolved measurements of the ion velocity $v_z = w(r)$. The ion velocity is of order 1 km/s and thus larger than the ion thermal velocity, comparable to the ion acoustic speed, and a significant fraction of the Alfvén velocity. Due to the large mass of the argon ions, $m_i\beta_e/m_e \sim 1$, there are significant electromagnetic fluctuations that are measured with magnetic probes. The reference time is L/v_A of order one microsecond.

6.1.4 *Stability analysis for the Helimak*

Two coordinate systems are needed for the analysis. First the machine geometry coordinates given in Fig. 6.11. Secondly, the tilted field line following coordinate. Now we explore the dynamics of a magnetofluid contained in a channel which corresponds to a slab model of the Helimak geometry. Since most of the calculations are performed in a rotated frame of reference aligned with magnetic field, in the following we use primes to denote the laboratory frame of reference. A geometry of the Helimak and the slab coordinate systems is given in Fig. 6.1. In the laboratory frame of reference, we let x correspond to the radial direction $\widehat{\boldsymbol{e}}_r$; y' correspond to the toroidal direction $\widehat{\boldsymbol{e}}_\phi$; and z' correspond to the axial direction $\widehat{\boldsymbol{e}}_z$. We reserve y and z without primes for the field line following coordinates.

Periodic boundary conditions are imposed in the y' and z' directions and no-slip boundary conditions in the radial direction at the walls of the chamber.

In the simulations we parameterize the radius r with dimensionless x such that

$$r = [a + b + (b - a)x]/2 \tag{6.4}$$

so that $x = -1$ is $r = a$ and $x = +1$ is $r = b$. Thus, $dx = 2dr/(b - a) = dr/L$ with L given in Table 6.1.

The natural time units and velocity units $\boldsymbol{v} = \boldsymbol{V}\, v_A$ and $t = L/v_A$ where L is the half width of the plasma annulus. The field lines spiral upward (downward) so that modes with a helical pitch close to that of the magnetic field line have very long ($> 100\,\mathrm{m}$) parallel wavelength. Thus $\omega_A = k_\parallel v_A \lesssim 100\,\mathrm{kHz}$ is easily achieved. Magnetic probes show that there is a significant δB_x produced in these plasmas with $\omega_{*e} \sim |k_\parallel| v_A \leq 100\,\mathrm{kHz}$ [Mays, *et al.* (2007)].

The boundary conditions are that the velocity vanish at the metal walls

$$u'(x = \pm 1) = v'(x = \pm 1) = w'(x = \pm 1) = 0 \tag{6.5}$$

where u' is the perturbed flow in the x-direction, v' is the perturbed flow in the y'-direction, and w' is the perturbed flow in the z'-direction in the laboratory frame.

In the Helimak, most of the magnetic field is due to the currents flowing through the sixteen external poloidal coils, which gives $B\phi(r)$ that decreases with radius as

$$B_\phi = \frac{B_0 R_0}{r} = \frac{2aB_0}{b + a + (b - a)x} \tag{6.6}$$

where the x coordinate corresponds to the scaled radial variable with $x = 0$ located at the center of the channel and $x = \pm 1$ to the walls. The magnetic field scaled to the inner wall ($x = -1$) value gives

$$\mathbf{B}'_{0y'}(x) = \frac{1.2}{2.2 + x}\widehat{\boldsymbol{e}}_{y'}, \qquad \mathbf{B}'_{0z'} = 0.1\widehat{\boldsymbol{e}}_{z'}. \tag{6.7}$$

Probe measurements of the radial electric field show a localized radial E_r well which means that there is a localized high speed vertical plasma stream in the vertical direction $W' = E_r/B_\phi = W'_0(x)$. The peaked vertical flows $W(x)$ in Fig. 6.2 are modeled with the hyperbolic function $\mathrm{sech}(r - r_0) \rightarrow \mathrm{sech}(3x)$. The baseline model has the toroidal flow $\boldsymbol{V}'_0 = 0$ and $\boldsymbol{V}'_0$ and the vertical flow

$$\boldsymbol{W}'_0(x) = 0.1\frac{\mathrm{sech}(3x) - \mathrm{sech}(3)}{\mathrm{sech}(0) - \mathrm{sech}(3)}\widehat{\boldsymbol{e}}_{z'} \tag{6.8}$$

where $\boldsymbol{V}'_0$ is the basic flow in the y'-direction and $\boldsymbol{W}'_0$ is the basic flow in the vertical z'-direction created by the applied radial electric field. Here in the lab coordinates y' is toroidal direction and z' is vertical in Figs. 6.1 and 6.2.

In the fusion literature the wave numbers α and β used here are typically written as $k_y = \alpha$ and $k_2 = \beta$ or ℓ/R.

In most toroidal plasma theories and codes a coordinate transformation is introduced to follow magnetic field line at each radius $r = r(x)$. Here we introduce this coordinate system by a rotation of the laboratory y', z' coordinates through the angle $\theta(x)$ given by $\tan\theta = -B_\phi(r)/B_z$ that is a tipping of the z' axis by about $80°$ to bring the new z coordinate parallel to the local $\mathbf{B}(x)$. Locally the plasma has translational invariance in the azimuthal direction $\widehat{e}_\phi$ and vertical direction $\widehat{e}_z$. Thus, we are able to define wavenumbers α and β corresponding to the bi-normal and parallel direction $(\mathbf{e}_\perp, \mathbf{b})$ as used in the fusion literature in the hydrodynamic literature are called the stream-wise (α) and the span-wise (β).

In the reduced resistive MHD equations have the two key functions have

$$F(\boldsymbol{k}, x) = \vec{k}\cdot\vec{V}_0(x) = \alpha V_0(x) + \beta W_0(x) \tag{6.9}$$
$$G(\boldsymbol{k}, x) = \vec{k}\cdot\vec{B}(x) = \alpha B_y(x) + \beta B_z(x). \tag{6.10}$$

in the local tilting coordinates.

There are rigid radial boundaries in the channel problem and in the Helimak. Boundaries can have both a stabilizing and a destabilizing influence. Another significant difference between the unbounded magnetized jet and the present bounded plasma problem is the lack of symmetry in the magnetic field. In particular, while the stream-wise flow field is symmetric in x, the magnetic field is about three times larger at $x = -1$ than at $x = 1$. This broken symmetry influences the details of the numerical method, in that one must retain all of the Chebyshev polynomials to account for the asymmetry. The numerical method is described in Section 6.1.5.

6.1.5 *Linearized resistive-viscous magnetohydrodynamic equations*

The vector fields for the incompressible plasma dynamics are

$$\boldsymbol{v}(\boldsymbol{x}, t) \equiv \text{flow velocity} \tag{6.11}$$

$$\boldsymbol{\zeta}(\boldsymbol{x},t) = \nabla \times \boldsymbol{v} \equiv \text{vorticity} \tag{6.12}$$

$$\mathbf{B}(\boldsymbol{x},t) \equiv \text{magnetic field} \tag{6.13}$$

$$\boldsymbol{j}(\boldsymbol{x},t) = \nabla \times \mathbf{B} \equiv \text{electric current density} \tag{6.14}$$

$$\Pi(\boldsymbol{x},t) \equiv \text{mechanical kinetic pressure} \tag{6.15}$$

$$+ \text{ thermal pressure per unit mass} \tag{6.16}$$

$$S \equiv \text{Lundquist number } = \text{av}_{\text{A}}/\eta \tag{6.17}$$

$$S_v \equiv \text{viscous Lundquist number } = \text{av}_{\text{A}}/\mu \tag{6.18}$$

The resistivity and viscosity are assumed to be constant and uniform. The velocities are measured in units of the Alfvén speed (v_A), and time is measured in units of the Alfvén transit time, L/v_A, where the characteristic distance, L, is defined by the channel half-width. In Table 6.1 both S and $S_v \simeq 10^4$.

The nonlinear magnetohydrodynamic (MHD) equations are

$$\frac{\partial \boldsymbol{v}}{\partial t} = \boldsymbol{v} \times \boldsymbol{\zeta} - \nabla \cdot \Pi + \boldsymbol{j} \times B + \frac{1}{S_v}\nabla^2 \boldsymbol{v} \tag{6.19}$$

and

$$\frac{\partial \mathbf{B}}{\partial t} = \nabla \times (\boldsymbol{v} \times \mathbf{B}) + \frac{1}{S}\nabla^2 \mathbf{B} \tag{6.20}$$

together with the constraints: $\nabla \cdot \boldsymbol{v} = 0$ and $\nabla \cdot \mathbf{B} = 0$.

After linearizing Eqs. (6.19) and (6.20), we decompose the first-order terms in the following manner:

$$a_1(x,y,z,t) = a(x)\; e^{i\alpha y + i\beta z - i\omega t} + \text{ c.c.} \tag{6.21}$$

where α is the stream-wise wavenumber, β is the span-wise wavenumber, and ω is the complex frequency and growth rate. Here ∇y is perpendicular to the radial gradient and the mid-radius magnetic field $\mathbf{B}_0$. This direction is called bi-normal or poloidal direction in the fusion literature. In the locally twisted field-line coordinates $\alpha \to k_y, \beta \to k_z$. Flute modes have $k_z = 0$.

After eliminating the pressure, the differential equations for the linear modes are found. To write the equations compactly we define the operators $D \equiv d/dx$ and $\widehat{\xi} = D^2 - (\alpha^2 + \beta^2)$, then the cross-stream radial velocity (u) and radial magnetic induction (b_x) equations reduce to:

$$\widehat{\xi}^2 u - iS_v F\widehat{\xi} u + iS_v(D^2F)u = -i\omega S_v \widehat{\xi} u + iS_v\left[(D^2G)b_x - G\widehat{\xi} b_x\right] \tag{6.22}$$

and

$$\left(\widehat{\xi} - iSF + i\omega S\right) b_x = -iSGu \tag{6.23}$$

with the boundary conditions: $u(x = \pm 1) = Du(x = \pm 1) = b_x(x = \pm 1) = 0$. In ideal MHD S and S_v are obtained by dividing Eq. (6.22) by S_v and Eq. (6.23) by S and taking the limit S_v and $S \to \infty$. Here u and b_x are the x (radial) components of the perturbed velocity and magnetic field, respectively. The steady plasma profiles enter the mode equations through

$$F = \boldsymbol{k} \cdot \boldsymbol{V}_0 = \alpha V_0 + \beta W_0 \tag{6.24}$$

and the local parallel wavenumber $G(x, \alpha, \beta) = k_{\|} B$ with

$$G = \boldsymbol{k} \cdot \mathbf{B} = \alpha B_{0y} + \beta B_{0z}. \tag{6.25}$$

α is the local k_y with respect to L in the local slab model for the helical magnetic field $\alpha = k_y$ and $\beta = k_{\|} = k_z$ at the mid-radius of the Helimak.

The fundamental equations (6.22) and (6.23) form a sixth-order system of coupled first-order ordinary differential equations with six boundary conditions.

Alfvén wave resonances occur from Eqs. (6.22) and (6.23) with $\omega^2 = G^2(x) = k_{\|}^2 v_A^2$ in the local limit with flow $F = 0$ and no dissipation $1/S = 1/S_v \to 0$. To get the 3D eigenfunction, one must solve perturbed vorticity ($\zeta = \beta v - \alpha w$) and electric current ($\chi = \beta b_y - \alpha b_z$) equations:

$$\left(\widehat{\xi} - i\alpha S_{\mathrm{v}} F\right) \zeta + i\alpha S_{\mathrm{v}} G \chi = -i\omega S_{\mathrm{v}} \zeta - \beta S_{\mathrm{v}} \left[(DQ)v - (DP)b_y\right] \tag{6.26}$$

and

$$\left(\widehat{\xi} - i\alpha S F\right) \chi + i\alpha S G \zeta = -i\omega S \chi - \beta S \left[(DQ)b_y - (DP)v\right]. \tag{6.27}$$

with the boundary conditions $\beta v(x = \pm 1) - \alpha w(x = \pm 1) = \beta b_y(x = \pm 1) - \alpha b_z(x = \pm 1) = 0$. On the RHS of Eqs. (6.26) and (6.27) $Df = \alpha f / \alpha x$ and on the LHS $\hat{\xi} = D^2 - (\alpha^2 + \beta^2)$. Here v and w denote the stream-wise and span-wise perturbed velocities, b_y and b_z denote the stream-wise and span-wise components of the perturbed magnetic field, and

$$P = \beta B_{0y} - \alpha B_{0z} \tag{6.28}$$

and

$$Q = \beta V_0 - \alpha W_0. \tag{6.29}$$

In the ideal inviscid limit the eigenmode equations reduced to the pair

$$\left[\omega \left(\partial_x^2 - k_\perp^2\right) + F'' + F\partial_x^2\right] u + \left[G\left(\partial_x^2 - k_\perp^2\right) - G''\right] b_x = 0,$$

$$G(x)u + (\omega - F(x))b_x = 0$$

giving the standard MHD Alfvén wave eigenmode equation (e.g. [Chen and Morrison (1990), Chen and Morrison (1991)]) $\{\partial_x[(\omega - F)^2 - G^2]\partial_x - k_\perp^2[(\omega - F)^2 - G^2]\}\xi = 0$ which is now expressed in the MHD Lagrangian displacement field $\xi(\omega - F) = u$.

Worked Problem 6.1: Decomposition in Chebyshev-τ Method

Express the linear equations with boundary conditions with the Chebyshev-τ method [Orszag (1971)] and find the equilibrium and the perturbed fields that are expanded in truncated Chebyshev series.

Solution:

One makes the expansions in the orthogonal polynomials $T_n(x)$ defined by

$$F(x) = \sum_{n=0}^{N} \widetilde{F}_n T_n(x) \quad \text{and} \quad G(x) = \sum_{n=0}^{N} \widetilde{G}_n T_n(x) \tag{6.30}$$

for the equilibrium fields and by the perturbed fields:

$$u(x) = \sum_{n=0}^{N} \widetilde{v}_n T_n(x) \quad \text{and} \quad b_x(x)) = \sum_{n=0}^{N} \widetilde{b}_n T_n(x). \tag{6.31}$$

The perturbed velocity field $\widetilde{u}_n$ is the expansion coefficient, and $T_n(x)$ is the nth Chebyshev polynomial of the first kind. The method is described by Goedbloed, *et al.* (2010) and Dahlburg and Einaudi (2000).

The equations satisfied by the unknown expansion coefficients, $\widetilde{F}_n, \widetilde{G}_n, \widetilde{v}_n, \widetilde{b}_n$, are obtained by substituting the $N \to \infty$ expansions of Eqs. (6.30) and (6.31) into (6.22) and (6.23), each of which produces a countable infinite number of equations in the expansion coefficients for $n = 0, 1, 2, \ldots$ when the orthogonality and recursion relations are used [Orszag (1971)]. All coefficients are set beyond $n = N$ to zero and use the $n = 0$ to $N-4$ equations from Eq. (6.22) and the $n = 0$ to $N-2$ equations from Eq. (6.23). The boundary conditions using $G_n(\pm 1) = (\pm 1)^n$ give the conditions

No-slip Boundaries

$$\sum_{n=0}^{N} \widetilde{v}_n = \sum_{n=0}^{N} (-1)^n \widetilde{v}_n = \sum_{n=0}^{N} n^2 \widetilde{v}_n = \sum_{n=0}^{N} (-1)^n n^2 \widetilde{v}_n = 0 \tag{6.32}$$

and for δB_x the boundary condition is

$$\sum_{n=0}^{N} \widetilde{b}_n = \sum_{n=0}^{N} (-1)^n \widetilde{b}_n = 0. \tag{6.33}$$

This method of truncation is called the τ-approximation [Lanczos (1956)].

The spectral discretization process yields a generalized eigenvalue problem that can be written as $AX = \omega BX$ where the vector $X = (\widetilde{v}_0, \widetilde{v}_1 \ldots \widetilde{v}_N, \widetilde{b}_0, \widetilde{b}_1 \ldots \widetilde{b}_N)$ and A and B are nonsymmetric $(2N + 2)$ by $(2N + 2)$ square matrices.

As is customary for this type of stability problem, either global or local methods are used to determine the eigenvalues.

For the global method we first transform the problem into the standard eigenvalue problem $CX = \omega X$ and then use the QR algorithm [Gary and Helgason (1970)]. This produces a full spectrum of complex eigenvalues $\omega_n(\alpha, \beta)$.

For the local method a good approximation or guess is required to start the iteration procedure using the Rayleigh procedure to find the nearby eigenvalue (and its associated eigenfunction) closest to the initial guess for the eigenvalue.

Generally we use the global method to identify the eigenvalue with the largest imaginary part (this is the most unstable mode or the least stable mode), and then use the local method to refine the eigenvalue and to compute the eigenfunction. The local method is also used for examining how the solution changes as a parameter, for example the Lundquist numbers is varied (Fig. 6.4).

Solutions of the linearized equations Eqs. (6.22) and (6.23) are given in Dalhburg, *et al.* (2001, 2002). For the baseline Helimak case we use the following parameters: the dimensionless stream-wise wavenumber $\alpha = k_\perp L = 2.3$, the span-wise wavenumber $\beta = 0.0$, the resistive Lundquist number $S = 5000$ and the viscous Lundquist number $S_v = 5000$. The corresponding drift wave normalized wavenumber is $k_\perp \rho_s = 0.1$ and growth rate is $\gamma L/c_s \simeq 0.8$.

This choice of α and β produces the largest growth rate for the specified values of S and S_v. There are a large number of damped, traveling modes for these particular parameters. There is also one unstable traveling mode. The form of the perturbed fields show that this is a global instability with comparable magnetic and flow velocity amplitudes.

6.1.6 *Sinuous, varicose and resistive varicose modes*

The stability analysis shows that there are three types of unstable modes and are identified for the magnetized jet: (1) an ideal sinuous mode, (2) an ideal varicose

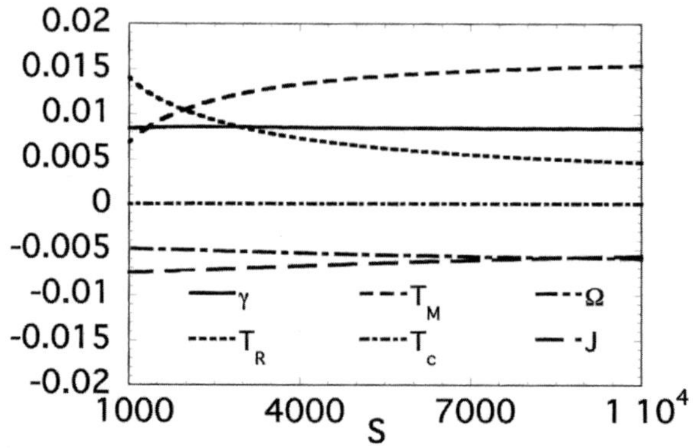

Fig. 6.4 Variation of perturbation energy balance terms with respect to Lundquist number (S). Here γ is the growth rate and T_R, T_M, T_C, Ω, and J represent normalized versions of the perturbed energy contributions due to the Reynolds stress, Maxwell stress, cross stress, viscous dissipation, and Ohmic dissipation, respectively. For these calculations $\alpha = 2.3, \beta = 0.0$ and $S_v = 5000$. The mode at mid-radius has $k_\perp = 4.6\,\mathrm{m}^{-1}$ and $k_\parallel = 0$.

mode, and (3) a resistive varicose mode. The sinuous mode magnetic field is kink-like in appearance, while for the varicose modes the magnetic field is sausage-like (ideal) or tearing-like (resistive).

For the Helimak fields the unstable mode is a distorted version of the sinuous mode reported for the magnetized jet [Dahlburg, *et al.* (1998)]. In contrast with the magnetized jet sinuous mode, the Helimak mode is suppressed near the inner wall at $x = -1$, where the magnetic field is enhanced. There are no unstable ideal or resistive varicose unstable modes, as are seen in the magnetized jet, found for the Helimak fields where the resistive varicose mode is the tearing-like mode in the fusion literature.

The dispersion relation and the variation of the growth rates with respect to α, parameterized by the span-wise wavenumber (β) are now considered. Calculations are shown for $S_v = S = 5000$. A nested family of curves is produced as the β is increased. All of the curves have the same form. This dispersion relation shows evidence of both low- and high-α cutoffs, as is typical for dissipative fluids confined to channels [Drazin and Reid (1981)]. The lower cutoff is due to the finite channel width and the upper cutoff is due to the finite dissipation. The variation of $\beta = k_{\|}L$ has several effects on the α dispersion relation. First, an increase in β generally leads to a decrease in the growth rate $\gamma = \mathrm{Im}(\Omega)$. Second, increasing β moves the maximum growth rate value to a larger value of $\alpha = k_{\perp}L$.

The variation of the growth rate $\gamma(\beta) = \mathrm{Im}(\omega)$, with respect to the span-wise wavenumber β, is parameterized by α. For all cases there is a cutoff value (β_c) above which the instability is stabilized. There is a decrease in the magnitude of the growth rate as β is increased. This implies that a Squires theorem [Squire (1933), Dahlburg and Karpen (1995), Seehafer and Schumacher (1998), Hughes and Tobias, (2001)] might be proved for the Helimak-like configuration, which we now proceed to consider.

6.1.7 *A magnetohydrodynamic Squire Theorem*

A comparison theorem is proven that states that for the same wavenumbers α and β and the same dissipation parameters S and S_v, the 2D system is more unstable than the 3D system. In hydrodynamics this theorem is known as the Squires theorem [Drazin and Reid (1981)]. Thus, we show how the Squires theorem is generalized to a dissipative magnetohydrodynamical system. Hence, for the same S and S_v, the growth rate of the two-dimensional unstable mode will be greater than that of the three-dimensional unstable eigenmode. A proof by the Squire Theorem is given in Dahlburg, *et al.* (2009) and requires detailed formulation of 3D system.

6.2 Turbulence in Toroidal Geometry with Helical Magnetic Fields

In the case of open field lines as in the mirror machine or a dipolar magnetosphere geometry, the calculation of the particle bounce averages of the wave fields is straightforward once the equilibrium $B(s)$ and $\Phi(\mathrm{s})$ fields are parameterized by the distance s along the field line. For the toroidal system with both toroidal and poloidal field components the calculation becomes more subtle. Namely, in the toroidal systems there are two types of magnetic field lines: those that close on themselves after one or several passages around the major axis and those that are ergodic and never close precisely on themselves. These field lines have an irrational value for their winding number or "safety factor" q. Electrons following these irrational winding number field lines tend to have a small, out-of-phase density response to the fluctuating potential.

The magnetic field component the long way around the torus called the toroidal B_t magnetic field and there is the magnetic field component the short way around the torus called the poloidal B_p magnetic field. The result of the vector sum of the two field (poloidal and toroidal) components produces the helical magnetic field structure. The rate of change of the pitch of the helical structure is defined as the magnetic shear in the field lines.

On the special **rational** surfaces the field line returns on its self after a given number n of transits around the major symmetry axis. On these rational surfaces the calculation proceeds as in the mirror geometry in each section of the field line between the $B_{\rm min}$ and $B_{\rm max}$. These closing field lines are called the rational field lines or field lines with rational rotational transforms or safety factor $q(r_{m,n}) = n/m$.

Charged particles with small pitch angles move over the whole field line. In the presence of an drift wave with mode numbers m and n, that is varying on the magnetic surface as $\cos(m\theta - n\phi)$, the result is a radial motion that is proportional to $1/(m - nq)$. For other radial positions a field line covers the magnetic surface ergodically and the integration of the fluctuation along the infinite field line length returns zero for the average as the motion ergodically covers the magnetic surface. The only exception is for modes that are strictly constant on the magnetic surface, that is for the $m = n = 0$ modes. Thus, charged particles on these ergodic field lines behave very differently depending on whether they are mirror trapped in regions of the minimum magnetic field $B_{\rm min}$ or if they are at small pitch angles to the magnetic field and move over $B_{\rm max}$ to average the wave field over the entire magnetic surface.

In the literature on mirroring particles the pitch angle is typically designated by α with $\cos\alpha = \vec{v}\cdot\vec{B}/vB$ and not to be confused with the bi-normal wavenumber $\alpha = k_\perp L$ used in Sec. 6.1.4. For circulating over $B_{\rm max}$ the condition on the pitch angle is $\sin\alpha < \sqrt{(1 - B_{\rm min}/B_{\rm max})}$. Thus the larger pitch angle particles are trapped in the region between $B_{\rm max}$ and $B_{\rm min}$ and respond more strongly to the perturbing fields than do the passing particles with small $\sin\alpha$ which average the wave fields over the entire magnetic surface. Here we define the critical pitch angle α_c at the local minimum of the magnetic field $B_{\rm min}$ by

$$\sin^2\alpha_c = \frac{B_{\rm min}}{B_{\rm max}}. \tag{6.34}$$

When the ion or electron pitch angle α is greater than α_c defined by Eq. (6.34) the particle is trapped in the weak field region making bounces between the mirroring points. When the condition $\alpha > \alpha_c$ from Eq. (6.34) is satisfied on the pitch angle, then the particle is reflected by the mirror force $F_{\|} = -\mu\nabla_{\|}B$ along the field line. For smaller pitch angles the particle passes over the magnetic field maxima and continually averages the wave perturbation over the lifetime of the fluctuation. If the fluctuation is frozen over this time, the average value is zero. If the fluctuation begins to change phase over the transit time of the particle then a nonzero average is returned. Clearly, the passing electrons will rapidly move along the field and average the perturbation to zero while, in contrast, the trapped electron respond with a value of the potential fluctuation characteristic of that at the $B_{\min}$.

This difference in the response is the origin of the trapped electron instability, where the passing electrons behave one way and the trapped electron behave very differently. This difference in response will also mean that in the trapped electron driven turbulence only the trapped electrons will undergo the turbulent transport in the system.

This trapped particle response situation is investigated through the study of electron orbits in the presence of models for the drift wave fields. In the limit of vanishing amplitude for the fields the collisional scattering of the guiding centers of the trapped and passing electrons make a random walk out of the system. This collisional random walk is called the neoclassical transport [Helander and Sigmar (2002)].

The mirror force trapped electron component of the plasma dominates the loss since (1) the trapped particles are more sensitive to the collisions and (2) the trapped particles have a larger radial excursion called the banana orbit. These characteristics apply both to the trapped ions and trapped electrons. In many experiments only the electrons are in the trapped particle regime. The trapped particle orbits are called banana orbits owing to their shape when displayed over one full bounce period in a plane perpendicular to the toroidal magnetic field. The radial width of the banana orbit vanishes as $1/B_p$ for the large magnetic field strength limit. Thus, it is the weaker poloidal magnetic field rather than the strong toroidal field component that confines the particle orbits in the tokamak. High plasma currents I_p are required for good plasma confinement in tokamaks. For example, $I_p \leq 7\,\text{MA}$ in JET and $I_p \leq 5\,\text{MA}$ in JT-60U compares with ITER designed to carry 15 MA of plasma current. In contrast, the Large Helical Device, LHD, has plasma currents of order 50 kA. The external magnetic field coils are shaped to produce a B_p designed for the required rotational transform $\iota(r) = 1/q(r)$.

In the presence of the drift waves with amplitudes greater than a small critical level the trapped electrons begin diffusing out of the system while the passing electrons have only small radial oscillations about their flux surface position. Calculations for the neoclassical transport in the presence of fluctuations have been unified with the drift wave induced fluxes for both the electrostatic and electromagnetic turbulence.

The attentive reader will now have realized that there is even a more fundamental problem (issue) for this toroidal plasma with alternating surfaces with

rational and **irrational** rotational transforms for the winding of the magnetic field line around the toroidal surfaces. Both tokamaks and stellarators with helical magnetic fields systems have this fundamental problem. Discussion on this issue is in the following section. An advantage of the mirror confinement and the field-reversed mirror called the FRC is the absence of the sheared helical magnetic field structure.

6.3 Density of Rational Surfaces and the KAM Tori

Consider a fast electron with zero pitch angle that maps out the magnetic field lines in the torus. If the electron is launched on a surface that has a rational value m/n for the rotational transform then the orbit is periodic: the electron returns to its initial position after some time passes for n revolutions around the torus. After n turns around the major axis the electron has made m turns around the minor axis. If, instead, the electron is launched on a neighboring surface with an **irrational** rotational transform it comes close, but never, returns precisely to its initial position. This orbit is called a Moser orbit or a quasi-period orbit. The effect of the perturbation to these two orbits is very different. In the first case there arises a vanishing denominator describing a secular change in the particle's orbit with time. From perturbation theory the original orbit appears unstable to the presence the perturbation [Jose and Saletan (2002)].

The KAM (Kolmogorov-Arnold-Moser) theory resolves this problem for the effect of these resonant orbits in the problem of celestial mechanics and then in Hamiltonian systems in general. The result is that there is the opening up of an "island" in the $r - \theta$ plane of radial width δr proportional to the square root of the amplitude of the perturbation strength. The island is constructed from the Poincare map by recording the puncture or crossing point on a reference surface for each passage of the electron around the torus. The perturbation varying with the fluctuation $\cos(m\theta - n\phi)$ creates magnetic islands with m-fold fixed-points inside islands formed on the rational surface $r(m/n)$ defined by the roots of $q(r) = m/n$. The orbit at the center of the islands is then a stable fixed point of the nonlinear map created by the electron equations of motion. Between two stable fixed points there is an unstable fixed point. Two particles approaching the unstable fixed points can leave the region with very different paths.

The fixed points $r(m/n)$ have the periodicity of the rational surface. The orbital stability question is now a famous result [Jose and Saletan (2002), Frisch (1996)] and numerous monographs at different levels of mathematical sophistication are available. The effect of the island is most easily computed if one substitutes an area preserving nonlinear map $M(x, y) \to x^*, y^*$ for original nonlinear flow generated by the orbital differential equations. The orbital motion is that from a Hamiltonian system of equations and thus is equivalent to a symplectic area preserving map on the surface of section. Generally, this map must be computed by numerical integration of the differential equations for the orbits. The topology and general properties, however, can often be seen by replacing the flow with a suitable nonlinear map. Many examples are given in detail in Chirikov (1983) and Lichtenberg and Lieberman (1992). The original mathematical solutions and theorems are clearly presented in the lectures of Moser (1973) in a Princeton University Monograph, Stable and Random Motions in Dynamical Systems.

The density and size of the islands around the rational points in the dynamical system is controlled by the twist in the rotational transform profile. For the simple example given above of a zero pitch angle electron the orbital twist is the same as the twist of the magnetic field lines which is given by the magnetic shear parameter $s(r)$ given by $s = rq'(r)/(q(r)$.

This magnetic shear profile $S(r)$ is of fundamental importance in the tokamak. In the following chapters many uses of both the rotational rate $1/q(r)$ and the magnetic shear $s = rq'(r)/q(r)$ profiles will be uncovered.

For ion orbits the phase space twist rotational transform can be dominated by the shear in the $\mathbf{E} \times \mathbf{B}$ drift velocity. The radial electric field E_r and since the parallel velocity determine the surface of section plots in general. For low values of $v_{\|}/q(r)$ however the shear in the E_r field often is the dominant influence on the ion orbits. The plasma fluid will rotate from the E_r field and thus the drift waves themselves be propagating in a sheared plasma flow velocity $v_\theta = -E_r(r)/B(r)$. In general this E_r shearing can add to or partially cancel the magnetic shear influence on the ions and ion fluid dynamics. The role of the fluid shearing is developed in Chapters 12 and 16. Sufficiently strong E_r shear can produce a drift wave version of the Kelvin-Helmholtz instability in magnetically confined plasmas. In other cases a localized shear flow region can decouple the edge from the core plasma giving rise to a transport barrier and the high-confinement mode called the H-mode regime. The H-mode confinement is analyzed in Chapters 12 and 13.

Sheared $\mathbf{E} \times \mathbf{B}$ flows in plasmas can strongly change the nature of the instabilities by creating smaller cross-field variations as the different elements of the plasma gas flow at different speeds $\delta v = \delta x \times dv_E/dx$ where the width $\delta x = \ell_c$ is the radial correlation width of the mode in the absence of sheared flows. A table of the effects for the level of the shear flowed required to effect a variety of plasma instabilities is given in Tajima, *et al.* (1991). The strong instabilities like the Rayleigh-Taylor are less disrupted by the sheared flows than the weaker instabilities like the ITG modes. The effects are well known in laser plasma physics as well as magnetic confinement plasma physics.

6.3.1 *Overlapping radial eigenmodes*

Determining the number of and the density of the rational surfaces is a famous problem in number theory. The solution for high-order n/m rational numbers is expressed in terms of the Euler quotient function as described in Beklemishev and Horton (1992). Here we give the approximation for the density of the rational surfaces for the limit of large n and m values. This density of rational surfaces is

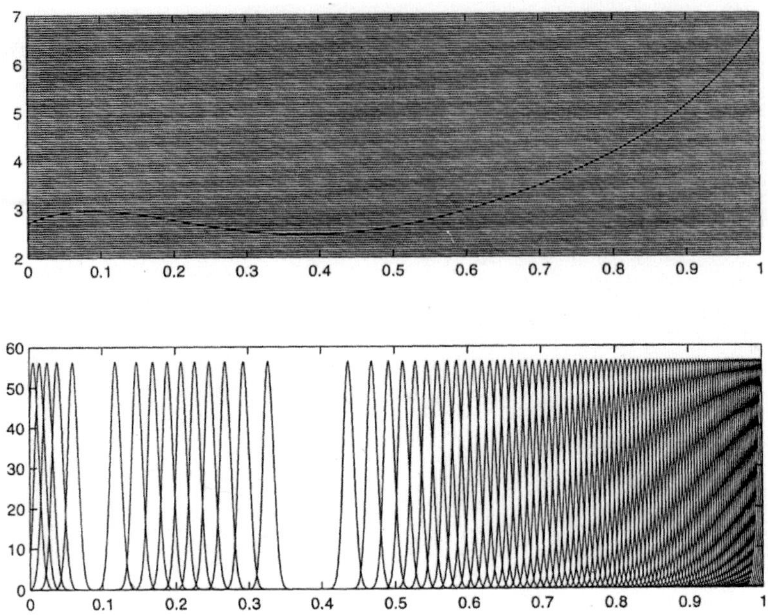

Fig. 6.5 The density of states for a reversed shear (RS) profile. The $q(r)$ profile is shown by the underlying solid curve in the top panel with resonant lines for toroidal mode numbers up to 20. The lower panel shows the density of rational surfaces with each given the same gaussian localized wave function. The local maximum at $q = 3$ gives a gap at $\rho = 0.1$ and the local minimum at $q = 5/2$ gives a large gap centered at $\rho = 0.37$ [Beklemishev and Horton (1992)].

given by

$$\rho_m(\text{rational surface density}) = m\,\frac{6|q'|}{\pi^2 q^2} \tag{6.35}$$

for all modes up to mode number m. Thus we see that the density increases rapidly and the total number of rational surfaces becomes very large. The spacing between the rational surfaces of order n is $dr = 1/nq'$ for non-vanishing shear. At low values of the magnetic shear $rq'/q \to 0$ the density of rational surfaces vanishes and there maybe only one or no rational surfaces in a given region. Given a measure for the radial width of the eigenmodes, or magnetic islands widths, on each rational surface, it is useful to construct the pictures shown in Fig. 6.5 from the summation over all rational surfaces. From Fig. 6.5 the outer regions $\rho > 0.6$ have a high density of strongly overlapping drift wave functions. In this region the ballooning mode theory applies with the summation over the outer regions and the rational density is such that the modes overlap significantly, as can be determined by counting the number of dark stripes appearing above a given radial position.

In Fig. 6.5 with reversed magnetic shear profile, we see immediately that in the reversal layer the gaps between the mode rational surfaces are large and the overlap is weak. Chapters 13, 14, and 15 shows experiments where there are sharp changes in the plasma transport across regions of reversed magnetic shear. These are called RS discharges for reversed magnetic shear.

Even before making the mode calculations or the orbital calculations of particles in these system one senses that by having a reversed magnetic shear RS system there is a region of reduced transport. In the confinement literature this effect is referred to as a transport barrier produced by reversed shear. In reality there is still transport across this region but the transport flux is greatly reduced as borne out by numerous experiments and simulations.

The motivation for the MHD stability problem is clear as recognized at the beginning of the thermonuclear fusion program. Briefly, there is an MHD potential energy $\delta W_{\rm MHD}$ from which it is clear that to not have the stabilizing perturbed magnetic energy overwhelm the destabilizing interchange energy that is released by convective motion, we must have an almost vanishing $k_\parallel$ on each toroidal surface. This vanishing $k_\parallel$ then conflicts with the condition for the wave function being doubly period on the torus except for the uninteresting case of the perturbation being constant on the surface. To see this more clearly suppose the function is doubly periodic with given m and n. Then one sees that the parallel wavenumber cannot stay small but increases rapidly as one deviates from the mode rational surface where $q = m/n$. In the neighborhood of the rational surface $q = m/n$, the $k_\parallel$ increase as $k_\parallel = (s/qR)(r - r_{m,n})$ where $q(r_{m,n}) = m/n$ and $s = rdq/dr$ is the magnetic shear parameter.

The magnetic shear is the rate of change of the twist in the helical magnetic field structure as the position moves across the magnetic surfaces. Later in this chapter and Chapters 13 and 14 we see that special modes occur on the reversed shear (RS) surfaces where the shearing of the field changes direction. Special considerations apply at the reversed shear surface and there can be a partial barrier to the anomalous transport set up at the shear reversal layer. Clear data for the change in the electron transport at the shear reversal layer is described in Chapter 15.

The ballooning mode solution to this problem is given in Connor, *et al.* (1979) and is explained and amplified in the works of Dewar and Glasser (1983) and Hazeltine and Waelbroeck (1988).

Later the method for the drift waves, which is somewhat more intricate than for the linear MHD stability problem, was developed. The drift wave literature since 1980 abounds with calculations of the surface eigenvalue problem for the wave functions expressed in the extended poloidal angle of the ballooning mode representation. In some of the flux tube gyrokinetic simulations the local twisted field line following coordinates are used in the ballooning mode formulation to return the information back to the (global) laboratory coordinates. Some of the early nonlinear calculations were also done in ballooning mode coordinates although it is now recognized that the severe truncations made in some of those calculations are not justified in view of the high-resolution numerical solutions needed for the nonlinear transport problem.

6.3.2 *Limitations of the MHD model and the fluid moments closure problem*

Serious problems arise in attempts to use the MHD model for drift waves, tearing modes and the mirror mode instability. The common origins of these problems is the failure of two key assumptions in the MHD model. First, the motions in these instabilities do not satisfy the condition of being FAST so that the adiabatic closure assumption that

$$\nabla \cdot \boldsymbol{q} = 0 \tag{6.36}$$

is not satisfied. Secondly, the drift waves and the tearing modes both contain, in essential ways, a finite parallel electric field. A more general way to express the failure of the MHD model for these problems, is to state that many plasma waves require a **generalized Ohm's law**.

The MHD model for the electric field $\mathbf{E}(\boldsymbol{x}, t)$ in the plasma is much too simple. As the field lines move in the plasma wave oscillation, the electrons rapidly move along the field line so as to set up a parallel electric potential near to a local Boltzmann distribution of the density and pressure variations along the magnetic field that follows the wave oscillations. In this case the parallel electric field follows the electron density variation and is given by $E_{\|} = (k_{\|}T_e/e)(\delta n_e/n_e)$. The parallel current is largely nondissipative and given by $j_{\|} = \sigma_{\|}E_{\|}$ with $\sigma_{\|} = in_e e^2(\omega - \omega_{*e})/m_e k_{\|}^2 v_e^2$. The parallel current-voltage phase relation changes sign at $\omega = \omega_*$. At $\omega = \omega_{*e}$ the approximate wave polarization is $\delta j_{\|} = 0$ rather than $\delta E_{\|} = 0$.

More generally for tearing modes, resistive-g modes, and shorter wavelength electron temperature gradient driven drift waves the Ohm's law required is of the form

Generalized Ohm's Law

$$E_{\|} + \frac{\nabla_{\|} p_e}{en_e} = \eta_{\|} j_{\|} \tag{6.37}$$

where $\eta_{\|}$ is the electrical resistivity along the magnetic field line. Note that electromagnetic waves at the gradient $\nabla_{\|} = (\widehat{b} + \delta\vec{B}/B)\cdot\nabla$ has a wave-induced component from $\delta\mathbf{B}$.

The parallel electron pressure gradient brings in the ion inertial scale length term

Ion-Sound Gyroradius

$$\rho_s = \frac{(m_i T_e)^{1/2}}{eB}. \tag{6.38}$$

The $\boldsymbol{j} \times \mathbf{B}$ Hall term brings in the scale length of the collisionless ion skin depth

Ion Inertial Scale Length

$$\delta_i = \frac{c}{\omega_{pi}}. \tag{6.39}$$

This δ_i-term eliminates the singularity of the electromagnetic waves at the ion cyclotron frequency ($\omega = \omega_{ci}$) and is important for the collisionless reconnection process [Galeev (1984), Galeev and Zeleny (1979), Drake, *et al.* (1992)].

For sufficiently small scale waves the electron inertial $\delta_e = c/\omega_{pe}$ becomes critical and is given by the keeping the electron mass in the parallel acceleration equation. This term plays a key role in the electron physics of magnetic reconnection and is described in the Alfvén wave experiments of Gekelman (1999) in the laboratory simulations of the Alfvén waves as they occur in the ionosphere and magnetosphere.

Partly because of the very large convergence of the magnetic field lines in the dipole geometry, the cross-field scale lengths due to the large scale Alfvén waves generated in the equatorial region of the magnetosphere converge to cross-field wavelengths of order 10 km which is comparable to the electron skin depth δ_e. These Alfvén waves where $k_\perp \delta_e$ is order unity are slowed down and dispersed due to the electron mass effect. Another source of Alfvén wave dispersion is the ion mass effect occurring through the parallel electron pressure gradient fluctuations. These so-called inertial Alfvén waves [Lysak and Lotko (1996)] have the dispersion relation

Inertial Alfvén Waves

$$\omega^2 = \frac{k_\parallel^2 v_A^2}{1 + k_\perp^2 c^2/\omega_{pe}^2}. \tag{6.40}$$

There are new types of large-scale vortex structures for the dispersive Alfvén waves in collisionless plasma. Alfvén waves can propagate in the form of Alfvén vortices of finite characteristic radius and characterized by magnetic flux ropes carrying orbital angular momentum. The structure of the toroidal and radial velocity, fluid and magnetic field vorticity, the longitudinal electric current in the plane orthogonal to the external magnetic field is important.

In the magnetic reconnection simulations in Chapter 14 the electrons stay attached to the magnetic field lines until the spatial gradient scale of the electron skin depth is reached. Then the electrons separate from the field lines forming high-current densities that produce lower-hybrid drift waves and in strong bursty events including the Buneman two-stream instability [Buneman (1958)]. The two-stream instability heats the electrons very rapidly and the plasma then rapidly evolves to two-stream stable with $T_e \gg T_i$ states. The hot electron plasma with a plasma current or electron drift velocity leaves the ion acoustic and current driven drift wave turbulence producing the anomalous resistivity.

The details of the magnetic reconnection processes are an active area of research. High-resolution computer simulations and laboratory experiments are clarifying the complexity of these events. Tearing mode theory in Chapter 14 describes well many aspects of the dynamics but the full complexity of the multiscale nonlinear dynamics requires computer simulations.

There are, in fact, many different regimes to this process of magnetic field line reconnection. Tools to study the microturbulence used to analyze these reconnection processes and their computer simulations are in Chapters 14. Resources for the physics of magnetic reconnection are *Nonlinear Magnetohydrodynamics* by Biskamp (1993) and for plasma shocks [Burgess (1995)]. Slow shocks are used to construct magnetic reconnection models.

6.4 Reduction of the Toroidal Electric Field Driven Dynamics

For the slow (subsonic and sub-Alfvénic) evolution of the confined plasma, the perpendicular part of the momentum equation should not be solved forward in time as performed in deriving Eq. (6.40) since the inertia acceleration term, $Mn\, d\boldsymbol{V}/dt$ is the small difference between the large, nearly balancing forces of the pressure gradient ∇p and $\boldsymbol{j} \times \mathbf{B}$ forces. Instead we solve the cross-field momentum equation for the current density $\boldsymbol{j}_\perp$ necessary to balance the perpendicular forces as

$$\boldsymbol{j}_\perp = \frac{\mathbf{B}}{B^2} \times \left[Mn \left(\frac{\partial \boldsymbol{v}}{\partial t} + \boldsymbol{v} \cdot \nabla \boldsymbol{v} \right) + \nabla p \right]. \tag{6.41}$$

Likewise from Ohm's law we solve for the cross-field velocity field $\boldsymbol{v}_\perp$ necessary to satisfy the perpendicular part of Ohm's law

$$\boldsymbol{v}_\perp = \frac{\mathbf{B}}{B^2} \times (-\mathbf{E} + \eta_\perp \boldsymbol{j}_\perp) \tag{6.42}$$

which is the $\mathbf{E} \times \mathbf{B}$ motion plus a resistive pressure gradient driven flux

$$\boldsymbol{v}_{\perp\eta} = -\frac{\eta_\perp}{B^2} \nabla_\perp p \tag{6.43}$$

where one uses $\boldsymbol{J} \times \mathbf{B} = \nabla_\perp p$.

At high temperature and strong magnetic field the flow from Eq. (6.43) is small compared to the radial flux from $\mathbf{E} \times \mathbf{B}$ convection that arises from the electric force due to $\eta_\parallel$. This $\eta_\parallel$ driven particle flux is called the Pfirsch-Schlüter "diffusion" even though the flux is from $\mathbf{E} \times \mathbf{B}$ convection. Thus, we have the following three equations to advance the system in time:

$$\frac{\partial n}{\partial t} = -\nabla \cdot (n\boldsymbol{v}) \tag{6.44}$$

$$\frac{\partial v_\parallel}{\partial t} = -S^2 \frac{B_\theta}{Br}\frac{\partial}{\partial \theta}\ell n(n) - \widehat{\boldsymbol{e}}_\parallel \cdot (\boldsymbol{v}\cdot\nabla)\boldsymbol{v} \tag{6.45}$$

$$\nabla \cdot \left[\frac{Mn}{B}\widehat{\boldsymbol{e}}_\parallel \times \left(\frac{\partial \boldsymbol{v}_\perp}{\partial t} + (\boldsymbol{v}\cdot\nabla)\boldsymbol{v}\right)\right] + \nabla\cdot\left(\frac{\mathbf{B}\times\nabla p}{B^2}\right) \tag{6.46}$$

$$+\nabla\cdot\left(\widehat{\boldsymbol{e}}_\parallel j_\parallel\right) = 0 \tag{6.47}$$

where for the toroidal geometry with $\mathbf{B} = B_\theta \widehat{\boldsymbol{e}}_\theta + B_\phi \widehat{\boldsymbol{e}}_\phi$, the flow fields are

$$\boldsymbol{v}_\perp = \frac{\widehat{\boldsymbol{e}}_\parallel}{B} \times \left(\nabla\varphi - \frac{R_0 E_0(t)}{R}\widehat{\boldsymbol{e}}_\phi + \eta_\perp \boldsymbol{j}_\perp + \eta_\perp j_r \widehat{\boldsymbol{e}}_r\right) \tag{6.48}$$

and

$$j_\parallel = \frac{1}{\eta_\parallel}\left(-\frac{B_\theta}{Br}\frac{\partial\phi}{\partial\theta} + \frac{B_\phi R_0}{BR}E_0(t)\right). \tag{6.49}$$

The sound speed squared is $S^2 = k_B(T_i + T_e)/m_i$ in Eq. (6.45). The system is coupled in that $\boldsymbol{j}$ in Eq. (6.46) depends on $\boldsymbol{v}$ which is being solved for in Eqs. (6.41) and (6.49). For small η this term is small in Eq. (6.48) and it is sufficient to use the previous value of $\boldsymbol{j}$ in Eq. (6.41) to compute $\boldsymbol{v}_\perp$ for the resistive component flows.

A fair amount of vector manipulation is required to write Eqs. (6.41)–(6.49) out in component form. Here we only give the reduction of one term. The unit vectors are defined as

$$\widehat{\boldsymbol{e}}_\parallel = \frac{\mathbf{B}}{B} = \frac{B_\theta}{B}\widehat{\boldsymbol{e}}_\theta + \frac{B_\phi}{B}\widehat{\boldsymbol{e}}_\phi \tag{6.50}$$

$$\widehat{\boldsymbol{e}}_\perp = \widehat{\boldsymbol{e}}_\parallel \times \widehat{\boldsymbol{e}}_r = \frac{B_\phi}{B}\widehat{\boldsymbol{e}}_\theta - \frac{B_\theta}{B}\widehat{\boldsymbol{e}}_\phi \tag{6.51}$$

and note that B_ϕ/B, B_θ/B are only functions of r. Then we have the following velocity and acceleration from partial differential equations results for Eq. (6.42) and Eq. (6.41)

$$\boldsymbol{v}_\perp = \frac{\widehat{\boldsymbol{e}}_\perp}{B}\frac{\partial\varphi}{\partial r} - \frac{\widehat{\boldsymbol{e}}_r}{B^2}\frac{B_\phi}{r}\frac{\partial\varphi}{\partial\theta} - \frac{R_0}{BR}E_0(t)\frac{B_\theta}{B}\widehat{\boldsymbol{e}}_r + \frac{\widehat{\boldsymbol{e}}_\perp}{B}\eta j_r - \widehat{\boldsymbol{e}}_r\frac{\eta j_\perp}{B} \tag{6.52}$$

and the components in Eq. (6.46)

$$\widehat{\boldsymbol{e}}_\parallel \times \frac{\partial \boldsymbol{v}_\perp}{\partial t} = -\frac{\widehat{\boldsymbol{e}}_r}{B}\frac{\partial^2\varphi}{\partial r\partial t} - \frac{\widehat{\boldsymbol{e}}_\perp}{B}\frac{B_\phi}{Br}\frac{\partial^2\varphi}{\partial\theta\partial t} - \frac{R_0\dot{E}_0(t)}{BR}\cdot\frac{B_\theta}{B}\widehat{\boldsymbol{e}}_\perp \tag{6.53}$$

$$-\frac{\eta_\perp}{B}\frac{\partial j_\perp}{\partial t}\widehat{\boldsymbol{e}}_\perp - \frac{\eta_\perp}{B}\frac{\partial j_r}{\partial t}\widehat{\boldsymbol{e}}_r.$$

The equation for $\nabla \cdot \boldsymbol{j} = 0$ for charge neutrality reduces to

$$
\begin{aligned}
&-\frac{1}{r}\frac{\partial}{\partial r}\left(\frac{Mn}{B^2}\frac{\partial^2\varphi}{\partial r\partial t}\right) - \frac{1}{r}\frac{\partial}{\partial\theta}\left(\frac{R_0\dot{E}_0}{BR}\cdot\frac{B_\theta B_\phi}{B^2}\right) \\
&-\frac{1}{r}\frac{\partial}{\partial r}\left(\frac{r\eta_\perp \dot{j}_r}{B^2}\right) - \frac{1}{r}\frac{\partial}{\partial\theta}\left(\frac{\eta_\perp \dot{j}_\perp B_\phi}{B^3}\right) + \nabla\cdot\frac{Mn}{B^2}\left(\widehat{\boldsymbol{e}}_\parallel \times (\boldsymbol{v}\cdot\nabla)\boldsymbol{v}\right) \\
&+\nabla\cdot\left(\frac{\mathbf{B}\times\nabla p}{B^2}\right) - \frac{1}{\eta_\parallel}\left(\frac{B_\theta}{Br}\right)^2\frac{\partial}{\partial\theta}\left(\frac{1}{B}\frac{\partial\varphi}{\partial\theta}\right) = 0
\end{aligned}
\tag{6.54}
$$

where

$$
j_r = -\frac{Mn}{B}\frac{\partial v_\perp}{\partial t} + \frac{Mn}{B}\left(\widehat{\boldsymbol{e}}_r\cdot\widehat{\boldsymbol{e}}_\parallel \times (\boldsymbol{v}\cdot\nabla)\boldsymbol{v}\right) - \frac{B_\phi}{B^2 r}\frac{\partial p}{\partial\theta} \tag{6.55}
$$

$$
j_\perp = \frac{Mn}{B}\frac{\partial v_r}{\partial t} + \frac{Mn}{B}\left(\widehat{\boldsymbol{e}}_\parallel \times (\boldsymbol{v}\cdot\nabla)\boldsymbol{v}\cdot\widehat{\boldsymbol{e}}_\perp\right) + \frac{1}{B}\frac{\partial p}{\partial r}. \tag{6.56}
$$

An important remaining task is to compute the components of $(\boldsymbol{v}\cdot\nabla)\boldsymbol{v}$ for the nonlinear inertial acceleration from fast flows.

All functions must be single valued in θ, φ and thus periodic in θ, φ. Several constraints are derived from Eqs. (6.54)–(6.56) by integrating over θ. Hazeltine, *et al.* (1971) analyze the equations by solving in the limit of small r/R_0 for the first order θ-variation and then computing the rate of change $(\partial/\partial t)$ in $\overline{n}(r,t), \overline{\phi}(r,t), \overline{v}_\perp$ where the bar denotes an average over θ $[\int(d\theta/2\pi)\times(R(\theta)/R_0).]$ Another approximation made in the Hazeltine, *et al.* (1971) analysis is that the radial gradients are sufficiently weak.

The electric field is

$$
\mathbf{E} = -\nabla\varphi(r,\theta,t) + \frac{R_0}{R}E_0(t)\widehat{\boldsymbol{e}}_\phi. \tag{6.57}
$$

The plasma is ohmically-heated by the applied external toroidal electric field $E_0(t)\widehat{\boldsymbol{e}}_\phi$ from increasing magnetic flux in the central solenoid.

The electric field is given as curl-free, and the magnetic field is taken to be the model tokamak field

$$
\mathbf{B} = \frac{R_0}{R}\left(0, B_\theta(r),\ B_\phi^{(0)}\right) \tag{6.58}
$$

where $R = R_0 + r\cos(\theta)$, $B_\phi^{(0)}$, R_0 are constants, and a toroidal current is specified by the form of B_θ by Amperé's law

$$
\frac{1}{r}\frac{\partial}{\partial r}(rB_\theta(r)) = \mu_0 j_\phi(r). \tag{6.59}
$$

There are eight field variables determined by the eight equations (6.52), (6.53), (6.54), and (6.55). Equation (6.55) determines n, Eq. (6.53) determines φ, Eq. (6.57) determines $\boldsymbol{j}_\parallel$, and $\boldsymbol{u}_\perp$.

The system of equations is third order in time from the time dependence of n and $\boldsymbol{v} = (v_r, v_\perp)$.

The time scales associated with the dynamics in the equations are as follows:

(1) Fast ion-sound oscillations with

$$\omega \simeq \pm \frac{m}{r} \frac{B_\theta}{B} c_s \tag{6.60}$$

where $c_s = [(T_i + T_e)/m_i]^{1/2}$ is the speed of sound in the isothermal equations and m is the approximate mode number of the θ-variation. There are density variations with $n = n_0 + n_1 \cos\theta$ and potentials $\varphi = \varphi_1 \sin\theta$ from the toroidal currents in the $\nabla B \neq 0$ toroidal magnetic field. These $m = 1$ ion acoustic waves are called geoacoustic modes (GAMS). In divertor geometry the same physics applies but a spectrum of m-modes develops from magnetic separatrix.

(2) Unstable axisymmetric ($m = n = 0$) rotation arises with the slow exponential growth rate

$$\gamma = \frac{\eta \beta c^2}{\mu_0 r^2} \cdot \frac{R_0^2}{r^2} \tag{6.61}$$

where $\beta = 8\pi p/B^2(\text{cgs}) = 2\mu_0 p/B^2(\text{mks})$. Large $m = 0$ potential oscillations are measured in Hamada, *et al.* (2005) where the fluctuations are associated with flows and the GAMS. The GAMS can reduce the turbulent plasma flux and are strong in rotating plasmas [Casson, *et al.* (2010), Kendl and Scott (2005)].

(3) Diffusion time scale τ_D in the rotating state is given by

$$\frac{1}{\tau_D} = \frac{1}{n} \frac{\partial}{\partial r} \langle n v_r \rangle \approx \frac{8\pi^2}{L^2} \frac{\eta \beta c^2}{r^2} \approx \left(\frac{B_\phi r}{B_\theta R} \right)^2 \frac{\eta_\| \beta c^2}{r^2} = \frac{q^2 \eta_\| \beta}{r^2}. \tag{6.62}$$

The theoretical analysis can be done in the limit of small resistivity η where the diffusion time $\tau_D = 4\pi L^2/c^2\eta$ is long compared to sound wave period qR/c_s.

Pfirsch-Schlüter transport drives a poloidal spin-up of the plasma [Stringer (1969), Hassam, *et al.* (1991)]. Hassam, *et al.* (1991) show that anomalous transport in the toroidal geometry with $\cos\theta$-variations on the flux surfaces also has a rotational spin-up. Hassam and Antonsen (1994) show the details of the spontaneous rotation in tokamaks. The competition between the sheared rotation created by the Reynolds stress tensor from the drift wave turbulence and the Stringer spin-up mechanism is analyzed by McCarthy, *et al.* (1993). The analysis becomes intricate with the scrape-off layer SOL becoming important as known from the confinement experiments. A review of the issues is given by Connor and Wilson (2000) in an article on the L-H transition dynamics.

6.5 Effect of Dust on Drift Instability in Edge Fusion Plasmas

Dusty plasmas occur in a large variety of natural plasmas both in the space environment and in high-temperature fusion plasmas. Dust particles measured in tokamaks and helical/stellarator machines are highly negatively-charged ($Z_d \cong 10^3 - 10^4$)

aggregates of atoms from the wall materials eroded by contact with the high-energy plasma ions. The dust particles are observed to reach sizes up to microns and to gain energy from collisions with hot plasma ions. Measurements in Tore Supra, DIII-D and the Large Helical Device show that some of the higher-energy dust particle cross the magnetic separatrix entering into the core plasma. These dust particles are then accelerated to high energies ultimately causing degradation of fusion plasma. Dust particles accelerated to high energies damage the plasma vessel walls. The dust particles are smaller than the plasma Debye length and when at sufficient density modify the properties of the edge plasma turbulence. The dust particles dynamically change their $Z_d(t)$ by their sheath plasma currents producing a probability distribution of charges and masses of the dust particles [Arnas, *et al.* (2010)].

The plasma-wall interaction produces plasma dust particles. Theoretical descriptions of the plasma dust transport are developed in detailed theory-experimental models that explain the observed dust profiles. The electron temperature profile $T_e(r,t)$ evolution in power balance transport studies in Tore Supra [Hoang, *et al.* (2003, 2004)], NSTX [Kaye, *et al.* (2007)], TCV [Asp, *et al.* (2008)] and DIII-D [Doyle (2007)] is expanded to include an additional kinetic equation $f_d(\boldsymbol{x}, \boldsymbol{v}, q_d, t)$ that describes the dynamics of the dust particles in a seven-dimensional $\boldsymbol{x}, \boldsymbol{v}, q_d$ phase space where q_d is the charge on the micron-sized dust particle in the plasma. The inclusion of the phase-space distribution of the dust particles was reported to be essential in impurity data from DIII-D and LHD where infrared and spectroscopic measurements show the motion of 100 nm to micro-sized clumps of atoms with high negative charge states $q_d(t) = -eZ_d(t)$ where values of Z_d range from 10^3 to 10^4. These dust particles have an electrical capacitance proportional to their radius a_d and a charge proportional to the dust density $n_d(t)$. Owing to the kilovolt per meter electric fields in the plasma a few of the dust particles are observed to be accelerated to kilovolt energies and penetrate into the core plasma.

Here we describe the dynamical charging and energy gain equations and consider the change in the drift wave turbulence produced by the dust particles. The drift wave turbulence in the edge plasma can be strongly increased by the dust at sufficient density. This effect may account for the well-known "short-fall" found when comparing drift simulations with the measured level of edge fluctuations and edge transport in tokamaks without accounting for dust particles.

The dust charge $Z_d(t)$ distribution results from a balance of competing processes described by Tsytovich and Havnes (1993). Dust densities n_d exceeding some critical value give rise to new collective drift and ion acoustic waves modes with $\widetilde{n}_i = \widetilde{n}_e + Z_d \widetilde{n}_d$ where the dust particle charge is $q_d = -eZ_d$. In particular, the modes with frequencies much less than the characteristic dust charging frequency $\nu_{\rm ch}$ are substantially changed when $Z_d \widetilde{n}_d = |\widetilde{n}_i - \widetilde{n}_e|$ is significant. Therefore, drift waves and turbulent transport is actively affected by the dust particles in the plasma.

In the case that there are slow perturbations of the electron and ion distribution functions $f_{e,i}$, the dust charge remains close to $Z_d^{\rm eq}$ which relaxes to this equilibrium value on the time scale determined by the charging frequency $\nu_{\rm ch}$ given by

$$\frac{d\delta Z_d}{dt} = \frac{dI}{dZ_d}\delta Z_d \equiv -\nu_{\rm ch}\delta Z_d, \tag{6.63}$$

where one derives that for dust particles with radius a and velocity v_n the charging rate is

$$\nu_{\rm ch} \equiv \frac{\omega_{pi}^2 a}{\sqrt{2\pi}\, v_n}(1+\tau+Z_d). \tag{6.64}$$

For an edge electron temperature $T_e \approx 50\,\mathrm{eV}$, density $n_e \approx 10^{13}\mathrm{cm}^{-3}$, and for dust particles of one micron size, the dust charge is $Z_d \approx 10^4$. Using these estimates one finds the charging rate $\nu_{\rm ch} \approx 10^8\mathrm{Hz}$.

The charging process described by Eq. (6.63) is a dissipative process and can drive negative energy waves unstable. When the electron reaches the surface of a dust particle, then it must overtake the Coulomb barrier to accumulate on the dust grain. The time scale of the capture of electrons is given by the frequency $\nu_{\rm capture}$

$$\nu_{\rm capture} = Z_d \frac{v_{Te}}{a}. \tag{6.65}$$

The condition that the capture frequency is greater than the electron cyclotron frequency is $\nu_{\rm capture} \gg \omega_{ce}$ with $\omega_{ce} = eB/m_e c$, giving that $a \ll \rho_e$ which is typically satisfied.

The drift wave frequency given by

$$\omega_{*e,i} = \mp\frac{k_y T_e c}{eBn_{0e,i}} \cdot \frac{\partial n_{0e,i}}{\partial x} \tag{6.66}$$

which is approximately $10^5 - 10^6\mathrm{Hz}$ for the SOL plasma parameters and thus $|\omega_*| \ll \nu_{\rm ch}$. The estimate for the charging rate is $\nu_{\rm ch} > 10^8$ Hz for one micron dust from Eq. (6.64). Numerous dusty plasma experiments and simulations have shown that there are ion-acoustic and drift waves in the dusty plasmas [Shukla and Mamun (2002)].

Chapter 7

Turbulence Theory for Drift and Alfvén Waves

Chapters 2–4 describe the physical processes that give rise to the low-frequency drift waves and temperature-gradient-driven instabilities in magnetized plasma. In those descriptions we used simple descriptions of the plasma motions to derive the waves and vortices. Now to understand these drift instability processes in laboratory and space plasmas, we must develop two lines of investigation: one is the FLR-fluid description which is a two-component set of fluid equations that are to be solved in limit that $m_a n_a \partial_t \boldsymbol{u}_a \ll e_a n_a \boldsymbol{u}_a \times \mathbf{B}$ so that the inertial acceleration is a small, but critical component of the wave vortex dynamics. To carry out this procedure is complicated as is also true for the corresponding geophysical fluid dynamics problem. In fact, it will be useful to see how the geostrophic limit of fluids on rapidly-rotating planets corresponds to the magnetized plasma limit where the dominant flow is

$$n_a \boldsymbol{u}_a = \frac{n_a \mathbf{E} \times \mathbf{B}}{B^2} + \frac{\mathbf{B} \times \nabla p_a}{e_a B^2}. \tag{7.1}$$

The books of Pedlosky (1987) and Gill (1982) on geophysical fluid dynamics define the geostrophic flow as follows. The basic idea is that the pressure gradient force balance is balanced by the Coriolis force

$$-\nabla p + 2\rho \boldsymbol{u} \times \boldsymbol{\Omega} = 0 \tag{7.2}$$

where p is the pressure of column, ρ mass density, and $\boldsymbol{\Omega}$ the angular velocity of the planet. Solving Eq. (7.2) for $\boldsymbol{u}$ gives the geostrophic flow velocity which is the analog of Eq. (7.1).

7.1 General Considerations: Analogs with Geophysical Fluids

In the next level of geophysical fluid dynamics where the inertial acceleration $\rho_m\, d\boldsymbol{u}/dt$ is retained for a finite value of ω/Ω is called the quasi-geostrophic fluid dynamics. This is the level that gives the all-important reduced dynamical equations of the Rossby waves and vortices. In plasma physics the corresponding flows are those arising from

$$n_a \boldsymbol{u}_a = \frac{\mathbf{B}}{e_a B^2} \times \left(m_a n_a \frac{d\boldsymbol{u}_a}{dt} + \nabla \cdot \boldsymbol{\pi}_a \right). \tag{7.3}$$

In Eq. (7.3) the accelerating forces on the right side are evaluated using the first-order flows in Eq. (7.1). That this iteration is valid is seen by comparing the two $n_a \boldsymbol{u}_a$-terms and finding their ratio is $\omega/\Omega_a = |(m_a/e_a B)d/dt| \ll 1$. In the general case the force from the divergence of the collisionless FLR momentum stress tensor has contributions that cancel the terms in

$$\frac{d\boldsymbol{u}_a}{dt} = \frac{\partial \boldsymbol{u}_a}{\partial t} + \boldsymbol{u}_a \cdot \nabla \boldsymbol{u}_a \tag{7.4}$$

that arise from the part of $\boldsymbol{u}_a \cdot \nabla \boldsymbol{u}_a$ from Eq. (7.1) due to the pressure gradient drift.

These cancellations are complicated and are described in Chapter 3 for the case of electrostatic waves in a constant **B** field. The cancellation, in a way, is a clue that the kinetic description may yield a compact, complementary method of deriving the dynamical equations. This, in fact, is found to be the case. Thus, the second major approach that is widely used is to reduce the full Vlasov or kinetic equation to its low-frequency limit in the form called the gyrokinetic equation. The gyrokinetic analysis of drift waves is developed in Chapters 10 and 11.

Now, the reader is wondering why not learn and use one description or the other? What happens is that each approach clearly reveals certain features, such as the conservation laws in the fluid description, and the drift-orbital motions and the critical conditions for marginal stability in the kinetic description, but tends to obscure other features. For this reason it seems that researchers wanting a full understanding of the low-frequency transport processes must be willing to use both descriptions. The comparisons and contrasts between the two descriptions is very revealing, and we will attempt to draw attention to these features as we develop the parallel formulations of the problem with both the kinetic and the FLR fluid mathematical descriptions.

One question may arise in the minds of those familiar with the stability analysis in fluid systems from what has already been stated: namely why is the kinetic analysis used for finding the conditions of marginal stability. Clearly, the dispersion relations derived from the fluid systems also give critical conditions for instability. Neutral fluid instability threshold formulas are found in Chandrasekhar (1961), for example, and are verified in neutral fluids experiments. The problem is that for the plasma fluid description the truncations that occur in the thermal energy or pressure equation that neglects the divergence of the flow of heat flux $q_\parallel$ along the magnetic field lines given by $\nabla_\parallel q_\parallel$. This is called the "closure problem." While for sufficiently fast modes this neglect of $\nabla_\parallel q_\parallel$ is a tolerable approximation and will even lead to reasonable approximations for the critical condition for an intrinsically fast MHD instability, the approximation will completely fail for slower instabilities. One famous example in the MHD context is explained in detail by Kulsrud (1983) is the mirror mode instability important in space plasmas. The threshold in the temperature anisotropy $T_\perp/T_\parallel$ for instability changes by a factor of six between the MHD [Chew, *et al.* (1956)] dissipationless and the Vlasov-Maxwell description. The problem is severe in general for drift waves since they are low frequency ($\omega \ll \omega_{ci}$), slow growing ($\gamma \sim v_i/L$) modes with relatively small wave lengths. Thus, while we want to know the fluid description's prediction for the stability threshold, we will generally not want to rely on this fluid-based critical condition until the corresponding kinetic theory answer is also known.

In both the fluid and kinetic theory cases the problem of determining the critical condition and the number of roots and even their values is greatly aided by knowledge of the Cauchy theorem for analytic functions. The dispersion relations are always of the form of analytic function

$$D(\omega, \boldsymbol{k}, \mu) = 0 \tag{7.5}$$

where μ denotes the parameters of the problem. In Eq. (7.5) the fluctuations of f are assumed to vary as

$$f = \delta f_{\boldsymbol{k}\omega} e^{i\boldsymbol{k}\cdot\boldsymbol{x} - i\omega t} + \delta f^*_{\boldsymbol{k}\omega} e^{-i\boldsymbol{k}\cdot\boldsymbol{x} + i\omega^* t} \tag{7.6}$$

so that the response functions, such as the polarization function,

$$P(\omega, \boldsymbol{k}, \mu) = \left\langle \left[\frac{\omega - \omega_* \left(1 + \eta(v^2 - 3/2)\right)}{\omega - \omega_D \left(v_\perp^2, v_\parallel^2\right) - k_\parallel v_\parallel} \right] J_0^2 \right\rangle \tag{7.7}$$

are made analytic by finding their analytic condition from the $\mathrm{Im}(\omega) > 0$ region of the complex ω-plane. The analytic continuation of Eq. (7.7) is made by deforming the real $v_\perp$ and $v_\parallel$-integrals into the complex plane while keeping the singular denominator away from the path of integration.

Once the analyticity of $D(\omega, \boldsymbol{k}, \mu)$ is in hand, then the Nyquist theorem is an important tool for finding the critical conditions and then for locating the roots of Eq. (7.5). The Nyquist theorem and the application to plasma waves is a topic developed in every textbook on plasma physics so we do not develop the procedure here. Since we will frequently use or quote the result, the reader may wish to review the procedure.

A typical linear spectrum of frequencies $\omega(k_x, k_y)$ for the drift waves is shown in Fig. 7.1. There is a maximum of the frequency in the $k_x - k_y$-plane shown in Fig. 7.1(a). The circle in the figure is contour of constant frequency at about one-half the maximum frequency. The units of k_x and k_y are the ion inertial scale length ρ_s. The imaginary part of the complex variable ω is defined as γ and the corresponding contours of growth and damping γ are shown in part (b) of Fig. 7.1.

The ring at $\gamma^\ell(k_x, k_y, \mu) = 0$ divides the growing modes from the damped modes. In the nonlinear mode coupling equations that follow, the growing modes feed energy into the turbulence and the damped modes absorb the energy. Here μ is a list (vector) of the system parameters. Thus, the amplitude and phases of each fluctuation component will evolve to a final turbulent spectrum. The condition for there to exists a well-defined turbulent spectrum are given in detail in Horton and Ichikawa (1996). Qualitatively, the condition is that there is sufficient damping

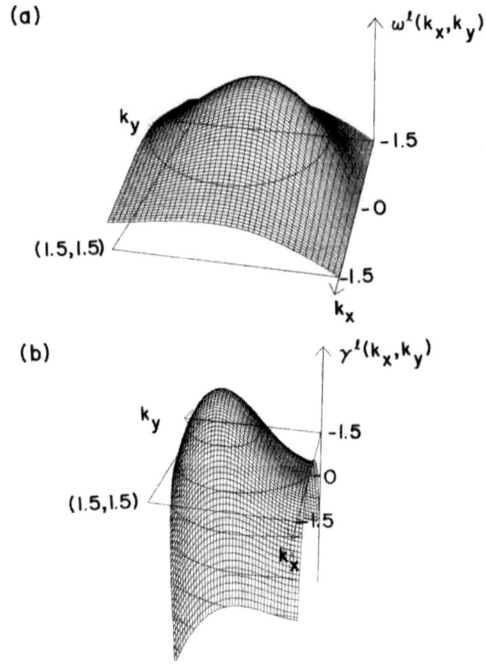

Fig. 7.1 Spectrum of typical drift wave frequencies $\omega(\mathbf{k})$ and growth-damping rate $\omega(\mathbf{k})$ for density gradient or trapped electron mode drift wave turbulence. The wavenumbers are normalized as $k\rho_s$ and the frequency normalized by $\omega L_n/c_s$. There is a well-defined maximum for the angular frequency and the growth rate at different points in the **k**-space. The lower panel (b) shows that there are a large number of strongly-damped eigenmodes that will absorb wave energy scattered by mode coupling or wave-particle interactions to high **k**-regions.

from negative $\gamma(\mathbf{k})$ modes at high wavenumbers from the small scale fluctuations. In some simulation problems this requires introducing a hyperviscosity in the vorticity equation. The energy absorbed at these smallest scales is then associated with other higher frequency modes such as lower-hybrid waves and ETG modes in Chapter 13.

In fully developed fluid turbulence, universal wavenumber spectra of the energy density $E(k)$ is of the power-law form $E(k) = \varepsilon k^{-m}$, developed from scale invariance and observed in fluid simulations, provided the stirring force is sufficient to prevent the formation of large-scale vortices. For 3D neutral fluid turbulence the Kolmogorov energy spectrum widely applies in what is called the "inertial range of k" where both $\omega_k = \gamma_k = 0$ [Frisch (1995)].

The presence of a significant number of vortex structures changes the energy spectrum as shown in Chapter 6 of Horton and Ichikawa (1996). In plasmas, however, there is no true inertial region with scale invariance due to the various dispersion scales e.g. $\rho_s = c(m_i T_e)^{1/2}/eB$ and the wave emission/absorption processes. The growth rates γ_k driving turbulence are typically peaked at $k_y\,\rho_s \approx 0.5-0.7$ and $k_x\,\rho_s \approx 0$. Thus, we investigate two drift wave models to determine the dependence of the spectral indices m_x, m_y defined by $E(k_x) = E_1\, k_x^{-m_x}$ and $E(k_y) = E_2\, k_y^{-m_y}$ on the system parameters and to examine the conditions for weak turbulence to be valid. For the weak turbulence measure we use the kurtosis $K = \langle\varphi^4\rangle/\langle\varphi^2\rangle^2$ which must be near three for homogeneous, space filling turbulence. A larger kurtosis $K \sim$ 6-10 in the long-time steady state is often found, which means that there are strong vortex structures characterized by radius r_0 and line density $n_v = N_v/L_x\, L_y$. The kurtosis is related to the packing fraction $f_p = n_v \pi\, r_0^2$ by $K \simeq 1/f_p$ where N_v is the number of well-defined vortices in the area $L_x L_y$ across the field $B\,\widehat{e}_z$.

Some aspects of drift wave turbulence related to the turbulence in two-dimensional fluid dynamics are collected in the volume edited by Moreau (1983) on the experiments, theory and simulations of 2D neutral fluid turbulence. The Terry-Horton (1982), Hasegawa-Mima (1977) and Novakovskii, *et al.* (1988) plasma works are related to the 2D neutral fluid turbulence. The relationship to coherent vortex structures is developed in Horton and Ichikawa (1996). An example of

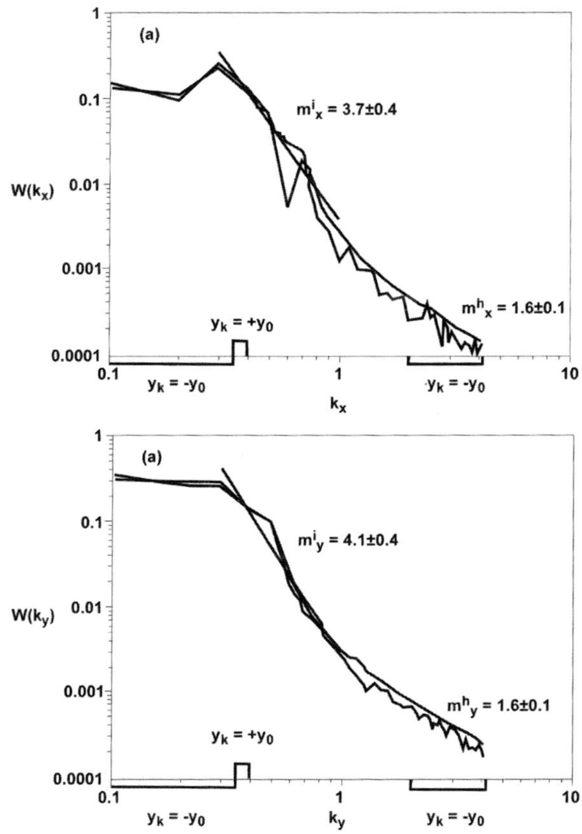

Fig. 7.2 Anisotropic wavenumber spectrum from the Novakovski model electron drift wave spectrum. In this model the isotropic rings of a growth γ_0 and damping $-\gamma_0$ are added as shown along the k-axis.

the 2D anisotropic turbulence spectrum using the Novakovskii model is shown in Fig. 7.2. The linear wave frequencies ω_k are less than the $\langle(\boldsymbol{k}_0\boldsymbol{v}_E)^2\rangle^{1/2}$ circulation frequency, and behave as a gas of vortices rather than as linear, dispersive waves. The dissipative Reynolds number $R_{en} = \widetilde{v}_E L/\nu \simeq (2)(176)/(.15) = 2.3 \times 10^3$, the kurtosis is $\langle\varphi^4\rangle / \langle\varphi^2\rangle^2 \simeq 6$ and the skewness near zero. The dissipationless $\mathbf{E}\times\mathbf{B}$ Reynolds number is $R_E \simeq 2$.

7.2 Consequences of Sheared Flows on the Drift Wave Spectrum

Sheared flows strongly influence the three wave coupling interactions by a nonlinear transverse redistribution of the fluctuation energy in k_x, k_y-space. The transverse sheared flow $v_y(x) = (dv_y/dx)(x - x_0)$ defines preferred directions. So the turbulent k-space is intrinsically anisotropic. An initial clump of fluctuation energy released at $t = 0$ from the region (K_x, K_y) will propagate in time to the region $K_x - K_y(dv_y/dx)t$ to a new region as shown in Fig. 7.3.

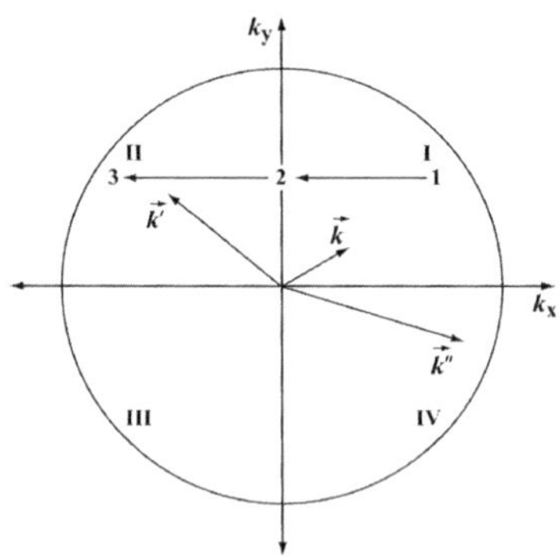

Fig. 7.3 The initial wave packet localized at high k_x at position 1 is swept by the sheared flow through the region of low k_x, region 2, and then moves to position 3. The amplitude increases sufficiently for nonlinear three-wave coupling to become strong in region 2 with $\boldsymbol{k} = \boldsymbol{k}' + \boldsymbol{k}''$ [Kim, *et al.* (1996)].

This process can allow an initial small amplitude signal to gain amplitude and under certain conditions, obtain sufficient amplitude to become a self-organized coherent structure as discussed in Chapter 9.

The initial wave packet localized at high k_x is swept by the sheared flow through the region of low-k_x where the amplitude increases sufficiently for nonlinear three-wave coupling to become strong. In some cases, as shown in Horton, *et al.* (2010), the wave packet becomes a coherent vortex structure.

7.3 Ion Temperature Gradient Induced Transport

In the presence of a strong temperature gradient the toroidal drift waves change character and propagate in the ion diamagnetic direction. Extensive simulations of

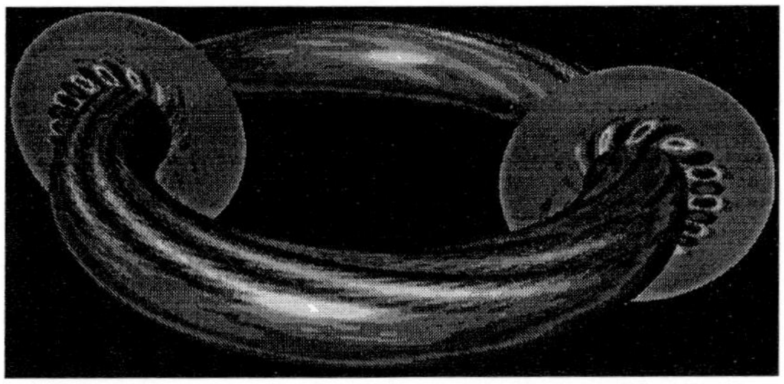

Fig. 7.4 Contours of the potential filaments aligned along the helical magnetic field lines in a simulation of the tokamak [Parker, et al. (1996)] driven by the ion temperature gradient.

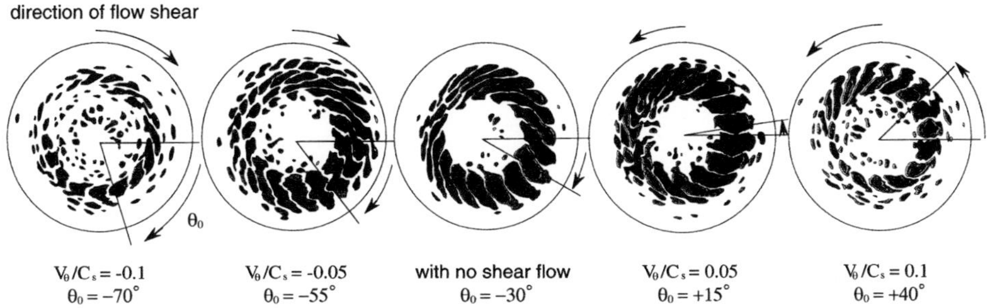

Fig. 7.5 The effect of the sheared poloidal flow $v_E(r)$ on the ITG turbulence. Contours of the toroidal drift-wave eigenvalues in the cross-sectional plane showing the radial scales for $a/\rho_s = 100$. The angle θ_0 of the Bloch-like wave function where $k_r = 0$ is clearly evident [Kim, et al. (1996)].

this toroidal ITG mode are available in the literature. Here we show in Fig. 7.4 a typical result for the vortices created by the turbulent ITG from Parker, *et al.* (1996). Further examples with a series of specific poloidal rotations v_θ is shown in Fig. 7.5 for five values of v_θ/c_s.

The cover of this book shows a state of ion temperature gradient driven turbulence in the complex 3D geometry of the Large Helical Torus (LHD) machine in NIFS in Toki, Japan.

An earlier example for the helical toroidal system (Heliotron-E) with electromagnetic ITG and MHD modes is developed in Hong, *et al.* (1991). In the top panel of Fig. 7.6 the electrostatic mode for $\beta/\beta_{\text{crit}} \to 0$ is shown in contrast to the wider mode in the bottom panel where the ion plasma pressure is high at $\beta = 0.015$ just below the critical beta for the ideal MHD instability, the complex electrostatic field and the fluctuating magnetic vector potential ψ are shown. The magnetic fluctuations are given by $\delta \mathbf{B}_\perp = (-ik_y\psi_k, \partial\psi_k/\partial x)$. The peak of the wave functions is at the rational transform $\iota\!\!\!-$ magnetic surface.

The eigenfunction ψ increases from zero at $\beta_e = 0$ to $\psi_{1,1} \sim \frac{1}{2}\,\phi_{1,1}$ at $\beta_e = 0.01$ to the MHD polarization with $\psi_{1,1}(r) \simeq \phi_{1,1}(r)$ at $\beta_e = 0.02$. The eigenfunctions

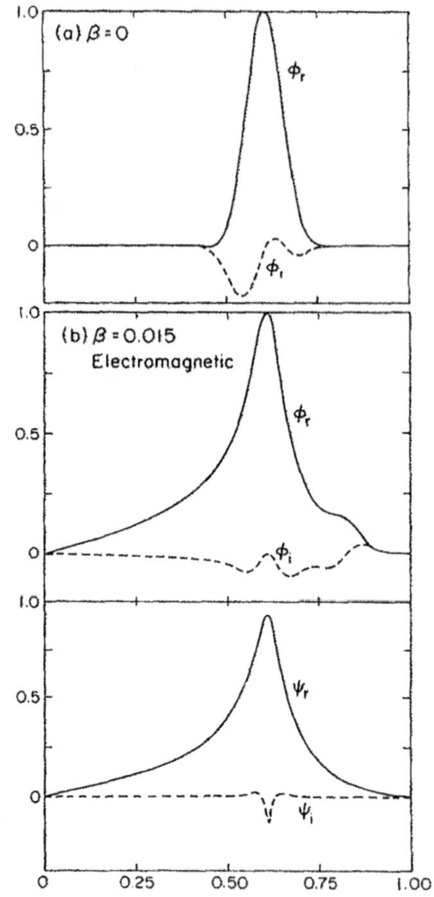

Fig. 7.6 Change in the eigenmode structure with increasing plasma pressure β in the Heliotron. (a) Electrostatic mode at $\beta = 0$. (b) Electromagnetic mode at $\beta = 0.015$ [Hong, *et al.* (1991)] in the Heliotron.

for the electrostatic limit and the transitional finite β_c value of $\beta_c = 0.015$ are shown in Fig. 7.6 for $\nu_e = 0$. As the growth rate increases with β_e, the wave functions broaden, approaching the MHD function at sufficiently high growth rates and β_e. A nonlocal transport may be computed from the perturbed fields ϕ, ψ, $\widetilde{n}$, $v_{\parallel}$, and $\widetilde{p}_i$. Here, we limit consideration for the localized resistive-g and η_i modes in the low-β_e limit. One finds that the compressibility effect on these η_i modes is not important, except near marginal stability for the finite g case, as demonstrated in Yagi, *et al.* (1989).

The small-scale turbulence from drift waves and ion acoustic waves couples to Alfvén waves and creates zonal flows on the large scale. These fluctuations propagate in the ion diamagnetic direction [Mikhailovskii, *et al.* (2006)]. The zonal flows merge with flows driven by the neutral beam heating and other sources of hydrodynamic flows in the plasma. The flows are also generated by the geodesic mode interactions [Zhang and Lin (2010), Kendl and Scott (2005)].

7.4 Drift-Wave Turbulence in Edge Plasma

The three-wave coupling is a classic problem in drift wave turbulence [Horton and Rowan (1994), Horton and Ichikawa (1996)]. The source and the sink regions of the turbulence are shown in Fig. 7.2 (the source region) and decay in the inertial range $k > 1$ is shown in Fig. 7.2. Parameterizing the numerically computed one-dimensional projections of the two-dimensional $\boldsymbol{k}$-spectra gives

$$\begin{aligned} W^{\text{sim}}(k_x) &= \sum_{k_y} W(k_x, k_y) \leftrightarrow \frac{1}{|k_x|^{m_x}} \\ W^{\text{sim}}(k_y) &= \sum_{k_x} W(k_x, k_y) \leftrightarrow \frac{1}{|k_y|^{m_y}}. \end{aligned} \tag{7.8}$$

Regression fits of Eq. (7.8) to the simulation turbulence data determine the values of $m_x \pm \delta m_x$ and $m_y \pm \delta m_y$ given in Fig. 7.2. More spectral experiment data is given in Terry (2000).

Exact solutions of three-wave ITG gyrofluid equations are used to compare with the exact kinetic theory three-wave solutions to attempt to verify closure models used in the fluid description. The system is reduced to set of three ODEs that have a structure similar to that of the famous Lorenz attractor problem in fluid dynamics. However, in the plasma case there is the "pump depletion" energy conserving behavior giving rise to periodic oscillations of the energy between the drift waves and the background profile gradient and the $k_y = 0$ mass flows. The reduced slab drift wave model driven by ∇T_e has been analyzed by Nakata, *et al.* (2010, 2016) using these methods. These interactions are described further in Chapter 13.

The simulations of the type shown in Fig. 7.2 give a typical form of the drift wave spectrum showing the change in the spectral indices m_x, m_y in the high-(m^h) and low-(m^l) wavenumber regimes as expected from the scaling theory. The exponents change as shown in Fig. 7.2 from $k\rho_s \gtrless 1$. In this simple example one finds that

$$\begin{aligned} m_x^h &= 1.6 \pm 0.1, \quad m_y^h = 1.6 \pm 0.1, \quad \text{for} \quad k\rho_s > 1 \\ m_x^l &= 3.7 \pm 0.4, \quad m_y^l = 4.1 \pm 0.4, \quad \text{for} \quad k\rho_s < 1. \end{aligned} \tag{7.9}$$

In the steady-state turbulence shown in Fig. 7.2, the total turbulent energy density and enstrophy are $W = 0.8$ and $U = 0.3$ in units of $(\rho/L_n)^2 n_0 T_e$. The space-time-averaged kurtosis and skewness of $\phi(\mathbf{x}, t)$ are 2.9 and -0.1, respectively. The system appears to be within the state of weak turbulence for these small values of γ_0.

Related simulations of drift wave turbulence driven by the parallel velocity ($\gamma_{\max} \simeq |du_{\|}/dx|$) shear flow in the ion-acoustic wave equation gives the spectra indices $m_x = 2$ for $0.2 < k_x\rho < 2$ and $m_x \cong 3$ for $2 < k_x\rho < 7$ [Horton and Rowan

(1994), Dong and Horton (1993)]. Huld, *et al.* (1991) show vortices and vortex transport produced by the KH instability in the edge of low-temperature density plasma.

Even in the isotropic, 2-D Euler limit of the Navier-Stokes equation, the question of the value of spectral indices remains an active area of research. High resolution simulations of the forced-damped 2-D Euler equation persistently show energy spectral decays faster than the rate $k^{-3}[\ln(k/k_1)]^{-1/3}$ for $k \gg k_1$ for enstropy transfer in space filling, isotropic, homogeneous 2-D turbulence theory [Kraichnan (1971)]. The reason for the larger spectra decay exponents has been traced to the emergence of long-lived, coherent vortex structures [McWilliams (1984), Legras, *et al.* (1988)]. The high resolution simulations of Legras, *et al.* (1988) show clearly the correlation of larger decay indices ($m \simeq 4$) with the appearance of monopole, dipole, and tripolar vortices. Legras, *et al.* (1988) compare the spectral indices and the associated vortex structures for $(512)^2$ simulations obtained with the three types of drivers (i) constant, single mode $\boldsymbol{k} = (k_l, 0)$, forcing; (ii) stochastic, narrow-band external forcing $\widetilde{F}(k_1 < |\boldsymbol{k}| < k_2)$ taken with news random phases at each time step; and (iii) narrow-band linear growth rate. A large-scale damping rate $-\nu_L \phi(\boldsymbol{k})$ and a hyperviscosity $-\nu_s (k_\perp^2)^p \phi(\boldsymbol{k})$ are used to produce the steady state. The resulting turbulence is characterized by the spectral index m and its vortex content as follows: (i) $m = 3.6$ and a low-packing fraction of relatively large vortices; (ii) $m = 3.5$ and a denser packing of smaller vortices; and (iii) $m = 4.2$ and the prominence emergence of several dipolar and one tripolar vortex. Thus, it appears that for 2-D turbulence the space filling, inverse cascade spectrum of $k^{-3}[\ln(k/k_l)]^{-1/3}$ is an idealization that is not achieved in practice due to the strong, self-organization of the coherent structures, described in Chapter 9, suppressing the fine-scale turbulent flows. A demonstration of the suppression of turbulent cascades by coherent vortices in 2-D turbulence is given by McWilliams (1990a,b).

Owing to the large ion orbits, excessive electrons can accumulate at tokamak edge. The non-neutralized electrons at tokamak edge contributes an electric compressive stress in the direction parallel to magnetic field by their mutual repulsive force [Zheng, *et al.* (2010, 2016)]. By extending the Chew-Goldberger-Low theory, it is shown that this newly recognized compressive stress can significantly change the plasma average magnetic well, so that a stabilization of magnetohydrodynamic modes in the pedestal can result. This linear stability regime helps to explain why in certain parameter regimes the tokamak high confinement can be rather quiet as observed experimentally.

7.5 Inertial Spectral Ranges in 2D and 3D Turbulence

Drift wave simulations have been carried out in models where the linear growth and damping is suppressed to study the transfer of energy between modes $\boldsymbol{k}$ from the nonlinear interactions. Such regions of $\boldsymbol{k}$ space are called inertial regions stemming from the origin of the nonlinear transfers in the dissipationless neutral fluid equations coming from the inertial terms $\rho \boldsymbol{v} \cdot \nabla \boldsymbol{v}$ alone. Kolmogorov used physical reasoning on the nature of interacting eddies and mathematical invariance scaling arguments to show that for 3D fluid turbulence the spectral energy density $\epsilon(\boldsymbol{k})$ must scale as

$$\epsilon(\boldsymbol{k}) = Ck^{-5/3} \ln\left(\frac{k}{k_0}\right). \tag{7.10}$$

The derivation of this spectrum is straight forward and given in text on fluid turbulence [Frisch (1995), Lesieur (2008)].

The situation is less straightforward in 2D turbulence owing to the existence of many invariants of the flows in addition to energy in the region of no growth or damping of the fluctuations. Kraichnan (1971) gave a set of definitive arguments to show that there are two inertial ranges in 2D neutral fluid turbulence. The first range is the energy transfer range also of the form of the Kolmogorov spectrum in Eq. (7.10). The second region is one in which there is a cascade of enstrophy rather than energy. The energy has an inverse cascade in 2D turbulence transferring energy from high-k to low-k. A striking example is in the ETG turbulence reported in Fu, *et al.* (2012). This is the range of greatest relevance to 2D fluid turbulence and continues down to the smallest scales where the viscous dissipation finally absorbs the vorticity of the fluctuations. The 2D enstrophy cascade given by Kraichnan (1971) is

$$\epsilon(\boldsymbol{k}) = C\eta^{2/3}k^{-3}\left[\ln\left(\frac{k}{k_1}\right)\right]^{-1/3}. \tag{7.11}$$

where η is the rate of enstropy transfer per unit mass. The wavenumber k_1 marks the bottom of the range where enstropy is pumped into the turbulence. Transfer in the energy and enstrophy inertial ranges is computed according to an almost-Markovian Galilean-invariant turbulence model. Transfer in the two-dimensional energy inertial range is found to be much less local than in three dimensions, with 60% of the transfer coming from wavenumber triads, where the smallest wavenumber is less than one-fifth the middle wavenumber.

Batchelor (1969), Leith (1968), and Kraichnan (1967, 1971) show that two-dimensional neutral fluid turbulence can exhibit two kinds of inertial range: an energy transfer range of the form

$$E(k) = C_E k^{-3} \tag{7.12}$$

and an enstrophy transfer range as in Eq. (7.11)

$$E(k) = C'k^{-5} \tag{7.13}$$

in isotropic 2D systems. In these equations, $E(k)$ is the isotropic energy spectrum, normalized so that $\int E(k)dk$ is the average kinetic energy per unit mass. The inviscid constants of motion in two-dimensional flow makes the universality of C and C_E, more questionable than for the inertial range in three dimensions. If Eq. (7.12) exists, the energy transfer is from higher to lower wavenumbers, while in Eq. (7.13) the enstrophy transfer is from lower to higher wavenumbers. There is zero enstrophy transfer in the lower wavenumber regime and zero energy transfer in the higher wavenumber range.

Worked Problem 7.1: Inertial Turbulence Spectra

Problem:

Derive the energy spectrum $E(k)$ for (1) 3-D isotropic turbulence in the inertial range $k_0 < k < k_d$ from the constant transfer of a turbulent energy flux $F(k)$ from the large-scale source $k_0 = 1/L_0$ to the small-scale dissipation $k_d = (\varepsilon/\nu^3)^{1/4}$ where ε is the turbulent energy per unit of mass and ν is the viscosity. (2) For 2-D isotropic turbulence, apply the same arguments to the flux of η conserved vorticity to derive the 2-D high-k spectrum.

Solution:

(1) The nonlinear convection rate at scale k is $\nu(k) = kv(k) = k(kE(k))^{1/2}$ (which is also known as the eddy-turnover-rate and the eddy viscosity rate). The flux of energy (per unit mass) is then

$$F(k) = \nu(k)v^2(k) = k^{5/2}\, E^{3/2}(k). \tag{7.14}$$

Taking the energy flux $F(k) =$ constant one finds that

$$E(k) = \frac{F^{2/3}}{k^{5/3}} \qquad \text{for} \qquad k < k_\nu = \left(\frac{\varepsilon}{\nu^3}\right)^{1/4}. \tag{7.15}$$

For $k > k_d$ the viscous dissipation of energy gives $E(k) = (\varepsilon/\nu^3)^2\, k^{-7}$, and the Heisenberg interpolation formula is

$$E(k) = E_0 \left(\frac{k_0}{k}\right)^{5/3} \left(1 + \left(\frac{k}{k_\nu}\right)^{8/3}\right)^{-2}. \tag{7.16}$$

(2) For 2-D turbulence the mean-squared-vorticity is conserved so that the vorticity flux

$$k(kE(k))^{1/2} k^2 E(k) k = \eta \tag{7.17}$$

is conserved in the nonlinear transfer to $k = k_D$. This leads to the steeper spectral exponent

$$E(k) = \frac{\eta^{2/3}}{k^3} \qquad \text{for} \qquad k_1 < k < k_\nu \tag{7.18}$$

where k_1 is the region of creation of vorticity. For $k_0 < k < k_1$ there is an inverse cascade of energy that is given by Eq. (7.15).

The subject of spectral distribution in turbulence is a rich field of research developed in numerous large works including Monin and Yaglom (1965), Kraichnan (1964), Krommes (1999, 2002), Leslie (1973), Stanisic (1985), and Frisch (1995).

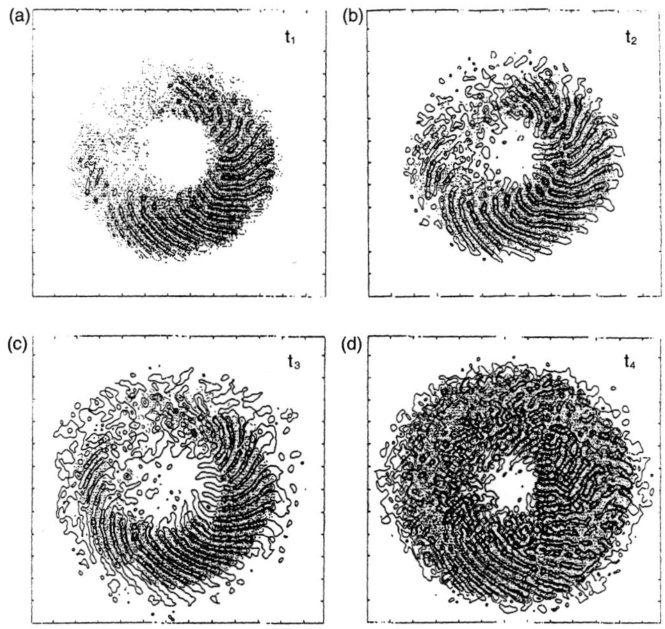

Fig. 7.7 Evolution of the isopotential contours for the ITG turbulence for $a/\rho_i = 28$, $L_T/R = 0.2$, $T_i = T_e$ with hyperbolic tangent profiles for the temperature. The density profile is flat [Sydora, *et al.* (1996)].

Experiments with various types of electromagnetic wave scattering diagnostics have reported the shape of the turbulent spectrum $E(k_\perp)$ for a variety of plasmas. Chapter 2 of the 2012 edition reviews these plasma experiments. A turbulence spectrum with two slopes, of the type described in Eq. (7.8), is reported for Tore Supra drift wave turbulence by Hennequin, *et al.* (2004).

In magnetized plasma, the turbulence is called quasi-2D turbulence. A review article [Horton and Choi (1979)] develops the drift wave turbulence spectral problem in some detail. In plasma confinement systems the turbulence is anisotropic, as shown in Fig. 7.7 for the turbulence by Sydora, *et al.* (1996) for ∇T_i driven drift wave turbulence. Coherent structures develop as shown by the spiraling radial fingers in the potential contours in Fig. 7.7. The temperature-to-density gradient parameter defined by $\eta_i = d\ell n\, T_i / d\ell n\, n_i$ is infinite in the Sydora simulation producing a strong turbulence. Large values of η_i are relevant to H-mode plasma confinement regimes which have flat density profiles in high ion temperature plasmas. Next we consider the role of a sheared plasma rotation.

The drift wave turbulence generates a sheared background flow velocity $v_E(x) = d\Phi(x)/B\, dx = d\Phi/d\Psi$ where Ψ is the magnetic flux function, and $\Phi(\Psi, t)$ is the electrostatic potential. The radial electric field $E_r(r,t) = -d\Phi/dx$, which is created by numerous processes, including the turbulence itself through the Reynolds' stress tensor.

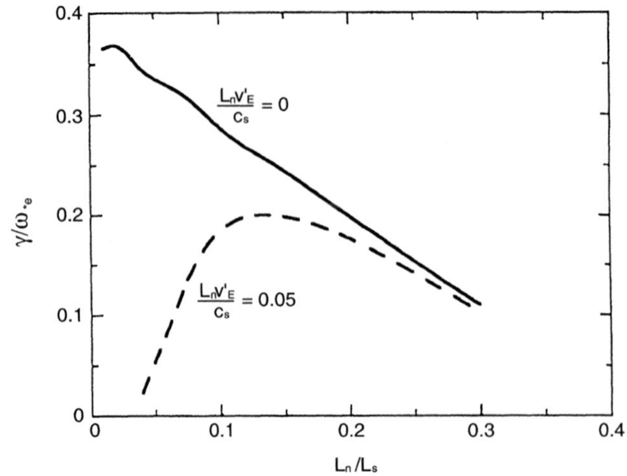

Fig. 7.8 The gyrokinetic growth rate as a function of magnetic shear $S = L_n/L_s$ with and without $\mathbf{E} \times \mathbf{B}$ shear flow [Hamaguchi and Horton (1992)]. Some mode growth rates are reduced by $\mathbf{E} \times \mathbf{B}$ shear and others are not.

Figure 7.8 shows how the flow shear dv_E/dx strongly changes the linear growth rate γ of the drift waves. In the high growth rate (solid) curve there is no shear flow in the system, whereas in the lower growth rate (dashed) curve there is a constant shear flow given by $dv_E/dx = 0.05(c_s/L_n)$. The shear flow combines with the magnetic shear S (on the x-axis) to shift the waves off the mode rational surface creating a $\langle k_{\|} \rangle \neq 0$ which breaks the symmetry of turbulence and produces a Reynolds stress transporting plasma momentum $\pi = \rho \langle v_x v_y \rangle$ across the magnetic flux surfaces as observed in a variety of laboratory plasmas.

Chapter 8

Impurity Transport Studies

This chapter analyzes the transport of impurity ions of charge Z and mass m_z in magnetized plasma. In small concentration $\alpha_z = n_z Z^2/n_e \to 0$ the impurities respond as test particles in the drift wave turbulence. Often, however, the concentration of the low-Z impurities is sufficient to make the impurities an active component in the lowest frequency part of the drift wave spectrum.

Turbulent particle transport experiments in tokamaks and stellarators have given detailed data on the transport of low-Z impurities including carbon, boron, beryllium, lithium, and helium. In this chapter we introduce partial differential equations (PDEs) for the multi-component plasmas. The PDEs are solved for the fluctuation fields as vectors $\boldsymbol{X_k}$ composed of the electric potential φ, hydrogenic density δn_i and the impurity densities δn_z. A typical set of impurities are the ions from the material coating the plasma facing metal walls, which would be beryllium with $Z = 4$ for JET and boron with $Z = 5$ for Alcator from the walls and the RF antennas along with metallic ions from chamber walls and divertors.

Turbulent particle transport is investigated with quasilinear theory that is motivated by the extensive beryllium transport experiments in ASDEX-U. Eigenvalue problems for sets of reduced fluid equations for multi-component plasmas are solved for the self-consistent fluctuating field vectors composed of the electric potential φ, the main ion density δn_i, the impurity density δn_z and the ion temperature fluctuation δT_i. We explore two drift wave models: (1) the density-gradient-driven impurity drift wave and (2) the ion-temperature-gradient-driven ion temperature gradient (ITG) mode. Analytic and numerical results for particle transport coefficients are derived and compared with the transport data and with the neoclassical transport. We explore the ability of the models to explain impurity density profiles in three confinement regimes: H-mode, I-mode and internal transport barrier (ITB) regime in impurity transport experiments for the high-Z metallic elements in the Large Helical Device and in JET. An example of the rapid ionization and inward transport flux of metallic ions, in this case, cobalt, is shown in Fig. 8.1.

In Sec. 8.9 the inward transport of tungsten ions is analyzed for H-modes in ASDEX-U and JET where the divertor chamber is changed to model that planned

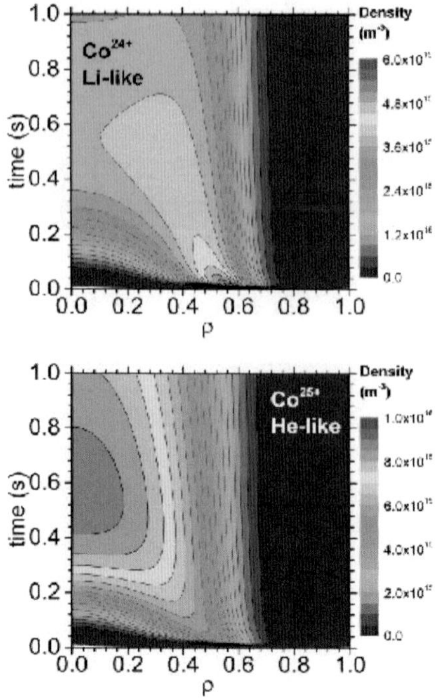

Fig. 8.1 Data showing the accumulation from inward transport of trace amount of high-Z ions from injected cobalt atoms placed in a small pellet from the impurity experiments on the Large Helical Device. (a) The lithium-like ($Z = +3$) and helium-like ($Z = +2$) ionization states of the injected cobalt atoms appear in the core plasma with time versus radius axes [Sudo, *et al.* (2013)]. (b) There is a lower concentration of highly-ionized cobalt ions in the core. Cobalt atoms are chemical neighbors to the Fe and Ni atoms of the steel vessel walls.

for ITER. Section 9.11 concludes Chapter 8 with the analysis transport of plasma dust from the erosion of the confinement vessel walls surfaces.

8.1 Drift Waves with Impurity Ions

Linear eigenmode analysis shows that there are the usual drift waves with hydrogen, and there are impurity drift waves supported by the impurity density gradients. The impurity drift waves add new degrees of freedom to the plasma allowing the hydrogen flux to be inward or outward. In these impurity drift wave modes the working gas ions behave quasi-adiabatically since the parallel phase velocity is now lower than in the hydrogen drift waves. The interaction between the hydrogen and the impurity drift waves makes the transport matrix large. Examples are given for the impurity particle fluxes in the L-, H-, ITB- and I-mode confinement regimes in this chapter.

The probability density for low-concentration tracer of impurity ions in a spectrum of hydrogenic drift waves is found to have non-Gaussian or non-normal spatial distributions owing to the large-scale convection cells in the turbulence. An example for the tracer-charged particle distribution in a spectrum computed from the resistive-g turbulence is given by Del-Castillo-Negrete (2000) and for drift waves in Dubuit, *et al.* (2007), and in Fu, *et al.* (2013). The authors show how to analyze the long-tailed PDFs of the test particles with fractional derivatives. The authors show that such non-diffusive transport is common for tracer particles. Thus, for trace amounts of impurities in a hydrogenic plasma, one may expect to see faster particle transport than given by a diffusion coefficient. Thus, the description of the particle data requires both a diffusivity D and a convection velocity V [Horton and Rowan (1994)]. Neoclassical theory gives a base level of the impurity D and V from Coulomb collisions in the absence of waves.

In the limit of a small concentration of impurities these charged molecules or ions are passively convected by the drift waves. Figure 8.2 shows a snapshot of the evolution of the impurity concentration for passive convection by the $\mathbf{E} \times \mathbf{B}$

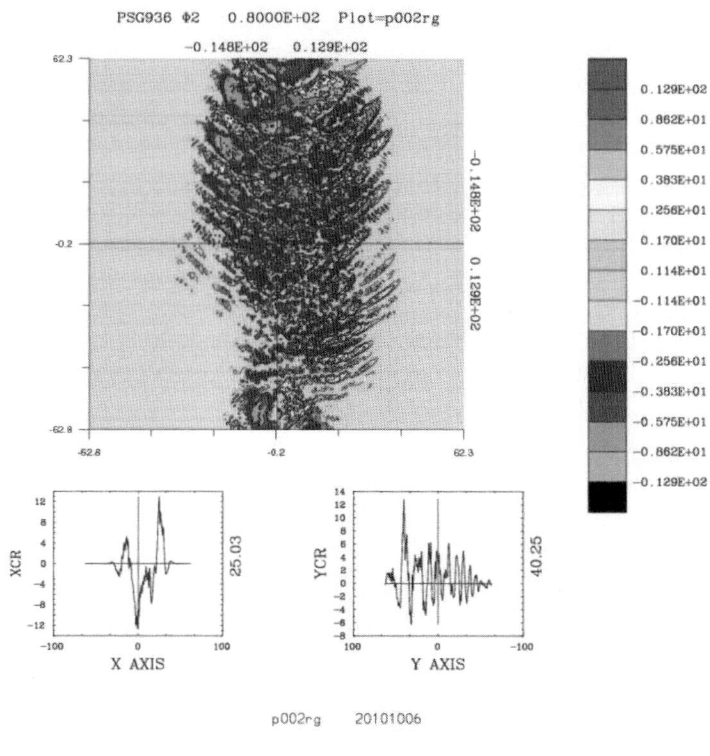

Fig. 8.2 Dispersion and diffusion of passively-convected impurities injected as a localized Gaussian puff at $t = 0$ in plasma in a steady turbulent state of drift wave turbulence. For the passive convection all impurities behave the same from the $\mathbf{E} \times \mathbf{B}$ motion provided $k_\perp \rho_s \ll 1$. The motion of the center-of-mass of the injected cloud is down the density gradient of the background plasmas however there is large dispersion with long fingers of impurity going toward the high density core plasma on the left. The lower two panels are cuts through the maximum of δn_z at $x = 25, y = 40$ showing the corregation in y and fingers in x.

motion in a drift wave turbulence. The waves produced a faster diffusive transport in the direction of the wave propagation, often called the zonal directions in the terminology taken from geophysical fluid dynamics, and a slower diffusion in the direction of the density gradient called the meridional direction in geophysical fluid dynamics. This transport often deviates some from the normal Gaussian diffusion of the particles as revealed by constructing the probability distributions of the particles at different stages of the evolution.

This passive convection is a famous and difficult problem in the dispersal of pollutants in the atmospheric flows and thus has a rich literature. No attempt is made to review this area, but we note that the early work of Taylor gives an important formula for the average of the particle displacement in homogeneous turbulent flows as

$$R^2(t) = 2\langle V_E^2 \rangle \int_0^t d\tau (t - \tau) C_L(\tau) \tag{8.1}$$

where $C_L(\tau)$ is the Lagrangian autocorrelation function for the $\mathbf{E} \times \mathbf{B}$ velocity of the particle. The function $C_L(\tau)$ generically goes as τ^2 for τ smaller than τ_c the correlation time of the turbulence and linearly as $C_L(\tau) \sim \tau$ for times long compared with the correlation time of the turbulence. This leads to diffusion at the rate

$$D = \langle V_E^2 \rangle \tau_L \tag{8.2}$$

for times long compare with the Lagrangian autocorrelations time τ_L.

The impurity transport problem has been investigated for many different models of drift wave turbulence. In work by Basu, *et al.* (2003) and Hasegawa and Wakatani (1983), the collisional turbulence model is used with a constant $C_\nu = \left\langle k_\parallel^2 v_e^2 \right\rangle / \langle \omega_{*e} \nu_e \rangle$. The collisionality parameter C_ν varies between 0.1 for the strongly collisional regime to the value of $C_\nu = 5$ for the weakly collisional regime.

The transport across the magnetic field is characterized as diffusivity while the transport in the direction of the wave propagation which is the symmetry direction in the equilibrium (zonal direction in geophysics for example) is super diffusivity meaning the variance of the spread of the particles in the direction of wave propagation is faster that $2D_y t$. In the example from Basu, *et al.* (2003) the waves produce a super-diffusion transport with $\langle Y^2 \rangle = D_{\text{sup}} t^{1.4}$. Figure 8.2 shows the rapid dispersal of the impurities in the direction of the wave propagation in the y-direction. When the motion is examined in detail we see that the impurities are trapped and detrapped in the wave potentials and carried as trapped particles with the wave propagation sufficiently often and far to produce a super-diffusion transport.

Turbulence driven by the ion temperature gradient, with mass shear flow parallel to the magnetic field, and the impurity ion density gradient is studied in a sheared slab magnetic configuration. The particle diffusion induced by the turbulence is obtained within the framework of quasilinear fluid theory and simulations. The effective diffusion coefficients that include the pinch effects show correlations with tokamak experimental observations such as an isotope scaling of plasma confinement time described in Casson, *et al.* (2015) for JET and ASDEX-U plasmas.

8.2 Drift Wave Eigenmodes with Active Impurity Components

Turbulent transport from drift waves is a key problem for fusion physics across all magnetic confinement geometries. To analyze plasmas with impurities three-component fluid equations are needed to find the eigenmodes and eigenfrequencies of a nonuniform, magnetized plasma with a four dimensional fluctuation vector $\boldsymbol{X}_k$ composed of fluctuations of the electron density, the working gas ion density, the impurity density and the electrostatic plasma potential. This structure of the eigenmodes and eigenvectors have two collisionality regimes: (i) the collisional drift waves appropriate for the scrape-off layer (SOL) and the edge plasma in limiter discharges and (ii) the trapped electron modes in the core plasma. From the eigenmodes and eigenvectors, the out-of-phase part of the density to potential fluctuations are derived which determine the quasilinear particle fluxes as a function of the power spectrum of the plasma potential fluctuations and the gradient parameters characterizing the L-, H-, I and ITB-confinement modes.

First we look at data from C-Mod with boronized walls and with boron injection to obtain a perspective on the wide range of plasma gradients and confinement regimes. Figure 8.3 shows the fully-ionized boron density profile evolution in a high-confinement C-Mod discharge that develops an internal transport barrier. Multiple types of drift wave modes exist in such plasmas. In the nonlinear state the modes are coupled in a complex manner. For weakly turbulent states the quasilinear particle fluxes for hydrogenic and impurity gases are calculated from the out-of-phase φ and δn_i and δn_z fluctuations. The qualitative changes of the particle fluxes with variation of the sign and strength of the density gradients are compared with C-Mod plasmas in the L-, H-, ITB- and I-mode regimes. The I-mode has its own low-frequency instability under conditions described in Coppi and Zhou (2012).

Following boronization, tokamak discharges in DIII-D have been obtained with confinement times up to a factor of 3.5 above the ITER89-P L-mode scaling and 1.8 times greater than the DIII-D/JET H-mode scaling relation. Very high confinement phases are characterized by relatively high central density with $n_e(0) \approx 1\times10^{20}\text{m}^{-3}$, and central ion temperatures up to 13 KeV at moderate plasma currents (1.6 MA) and heating powers (12.5-15.3 MW). These discharges exhibit a low fraction of radiated power, $P \leq 25\%$, with a $Z_{\text{eff}}(0)$ close to unity, and with lower impurity influxes than comparable DIII-D discharges before boronization of the walls.

Nonlinear simulations show a modest change in the amplitude of the turbulence for a change from boron to argon in the fully-ionized hydrogenic plasma. The change of plasma confinement regimes arises from the balancing of the hydrogen and impurity particle fluxes. The phase shifts in the fluctuations change the hydrogen-to-impurity flux ratios.

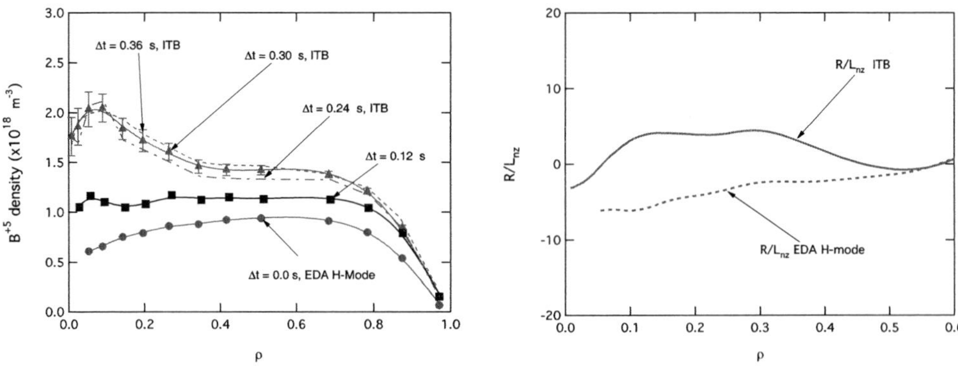

(a) (b)

Fig. 8.3 (a) Time evolution of boron density profile in the C-Mod discharge with 3.5 MW ICRF heating, from the H-mode (circles), transition (squares), to ITB (triangles in the error bars); (b) boron density gradient profile in H-mode (dash) and in the ITB (solid) at time $t = 0.30\,\mathrm{s}$ [Rowan, *et al.* (2008)].

From gyrokinetic equations one derives the Vlasov-Poisson dispersion relation with the impurities. There are three types of drift-wave instabilities that produce turbulent particle and impurity fluxes:

(1) the drift wave with hydrodynamic ion response giving ITG modes with quasi-adiabatic electrons,

(2) the trapped electron-ITG mode (TEM-ITG) that has trapped electron resonances and FLR-hydrodynamic ions and impurities, and

(3) the impurity drift wave where the impurity has FLR-hydrodynamic response and the hydrogenic working gas has adiabatic response. The impurity mode is strongly unstable when the hydrogenic ions and the impurity ions have reversed (opposite direction) radial gradients and the turbulent flux then transports each ion type down its respective density gradient leading to a thorough mixing of the two ion types. Simulations show turbulence growth from initial plasma states with $\delta n_e = \varphi = 0$ with only a tiny injection of the impurity.

Onset of the impurity mode requires that the impurity density contribution to effective charge $Z_{\rm eff}$ be of order unity. The impurity contribution is specified by the value of $\alpha_z = n_z Z^2/n_e$. For a small impurity contribution $\alpha_z \ll 1$ to the charge $Z_{\rm eff}$ the impurity mode is stable. Owing to the numerous system parameters, the exact stability condition is complicated and must be determined numerically [Futatani, *et al.* (2010)].

8.2.1 *Eigenmodes and the quasilinear fluxes with impurities*

The eigenvectors of the impurity drift wave matrix give the polarization of the fluctuations that determine the relative strength and direction of the impurity and hydrogenic ion transport.

For example, the trapped electron drift eigenvalue problem with impurity ions δn_z is given by the matrix

$$\begin{bmatrix} 1 & Z & -1 & -f_p \\ -i\omega & 0 & 0 & i\omega_{*i} - i\left(\omega + i\mu_i k_\perp^2\right) k_\perp^2 \rho_{s,i}^2 \\ 0 & -i\omega & 0 & i\omega_{*z} - i\left(\omega + i\mu_z k_\perp^2\right) k_\perp^2 \rho_{s,z}^2 \\ 0 & 0 & -i\omega + i\omega_{De} + \nu_{\text{eff}} & i\omega_{*e} f_t + i\omega_{De} f_t \end{bmatrix} \begin{bmatrix} \dfrac{\delta n_i}{n_e} \\ \dfrac{\delta n_z}{n_e} \\ \dfrac{\delta n_e}{n_e} \\ \dfrac{e\phi}{T_e} \end{bmatrix} = \begin{bmatrix} 0 \\ 0 \\ 0 \\ 0 \end{bmatrix} \tag{8.3}$$

where f_p is the passing and f_t the trapped fractions of the electrons. The matrix in Eq. (8.3) is of the form $\omega \mathbf{B}_{\boldsymbol{k}} \boldsymbol{X}_k = \boldsymbol{A} \boldsymbol{X}_{\boldsymbol{k}}$ where ω is the complex eigenvalue.

Due to the complex dependence of the dynamics on the fluctuation frequency ω the linear equations of matrix by the $\boldsymbol{A}$ and $\mathbf{B}$ matrices in Eq. (8.3) lack the symmetries of the $\boldsymbol{A}$ and $\mathbf{B}$ matrices in the standard orthogonal form of Hamiltonian systems. Thus we introduce the left $\boldsymbol{y}_i^* = \langle i \mid$ and the right $\boldsymbol{x}_j = \mid j\rangle$ eigenvectors of the auxiliary matrix $\boldsymbol{M} = \mathbf{B}^{-1}\boldsymbol{A}$.

The left eigenvectors are taken from

$$\boldsymbol{y}_i^* \boldsymbol{M} = \lambda_i \boldsymbol{y}_i^* \tag{8.4}$$

and the right eigenvectors are taken from

$$\boldsymbol{M} \boldsymbol{x}_j = \lambda_j \boldsymbol{x}_j. \tag{8.5}$$

Then post-multiply Eq. (8.4) by $\boldsymbol{x}_j$ and subtracting from pre-multiply Eq. (8.5) by $\boldsymbol{y}_i^*$ leads to

$$\left(\lambda_j - \lambda_i\right) \boldsymbol{y}_i^* \boldsymbol{x}_j = 0 \tag{8.6}$$

from which we derived that $\langle i \mid j\rangle = \boldsymbol{y}_i^* \boldsymbol{x}_j = 0$ for nondegenerate $\lambda_i \neq \lambda_j$ eigenvalues. Now we can normalize $\langle i \mid i\rangle = 1$ and proceed as in quantum mechanical perturbation theory.

To get the eigenvectors X_α for each root ω_α of the dispersion relation $D(\omega, k_x, k_y, k_\parallel, \mu) = \det(\boldsymbol{M} - \lambda \boldsymbol{I}) = 0$ where the parameters of the system are listed in the vector μ. We factor out the value of the X_4 component $(e\phi_k/T_e)$ and write the eigenvector as the independent $\boldsymbol{X} = (e\phi_k/T_e)\widehat{\boldsymbol{X}}$. Then the components of the drift wave polarization vector $\widehat{\boldsymbol{X}}$ are

$$\widehat{\boldsymbol{X}} = \begin{bmatrix} \dfrac{\delta n_i}{n_e} / \dfrac{e\phi}{T_e} \\ \dfrac{\delta n_z}{n_e} / \dfrac{e\phi}{T_e} \\ \dfrac{\delta n_e}{n_e} / \dfrac{e\phi}{T_e} \\ 1 \end{bmatrix}. \tag{8.7}$$

Physically, the $\widehat{\boldsymbol{X}}$ vector describes the "polarization" of density fluctuations relative to the electrostatic potential fluctuation. Examples of the polarization diagram is shown in Fig. 8.4 for two eigenmodes.

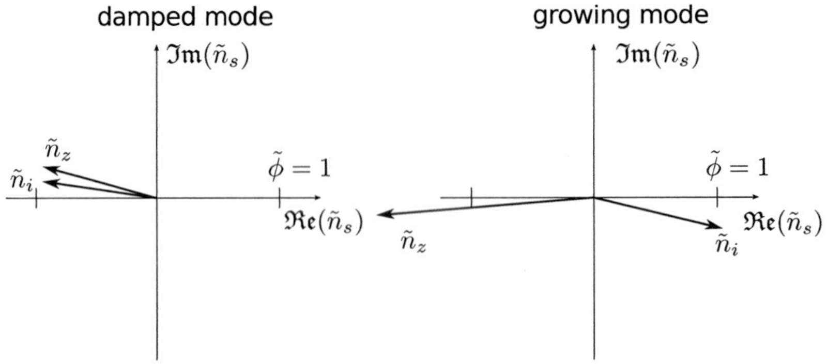

Fig. 8.4 Polarization diagram of eigenmodes of the usual drift wave with impurities.

Now the particle fluxes are given by the polarization vector as

$$\begin{aligned}
\Gamma_i &= \mathrm{Re}\sum_k \frac{ik_y\phi_k^*}{B}\delta n_i = -n_e\frac{T_e}{eB}\sum_k \frac{k_y|e\phi_k|^2}{T_e^2}\mathrm{Im}\widehat{X}_1(k) \\
\Gamma_z &= \mathrm{Re}\sum_k \frac{ik_y\phi_k^*}{B}\delta n_z = -n_e\frac{T_e}{eB}\sum_k \frac{k_y|e\phi_k|^2}{T_e^2}\mathrm{Im}\widehat{X}_2(k) \\
\Gamma_e &= \mathrm{Re}\sum_k \frac{ik_y\phi_k^*}{B}\delta n_e = -n_e\frac{T_e}{eB}\sum_k \frac{k_y|e\phi_k|^2}{T_e^2}\mathrm{Im}\widehat{X}_3(k).
\end{aligned} \tag{8.8}$$

Analytically we know that the formula in the first row of Eq. (8.3) gives that $\Gamma_e = \Gamma_i + Z\Gamma_z$. Figure 8.5 shows an example of the boron impurity flux in a deuterium plasma and the deuterium flux computed for Alcator C-Mod data as a function of the boron impurity gradient R/L_{nz}.

The quasilinear formulas in Eq. (8.8) are valid when there are overlapping resonances in the Hamiltonian motion of the particles. The conditions are given in Horton and Ichikawa (1996). In the nonlinear state the fastest growing modes couple to the damped eigenmodes driving them up to the level required for the power flow from the unstable waves to balance the damping from the stable modes.

The nonlinear waves will saturate when the finite k_x part of the spectrum grows. The rms values of the density fluctuation is then

$$\left(\frac{\delta n_i}{n_i}\right)_{\mathrm{rms}} \sim Z\left(\frac{\delta n_z}{n_e}\right)_{\mathrm{rms}} \sim \frac{1}{\langle k_x^2 L_n^2\rangle^{0.5}}. \tag{8.9}$$

The saturation level in Eq. (8.9) has the physical meaning that the average gradient of the rms impurity fluctuations $\langle k_x^2\rangle^{1/2}\delta n_{z,\mathrm{rms}}$ reaches the level of the background density gradient.

The $\boldsymbol{k}$ spectrum has an isotropic part at higher $k_\perp$ and an anisotropic part from zonal flow and density flattening part at low k_y values. There are coherent vortices that come and go in the turbulence and are strong in the limit of small

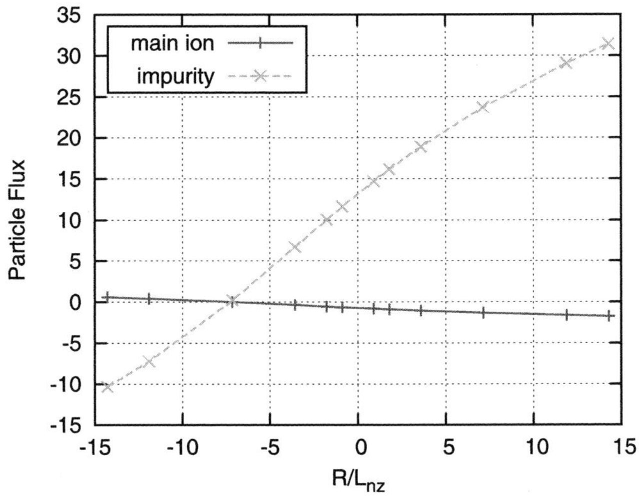

Fig. 8.5 Quasilinear fluxes of the main ion and the impurity, in this case boron (B^{5+}) as function of the impurity density gradient. Transport coefficients D and V are extracted from the slope and the intercept of the flux versus gradient curve.

parallel diffusivity $\nu_{\|} \ll |\omega_k|$. The turbulent wavenumber spectrum that enters Eq. (8.8) is modeled as

$$\left|\frac{e\phi_k}{T_e}\right|^2 = \frac{\rho_s^2}{L_n^2}\frac{1}{(1+k_\perp^2\rho_s^2)^{n/2}}. \tag{8.10}$$

with the spectral index $n = 3$ to 5 from simulations and scattering experiments.

Figure 8.5 shows the quasilinear flux of ion density and impurity density. The quasilinear fluxes are calculated for $0.1 < k_\perp\rho_s < 1.0$ with Eq. (8.10). The quasilinear flux of impurity Γ_z is a function of the impurity species. There is no significant phase shift between electron density and the potential. In other words, the adiabaticity of electron motion is high enough for $\Gamma_e \cong 0$.

The ion density flux is interesting. The direction of the convective part of the flux depends on the phase velocity. There are reversals of the fluxes under certain conditions. Obviously, reversal of a flux of ions seen in Fig. 8.5 is of critical importance. The flux of impurities has a dependence on the species. Higher mass impurities have smaller quasilinear fluxes, but their accumulation seriously degrades the fusion plasma performance.

An assumption on the slope of the power spectrum of $\int I(k)dk = \langle|e\phi/T_e|^2\rangle$ is needed to compute the turbulent diffusivities. A typical drift wave spectrum is

$$I(k_\perp) = \frac{I_1}{\left(k_x^2 + k_y^2\right)^{3/2}} \tag{8.11}$$

for $k_x^2 \geq \langle k_x^2 \rangle$ where the turbulent mixing radial correlation length $l_c \approx 1/\langle k_x^2 \rangle^{1/2}$. The associated mixing length amplitude is

$$\left\langle \left(\frac{e\phi}{T_e} \right)^2 \right\rangle^{1/2} \sim \frac{l_c}{L_{n_\alpha}} \tag{8.12}$$

where $1/L_{n_\alpha}$ is the dominant driving gradient ($\alpha = i, z, e$) for the growth rate.

The turbulent diffusivity is then

$$\begin{aligned} D_z &= \langle k_y \rho_s \rangle \left\langle \left(\frac{e\phi}{T_e} \right)^2 \right\rangle \frac{T_e}{eB} \\ &\approx \frac{\rho_s}{l_c} \left(\frac{l_c}{L_{n_z}} \right)^2 \frac{T_e}{eB} = \frac{\rho_s l_c}{L_{n_z}^2} \frac{T_e}{eB} \end{aligned} \tag{8.13}$$

when the impurity gradient $1/L_z$ determines the growth rate. Both magnetic shear $1/L_s$ and the E_r-shear control the radial correlation length l_c.

8.3 Three-Component Fluid Equations

The Braginskii equations, which describe the evolution of fluid moments, are obtained by taking moments of the Fokker-Planck equations and using a heat flux closure model in the thermal balance equation. Here we use the low-ion temperatures ($T_i, T_z \ll T_e$) and the isothermal electron models to close the system of fluid equations [Futatani, *et al.* (2010)].

The equations for the evolution of densities are

$$\frac{\partial n_s}{\partial t} + \nabla \cdot (n_s \boldsymbol{v}_s) = 0 \tag{8.14}$$

where n_s and v_s are the density and the fluid velocity of the species $s = e$, i, and z for electrons, hydrogenic ions and impurities ions.

The conservation of charge follows from Eq. (8.14) by computing $\partial_t \rho_{ch} = 0$ with $\rho_{ch} = \sum_s e_s n_s$. In the initial evolution following the impurity injection there is a rapid evolution ($\Delta t \sim qR/v_e$) of $\rho_{ch}(x,t)$ to zero from the parallel electron motion taking the system to $\nabla \cdot \boldsymbol{J} = 0$ where $\boldsymbol{J} = \sum_s e_s n_s \boldsymbol{v}_s$.

The evolution of fluid velocities is given by

$$n_s m_s \frac{d\boldsymbol{v}_s}{dt} = n_s e_s (\mathbf{E} + \boldsymbol{v}_s \times \mathbf{B}) - \nabla p_s - \nabla \cdot \mathbf{\Pi}_s \tag{8.15}$$

where p_s and $\mathbf{\Pi}_s$ are the pressure and the off-diagonal momentum stress tensor. For low ion temperatures a small ion viscosity is kept in the ion momentum stress tensors.

8.3.1 Reduction of three-component fluid equations

An outer product of **B** with Eq. (8.15) is taken to derive the perpendicular components of drift velocity. The result is

$$\begin{aligned} \boldsymbol{v}_{s\perp} &= \frac{\mathbf{E}\times\mathbf{B}}{B^2} + \frac{1}{e_s n_s}\frac{\mathbf{B}}{B^2}\times\nabla p_s + \frac{m_s}{e_s}\frac{\mathbf{B}}{B^2}\times\frac{d\boldsymbol{v}_{s\perp}}{dt} \\ &= \boldsymbol{v}_E + \boldsymbol{v}_* + \boldsymbol{v}_{\text{pol}} \end{aligned} \tag{8.16}$$

where the second line of Eq. (8.16) defines the three components of the velocity field; the $\mathbf{E}\times\mathbf{B}$ drift $\boldsymbol{v}_E = \mathbf{E}\times\mathbf{B}/B^2$, the diamagnetic drift $\boldsymbol{v}_* = \frac{1}{e_s n_s}\frac{\mathbf{B}}{B^2}\times\nabla p_s$ and the polarization drift $\boldsymbol{v}_{\text{pol}} = \frac{m_s}{e_s}\frac{\mathbf{B}}{B^2}\times\frac{d\boldsymbol{v}_{s\perp}}{dt}$. The electron polarization drift is negligible owing to the proportionality to the electron mass. The polarization drifts for the hydrodynamic and impurity ions reduce to

$$\boldsymbol{v}_{\text{pol,s}} = -\frac{m_s}{e_s B^2}\frac{d}{dt}\nabla\phi \tag{8.17}$$

and the polarization current is

$$\boldsymbol{j}_{\text{pol}} = -\frac{n_i m_i + n_z m_z}{B^2}\frac{d}{dt}\nabla\phi. \tag{8.18}$$

In the simplest description of drift waves the divergence of the cross field polarization current $\boldsymbol{j}_{\text{pol}}$ balances the divergence of the parallel electron currents.

The advection due to the diamagnetic drift is often negligible. For the polarization drift,

$$|v_{\text{pol}}| = \frac{m_s}{eB}\frac{d\boldsymbol{v}_E}{dt} \sim \frac{\rho_s\omega}{c_s}v_E \tag{8.19}$$

where $1/\omega$ is a typical time scale of the phenomena. In toroidal fusion plasmas, the typical plasma of core tokamak, has the space-time scales such that $\rho_s \sim 10^{-3}$m, sound speed $c_s \sim 10^5\ \text{ms}^{-1}$ and the drift wave-ITG turbulence has a frequency of the order $\omega \sim 10^5\text{s}^{-1}$. Therefore, the order of polarization drift becomes $v_{\text{pol}} \sim 10^{-3}v_E$. Therefore, v_{pol} can be neglected compared to other drift velocity, when compared directly with the drift velocity from $\mathbf{E}\times\mathbf{B}$ and diamagnetic pressure gradient terms v_{*s}. However, the divergence of the plasma current from $\boldsymbol{v}_{\text{pol}}$ from Eq. (8.18) is comparable to the other plasma current charge density contributions.

The characteristic dispersion scale of the ion gyroradius with the electron temperature in the three component plasma becomes

$$\rho_\tau^2 = \rho_{s,i}^2\frac{n_i}{n_e} + Z^2\rho_{s,z}^2\frac{n_z}{n_e}. \tag{8.20}$$

The definition of effective charge Z_{eff} that determines the resistivity η and the impurity dilution of the hydrogenic fusion fuel component is

$$Z_{\text{eff}} = \frac{n_i + Z^2 n_z}{n_i + Z n_z}, \qquad 1 < Z_{\text{eff}} < Z. \tag{8.21}$$

For tungsten $W(Z = 74)$, the $Z_{\text{eff}} = 3$ limit corresponds to the small concentration $n_z/n_e = 4\times10^{-4}$ from Eq. (8.21).

There is an enhancement in the dispersion radius ρ_s^2 from the impurities. The ratio of the impurity density n_z to electron density n_e follows from the effective charge Z_{eff} relative to the impurity charge Z through

$$\frac{n_z}{n_e} = \frac{Z_{\text{eff}} - 1}{Z(Z-1)}. \tag{8.22}$$

For the case $Z_{\text{eff}} = 3$ from fully ionized beryllium Be^{4+}, the fraction is $n_z/n_e = (3-1)/(4 \times 3) = 1/6$. Therefore, the enhancement term $1 + (A - Z)n_z/n_e$ to ρ_s^2 is $1 + (5 - 2) \times 1/6 = 1.5$. Here, the upper limit of $Z_{\text{eff}} = 3$ was taken. In the same limit for fully ionized, C^{6+} for $n_z/n_e = (3-1)/(6 \times 5) = 1/15$. Therefore, the term $1 + (A - Z)n_z/n_e$ for fully ionized carbon is $1 + (12 - 6) \times 1/15 = 1.4$. For a partially stripped heavy metal, for example W^{30+}, the concentration is $n_z/n_e = (3-1)/(30 \times 29) = 1/435$. Therefore, the term $1+(A-Z)n_z/n_e$ for tungsten W^{30+} is then $1 + (184 - 74) \times 1/435 = 1.253$.

For the drift frequencies with the impurities, one defines

$$\begin{aligned}
\omega_{*i} &= k_y \frac{T_e}{eB} \frac{1}{n_e} \frac{dn_i}{dx} \\
\omega_{*z} &= k_y \frac{T_e}{ZeB} \frac{1}{n_e} \frac{dn_z}{dx} \\
\omega_{*e} &= -k_y \frac{T_e}{eB} \frac{1}{n_e} \frac{dn_e}{dx} \\
\omega_{De} &= k_y \frac{2T_e}{eBR} \cos\theta.
\end{aligned} \tag{8.23}$$

The charge neutrality condition and Eq. (8.23) give

$$\omega_{*e} = \omega_{*i} + Z^2 \omega_{*z}. \tag{8.24}$$

Other impurity system parameters are

$$\rho_{s,i}^2 = \frac{m_i T_e}{e^2 B^2} \quad \text{and} \quad \rho_{s,z}^2 = \frac{m_z T_e}{Ze^2B^2}, \quad D_\parallel = \frac{\sigma_\parallel T_e}{n_e e^2} \quad \text{and} \quad \sigma_\parallel = \frac{n_e e^2}{m_e \nu_e}. \tag{8.25}$$

The determinant of the fluctuation matrix Eq. (8.3) gives a cubic polynomial dispersion equation

$$\begin{aligned}
&ik_\perp^2 \rho_T^2 \omega^2 (\omega - \omega_{De}) \\
&\quad - \nu_\parallel \left\{ 1 + k_\perp^2 \rho_T^2 + \left(\mu_i \rho_{s,i}^2 + \mu_z Z \rho_{s,z}^2\right) \frac{k_\perp^4}{\nu_\parallel} + i(Z^2 - Z) \frac{\omega_{*z}}{\nu_\parallel} \right\} \omega^2 \\
&\quad + (\nu_\parallel + i\omega_{De}) \left\{ \omega_{*i} + Z\omega_{*z} - i\left(\mu_i \rho_{s,i}^2 + \mu_z Z \rho_{s,z}^2\right) k_\perp^4 \right\} \omega = 0
\end{aligned} \tag{8.26}$$

where $\nu_\parallel = k_\parallel^2 D_\parallel$ and $\rho_T^2 = \rho_{s,i}^2 + \rho_{s,z}^2 Z$. For all k values, the cubic equation factors into the modes $\omega = 0$ and two finite ω roots of a equivalent quadratic equation.

8.3.2 *Impurity transport from trapped electron modes inside the SOL*

Inside the scrape-off layer (SOL) the plasma collisionality is low so the collisional drift wave evolves into the trapped electron mode. In the core plasma the few collisions transform the instability to the collisionless trapped electron mode (CTEM). The modulation of the toroidal magnetic field strength is modeled as

$$B = B_0 \left(1 - \frac{r}{R}\cos\theta\right). \tag{8.27}$$

The small aspect ratio $\epsilon = r/R \ll 1$ is considered here, then the fraction of trapped particles, f_t, can be defined by

$$f_t = \frac{\text{trapped electrons}}{\text{total electrons}}. \tag{8.28}$$

The trapped particle fraction f_t is estimated by

$$f_t = \left(\frac{B_{\max}}{B_{\min}} - 1\right)^{1/2} = \left(\frac{1+\epsilon}{1-\epsilon} - 1\right)^{1/2} \tag{8.29}$$

$$= 1 - f_p \tag{8.30}$$

where f_p is the fraction of passing electrons.

From the electron kinetic equation, we derive the fluctuating δf_e given by

$$\delta f_e = f_M \left\{\frac{e\varphi}{T_e} - \frac{1}{\tau}\int \frac{ds'}{v_\parallel}\varphi(s')\frac{\omega - \omega_{*e}}{\omega - \omega_{De} + i\nu_{\text{eff}}}\right\} \tag{8.31}$$

where the orbit integral is discontinuous with

$$\begin{aligned} \frac{1}{\tau}\int \frac{ds'}{v_\parallel}\varphi &= 0 \qquad \text{(for passing electrons)} \\ &\simeq \varphi \qquad \text{(for trapped electrons).} \end{aligned} \tag{8.32}$$

The passing electron integral in Eq. (8.32) vanishes due to phase mixing of φ averaged over the irrational toroidal field lines. From Eqs. (8.31) and (8.32) one derives

$$\delta n_p = \int d^3v \delta f_p = f_p \frac{n_e e\varphi}{T_e} \tag{8.33}$$

$$\delta n_t = \int d^3v \delta f_t = f_t \frac{n_e e\varphi}{T_e}\left[1 - \frac{\omega - \omega_{*e}}{\omega - \omega_{De} + i\nu_{\text{eff}}}\right]. \tag{8.34}$$

There are two regimes of the trapped electron mode: the dissipative (DTEM) and the collisionless (CTEM) contribution. For the DTEM, $\nu_{\text{eff}} \gg |\omega| \sim \omega_{De}$ we have

$$\frac{\delta n_e}{n_e} = \left[1 + i f_t \left(\frac{\omega - \omega_*}{\nu_{\text{eff}}}\right)\right]\left(\frac{e\delta\varphi}{T_e}\right).$$

For the CTEM, $\nu_{\text{eff}} \ll \omega_{De}$, the electron density response is

$$\frac{\delta n_e}{n_e} = [1 + i\pi f_t \langle \delta(\omega - \omega_{De}) \rangle (\omega - \omega_{*e})] \left(\frac{e\delta\varphi}{T_e}\right)$$

for the collisionless (CTEM) mode.

The equation for the passing electron density fluctuation is $\delta n_p = (en_p/T_e)\varphi_k$ where $\delta n_p = n_p - n_{p0}$ is linear and in phase with the potential φ_k. The trapped electron density in the dissipative limit may be modeled with the fluid equation

$$\frac{\partial n_t}{\partial t} + \boldsymbol{v}_E \cdot \nabla n_t = -\nu_{\text{eff}} (n_t - n_{t0}) \tag{8.35}$$

with the steady-state trapped density given by $n_{t0} = f_t n_e$. The fluctuating trapped electron density δn_t is given by

$$(-i\omega + \nu_{\text{eff}})\, \delta n_t + \frac{-ik_y\phi}{B}\frac{dn_t}{dr} = 0. \tag{8.36}$$

For the transition from the collisional regime to the hot core plasma the dissipation is replaced with $\nu_{\text{eff}} \to c_D|\omega_{De}|$ with the constant c_D chosen to get agreement with collisionless gyrokinetic limit. For this system with impurity ions, the dynamics reduces to

$$\begin{aligned}
&\frac{dn_i}{dt} + \mu_i \frac{m_i}{eB^2}\nabla^4\phi = \nabla \cdot \left(\frac{m_i n_i}{eB^2}\frac{d}{dt}\nabla\phi\right) \\
&\frac{dn_z}{dt} + \mu_z \frac{m_z}{eB^2}\nabla^4\phi = \nabla \cdot \left(\frac{m_z n_z}{ZeB^2}\frac{d}{dt}\nabla\phi\right) \\
&\frac{d}{dt}\delta n_t = -\nu_{\text{eff}}\delta n_t = n_e f_p \frac{e\varphi}{T_e} + \delta n_z + n_{t0}
\end{aligned} \tag{8.37}$$

The linear fluctuations are given by

$$\begin{aligned}
&-i\omega\delta n_i - \frac{ik_y\phi_k}{B}\frac{dn_i}{dr} = i\frac{k_\perp^2 m_i n_i}{eB^2}\omega\phi_k = i\omega k_\perp^2 \frac{m_i n_i T_e}{e^2B^2}\left(\frac{e\varphi_k}{T_e}\right) \\
&-i\omega\delta n_z - \frac{ik_y\phi_k}{B}\frac{dn_z}{dr} = i\omega\frac{k_\perp^2 m_z n_z T_e}{e^2B^2}\left(\frac{e\varphi_k}{T_e}\right) \\
&\delta n_e = \left(n_e f_p + \frac{k_y \frac{dn_t}{dr}\frac{T_e}{e}}{B(\omega + i\nu_{\text{eff}})}\right)\left(\frac{e\varphi_k}{T_e}\right) \\
&\delta n_i + Z\delta n_z = \delta n_e + n_e f_p \frac{e\varphi_k}{T_e}.
\end{aligned} \tag{8.38}$$

The system of linear Eq. (8.38) determines the polarization of the fluctuations for the trapped electron modes with impurity ions.

8.3.3 *Spectral expansion of the nonlinear fields for turbulent impurity transport*

To investigate the nonlinear fluctuations we expand the turbulent fluctuations in terms of the eigenvectors of the linear system. The advantage of the eigenmode expansion over that of Fourier modes is that the complex linear couplings are taken into account exactly in each eigenvector. The time evolution of the expansion coefficient $C_{k,i}(t)$ is then due entirely to the nonlinear interactions.

The vector field $\boldsymbol{X}(t)$ expressed as the sum over a set of k_x, k_y, k_z modes there is a sum over the three distinct eigenmodes, $j = 1, 2, 3$ for each $\boldsymbol{k}$ vector. The time-dependent expansion coefficients are defined as

$$\boldsymbol{X}(t) = \sum_{j,\boldsymbol{k}} C_{\boldsymbol{k},j}(t)\boldsymbol{X}_j e^{i\boldsymbol{k}\cdot\boldsymbol{x}-i\omega_{\boldsymbol{k},j}t} + \text{c.c.} \tag{8.39}$$

where the dynamics $dC_{\boldsymbol{k},j}/dt$ is due to the nonlinear terms in Eq. (8.40) and $X_j(\boldsymbol{k})$ is the eigenvector for $j = 1, 2, 3$ from the eigenmodes.

Substituting Eq. (8.39) into the dynamical equations we find

$$\begin{aligned}&\sum_{\boldsymbol{k},j}\left(\omega C_{\boldsymbol{k},j} + i\frac{dC_{\boldsymbol{k},j}}{dt}\right) BX_{\boldsymbol{k},j}e^{i\boldsymbol{k}\cdot\boldsymbol{x}-i\omega_j t}\\ &\quad = \sum_{\boldsymbol{k},j} A(\boldsymbol{k})C_{\boldsymbol{k},j}X_j + \sum_{\boldsymbol{k}',j'}\sum_{\boldsymbol{k}'',j''} N(X_{\boldsymbol{k}'}, X_{\boldsymbol{k}''})C_{\boldsymbol{k}',j'}C_{\boldsymbol{k}'',j''}.\end{aligned} \tag{8.40}$$

Now multiplying through Eq. (8.40) by $\mathbf{B}^{-1}$ and contracting, or projecting, the equation onto the left eigenvector $\boldsymbol{y}_{l,\boldsymbol{k}''} = \langle l, k'' \mid$ yields

$$\begin{aligned}&\sum i\langle y_{l,\boldsymbol{k}''}|x_{l,\boldsymbol{k}}\rangle\frac{dC_{l,\boldsymbol{k}}}{dt}\\ &\quad = \sum_{m,\boldsymbol{k}_1,n,\boldsymbol{k}_2}\langle y_{l,\boldsymbol{k}''}|N\left(x_{m,\boldsymbol{k}_1}, x_{n,\boldsymbol{k}_2}\right)\rangle C_{m,\boldsymbol{k}_1}(t)C_{n,\boldsymbol{k}_2}(t)\exp\left\{i\left(\omega_{\boldsymbol{k}} - \omega_{\boldsymbol{k}_1} - \omega_{\boldsymbol{k}_2}\right)t\right\}.\end{aligned} \tag{8.41}$$

The diagonal terms are normalized such that $\langle y_{l,\boldsymbol{k}''} \mid x_{l,\boldsymbol{k}}\rangle = \delta_{l,\boldsymbol{k}}$. Then we arrive at the nonlinear dynamical equation

$$i\frac{dC_{l,\boldsymbol{k}}}{dt} = \sum_{\boldsymbol{k}}\langle y_{l,\boldsymbol{k}}|N\left(x_{m,\boldsymbol{k}_1}, x_{n,\boldsymbol{k}_2}\right)\rangle C_{m,\boldsymbol{k}_1}(t)C_{l,\boldsymbol{k}_2}(t). \tag{8.42}$$

Equation (8.42) is of the form that occurs in quantum mechanics for the phonon-phonon interactions with a summation over $\boldsymbol{k}$-space lattice and a second summation over the quantum states j. In the plasma problem the matrices are not Hermitian. The interactions come in resonant $\boldsymbol{k}$-vector triads but contain the additional complexity that each triad has 3 degrees of freedom for the interaction within each triad.

When the wave fields are arbitrarily restricted to one vector triad the system becomes integrable. The single triad has a reversible "pump depletion" behavior. As more interacting vector triads are included, the dynamics become chaotic and the fields diverge and mix in their phase space.

8.4 Impurity Transport in High-Density Regimes

High-purity plasmas are associated with high-plasma density operation. The self-cleaning ability of high-density plasmas was documented by Alcator C-Mod [Greenwald, *et al.* (1995)] and the Frascati Tokamak Upgrade (FTU) [Apicella, *et al.* (1997)] experiments. A scaling law relating plasma purity, radiated power, and machine dimensions was derived from a number of experiments [Marmar, *et al.* (1982)]. Based on these experiments, average Z_{eff} values of around 1.2 to 1.5 should be possible in a high-density fusion reactor. The problem of plasma purity is associated with the problem of reducing the power load on the first wall where the wall sputtering and evaporation produce impurities from the plasma-wall interactions.

One solution to reduce the impurity concentration is to use a cold radiating plasma mantle. The radiating plasma mantle allows the use of molybdenum tiles as the first wall material. The high density regimes operating with strong screening of the main body of the plasma column from wall-produced impurities have been produced. Experimental results [Weynants, *et al.* (1999)] indicate the possibility of operating with a radiating plasma mantle to dissipate up to 90% of the total power lost from the core plasma without energy confinement degradation. Thermal loads have been calculated for an ideal continuous first wall under the assumption that 70% of the input power is radiated, which would reduce the maximum thermal heat flux to the plasma-facing components to $0.5\,\text{MW/m}^2$ or less in compact-high-field tokamaks.

8.5 Trace Impurity Transport Studies

Early impurity transport experiments in the Texas Experimental Tokamak (TEXT) [Gentle (1981)] were compared with the predictions of turbulence transport models in Horton and Rowan (1994). In these experiments, scandium Sc was injected into the plasma using laser ablation and the time resolved profiles of the Sc ionization stages were measured along with the radial potential fluctuation profile using the heavy ion beam probe (HIBP). The experiment was simulated using a one-dimensional (1D) radial transport code with the standard density transport flux

$$\Gamma = -D\frac{\partial n}{\partial r} + nV. \tag{8.43}$$

The diffusion coefficient D and convective velocity V parameters were varied until the time-dependent transport simulations reproduced the data. The HIBP data gave the radially-resolved electrostatic potential ϕ. Representation in Eq. (8.43) for the impurity transport is compared with the $\mathbf{E} \times \mathbf{B}$ turbulent diffusivities and mobility based on the fluctuation data and the measured radial electric field. The electron energies for the first ionization potentials and the corresponding ionization rates $\langle \sigma v \rangle$ are given in Table 8.1 for the dominant impurities present in the confinement vessels.

Table 8.1 Ionization rates $\langle \sigma v \rangle$ for atoms by electron impact [Voronov (1997)].

Atom	1st Ionization Energy	$\langle \sigma v \rangle$ at $E = T_e$
H	13.6 eV	8.7×10^{-9} cm^3/s
He	24.6 eV	5.5×10^{-9} cm^3/s
Li	5.4 eV	3.6×10^{-8} cm^3/s
Be	9.3 eV	2.7×10^{-8} cm^3/s
B	8.3 eV	2.0×10^{-8} cm^3/s
C	11.3 eV	2.1×10^{-8} cm^3/s
N	14.5 eV	1.7×10^{-8} cm^3/s
O	13.6 eV	1.2×10^{-8} cm^3/s

The agreement with the impurity transport data was best with the $\mathbf{E} \times \mathbf{B}$ diffusivity taken in the strong turbulence regime where

$$D = c_D \left(\frac{\widetilde{\phi}}{B_T} \right) \tag{8.44}$$

while the comparison with the weak turbulence diffusivity and the collisional (no fluctuations) diffusivity results were in qualitative disagreement with the Sc transport data. Here $\widetilde{\phi}$ is the local rms value of the electrostatic potential. The percolation theory gives the diffusivity

$$D_{\text{perc}} = \left(\frac{\widetilde{\phi}}{B_T} \right)^{7/10} \left(\frac{\Delta\omega}{k_\perp^2} \right)^{3/10} \tag{8.45}$$

derived by Isichenko, *et al.* (1992) with a weaker dependence on $\widetilde{\phi}$.

Predicting the transport of impurities in tokamak confinement systems is of key importance for projections of the operation to fusion reactors. In addition, the analysis of the transport of impurity ions injected at sufficiently low levels of concentration that the collective modes responsible for the fluctuations are not materially altered provides important information for testing impurity transport theory and modelling.

Now we review the results of studies of the impurity transport experiments in which trace amounts of impurity ions are injected by laser ablation and the observed transport is compared with the $\mathbf{E} \times \mathbf{B}$ turbulent diffusivities for test particles. For this purpose experiments were performed to provide both impurity profile evolution data and fluctuation data in the same, and in similar discharges. For injection experiments where $n_z Z^2 \ll n_e$ the effect of the impurity on the collective modes is negligible. Thus, one may take the impurity ions convected by the $\mathbf{E}\times\mathbf{B}$ fluctuations velocity as test particles in the turbulence.

Impurity injection experiments in the large helical device (LHD) with traces of vanadium (V), manganese (Mn), and cobalt (CO) show that the turbulence and plasma are not appreciably disturbed. These three elements are neighbors to the intrinsic impurities of iron Fe and chromium Cr from the stainless steel vessel walls [Sudo, *et al.* (2012)].

The study of test particle (ion) transport by the turbulent $\mathbf{E} \times \mathbf{B}$ drifts in electric fields from drift wave-type fluctuations has been an active area of theoretical research [Horton, *et al.* (1981), Horton (1985), Kleva and Drake (1984), Isichenko, *et al.* (1992), Misguich, *et al.* (1988), Gruzinov, *et al.* (1990), Isichenko and Horton (1991)].

Rather well understood test particle diffusivities from the $\mathbf{E} \times \mathbf{B}$ drifts are reported in the extensive computer experiments. The results show that once the fluctuation level exceeds a relatively low level, the diffusion leaves the weak turbulence regime in which

$$D_\perp = \frac{\widetilde{E}_\perp^2 \tau_c}{B^2} \tag{8.46}$$

and enters the strong turbulence, or renormalized turbulence regime, in which $D_\perp \propto \widetilde{E}_\perp \lambda_\perp / B \sim \widetilde{\phi}/B$. At still higher fluctuation amplitudes, the diffusion increases even more slowly with the fluctuating potential. In the percolation theory limit [Isichenko, *et al.* (1992), Gruzinov, *et al.* (1990), Isichenko and Horton (1991)] the effective test particle diffusivity increases as $D_\perp \propto (\widetilde{\phi}/B)^{7/10}(\Delta\omega/k_\perp^2)^{3/10}$ where $\Delta\omega, k_\perp$ are the characteristic frequency band and mean wavenumber of the fluctuations. These predictions are tested numerically in Isichenko, *et al.* (1992) and compared with earlier works [Kleva and Drake (1984), Horton (1985), Misguich, *et al.* (1988), Gruzinov, *et al.* (1990)]. For the ITER parameters the turbulent diffusivity parameters are summarized in Table 8.2 for the ITER parameters.

Table 8.2 Typical ITER drift wave frequencies and plasma parameters.

	Core Plasma	SOL Plasma
Major Radius	$R = 6\,\mathrm{m}$	$R = 6\,\mathrm{m}$
Minor Radius	$a = 2\,\mathrm{m}$	$a = 2\,\mathrm{m}$
Magnetic field	$B = 5\,\mathrm{T}$	$B = 5\,\mathrm{T}$
Safety factor	$q = 2$	$q = 5$
Electron Temperature	$T_e = 10\,\mathrm{KeV}$	$T_e = 1\,\mathrm{KeV}$
Electron Density	$n_e = 10^{20}\,\mathrm{m}^{-3}$	$n_e = 10^{19}\,\mathrm{m}^{-3}$
Impurity Density (Be^{+4})	$n_z/n_e = 1\%$	$n_z/n_e = 3\%$
Parallel Wavenumber	$k_\parallel = 0.5q/R \sim 0.17\,\mathrm{m}^{-1}$	$k_\parallel \sim 0.42\,\mathrm{m}^{-1}$
Electron Thermal Velocity	$v_e = 1.5 \times 10^9\,\mathrm{m/s}$	$v_e = 4.7 \times 10^8\,\mathrm{m/s}$
Sound Speed	$c_s = 3.4 \times 10^7\,\mathrm{m/s}$	$c_s = 1.1 \times 10^7\,\mathrm{m/s}$
Ion Sound Gyroradius	$\rho_s = c_s/\omega_{ci} = 7.3 \times 10^{-3}\,\mathrm{m}$	$\rho_s = 3 \times 10^{-3}\,\mathrm{m}$
Electron Parallel Diffusion	$D_\parallel = \frac{v_e}{k_\parallel} = 8.8 \times 10^9\,\mathrm{m^2s^{-1}}$	$D_\parallel = \frac{v_e}{k_\parallel} = 1.1 \times 10^9\,\mathrm{m^2s^{-1}}$
Electron Collision Rate	$\nu_e = 1.1 \times 10^5\,\mathrm{s}^{-1}$	$\nu_e = 1.6 \times 10^5\,\mathrm{s}^{-1}$
Ion Collision Rate	$\nu_i = 1.6 \times 10^3\,\mathrm{s}^{-1}$	$\nu_i = 2.3 \times 10^3\,\mathrm{s}^{-1}$
Parallel Diffusion Rate	$\nu_\parallel = k_\parallel^2 D_\parallel = 6.6 \times 10^5\,\mathrm{s}^{-1}$	$\nu_e/(a/2R) = 2.2 \times 10^5\,\mathrm{s}^{-1}$
Electron Density Gradient Length	$L_{n_e} = 2.0\,\mathrm{m}$	$L_{n_e} = 5.0\,\mathrm{m}$
Impurity Density Gradient Length	$L_{n_z} = 2.0\,\mathrm{m}$	$L_{n_z} = -2.0\,\mathrm{m}$
Trapped Fraction	$f_t < 0.1$	$f_t = \sqrt{r/R} = \sqrt{a/2R} = 0.41$

In addition to the diffusivity the inward radial electric field E_r produces a mobility μ component to the impurity flux that is also proportional to the fluctuation level. In collisional transport theory the mobility μ_z and diffusivity D_z are related by the Einstein relationship

$$D_z = \mu_z T_z. \tag{8.47}$$

Due to the inward radial electric field $E_r = -d\Phi/dr < 0$, this value of the mobility produces a strong inward impurity flux such that the Boltzmann relationship

$$n_z(r) = n_0 \exp\left(-\frac{Ze\Phi(r)}{T_z}\right) \tag{8.48}$$

is an equilibrium state at zero impurity flux $\Gamma_z = 0$.

In the turbulent $\mathbf{E}\times\mathbf{B}$ transport the mobility-to-diffusivity ratio can be much less than the value given by the Einstein relation. Since the theoretical studies for the effective mobility are not as complete as those for the diffusivity we may introduce the effective coefficient c_E (with E for Einstein relation) in the mobility relationship $\mu_z = c_E(D_z/T_z)$ that gives the best agreement with the evolution of the impurity density profiles.

The scandium test particle injection experiments in the Texas Experimental Tokamak (TEXT) [Gentle (1981)] were explained within a factor of 2 by the formulas

$$\Gamma_z = -D_z \frac{dn_z}{dr} + \mu_z n_z Ze E_r \tag{8.49}$$

with $D_z = c_D(\widetilde{\phi}/B_T)$ and $\mu_z = c_\mu(\widetilde{\phi}/B_T T_z)$ with $c_D = 0.28 \pm 0.09$ and $c_\mu = 0.043 \pm 0.007$ for hydrogen as a working gas. For these values of the coefficients the deviation from the Einstein formula is given by $c_E = \mu T_z/D_z = c_\mu/c_D \simeq 0.15$ compared with unity. While with He for the working gas, $c_D = 0.22 \pm 0.09$, $c_\mu = 0.16 \pm 0.07$, $c_E \simeq 0.73$. Thus, for helium as the working gas the mobility is closer to the limiting Einstein value. While the domains of applicability of the turbulent transport formulas are an active area of theoretical research, the formula (8.49) is consistent with the scandium injection experiments. Theoretical work with computer experiments performed for the nonlinear $\mathbf{E}\times\mathbf{B}$ drift motion of test particles in drift wave-type fluctuating electric fields interpret the impurity data well.

The fluctuation data from the heavy ion beam probe in the Texas Experimental Tokamak (TEXT) with respect to the regime of the $\mathbf{E}\times\mathbf{B}$ transport showed that the strong nonlinearity condition $R_E = \Omega_E/|\omega_k| \gtrsim 1$ was satisfied. Here Ω_E is the rotation rate around the maximum and minimum of $\widetilde{\phi}$ which for a fluctuation with $\boldsymbol{k} = (k_x, k_y)$ is approximately given by $\Omega_E = k_x k_y \widetilde{\phi}/B \sim k_\perp^2 D$. The measured value of the dimensionless nonlinearity parameter R_E is $R_E \simeq 1.5$ to 2.5. This measure R_E is named the Kubo number.

In the regime $R_E > 1$, the nonlinearity determines the correlation time between the ion and the fluctuation. This shortening of the correlation time with increasing amplitude is in contrast with quasilinear theory where the correlation time is independent of the amplitude. Thus, the structure of the transport of trace

Table 8.3 Scandium and titanium transport experiments in TEXT

$m_z = 45\,\text{amu} \qquad Z = 21$
$Z = 12, 16, 17, 18$
$n_z \le 2 \times 10^7\,\text{cm}^{-3}$
For $Z = 18$ and $m_z = 45$ rm the experiments have
$\nu_{zi} = 10^4/\text{s}$
$\rho_z = 0.03\,\text{cm}= 300\,\mu\text{m}$
$v_z = 3 \times 10^6\,\text{cm/s}$
The working gas is hydrogen at
$T_i = 300\,\text{eV}$ and the impurities are $45\,\text{Sc}^{21}$ $(3d^4\,\text{s}^2)$
and 48 Ti^{22} $(3d^2 4\,\text{s}^2)$.

amounts of impurity elements provides an important comparison with the nonlinear regime of $\mathbf{E} \times \mathbf{B}$ turbulent transport.

The theoretical value of c_D determined from the regression fits to the ensemble averaged diffusivity is a functional of the fluctuation spectrum. Some results for the sensitivity of c_D to changes in modeling of the fluctuation spectrum are shown in Horton, *et al.* (1987) and Kim, *et al.* (1990). The effect of Coulomb pitch angle scattering along with the turbulent $\mathbf{E}\times\mathbf{B}$ flow shows that the neoclassical diffusivity is recovered at a sufficiently low fluctuation level.

8.6 Thermalization of Impurities and their Collisional Fluxes

The laser ablation injection method creates a burst of impurity ions with parameters $n_z, m_z, q_z = Z_z, T_z$ in a localized radial region. The impurity ions stream over the magnetic surface in time $qR/v_{zi} \simeq 10^{-4}\,\text{s}$ and scatter in velocity space by Coulomb collisions with the working gas ions n_i, m_i, Z_i, T_i. This initial relaxation is given by the kinetic equation

$$\frac{\partial f_z}{\partial t} + \frac{v_\parallel}{qR}\frac{\partial f_z}{\partial \theta} = C_{zi}(f_z) = \frac{\partial}{\partial \boldsymbol{v}} \cdot \left(\frac{\boldsymbol{v} - \boldsymbol{u}_i}{\tau_s} f_z + D_v \frac{\partial f_z}{\partial \boldsymbol{v}} \right) \tag{8.50}$$

where the slowing-down time τ_s from the collisions with the thermal ions is given by

$$\frac{1}{\tau_s} = \frac{16\pi^{1/2}}{3} \frac{n_i\, e^4\, Z^2\, Z_i^2}{m_z\, m_i\, v_i^3}\, \ell n\, \Lambda. \tag{8.51}$$

In Braginskii (1965) the impurity-ion collision frequency $\nu_{zi} = 1/\tau_s$ occurs with

$$\nu_{zi} = \frac{4(2\pi)^{1/2}}{3} \frac{n_i\, e^4\, Z^2\, Z_i^2\, \ell n\, \Lambda}{m_z\, m_i^{-1/2}\, T_i^{3/2}} \tag{8.52}$$

where Eq. (8.52) is equivalent to Eq. (8.51) with $v_i = (2T_i/m_i)^{1/2}$. For a hydrogen discharge at $n_i = 3 \times 10^{13}\,\text{cm}^{-3}$, $T_i = 300\,\text{eV}$, this time is $\tau_s \simeq 10^{-4}\,\text{s}$ and for the helium discharge the time is approximately twice as long. The velocity diffusion coefficient in Eq. (8.50) is given by $D_v = T_i/(m_z\, \tau_s)$ so that in the absence of driving forces the impurity-ion collisions force $f_z(\boldsymbol{x}, \boldsymbol{v}, t)$ to $f_z = n_z(r,t)\exp[-m_z(\boldsymbol{v} - \boldsymbol{u}_i)^2/2T_i]$ for times long compared with τ_s. Here $\boldsymbol{u}_i(\boldsymbol{x},t)$ and $T_i(\boldsymbol{x},t)$ are the center of mass flow velocity and temperature of the working gas.

Note that collisional impurity transport formulas are often presented in terms of the ion-impurity collision frequency $\nu_{iz} = (n_z\, m_z/n_i\, m_i)\nu_{zi}$ and the thermal ion gyroradius $\rho_i = (2m_i\, T_i)^{1/2}/eZ_i\, B_T$ (rather than the gyroradius ρ_z of the impurity ions) which changes the appearance of the formulas.

In regimes where the impurity-ion collision frequency is high enough that the impurity mean free path $v_z\,\tau_s$ is less than qR the collisional transport is in the Pfirsch-Schlüter regime. Typical values for these experiments with scandium (Sc) impurity ions are $\rho_z(Z = 18) \simeq 0.025\,\mathrm{cm}$, $\nu_{zi} = 1/\tau_s = 10^4\,/\mathrm{s}$ and $v_z\,\tau_s = 3\times 10^6\,\mathrm{m/s}\cdot 10^{-4}\,\mathrm{s} = 300\,\mathrm{cm}$. For $q = 3$ the P-S diffusivity is then

$$D_z^{\mathrm{PS}} = (1+2q^2)\nu_{zi}\,\rho_z^2 = (1+18)(0.025\,\mathrm{cm})^2 10^4\,\mathrm{s}^{-1} \cong 120\,\mathrm{cm}^2/\mathrm{s} \tag{8.53}$$

which is about one order of magnitude smaller than the measured rate of diffusion. Early experiments [Marmar, *et al.* (1982), Synakowski, *et al.* (1989), Kim, *et al.* (1990)] attempted to use collisional transport theory to explain impurity transport and met with difficulty. Note that $\nu_{zi}\,\rho_z^2$ is the same for all charge states Z and independent of m_z. The Alcator impurity study of Marmar, *et al.* (1982) reports that the diffusivities of order $2.6\times 10^3\,\mathrm{cm}^2/\mathrm{s}$ are required to explain the transport of the trace silicon ions.

For a general $\mathbf{E}_\perp(\boldsymbol{x},t)$ field the exact motion of the impurity particles has been studied extensively in theoretical works. The studies show that the initial distribution

$$n_z(\boldsymbol{x},t) = \sum_{j=1}^{N} \delta(\boldsymbol{x} - \boldsymbol{r}_j(t)) \tag{8.54}$$

evolving with

$$\frac{d\boldsymbol{r}_j}{dt} = \frac{\mathbf{E}(\boldsymbol{r}_j,t)\times\mathbf{B}}{B^2} \tag{8.55}$$

is described by an effective diffusion coefficient D after the convection over several spatial correlation lengths $\ell_c = 1/\Delta k_\perp$. After this transition the impurity density spreads, in the absence of ionization and recombination, according to

$$\frac{\partial n_z}{\partial t} = \frac{-1}{r}\frac{\partial}{\partial r}(r\,\Gamma_z) \tag{8.56}$$

with

$$\Gamma_z = -D\,\frac{dn_z}{dr} + Ze\,n_z\,\mu\,E_r. \tag{8.57}$$

Now the precise determination of the effective diffusion coefficient D and mobility μ are difficult problems in nonlinear dynamics and depend on the spectral features of the fluctuations [Isichenko, *et al.* (1992), Misguich, *et al.* (1988), Gruzinov, *et al.* (1990), Isichenko and Horton (1991)].

The estimates for ITER-like parameters for D are given in Table 8.2. The main point is that for the measured fluctuations the impurity dynamics is nonlinear with the decorrelation rate ν determined by the turbulent $\mathbf{E}\times\mathbf{B}$ motion itself rather than parallel transit motions.

The associated mobility of the impurity ions is then

$$\mu = c_E \frac{D}{T_z} \tag{8.58}$$

where c_E is a constant between zero and unity. For $c_E = 1$ the mobility is given by the Einstein relation so that the $\Gamma_z = 0$ state corresponds to the Boltzmann impurity density profile $n_z = n_0 \exp(-Ze\,\Phi(r)/T_z)$ when $E_r = -d\Phi/dr$. Some $\mathbf{E}\times\mathbf{B}$ nonlinear dynamics studies [Aydemir (1990)] indicate that the mobility may fall below this value so that $c_E \ll 1$. Since c_E has not been accurately determined in the literature we leave the coefficient $c_E = c_\mu/c_D$ to be determined by the impurity transport data.

In a theoretical simulation value of c_D a study of the stochastic $\mathbf{E}\times\mathbf{B}$ diffusion with a six-wave model for the fluctuation spectrum gave the nonlinear regression fit of $D_z = 0.3\widetilde{\phi}^{0.92}$ in Fig. 4 of Isichenko and Horton (1991) for the normalized Reynolds number R_E in the range [1,100]. The corresponding Kolmogorov entropy K calculated from the exponential rate of separation of neighboring $\mathbf{E}\times\mathbf{B}$ trajectories is $K = 0.2\widetilde{\phi}^{0.56}$. The Kolmogorov entropy defines the exponential rate of separation of neighboring impurity ions.

8.7 Scandium and Titanium Transport with Ionization and Recombination

High-Z impurity transport was researched extensively using the injection scandium ions. Scandium and titanium ions are readily followed and are representative of high-Z impurities. The transport of Sc atoms in a series of ionization states was measured in H and He discharges with common external control parameters. In the experiments, a small quantity of Sc was injected into the plasma using laser ablation. The electron density perturbation was less than 1%.

After injection, the trace impurity ions are transported and ionized in their motion through the plasma. On encountering a vessel surface outside of the plasma, the Sc ions stick to the wall surface and are not recycled in significant quantity. For measurements of transport, the impurity spatial distributions are as important as the temporal data. Because Sc($m = 45, Z = 21$) must be very close in its transport properties to titanium transport Ti($m = 48, Z = 22$), an intrinsic chamber wall impurity, the spatial distributions for Ti can be predicted with the temporal data for Sc to provide the transport coefficients for the metallic species from the vessel walls. The analysis for the small atomic differences in the two elements, Sc and Ti is taken into account. The impurity transport of both Sc and Ti were simulated with a fixed background plasma model using the measured profiles of T_e and n_e.

A high-energy (500 KeV) accelerator called the heavy ion beam probe (HIBP) as used in the TEXT tokamak is currently used in the Large Helical Device (LHD) for confinement. The mean electrostatic potential $\overline{\phi}(\boldsymbol{r})$, and the root-mean-square fluctuating potential $\widetilde{\phi}(\boldsymbol{r})$ are measured by the charge exchange collisions products [Schoch, *et al.* (1986)].

The impurity density profiles are simulated using the set of coupled impurity transport equations

$$\frac{\partial n_z}{\partial t} + \nabla \cdot \Gamma_z = -n_e\, n_z (S_z + \alpha_z) + n_e\, n_{z-1} S_{z-1} + n_e\, n_{z+1} \alpha_{z+1} \tag{8.59}$$

where the S_z is the ionization rate coefficient for the zth impurity ionization stage and α_z is the recombination rate coefficient. The functions S_z and α_z depend principally on the $T_e(r)$ profile which is known from Thomson scattering and electron cyclotron emission (ECE) data. The diagnostic is called the heavy ion beam probe (HIBP). For each ion species the impurity flux, Γ, is represented with the transport formula is

$$\Gamma_z = -D\,\frac{\partial n_z}{\partial r} + n_z\, V \tag{8.60}$$

where both D and V are dependent on radius and the ionization species.

There are two sets of transport equations; one for titanium Ti and one for scandium Sc. The diffusivity $D(r)$ and the convection $V(r)$ are common. There is good agreement between the experiments and the simulations for these two heavy impurities at similar stages of ionization.

The predicted turbulent impurity flux is then where

$$\Gamma_j = -D_j \left(\frac{\partial n_j}{\partial r} + c_E\, \frac{n_j\, e_j}{T_j}\, \frac{\partial \Phi}{\partial r} \right) \tag{8.61}$$

renormalized turbulence theory transport formulas give

$$D = c_D\, \frac{\widetilde{\phi}}{B_T} \tag{8.62}$$

$$V_j = c_\mu\, \frac{\widetilde{\phi}}{B_T}\, \frac{e_j}{T_j}\, E_r \tag{8.63}$$

with the constant dimensionless ratio

$$c_E = \frac{c_\mu}{c_D} \tag{8.64}$$

less than or equal to unity.

8.8 Divertor Designs for Impurity Control

The design of the divertor in tokamaks strongly influences the impurity transport. Numerous experimental studies were made with variations of the divertor design. A comparison of the particle flux control from the divertor in the ASDEX Upgrade (AUG) [Gruber, *et al.* (1999)] with the fluxes measured in the JET tokamak is given by the ASDEX team in Angioni, *et al.* (2007). The comparison was aimed at optimizing the design of the divertor for the ITER machine.

For stellarators and the LHD steady-state machine, there are high spatial resolution transport measurements of trace amounts of the high-Z elements. Small pellets with small amounts of vandium, manganese and cobalt that neighbor the elements of the stainless steel in the vacuum vessel walls are injected in steady-state LHD plasmas [Sudo, *et al.* (2012)]. These experiments used the UV emission from the ionization stages and the K_α line radiation for following the motion of the heavy elements in the KeV core plasma. These LHD experiments show that there was a transport barrier that kept the injected trace elements in the core and prevented the intrinsic iron and chromium elements from the wall from reaching the core plasma.

An ion cyclotron range of frequencies (ICRF) system with power up to 6.0 MW and a lower-hybrid current drive (LHCD) system up to 4 MW were applied for heating and current drive experiments in Experimental Advanced Superconducting Tokamak (EAST). Significant progress was made with ICRF heating and LHCD for reaching the H-mode plasma operation in EAST. During 2010 and 2012 experimental campaigns, ICRF heating experiments were carried out at the fixed frequency of 27 MHz, achieving effective ions and electrons heating with the H minority heating (H-MH) mode. The H-MH mode produced good plasma performance, and realized H-mode using ICRF power alone in 2012. In 2010, H-modes were generated and sustained by LHCD alone, where lithium coating and gas puffing near the mouth of the LH launcher were applied to improve the LHCD power coupling and penetration into the core plasmas of H-modes. In 2012, the combination of LHCD and ICRH power extended the H-mode duration to over 30 s. H-modes with various types of edge-localized modes (ELMs) were achieved with $\mathrm{H}_{\mathrm{IPB98}}(y, 2)$ ranging from 0.7 to over unity. An overview of LHCD and ICRF Heating experiment and their application in achieving H-mode operation during these two EAST campaigns is presented in Zhang, *et al.* (2014).

8.9 Impurity Fluxes in JET and ASDEX-U Divertor Plasmas

Single null divertor geometry as in JET, ASDEX-Upgrade, and WEST produce a broken symmetry that lowers the power required for the L- to H-mode bifurcation. The bifurcation modifies the plasma transport properties. The transport of the minority species is known from theory and experiments in JET and ASDEX Upgrade (AUG) [Casson, *et al.* (2010)]. The data for the tungsten (W) accumulation is discussed from JET in Fig. 8.6 and from ASDEX Upgrade in Fig. 8.7 shows that the heavy impurity ions profiles are strongly changed by the broken poloidal symmetry and fast plasma rotation. Background electrostatic fields in the plasma are generated by the anisotropy created by external heating beams and RF antennas. Significant poloidal variations in the densities of the heavy impurities are measured. These asymmetries increase both the neoclassical and turbulent transport fluxes.

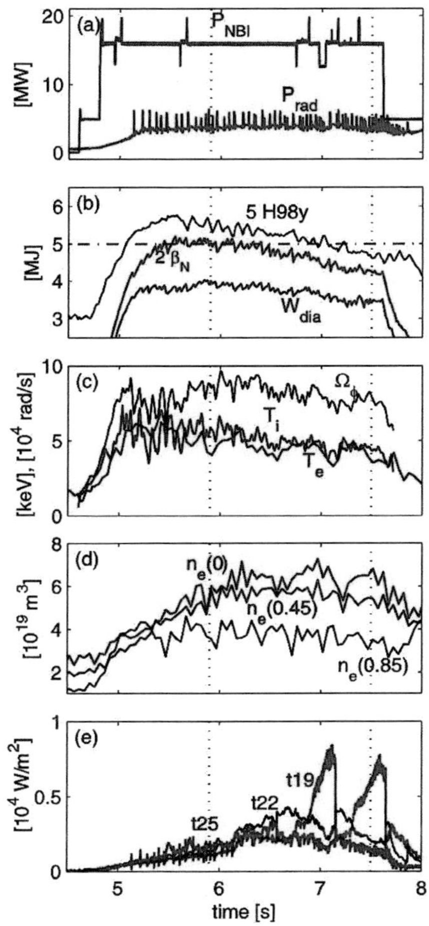

Fig. 8.6 Overview of a high-power NBI driven JET discharge used to analyze the inward transport of tungsten impurity ions. (a) shows the 17 MW of NBI power and the 3 MW radiated power. (b) gives the stored plasma energy, the plasma β, and the H-mode factor (c) shows the high rotation frequency Ω and the temperature drops due to ELMS (d) three positions $\rho = 0$, 0.45, and .85 for the density. (e) the radiated x-ray power from the tungsten ions. The first sawtooth crash at $t = 6.6$ s produces a small drop in T_e from the SXR diagnostic with the tungsten ions W off axis. The second sawtooth crash at $t = 7.5$ produces a large peak in SXR emission of $10^4 \mathrm{W/m^2}$ (frame e). The analysis shows that the tungsten ions W44+ and W 46+ are in the plasma center at $t = 7.5$ s. Earlier the tungsten ions are W41+ and W46 are at mid-radius and the core the ions are W38+ and W41+ [Casson, *et al.* (2014)].

The measured increases are found to increase the collisional tungsten transport fluxes by an order of magnitude.

Using tomography from soft x-ray diagnostics, Casson, *et al.* (2010) compared the measured tungsten impurity distributions with the theoretical-modeling predictions for the steady-state, 2-D transport measured in JET. The modeling with the enhanced neoclassical transport from the poloidal asymmetries increases the

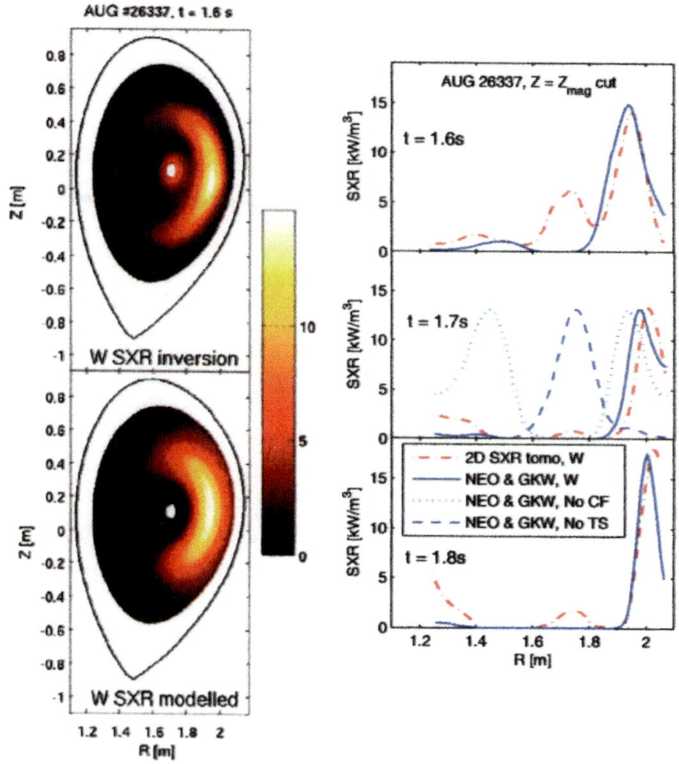

Fig. 8.7 Comparison of the predicted and soft x-ray measured tungsten impurity in the core of AUG. Tomographic inversions of soft x-ray (SXR) emission with Brehmstrahlung subtracted (top left) and W SXR emission forward modeled from the predicted W distribution (bottom left) for $t = 1.6$ s of AUG 26337. The maximum emission value is used to normalize the modeled result to the tomography (right). The same data is cut horizontally through the magnetic axis (all three time slices). For $t = 1.7$ s, the modeling excluding centrifugal force (CF) effects and temperature screening (TS) is also shown [Casson, *et al.* (2015)].

collisional neoclassical transport rates by an order of magnitude over the rates for symmetric circular poloidal flux surfaces.

The fluxes describe the tungsten accumulation in the central core of the plasma that depends on the density and temperature profiles and the toroidal rotational velocity of the plasma. The turbulent diffusion and collisional fluxes can increase or reduce the high-Z impurity accumulation in the core plasma. Ion cyclotron RF heating of the minority species significantly changes the radial tungsten fluxes [Liao, *et al.* (2011)].

Casson, *et al.* (2010) model the turbulent and neoclassical heavy impurity transport for the single null divertor geometry of JET using the numerical tools GKW and NEO which include the poloidal asymmetries from the single null divertor configuration. The high-Z transport predictions agree with the soft x-ray data from the core tungsten W distributions measured in the JET

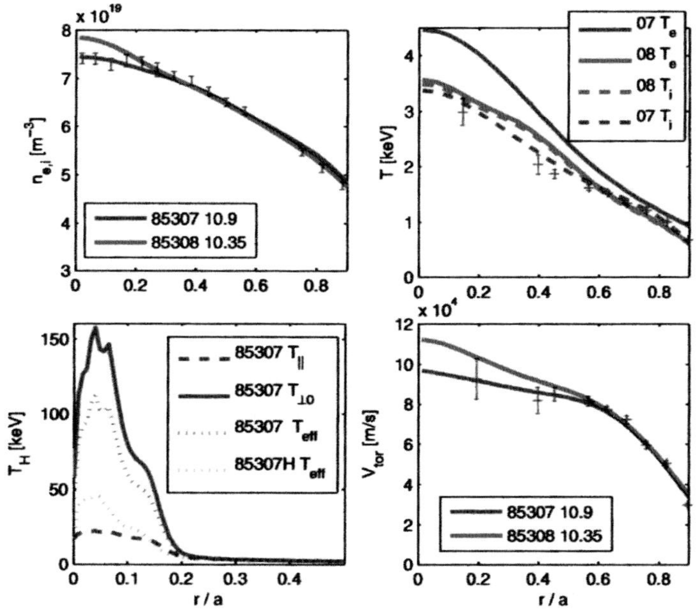

Fig. 8.8 Profiles for the JET modeling, with selected data and error bars with high power ICRH heating and $T_{\perp} \gg T_{\|}$. The density n_e and temperature T_e are profiles measured by Thompson scattering (LIDAR), T_i and V_{tor} are measured by charge exchange. (Bottom left) H minority temperatures produced by TORIC-SSFPQL for the ICRH case, inner radii only (the case 85307H is with half the ICRH power).

and the ASDEX-Upgrade experiments. The modeling showed that the effect of the poloidal asymmetries in neoclassical convection depends strongly on the collisionality regime. The asymmetries are needed to describe the measured heavy impurity distributions.

In the ASDEX-Upgrade H-mode with toroidal current overshoot, the flat density profiles measured confirm that neoclassical temperature screening effect is sufficiently strong to reduce the heavy impurity ion accumulation in the core plasma. The experiments show the trapped tungsten ions are in the outer half of the plasma, as shown in Fig. 8.7. Here, the centrifugal force effects decrease the tungsten peaking and enhances the neoclassical transport by an order of magnitude such that it dominates the impurity turbulent transport over most of the plasma radius.

In JET baseline H-modes, the centrifugal force effects increase the W peaking, in contrast to the hybrid scenario in which their effect on the overall peaking is small [Valisa, *et al.* (2011)]. Strong minority ICRH, heating enhances neoclassical impurity temperature screening and reverses the convection in the region of the ICRH in agreement with Angioni, *et al.* (2014). The anisotropy-induced poloidal asymmetry reduces neoclassical impurity diffusivity and the minority-impurity temperature screening is enhanced when both species are localized at the last closed flux surface. These effects are complementary to flatter density profiles and a wider region of turbulent diffusion in the ICRH plasmas which works to reduce the central W accumulation.

The impurity profiles are derived from tomographic inversions of the soft x-ray measurements and compared with the impurity profiles from the modeling codes. Satisfactory agreement is reported over a range of plasma conditions. This validation re-emphasizes that poloidal asymmetries are an essential ingredient for accurate modeling of heavy ion impurity transport. The modeling shows that the temperature gradients from externally-heated ion species significantly modifies the impurity temperature screening in the neoclassical fluxes.

8.9.1 *Tungsten transport in JET with ITER-like walls*

A typical JET plasma in the ILW regime is shot #82722 with $B_\phi = 2\,\mathrm{T}$, plasma current $I = 1.7\,\mathrm{MA}$ and low triangularity similar to Fig. 8.6. The neutral beam heating is $P_{\mathrm{nbi}} \sim 18\,\mathrm{MW}$ and the radiated power is $P_{\mathrm{rad}} \sim 4\,\mathrm{MW}$. The plasma energy was $W \sim 5\,\mathrm{MJ}$ during the flat-top of the discharge. The plasma in the quasi-steady state reached $T_i \sim T_e \sim 5\,\mathrm{KeV}$ and $n_e = 4.5 \times 10^{19}\mathrm{m}^{-3}$.

The simulation codes NEO and GKW used W^{+35} at edge and W^{+45} in core at time $t = 5.9\,\mathrm{s}$ with the fractional W^{+Z} densities of $n_W/n_e \cong 10^{-5}$ to 10^{-4}. These tungsten contributions are negligible in the quasi-neutrality equation but critical in the power balance and heat transport equations [Tokar, *et al.* (1997)] and JETTO/SANCO [Taroni, *et al.* (1994)]. Neoclassical transport is increased by the rotational forces by an order of magnitude in the core for typical NBI heated plasmas [Wong (1987), Romanelli and Ottaviani (1998), Fulop, *et al.* (1999)]. Electron cyclotron heating (ECH) is a tool for controlling the W transport as shown in ASDEX Uprade experiments [Neu, *et al.* (2003), McDermott, *et al.* (2011)]. For reference the tungsten atom has $Z = 74$ and $A = 184$.

8.10 ICRH Control of Turbulent Impurity Pinch in Divertor Tokamaks

Experiments in various tokamaks [ASDEX Upgrade, DIII-D, JET, TCV, TFTR, Tore Supra and EAST] show evidence for a turbulent particle pinch as reviewed by Bourdelle (2005). The evidence for the turbulent particle pinch is clearly present in the studies of particle transport in Tore Supra during long steady-state discharges Hoang, *et al.* (2003)]. The turbulent particle pinch was derived in Weiland, *et al.* (1989) and confirmed in nonlinear simulations by Garbet, *et al.* (2003). Experimentally, the particle pinch is observed to depend on the gradient of the magnetic field as well as the gradients of the density and temperature as described in Hoang, *et al.* (2004) and Angioni, *et al.* (2004). The observed variations with plasma parameters is shown by Bourdelle to agree sufficiently with the theory and simulations to give an understanding of the inward particle transport, often called the particle "pinch" in tokamaks. The particle pinch occurs when the temperature gradient is sufficiently greater than the density gradient. The inward particle flux with an outward thermal flux follow from the quasilinear formulas and is called the thermal pinch effect.

Plasmas with a particle-pinch term will evolve to a sufficiently-peaked density profile brings the net particle flux to vanish in the core and balance the particle sources from in the edge or scrape-off layer plasma.

This behavior is documented in many different tokamaks with different methods of density fueling from the edge plasma including NBI, gas puffing and laser blow-off experiments. The Tore Supra Equipe used advanced simulation codes and the collisional transport codes – particularly NCLASS – to model the density evolution and validate the codes used to predict the density profiles in a variety of tokamak experiments.

8.11 Dust Transport from Vessel Wall Erosion

Extensive dust dynamics data is available for the steady-state hydrogen plasmas in the Large Helical Device (LHD) from using high-speed infrared cameras. Figure 8.9 shows the typical paths of the dust particles measured with the infrared cameras in the Large Helical Device experiments between the outboard chamber wall and the plasma in the divertor chamber. Some dust particle trajectories enter the core plasma crossing the magnetic separatrix while most dust particles remain between the magnetic separatrix and metallic chamber walls. The dust particles are often reflected at the walls and remain in these plasmas for long periods. Some of the dust

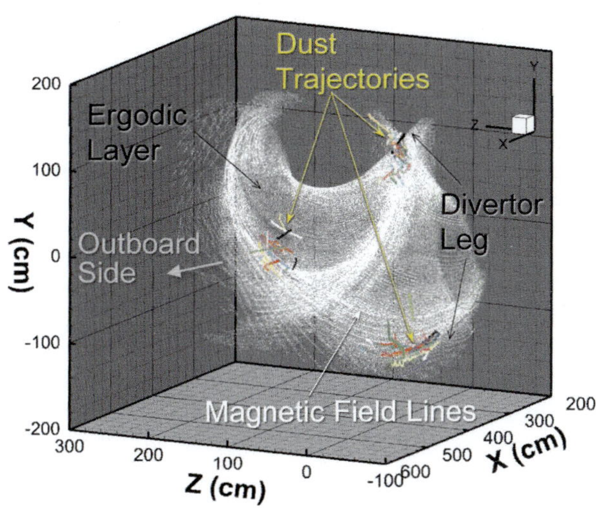

Fig. 8.9 Perspective showing the three-dimensional trajectories of dust particles from a bubble on the chamber wall in the divertor chamber of the Large Helical Device (LHD) [Shoji, *et al.* (2015)].

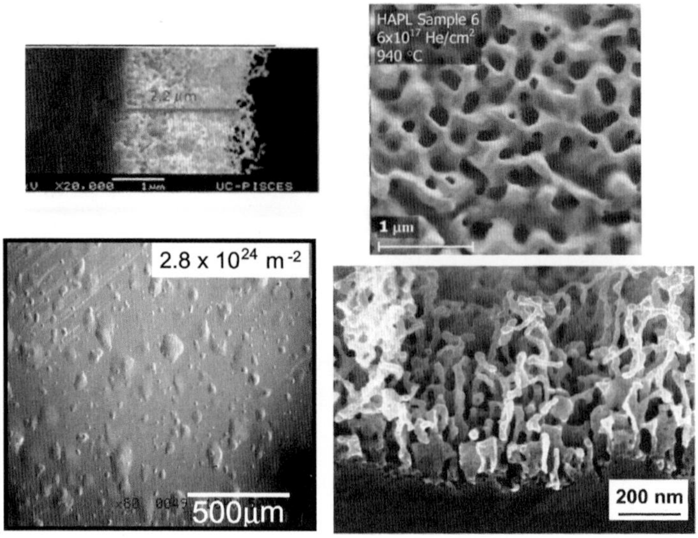

Fig. 8.10 The formation of dust bubbles on the chamber wall. The top left shows 2.2μ m tungsten fuzz surface layer observed at the PISCES-A facility [Doerner, *et al.* (2011)]. The top right pores formed from ruptured bubbles absorbed in the high-average power laser (HAPL) project [Cipiti and Kulcinski (2005)] with implantation of $6 \times 10^7 \text{He/cm}^2$ at 1213 K. The bottom left shows the blistering found from HiFIT irradiated tungsten [Ueda, *et al.* (2004)]. And the bottom right are experiments performed in the NAGDIS-Idevice showing the resulting morphology from subsurface pinholes leading to nanometer-sized fiber-form structures in irradiated tungsten [Kajita, *et al.* (2009)].

particles enter the core plasma. Their velocities are measured to be in the ranges 5-20 m/s. The dominant dust density is between the magnetic separatrix and the vessel wall with a significant fraction of the dust crossing the magnetic separatrix to enter the confined plasma.

The dominant forces on the dust are the plasma electric field, the drag force from the Coulomb collisions with plasma ions and the drag force from the electrons. The diameter of the carbon dust is 3-5 μm dust and about 1 μm for the metallic ion dust. The charged dust particles are reflecting from the walls thus remaining in the plasma for long periods.

The transport of the dust particles implies wall erosion due to evaporation of the surface material where even below the melting point of the wall material, there is sufficient kinetic energy in the collisions to overcome the surface chemical binding energies to induce dust cloud formations. This erosion in plasma facing components occurs in concert with changes in the surface and subsurface morphology of the material interface along chamber walls as shown in Fig. 8.10. For example, in tungsten wall surfaces nanometer-sized microstructures are shown in Kajita, *et al.* (2009), tungsten fuzz by Petty, *et al.* (2015), pore formation by Cipiti and Kulcinski (2005), and bubbles/blisters by Enomoto, *et al.* (2009), as shown in the four frames in Fig. 8.10.

In conditions commensurate to those expected in ITER, the tungsten armor on the first wall and divertor plate will accumulate gas molecules of H_2, D_2, and He from energetic plasma bombardment at the surface. The implanted molecules are trapped in lattice defects, even though W demonstrates low hydrogen solubility. The resulting gas pressures and flow stresses develop into blisters along the W surface, typically from a few microns (μ m) to several tens of μ m. These blisters dynamically change in size and shape, frequently leading to irregular bulging and rupturing as in the two lower frames in Fig. 8.10. Moreover, the irradiation from D^+ leads to more extreme blistering than H^+, and the ultimate blister size and behavior determines the type and amount of gas retention [Enomoto, *et al.* (2009)].

Worked Problem 8.1: Collisional mass flows and impurity transport (with temperature gradient screening) tokamaks

Once the impurities are released, they flow on the magnetic surface to set upon equilibrium poloidal flow patterns that gives rise to the Pfirsch-Schlüter transport enhancement factor of $1+2q^2$. The flows are fast and of fundamental importance. The key equations for the establishment of these flows in the absence of rotation are

$$e_1\, n_1 \left(\mathbf{E} + \frac{\boldsymbol{v}_1}{c} \times \mathbf{B}\right) - \nabla p_1 - m_1\, n_1\, \nu_{12}(\boldsymbol{u}_1 - \boldsymbol{u}_2) = 0 \quad (8.65)$$

$$e_2\, n_2 \left(\mathbf{E} + \frac{\boldsymbol{v}_2}{c} \times \mathbf{B}\right) - \nabla p_2 - m_2\, n_2\, \nu_{21}(\boldsymbol{u}_2 - \boldsymbol{u}_1) = 0. \quad (8.66)$$

From Newton's third law we have where the working gas is designated by 1 and the impurity ions by 2 with the equal and opposite frictional forces related through $m_1\, n_1\, \nu_{12} = m_2\, n_2\, \nu_{21}$.

Calculating the perpendicular flux and the parallel components we have

$$n_1\, \boldsymbol{u}_1 = \frac{n_1\, \mathbf{B}}{B^2} \times \left[-\mathbf{E} + \frac{\nabla p_1}{e_1\, n_1} + \frac{m_1\, n_1\, \nu_{12}}{e_1\, n_2}\,(\boldsymbol{u}_1 - \boldsymbol{u}_2)\right] \quad (8.67)$$

and

$$e_1\, n_1\, E_\parallel - \nabla_\parallel\, p_1 = m_1\, n_1\, \nu_{12}(u_{1\parallel} - u_{2\parallel}) \quad (8.68)$$

with the appropriate interchange of 1 and 2 for the second ion species.

Now the continuity equations $\nabla \cdot n\boldsymbol{u}_i = 0$ in the toroidal magnetic field forces the parallel flow $u_\parallel$ to be established on the time scale qR/v_T such that

$$\frac{B_\theta}{rB}\frac{\partial}{\partial\theta}\,(n_1\, u_{\parallel 1}) = -\nabla \cdot \left[\frac{n_1\, \mathbf{B}}{B^2} \times \left(-\mathbf{E} + \frac{\nabla p_1}{e_1\, n_1}\right)\right] \quad (8.69)$$

$$= \frac{2\sin\theta}{BR}\left(n_1\, E_r - \frac{1}{e_1}\frac{\partial p_1}{\partial r}\right) \quad (8.70)$$

giving

$$n_1\, u_{\parallel 1} = -\frac{2r\cos\theta}{B_\theta\, R}\left(n_1\, E_r - \frac{1}{e_1}\frac{\partial p_1}{\partial r}\right). \quad (8.71)$$

Computing $u_{\|1} - u_{\|2}$ from Eq. (8.71) and substituting into Eqs. (8.69) and (8.70) gives the poloidal angle variations on the magnetic surface

$$\frac{1}{Br}\frac{\partial}{\partial\theta}\left(e_1\, n_1\, \Phi + p_1\right) = \frac{m_1\, n_1\, \nu_{12}}{B_\theta^2\, R}\left(2r\cos\theta\right)\left(\frac{p_1'}{e_1\, n_1} - \frac{p_2'}{e_2\, n_2}\right) \quad (8.72)$$

and

$$\frac{1}{Br}\frac{\partial}{\partial\theta}\left(e_2\, n_2\, \Phi + p_2\right) = \frac{m_2\, n_2\, \nu_{21}}{B_\theta^2\, R}\left(2r\cos\theta\right)\left(\frac{p_1'}{e_1\, n_1} - \frac{p_2'}{e_2\, n_2}\right). \quad (8.73)$$

Now the poloidal $e_{1,2}E_\theta$ and $\partial p_{1,2}/\partial\theta$ forces crossed with the toroidal magnetic field generate the radial convective fluxes given by the radial component of Eq. (8.67). From Eqs. (8.67) and (8.72)–(8.73) we derive the radial fluxes

$$(n_1\boldsymbol{u}_1)_r = -\frac{B_\phi}{B^2}\left(1+\frac{2r\,B^2\cos\theta}{R\,B_\theta^2}\right)\left(\frac{m_1\, n_1\, \nu_{12}}{e_1\, B}\right)\left(\frac{p_1'}{e_1\, n_1} - \frac{p_2'}{e_2\, n_2}\right) \quad (8.74)$$

$$(n_2\boldsymbol{u}_2)_r = -\frac{B_\phi}{B^2}\left(1+\frac{2r\,B^2\cos\theta}{R\,B_\theta^2}\right)\left(\frac{m_1\, n_1\, \nu_{12}}{e_2\, B}\right)\left(\frac{p_2'}{e_2\, n_2} - \frac{p_1'}{e_1\, n_1}\right). \quad (8.75)$$

These formulas show that there is ambipolarity in the ion-impurity fluxes before making the surface average.

The formulas (8.74) and (8.75) also give the full r-θ pattern of the impurity convection from E_θ and $\partial p_{1,2}/\partial\theta$ before making the flux surface average. One sees that the broken symmetry from a single null divertor will enhance the resulting fluxes. Once the surface average $dS = r d\theta(R_0 + r\cos\theta)d\phi$ is carried out by integrating over dS, the direct collisional random walk flux and the toroidal-poloidal flow enhancement combine to give

$$D_z^{ps} = (1 + 2q^2)\nu_{zi}\,\rho_z^2 \quad (8.76)$$

with $\Gamma_i = 2\pi\int_{-\pi}^{\pi} R(\theta) r d\theta\,(n_i\boldsymbol{u}_i)_r$, where D_z^{ps} is defined such that

$$\Gamma_z = -D_z^{ps}\,\frac{1}{T_z}\left(\frac{\partial p_z}{\partial r} - \frac{n_z\, Z}{n_i\, Z_i}\frac{\partial p_i}{\partial r}\right).$$

For an isothermal plasma the condition $\Gamma_z = 0$ is achieved by the inward pinch velocity v_z when

$$n_z = n_i^{(ZT_i/T_z)} \simeq (n_i)^Z \quad (8.77)$$

when $T_z = T_i$. This strong exponential dependence on Z indicates the potential strength of high-Z impurity peaking in the core plasma from this an analytic analysis.

The Rutherford (1974) generalization of Γ_z to include the thermoelectric effect in the frictional force in Eq. (8.66) which includes $\boldsymbol{R}_{iz} = -c_1(\frac{m_i\, n_i}{\tau_{iz}})(\boldsymbol{v}_i - \boldsymbol{v}_z) - c_2\, n_i\,\nabla_\| T_i$ with the values of c_1, c_2 given by first order rational polynomials in $\alpha = n_z\, Z^2/n_e$. The Rutherford formula for the impurity particle flux is

$$\Gamma_z = -\frac{n_i\,\rho_i^2\,q^2}{T_i\,Z^2\,n_z\,\tau_{iz}}\left[(c_1 + c_2^2/c_3)\left(\frac{\partial p_z}{\partial r} - \frac{n_z\,Z}{n_i}\frac{\partial p_i}{\partial r}\right) - \frac{5}{2}\frac{c_2}{c_3}\,(n_z\,Z)\frac{\partial T_i}{\partial r}\right] \tag{8.78}$$

where c_2 and c_3 arise from the parallel ion thermal conductivity

$$q_{i\parallel} = c_2\,n_i\,T_i\,u_\parallel - c_3\left(\frac{n_i T_i \tau_{iz}}{m_i}\right)\nabla_\parallel\,T_i. \tag{8.79}$$

The outward Γ_z from ion temperature gradient terms in Eq. (8.78) are important (neoclassical) temperature gradient screening terms. In the trace impurity experiments the terms proportional to n_z/n_i in Eq. (8.78) can be dropped.

The question arises as to how this impurity flux depends on the charge state and the mass of the impurity ions. The $\mathbf{E}\times\mathbf{B}$ turbulent flux is independent of Z and m_z provided $k_\perp\,\rho_z \ll 1$ and $k_\parallel\,v_z/\omega \ll 1$. Now using Eq. (8.52) we find that

$$\rho_z^2\,\nu_{zi} = \frac{m_z\,T_z}{e^2\,Z^2\,B^2}\cdot\frac{n_i\,e^4\,Z_i^2\,Z^2\,\ell n\,\Lambda}{m_z\,m_i^{-1/2}\,T_i^{3/2}} \tag{8.80}$$

so that both m_z and Z^2 drop out. Thus, all charge states of Sc, Ti, and W etc. have the same D_z^{ps} and are transported at the same rate from the collisions with the working gas ions in this collisional model.

8.12 Plasma-Wall Interactions in Long-Pulse Fusion Plasmas

The plasma-wall interactions include the large metal surface of the confinement vessel coated with beryllium (Be), the limiter surfaces and surfaces in the divertor chambers covered with tungsten (W), with special emphasis on the metallic surfaces of the high-powered RF antennas. The outboard vessel surfaces are subjected to $10\,\mathrm{MW/m^2}$ transients from the edge-localized modes (ELMs). Fusion plasma are maintained for quasi-steady operation by antennas that launch radio frequency (RF) waves with powers of up to 20 MW for heating the hydrogenic ions (ICRF) and driving the plasma current and heating the electrons (LHCD) for minutes.

The W surfaces in Tore Supra contained 6-min plasmas in 3-5 KeV temperature plasma with one gigajoule energy passing from the RF antennas to the electrons and then exhausted through a divertor chamber. These same high-power RF antennas are driving the WEST plasma-wall interactions. WEST – in a collaboration with EAST – is at the forefront of research for plasma-surface interaction research. The EAST and WEST machines are delivering definitive experiments on plasma-wall interactions.

A key issue for a magnetic fusion confinement system is the accumulation of plasma impurity elements from the confinement vessel walls. The wall contact with the hot plasma is called the plasma-facing component (PFC) problem. The inward impurity particle flux is single ions W^+, Mo^+, Fe^+ and clumps of partially ionized impurity atoms called dust. In this dusty plasma drift wave turbulence with certain phase relations between the electric field and the dust particle density fluctuations give the net dust transport. Increase of the wall temperature, still well below the melting temperature, creates blistering that adds to the sputtering of the surface coatings from the fast particle impacts on the hot vessel walls. Both the turbulent impurity transport and the neoclassical collisional transport predict an accumulation of the high-Z materials like Ni, Ti and W from the chamber walls in the core plasma. Extensive measurements of impurity profiles are now in the literature from Alcator C-Mod data [Rowan, *et al.* (2008)] and the Large Helical Device [Sudo, *et al.* (2013)] applying drift wave turbulence models for impurity ion transport and ionization data with transport codes.

Experiments from the LHD team inject 500 micron diameter plastic pellets that are filled with atoms of V23, Mn25, Co27 which are the neighboring elements of the vacuum vessel walls Ni28, Fe26 and Cr24. The injected pellets release along the observed path the high-Z impurities which are then tracked by measuring the soft X-ray spectra from their K-alpha line emissions. The experiments have given unprecedented detail information on the transport in the plasma of the metallic wall impurities in the KeV electron temperature plasmas. Initial indication from the collaboration with the LHD-team is reported in the 2013 impurity transport publication [Fu, *et al.* (2013)]. While first results from the TESPEL experiments are presented in Sudo, *et al.* (2012). Bursts of turbulence from the scrape-off layer are described by Meyerson, *et al.* (2014) that include the high-Z impurity ion transport. There is experimental evidence of regimes with inward impurity transport leading to the injection of the high-Z wall atoms into the core plasma.

Dense sets of flush-mounted Langmuir probes are built into the plasma-facing wall components to provide detailed data for understanding the stresses and temperature limits for blistering and bubble formation on the PFC surfaces. Those components in fusion machines will be subjected to intense plasma bombardment including the surfaces and parts in the divertor chamber, the divertor dome, and the RF antennas. These components must be actively cooled by forcing water through the core of large arrays of rectangular tungsten blocks as shown in Fig. 8.11. These PFCs are long rectangular tubes with cross-sections of about $3\,\mathrm{cm} \times 3\,\mathrm{cm}$ are called "monoblocks". The power handling limits of the tungsten monoblocks are being determined on the WEST tokamak. The name WEST derives from these features: Tungsten (W) Environment (E) Steady state (S) Tokamak (T).

WEST construction was completed at the time of writing (2016) and began steady-state operation in 2017 to model of the $500\,\mathrm{m}^3$ ITER machine in a $15\,\mathrm{m}^3$ superconducting steady-state tokamak. The principal heating and current drive for the steady-state operation is provided by the two high-power (5 GW) lower-hybrid RF antennas. These RF antennas were successful in producing steady fusion-grade plasmas for periods from several minutes up to six minutes in the predecessor –

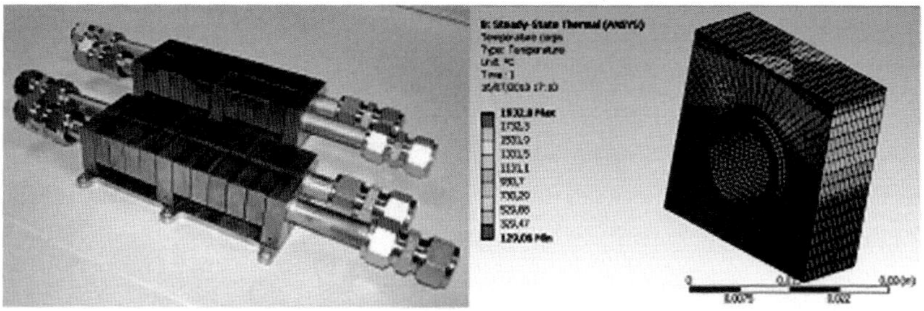

Fig. 8.11 Plasma wall architecture for WEST. On the left, the assembly of tungsten monoblocks with water pipes for cooling the first wall of the superconducting steady-state machine. On the right, simulation of the temperature distributions on a monoblock approaching the critical condition for blistering (M. Grossman, CEA, 2014).

Tore Supra – with circular cross-section and conventional plasma limiters. WEST has the same single-null divertor structure as planned for ITER.

8.13 Heat Load Limits on the Metal Surfaces of the Confinement Vessel

One of the severest problems of a fusion power system is the plasma-surface interaction problem. The flux of plasma on the plasma walls and divertor plates in the standard toroidal systems exceeds the limits for long-time (years) survival of the metallic chamber surfaces. Current estimates are that the walls may need replacing after a period of a few years. Their lifetime is unknown. The problem is that the plasma flux to the machine surfaces is sufficient to sputter the surfaces severely producing impurity contaminated plasmas that would have a low nuclear fusion reactivity due to the lower $n_D + n_T = n_e - Z_{\text{eff}} n_Z$ nuclear fuel density. The current solution for the ITER machine is to coat the walls with beryllium since this is the lowest-Z material that can withstand the estimated plasma flux to the chamber walls. The JET machine showed that replacing the carbon/graphite coating with beryllium lowered the plasma impurity content by an order of magnitude.

Chapter 9

Coherent Structures in Plasmas

Vortices are created in fluids by relative motions of parts of the fluids with different speeds and thus occur universally in nature as seen in rivers and the atmosphere. The most efficient way to mix inhomogeneous materials is to create vortices in the material. When one mixes the various ingredients in a cake recipe the process is quick by using an appropriate spoon in circular motion to create the vortical flows.

Mixing machines of many types create homogeneous fluids from the inhomogeneous initial materials. For gases the processes is widely used to create a homogeneous mixture. One popular car advertised that their compound vortex combustion chamber gave their engines higher efficiency than that of its competitors. The same situation applies to plasmas. A well-documented plasma example is that from the flow of the solar wind plasma around the Earth's magnetosphere. Both spacecraft data and computer simulations show that the Kelvin–Helmholtz instability grows from the subsolar apex of the magnetopause along the flanks forming vortex convection cells which mixes the solar wind plasma with its higher helium concentration with the plasma trapped in the Earth's magnetic field with its higher oxygen concentration.

9.1 Kelvin–Helmholtz Instability and Vortices in Magnetized Plasma

In this chapter we investigate the nature of the vortices in magnetized, inhomogeneous plasmas. The theoretical structure of the problem and the numerical simulations of the resulting nonlinear partial differential equations are examined. The data from controlled plasma experiments measuring precisely the vorticity of the electrically conducting plasma with a probe expressly designed to measure the plasma vorticity.

Many of the partial differential equations in plasma physics that describe the nonlinear evolution of the plasma have the vorticity dynamics as their central component. This statement applies to such diverse phenomenon as the magnetized Rayleigh-Taylor instability, the dynamics of magnetic reconnection and the propagation of drift waves. For shear Alfvén waves the two essential partial differential equations are one for the evolution of the plasma vorticity forced by the divergence of the parallel plasma electrical currents and the second equation for the dynamics of the magnetic flux function associate with the bending of the magnetic field lines.

These plasma vortices are quasi-two-dimensional since the ambient magnetic field $B\boldsymbol{e}_z$ makes the spatial variation slow along the magnetic field. Mathematically, this appears more concisely expressed in the Fourier transformed space derived from the 3D Euclidean space. In Fourier space each fluctuation is represented by a wavenumber vector $\boldsymbol{k}$. The slow variation along the field is contained in the energy spectrum of the fluctuations as concentrated in the regions where $k_\perp \gg k_\parallel$ where $k_\parallel = \boldsymbol{b} \cdot \boldsymbol{k}$ and $\boldsymbol{k}_\perp = \boldsymbol{b} \times (\boldsymbol{k} \times \boldsymbol{b})$. Here $\boldsymbol{b}$ is the unit vector along the magnetic field. Often the nonlinear dynamics is most accurately described by computer simulations that advance the fields in time in Fourier transform space.

To model the Kelvin–Helmholtz instability we consider the electrostatic response of the plasma in a two-fluid model, where the dynamics for the electrostatic potential $\varphi(\boldsymbol{x}, t)$ is obtained from the conservation of charge under the assumption of quasineutrality, i.e.,

$$\nabla \cdot \boldsymbol{j} = 0. \tag{9.1}$$

For the 2D model with $\boldsymbol{j} = \boldsymbol{j}_p$, the polarization current, the dynamics reduces as follows, with the nonlinear polarization current

$$\boldsymbol{j}_p = -\frac{m_i n_i}{B^2} \frac{d}{dt} \nabla \varphi = -\frac{m_i n_i}{B^2} \left(\frac{\partial}{\partial t} + \boldsymbol{v}_E \cdot \nabla \right) \nabla \varphi \tag{9.2}$$

where

$$\boldsymbol{v}_E = \frac{\mathbf{E} \times \mathbf{B}}{B^2} = \frac{1}{B} \widehat{\mathbf{z}} \times \nabla \varphi. \tag{9.3}$$

Substituting Eq. (9.2) into the current closure statement of Eq. (9.1) yields the dynamical equation for the vorticity

$$\left(\frac{\partial}{\partial t} + \vec{v}_E \cdot \nabla \right) \nabla^2 \varphi = \frac{\partial}{\partial t} \nabla^2 \varphi + [\varphi, \nabla^2 \varphi] = 0 \tag{9.4}$$

where we have introduced the Poisson bracket $[f, g]$ between the two 2D functions f, g:

$$[f, g] \equiv \frac{\partial f}{\partial x} \frac{\partial g}{\partial y} - \frac{\partial f}{\partial y} \frac{\partial g}{\partial x}. \tag{9.5}$$

In the vorticity equation the Poisson bracket arises from the directional derivative of the vorticity along the $\mathbf{E} \times \mathbf{B}$ flow velocity. Now we consider the shear flows and vorticities in the LAPD experiments.

In the velocity shear layer $v_E(x)$ the density scale length L_{n_e} is calculated from the spatial profile of the mean ion saturation current. For the plasma in Sec. 9.2 this gives the density gradient scale length $L_{n_e} \sim 15\,\mathrm{cm}$ which is large enough to justify neglect the density gradients in the shear flow modeling. The velocity scale length L_v is approximately $15\,\rho_i \simeq 4.5\,\mathrm{cm}$ where $\rho_i = 3\,\mathrm{mm}$.

The simple KH model in Eq. (9.4) has canceling electron and ion currents from the $\vec{E} \times \vec{B}$ motion. The vorticity equation, (9.4), is the statement that the divergence of the cross-field electric current from ion inertia vanishes.

Considering a slab approximation for this plasma we take x as the local radial coordinate and y to be the periodic polar coordinate. Equation (9.4) is linearized around a steady-state poloidal flow $v_y(x)$ like that in the LAPD plasma rotation. By doing so we obtain Rayleigh's eigenvalue equation given by

Rayleigh Equation

$$\varphi_k''(x) - k^2\varphi_k(x) - \frac{kv_y''(x)}{kv_y(x) - \omega}\varphi_k(x) = 0. \tag{9.6}$$

With boundary conditions on $\varphi(x)$ at x_1 and x_2, the Rayleigh equation defines a non-normal eigenvalue problem with non-orthogonal basis vectors. The shear flow problem then has numerous special features given in Antar (2003), Burrell (1997), Kim, *et al.* (2006), and Horton, *et al.* (1987, 2010).

9.2 Drift Wave Models for LAPD

During the plasma rotation from the radial electric bias pulse, the maximum velocity in the flow corresponds to an electric field $E \approx 1.5\,\mathrm{kV/m}$ which occurs approximately at $r \approx 0.4\,\mathrm{m}$. Taking the magnetic field to be $B = 0.075\,\mathrm{T}$ along the z direction we obtain

$$v_{\max} = \mathbf{E} \times \mathbf{B} \approx 2 \times 10^4\,\mathrm{m/s}. \tag{9.7}$$

The eigenvalue problem for an idealized triangular flow profile that models the equilibrium electric field profile is solved and the flow profile defined by

$$v_y(x) = v_{\max} \begin{cases} 0 & x \leq -1 \\ (1 - |x|) & |x| \leq 1 \\ 0 & x \geq 1 \end{cases} . \tag{9.8}$$

Here the velocity is normalized to $v_{\max}$, and all lengths to $a = 0.1\,\mathrm{m}$, corresponding to the half width of the shear layer. The flow in Eq. (9.8) is called a jet in Chapter 6 with maximum velocity $v_{\max}$. Dimensionless variables are used with $x = \pm 1$ measuring the width $a = 0.1\,\mathrm{m}$ of the stream and $v_y = 1$ measuring the maximum velocity $v_{\max}$ of the stream.

Differentiation with respect to x gives

$$v_y''(x) = \delta(x+1) - 2\delta(x) + \delta(x-1). \tag{9.9}$$

Therefore the eigenfunctions obey the simple equation

$$\varphi_k''(x) - k^2\varphi_k(x) = 0 \tag{9.10}$$

except at the points $x = 0, \pm 1$. The jet flow has a steep, localized vorticity gradient, modeled by Eq. (9.9) giving secondary connection conditions, which are derived by integrating Eq. (9.6) across the singular points. Doing so results in the three continuity conditions

$$\Delta[\varphi_k'(-1)] = -\frac{k}{\omega}\varphi_k(-1) \tag{9.11}$$

$$\Delta[\varphi_k'(0)] = -\frac{2k}{kv_0 - \omega}\varphi_k(0) \tag{9.12}$$

$$\Delta[\varphi_k'(1)] = -\frac{k}{\omega}\varphi_k(1) \tag{9.13}$$

where we have defined the difference operator

$$\Delta[f(x_0)] \equiv f(x_0^+) - f(x_0^-). \tag{9.14}$$

In practice, the profiles are continuous, but vary rapidly over a small scale δ. The jump conditions are well satisfied if $k\delta \ll 1$; for $k = m/r$ this is equivalent to $m \ll r/\delta$.

Equation (9.6) is invariant under the transformation $x \to -x$, which means that if $\varphi_k(x)$ is a solution to Eq. (9.6) with eigenvalue ω, then $\varphi(-x)$ is also a solution with the same eigenvalue. This fact allows us to find eigenfunctions with a definite parity. Let us first look for odd eigenfunctions, i.e., eigenfunctions such that $\varphi_k(x) = -\varphi_k(-x)$. The general solution to the Rayleigh equation, Eq. (9.6), in this case is of the form

$$\varphi_k^{\text{odd}}(x) = \begin{cases} Ae^{-k|x|}\text{sgn}(x) & |x| > 1 \\ B\cosh(kx)\text{sgn}(x) + C\sinh(kx) & |x| \le 1 \end{cases} \tag{9.15}$$

where we have chosen the solution to decay at $x \to \pm\infty$. By using the matching conditions at the singular points, we obtain the dispersion relation

$$\omega_k = \frac{1}{2}\left(1 - e^{-2k}\right). \tag{9.16}$$

All the odd modes are neutrally stable and are called varicose modes in the neutral fluid literature and in Chapter 6. These modes, however, are important in the non-linear analysis and magnetic reconnection in Chapter 14.

Even Parity Modes

Now we proceed to consider the even modes, which satisfy $\varphi_k(x) = \varphi_k(-x)$ which are called sinuous modes in fluids. The general solution to the Rayleigh equation in this case is

$$\varphi_k^{\text{even}}(x) = \begin{cases} Ae^{-k|x|} & |x| > 1 \\ B\cosh(kx) + C\sinh(k|x|) & |x| \le 1. \end{cases} \tag{9.17}$$

Applying the conditions at the singular point we obtain

$$\omega_k = \frac{1}{4}\left[(2k + e^{2k} - 1) \pm \sqrt{G(k)}\right] \tag{9.18}$$

where

$$G(k) \equiv 9 - 10e^{-2k} - 12k + e^{-4k} - 4e^{-2k}k + 4k^2. \tag{9.19}$$

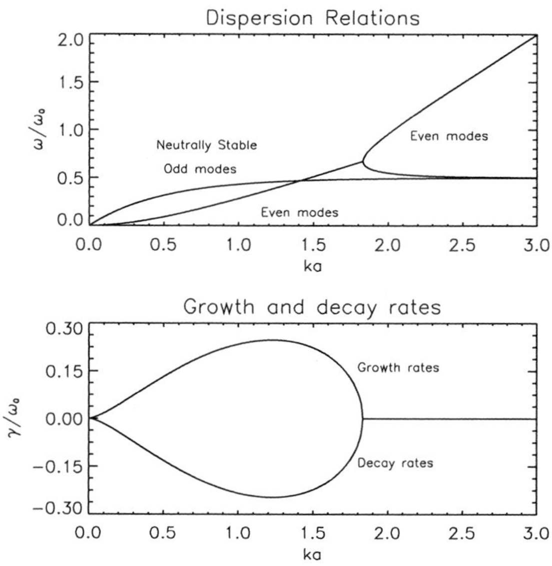

Fig. 9.1 Frequency and growth rate as a function of $k_\theta a$ for the model of the equilibrium radial electric field measured in the plasma. There are two modes: the unstable Kelvin–Helmholtz mode and the neutrally stable modes with odd symmetry corresponding to a wavy motion of the jet. The maximum velocity v_0 drops to zero at $\Delta r = \pm a$ and the reference frequency is $\omega_0 = v_0/a$.

The function $G(k)$ is negative for $0 \leq k \leq k_c$, with $k_c \approx 1.833$. Hence, in this range, the eigenvalues are

$$\omega_{k\pm} = \omega_R \pm i\gamma_k \tag{9.20}$$

where

$$\omega_R(k) = \frac{1}{4}\left(2k + e^{2k} - 1\right) \tag{9.21}$$

$$\gamma_k = \frac{1}{4}\sqrt{-G(k)}. \tag{9.22}$$

In this case modes with eigenvalue ω_{k+} are unstable and grow according to the values in Fig. 9.1 and Table 9.1

$$|\varphi_k(x,t)| = |\varphi_k(x)|\, e^{\gamma_k t}. \tag{9.23}$$

On the other hand we see that for every growing mode there corresponds a damped mode, due to the Hamiltonian structure of the system [Morrison (1998)], Balmforth and Morrison (2002)], namely, modes with eigenvalue ω_k. If we include dissipation from viscosity and resistivity, the Hamiltonian symplectic structure is broken. However, dissipation is more important for the high-k modes, while the nonlinearity transfers energy to low-k modes.

The general wave function is of the form

$$\varphi(x,y,t) = \varphi_0(x) + \mathrm{Re}\left[\sum_k e^{\gamma_k t}\varphi_k(x)e^{i(ky-\omega_k t)}\right]. \tag{9.24}$$

Table 9.1 Linear KH eigenfrequencies for the triangular flow profile model in Eq. (9.8) with frequency and growth rate given for the LAPD experiment.

Mode Number m	ω_0 (KHz)	$\gamma\,(ks^{-1})$
1	5	13.6
2	18	33.3
3	36.8	52.8
4	59.7	70.1
5	85.6	84
6	113.5	93.7
7	143	98.4
8	173.6	97.4
9	205	89.1
10	237	69.4
11	273.7	0

Within the range $0 \leq k \leq k_c$ we have the frequencies and growth rates given in Eqs. (9.21), (9.22) and Fig. 9.1.

The fastest growing mode corresponds to $m = 7$. Figure 9.2 shows isolines of the total potential

$$\varphi(x, y, t) = \varphi_0(x) + \varphi_k(x, y, t), \tag{9.25}$$

from a simulation with the step-up flow profile where

$$\varphi_0(x) = \begin{cases} 0 & x \leq -1 \\ \frac{1}{2}x^2 + x + \frac{1}{2} & -1 \leq x \leq 0 \\ -\frac{1}{2}x^2 + x + \frac{1}{2} & 0 \leq x \leq 1 \\ 1 & x > 1 \end{cases} \tag{9.26}$$

is the basic flow-stream function from a vee-shaped electric field profile.

In the turbulent flow layer there is a broad band of fluctuation frequencies that is difficult to associate with the linear eigenmode frequencies shown in Table 9.1. The power spectrum of the floating potential and the vorticity for different time series $\Delta T = N_T \Delta t$ was computed for time spans from $N_T = 256$ to 2048 and the results of the spectra over the M ensembles have been averaged and defined by dividing the stationary section of the bias pulse into M records of length $N_T \Delta t$. The result shows a broadband frequency power spectrum where the power decreases as f^{-3} from $f_1 = 1/N_T \Delta t \sim 1\,\text{kHz}$ to $f_{\max} = \pi/\Delta t = 1.6\,\text{MHz}$ in the case where $N_T = 1024$. Thus, the bulk of the turbulence power is at the low-frequency end of the spectrum.

The swirling motion of neutral fluids and plasmas is quantified by the vorticity, which can be viewed as the source function for the velocity field, in the same sense that an electric current is the source for a magnetic field. Vorticity is thus a primary physical quantity in equations for fluids and plasma dynamics. For quasi two-dimensional velocity flow fields that occur in geophysical fluid dynamics (GFD)

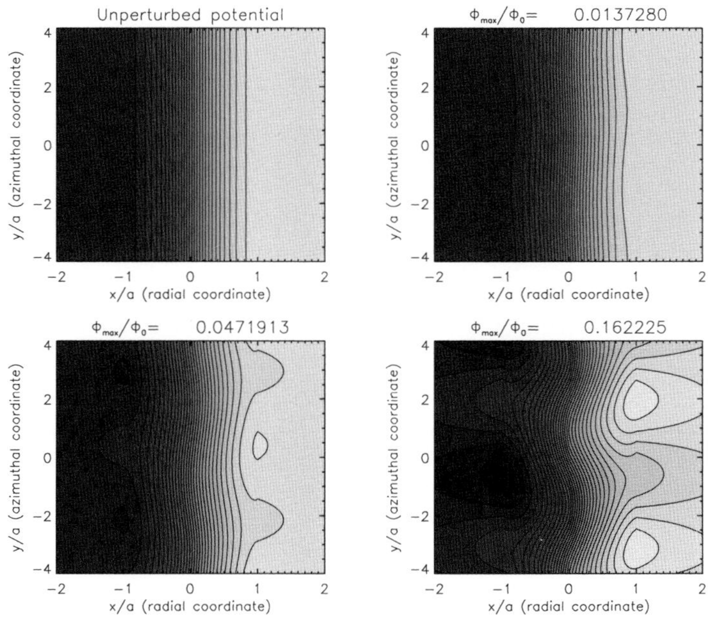

Fig. 9.2 Simulation data for the stream function $\psi = \phi/B$ from the isopotentials of the unstable eigenmode of Eq. (9.6). The last frame in the saturated state shows the alternation of the vortex rotational directions across the jet.

and in magnetized plasmas, it is essential to measure the vorticity field in order to understand quantitatively the dynamics of vortices and other coherent structures in the flow [Maurel and Petitjeans (1999), Horton, *et al.* (2000)].

For shallow water and atmospheric flows of depth H in rotating vessels and on planets with Coriolis frequency f there is a stream function analogous to the plasma stream function $\varphi(x, y, t)/B$. For quasi two-dimensional flows there is a scalar stream function whose Laplacian is the vorticity. In GFD the stream function is $\psi = gH(x, y, t)/f$, where $gH(x, y, t)$ is the gravitational potential energy in the height H of a column of neutral fluid and f is the Coriolis parameter from the rotation of the planet. In magnetized plasmas the stream function is $\psi = \varphi(x, y, t)/B$, where φ is the electrostatic potential and B is the ambient magnetic field strength.

The fact that the plasma stream function is the electrostatic potential, which can be measured directly by Langmuir probes, gives an unique advantage to the measurement of vorticity in plasmas over the measurement of vorticity in neutral fluids. In neutral fluids, vorticity is constructed from velocity field measurements using particle image velocimetry (PIV), in which digital images of neutrally buoyant test particles (or seeds) in the fluid are processed with sophisticated computer algorithms to infer the flow velocity, an expensive and somewhat complicated procedure that leads to significant levels of noise.

The coalescence and scattering of monopolar and dipolar drift wave vortices are analyzed with a background of drift waves with and without the KdV nonlinearity and compared with fluid Rossby wave vortices and waves [Horton, *et al.* (1990)] using theory and simulations.

A plasma vorticity probe is designed for measuring directly the vorticity of the plasma motion. Fast digital measurements of the field-aligned vorticity $\omega(x, y, t) = \nabla^2\varphi/B$ can be measured with the vorticity probe. Here we describe the design of the vorticity probe and its use in large steady-state cylindrical plasma in the large plasma device (LAPD). The experimental situation chosen is one in which the Kelvin–Helmholtz instability is present due to a strongly localized radial electric field at the mid radius of the confinement vessel.

The vorticity probe measures both the DC and AC components of the vorticity with appropriate time filter. The principle of the vorticity probe is to use Langmuir probes in the stencil of the discretized Laplacian operator in the layout as shown in Fig. 9.4.

The experiments are performed in the upgraded Large Plasma Device shown in Fig. 9.3 at UCLA, which is an 18 m long, $\lesssim 1$ m diameter magnetized plasma column created by a pulsed discharge from a Barium Oxide coated emissive cathode. The typical plasma parameters in LAPD are $n_e \sim 5 \times 10^{12}\,\text{cm}^{-3}$, $T_e \lesssim 15\,\text{eV}$, $T_i \lesssim 1\,\text{eV}$, $B < 0.2\,\text{T}$ with helium, argon, and neon as the working gas ($P_{\text{fill}} \sim 5 \times 10^{-5}$Torr). Discharges in LAPD are typically fully ionized, but some residual neutrals are present. A conservative estimate of ion-neutral collisionality is obtained by assuming a neutral density equal to the gas density before ionization (consistent with the fill pressure at room temperature). This assumption yields a small ion-neutral collision frequency $\nu_{\text{in}} = n_0 v_i \sigma_{\text{in}} \sim 100/\text{s}$. These collisions produce a small background viscosity for the plasma that is negligible for the wavenumbers relevant to the fluctuations considered here. A drawing of the apparatus with a schematic of the discharge circuit and rotation biasing circuit is shown in Fig. 9.3.

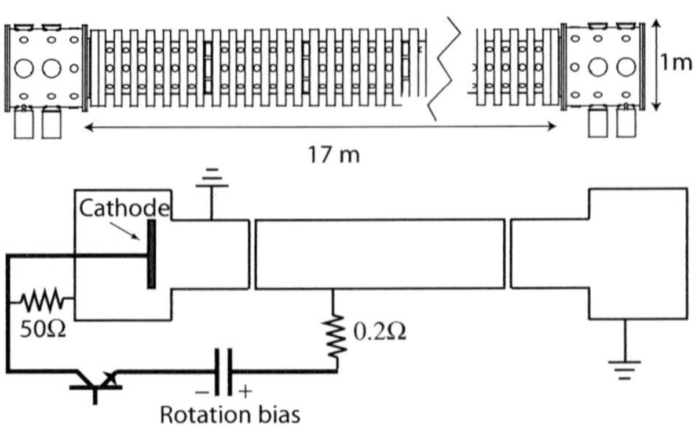

Fig. 9.3 Schematic of the large plasma device (LAPD), including a diagram of the circuit used for plasma column biasing.

9.2.1 *The vorticity probe*

It is well known that the most natural quantity for describing eddies or vortices in neutral fluids and in plasmas is the fluid vorticity, defined as

$$\boldsymbol{\omega} = \nabla \times \boldsymbol{v}. \tag{9.27}$$

This gives a local measure of the circulation of the velocity field at every point in the fluid plasma.

In the limit of a uniform magnetic field $B\hat{z}$ and with an electrostatic field $\mathbf{E} = -\nabla\varphi(\boldsymbol{x}, \boldsymbol{t})$, the ions and electrons move across the $\mathbf{B}$-field with velocities

$$v_x = -\frac{1}{B}\frac{\partial\varphi}{\partial y}(x, y, t) \tag{9.28}$$

$$v_y = \frac{1}{B}\frac{\partial\varphi}{\partial x}(x, y, z, t). \tag{9.29}$$

The component of the vorticity vector $\boldsymbol{\omega} = \boldsymbol{\nabla} \times \boldsymbol{v}$ along $\mathbf{B}$ is given by

$$\omega = \frac{\partial v_y}{\partial x} - \frac{\partial v_x}{\partial y} = \frac{1}{B}\nabla_\perp^2\varphi(x, y, z, t). \tag{9.30}$$

The vorticity is measured with the vorticity probe shown in Fig. 9.4. From the known spacing $\Delta x = \Delta y = h$ of the four corners of the probe with respect to the center of the probe, we use the sum of the floating potentials of the corner, minus four times the floating potential at the center, $\varphi_1 + \varphi_2 + \varphi_3 + \varphi_4 - 4\varphi_0$ to obtain the vorticity ω.

The vorticity calculated in this way from the floating potential is valid as long as one assumes that the number of primary electrons at probe positions is small and that the plasma temperature is the same at all probe tips. For the measurements in the LAPDs, the condition is satisfied. Measurements of temperature profiles show

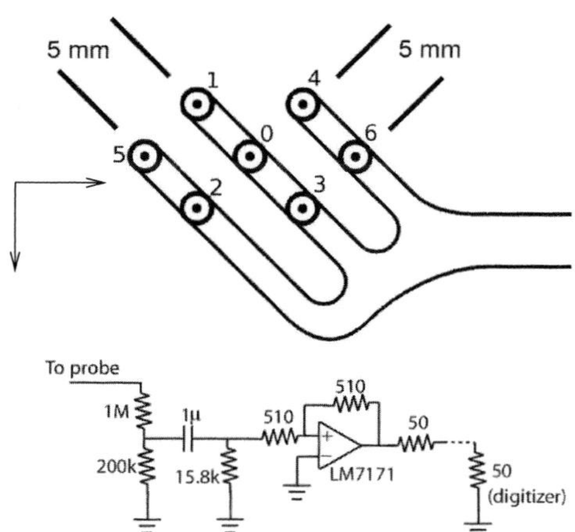

Fig. 9.4 Vorticity probe design. The probe is inserted into the LAPD plasma radially, so that the magnetic field is perpendicular to the surface of the probe tips.

typical temperature scale length of the order of $L_{T_e} \sim 100\,\text{mm}$ which is sufficiently greater than the probe tip distance $h = 5\,\text{mm}$. If this were not the case we would need to measure the temperature profile at each probe tip in order to get the plasma potential φ_p, from the probe floating potential φ_f

$$\varphi_p(eV) = \varphi_f(eV) - \frac{T_e}{2e}\ln\left(2\pi\frac{m_e}{m_i}\right) \tag{9.31}$$

$$= \varphi_f(eV) + 3.5T_e \quad \text{for Helium.} \tag{9.32}$$

The floating potential φ_f of the probe is discussed further in Section 10.2.3.

This probe stencil is based on the finite difference approximation to $\nabla^2\varphi$ given by

$$\nabla^2\varphi = \frac{(\varphi_{1,0} + \varphi_{0,1} + \varphi_{-1,0} + \varphi_{0,-1} - 4\varphi_{0,0})}{h^2}. \tag{9.33}$$

The vorticity is then given by dividing by the magnetic field B as in Eq. (9.30). The velocities into and out of the four sides of the square are also of interest. For the left side, for example, the velocity is given by

$$v_x(\text{left}) = \frac{1}{B}\frac{(\varphi_1 - \varphi_2)}{2h}. \tag{9.34}$$

Clearly, the vorticity in Eq. (9.33) vanishes for linearly potential with $\varphi = C + Ax + By$. Only quadratic variations over the probe yield nonzero vorticity.

Just as in numerical simulations, the five-point sample of $\varphi(x, y, z, t)$ yields the true vorticity only for that part of the spectrum $\varphi(k_x, k_y, z, t)$ that has $k_x h < 1$ and $k_y h < 1$. Higher $\boldsymbol{k}$-modes yield an anomalous value for ω, although the value can still be interpreted unless $k_x h \gg 1$. For example, a mode with $k_x h = \pi$ and $k_y h = \pi$ leads to the value $-8\varphi_{0,0}/h^2$ for the right-hand side of Eq. (9.33). For larger $k_x h$ values the vorticity is small from the simulations due to inverse cascade in quasi-2D turbulence.

The rate of convective transport of ω is given by

$$\frac{\partial\omega}{\partial t} + \boldsymbol{v}\cdot\nabla\omega = S_\omega(\boldsymbol{x}, \boldsymbol{t}) \tag{9.35}$$

where the source/sink function S_ω includes various mechanisms that produce and absorb vorticity.

Since the $\mathbf{E}\times\mathbf{B}$ drift is incompressible for a uniform $\mathbf{B}$-field, Eq. (9.35) also has the conservation form

$$\frac{\partial\omega}{\partial t} + \nabla\cdot(\omega\boldsymbol{v}) = S_\omega. \tag{9.36}$$

Averaging using Eq. (9.36) over the symmetry direction y (or θ) yields the transport equation,

$$\frac{\partial}{\partial t}v_y'(x,t) + \frac{\partial}{\partial x}\langle\omega v_x\rangle = \overline{S}_\omega(x,t) \tag{9.37}$$

for the generation of sheared zonal flows $v_y' = d\langle v_y\rangle/dx$.

An important theorem is that the average flux of vorticity $\langle v_x \omega \rangle$ equals the divergence of the Reynolds stress $\langle v_x v_y \rangle$ per unit mass. Both the vorticity flux and the Reynolds stress are measured by the vorticity probe.

In the LAPD experiment, the differential axial confinement of the electrons at the radius of the cathode controls the source of vorticity S_ω through the biasing of the cathode/anode plasma source assembly with respect to the confinement chamber walls. For tokamaks a similar E_r well structure is formed by the differential confinement of the ions and the electrons at the edge of the plasma. This E_r well becomes a strong feature after the transition to the H-mode [Burrell (1997)].

The vorticity probe was constructed from seven Tantalum tips, as shown in Fig. 9.4. The tips are cylindrical, 0.02 inch (0.5 mm) in diameter and 1 mm long. The tips are oriented along the direction of the magnetic field (into the page in Fig. 9.4). The Tantalum tips are held in alumina ceramics, which in turn are held by a stainless steel structure that is slotted in order to minimize perturbation to the plasma. The central five tips are arranged in a diamond pattern, with the outer four tips separated from the central tip by 5 mm. The five inner tips are used to measure the floating potential, which is then used to compute a finite difference estimate of the vorticity, while the two outer tips allow measurement of the ion saturation current, in order to obtain the plasma density and radial particle flux $\langle nE_\theta \rangle / B$, as well as the Reynolds stress $nm_i \langle v_r v_\theta \rangle = (\rho/B^2)\langle E_r E_\theta \rangle$ [Lochte-Holtgreven (1968)].

In order to compute the finite-difference estimate of the vorticity, four times the floating potential on the central tip must be subtracted from the floating potential on the surrounding four tips. During biased rotation experiments in LAPD, the DC floating potential can reach values of order 20 V, substantially larger than the observed fluctuation amplitude ($\sim$ 1 V). The floating potential measurements are therefore performed using AC coupled amplifiers in order to reject the large low-frequency floating potential signal and to maximize the use of the dynamic range of the available digitizers. The amplifiers are constructed using wide band operational amplifiers (National Semiconductor LM7171) with a flat response from 1 KHz to 10 MHz. A schematic of the measurement circuit is shown in Fig. 9.3. Data is acquired using 100 MS/s, 14 bit VME-based digitizers (eight channels per board, four available boards).

The origin of the peaking of the electron temperature peak in the shear flow layer is not well understood and suggests further research studies on the anomalous electron viscous heating and magnetic fluctuations.

9.2.2 *Vorticity probe measurements in the Kelvin–Helmholtz instability*

The Kelvin–Helmholtz instability is excited in the LAPD experiment when biasing the floating anode-cathode plasma source with respect to the chamber walls. This results in a sharply localized radial electric field as shown in Fig. 9.5. This electric field, along with the axial magnetic field, creates a sheared poloidal $\mathbf{E} \times \mathbf{B}$ jet stream

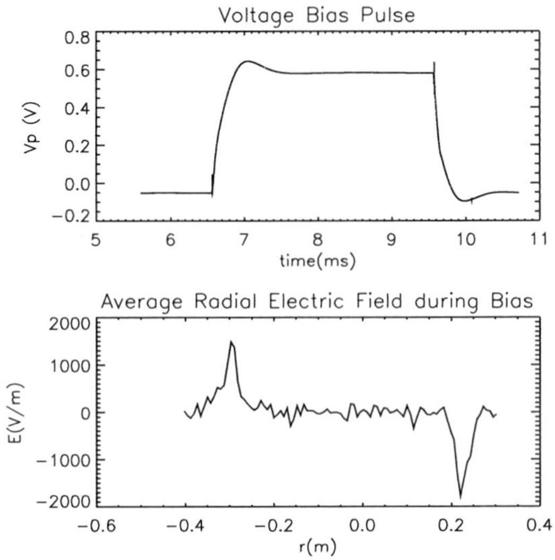

Fig. 9.5 Top: Typical cathode/anode bias voltage pulse with respect to the chamber wall for establishing the rotational jet. Bottom: Radial electric field measured in the stationary section of the bias pulse.

at the edge of the plasma column. This jet stream flow forms as one of the natural self-organized states after the K-H instability in which the vorticity ω as function of the potential φ is a steady-state solution of the vorticity equation

$$\vec{v}_E \cdot \nabla\omega = 0. \tag{9.38}$$

The nonlinear equation, Eq. (9.38), has vortex solutions of the $\omega = F(\varphi - ux)$ moving with speed u in the symmetry direction $y = r\theta$ of the cylinder. Various types of solutions exist with or without embedded vortices. The simplest solution is a jet stream with localized $v_\theta(r)$. In Section 9.2.3 we show a nonlinear simulation that illustrates this feature.

In Fig. 9.6 we show vorticity probability distribution functions or PDFs at representative points of the shear layer. In each plot we combine vorticity data from 25 similar shots, during 2 ms of the bias pulse between $t = 7.5$ ms and $t = 9.5$ ms, see Fig. 9.5. This corresponds to a total of 77525 data points divided into 50 equally spaced bins of fluctuating vorticity amplitudes normalized to the standard deviation.

The heavy tails of the vorticity probability distributions functions shown in Fig. 9.6 signify that large values of the vorticity occur frequently. The situation here for vorticity fluctuations driven by the background plasma jet is the plasma has PDFs of the vorticity with heavy tails as do the fluctuations $\langle v_x \delta n\rangle$ from the density fluctuations and density fluxes $\langle v_x \delta n\rangle$ from coherent intermittent structures. Similar measurements are found in the linear machine PISCES-A [Antar (2003)] and the ADITYA tokamak machine [Jha, *et al.* (1992)].

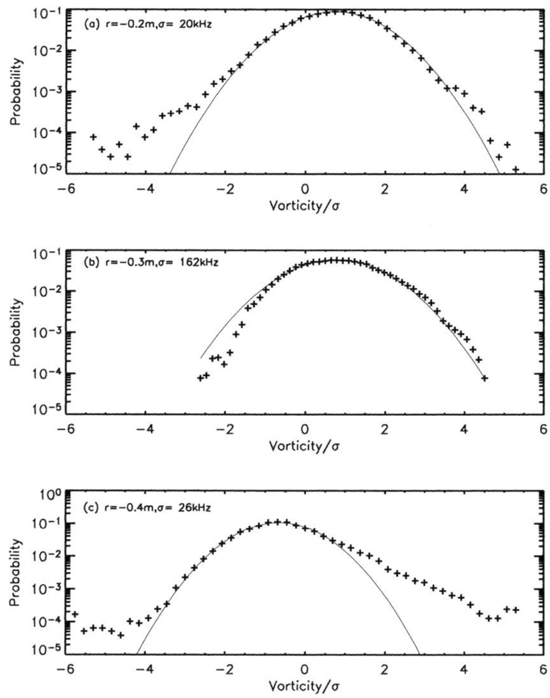

Fig. 9.6 Vorticity PDF for four radial position across the shear-flow layer for the left side ($r < 0$) of the plasma column. The sign of the excess vorticity counts over the Gaussian value changes from negative (counterclockwise rotation) to positive (clockwise rotation) in moving across the shear-flow layer.

Figure 9.7 shows the relevant statistical quantities for an ensemble of 25 experimental runs under identical conditions. The measurements of vorticity show a high degree of reproducibility over the experiments. First we consider the degree of correlation of the floating potential difference between the outermost probe pins 5 and 6 as the probe is scanned across the shear-flow layer. The spacing between these probes is $d = 2\sqrt{2}h\,\mathrm{mm} = 14\,\mathrm{mm}$. Frame (a) shows that the normalized correlation function $\langle \varphi_5 \varphi_6 \rangle$ is near unity except in the shear layer where it drops to about 0.7. Thus, we infer that the radial correlation length of the fluctuations drops from greater than 14 mm to about 10 mm in the center of the shear-flow layer. Next we consider the vorticity mean values, as the probe is scanned along the radial direction, by averaging vorticity over time at each point in the shear layer and later averaging over the ensemble of the 25 experimental samples. Frame (b) of Fig. 9.7 shows the mean vorticity $\langle \omega \rangle$ as a function of the radius. From this plot we see that the vorticity reaches a maximum value $\omega_{\max} = 150\,\mathrm{kHz}$ which is comparable to $v_{\max}/a \sim 10^4\,(\mathrm{m/s})/0.05\,\mathrm{m} \sim 200\,\mathrm{kHz}$ from the equilibrium radial electric field profile at $B = 0.075\,\mathrm{T}$. The positive core value agrees with sign of chain of vortices expected from the KH instability [Horton, *et al.* (1987)].

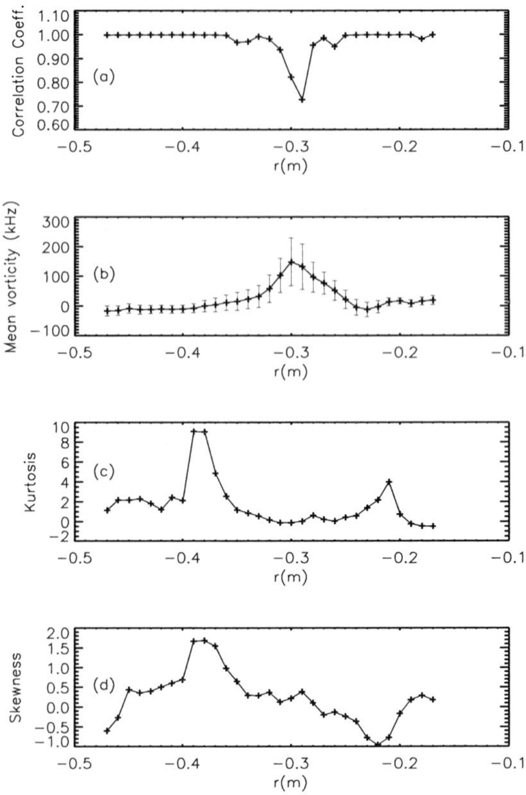

Fig. 9.7 LAPD data analysis with the vorticity probe. Radial variation on the left side of the cylinder for the mean, the kurtosis [frame (c)] and skewness [frame (d)]. The antisymmetry of the skewness agrees with the change if vortex rotation direction across the layer and coincides with the larger values of kurtosis in frame (c) that occurs in for a field of vortices.

Frames (c) and (d) show that outside the core of the jet there is an intermittency of the vorticity probability distribution function consistent with the heavy tails shown in Fig. 9.6. While the error bars on the mean vorticity suggest that it may be consistent to take the vorticity as vanishing outside the core of the jet stream, the small relative bias that is about equal to the width of the error bar is such as to give vortices rotating counter-clockwise outside the stream $r < -0.4\,\mathrm{m}$ and clockwise inside the stream $-0.4\,\mathrm{m} < \mathrm{r} < 0$ on the left side of the cylinder. The computer simulations show such a pattern of counter-rotating vortices on the flanks of the jet stream.

There are two types of aliasing errors: (i) the usual single probe sampling error for signal frequencies higher than the Nyquist $f_N = 1/2\delta t$ and (ii) spatial aliasing errors for short wavelength signals. For each pin there are anti-aliasing filters in frequency. This helps reduce anti-aliasing from $k_N = \pi/h$ under the assumption that small scale structures exist at higher frequency. The sample rate (and anti-aliasing filter) is at much higher frequency than the frequency of the observed fluctuations. The probe sampling rate is at few MHz, and the fluctuations are below 5 kHz.

For spatial aliasing problem we consider the effect of the finite probe space h on a single wavelength signal, the ratio of vorticity measured with the probe to the actual value of form factor vorticity is given by

$$R(\vec{k}, h) = 4\frac{\sin^2\left(\frac{k_x h}{2}\right) + \sin^2\left(\frac{k_y h}{2}\right)}{h^2(k_x^2 + k_y^2)}. \tag{9.39}$$

For $kh < 1$ we have $R \approx 1$. For $k'_x = k_x + 2\pi n$, $k'_y = k_y + 2\pi m$, the probe picks up the same signal for $\vec{k}$ and $\vec{k}'$ which is the aliasing error. However, the probe value is smaller as given by $R(\vec{k}, h)$ at the spatial Nyquist frequency $k_N = \pi/h$ where the ratio function is $R = 4/\pi^2 \approx 0.444$. So, keeping probe data for $k < k_N \approx 628\,\mathrm{m}^{-1}$ largely suppresses the aliasing errors. Using a range of probe sizes h would allow an accurate assessment of the power spectrum in the high-k components.

9.2.3 *Kelvin–Helmholtz turbulence with drift waves*

Now we examine the nonlinear simulations of the LAPD shear layer during the rotation bias pulse. A nonlinear pseudo-spectral code is used with Fourier decomposition in space to compute gradients and Poisson brackets in k-space and adaptive fifth/sixth order Runge-Kutta method to advance in time the solution in k-space. Then we transform the advanced solution back to configuration space where the quadratic nonlinear terms are computed.

There are two versions of the code with different boundary conditions. One code is implemented with periodic boundary conditions in both x (radial) and y (poloidal) coordinates [Horton (1989)]. There are two disadvantages to using the local periodic slab model for the simulation. First, the DC rotation through

$$\langle\omega\rangle = \frac{1}{r}\frac{\partial(rv_\theta)}{\partial r} = \frac{\partial v_\theta}{\partial r} + \frac{v_\theta}{r}$$

has two parts. The first gradient component is dominant, for the profile in Perez (2006), and is about 200 kHz while the second rotational frequency Ω, which is symmetric about the peak of the jet, is $\Omega = v_{\theta,\max}/r \sim 60\,\mathrm{kHz}$. The second disadvantage is that the radial boundary conditions are not periodic. However, we can still receive some qualitatively important results by using the local potential jet $\varphi(x, y, t)$ shown in Fig. 9.8.

The top frame of Fig. 9.8 shows the initial state in which opposite vortices are shed from the right and left sides of the jet, so that there is a definite skewness $\langle\omega^3\rangle$ that changes sign across the layer as measured with the vorticity probe. After some time, the simulation shows that the average value of vorticity along the y-direction of Fig. 9.8 tends to decrease at each x-point due to the alternate counter rotating vortices lining up along the y-direction. The nonlinear simulations use a slab of the rotating shear layer and therefore which ignores the fact that the plasma is rotating

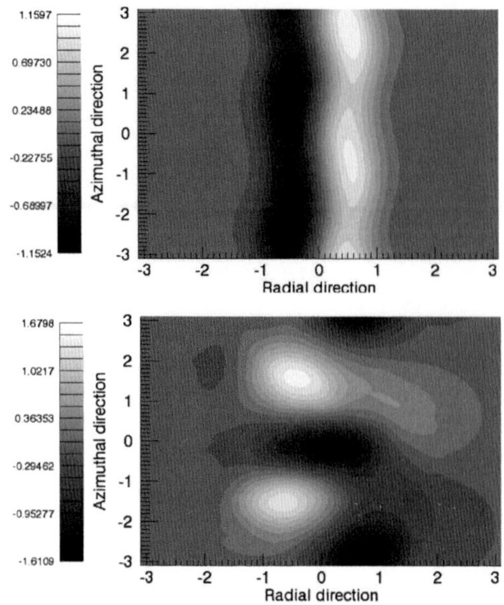

Fig. 9.8 Top panel is the linear stage and bottom panel is the nonlinear stage of the jets. In the nonlinear state, the jet disassembles into positive and negative vortices. Afterwards there is a coalescence of the vortices and a separation of the direction of rotations according to the direction of the jet. In the experiment the momentum is transported away by the divergence of the Reynolds stress tensor.

as a whole, giving rise to a net background positive (clockwise when viewed into the magnetic field vector) vorticity associated to the equilibrium rotation. This rotational vorticity explains the dominance of positive mean values in Fig. 9.8 and a peak at the center of about 150 kHz, where the flow velocity is maximum. Analysis of the cylindrically-rotating plasmas is given in Chapter 16.

9.3 Experimental Measurement of Vorticity Dynamics Studies and the Reynolds Stress

The analytic model used in simulations takes the plasma density as constant over the shear layer. From the ion saturation current $I_{\rm sat}(r)$ measured taken through a low-pass filter we know just inside the shear-flow layer there is a sharp increase of the electron density. In and outside the layer we find that density gradient scale length $L_{ne} \sim 15\,\text{cm}$ which gives a sufficiently low-gradient frequency ω_{*e} to neglect the $\delta n_e/n_e$ from $\vec{E} \times \vec{B}$ convection across the density gradient. Thus, the shear flow of the jet reduces the particle flux $\langle n_e v_x \rangle$ across the layer.

The cross-correlation signals between the widest pin spacing in the probe shows that the pin signals are correlated and thus contain useful wave-vortex information rather than noise. The cross correlation drops from near unity to about 70%

in the shear-flow layer for the outermost pin tips separated by 14 mm which is approximately 20 ion gyro-radii. The mean vorticity shows a large peak of 1.5×10^5 /s at the center and small residual positive and negative values of order $\pm 1 \times 10^4$ /s just outside the shear-flow layer. The small residual values are just at the estimate of one standard deviation $\sigma_\omega(r)$ in the spread (variance of ω) about the mean of the vorticity PDF. These residual values of ω are statistically significant and thus we see a confirmation of the odd symmetry of the sign of vorticity shed from the plasma jet as shown in the computer simulations.

A set of vorticity of probes with different tip spacing would allow the collection of a relatively complete picture of the vorticity field from $\mathbf{E}\times\mathbf{B}$ flows in a magnetized plasma. Future research would integrate test particle motion and compare with tagged ions followed by the laser induced fluorescence technique to investigate more details of the transport across the plasma jet from the E_r well.

The probability distribution functions (PDF) of the vorticity were computed across the shear layer, as sampled at $\delta t = 0.3\mu$s over a steady-state period of $T = 2$ ms. The PDF shows strong tails in the regimes of maximum vorticity $v_\theta' = -E_r'/B$ near $r = 0.4$ m, where E_r' is a maximum. These results are interpreted as showing the existence of a string of vortices of opposite signs on either side of the $E_r' = 0$ layer. The antisymmetric form of the skewness moment of the vorticity in Fig. 9.7, as well as the local peaks of the kurtosis function, are consistent with a string of oppositely signed vortex chains on either side of the maximum of the $E_r(r)$ profile. This antisymmetric double chain of vortices was studied with pseudo-spectral code simulations of the Kelvin–Helmholtz instability for the experimental system parameters.

Other quantities available from the probe are the ion saturation current, for a measure of the plasma density and particle flux $\Gamma = \langle nE_\theta \rangle / B$ and the Reynolds stress $\Pi_R = nm_i \langle v_r v_\theta \rangle = (\rho/B^2)\langle E_r E_\theta \rangle$. Analysis of these fluxes are given in Perez (2006), along with a comparison to test particle transport in models for the measured fluctuations.

Plasma has the unique advantage over a neutral fluid in that the stream function for quasi-two-dimensional flow is the electric potential, which can be sensitively measured directly by electrical probes with standard Langmuir probe techniques. Vorticity in neutral fluids is inferred from particle image velocimetry of neutrally buoyant test particles in the flow. This measurement is a relatively complex procedure compared with the measurement of the vorticity probe and is subject to high noise level when the vorticity is constructed from the measured velocity field.

The measurement of vorticity is of fundamental importance in theory of turbulent fluids and plasmas. Closure models lead to various relations for the probability distributions and moments of the vorticity. A classical neutral fluid example is the Proudman-Reid formula $d^2\overline{\omega}^2/dt = \frac{1}{3}(\overline{\omega}^2)^2$ discussed in detail in Monin and Yaglom (1965). Developing the vorticity probe for plasma provides new insight into plasma turbulence.

The nonlinear stability of the plasma is intimately connected to the formation of vortex structures. Kim, *et al.* (2006) show had the presence of sheared cross-field plasma flow that is linearly stable by the Rayleigh condition produces convective amplification of an initial wave packet disturbance. The sheared flow transforms as a motion $\boldsymbol{k}$-space from the initial region of the wave packet. In the process of $\boldsymbol{k}$-space convection the wave packet is amplified becoming a nonlinear coherent vortex under the conditions given by Kim, *et al.* (2006) and Horton (1989).

In tokamaks, scrape-off layers are formed by applying a horizontal magnetic field that creates an X-point in the poloidal magnetic field structure. The magnetic separatrix then become a boundary between the core plasma and the scrape-off layer plasma. These driven plasmas are found to have large, intermittent coherent structures often called "blobs" expelled from inside the separatrix into the outer scrape-off layer plasma. In the scrape-off layer the plasma streams along the magnetic field line to the carbon or beryllium coated metal plates placed in a small volume called the divertor chamber. Antar, *et al.* (2001a,b, 2010) have made images of these structures and constructed the probability distributions for these events in the spherical tokamak called MAST standing for Mega Ampere Spherical Tokamak.

The electron Bernstein wave (EBW) [Laqua (2007)] is typically the only wave in the electron cyclotron (EC) range that can be applied in spherical tokamaks for heating and current drive. Spherical tokamaks (STs) operate generally in high-β regimes, in which the usual electron cyclotron called EC O- and X-modes are cut off. In this case, EBWs seem to be the only option that can provide features similar to the EC waves that are used for core plasma heating as well as for accurate external plasma stabilization control.

The electron Bernstein wave (EBW) [Laqua (2007)] is a quasi-electrostatic wave excited by mode conversion from a suitably launched X- or O-mode high-frequency transverse wave. The EBW develops a short wavelength which heats the electrons at a series of cyclotron harmonics. Numerical study of EBW electron plasma heating is measured in four typical ST plasmas (NSTX L- and H-mode, MAST Upgrade and NHTX). Coupled ray-tracing and Fokker-Planck (LUKE) codes are employed to simulate EBWs for varying frequencies and launch conditions. The ECH waves through this mode conversion to the Bernstein wave resonant with many cyclotron harmonics heats or energizes the electrons to KeV energies as shown in W7-X data in Urban, *et al.* (2011) and Laqua (2016).

9.4 Electromagnetic Plasma Vortices

Drift wave-Alfvénic vortex solutions are known for the electromagnetic structures in the finite pressure magnetized plasma. The generalization follows from the continuity of the plasma currents, which leads to fourth order differential equations for the electromagnetic fields. Special solutions for two types of boundary value problems are known for the fourth-order partial differential equations governing the vortex structures. There are several types of vortex structures: large-scale vortex structures called the intrinsic electromagnetic drift-Alfvén vortex and the small electron scale Alfvén vortex structures.

The electron scale electromagnetic vortices have more complicated radial structures with inner and outer regions. The large-scale Alfvén drift-wave vortex [Onishchenko, *et al.* (2015)] have relatively simple $\delta\mathbf{E}$ and $\delta\mathbf{B}$ fields. The complex vortex structures guarantee the continuity of the perturbed magnetic field $\delta\mathbf{B}_\perp$ with a parallel current $j_\parallel$ on the boundary between the inner and outer regions of the vortex. The allowed regions of propagation speeds and the amplitude for these vortices are now analyzed.

Fast imaging cameras with 10μs time resolutions and fast reciprocating probes are used to study the characteristics of the "blobs" ejected from the confinement region in the scrape-off layer. The structures are larger and more frequent on the out-board side where the toroidal magnetic field curvature is such as to accelerate the higher pressure blob plasma out of the confinement region. The outboard side is also called the low-field side of the torus and the magnetic separatrix is called the LCFS for last closed flux surface.

Antar, *et al.* (2001a,b, 2010) report measuring these structures with diameters of from 5 to 10 cm across the magnetic field. Once in the open magnetic field region outside the magnetic separatrix they move with speeds of 10^3 m/s along the field lines to the strike points on the divertor plates. The accumulative distribution of the blobs makes a probability distribution function (PDF) that has a skewness ~ 3 and a flat (kurtosis ~ 20) structure. In contrast, on the high-field inboard side the intermittency is not significant and the distributions are approximately Gaussian. Krasheninnikov (2001) anticipated the likely features of the ejected coherent plasma structures and chose to call them "blobs".

9.4.1 *Large-scale Alfvén vortex structures*

Three-dimensional nonlinear structures of Alfvén waves in the form of vortices extended along the external magnetic field are investigated in Antar, *et al.* (2001ab, 2003, 2006, 2010), and Balmforth and Morrison (2002). In these vortices, the dispersion effects play a key role. The condition of the existence of these quasi-stationary structures is the compensation of the wave dispersion by the "vector" nonlinearity effects. The dispersion effects in shear Alfvén waves depend on values of plasma β, namely $\beta_e = 2\mu_0 n T_e/B_0^2$ and $\beta_i = 2\mu_0 n T_i/B_0^2$, where μ_0 is the permeability of free space, n is the plasma number density, T_e and T_i are the electron and ion temperature, respectively, and B_0 is the modulus of the external magnetic field $\mathbf{B}_0$.

The dispersion relation for shear Alfvén waves in low-β plasma, $\beta_e \ll m_e/m_i$, where m_e and m_i are the electron and ion masses, respectively, takes the form

$$\omega^2 = k_z^2 v_A^2 / \left(1 + k_\perp^2 \delta_e^2\right), \tag{9.40}$$

where ω is the wave frequency, k_z and $k_\perp$ are longitudinal and transverse wave numbers, respectively, $v_A = B_0/(\mu_0 m_i n)^{1/2}$ is the Alfvén velocity, $\delta_e = c/\omega_{pe}$ is the electron collisionless skin depth, c is the speed of light, and ω_{pe} is the electron plasma frequency. These types of waves are called inertial Alfvén waves.

The dispersion relation of the shear Alfvén wave in intermediate-β plasma, $m_e/m_i \ll \beta_e \ll 1$, changes form to be

$$\omega^2 = k_z^2 v_A^2 \left[1 + k_\perp^2 \rho_s^2 + (3/4) k_\perp^2 \rho_i^2\right] \tag{9.41}$$

where $\rho_i = (T_i/m_i)^{1/2}/\omega_{ci}$ is the ion Larmor radius and $\rho_s = (T_e/m_i)^{1/2}/\omega_{ci}$ is the ion acoustic radius or the ion Larmor radius calculated at the electron temperature. These waves are referred to as the kinetic Alfvén waves. The dispersion relation gives

$$\omega^2 = k_z^2 v_A^2 \left[1 + k_\perp^2 \rho_*^2 + (3/4) k_\perp^2 \rho_i^2\right], \tag{9.42}$$

where $\rho_*^2 = \rho_s^2/(1 + T_e/T_i)$.

In Antar (2003, 2006, 2010) and Balmforth and Morrison (2002) the nonlinear Alfvén wave structures are derived in the form of dipole vortices. The characteristic transverse scales of these vortices are comparable with the dispersion scales δ_e, ρ_i, ρ_s, or ρ_*.

Nonlinear Alfvén waves play an important role in the coupling between the ionosphere and magnetosphere. They determine some of the fine structure in the many auroral forms. This was demonstrated by observations on the Intercosmos-Bulgaria-1300 [Banerjee, *et al.* (2000)] satellite. Several new nonlinear Alfvén wave structures, such as vortex streets, have also been found.

In contrast, a number of problems of solar physics, space physics, geophysics, and astrophysics, [Burrell (1997), Chmyrev, *et al.* (1988), Fedun, *et al.* (2011), Hazeltine, *et al.* (1985)] are of interest to the study larger-scale vortex structures with characteristic transverse scale many orders greater than the ion Larmor radius (or collisionless electron skin-depth). Onishchenko, *et al.* (2015) investigate vortices in large-scale nonlinear structures of dispersionless Alfvén waves.

In order to describe the vortex motion in Alfvén waves, we make use of a cylindrical coordinate system (r, ϕ, z). The magnetic field is considered as $\mathbf{B} = \mathbf{B}_0 + \mathbf{B}_\perp$, where $\mathbf{B}_0 = \boldsymbol{e}_z B_0$ is the external magnetic field which, for simplicity, is assumed to be uniform and $\mathbf{B}_\perp = (B_r, B_\phi, 0)$ is the perpendicular part of the wave magnetic field from the vortex. The field-aligned magnetic field perturbation δB_z in the Alfvén waves is assumed to be negligibly small. Furthermore, $\boldsymbol{e}_z$ is the unit vector along the z-axis. The electric field is given as $\mathbf{E} = \mathbf{E}_\perp + \boldsymbol{e}_z E_z = (E_r, E_\phi, E_z)$. For the analysis we introduce the scalar potential φ and the z-component of the vector potential A, so that $E_z = -\partial_t A - \partial_z \varphi$, $\mathbf{E}_\perp = -\nabla_\perp \varphi$, and $\mathbf{B}_\perp = \nabla A \times \boldsymbol{e}_z$. Here, $\partial_t \equiv \partial/\partial t$ and $\partial_z \equiv \partial/\partial z$; the subscripts z and $\perp$ denote the components along and perpendicular to B_0, respectively.

If one neglects the electron density compressibility and the inertial effects $(\delta_e/r_0 \to 0)$ in the electron motion along the total magnetic field the field-aligned electric field vanishes, i.e., $E_\parallel = E_z + B_0^{-1}\mathbf{B}_\perp \cdot \mathbf{E}_\perp = 0$, or

$$\frac{\partial A}{\partial t} + \nabla_\parallel \varphi = 0, \tag{9.43}$$

where the operator $\nabla_{\|}$ is given by

$$\nabla_{\|} = \frac{\mathbf{B}\cdot\nabla}{B_0} = \frac{\partial}{\partial z} + \frac{1}{B_0 r}\frac{\partial A}{\partial \phi}\frac{\partial}{\partial r} - \frac{1}{B_0 r}\frac{\partial A}{\partial r}\frac{\partial}{\partial \phi}. \tag{9.44}$$

The second equation connecting φ and A is the current closure equation

$$\nabla_{\perp}\cdot\boldsymbol{j}_{\perp} + \nabla_{\|} j_z = 0, \tag{9.45}$$

where $\boldsymbol{j}_{\perp}$ and j_z are the perpendicular and parallel components of the plasma current density, respectively. If the finite ion Larmor radius effects are neglected the perpendicular current is determined by the inertial (polarization) part of the ion velocity

$$\boldsymbol{j}_{\perp} = qn\boldsymbol{V}_i \tag{9.46}$$

where

$$\boldsymbol{V}_i = -\frac{1}{B_0\omega_{ci}}\frac{d}{dt}\nabla_{\perp}\varphi, \tag{9.47}$$

and

$$\frac{d}{dt} = \frac{\partial}{\partial t} + \boldsymbol{v}_E\cdot\nabla = \frac{\partial}{\partial t} + \frac{1}{B_0}(\nabla\varphi\times\nabla)_z, \tag{9.48}$$

where $\boldsymbol{v}_E = B_0^{-1}\boldsymbol{e}_z\times\nabla\varphi$ is the $\mathbf{E}\times\mathbf{B}$ drift velocity. From the z-component of the Ampere's law, we have

$$j_z = -\left(\frac{1}{\mu_0}\right)\nabla_{\perp}^2 A \tag{9.49}$$

where the operator $\nabla_{\perp}^2$ is denoted by

$$\nabla_{\perp}^2 = \frac{1}{r}\frac{\partial}{\partial r}r\frac{\partial}{\partial r} + \frac{1}{r^2}\frac{\partial^2}{\partial\phi^2}. \tag{9.50}$$

Making use of (9.46) and (9.51), then the $\nabla\cdot\mathbf{j} = 0$ reduces to

$$\frac{d}{dt}\nabla_{\perp}^2\varphi + v_A^2\nabla_{\|}\nabla_{\perp}^2 A = 0. \tag{9.51}$$

The pair of Eqs. (9.43) and (9.51) coincide with known equations for dispersionless shear Alfvén waves [Horton, *et al.* (1987), Horton (1989)]. The set of these equations, Eqs. (9.43) and (9.51) possesses the integral of energy conservation

$$W = \frac{1}{2}\int\left[(\nabla_{\perp}\varphi)^2 + v_A^2(\nabla_{\perp}A)^2\right]d\boldsymbol{r}, \tag{9.52}$$

with $dW/dt = 0$ where W is the wave energy per unit length in the z-direction. In Horton, *et al.* (1990) a generalization of Eqs. (9.43) and (9.51) is given by taking the gravitational force into consideration for the solar plasmas.

9.4.2 *Emission of plasma filaments and vortices in the scrape-off layer*

Extensive measurements and simulations are available describing the shedding of plasma blobs, better described as long-thin filaments from the steep gradients of the plasma at and near the magnetic separatrix. The plasma region at and just beyond the magnetic separatrix is called the scrape-off layer (SOL). Many publications describe the observation of structures variously called blobs, vortices, or plasma filaments that move rapidly from the hot core plasma into the outer regions called the SOL plasma and then to the plasma vessel walls. These localized coherent structures are best described as plasma filaments. Since the data is extensive for these filaments, there are computer simulations using different algorithms for the same data. In particular, Militello, *et al.* (2016) have reported the use of the plasma data from the double-null divertor plasmas in MAST to search for the agreement between the filament data in the scrape-off layer with four simulation codes: (i) STORM module of BOUT++ [Sykes, *et al.* (2001), Harrison, *et al.* (2011)], with (ii) GBS [Scannell, *et al.* (2010)], (iii) HESEL [Yang, *et al.* (2003), Lao, *et al.* (1985)] and (iv) 10 KAM3X [Elmore, *et al.* (2012)] simulation codes.

The MAST name is for the mega-ampére spherical torus which is a machine similar to NSTX in having a low-toroidal magnetic field, ($B_T < 1\,\mathrm{T}$) and large plasma current with small aspect ratio R/a.

The MAST and NSTXU parameters are given in Table X. The MAST experiment used for the comparison of filament data and codes used the double-null separatrix with $B_T = 0.4\,\mathrm{T}$, $R = 0.66\,\mathrm{m}$, and $B_p = 0.12\,\mathrm{T}$. The neutral beam (NBI) heating (1.7 MW) to (0.7 MW) produced L-mode plasmas with $I_t = 0.56\,\mathrm{MA}$, n_e few times $10^{19}\mathrm{m}^{-3}$ and $T_e = 1 - 2\,\mathrm{KeV}$. The plasma density and electron temperature were measured with high-resolution Thomson scattering (HRTS) [Huang, *et al.* (2010)] during the time the filaments were observed with a 100 kHz camera observing the light emission from the D_α spectral lines. The camera resolution is sufficient to measure filaments at their early stage of development from milliliters to their final stage of centimeters.

With the 3μ sec camera exposure the resolution of the filaments growth in size and shape is adequate for differentiating the structures given by the four computer simulation models. The plasma density and electron temperature profiles show a sharp increase of radial gradients beyond the magnetic separatrix. The L-mode plasmas are described in detail in Allan, *et al.* (2016). The magnetic equilibrium is simulated using EFIT++ [Lipschultz, *et al.* (2005)] with data from the magnetic probes and high-resolution Thomson scattering.

The authors give some details in the 3D-STORM BOUT++ code, the plasma-wall sheath boundary conditions where the field lines strike the divertor end-plates called the "targets" in tokamak literature. There is an exponentially increasing density source $S_n(z)$ from the core to the target plates and an exponential decaying energy sink $S_E(z)$ from the core to the target plates while keeping plasma pressure constant along the field lines. The system of five nonlinear partial differential equations for $d\Omega/dt, dn/dt, dv_{\|}/dt, dv_{\|e}/dt$ and dT_e/dt are solved with a complex set of collisional transport rates following the Braginskii equations in Chapter 3.

The results of the comparison of the four simulation codes and the MAST plasma data gives information showing agreement in general, while revealing important differences in the codes themselves. The differences include both the filament trajectories and the shapes of the vortex structures created by the models. In a related simulation Zhu, *et al.* (2015) show the development of blobs or vortex structures in the nonlinear stage of interchange dynamics.

Steady-state traveling vortex structures along the external magnetic field with velocity v_A are found with a transform to the Alfvén structure frame with the variable $\eta = z - v_A t$. In this frame, the equations (9.43) and (9.51) reduce to

$$-v_A \frac{\partial}{\partial \eta} A + \frac{\partial}{\partial \eta} \varphi - J(A, \varphi) = 0, \tag{9.53}$$

and

$$-v_A \frac{\partial}{\partial \eta} \nabla_{\perp} \varphi + \frac{1}{B_0} J(\varphi, \nabla_{\perp}^2 \varphi) + v_A^2 \frac{\partial}{\partial \eta} \nabla_{\perp}^2 A - \frac{v_A^2}{B_0} J(A, \nabla_{\perp}^2 A) = 0, \tag{9.54}$$

where

$$J(a,b) \equiv \frac{\partial a}{\partial r} \frac{1}{r} \frac{\partial b}{\partial \phi} - \frac{\partial b}{\partial r} \frac{1}{r} \frac{\partial a}{\partial \phi}$$

is the Jacobian. This system of $A - \varphi$ equations has the solution

$$\varphi = v_A A(r, \phi, \eta). \tag{9.55}$$

In this frame the waves, not only have $E_{\|} = 0$, but also $E_z = 0$.

Now one looks for solutions of the scalar potential $\varphi(r, \phi, \eta)$ in the form

$$\varphi = F(r, \phi) \Phi(\eta). \tag{9.56}$$

For the function $\Phi(\eta = z - v_A t)$, one can choose the periodic function $\exp(i\eta/L_z)$, where L_z is the characteristic longitudinal wave scale. The function F describes the vortex structure in the (r, ϕ) plane in the moving coordinate system with constant η. Therefore, it is important to find self-similar solutions in the moving coordinate system.

The wave structure with such a function F should satisfy the two necessary conditions: (1) the limited energy W, defined in Eq. (9.52), and (2) the condition of global electroneutrality $\int (\nabla \cdot \mathbf{E}) d\boldsymbol{r} = 0$, or

$$\int \nabla_{\perp}^2 \varphi d\boldsymbol{r} = 0. \tag{9.57}$$

Furthermore, the function $F(r, \phi)$ is a periodic function of the angle ϕ around $\mathbf{B} = B_0 \widehat{\mathbf{e}}_z$.

These conditions are satisfied for

$$F(r, \phi) = F_1 = \alpha B_0 v_A \frac{r^{2n}}{2 r_0^{2n-1}} \exp\left(-\frac{r^2}{r_0^2} \pm i l \phi \right). \tag{9.58}$$

Here, $0 < |\alpha| < 1$ is the amplitude constant and l is the structure integer. Taking into account that the integral of energy W reaches a minimum value when $n = 1$, we first take

$$F(r, \phi) = \alpha B_0 v_A \frac{r^2}{2r_0} \exp\left(-\frac{r^2}{r_0^2} \pm il\phi\right). \quad (9.59)$$

Making use of Eqs. (9.55) and (9.59) for W in Eq. (9.52), we derive for the vortex energy in Eq. (9.52)

$$W_l = \frac{\alpha^2 \pi r_0^2 v_A^2 B_0^2}{4}\left(1 + \frac{l^2}{2}\right). \quad (9.60)$$

From Eq. (9.59)) we derive the plasma flow velocity components

$$\frac{v_\phi}{v_A} = \frac{1}{v_A B_0}\frac{\partial \varphi}{\partial r} = \alpha \frac{r}{r_0}\left(1 - \frac{r^2}{r_0^2}\right)\exp\left(-\frac{r^2}{r_0^2}\right)\cos(l\phi), \quad (9.61)$$

and

$$\frac{v_r}{v_A} = -\frac{1}{v_A \mathbf{B}_0 r}\frac{\partial \varphi}{\partial \phi} = \alpha l \frac{r}{2r_0}\exp\left(-\frac{r^2}{r_0^2}\right)\sin(l\phi), \quad (9.62)$$

From the expression $\mathbf{B}_\perp = \nabla A \times \boldsymbol{e}_z$, the magnetic field components are

$$B_r = \frac{1}{r}\frac{\partial A}{\partial \phi} = -B_0 \frac{v_r}{v_A}, \quad (9.63)$$

and

$$B_\phi = \frac{\partial}{\partial r} A = -B_0 \frac{v_\phi}{v_A}, \quad (9.64)$$

The vertical fluid vorticity is $\omega_z = (\nabla \times \boldsymbol{v})_z = \nabla_\perp^2 F$. The normalized magnetic field vorticity is $\Omega = (v_A/B_0)(\nabla \times \mathbf{B}_\perp)_z$, and the normalized longitudinal electric current $J_z = (\mu_0 j_z v_A/B_0)$ are related to each other through the equations

$$\omega_z = -\Omega = -J_z = 2\alpha \frac{v_A}{r_0}\left[\left(1 - \frac{r^2}{r_0^2}\right)^2 - \frac{r^2}{r_0^2} - \frac{l^2}{4}\right]\exp\left(-\frac{r^2}{r_0^2}\right)\cos(l\phi). \quad (9.65)$$

When $r = r_0$ and $l = 0$, the vertical vorticity is $\omega_z = -(2\alpha v_A/r_0)e^{-1}$.

For $l = 0$, such structures correspond to axially-symmetric vortices with $v_r = 0$ and $B_r = 0$. Note that for this case the nonlinear term in Eq. (9.53) and both nonlinear terms in Eq. (9.54) are equal to zero for these coherent vortex structures.

When $l = 1, l = 2$, the Alfvén vortices form dipole and quadrupole structures.

Figures 9.9 and 9.10 show the radial structure for the normalized values of the toroidal speed v_ϕ and vertical vorticity ω_z or the longitudinal electric current j_z in an axially symmetrical vortex as a function on the normalized radius r/r_0.

Figure 9.9 shows that the toroidal speed v_ϕ changes sign at $r = r_0$ and reaches a maximum and a reversed flow minimum value when $r \simeq 0.5r_0$ and $r \simeq 1.5r_0$, respectively. Figure 9.10 shows that the longitudinal plasma current j_z and vorticity ω_z change the sign when $r \simeq 0.6r_0$ and attain the maximum value at the center of the vortex and for $r \simeq 1.6r_0$.

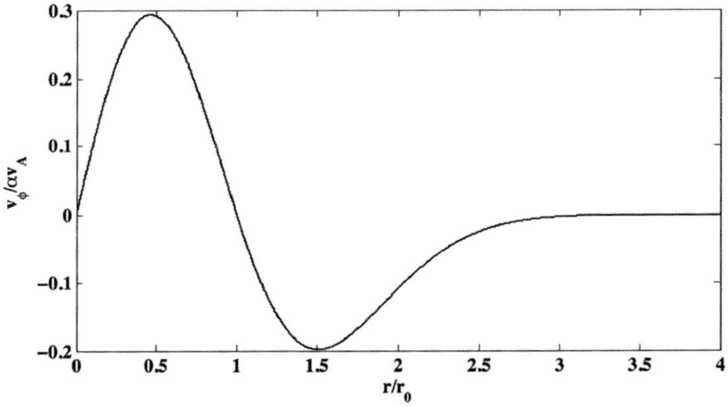

Fig. 9.9 The normalized value of the toroidal speed v_ϕ as a function of distance r/r_0 from the center of Alfvén vortex from Eq. (9.65).

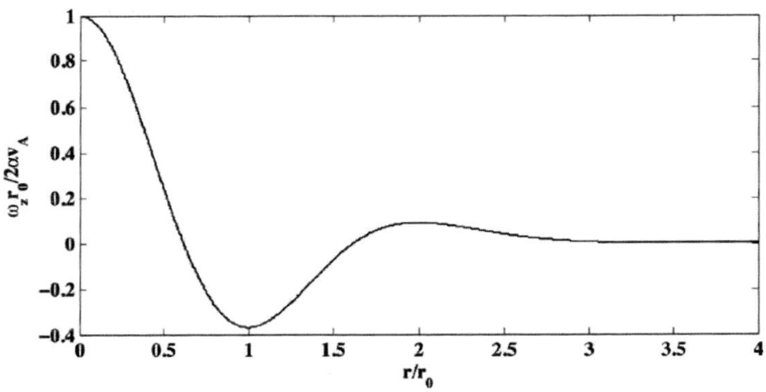

Fig. 9.10 The normalized value of the vertical vorticity ω_z as a function of distance from the center of vortex r/r_0.

Figure 9.11 gives a 2D visualization of the normalized vortex velocity field for axially symmetric ($l = 0$), dipole ($l = 1$), and quadrupole ($l = 2$) structures. Accordingly to Eqs. (9.63) and (9.64), these plots also give the magnetic field distribution, but magnetic vector field direction is opposite to $\boldsymbol{v}$.

These nonlinear Alfvén structures propagate in the form of the magnetic flux ropes with vorticity in the plasma motion [Fedun, *et al.* (2011)]. The characteristic radial size of these structures r_0 can be many orders of magnitude larger than the ion Larmor radius or the electron collisionless skin depth. These large-scale vortices are complementary to the small vortices that have exterior fields that have exterior fields that decrease at $\exp(-r^2/r_0^2)$.

Large-scale Alfvén structures are observed in space and solar physics. Understanding the scalable nature of vortex motions in space plasmas provides unique insights into the association of large- and small-scale structures in the atmosphere

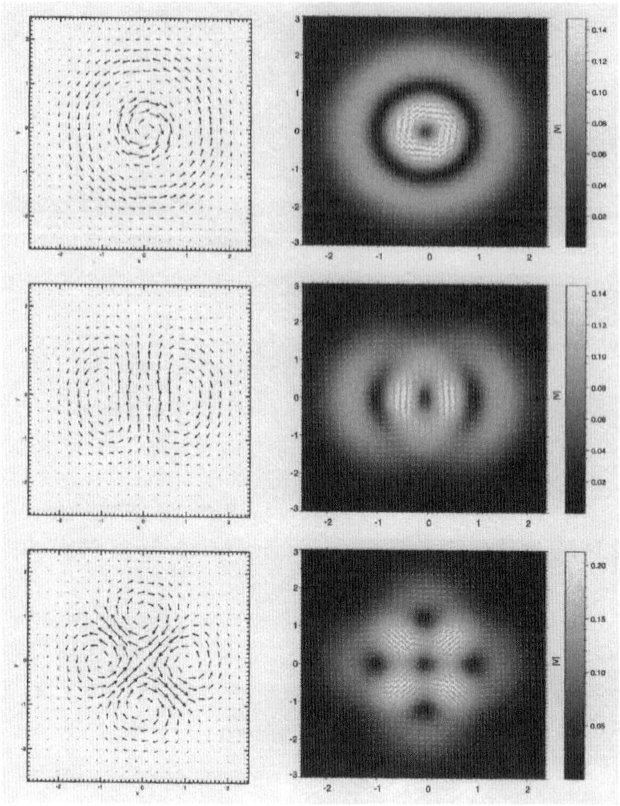

Fig. 9.11 The horizontal structure of the three different vortex velocity field patterns calculated for $l = 0, 1$, and 2 (from the top to the bottom, correspondingly) [Onischenko, *et al.* (2015)].

of the Sun. Giant tornadoes [Wedemeyer-Bohm, *et al.* (2013)] – the observed analogue to large-scale vortex motions in the solar atmosphere – are recently observed as the rotating legs of large-scale solar prominence structures. These plasma tornadoes are thought to be formed in a similar way as with vortex-driven, small-scale magnetic tornadoes and they are often associated with triggering the eruption of prominences into the heliosphere.

Large-scale vortex motions appear to play an important role in the dynamics of the solar transition region [Horton, *et al.* (2010), Jha, *et al.* (1992)]. Similarities with small-scale vortex motions are present in their measured Doppler velocities (i.e., $\pm 6\,\mathrm{m\,km/s}$) within the adjacent legs of a quiescent hedge-row prominence [Kadomtsev and Pogutse (1974)]. Coordinated ground and space-based observations of another giant tornado also exhibited similar opposing Doppler velocity signatures (indicative of rotation of the structure) within the legs of a prominence [Kim, *et al.* (2006)], although with larger Doppler velocities of $\pm 25\,\mathrm{km/s}$. However, it is not yet clear whether giant tornadoes actually rotate as a vortex with a central axis or whether this motion is a representation of flows along a stable helical magnetic structure [Krasheninnikov (2001)]. The different oscillatory modes within giant solar tornadoes may be identified using current and future solar observational high-resolution imaging capabilities, in order to confirm the presence of the dispersionless Alfvén wave vortex structures.

9.5 Formation of Blobs from a Pure Interchange Process

In a simplified geometry the formation of the hot plasma vortex structures propagating out of the confinement region called "blobs" was examined in Zhu, *et al.* (2015) with a set up that focused on the pure interchange dynamics without magnetic shear. To simplify the problem slab geometry approximation was made in the magnetohydrodynamic simulations. The simulations show how the blob-like or vortex plasma structures emerge out of the nonlinear interchange instability. Such structures are observed with Phase Contrast Imagining in Allocator C-Mod [Terry, *et al.* (2005)] originating in a pedestal-like transition region with steep gradients. In the early nonlinear stage, filamentary structures develop and extend in the direction of the effective gravity. The blob structures appear when the radially extending filaments break off and disconnect from the core plasma. The morphology and the dynamics of these filaments and blobs vary dramatically with a sensitive dependence on the dissipation mechanisms in the system and the initial perturbation. Despite the complexity in morphology and dynamics, the nature of the entire blob formation process remains strictly interchange dynamics without involving a change in magnetic topology.

The mathematical structure of the plasma vortices is complicated and we refer the reader to the First edition of Turbulent Transport in Magnetized Plasma, ISBN 978-981-4383-53-0 for solutions of several types of three-field nonlinear plasma vortex equations. Here we give one Worked Problem.

Worked Problem 9.1:

Find the localized solutions of radius $r = a$ for the Type 1 and Type 2 vortex partial differential equations.

Type 1 drift Alfvénic vortex equation:

Find localized solutions of the fourth-order pdes given by

$$\nabla_\perp^4 \phi(r,\theta) - \alpha_1 \nabla_\perp^2 \phi(r,\theta) = 0, \qquad (r > a) \tag{9.66}$$

$$\nabla_\perp^4 \phi(r,\theta) + \alpha_2 \nabla_\perp^2 \phi(r,\theta) + \alpha_3[\phi(r,\theta) - \alpha_4 r\cos\theta] = 0, \quad (r < a) \tag{9.67}$$

where

$$\nabla_\perp^2 = \frac{1}{r}\frac{\partial}{\partial r}\left(r\frac{\partial}{\partial r}\right) + \frac{1}{r^2}\frac{\partial^2}{\partial \theta^2}$$

and $\alpha_1, \alpha_2, \alpha_3, \alpha_4$ are constants determined by parameters of the large scale drift-Alfvén wave equations. [The values of α_n are given in the first edition with Eqs. (10.81) and (10.82).] Find the localized, continuous solution of ϕ along with the conditions on α_i $(i = 1, 2, 3, 4)$.

Type 2 electron inertial scale drift Alfvénic vortices:

Find the localized solutions of the equations

$$\nabla_\perp^4 \phi - \beta_1 \nabla_\perp^2 \phi + \beta_2 \phi = 0 \qquad (r > a) \tag{9.68}$$

$$\nabla_\perp^4 \phi + \beta_3 \nabla_\perp^2 \phi + \beta_4 \phi + \beta_5 r \cos\theta = 0 \qquad (r < a) \tag{9.69}$$

where β_1, β_2, β_3, β_4, and β_5 are constants determined by small space scale electron skin depth vortices of radius a. Find localized, continuous solution of ϕ along with the conditions on β_i $(i = 1, 2, 3, 4, 5)$ which are specified in Eqs. (10.108) and (10.109).

The study is limited to solutions of the form

$$\phi(r, \theta) = \Phi(r) \cos\theta \tag{9.70}$$

for Type 1 and Type 2 problems with the boundary conditions

$$\phi(r, \theta) \to 0 \text{ as } r \to \infty, \text{ and } \phi \text{ is regular when } \quad r = 0. \tag{9.71}$$

Now we give some details of the procedure for solving Problem 1; while for Problem 2, the procedure is similar and we only give the solutions. Substituting Eq. (9.70) into (9.66) and (9.67) gives

Outside Vortex Core Equation

$$\left(\frac{d^2}{dr^2} + \frac{1}{r}\frac{d}{dr} - \frac{1}{r^2}\right)\left[\left(\frac{d^2}{dr^2} + \frac{1}{r}\frac{d}{dr} - \frac{1}{r^2} - \alpha_1\right)\Phi(r)\right] = 0 \qquad (r > a) \tag{9.72}$$

and

Inside Vortex Core Equation

$$\left(\frac{d^2}{dr^2} + \frac{1}{r}\frac{d}{dr} - \frac{1}{r^2}\right)\left[\left(\frac{d^2}{dr^2} + \frac{1}{r}\frac{d}{dr} - \frac{1}{r^2} + \alpha_2\right)\Phi(r)\right] + \alpha_3 [\Phi(r) - \alpha_4 r] = 0. \qquad (r < a). \tag{9.73}$$

Outside Solution

Due to the property of the operator $d^2/dr^2 + d/rdr - 1/r^2$, the general solution of Eq. (9.72) must satisfy the second order equation

$$\left(\frac{d^2}{dr^2} + \frac{1}{r}\frac{d}{dr} - \frac{1}{r^2} - \alpha_1\right)\Phi(r) - f(r) = 0 \tag{9.74}$$

where

$$f(r) = Ar + Br^{-1}; \quad A, B = \text{ constants.} \tag{9.75}$$

The solution of Eq. (9.74) which vanishes at infinity is

$$\Phi_{\text{out}}(r) = A_1 K_1(\sqrt{\alpha_1} r) + A_2 r^{-1} \tag{9.76}$$

where K_1 is the first order McDonald function and A_1, A_2 are integration constants. From the reality requirement of $\sqrt{\alpha_1}$ we have

$$\alpha_1 > 0. \tag{9.77}$$

The solutions are exponentially localized when $A_2 = 0$ and are extended, or global when $A_2 \neq 0$

Inner Solutions

From the form of Eq. (9.73) and the boundary condition at the origin, the solution of Eq. (9.73) can be any one of the following three expressions, depending on the relations between the coefficients α_2 and α_3 in Eq. (9.67):

$$\Phi_{\text{in}}(r) = A_3 J_1(\lambda r) + A_4 I_1(\mu r) + \alpha_4 r \tag{9.78a}$$

$$\Phi_{\text{in}}(r) = A_3 J_1(\lambda^{(1)} r) + A_4 J_1(\lambda^{(2)} r) + \alpha_4 r \tag{9.78b}$$

$$\Phi_{\text{in}}(r) = A_3 I_1(\mu_1 r) + A_4 I_1(\mu_2 r) + \alpha_4 r \tag{9.78c}$$

where J_1, I_1 are the first order Bessel and modified Bessel functions, respectively, and A_3, A_4 are integration constants. Substituting Eqs. (9.78a)–(9.78c) into Eq. (9.73) yields the following conditions corresponding to the different solutions:
Solution Eq. (9.78a) in which

$$\lambda^2 = \frac{1}{2}\left(\sqrt{\alpha_2^2 - 4\alpha_3} + \alpha_2\right) \tag{9.79a}$$

$$\mu^2 = \frac{1}{2}\left(\sqrt{\alpha_2^2 - 4\alpha_3} - \alpha_2\right) \tag{9.79b}$$

corresponds to

$$\alpha_3 < 0. \tag{9.80}$$

For positive α_3 the solution of Eq. (9.78b) is

$$\left(\lambda^{(1),(2)}\right)^2 = \frac{1}{2}\left(\alpha_2 \pm \sqrt{\alpha_2^2 - 4\alpha_3}\right) \tag{9.81}$$

with the conditions

$$\alpha_2 > 0 \quad \text{and} \quad \alpha_2^2 > 4\alpha_3 > 0. \tag{9.82}$$

Solution of Eq. (9.78c) is

$$\mu_{1,2}^2 = \frac{1}{2}\left(-\alpha_2 \pm \sqrt{\alpha_2^2 - 4\alpha_3}\right) \tag{9.83}$$

with the conditions

$$\alpha_2 < 0 \quad \text{and} \quad \alpha_2^2 > 4\alpha_3 > 0. \tag{9.84}$$

It is easy to show that Eqs. (9.78a)–(9.78c) with corresponding conditions (9.80), (9.82), and (9.84), give inner solutions of Φ which are regular at the origin. To further determine the solution, we must impose some matching conditions to the electromagnetic fields Φ_{out} and Φ_{in} on the common boundary $r = a$.

9.6 Further Comments on Alfvén-Drift Wave Vortices

Solitary drift wave vortices are well known and are the subject of Chapters 6 in Horton and Ichikawa (2000). Such vortex solutions were first given by Larichev, *et al.* (1976) in the course of study of nonlinear Rossby waves in rotating fluids. Solitary vortex solutions which seem ubiquitous in nature have been discussed by Mikhailovskii, *et al.* (1984, 1985), Shukla, *et al.* (1985a), Su, *et al.* (1991, 1992) and Hazeltine, *et al.* (1985). In Shukla, *et al.* (1985b) the drift-Alfvén vortex is analyzed. In a work by Mikhailovskii, *et al.* (1985) two other nonlinear equations of electromagnetic perturbations in magnetized plasma are derived, and the possibility of the existence of solitary vortex solutions for these equations was discussed without presentation of explicit solutions. In work by Hazeltine, *et al.* (1985) a simple electromagnetic generalization of drift-solitary wave for certain nonlinear equations was suggested.

A direct application of the Rossby-type vortex solutions to the electromagnetic Alfvén vortex solutions, as given by Mikhailovskii, *et al.* (1984) and Shukla, *et al.* (1985a), has the problem that the perturbed magnetized field $\delta\mathbf{B}_\perp$ and the parallel current $j_\parallel$ are discontinuous on the boundary between the inner and outer regions of the vortex. These discontinuities do not appear in the solution given by Shukla, *et al.* (1985b), but only because that solution is an approximate one. Liu and Horton (1986a,b) derive continuous vortices in rapidly rotating cylindrical plasmas.

These continuous electromagnetic vortex solutions are important for space plasmas where the CLUSTER spacecraft mission found evidence for such magnetic vortex structures. The small-scale vortex structures are also found in laboratory plasmas where $\beta = 2\mu_0\langle p\rangle/B^2$ is finite. The analysis in this chapter of the corresponding nonlinear dispersion relations is intrinsic to the electromagnetic vortices. As the plasma beta vanishes, one of the solutions reduces to the electrostatic vortices given by Meiss and Horton (1983) and Liu and Horton (1986a,b). More general vortex structures are given by Schepp, *et al.* (1994) with FLR terms in the vorticity equation for nonlinear Alfvénic drift wave equations. The solutions of Eqs. (9.68) and (9.69) are developed in Liu and Horton (1986b).

Chapter 10

Fluctuating Magnetic Fields and Transport

In this chapter we develop the kinetic theory for the response of the plasma to fluctuating magnetic fields. Fluctuating magnetic fields often lead to the direct loss of electrons and electron thermal energy by the motion of the fast electrons along the magnetic field lines. The kinetic theory particle response $\delta f(\boldsymbol{x}, \boldsymbol{v}, t)$ is derived to general electromagnetic field fluctuations.

In Section 10.1–Section 10.2 we derive the electromagnetic field equations governing the self-consistent dynamics of the fluctuating magnetized particle orbits and the fluctuating electric field $\delta\mathbf{E}(\boldsymbol{x}, t)$ and fluctuating magnetic field $\delta\mathbf{B}(\boldsymbol{x}, t)$. In Section 10.3 we derive the ion losses and in Section 10.4 the electron losses from turbulent transport in these fluctuating fields.

Anisotropic electron phase space densities are required to maintain the plasma current density in the tokamak fusion system. For steady-state fusion plasmas the toroidal plasma current maintenance and control is required for stability and performance optimization. Enhanced confinement regimes require specific current profiles. The two methods for driving the current profiles use the injection of multi-megawatts of high-frequency radio frequency (RF) waves. The first most thoroughly validated method uses gigahertz lower-hybrid waves launched with complex antennas designed to send the waves in the direction opposite to the plasma current. The wave-electron Landau resonances then transfer the wave momentum to the electrons producing the local current density $j_{\|} = -en_e v_{\|e}$. The method is effective and was used to produce the record-long steady-state discharges over several minutes in time in Tore Supra and EAST. These discharges pass gigajoules of energy through the plasma to the specially designed axisymmetric divertor chamber, giving the steady-state fusion-grade tokamak plasma operation tests in the period from 2000 to the present. The second method to drive and control the plasma current is the use of electron cyclotron resonance (ECRH) waves at over one hundred gigahertz frequency. These waves resonate with the electrons at the upper hybrid resonance layer and drive up the perpendicular electron kinetic energy. They have been used successfully in the DIII-D, NSTXU, EAST, Tore Supra, and WEST tokamaks to heat, drive current, and control the size of the lower-order magnetic islands that appear particularly at the $q = 3/2$ and $q = 2$ magnetic surfaces. Both the LHCD and ECRH will be used in ITER to control the plasma.

These high-powered radio frequency (RF) waves drive the electron velocity distributions away from the Maxwellian in the higher than thermal energy region called superthermal energy distributions. Ion cyclotron resonance (ICRF) and neutral beam injection (NBI) have a similar action on the ion phase space distributions.

10.1 Kinetic Theory Formulas for the Fluctuating Electromagnetic Fields

The fluctuating phase space density $\delta f(\boldsymbol{x}, \boldsymbol{v}, t)$ in a magnetized plasma is given by the time-history integral specifying the effect of the fluctuation fields on the background distribution function

$$\left(\frac{\partial}{\partial t} + \boldsymbol{v} \cdot \boldsymbol{\nabla} + \frac{q}{m}\boldsymbol{v} \times \mathbf{B}_0 \cdot \frac{\partial}{\partial \boldsymbol{v}}\right) \delta f(\boldsymbol{x}, \boldsymbol{v}, t) = -\frac{q}{m}(\delta\mathbf{E} + \boldsymbol{v} \times \boldsymbol{\delta}\mathbf{B}) \cdot \frac{\partial F(\boldsymbol{v})}{\partial \boldsymbol{v}}. \quad (10.1)$$

The plasma electric current and charge density are given by

$$\delta \boldsymbol{j} = \sum_s q_s \int d^3v \boldsymbol{v} \delta f_s \quad (10.2)$$

$$\delta \rho = \sum_s q_s \int d^3v \delta f_s \quad (10.3)$$

where s is the species index of the charged particles. The solution of Eq. (10.1) is

$$\delta f(\boldsymbol{x}, \boldsymbol{v}, t) = -\frac{q}{m} \int_{-\infty}^{t} dt' \left[\delta\mathbf{E}(\boldsymbol{x}', \boldsymbol{v}', t') + \boldsymbol{v}' \times \delta\mathbf{B}(\boldsymbol{x}', \boldsymbol{v}', t')\right] \cdot \frac{\partial F(\boldsymbol{v}')}{\partial \boldsymbol{v}'} \quad (10.4)$$

with $\boldsymbol{x}' = \boldsymbol{x}(t'), \boldsymbol{v}' = \boldsymbol{v}(t')$ being the orbits in the absence of the fluctuations specified by the derivatives on the left-hand side of Eq. (10.1).

Let the fluctuating fields vary as $\exp(i\boldsymbol{k} \cdot \boldsymbol{x}' - i\omega t')$. Maxwell's equations give $\delta\mathbf{B} = \boldsymbol{k}/\omega \times \delta\mathbf{E}$. Eliminating $\delta\mathbf{B}$ in Eq. (10.4) gives

$$\begin{aligned}\delta f(\boldsymbol{x}, \boldsymbol{v}, t) &= -\frac{q}{m} \int_{-\infty}^{t} dt' \exp(i\boldsymbol{k} \cdot \boldsymbol{x}' - i\omega t') \left[\delta\mathbf{E}(\boldsymbol{k}, \omega) + \boldsymbol{v}' \times \left(\frac{\boldsymbol{k}}{\omega} \times \delta\mathbf{E}\right)\right] \\ &\quad \cdot \frac{\partial F(\boldsymbol{v}')}{\partial \boldsymbol{v}'} \\ &= -\frac{q}{m\omega}\delta\mathbf{E} \cdot \int_{-\infty}^{t} dt' \exp(i\boldsymbol{k} \cdot \boldsymbol{x}' - i\omega t') \left[\boldsymbol{I}(\omega - \boldsymbol{v}' \cdot \boldsymbol{k}) + \boldsymbol{v}'\boldsymbol{k}\right] \cdot \frac{\partial F(\boldsymbol{v}')}{\partial \boldsymbol{v}'}.\end{aligned} \quad (10.5)$$

The gyro-orbits expressed such that $\boldsymbol{x}' = \boldsymbol{x}$ and $\boldsymbol{v}' = \boldsymbol{v}$ at $t' = t$ are given in terms of $\tau = t - t'$ by

$$v_x' = v_\perp \cos(\phi + \Omega\tau) \quad (10.6)$$

$$v_y' = v_\perp \sin(\phi + \Omega\tau) \quad (10.7)$$

$$v_z' = v_\parallel \tag{10.8}$$

$$x' = x - \frac{v_\perp}{\Omega}\left[\sin(\phi + \Omega\tau) - \sin(\phi)\right] \tag{10.9}$$

$$y' = y + \frac{v_\perp}{\Omega}\left[\cos(\phi + \Omega\tau) - \cos(\phi)\right] \tag{10.10}$$

$$z' = z - v_\parallel \tau. \tag{10.11}$$

Using the orbits in Eqs. (10.6)–(10.11) the fluctuating distribution function reduces to

$$\delta f_{\mathbf{k}} = -\frac{q}{m\omega}\delta\mathbf{E}_{\mathbf{k}} \cdot \int_0^\infty d\tau \, \exp\left(i\boldsymbol{k}\cdot\boldsymbol{x}' - i\omega t'\right)\left[\boldsymbol{I}(\omega - \boldsymbol{v}'\cdot\boldsymbol{k}) + \boldsymbol{v}'\boldsymbol{k}\right]\cdot\frac{\partial F(\boldsymbol{v}')}{\partial \boldsymbol{v}'}. \tag{10.12}$$

First we reduce the integrand of Eq. (10.12) to

$$\delta\mathbf{E} \cdot \left[\boldsymbol{I}(\omega - \boldsymbol{v}'\cdot\boldsymbol{k}) + \boldsymbol{v}'\boldsymbol{k}\right]\cdot\frac{\partial F(\boldsymbol{v}')}{\partial \boldsymbol{v}'} \tag{10.13}$$

$$= \delta E_z\left(\omega\frac{\partial F}{\partial v_z} - (k_x v_x + k_y v_y)\left(\frac{\partial F}{\partial v_z} - \frac{v_z}{v_\perp}\frac{\partial F}{\partial v_\perp}\right)\right). \tag{10.14}$$

The exponential phase function $\boldsymbol{k}\cdot\boldsymbol{x}' - \omega t'$ can be rewritten, using $k_x = k_\perp \cos\theta$ and $k_y = k_\perp \sin\theta$ as

$$\begin{aligned}
\boldsymbol{k}\cdot\boldsymbol{x}' - \omega t' &= \boldsymbol{k}\cdot\boldsymbol{x} - \omega t - \left\{k_\perp \cos\theta\frac{v_\perp}{\Omega}\left[\sin(\phi + \Omega\tau) - \sin\phi\right]\right\} \\
&\quad - \left\{k_\perp \sin\theta\frac{v_\perp}{\Omega}\left[\cos(\phi + \Omega\tau) - \cos\phi\right]\right\} - k_z v_\parallel \tau + \omega\tau \\
&= \boldsymbol{k}\cdot\boldsymbol{x} - \omega t - \frac{k_\perp v_\perp}{\Omega}\left[\sin(\phi - \theta + \Omega\tau) - \sin(\phi - \theta)\right] + (\omega - k_\parallel v_\parallel)\tau.
\end{aligned} \tag{10.15}$$

Let us define the wave-particle phase function

$$\beta_k(\tau) = -\frac{k_\perp v_\perp}{\Omega}\left[\sin(\phi - \theta + \Omega\tau) - \sin(\phi - \theta)\right] + (\omega - k_\parallel v_\parallel)\tau \tag{10.16}$$

from Eq. (10.15). Then the phase becomes $\exp[i(\boldsymbol{k}\cdot\boldsymbol{x}' - \omega t')] = \exp[i(\boldsymbol{k}\cdot\boldsymbol{x} - \omega t) + i\beta_k(\tau)]$. With $\theta = 0(k_x = k_\perp,\ k_y = 0)$ the fluctuation phase is

$$\beta_k(\tau) = -\frac{k_\perp v_\perp}{\Omega}\left[\sin(\phi + \Omega\tau) - \sin\phi\right] + \left(\omega - k_\parallel v_\parallel\right)\tau. \tag{10.17}$$

The following Bessel function identity is now used

$$\exp\left(iz\sin\phi\right) = \sum_{n=-\infty}^{\infty}\exp\left(in\phi\right)J_n(z) \tag{10.18}$$

to carry out the τ-integral in Eq. (10.12). Later we will use the identities

$$\int_0^{2\pi} \frac{d\phi e^{-iz[\sin(\phi+\Omega\tau)-\sin(\phi)]}}{2\pi} \begin{pmatrix} \sin\phi\sin(\phi+\Omega\tau) \\ \sin\phi\cos(\phi+\Omega\tau) \\ \cos\phi\sin(\phi+\Omega\tau) \\ \cos\phi\cos(\phi+\Omega\tau) \end{pmatrix} = \sum_{n=-\infty}^{\infty} e^{-in\Omega\tau} \begin{pmatrix} (J_n')^2 \\ -\frac{in}{z}J_nJ_n' \\ \frac{in}{z}J_nJ_n' \\ \frac{n^2}{z^2}J_n^2 \\ J_n^2 \end{pmatrix}. \tag{10.19}$$

The fluctuating plasma current is then

$$\delta \boldsymbol{j} = \sum_s q_s \int d^3v \boldsymbol{v} \delta f_s = -\sum_s \frac{\epsilon_0 \omega_{ps}^2}{n_0 \omega} \int_0^\infty \int_{-\infty}^\infty \int_0^{2\pi} v_\perp dv_\perp dv_\parallel d\phi$$

$$\times \boldsymbol{v} \int_0^\infty d\tau e^{-iz[\sin(\phi+\Omega\tau)-\sin(\phi)]} e^{-i(\omega-k_z v_z)\tau}$$

$$(\delta E_x v_\perp \cos(\phi+\Omega\tau) + \delta E_y v_\perp \sin(\phi+\Omega\tau))$$

$$\times \left(k_z \frac{\partial F}{\partial v_z} + \frac{1}{v_\perp}(\omega - k_z v_z)\frac{\partial F}{\partial v_\perp}\right). \tag{10.20}$$

The wave-particle gradient $\widehat{D}_k F$ is defined by

$$\widehat{D}_k(\boldsymbol{k},\omega,\boldsymbol{v})F = \left(k_z v_\perp \frac{\partial F}{\partial v_z} + (\omega - k_z v_z)\frac{\partial F}{\partial v_\perp}\right) \tag{10.21}$$

enters the transport tensor in Sec. 10.3. The components of the electric current reduce to

$$\delta j_x = \sum_s \frac{\epsilon_0 \omega_{ps}^2}{\omega} \int_0^\infty \int_{-\infty}^\infty \int_0^{2\pi} v_\perp dv_\perp dv_\parallel d\phi \int_0^\infty d\tau e^{-iz[\sin(\phi+\Omega\tau)-\sin(\phi)]} v_\perp \cos\theta$$

$$(\delta E_x v_\perp \cos(\phi+\Omega\tau) + \delta E_y v_\perp \sin(\phi+\Omega\tau))\, \widehat{D}_k F \tag{10.22}$$

$$= -\sum_s \frac{\epsilon_0 \omega_{ps}^2}{n_0 \omega} \int_0^\infty \int_{-\infty}^\infty v_\perp dv_\perp dv_\parallel$$

$$\frac{2\pi i\, \widehat{D}_k F}{\omega - k_z v_z - n\Omega} \sum_{n=-\infty}^{\infty} \left(\frac{n^2\Omega^2}{k_\perp^2 v_\perp^2} v_\perp J_n^2 \delta E_x + \frac{in\Omega}{k_\perp v_\perp} v_\perp J_n J_n' \delta E_y\right) \tag{10.23}$$

and

$$\delta j_y = -\sum_s \frac{\epsilon_0 \omega_{ps}^2}{n_0 \omega} \int_0^\infty \int_{-\infty}^\infty \int_0^{2\pi} v_\perp dv_\perp dv_\parallel d\phi$$

$$\int_0^\infty d\tau e^{-iz[\sin(\phi+\Omega\tau)-\sin(\phi)]} v_\perp \sin\phi\, (\delta E_x v_\perp \cos(\phi+\Omega\tau) + \delta E_y v_\perp \sin(\phi+\Omega\tau))\, \widehat{D}_k F$$

$$= -\sum_s \frac{\epsilon_0 \omega_{ps}^2}{n_0 \omega} \int_0^\infty \int_{-\infty}^\infty v_\perp dv_\perp dv_\parallel$$

$$\frac{2\pi i\, \widehat{D}_k F}{\omega - k_z v_z - n\Omega} \sum_{n=-\infty}^{\infty} \left(-\frac{in\Omega}{k_\perp v_\perp} v_\perp J_n J_n' \delta E_x + v_\perp (J_n')^2 \delta E_y \right). \tag{10.24}$$

In matrix form the two cross-field plasma currents are

$$\begin{pmatrix} \delta j_x \\ \delta j_y \end{pmatrix} = \sum_s \sum_{n=-\infty}^{\infty} \frac{\epsilon_0 \omega_{ps}^2}{n_0 \omega} \int_0^\infty \int_{-\infty}^{\infty} v_\perp dv_\perp dv_\| \frac{2\pi i\, \widehat{D}_k F}{\omega - k_z v_z - n\Omega}$$

$$\begin{pmatrix} \frac{n^2\Omega^2}{k_\perp^2 v_\perp^2} v_\perp J_n^2 & \frac{in\Omega}{k_\perp v_\perp} v_\perp J_n J_n' \\ -\frac{in\Omega}{k_\perp v_\perp} v_\perp J_n J_n' & v_\perp (J_n')^2 \end{pmatrix} \cdot \begin{pmatrix} \delta E_x \\ \delta E_y \end{pmatrix}. \tag{10.25}$$

The plasma conductivity defined by $\boldsymbol{j} = \sum_s \boldsymbol{\sigma}_s \cdot \delta \mathbf{E}$ is given by

$$\boldsymbol{\sigma}_s(\boldsymbol{k}, \omega) = \sum_{n=-\infty}^{\infty} \frac{ie_s^2}{m_s} 2\pi \int_0^\infty \int_{-\infty}^{\infty} v_\perp dv_\perp dv_\| \frac{\widehat{D}_k F}{\omega k_z v_z n\Omega}$$

$$\times \begin{pmatrix} \frac{n^2\Omega^2}{k_\perp^2 v_\perp^2} v_\perp J_n^2 & \frac{in\Omega}{k_\perp v_\perp} v_\perp J_n J_n' \\ -\frac{in\Omega}{k_\perp v_\perp} v_\perp J_n J_n' & v_\perp (J_n')^2 \end{pmatrix} \begin{pmatrix} \delta E_x \\ \delta E_y \end{pmatrix}. \tag{10.26}$$

The formulas for $\boldsymbol{\sigma}$ are required for computing the fluctuating plasma currents for arbitrary applied electric and magnetic fluctuations. External antennas on vessel walls are installed for launching the radio frequency (RF) waves to heat the plasma and to drive the plasma current. The RF waves are also used for control of edge localized modes (ELMs).

10.2 Requirements for Self-Consistent Fluctuating Fields

Now the electromagnetic fields are made self-consistent with the plasma currents through Maxwell field equations

$$i\boldsymbol{k} \times \mathbf{E} = i\omega \mathbf{B} \tag{10.27}$$

$$i\boldsymbol{k} \times \mathbf{B} = \mu_0 \boldsymbol{J} - \frac{i\omega}{c^2} \mathbf{E} \tag{10.28}$$

where the current density $\boldsymbol{J}(\boldsymbol{k}, \omega)$ is given by the kinetic calculation with $\boldsymbol{J} = \boldsymbol{\sigma} \cdot \mathbf{E}$. This substitution gives the linear homogeneous system of equations

$$\boldsymbol{k} \times (\boldsymbol{k} \times \mathbf{E}) + i\mu_0 \omega \boldsymbol{J} + \frac{\omega^2}{c^2} \mathbf{E} = 0. \tag{10.29}$$

In matrix or dyadic form Eq. (10.29) becomes

$$\left(\boldsymbol{kk} - k^2 \boldsymbol{I} + i\mu_0 \omega \boldsymbol{\sigma} + \frac{\omega^2}{c^2} \boldsymbol{I} \right) \cdot \mathbf{E} \tag{10.30}$$

or, equivalently, the matrix

$$\left[\left(k^2 - \frac{\omega^2}{c^2}\right)\boldsymbol{I} - \boldsymbol{kk} - i\omega\mu_0\boldsymbol{\sigma}\right]\cdot\mathbf{E} = 0 \tag{10.31}$$

that constrains the fields to be the naturally-occurring, collective self-consistent fluctuations in the plasma. For driven RF waves there is $\boldsymbol{J}_{\text{ant}}$ on the plasma wall.

For the linear, homogeneous system of Eq. (10.31) to have nontrivial solution with $\mathbf{E} \neq 0$, we must have

$$D(\boldsymbol{k},\omega) = \det\left[\left(k^2 - \frac{\omega^2}{c^2}\right)\boldsymbol{I} - \boldsymbol{kk} - i\omega\mu_0\boldsymbol{\sigma}\right] = 0. \tag{10.32}$$

The complex valued function $D(\boldsymbol{k},\omega)$ vanishes at numerous points in the complex ω-plane. These complex roots give multiple modes with $\omega_\alpha = \omega_\alpha(\boldsymbol{k})$ $\alpha = 1, 2, 3...n_{\text{modes}}$ and their electric field polarization vectors follow from Eq. (10.31) by computing ratios of the $\mathbf{E}_{\boldsymbol{k}}$ components with the roots $\omega = \omega_\alpha(\boldsymbol{k})$ substituted into the matrix.

In the cold plasma limit the conductivity reduces to

$$\boldsymbol{\sigma} = \sum_s \omega_{ps}^2 i\epsilon_0 \begin{bmatrix} \dfrac{\omega}{\omega^2-\Omega^2} & \dfrac{-i\Omega}{\omega^2-\Omega^2} & 0 \\ \dfrac{i\Omega}{\omega^2-\Omega^2} & \dfrac{\omega}{\omega^2-\Omega^2} & 0 \\ 0 & 0 & \dfrac{1}{\omega} \end{bmatrix}.$$

For weak electron-neutral collisions $1/\omega \to 1/(\omega + i\nu_{en})$ and $\sigma_\parallel = n_e e^2/m_e(\nu_{en} - i\omega)$. For $|\omega| < \nu_{en}$ the $1/\sigma_\parallel$ becomes the collisional resistivity. Using this $\boldsymbol{\sigma}$ in Eq. (10.31) gives the Stix matrix.

10.2.1 *Dispersion relation for two-temperature Maxwellian distribution*

Fusion plasmas are strongly heated with powerful neutral beams (NBI) and radio frequency (RF) antennas. The particle energies above the thermal energy given by the core plasma temperature are preferentially heated especially in the RF heating methods designed to drive the plasma current. Thus, the particle distribution functions have anisotropies in their high-energy parts. These anisotropies can drive instabilities by transferring energy from the particles to the waves. In one case called "alpha channeling", the antenna system is designed to have the fusion reaction produced alpha particles drive the waves up to significant amplitudes where they may then resonantly heat the thermal ions in the plasma. This mechanism has the potential of allowing a fast, direct energy transfer from the alpha particles produced by fusion reactions to the hydrogen isotopes of deuterium and tritium used for fuel. In the standard collisional scenario the alpha power goes into heating the electrons which on a long time scale of seconds transfer their energy to the deuterium and tritium through collisions.

Let us evaluate the dispersion relation for the bi-Maxwellian distribution function using the plasma dispersion function $Z(\alpha)$ derived in Worked Problem 2.2. The bi-Maxwellian distribution function is given by

$$F\left(v_\perp, v_\parallel\right) = \frac{1}{\sqrt{2\pi}} \frac{1}{v_{T\parallel}} \frac{1}{v_{T\perp}^2} \exp\left(-\frac{v_\perp^2}{2v_{T\perp}^2}\right) \exp\left(-\frac{v_\parallel^2}{2v_{T\parallel}^2}\right) \tag{10.33}$$

The velocity gradients of F are

$$\widehat{D}_k F = k_z v_\perp \frac{\partial F}{\partial v_z} + (\omega - k_z v_z) \frac{\partial F}{\partial v_\perp} \tag{10.34}$$

with

$$\frac{\partial F}{\partial v_z} = \frac{1}{\sqrt{2\pi}} \frac{1}{v_{T\parallel}} \frac{1}{v_{T\perp}^2} \frac{-v_z}{v_{T\parallel}^2} \exp\left(-\frac{v_\perp^2}{2v_{T\perp}^2}\right) \exp\left(-\frac{v_\parallel^2}{2v_{T\parallel}^2}\right) \tag{10.35}$$

$$\frac{\partial F}{\partial v_\perp} = \frac{1}{\sqrt{2\pi}} \frac{1}{v_{T\parallel}} \frac{1}{v_{T\perp}^2} \frac{-v_\perp}{v_{T\perp}^2} \exp\left(-\frac{v_\perp^2}{2v_{T\perp}^2}\right) \exp\left(-\frac{v_\parallel^2}{2v_{T\parallel}^2}\right). \tag{10.36}$$

The integration over all parallel velocities of the response function gives

$$\begin{aligned} \frac{1}{\sqrt{2\pi}} \int dv_z \frac{k_z v_z}{\omega - k_z v_z - n\Omega} \frac{1}{v_{T\parallel}} \frac{1}{v_{T\parallel}^2} \exp\left(-\frac{v_z^2}{2v_{T\parallel}^2}\right) \\ = -\frac{1}{v_{T\parallel}^2} \left(1 + \alpha_n Z(\alpha_n)\right) \end{aligned} \tag{10.37}$$

where $\alpha_n = (\omega - n\Omega)/\sqrt{2}\,|k_\parallel| v_T$. Here the $Z(\alpha_n)$ is the plasma dispersion function tabulated in Fried and Conte (1961) and introduced in Worked Problem 2.2 of Chapter 2.1. The graphs of Z and Z' are shown in Fig. 2.1 and 2.2. Note that in the low-temperature limit $|\alpha| \gg 1$ and from Worked Problem 2.2 the cold plasma response is $Z(\alpha) \simeq -1/\alpha$ and thus the integral reduces to the usual cyclotron response $\omega/(\omega - n\Omega)$ for low-thermal energy (cold) particles.

Worked Problem 10:1: Conductivities for Maxwell–Boltzmann Plasma

Carry out the integration over the perpendicular velocities in the response functions for the currents in Eq. (10.25) for a Maxwellian distribution $\exp(-mv_\perp^2/2T_\perp)$ of the perpendicular velocities defining $v_{\perp T_\perp}^2/m$ to derive the formulas for the kinetic RF electrical conductivities.

Solution:

The integral over the Maxwell distribution of $\left\langle J_0^2(k_\perp v_\perp/\Omega)\right\rangle_{\mathrm{M}} = \exp(-b) I_0(b)$ where $b = k_\perp^2 v_{T\perp}^2/2\Omega^2$. See Worked Problem 1 in Chapter 1. Following this analysis one derives for the upper σ (2×2) matrix elements D_{11}, $D_{12} = D_{21}^*$ and D_{22} given by

$$D_{11} = \frac{n^2\Omega^2}{k_\perp^2}\exp\left(-\frac{k_\perp^2 v_{T_\perp}^2}{\Omega^2}\right) I_n\left\{\left(\frac{v_{T\perp}^2}{v_{T\|}^2}+1\right)(1+\alpha_n Z(\alpha_n)) - \frac{\omega}{k_\perp v_{T\|}} Z(\alpha_n)\right\}$$

$$D_{12} = in\exp\left(-\frac{k_\perp^2 v_{T_\perp}^2}{\Omega^2}\right)(I_n - I_n')\left\{\left(\frac{v_{T\perp}^2}{v_{T\|}^2}+1\right)(1+\alpha_n Z(\alpha_n)) - \frac{\omega}{k_\perp v_{T\|}} Z(\alpha_n)\right\}$$

$$D_{22} = \exp\left\{\frac{n^2\Omega^2}{k_\perp^2} v_{T\perp}^2 I_n + 2\frac{k_\perp^2 v_{T_\perp}^2}{\Omega^2}(I_n - I_n')\right\}$$
$$\times\left\{\left(1-\frac{v_{T\perp}^2}{v_{T\|}^2}\right)(1+\alpha_n Z(\alpha_n)) - \frac{\omega}{k_\perp v_{T\|}} Z(\alpha_n)\right\}. \quad (10.38)$$

In the limit of low-perpendicular temperature, one has $I_{n=1} \to k_\perp^2 v_{T\perp}^2/2\Omega^2$ and $I_{n=1}' \to 1/2$. The matrix reduces to that of cold plasma Stix matrix.

10.2.2 *Electromagnetic dispersion relation for cross-field beam injected plasma*

Both neutral beam injection NBI and cyclotron resonance heating create strong anisotropies and beam populations in the plasma. Ion cyclotron waves are observed during perpendicular neutral beam injection in JET [Dendy, *et al.* (1995)].

For the parameterization of a cross-field particle beam with plasma $v_\perp = v_b$ phase space distribution, we take

$$F = g_1(v_\perp)\, g_2(v_z) \quad (10.39)$$

$$g_2(v_z) = \left(\frac{m}{2\pi T_\|}\right)^{1/2}\exp\left(-\frac{mv_z^2}{2T_\|}\right) = \frac{1}{\sqrt{\pi}\, v_{T_\|}}\exp\left(-\frac{v_z^2}{v_{T_\|}^2}\right) \quad (10.40)$$

$$g_1(\overline{v}_\perp) = \frac{v_\perp}{v_r}\frac{1}{V_0^2}\frac{2}{\sqrt{\pi}}\frac{1}{\overline{v}_{T_\perp}}\exp\left[-\left(\frac{v_\perp - v_b}{\sqrt{2}\, v_{T_\perp}}\right)^2\right]. \quad (10.41)$$

First we consider the low-perpendicular thermal spread limit. From the cyclotron harmonics $n = \pm 1$ we derive

$$\int_0^\infty\int_{-\infty}^\infty v_\perp dv_\perp dv_\| \frac{2\pi}{\omega - k_z v_z \mp \Omega}\left(v_\perp k_z\frac{\partial F}{\partial v_z} + (\omega - k_z v_z)\frac{\partial F}{\partial v_\perp}\right)\frac{1}{2}v_\perp J_0^2 \quad (10.42)$$

$$= -\frac{1}{v_{T\|}^2}(1+\alpha Z(\alpha))B_{k_\perp} + \left(1 - \frac{n\Omega}{k_z v_{T\|}}Z(\alpha)\right)A_{k_\perp} \quad (10.43)$$

where the key stability functions of $k_\perp$ and the beam distribution functions are

$$A_{k_\perp} = \int dv_\perp v_\perp \frac{v_\perp}{2}\frac{\partial g_1(v_\perp)}{\partial v_\perp}J_0^2 \quad (10.44)$$

$$B_{k_\perp} = \int dv_\perp v_\perp \frac{v_\perp}{2}J_0^2 g_1(v_\perp). \quad (10.45)$$

The dispersion relation at $k_\perp \rho_s \ll 1$ reduces to

$$\omega^2 - k^2c^2 + \sum_s \frac{1}{2}\omega_{ps}^2 \int_{-\infty}^{\infty} d^3v \frac{1}{\omega - k_z v_z \pm \Omega} v_\perp \left(v_\perp k_z \frac{\partial F}{\partial v_z} + (\omega - k_z v_z) \frac{\partial F}{\partial v_\perp} \right) = 0 \tag{10.46}$$

which gives electromagnetic ion cyclotron (EMICs) waves and whistlers. Let $\omega \to \omega_R + i\gamma$, $\gamma \ll \omega(k_\perp, k_z)$, where the subscript R is for the real part of the complex $\omega(k)$-function and $\gamma(k)$ is the imaginary part. Evaluating the resonant particle-wave interactions gives the imaginary terms:

$$2\gamma_{\mathbf{k}}\omega_R(k_\perp, k_z) + \sum_s \frac{1}{2}\omega_{ps}^2 \Bigg[-\langle v_\perp^2 \rangle \pi \frac{\partial f^{(0)}}{\partial v_z} + \int d^3v v_\perp \times \left(1 \mp \frac{\Omega_s / k_z}{v_z - (\omega \pm \Omega_s)/k_z} \right) \frac{\partial F}{\partial v_\perp} \Bigg] \tag{10.47}$$

$$2\gamma_{\mathbf{k}}\omega_k(k_\perp, k_z) + \sum_s \frac{1}{2}\omega_{ps}^2 \left[-\langle v_\perp^2 \rangle \pi \frac{\partial f^{(0)}}{\partial v_z} \mp \frac{2\Omega_s}{k_z} f^{(0)}(v_z) \right]_{v_z = (\omega \pm \Omega_s)/k_z} = 0 \tag{10.48}$$

$$\gamma_{\mathbf{k}} = \frac{\pi}{2\omega_k(k_\perp, k_z)} \sum_s \left[\frac{\langle v_\perp^2 \rangle}{2} \frac{\partial f^{(0)}}{\partial v_z} \pm \frac{\Omega_s}{k_z} f^{(0)}(v_z) \right]. \tag{10.49}$$

If there is a Maxwellian distribution in the parallel velocity v_z, then $\frac{\partial f}{\partial v_z} = -\frac{v_z}{v_{T_\parallel}^2} f$. The average $v_\perp^2$ is defined by $\langle v_\perp^2 \rangle = 2v_{T_\perp}^2 = T_\perp/m$. The growth rate function $\gamma(\mathbf{k})$ from Eq. (10.49) reduces to

$$\gamma_{\mathbf{k}} = \frac{\pi}{2\omega_R(k_\perp, k_z)} \sum_s \left[-\frac{(\omega_R(k_\perp, k_z) \pm \Omega_j)}{k_z} \left(\frac{T_\perp}{T_\parallel} \left(\frac{T_\perp}{T_\parallel} - 1 \right) \right) \right] \cdot \times \frac{1}{\sqrt{2\pi}\, v_{T_\parallel}} \exp\left(\frac{-(\omega_R(k_\perp, k_z) \pm \Omega_j)^2}{2k_z^2 v_{T_\parallel}^2} \right).$$

10.3 Quasilinear Fluctuation Diffusion Tensor

The reaction of the $\mathbf{E_k}, \delta\mathbf{B_k}$ fluctuations on the background phase space distribution $\langle f \rangle$ is given by

$$\frac{\partial \langle f \rangle}{\partial t} = \frac{e^2}{m^2} \sum_{\boldsymbol{k}} \widehat{D}_{\boldsymbol{k}} \mathbf{E}_{\boldsymbol{k}}^* \cdot \mathrm{Im}(\mathcal{X}_{\boldsymbol{k}}) \cdot \mathbf{E}_{\boldsymbol{k}} \widehat{D}_k \langle f \rangle \tag{10.50}$$

where the operator $\widehat{D}_k$ is defined in Eq. (10.21) and the resonant part of the conductivity tensor $\mathcal{X}_{\boldsymbol{k}}$ is the linear wave-particle resonance propagation

$$\omega_{\boldsymbol{k}} - n\Omega = k_\parallel v_\parallel \tag{10.51}$$

for all waves $\boldsymbol{k}, \omega_{\boldsymbol{k}}$ and cyclotron harmonics n in the fluctuation spectrum. Adding source terms from the NBI on the electromagnetic field from the RF antenna to Eq. (10.50) gives the evaluation of the high-energy particles.

10.4 Electron Transport from Magnetic Flutter

Magnetic flutter is a term introduced by Callen (1977) to describe losses from turbulent magnetic fields.

10.4.1 *Polarization and transport for low-frequency electromagnetic fluctuations*

At moderate plasma pressure, $\beta < 1$, there are general polarization features of the low-frequency drift wave fluctuations that simplify the transport analysis. In contrast, for $\beta \gtrsim 1$, all three electric field components are coupled in a complicated manner; however, the drift-type modes are then unstable only under rather special conditions.

In moderate pressure plasma the highly elongated fluctuations $k_\perp \gg k_\parallel$ of the drift wave and tearing-mode type decouple from the fast compressional motion with $\omega^2 \ll k^2 v_A^2$. The polarization of the slow convective modes is determined by the total pressure balance across the magnetic field $B\delta B_\parallel(k) + 4\pi\delta p_\perp(k) = 0$ that is maintained on the scale $k_\perp^{-1}$ during the slow oscillations. From this relation and the formula for $\delta p_\perp(k)$ the magnitude of the magnetic compression follows as

$$\frac{\delta B_\parallel(k)}{B} = \frac{1}{2}\beta\left[\frac{iekE_y(k)}{k_yT}\right] \cong \frac{1}{2}\beta\left[\frac{e\varphi(k)}{T}\right] \tag{10.52}$$

which is valid for modes rotating in either the ion or the electron diamagnetic direction. The change in the magnetic field strength $\delta B_\parallel(k)$ ensures that there is no compression of the plasma $\nabla\cdot\boldsymbol{v}_E = 0$ in the convection of the plasma across the finite beta equilibrium with $B^2(r) + 8\pi p_\perp(r) = \text{const}$.

The perpendicular magnetic perturbation $\delta B_\perp(k)$ considerably exceeds $\delta B_\parallel(k)$ and takes on the magnitude and phase required to short out the parallel electric field, $E_\parallel(k) = -ik_\parallel\varphi(k) + i(\omega/c)A_\parallel(k) \approx 0$. The important case where the inductive part of the parallel electric field is not able to cancel the parallel electric field occurs when the $\boldsymbol{k}\omega$ components are such as to incur in an anomalously low-parallel conductivity $\boldsymbol{\sigma}_\parallel(\boldsymbol{k},\omega)$. The fluctuations for which the parallel conductivity is low are $\omega \sim \omega_{*e}$ and $k_\parallel v_e \gtrsim \omega$ since from the electron current $j_\parallel(k)$,

$$\boldsymbol{\sigma}_\parallel(k) \cong \frac{im_e n_e e^2}{2k_\parallel^2 T_e}(\omega - \omega_{*e})\, Z'\left(\frac{\omega}{|k_\parallel|v_e}\right). \tag{10.53}$$

For fluctuations in this domain it follows from $\nabla\times\delta\mathbf{B}_\perp = 4\pi\delta j_\parallel/c$ that the perpendicular magnetic perturbation is

$$\frac{\delta B_\perp(k)}{B} \cong i\left(\frac{\beta}{2}\right)^{1/2}\left(\frac{\omega_{*e}}{k_\parallel v_A}\right)k_\perp\rho_i\left[\frac{e\varphi(k)}{T_e}\right] \tag{10.54}$$

where $E_y(k) \cong -ik_y\varphi(k)$ is used.

In characterizing the moderate β drift-tearing mode fluctuations one observes that the field line bending from $\delta B_\perp(k)$ couples the parallel transport into a radial motion giving cross-field transport. As an example, we observe that in the absence of line bending the electron temperature fluctuation determined for $\nu_e > k_\parallel v_e$ from the thermal balance equation with $q_{\parallel e} = -\chi_{\parallel e} \nabla_\parallel \delta T_e$ where $\chi_{\parallel e} = n_e T_e / m_e \nu_e$ yields

$$\frac{\delta T_e(k)}{T_e} = \left(\frac{i\nu_e}{k_\parallel^2 v_e^2}\right)\left[\omega - \omega_{*e}\left(1 - \frac{3}{2}\eta_e\right)\right]\left[\frac{e\varphi(k)}{T_e}\right] = \left[\frac{\delta T_e(k)}{T_e}\right]_{\text{conv}} \tag{10.55}$$

from the balance of convection and parallel diffusion. In the presence of line bending, however, the thermal transport exceeds that due to convection. At the magnetic amplitude $\delta B_\perp(k)$ when the parallel inductive field $(\omega/c)A_\parallel(k)$ due to $\delta B_\perp(k) = ik_y A_\parallel(k)$ is such that

$$E_\parallel^{(A)} = \frac{i\omega}{c} A_\parallel(k) \simeq \left(\frac{\omega}{\omega_{*e}}\right)\left[-ik_\parallel \varphi(k)\right].$$

The opposite parity in $k_\parallel$ of $A_\parallel(k)$ and $\varphi(k)$ is a general feature of the low-frequency fluctuations. The relationship of electromagnetic drift modes to magnetohydrodynamic modes is described further in Horton, *et al.* (1983).

The polarization relations for the low-frequency drift and tearing mode fluctuations at $\beta < 1$ allow the particle and energy flux formulas to be simplified. First consider the contribution of $\delta B_\parallel(k)$ to the Lorentz force $F_y^j(k, v_\perp, v_\parallel)$. Using Eq. (10.52) to relate $E_y(k)$ and $\delta B_\parallel(k)$ we observe that $F_y^j \simeq E_y(k)[J_0 + i\beta J_1/(k_\perp \rho_j)] \simeq E_y(k) J_0(k_\perp v_\perp/\omega_{cj})$for $\beta < 1$ and arbitrary $k_\perp \rho_j$. Now consider the contribution to F_y^j due to $\delta B_\perp(k)$. For the $\delta B_\perp(k)$ given in Eq. (10.54), it follows that $F_y^j \cong E_y J_0(1 + v_\parallel/v_A)$ when $\omega_{*e} \sim k_\parallel v_A$. For the plasma pressure $\beta < 1$ the Alfvén speed is intermediate to the ion and electron thermal velocities, so that the magnetic force is small on the ions with $F_y^i \cong J_0 E_y$ and dominant on the electrons $F_y^e \cong (v_\parallel/c) B_x(k)$. Taking into account these simplifications, we have

$$\Gamma_i = \sum_k \left|\frac{cE_y(k)}{B}\right|^2 \int d\boldsymbol{v} \left(\frac{\partial f_i}{\partial x} + \frac{\omega \Omega_i}{k_y}\frac{\partial f_i}{\partial \varepsilon}\right) J_0^2\left(\frac{k_\perp v_\perp}{\Omega_i}\right) \text{Im}\ g_k^i(\boldsymbol{v}) \tag{10.56}$$

$$\Gamma_e = \sum_k \left|\frac{B_x(k)}{B}\right|^2 \int d\boldsymbol{v} \left(\frac{\partial f_e}{\partial x} + \frac{\omega \Omega_e}{k_y}\frac{\partial f_e}{\partial \varepsilon}\right) v_\parallel^2 \,\text{Im}\ g_k^e(\boldsymbol{v}) \tag{10.57}$$

and the energy flux Q_j is obtained by including the kinetic energy $\frac{1}{2} m_j v^2$ under the velocity integrals. For fluctuations with sufficiently low frequencies, $|\omega| \ll k_y \langle v^2 \rangle / 2\Omega_j \langle d \ln f_j / dx \rangle$ the anomalous flux is driven directly by $\partial f_j / \partial x$. The finite $(\omega/k_y)(\partial f/\partial \varepsilon)$ terms give the convection velocity to the particle and thermal fluxes Γ_s.

10.4.2 *Anomalous transport due to magnetic fluctuations*

Low-frequency fluctuations near a neutral sheet such as in the geomagnetic tail plasma, e.g. [Galeev (1978)], or in the laboratory near the mode rational surfaces of the helical magnetic fields used for confinement, e.g. [Callen (1977)], produce parallel inductive electric fields $E_\parallel = -\partial A_\parallel / c\partial t$ and perpendicular magnetic fields $\delta \mathbf{B}_\perp = \nabla A_\parallel \times \widehat{\boldsymbol{z}}$. Such magnetic perturbations lead to anomalous transport. According to the quasilinear formula Eq. (10.57), the electron transport is proportional to $\left\langle |\delta B_x(k)|^2 \right\rangle$ and the particle-fluctuation interaction strength $\operatorname{Im} g_k^e(\boldsymbol{v})$. Callen (1977) describes the process as transport due to "magnetic flutter". Even in the case of static magnetic perturbations with $\left\langle \delta B_\perp(\boldsymbol{k})|^2 \right\rangle = \left\langle \delta B_\perp(\boldsymbol{k})|^2 \right\rangle \delta(\omega)$, there is an enhanced transport due to the particle motion along the magnetic field line with the interaction the Im $g_k^e = -\pi\delta(k_\parallel v_\parallel)$. The anomalous transport produced by the static magnetic field perturbations is described by Stix (1973) as transport due to "magnetic braiding."

For the electron flux in Eq. (10.57) the resonance $-$ Im $g_k^e(v_\parallel) = \pi / |k_\parallel v_\parallel|$ gives the simple formula

$$D_M = \frac{\pi v_e}{|k_\parallel|} \frac{\left\langle \delta B_x^2 \right\rangle}{B^2}$$

for the electron diffusion from the magnetic fluctuations as in Eq. (10.57).

Broadband kilohertz magnetic fluctuations with $\left\langle \delta B_x^2 \right\rangle^{1/2} / B \sim 10^{-5}$ to 10^{-4} are widely observed in tokamaks. The technique is given in Colas, *et al.* (1998) was that the $e\boldsymbol{v}_w \delta B_x$ Lorentz force changes the polarization of the incident electromagnetic wave from $X \leftrightarrow O$ polarization. This diamagnetic δE_z is called cross-polarization scattering. The X-mode with $\delta E_\perp$ is used for ECH heating.

The cross-polarization scattering (CPS) diagnostic was developed and used in Tore Supra to measure the spectrum of $\left\langle \delta B^2 \right\rangle_{\mathbf{k}}$ from $k = 6$ to $26\,\mathrm{cm}^{-1}$ showing a transition from a $k_\perp^{-3.5}$ to $k_\perp^{-7}$ spectrum with $\delta B_{\mathrm{rms}} / B \leq 7 \times 10^{-4}$. The fluctuation data from Sabot, *et al.* (2009) in Tore Supra supports the conclusion that the electromagnetic ETG turbulence gives the dominance mechanism for the turbulence and the turbulent transport in plasmas with $\beta_e > m_e/m_i$. A similar conclusion is reached from fluctuation measurements in the FT1 tokamak. The density fluctuation power spectrum is reported as $k_\perp^{-3}$ for $k\rho i \leq 1$ and $k_\parallel^{-6}$ for $k\rho i > 1$ in Tore Supra with measurable turbulence levels extending up to $k_\perp = 30\,\mathrm{cm}^{-1}$.

Stix (1973) points out that a magnetic perturbation $\delta \mathbf{B}_\perp(\boldsymbol{x})$ that is resonant with a helical magnetic field produces a chain of magnetic islands of width Δ_k about the resonant surface where $\mathbf{B} \cdot \nabla = ik_\parallel B = 0$. Writing $\mathbf{B} = B_y(x)\widehat{y} + B_0 \widehat{z}$ for the local magnetic field and $L_s^{-1} = dB_y / Bdx$ for the local shear, the width Δ_k of the magnetic island is given by

$$\Delta_k = \left(\frac{2L_s}{k_y} \frac{\delta B_x}{B_0} \right)^{1/2} . \tag{10.58}$$

When two or more of these chains of islands occur, as shown in Fig. 10.1, the magnetic field becomes ergodic first along the separatrix and then throughout the region between the islands with increasing amplitude of the magnetic perturbation. Owing to their high speed, the electrons follow the braided field lines to produce a transport across the ambient helical field. The magnetic perturbations arise from

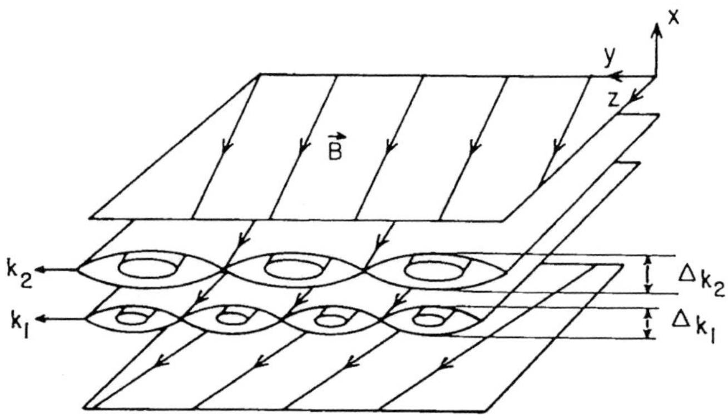

Fig. 10.1 The magnetic island chains by two nearby resonant perturbations $\boldsymbol{k}_1 \cdot \mathbf{B} = 0$ and $\boldsymbol{k}_2 \cdot \mathbf{B} = 0$ in the sheared magnetic field. The $\delta \mathbf{B}_\perp(\boldsymbol{k})$ amplitudes are small enough that overlapping has not yet occurred [Galeev (1978)].

a variety of reasons such as filamentation of an Ohmic heating current and electromagnetic fields from auxiliary heating methods.

Low-frequency drift waves and tearing modes which arise from the gradients in nonuniform plasmas, produce small-scale magnetic perturbations as measured in Tore Supra.

The radial gradient of the large plasma current density $j_\phi(r)$ in a tokamak that is driven either by (i) the ohmic inductive electromotive force, (ii) by radio frequency current drive or (iii) by the pressure gradient driving the bootstrap currents adds a symmetry breaking term to the ballooning interchange mode equation [Zheng and Furukawa (2010)]. With the current gradient there is coupling of the ballooning modes to the odd symmetry modes that evolve into the tearing mode or varicose mode in the nonlinear state [Muraglia, *et al.* (2009), Ishizawa and Nakajima (2010)]. This instability is a source of the magnetic turbulence giving the anomalous transport of electrons. Chapter 14 analyzes this form of the plasma turbulence.

For finite beta drift wave turbulence, $\delta B_x(x)$ vanishes at the resonant surface, rises to a maximum of δB_{xA} at $k_\| v_A \simeq \omega_{*e}$, and decreases at $\delta B_{xA}(x_A/x)$ for $k_\| v_A > \omega_{*e}$. Here, x_A is the distance from the resonant surface $k_\| = 0$ to the region where $k_\| v_A = k_y v_A x_A / L_s = \omega_{*e}$ which is $x_A = \rho_s \beta^{1/2}(L_s/r_n)$. For a drift wave magnetic island width $\Delta_k^{\mathrm{dw}} > x_A$, one finds the island width

$$\Delta_k^{\mathrm{dw}} \simeq \left(\frac{L_s x_A}{k_y} \frac{\delta B_{xA}}{B_0} \right)^{1/3}$$

and the distance L_k along the magnetic field to circumnavigate the magnetic island $L_k = L_s / k_y \Delta_k \gg L_s$. Since L_k is greater than L_s the drift wave frequency ω_k resonates with the electron transit frequency around the island $\omega_k \cong v_e / L_k$. Including the buildup of an ambipolar potential $\Phi(x)$ to equalize the ion and electron loss rates, Callen (1977) estimates that the electron diffusion and ambipolar potential are given by

$$\Gamma_e \simeq -\frac{n_e}{2}\sum_k \Delta_k^2(\nu_e+\gamma_k)\left[\frac{d\ln n_e}{dx}-\frac{e}{T_e}\frac{d\Phi}{dx}-\frac{\omega e B}{ck_yT_e}\right]\simeq 0,$$

$$\frac{e}{T_e}\frac{d\Phi}{dx}=\left(1-\frac{\omega_{\overline{k}}}{\omega_{*e}}\right)\frac{d\ln n_e}{dx}\approx\frac{d\ln n_e}{dx} \tag{10.59}$$

giving electrostatic confinement of the electrons with radially outward $E_r>0$ electric field. The electron thermal transport $Q_e=-n_e\chi_e(dT_e/dx)$ is estimated by

$$\chi_e=\frac{3}{4}\sum_k \Delta_k^2(\nu_e+\gamma_k) \tag{10.60}$$

where ν_e is the electron collision frequency and $1/\gamma_k$ is the correlation time or the lifetime of the magnetic island Δ_k. Equation (10.60) is a nonlinear generalization of the quasilinear formula in which the island width

$$\Delta_k \propto \begin{cases}\left[\dfrac{\delta B_{xA}}{B}\right]^{1/3} & \text{for drift waves}\\[2ex] \left[\dfrac{\delta B_x(0)}{B}\right]^{1/2} & \text{for tearing modes}\end{cases} \tag{10.61}$$

replaces the linear radial excursion $v_\parallel B_x(k)/B\Delta(\omega-\boldsymbol{k}\cdot\boldsymbol{v})=[B_x(k)/B]v_\parallel\tau_c$ that occurs in the quasilinear formula in Eq. (10.57).

The magnitude of the anomalous transport produced by magnetic braiding depends sensitively on the correlation time of the electron with the magnetic perturbation $\delta\mathbf{B}_\perp$, as well as the obvious amplitude dependence. Rechester and Rosenbluth (1978) and daSilva, *et al.* (2001) introduce the method of stochastic trajectories to obtain formulas describing the finite correlation time τ_c or length L_c along the field line.

First the statistical properties of the magnetic field lines in the presence of the stochastic perturbations $\delta\mathbf{B}_\perp=\sum_k\delta\mathbf{B}_k\exp(i\boldsymbol{k}\cdot\boldsymbol{x})$ are established. Rechester and Rosenbluth introduce the stochasticity parameter

$$s=\frac{1}{2}\frac{(\Delta_{k_1}+\Delta_{k_2})}{|(r_{k_1}-r_{k_2})|} \tag{10.62}$$

that compares the total width of two magnetic islands with their radial separation. When this parameter exceeds unity the magnetic surfaces are destroyed between the chains of islands. In this region the field lines wander ergodically and a small circle of radius l_0 is mapped, as shown in Fig. 10.2, by following the field line into a complex, elongated structures that have the same area as the initial circle owing to the constraint $\nabla\cdot\mathbf{B}=0$. The stretching of a typical side of this area is given by

$$I(z)=l_0\exp\left(\frac{z}{L_c}\right) \tag{10.63}$$

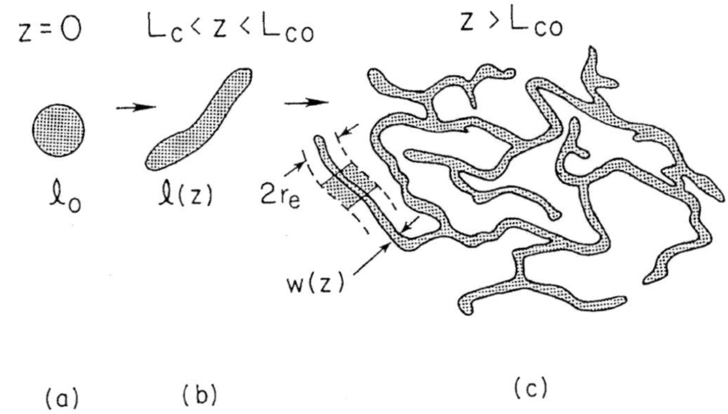

Fig. 10.2 In the stochastic regime of strong island overlapping $s \gg 1$, the small circle in (a) is mapped by the magnetic field to (b) showing the stochastic instability of trajectories. When the filament length in (b) exceeds the cross-field correlation distance, the mapping evolves as shown in (c) [Rechester and Rosenbluth (1978)].

where $L_c = \pi R/\ln(\pi s/2)$ from the stochastic instability of the field line trajectories. To conserve area the width $w(z)$ of the area must decrease according to $w(z) = l_0 \exp(-z/L_c)$. When the elongation $l(z)$ exceeds the cross-field correlation distance δ different regions separated by δ evolve independently. This condition, $l(z) = \delta = 1/\Delta k_\perp$, defines the correlation length

$$L_{c0} = L_c \ln\left(\frac{\delta}{l_0}\right) \tag{10.64}$$

for the area. For $L \gg L_{c0}$ the magnetic mapping is a random walk with the average squared displacement of the connected area, shown in Fig. 10.2, described by

$$\langle \Delta r^2 \rangle = 2D_F L \tag{10.65}$$

where

$$D_F = \pi \sum_k \left| \frac{B_x(\boldsymbol{k})}{B} \right|^2 \delta(k_\parallel) \tag{10.66}$$

is the field (D_F) diffusion coefficient.

Now consider the motion of a small patch of electrons whose guiding centers follow the magnetic field lines. The small patch of electrons of radius r_e move along the field mapping into the complicated thin filaments of width $r_e \exp(-z/L_c)$ until they suffer a collision at the average distance $z = \lambda_e = v_e/\nu_e$. At the time of the collision the initial patch of electrons has spread the average squared perpendicular distance $\langle \Delta r^2 \rangle = 2D_F \lambda_e$ provided $\lambda_e > L_{c0}$. After the collision the thin filaments expand to r_e, and subsequently the new patch of radius r_e repeats the process. These random steps give rise to the thermal diffusion

$$\chi_e = \frac{1}{2}\langle \Delta r^2 \rangle \nu_e = v_e D_F \tag{10.67}$$

valid for $\lambda_e \gg L_c \sim 1/\Delta k$. Rechester and Rosenbluth (1978) note that this transport applies to the electron thermal energy since the particle transport must be limited by the ambipolar potential.

These stochastic magnetic fields may also be created with external magnetic coils on the vessel walls to produce what is called an ergodic magnetic limiter in some tokamaks. The stochastic magnetic fields occur automatically in the helical/stellarator geometries owing to the broken toroidal symmetry. For example, in the Large Helical Device (LHD) there are $l = 2/n = 10$ external coils creating an ergodic magnetic limiter.

10.4.3 *Transport from static magnetic fluctuations*

Kadomtsev and Pogutse (1978) formulated the problem of calculating the anomalous electron transport from stochastic magnetic perturbations within the electron fluid approximation for thermal conduction. The dimensionless strength of the perturbation is $R_M = bL_c/i$. The basic transport law is taken as

$$\nabla \cdot \boldsymbol{q} = 0 \tag{10.68}$$

with the collisional formula for the local thermal flux

$$\boldsymbol{q} = -\chi_{\parallel}\boldsymbol{h}(\boldsymbol{h} \cdot \nabla T_e) - \chi_{\perp}\nabla T_e \tag{10.69}$$

where $\chi_{\parallel}$ exceeds $\chi_{\perp}$ by several orders of magnitude. The collisional values of approximately $\chi_{\parallel} = nT_e/m_e\nu_e$ and $\chi_{\perp} = nT_e\nu_e/m_e\omega_{ce}^2$ give the ratio $\chi_{\perp}/\chi_{\parallel} = (\nu_e/\omega_{ce})^2 = \gamma^2$ where γ is very small. The vector $\boldsymbol{h}$ along the magnetic field is written as the sum

$$\boldsymbol{h} = \widehat{\boldsymbol{e}}_z + \boldsymbol{b} \tag{10.70}$$

of the constant vector in the direction of the ambient magnetic field and small stochastic vector $\boldsymbol{b} = \delta\mathbf{B}_{\perp}/B$. From Eq. (10.69) we see that for $\langle b^2\rangle\chi_{\parallel} \sim \chi_{\perp}$ the contribution from the parallel transport is comparable to that from the cross-field transport.

The stochastic magnetic field $\boldsymbol{b}(\boldsymbol{r}_{\perp}, z)$ has $\nabla_{\perp} \cdot \boldsymbol{b} = 0$ since $\mathbf{B}_0 = \text{const}$. The stochastic field $\boldsymbol{b}$ is characterized by (i) the small root-mean-square value $b_0 = \langle b^2\rangle^{1/2}$, (ii) isotropic properties in the plane perpendicular to the ambient magnetic field, (iii) the cross-field correlation distance δ and (iv) the parallel field correlation distance L_c. In this section the logarithmic factors that distinguish L_c, L_{c0} and $L_{c\delta}$ are disregarded [Krommes, *et al.* (1983)].

The equation for the transverse coordinates $\boldsymbol{r}_{\perp}$ of a point following the field lines is

$$\boldsymbol{r}_{\perp}(z) = \int_0^z \boldsymbol{b}\left[r(z'), dz'\right] dz'. \tag{10.71}$$

In the limit of small $b_0 = \langle \boldsymbol{b}^2\rangle^{1/2}$ the quasilinear approximation for $\langle r_{\perp}^2(z)\rangle$ at large z increases as

$$\langle r_{\perp}^2\rangle = 4D_F z \tag{10.72}$$

where

$$D_F = \frac{1}{4}\int_{-\infty}^{+\infty} \langle \boldsymbol{b}(0,z)\boldsymbol{b}(0,0)\rangle dz. \tag{10.73}$$

The parallel correlation length L_c is defined by

$$D_F = \frac{1}{4} b_0^2 L_c \tag{10.74}$$

with which the quasilinear diffusion for Eq. (10.73) may be written as $\langle r_\perp^2 \rangle = b_0^2 L_c z$ for $z > L_0$. The quasilinear approximation of neglecting $\boldsymbol{r}(z)$ in Eq. (10.73) is valid when $\langle r_\perp^2 \rangle^{1/2} \simeq b_0 L_c \leq \delta$ where δ is the cross-field correlation distance. Thus, the amplitude parameter $R = b_0 L_c/\delta$, defines the quasilinear regime by $R < 1$.

To determine the field diffusion D_F in the regime $R > 1$, Kadomtsev and Pogutse (1978) use the statistical theory of plasma turbulence. With $N(\boldsymbol{r}, z)$ defined as the conserved density of magnetic field lines, the magnetic transport equation is

$$\frac{\partial N}{\partial z} + \boldsymbol{b}\cdot\nabla N(\boldsymbol{r}_\perp, z) = 0. \tag{10.75}$$

Splitting the density into its mean value $\overline{N}$ and its fluctuation N', gives the well-known coupled turbulence equations

$$\frac{\partial \overline{N}}{\partial z} = -\nabla\cdot\langle \boldsymbol{b}N'\rangle = D_F \nabla_\perp^2 \overline{N}, \qquad \frac{\partial N'}{\partial z} - D_F \nabla_\perp^2 N' = -\boldsymbol{b}\cdot\nabla N'. \tag{10.76}$$

Solving these equations we derive

$$D_F = \frac{1}{2}\int \frac{b^2(k)}{ik_z + k_\perp^2 D_F} dk \tag{10.77}$$

where the factor 1/2 results from the angle average of the isotropic spectrum. Here, the Fourier spectrum of the isotropic magnetic fluctuation correlation function is

$$b^2(\boldsymbol{k}) = \frac{1}{(2\pi)^3}\int d\boldsymbol{r}\langle \boldsymbol{b}(0)\boldsymbol{b}(\boldsymbol{r})\rangle \exp\left(-i\boldsymbol{k}\cdot\boldsymbol{r}\right).$$

The quasilinear formula is recovered from Eq. (10.77) when $\Delta k_z > k_\perp^2 D_F$ where $D_F = (\pi/2)\int d\boldsymbol{k} b^2(\boldsymbol{k})\delta(k_z) = b_0^2 L_c$. For large amplitudes $R = b_0 L_c/\delta > 1$ the cross-field decorrelation dominates, $k_\perp^2 D_F > \Delta k_z$, and Eq. (10.77) reduces to

$$D_F^2 = \frac{1}{2}\int \frac{b^2(\boldsymbol{k})}{k_\perp^2} d\boldsymbol{k} \cong b_0^2\delta^2. \tag{10.78}$$

Thus, the field diffusion varies according to

$$D_F = \begin{cases} b_0^2 L_c \text{ for } R = \dfrac{b_0 L_c}{\delta} < 1, \\[2ex] b_0\delta \quad \text{for } R = \dfrac{b_0 L_c}{\delta} > 1. \end{cases} \tag{10.79}$$

Returning to the solution of Eqs. (10.68) and (10.69), the temperature field is split into its mean T_e and fluctuating components T_e' where linearization gives

$$\left(k_z^2 + \gamma^2 k_\perp^2\right)\delta T_e(\boldsymbol{k}) = ik_z b_z(\boldsymbol{k})\frac{d\overline{T}_e}{dx}. \tag{10.80}$$

The mean value of the radial component of the heat flux is given by

$$q_e = -\chi_\parallel \langle b_x^2 \rangle \frac{d\overline{T}_e}{dx} - \chi_\parallel \left\langle b_x \frac{dT_e'}{dx} \right\rangle = -\frac{\chi_\parallel}{2} \int \frac{\gamma^2 k_\perp^2 b^2(k) d\boldsymbol{k}}{k_z^2 + \gamma^2 k_\perp^2} \frac{d\overline{T}_e}{dx}. \tag{10.81}$$

When we restrict consideration to $b_0 > \gamma = \nu_e^e/\omega_{ce}$ and take $R \lesssim 1$, the approximation for the effective transport resulting from Eq. (10.81) is

$$\chi_e = \left(\chi_\parallel \chi_\perp\right)^{1/2} \left(\frac{b_0^2 L_c}{\delta}\right) = \left(\frac{cT_e}{eB}\right)\left(\frac{b_0^2 L_c}{\delta}\right) \tag{10.82}$$

where $(\chi_\parallel \chi_\perp)^{1/2} = D_B = cT_e/eB$ and $b_0 L_c \leq \delta$ is the limit of the quasilinear regime.

The rms fluctuation level $b_0 = \left\langle \delta B_\perp^2 / B^2 \right\rangle^{1/2}$ in a tokamak ranges from 10^{-5} to 10^{-4} which is well above ν_e/ω_{ce}.

In both parameter ranges of the ratio $\chi_\parallel/\chi_\perp$ the anomalous thermal conductivity is related to the field diffusion by

$$\chi_e = D_f \overline{v}. \tag{10.83}$$

In a plasma with high conductivity the $\delta B_\perp$ arises from displacements with $\xi \sim \delta$ which bend the field according to $\delta B \approx B_0 \delta / L_c$. Such plasma displacements put the magnetic turbulence in the $R \sim 1$ regime where χ_e is only weakly dependent on $\gamma = (\chi_\perp/\chi_\parallel)^{1/2}$. The value of χ_e is then

$$\chi_e = \begin{cases} \chi_\parallel \left(\dfrac{\delta^2}{L_c^2}\right) & \text{for } \lambda_e < L_c, \\ v_e \left(\dfrac{\delta^2}{L_c}\right) & \text{for } \lambda_e > L_c \end{cases} . \tag{10.84}$$

For low-frequency fluctuations of the drift-mode type, the natural scale for the reconnection of the magnetic field lines is the collisionless skin depth $\delta = c/\omega_{pe}$. The argument is that at these low frequencies the magnetic surfaces are frozen into the electrons on this scale. With this assumption the formula Eq. (10.84) for the anomalous electron thermal conductivity becomes

$$\chi_e = \frac{c^2}{\omega_{pe}^2} \frac{v_e}{L_c} \quad \text{for} \quad \nu_e < \frac{v_e}{L_c} \tag{10.85}$$

where the parallel correlation length L_c is now the effective length of the system. In a toroidal confinement device with major radius R and inverse rotational transform or safety factor q, the effective length is $L_c = qR$. In this case the formula agrees with the empirical formula given by Ohkawa (1978) to explain the parametric dependence of the electron thermal transport in toroidal confinement experiments.

There are two effects that arise from the finite radial correlation distance. First, there is a further loss of the phase coherence due to the perpendicular diffusion over δ_x, which is described by $\nu_{\text{eff}} \to \nu_{\text{eff}} + k^2 \chi_e$; and secondly, there is an additional decorrelation from the parallel motion of the electron along the radially tilted field line which is described by the nonlinear $\boldsymbol{k} \cdot \mathbf{B} = k_{\parallel 0} B + k_x \delta B_x$. For the second effect it is adequate to take into account that the mean value of the stochastic $k_\parallel$ is given by

$$\left\langle k_\parallel^2 \right\rangle = k_\parallel^2 + k_x^2 b_0^2 \tag{10.86}$$

from which it is evident that the nonlinear broadening of the parallel phase dominates when

$$b_0 > \frac{\max(k_\parallel)}{\Delta k_x} = \frac{\delta_x}{L_s}. \tag{10.87}$$

Taking these two effects into account generalizes correlation function so that the quasistatic thermal diffusivity becomes

$$\chi_e = \chi_{\parallel e} \sum_{\boldsymbol{k}} \left| \frac{B_x(\boldsymbol{k})}{B} \right|^2 \frac{\nu_{\text{eff}} + k_x^2 \chi_e}{\left(k_\parallel^2 + k_x^2 b_0^2\right) \chi_{\parallel e} + \nu_{\text{eff}} + k_x^2 \chi_e}. \tag{10.88}$$

The approximation of replacing $k_\parallel^2$ the average nonlinear $\left\langle k_\parallel^2 \right\rangle$ given by Eq. (10.86) is adequate to recover the essential results of the complex nonlinear dynamics.

In the nonlinear regime $b_0 > \delta_x/L_s$ the stochastization of the parallel phase is unimportant according to Eq. (10.88). Clearly, there remain two subdomains depending on the magnitude of the nonlinear broadening $k_x^2 b_0^2 \chi_{\parallel e}$. In the lower-amplitude subdomain clearly Eq. (10.88) reduces simply to

$$\chi_e = \chi_{\parallel e} b_0^2 \tag{10.89}$$

where now the condition for applicability becomes $\delta_x/L_x < b_0 < \delta/\lambda_e$. When the nonlinear parallel decorrelation is substantial, the only consistent solution of Eq. (10.88) has $k_x^2 b_0^2 \chi_{\parallel e} \sim k_x^2 \chi_e$ and consequently $\chi_e \cong \chi_{\parallel e} b_0^2$, valid when the mean-free-path is less than the parallel correlation length. The parallel correlation length is given by $1/L_\parallel \sim k_x b_0 \sim b_0/\delta_x$. Thus, the well-known self-similar form $\chi_e \cong \chi_{\parallel e} b_0^2$ remains valid until $b_0 > \delta_x/\lambda_e$ whereupon

$$\chi_e = v_e D_F \cong v_e b_0 \delta_x \tag{10.90}$$

gives the nonlinear, collisionless thermal transport. The self-similar result $\chi_e = \chi_{\parallel e} \langle \delta B_x^2 / B^2 \rangle$ is universal in the sense that the formula applies for both strong collisions and high-level magnetic fluctuations.

Additional studies clarifying the assumptions [Rose (1982)] and relationships [Krommes, *et al.* (1983)] between the different regimes of anomalous magnetic transport have been developed. Closely-related formulas for χ_e from electromagnetic fluctuations are derived in Zhang and Mahajan (1988), Hong and Horton (1989-), and Horton, *et al.* (1987). For plasma $\beta_e \gtrsim 0$ of the magnetic fluctuations soften and lower the turbulent electrostatic fluxes. Citrin, *et al.* (2014) describes this reduction of the electron turbulent fluxes as reducing the stiffness and give examples for JET discharges described in Chapter 14.

In conclusion, there are different mechanisms for anomalous transport depending on the regime of the plasma. There are no simple rules for determining which of the formulas presented here govern transport in a specific plasma. One must evaluate the stability to the various drift waves to determine the dominant type of magnetic turbulent fluctuations. Three-dimensional codes are used to find the regimes of the magnetic turbulence and the associated thermal transport coefficients for the dominant instability. This method has been successful in explaining tokamak transport features for a variety of machines.

Now we focus on examining a pure interchange process in a shear-less slab configuration as a prototype mechanism for the formation of vortex and blob-like plasma structures. Full nonlinear magnetohydrodynamic simulations are required to demonstrate that the blob-like structures can emerge through the nonlinear development of interchange instabilities originating from a pedestal transition region. In the early nonlinear stage, filamentary structures develop and extend in the direction of the effective gravity. The blob-like structures appear when the radially extending filaments break off and disconnect from the core plasma. The morphology and the dynamics of these filaments and blobs vary dramatically with a sensitive dependence on the dissipation mechanisms in the system and the initial perturbation. Despite the complexity in morphology and dynamics, the nature of the entire blob formation process in the weak shear configuration remains interchange without involving any appreciable change in magnetic topology [Zhu, *et al.* (2015)].

In a pure plasma interchange dynamics in a shearless slab configuration a mechanism exists for the formation of plasma "blobs". Using full magnetohydrodynamic simulations [Zhu, *et al.* (2015)] demonstrated that the blob-like structures emerge through the nonlinear development of the pure interchange instability originating from a pedestal transition region. In the early nonlinear stage, filamentary structures develop and extend in the direction of the effective gravity. The blob-like structures appear when the radially extending filaments break off and disconnect from the core plasma. The morphology and the dynamics of these filaments and blobs vary dramatically with a sensitive dependence on the dissipation mechanisms in the system and the initial perturbation. Despite the complexity in morphology and dynamics, the nature of the entire blob formation process in the shear-less slab configuration remains strictly interchange without involving any change in magnetic topology [Aydemir (2005)].

10.5 RF wave propagation and scattering in turbulent plasmas

Drift wave turbulence driven by the steep electron and ion temperature gradients in H-mode divertor tokamaks produces scattering of the RF waves used for heating and current drive. The X-ray emission spectra produced by the fast electrons requires the turbulence broaden RF wave spectrum. Both the 2-5 GHz lower-hybrid waves and the 150-170 GHz electron cyclotron (EC) RF waves experience scattering and diffraction by the electron density fluctuations. With strong LHCD there are bifurcations in the RF coupled turbulent transport dynamics giving improved steady-state confinement states. The scattering of the RF rays makes the prediction of the distribution of the rays and the associated particle heating a statistical problem. Thus, one introduces a Fokker-Planck equation for the probably density of the RF rays in the turbulence [Horton, *et al.* (2015)]. The general framework of the coupled system of coupled high-frequency current driving rays with the low-frequency turbulent transport determines the profiles of the plasma density and temperatures [Peysson, *et al.* (2011)].

Nonlinear phenomena in the RF wave propagation has been observed from the earliest days in basic laboratory experiments going back to the 1960s, followed by observations of parametric decay instability (PDI) phenomena in RF heating experiments in magnetized fusion plasmas in the 1970s. Although not discussed further here, the importance of PDI phenomena may also be central to understanding anomalous absorption in laser-fusion experiments (ICF). The fundamentals of nonlinear interactions among waves and particles are discussed in Horton, *et al.* (2015) and in particular, their role in PDIs. This phenomenon is distinct from quasilinear phenomena that are often invoked in calculating absorption and compressional heating of RF power in wave heating experiments in the core of magnetically confined toroidal plasmas. Indeed, PDIs are most likely to occur in the edge of magnetized fusion plasmas where the electron temperature is modest and hence the oscillating quiver velocity of charged particles can be comparable to their thermal speeds. The PDI theory gives examples from past experiments in the ECH/EBW, lower-hybrid (LHCD) and ICRF/IBW frequency regimes are given in Porkolab (2015). Importantly, PDI plays a role in determining the so-called "density limit" in lower-hybrid driven toroidal experiments that has persisted in edge plasma may be central to understanding present day experiments.

Chapter 11

Toroidal Confinement Systems

Much of this book is concerned with the turbulent transport in the transient tokamak confinement system which has set the records for the highest achieved fusion power in the laboratory. Looking ahead to achieving a steady-state fusion power system with full external control of the plasma conditions, one is led to search for a toroidal magnetic confinement system without the large internal plasma current that must be maintained in the tokamak by an external central solenoid driver. To achieve a steady state with external magnetic coils leads one to research the large helical device (LHD) shown in Fig. 11.1 and the Wendelstein 7-X (`www.ipp.mpg.de/W7X`) [Laqua, *et al.* (2016)].

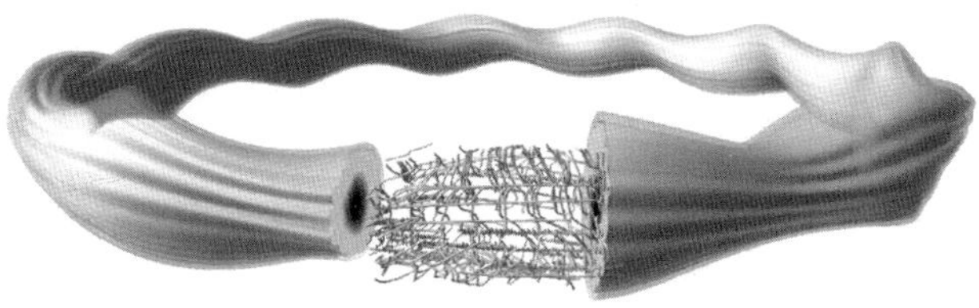

Fig. 11.1 Simulation of the Large Helical Device plasma with helical period periodicity $\ell = 2$ and toroidal periodicity $n = 10$ during an unstable motion when the plasma pressure exceeds that of ideal MHD stability. The motion is driven by the curvature of the toroidal system with line-bending magnetic field tension as the chief stabilizing force. The threads shown in the core of the cut-away section are drawn from a 3D rendering of the plasma mass flows generated by the Reynolds stresses from the MHD unstable motion. Owing to the $\ell = 2, n = 10$ variations of the equilibrium there is extensive mode coupling from the $\cos\theta$-driven toroidal curvature ballooning interchange motion in the MHD unstable motion [Miura and Nakajima (2010)].

11.1 Alternative Toroidal System with High-Temperature Plasmas

The alternative to providing the confining poloidal magnetic field by the large plasma current I_p is to design external coils that create the helical magnetic field along with the toroidal magnetic field. From a sketch of these two magnetic field structures one is led to consider helical coils around the torus. The large helical device (LHD) is a superconducting toroidal magnetic confinement machine designed from experience with the earlier series of Heliotron machines built in Kyoto University in the 1980-1990s. The physics of Heliotron E, the last machine in the series before construction of the LHD, is thoroughly documented in the book, *Stellarator and Heliotron Devices* by Wakatani (1998).

The stability analysis of Fu, *et al.* (2000) shows that both magnetic curvature and the Pfirsch-Schlüter current contribute to the pressure limit of the MHD stability of the helical/stellarator device. They report that the system is stable without a close conducting wall up to dimensionless plasma pressure $\beta = 0.05$.

Helical toroidal systems have numerous advantages over the tokamak for the design of a steady-state power reactor as discussed in Chapter 9.6 of Wakatani (1998). The external control of both the toroidal and poloidal magnetic fields is a large advantage in view of the unpredictable current disruption events known to occur in tokamaks operating near their peak performance. The helical system has a natural magnetic divertor with a large surface area from the stochasticity of the magnetic field lines readily understood from the bending of the straight $R = \infty$ helical coils with $\ell = 2$ and $n = 10$ with its regular magnetic separatrix into the toroidal shape which adds the standard $B_{\text{tor}} = B_0/[1 + r/R\cos(\theta)]$ toroidal field as a perturbation to the external superconducting coil-driven helical field $B_n = B_n(r/a)^2\cos(2\theta - 10\phi)$ producing a wide edge region of stochastic magnetic field lines. This combination of the helical and toroidal magnetic field produces what is called an ergodic limiter that has been studied in several tokamaks including TEXT, JFT-2M and Tore Supra. The ergodic divertor lowers the power delivered to the plasma facing components. Chapters 8 and 9 in Wakatani (1988) describe in some detail the favorable features of a steady-state fusion reactor based on the design of the Large Helical Device at NIFS.

The second new major helical system is the Wendelstein 7-X machine commission in 2015 at the Max-Planck Institute for Plasma Physics in Greifswald, Germany. Plasmas with $T_e \lesssim 7\,\text{KeV}$ and $T_i \lesssim 2\,\text{KeV}$ for several seconds were achieved in early 2016. Gyrokinetic simulations of the drift waves in the Wendelstein 7-X stellarator are given in Xanthopoulos and Jenko (2007).

For LHD the 3D simulations are given in Nunami, *et al.* (2012) and described in the first edition of *Turbulent Transport in Magnetized Plasmas* [Horton (2012)]. The cover of the book shows the 3D structures of the drift wave turbulence in the LHD computed for the ion temperature gradient driven turbulence by Nunami and the NIFS team.

The six rings of poloidal field coils shift the plasma in LHD to a smaller major radius in the left column of Fig. 11.2 and to a larger major radius in the right column. For the inward shift there is a magnetic hill with magnetic rotational transform $t(0) < 0.5$ and for the outward shift there is a magnetic well with $t(0) > 0.5$ in the right column of Fig. 11.2. For the outward shift plasma there are substantial density "ears" in the plasma density $n_e(\rho)$ profile from a transport barrier as shown in Fig. 11.2b.

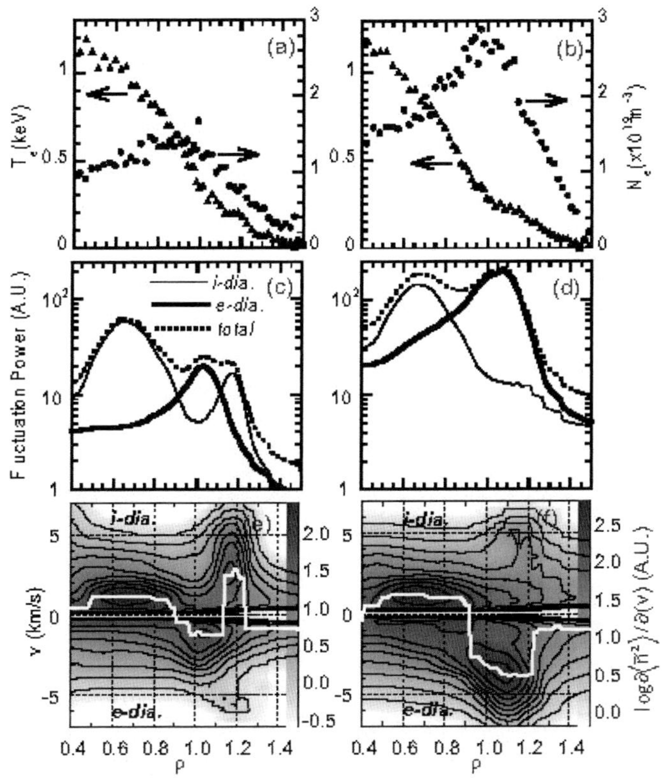

Fig. 11.2 The results of CO_2 laser phase contrast imaging (PCI) for an L-H transition in the LHD plasma from Tanaka, *et al.* (2008). The left column is data at $t = 2.5$ s before the transition and the right column is at $t = 2.7$ s after the transition. (a)-(b) show the YAG laser incoherent Thomson scattering derived electron density and temperature profiles. (c) and (d) show the profiles of fluctuation power. (e) and (f) show fluctuation components are poloidally dominated and Doppler shifted by $E_r \times B_t$ poloidal rotation.

The performance of net-current free heliotron plasmas has been developed by findings of innovative operational scenarios in conjunction with an upgrade of the heating power and the pumping/fueling capability in the large helical device (LHD). The operational regime has been extended, in particular with regard to high density, long pulse length and high beta. Diversified studies in LHD have elucidated the advantages of net-current free heliotron plasmas. In particular, an internal diffusion barrier (IDB) by a combination of efficient pumping of the local island divertor function and core fueling by pellet injection produced a super dense core as high as $5 \times 10^{20}\text{m}^{-3}$, which stimulates an attractive super dense core reactor. Achievements of a volume-averaged beta of 4.5% and a discharge duration of 54 min with a total input energy of 1.6 GJ (490 kW on average) are also highlighted. The progress of LHD experiments in this period is overviewed in Motojima, *et al.* (2007) describing the high-β and long pulse plasmas with the internal transport barrier.

11.2 Helical Toroidal Systems: LHD, Heliotrons, and Stellarators

The toroidal heliotrons like the Large Helical Device shown in Fig. 11.1 are relatively simple in design with respect to the current carrying coil design. The helical toroidal machines now have a large database. The transport is rather difficult to compute due to the nonaxisymmetry of the combined helical ripples $\epsilon_h \cos(2\theta - 10\phi)$ and smaller toroidal magnetic field strength modulation as $\epsilon_t \cos(\theta)$. The broken symmetry leads to the nonambipolar rate of ion and electron transport creating two distinct modes of operation. The tokamak-like regime is called the ion-root where the ions are lost faster than the electrons creating an inward radial electric field at the stable ion fixed point of matched ion and electron radial particle transport. This regime occurs in hot-ion plasmas created by either NBI or ICRF as the dominant plasma heating process. Ion dominant heating occurs with positive ion based NBI with acceleration voltage 40 KV and ICRF heating with large contents of minority ions. Ion heating case using positive ion neutral beam injection (NBI), the E_r is positive in edge region ($\rho > 0.7$) [Tanaka, *et al.* (2010), Ida, *et al.* (2001)]. With strong ion heating at high density where energy equipartition from electron to ion is fast, the E_r is negative.

Figure 11.2 shows two snapshots for the fluctuation spectrum across the entire cross-section of the plasma for wavenumbers in the range of the drift wave fluctuations in the L-mode at $t = 2.5$ s on the left and after a transition to the H-mode at $t = 2.7$ on the right for a Large Helical Device plasma. The top-left and top-right figure shows the electron density and temperature profiles. The middle left and right figures show the intensity of the electron density fluctuations in the wavenumber range $k = 10\,\mathrm{cm}^{-1}$ before and after the L-H transition. The lower row of figures gives the radial variation of the dominant power resolved by the direction of the phase velocity. The rotation in the ion-diamagnetic velocity arises from a positive radial electric field and rotation in the electron-diamagnetic direction arises from the negative radial electric field. Thus, the transition to the H-mode is consistent with a substantially increased inward (negative) pointing radial electric field. This is also typical for tokamak plasmas that have stronger negative radial electric fields required for higher confinement. From this data one sees the ability of the phase contrast imaging (PCI) method to give time-resolved fluctuation levels over large spatial regions [Tanaka, *et al.* (2008)].

The second stable root from the condition that the total ion radial loss rate balance in the electron loss rate occurs in hot electron plasmas driven by strong electron cyclotron resonance heating ECRH power. This regime is called the electron root and the radial electric field points outward or $E_r > 0$. In this state the plasma rotates in the ion diamagnetic direction with respect to the laboratory frame. The PCI infrared laser imaging system is able to see both these rotation regimes and confirms the general description of the radial electric field and the plasma rotation [Tanaka, *et al.* (2008)]. The LHD has a heavy ion beam probe HIBP that directly measures the local plasma electric potential in the plasma. Both regimes have their

Table 11.1 Plasma parameters achieved in the large helical device (LHD).

$T_i(0)$(KeV)	$T_e(0)$(KeV)	$n_e(0)(10^{19}\ \text{m}^{-3})$	$\beta(\%)$	$\tau_E(s)$	P_{abs} (MW)	$B(T)$
0.22	0.22	125	1.43	0.039	21.4	2.51
-	0.43	2.3	5.1	0.040	11.9	0.425
5.6	3.8	1.6	0.80	0.046	19.1	2.9
-	18.9	0.24	0.19	0.043	3.7	2.71

advantages. In Table 11.1 record parameters for plasma confinement in terms of plasma temperatures and densities and beta achieved in different regimes are shown as reported by the Yamada for the LHD group at the 23 IAEA Fusion Conference in 2010. The radial electric field is derived from a combination of the HIBP and the CXRS signals.

An example of the turbulent transport in the LHD device for the hot ion mode is shown in Fig. 11.3 from simulations by Nunami, *et al.* (2011). In this simulation the ion temperature gradient drives the drift wave turbulence and the electrons may be taken with the simple adiabatic response. The figure shows a snapshot in time of the nonlinear state of turbulence with shaded contours of the electric potential turbulence for annular volume sectors centered on flux surfaces locations at $r/a = 0.46$, 0.65, and 0.83. The ambient background plasma profiles are taken from codes for the magnetic configuration with reconstructed experimental data for a high ion-temperature discharge. The non-axisymmetric geometry requires the use of the GKV-X code for describing the drift wave turbulence in the Large Helical Device.

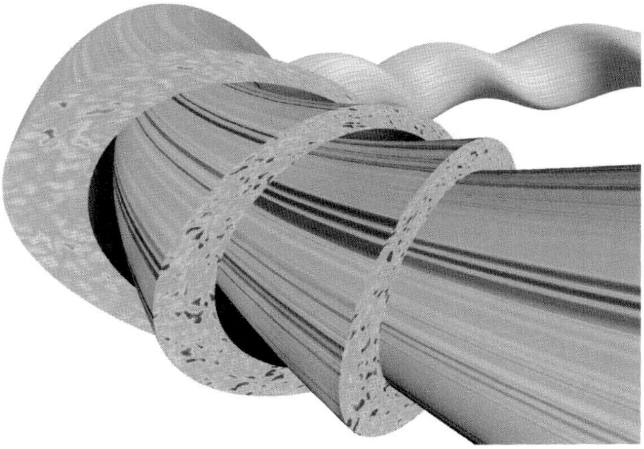

Fig. 11.3 Contours of the electric potential turbulence for a simulation of the ion temperature gradient driven instability in the Large Helical Device. The simulations from the GKV-X code are performed with simple adiabatic electron response for the electron dynamics at the three different radial flux surfaces locations at $r/a = 0.46$, 0.65, and 0.83 in the non-axisymmetric helical torus. The ambient background plasma profiles are taken from codes for the magnetic configuration reconstructed from the experimental data for the high ion-temperature discharge.

The continuum gyrokinetic code of Watanabe, *et al.* (2007) computes the ion temperature gradient driven drift wave turbulence with the adiabatic $\delta n_e = en_0\delta\varphi/T_e$ response for plasma profiles taken from experimental data. A typical result for the electrostatic potential is shown in Fig. 11.3 and on the book cover.

11.3 Neoclassical Dynamics and Transport in Toroidal Systems

Chapter 6.3.4 develops the resistive-viscous partial differential equations for the plasma flows and associated rotation with convective transport in Eqs. (7.53)–(7.58). In Eq. (7.76) the parallel resistivity $\eta_\parallel$ produces convective cells in the toroidal equilibrium with a $m = 1$ or $\cos\theta$ and $\sin\theta$ structures from the guiding-center drifts in the toroidal curvature of the toroidal system. The $\mathbf{E}\times\mathbf{B}$-convection of plasma at low collisionality

$$\frac{\nu_\alpha}{(\Delta B/B)} < \frac{v_\alpha}{L_c}\left(\frac{\Delta B}{B}\right)^{1/2} \tag{11.1}$$

becomes the dominant loss mechanism from the Coulomb collisional scattering rate ν_α giving rotations of the particle velocity vector $\mathbf{v}$ over 90° in the time

$$\tau_\alpha = \frac{1}{\nu_\alpha} = \frac{4\pi\epsilon_0^2 m_\alpha^2 \mathrm{v}_\alpha^3}{n_\beta e_\alpha^2 e_\beta^2 \ell n\,\Delta} \tag{11.2}$$

from collisions with particles n_β of charge e_β.

There are various definitions of the collisional scattering rate $\nu_{\alpha/\beta}$ depending on the calculation being performed. The definition given in Eq. (11.2) is from the Appendix of Helander and Sigmar (2002). See the NRL formulary at `http://nrlformulary.gov` for another commonly used definition of $\nu_{\alpha/\beta}$.

In the banana regime the locus of the guiding-center motion projected onto a poloidal cross-section of the torus traces out the shape of a banana. Other projections of the guiding-center orbits on the cross-section can give figures resembling potatoes. Clearly, these large orbital excursions create a fast loss of particles when the Coulomb scattering and collective modes of drift waves are taken into account. The Coulomb scattering loss rate is intrinsic to the system and is the subject of the books on tokamaks and helical toroidal confinement systems [Helander and Sigmar (2002), Wakatani (1998)]. The electric potential-parallel current system given in Eq. (7.76) is called the Pfirsch-Schlüter (PS) regime of plasma transport. As the temperature of the plasma increases this transport mechanism makes a transition from the PS-regime to first a "plateau" regime in which the dissipation from the Landau-particle resonance at

Plateau Regime

$$k_\parallel \mathrm{v}_\parallel + \omega_D = 0 \tag{11.3}$$

replaces the resistivity $\eta_\parallel$. Further increase in the temperature in the plasma enters the "banana" regime in which the trapped particles increase the transport. The collective drift-wave instabilities associated with these large trapped particle orbits are called trapped particle modes. There is a trapped electron mode

TEM

with frequency

$$\omega_{\boldsymbol{k}}(\text{TEM}) \cong \frac{\epsilon^{1/2}\omega_{*e}I_0(b)e^{-b}}{1+\frac{T_e}{T_i}\left(1-I_0(b)e^{-b}\right)} < \frac{v_e}{L_c}\left(\frac{\Delta B}{B}\right)^{1/2} \tag{11.4}$$

and a lower-frequency trapped ion mode

TIM

$$\omega_{\boldsymbol{k}}(\text{TIM}) = \frac{\epsilon^{1/2}\omega_{*i}F\left(k_\perp^2\rho_b^2\right)}{1+T_i/T_e} < \frac{v_i}{L_c}\left(\frac{\Delta B}{B}\right)^{1/2} \tag{11.5}$$

where $\Delta B/B$ is the depth of the magnetic well over the connection length L_c that traps the guiding-center particle motion.

These trapped particle instabilities and transport mechanisms are universal in that they occur in tokamaks, toroidal helical systems, stellarators and mirror machines. The transport fluxes are different for the banana regime and the plateau regime.

11.4 Large Helical System Fields and Transport

The structure for the fluctuations in the toroidal geometry is a more complicated problem than first appears. While the Fourier decomposition in the double periodicity in the two toroidal angles ϕ and θ is a natural choice for satisfying the single valued condition on the toroidal surface ψ, such a decomposition as $\delta n_e \cos(m\theta + n\phi - \omega t)$ does not describe the alignment of the fluctuation with the magnetic flux tube away from the rational surfaces. The rational surfaces are of special importance but have zero measure in terms of their density. Thus we need to introduce representations that allow the conflicting requirements of weak variation along the magnetic field lines while maintaining single valued functions in the toroidal magnetic geometry. The helical magnetic fields are described by $\mathbf{B} = (B_\theta, B_\phi)$. The fast motion and waves parallel to the helical magnetic field is the dominant effect in determining the mode structure in the long, thin fluctuation filaments.

11.4.1 *Fluctuations in the ballooning mode representation*

For fluctuations to be unstable in both MHD models and for the full kinetic theory dynamics the fluctuating potential φ must satisfy, to the lowest order, the magnetic differential equation

$$\mathbf{B}\cdot\nabla\varphi = 0 \tag{11.6}$$

for field lines that have an irrational value of the winding number or twist function

$$q(\psi) = \frac{rB_\phi}{RB_\theta} \neq \text{rational number}. \tag{11.7}$$

Then the condition Eq. (11.6) and the condition of double periodicity are in conflict for the fluctuations simply expanded in the double Fourier series $\exp(im\theta - in\phi)$ since the operator in the magnetic differential equation then gives

$$\mathbf{B} \cdot \nabla\varphi = ik_{\|}\varphi \neq 0. \tag{11.8}$$

The resolution of this difficulty is given by the ballooning mode representation developed in Connor, *et al.* (1978). The idea behind the method is to extend the flux tubes to infinity and use a new coordinate η on the infinite domain to describe the variation along the field line. The functions are Fourier decomposed in $\exp(im\eta)$ modes on the infinite domain $\eta = [-\infty, = \infty]$. These Fourier coefficients $\varphi_m = \int d\eta \exp(im\eta)\varphi(\eta)$ are then used to construct the periodic function in the poloidal angle θ through the Fourier summation

$$\varphi(\theta, \psi) = \sum_m \varphi_m \exp(-im\theta). \tag{11.9}$$

This decomposition is performed for each toroidal mode number n with the nonlinear fields then being the summation over all the ballooning modes with $\sum_n \exp(in\phi)\varphi(\theta, \psi)$.

The details of the theoretical construction are given [Connor, *et al.* (1978, 1991) and Hastie, *et al.* (1979)] and in the expanded article in the *Proceedings of the Royal Society.* The reasoning is intricate and the calculations are complicated. A small expansion parameter is introduced to solve the equations in the order: (i) the ballooning eigenmode structure for variation of the fluctuation along the flux tube given by the function $\varphi(\eta)$ on the infinite domain of the coordinate η. This is called the ballooning mode eigenfunction and typically extends well beyond the spatial domain of 2π of the poloidal coordinate. The eigenvalue for this 1D eigenmode equation is then a key function called $\omega_0(\psi, \theta_0, n)$; (ii) the first-order correction to the lowest-order equation is derived; (iii) the second order correction to the eigenvalue equation is derived. The effective potential of the second-order equation then determines the radial mode structure. This equation has a bounded solution when the variation of the lowest-order eigenvalue $\omega_0(\theta_0)$ with respect to the variable θ_0 vanishes. With this condition satisfied there is then a rather standard second-order radial Storm-Liouville eigenvalue problem to solve for the radial mode structure in the flux coordinate ψ.

There are only a limited number of cases where this full ballooning procedure has been carried out. In the case of MHD without flow the self-adjointness of the operators in the formula for δW gives considerable simplification and the effective radial potential for the radial eigenvalue problem can be derived from the radial variation of the $\omega^2(\psi$ radial coordinate). This MHD ω^2 functional is extremely important in itself since it defines what is a stable magnetic confinement system. So this class of problems has been solved extensively for a wide range of axisymmetric toroidal geometries.

11.4.2 *Drift waves in the ballooning representation*

For the case of drift wave fluctuations the operators in the linear partial differential equation are not self-adjoint and there is no useful variational form of the associated eigenvalue problem. Here the ballooning mode theory procedure is worked out by Choi and Horton (1980). The procedure is similar to that in MHD but the ordering of wave frequency and mode numbers is different. The theory is interesting and the results show that often the drift waves in the torus are strongly localized to the outer equatorial plane and in other cases to the top and bottom of the torus.

Energetically, we can understand these localizations by the destabilizing effect of the magnetic curvature vector $\boldsymbol{\kappa} = \boldsymbol{b} \cdot \boldsymbol{\nabla} \boldsymbol{b}$. At the outboard mid-plane or equatorial plane there is a large effective acceleration of the plasma given by $g_{\text{eff}} = 2v_{th}^2/R_c$ which produces a local ballooning interchange driving mechanism. For sufficiently strong localization the stabilizing role of the ion acoustic waves over comes to the destabilization from the g_{eff}. At the top and bottom of the torus where the poloidal angle $\theta = \pm\pi/2$ the effective g_{eff} acceleration is horizontal and thus tangential to the flux surface at $\theta = \pm\pi/2$. This also creates and interchange release of energy, but the wave function must be peaked in these regions just before the values $\theta = \pm\pi/2$ to use this effective force to move plasma down the ∇B.

In Fig. 11.4 the typical drift wave ballooning mode structure is shown and in Fig. 11.6 the corresponding eigenvalue shown versus the strength of the toroidicity parameter $\epsilon = r/R$. Note the sudden decrease in the damping rate of the eigenvalue for the toroidicity induced mode for $\epsilon \gtrsim 0.05$ for the symmetric eigenmode labeled with the filled circles in Fig. 11.4 [Chen and Cheng (1980)].

The ballooning mode representation gives the method of calculating the parallel wavenumber $k_\parallel$ for general flux surfaces with irrational magnetic winding numbers $q(\psi)$. The parallel wavenumber $k_\parallel$ has a spectrum due to the summation over the integer m values of the transform of the infinite domain ballooning mode. These means that the modes can now be radial extended structures with the wave functions in the linear mode being a coherent sum of localized ballooning modes. This behavior gives rise to a rich spectrum seen experimentally most clear in the Alfvén modes in the tokamak. Since the Alfvén wave frequency is $\omega^2 = k_\parallel^2 v_A^2$ the overlapping and coupling of the different poloidal harmonics is most critical. This result is that new modes called the Alfvén gap modes that are discrete localized modes in the gaps between the $k_\parallel$ lines drawn for different poloidal mode numbers. The details of the ITG eigenmode in the ballooning mode representation are given in Rewoldt, *et al.* (1982), Rewoldt and Tang (1983), Dong, *et al.* (1992), and Horton, *et al.* (1979).

The simulation codes use what is called the flux tube representation is a numerical version of the infinite domain coordinate η defined in Eq. (11.9). The

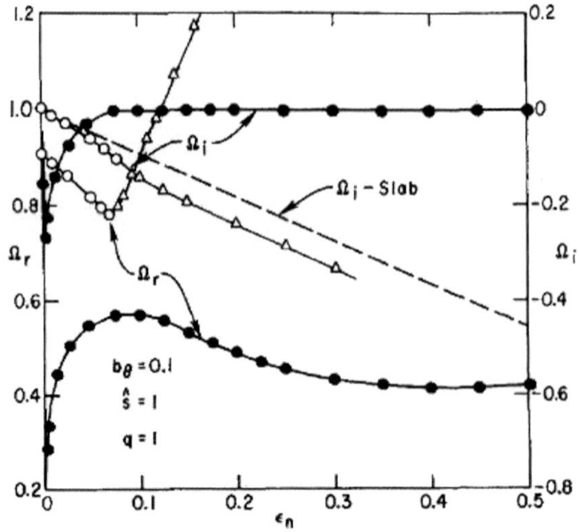

Fig. 11.4 Plot of eigenmode frequencies Ω vs. ϵ_n for $b_\theta = 0.1$, $q = 1$, and $\widehat{s} = 1$. ∘, • and Δ correspond, respectively, to the $n = 0$ slab-like, $n = 0$ and $n > 0$ toroidicity-induced eigenmodes [Chen and Cheng (1980)].

development of this numerically flux tube representation that is described in Cowley and Sudan (1997) and Beer and Hammett (1996). The work is rather technical so we leave this discussion here and refer the reader to these works and review by Brizard and Hahm (2007) for the details of how the flux tube method is implemented in the simulation and gyrokinetic equations. The method used by Rewoldt is rather simpler consisting of a set of harmonic oscillator functions as given in Chapter 3 on each rational surface up to the maximum number of rational surfaces kept the simulations.

The relationship between the kinetically-modified MHD ballooning mode and the finite-β ITG mode is explored in Hong, *et al.* (1989) for tokamaks and for a stellarator in Hong, *et al.* (1991) and Yagi, *et al.* (1989). Boozer, *et al.* (1990) compare the properties of stellarators and tokamaks.

Nonlinear simulations of the energetic particle destabilization and evolution of these TAE modes is described in Todo, *et al.* (2010). The saturation level is shown to require both nonlinear particle dynamics and nonlinear MHD dynamics when the amplitudes reach high levels with $\delta B/B \sim 10^{-2}$. There is a zonal flow and nonlinear coupling to high-n modes up to $n \sim 10$.

Other low-frequency MHD modes where the high energy injected ions create instabilities are called energetic particles modes or EP-modes for short. One of the more interesting and still incompletely understood of the energetic particle modes that is of concern as to how and when they would appear in the burning plasma driven by the fusion products of alpha particles in the "fish bone" mode. The name is derived from the plot of the fluctuating magnetic field versus time which has the structure resembling the skeleton with backbone of a large-flat bonita type of fish. Recent progress in the research of fishbones is given in Okabayashi, *et al.* (2011). The waves are driven by a population of high-energy ions resonating with the low-frequency modes through the bounce averaged gradient-B drift frequency with the transit motion frequency. The polarization of the magnetic field fluctuations indicate the mode has a compressional magnetic component in addition to the line-bending shear Alfvén wave component. The radial gradient of the fast ion distribution function is important. The mode ejects some fraction of the fast ions from the plasma and repeats in rapid succession, creating the "fish-bone" signature on the magnetic field probes. New results on DIII-D show that when the mode occurs sufficiently off-axis the magnetic activity destabilizes or drives the resistive-wall mode. Experiments showing the importance of the toroidal Alfvén eigenmodes were carried out on TFTR by Wong, *et al.* (1991).

11.5 Wendelstein 7-X Stellarator

Plasma configurations describing the stellarator experiment Wendelstein 7-X (W7-X) are modeling the geometry of the test-divertor unit shown in Fig. 11.5. A high-heat-flux divertor will be installed in the vacuum chamber of the device [Gasparotto, *et al.* (2014)] in 2017-2018.

Fig. 11.5 The current phase of operation of W7-X uses a test divertor unit (TDU) facing the engineer that is an inertially-cooled island divertor with the same shape as the that of the water cooled metallic heat flux divertor with carbon tile baffles. With the water cooling divertor the team expects to achieve 8 MW discharges for 10 seconds in 2017. Current results are available at `http://www.ipp.mpg.de/2285/en`.

Experiments from the first 1000 discharges in the new superconducting Wendelstein 7-X helical machine were reported at Exploratory Plasma Research (EPR) 1016, Auburn University. With plasmas produced by electron cyclotron heating in both the X-mode polarization of the ECH waves and then 2nd harmonic O-mode polarization for powers up to 10 MWs. Twenty plasma diagnostics were used to verify the stability and long-pulse duration of the plasmas. Plasmas reached 8 KeV core electron temperatures and 2 KeV core ion temperatures for densities of $n_e = 3 \times 10^{19}/\mathrm{m}^3$ in helium and deuterium discharges. The diagnostics included Thomson scattering of laser beams for the electron temperature and density and imaging with X-ray spectroscopy (XICS) for both the electron and ion temperature profiles [Bozhenkov, *et al.* (2016)].

11.5.1 *Toroidal Alfvén Eigenmodes*

Fast neutral beam injection that exceeds the stability limit of the internal $m = 1$ modes gives rise to nonlinear relaxation oscillations called "fishbones" due to the character of the times series from the poloidal magnetic field probes. The bursts of the $m = 1$ oscillations ejects a fraction of the injected energetic ions and limits their density and pressure. A simple description of the dynamics consistent with the data from Heidbrink, *et al.* (1986) and McGuire, *et al.* (1983) is given by Coppi and Porcelli (1986).

Alpha particle driven toroidal Alfvén eigenmodes (TAEs) were first observed in deuterium-tritium plasmas in TFTR [Nazikian, *et al.* (1997)]. The modes were in the core of the plasma with amplitudes $\delta B/B \sim 10^{-5}$ with toroidal mode numbers $n = 2-4$ in a low magnetic-shear regime. The dimensionless alpha particle pressure $\beta_\alpha \sim 2 \times 10^{-4}$ was below the value expected for onset of the modes from theory. A new interpretation of the alpha particle driven instabilities more consistent with the data from the TFTR experiments is given in Nazikian, *et al.* (2003). The Alfvén waves in the tokamak are coupled in their poloidal angle-θ variation in the $B_\phi(\theta) = B_0/(1 + r/R\cos\theta)$ toroidal magnetic field. The coupling results in the formation discrete eigenmodes where the frequencies $\omega_\pm(m,n)$ of the uncoupled modes derived in the $\epsilon = r/R \to 0$ limit intersect. The frequency of the gap eigenmode is derived by setting the frequencies of the two neighboring Alfvén wave continuum modes $\omega_1^2 = k_\|^2(m,n)v_A^2$ and $\omega_2^2 = k_\|^2(m+1,n)v_A^2$ equal to each other in the limit $\epsilon \to 0$. The finite ϵ-coupling of the two degenerate continuum ω_1, ω_2 is from the gap mode at the radial position $r = r_{\mathrm{TAE}}$.

The presence of the toroidal Alfvén eigenmodes, or TAE modes, is analogous to the discrete bound electron hole-modes in the semi-conductors in the band-gaps of solid state physics. Whereas the Alfvén wave continuum modes have strong electron Landau damping, the discrete Alfvén gap modes are weakly damped and easily destabilized by energetic electron or ion Landau wave-particle resonances. The Alfvén gap modes have been extensively developed since the fusion plasma has a high-energy population of alpha (He^{+4} helium ions) particles that destabilize the

Alfvén gap modes. Laboratory simulation experiments by Wong, *et al.* (1991) with high-energy neutral beam produced hydrogen nuclei show that the nonlinear TAE modes have a wide variety of frequency spectra that wave-particle produce a fast loss of the resonant ions through transport processes [Berk, *et al.* (1997), Todo, *et al.* (2010), Gorelenkov, *et al.* (2007), Fu and Van Dam (1989), Breizman and Sharapov (1995)].

In an homogeneous magnetized plasma, linear ideal-MHD arguments show the existence of a shear-Alfvén wave of frequency ω_A with the dispersion relation

$$\omega_A^2 = k_\parallel^2 v_A^2 \tag{11.10}$$

where v_A is the local Alfvén velocity and $k_\parallel$ is the wavenumber in the direction of the equilibrium magnetic field $\mathbf{B}_0$.

Let us consider axisymmetric toroidal plasmas. In the cylindrical limit, the periodicities of the system require that there exists two integers, a toroidal mode number n and a poloidal mode number m, such that

$$k_\parallel = \frac{n - m/q(r)}{R_0}, \tag{11.11}$$

where R_0 is the distance from the symmetry axis of the tokamak to the magnetic axis. In a non-homogeneous plasma in a sheared magnetic field, both $k_\parallel$ and v_A are functions of r where r is the local radial flux coordinate. The simple dispersion relation Eq. (11.10) is still valid in this configuration and it is called the Alfvén continuum. Since phase velocity is a function of radius, a wave packet with finite radial extent would suffer from phase-mixing, producing the so-called continuum damping. Except for energetic particle modes, resonant drive-by thermal particles is not enough to overcome this damping. However, a toroidal coupling of two successive poloidal modes m and $m+1$ breaks up the continuous spectrum. This is illustrated in Fig. 11.6, which shows the Alfvén continuum for $n = 1$, $m = 2$, and $n = 1$ and $m = 3$, in cylindrical geometry, where two poloidal continuum modes are decoupled, and in toroidal geometry, with a two-mode coupling model. The latter is obtained with equilibrium plasma parameters corresponding to JT-60U shot E32359 at $t = 4.2\,\text{s}$ assuming concentric circular magnetic flux surfaces, retaining toroidicity effects in the first order in inverse aspect ratio. Though we show only the $\omega > 0$ half-plane, the continuum is symmetric with respect to ω. Coupled modes are (n, m) and $(-n, -m-1)$ for $\omega > 0$, and $(n, m+1)$ and $(-n, -m)$ for $\omega < 0$. The gap is centered at a radius r_A such that $q(r_A) = (m + 1/2)/n$, where the two continuous spectra would cross in the absence of coupling, and where $|k_\parallel| = 1/2qR_0$. The resulting discrete eigenmode is a TAE, at a frequency $\omega_A = v_A/2qR_0$.

For a deuterium plasma with typical magnetic field $B_0 \sim 1\,\text{T}$ and density $n_i \sim 10^{20}\,\text{m}^{-3}$, the Alfvénic energy is $E_A \equiv m_i v_A^2/2 \sim 10\,\text{KeV}$, which is in the range of passing particles induced by NBI. For ITER parameters, $E_A \sim 1\,\text{MeV}$, which is in the range of passing α-particles born from fusion reactions. In both cases, TAEs can be driven unstable by resonance with energetic particles. For far-passing particles, the resonance condition is $\Omega = \omega_A$, where

$$\Omega = n\omega_\zeta + l\omega_\theta, \tag{11.12}$$

where $\omega_\zeta = v_\parallel/R_0$ and $\omega_\theta = v_\parallel/qR_0$ are the particle frequencies of toroidal motion and poloidal motion, respectively, and $l = -m$ for co-passing particles, $l = m$ for counter-passing particles. Since we analyze TAEs driven by co-injected ions, we can simplify the discussions by considering only co-passing particles. Then, the resonance condition is

$$\omega_A - n\frac{v_\parallel}{R_0} + m\frac{v_\parallel}{qR_0} = 0. \tag{11.13}$$

In JET the TAE modes are routinely observed and used to interpret the $q(r,t)$ profiles in the evolving plasma. An early example is given by Fasoli, *et al.* (1995) for the measurement of the damping of toroidicity-induced Alfvén eigenmodes.

An important low-order TAE mode example, as shown in Fig. 11.4 occurs for the coupling of the $m = 2$ and $m = 3$ forward and backward $\omega_(m+1,n,r)$ Alfvén wave continuum modes. Writing the condition for

$$\omega_+ = k_\parallel(m,n,r)v_A \quad \text{and} \quad \omega_- = -k_\parallel(m+1,n,r)v_A \tag{11.14}$$

to be degenerate $\omega_+ = \omega_-$ gives

$$\left(\frac{n}{R} - \frac{m}{qR}\right)v_A = -\left(\frac{n}{R} - \frac{m+1}{qR}\right)v_A \tag{11.15}$$

which reduces to

$$\frac{2n}{R} = (m+1+m)/qR \tag{11.16}$$

giving the resonant radius for the gap mode of

$$q\left(r_0, m, n\right) = \frac{m+1/2}{n}. \tag{11.17}$$

Extended radial TAE-modes occur for $m = 1, n = 1$ coupled to $m = 2, n = 1$ modes at $q(r_0, m, n) = (m+1/2)/n$ as shown in Fig. 11.6.

Nonlinear simulations of the energetic particle destabilization and evolution of these TAE modes is described in Todo, *et al.* (2010). The saturation level is shown to require both nonlinear particle dynamics and nonlinear MHD dynamics when the amplitudes reach high levels with $\delta B/B \sim 10^{-2}$. There is a zonal flow and nonlinear coupling to high-n modes up to $n \sim 10$.

Other low-frequency MHD modes where the high energy injected ions create instabilities are called energetic particles modes or EP-modes for short. One of the more interesting and still incompletely understood of the energetic particle modes that is of concern as to how the waves might appear in the burning plasma driven by the fusion products of alpha particles. The name is derived from the plot of the fluctuating magnetic field versus time that has the structure resembling the skeleton with backbone of a large-flat bonita type of fish. Recent progress in the research of fishbones is given in Okabayashi, *et al.* (2011). The waves are driven by a

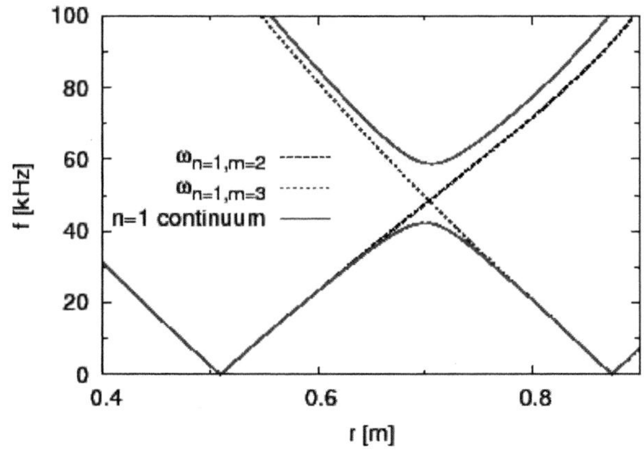

Fig. 11.6 Alfvén continuum for $n = 1$ with (solid line) and without (dashed line) coupling between $m = 2$ and $m = 3$ poloidal modes. Note that the discrepancy, relatively far from resonance, between the upper branch of coupled continuum and the uncoupled $m = 2$ branch, is accounted for by the terms of second-order in the inverse aspect ratio $\epsilon = r/R$ expansion.

population of high-energy ions resonating with the low-frequency modes through the bounce averaged gradient-B drift frequency with the transit motion frequency. The polarization of the magnetic field fluctuations indicate the mode has a compressional magnetic component in addition to the line-bending shear Alfvén wave component. The radial gradient of the fast ion distribution function is important. The mode ejects some fraction of the fast ions from the plasma and repeats in rapid succession, creating the "fish-bone" signature on the magnetic field probes. In addition, the new results on DIII-D show that when the mode occurs sufficiently off-axis the magnetic activity destabilizes or drives the resistive-wall mode. Experiments showing the importance of the toroidal Alfvén eigenmodes were carried out on TFTR by Wong, *et al.* (1991).

11.6 Gyrokinetic Simulation of Zonal Flows and Ion Temperature Gradient Turbulence in Helical Systems

The gyrokinetic-Vlasov (GKV) simulation code is applied to zonal flows and the ion temperature gradient (ITG) turbulence in helical systems with $L = 2$ and $M = 10$ like the large helical device in Fig. 11.1. Here L and M denote poloidal and toroidal periodicities of the main helical component of the confinement field, respectively) for the standard and inward-shifted model configurations. Because of the slower radial drift motion of helical-ripple-trapped particles, the inward-shifted case provides a higher zonal-flow response than that in the standard model with smaller side-band helical field components. The nonlinear GKV simulations show that the ITG turbulent transport in the inward-shifted regime has larger growth rates of the ITG stability which are regulated by the zonal flows.

The book cover shows the fine structure of the ITG turbulence from 3D simulations of the Large Helical Device by Nunami *et al.* (2012). There are complex multi-scale structures captured in the simulation that create the ion thermal transport measured in the high ion temperature LHD plasma. The ion temperature gradient driven turbulence is described further in the next chapter.

Chapter 12

Temperature Gradient Driven Instabilities

Magnetized plasmas with a significant temperature gradient across the magnetic field are a common situation owing to the thermal insulating properties of the magnetic field. Such temperature gradients occur in the magnetosphere, the solar corona and with particularly high gradients in the laboratory experiments aimed at magnetic confinement for nuclear fusion. In the large tokamaks experiments reviewed in Chapter 1 we see that the core ion temperatures have reached 40 KeV in the TFTR device (Sec. 1.7.2) and in the JT-60U (Sec. 1.7.3) machine with $R/a = 3.1\,\mathrm{m}/0.7\,\mathrm{m}$. Thus, we know that the magnetic field can provide strong thermal insulation withstanding gradients greater than 20 KeV/m under ideal conditions of the closed irrational magnetic flux tubes. Table 12.1 gives the parameters for a well-known JT-60U shot that has a core ion temperature of 38 KeV and core electron temperature of 12 KeV. The ion (deuterium) quantities T_i and v_ϕ are inferred from spectroscopic measurement of carbon ions.

The whole concept of magnetic confinement for thermonuclear fusion of plasmas depends on this thermal insulating principle. From the deuterium-tritium reactivity function shown in Fig. 1.1 one finds that the core ion plasma temperatures of order $10^7 - 10^8$ K are required for significant fusion nuclear reactivity. This high-temperature plasma must be partially insulated from the vessel walls which must remain below 5000 K to retain their integrity. Thus, steep temperature

Table 12.1 JT-60U high-β_p experiment 17110 with internal transport barrier: parameters in Phase I of shot.

R/a	$3.1\,\mathrm{m}/0.7\,\mathrm{m}$
B_ϕ	4.4 T
I_p	2 MA
P_{NBI}	27 M
$n_D(0)$	$4.1 \times 10^{19}\,\mathrm{m}^{-3}$
$T_i(0)/T_e(0)$	38 KeV/12 KeV
$n_D \tau_E T_i$	$1.1 \times 10^{21}\,\mathrm{m}^{-3} \cdot s \cdot \mathrm{KeV}$
$v_\phi(0)$	-100 km/s

gradients of order 20 KeV/m must be stably maintained in the thermonuclear fusion device. The associated ion thermal diffusivity χ_i is then of order or less than $1\,\mathrm{m}^2/\mathrm{s}$ in devices with minor radius a few meters.

Such magnetic thermal insulation is consistent with the results from the latest generation of tokamak and spherical tokamak experiments under favorable conditions. The thermal flux across the magnetic surfaces is large $q_i = -n_i \chi_i dT_i/dr$ reaching $0.1\,\mathrm{MW/m}^2$ and requiring large heating rates of many MW to maintain the plasma against turbulent losses. For the alpha particle decay products from the fusion reactions to provide the power to maintain the ion temperature, the core electron temperature T_e must also be sufficiently high – in the 10 KeV range. The next chapter, Chapter 13, is devoted to instabilities driven by the electron temperature gradient and the associated turbulent transport of electron thermal energy across the magnetic flux tubes. This electron thermal transport problem is more difficult to control in the magnetic confinement experiments than the ion thermal turbulent transport as emphasized by Kadomtsev (1992).

12.1 Ion Temperature Gradient Instabilities

In both laboratory plasmas and in astrophysical plasma where a strong local heating takes place there is clearly the potential source of plasma instability. There can be MHD (single conducting fluid in Chapter 5) instabilities controlled by the constraints of frozen-in magnetic flux and the adiabatic equation of state. These fast instabilities limit the plasma pressure gradients with respect to the confining magnetic pressure. These fast, single fluid instabilities have been eliminated in the magnetic confinement devices proposed for nuclear fusion. In this chapter the plasma is stable to ideal MHD motions and study the remaining smaller scale drift wave instabilities. The drift waves have dispersive frequencies and finite parallel electric fields which means the associated waves and vortices have resonant wave-particle interactions. Figure 12.1 shows the growth of the ion temperature gradient driven drift wave vortices starting on the outside of the torus in frame (a) and spreading throughout the torus in frames (b), (c), and (d).

The drift wave dispersion relation in a collisionless plasma does not assume a local thermal Maxwell-Boltzmann distribution function. In the high temperatures of fusion and geospace plasmas the particle distributions in energy typically have high energy tails and different effective temperatures across and parallel to the magnetic field. In low beta $\beta = p/(B^2/2\mu_0)$ plasmas the ion temperature gradient instability, known as the ITG mode, has nearly an electrostatic polarization for the coupled drift waves and the ion acoustic waves that gives rise to the collective instability. In higher pressure plasmas with $\beta > m_e/m_i$ the ITG mode couples with the shear Alfvén wave becoming a dispersive oscillation with an electromagnetic polarization. As the plasma pressure increases the inductive electric field from the time changing δB_x cancels part of the electrostatic component of the parallel

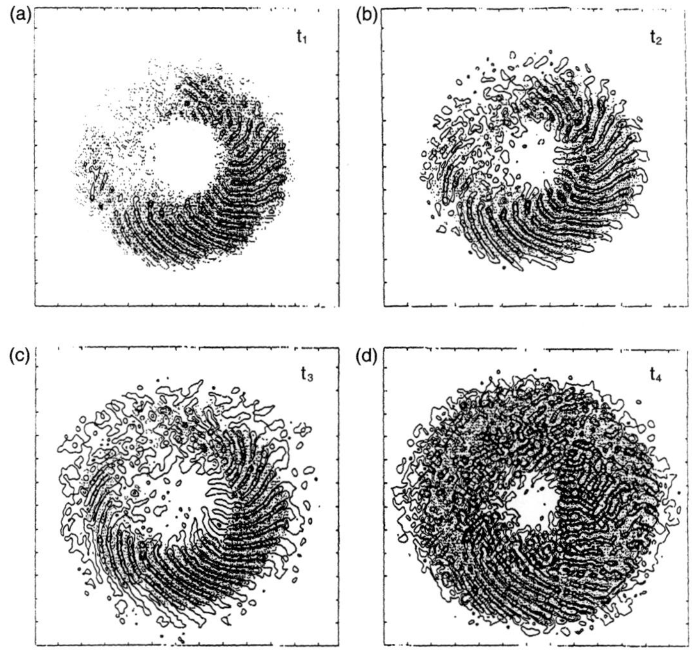

Fig. 12.1 Evolution of the isopotential contours for the ITG turbulence for $a/\rho_i = 28$, $L_T/R = 0.2$, $T_i = T_e$ with hyperbolic tangent profiles for the temperature. The density profile is flat [Sydora, *et al.* (1996)].

electric field. This cancellation reduces the energy transfer rate $\langle j_\parallel E_\parallel \rangle$ and reduces the growth rate of the instability. There are numerous studies of these weakly electromagnetic ITG modes including Kim, *et al.* (1993), Rewoldt, *et al.* (1998), and Hong, *et al.* (1989a,b) that detail how the ITG modes change with increasing plasma beta. At certain intermediate values of plasma beta both the ITG and the kinetic ballooning mode of the MHD polarization are present with different mode frequencies and growth rates.

We begin with a thermodynamic picture of the energy source for the turbulence from the temperature gradient. Then we consider the kinetic theory in the electrostatic limit of the ion temperature gradient driven stability. In Section 13.2 we consider the stability analysis for the long-tailed ion energy distribution functions characteristic of some neutral beam heated plasmas and astrophysical plasmas. For example, the dense plasma in the wake of the solar wind behind magnetized planets are typically parameterized by a power laws distribution that is called the "kappa" distribution [Kivelson and Russell (1995)] with power law energy E distributions parameterized as $f_\kappa(E) \simeq 1/E^\kappa$. The energy spectrum index parameter κ for heavy-tailed distributions varies from $\kappa = 2$ or 3 energy distributions with every large high-energy fluxes suitable for modeling shock accelerated protons to large values of kappa where the bulk of the distribution is close to the Maxwell-

Boltzmann distribution. The κ-distribution is used in the space physics [Kivelson and Russell (1995)] for the hot geotail plasma and the solar corona [Aschwanden (2005)] for the proton distributions. In both these astrophysical plasma the time for Coulomb collisions exceeds the dynamical times scales by orders of magnitudes.

12.2 Mechanism of the Ion Temperature Gradient Instabilities

For the Maxwellian phase space distributions the electrostatic ITG (ion temperature gradient) instability starts when the ratio of the ion temperature gradient $d\ln T_i/dr$ exceeds the density gradient $d\ln n_i/dr$ by a number that depends on the compressibility of the ion gas. This critical gradient is best understood by considering the maximum work that can be extracted from the plasma temperature profile with a Carnot cycle operating between the high- and low-temperature regions. The "engine" for the cycle is the drift wave convection between the high-temperature and low-temperature regions and the working gas is the ion component of the plasma with an adiabatic gas constant $\Gamma = (d+2)/d$ where $d =$ number of degrees of freedom active in the dynamics of the ion gas. The situation is shown in Figs. 3.3 and 12.1 where the radially-extended vortex cells, tilted to follow the twist of the magnetic field lines, cycle the hot interior plasma to the lower-temperature plasma [Sydora, *et al.* (1996)].

One uses the Carnot cycle to calculate the maximum energy W_{ITG} that can be extracted per cycle from the drift waves. The plasma is adiabatically compressed and expanded between the high and low ion-temperature regions so that the adiabatic gas constant controls the stability condition. We find that the condition for a net energy release from the temperature gradient is given by

$$\eta_i > \eta_{\text{crit}} = \frac{2}{3} \tag{12.1}$$

for the usual $d = 3$ degrees of freedom gas equation of state.

This temperature gradient η parameter is widely used in the plasma literature and defined for a charge species s by

$$\eta_s = \frac{d\ln T_s}{d\ln n_s}. \tag{12.2}$$

In the early plasma literature the instabilities driven by ion temperature gradient were often called the η_i modes [Horton and Varma (1972)].

The temperature gradient instability condition follows from a calculation of the work done $W = Q_1 - Q_2$ from the thermal energies Q taken from the hot ion region at $T_i(x)$ and adiabatically expanding the gas to the lower-temperature regions at $T_i(x + \Delta x)$. Then the convection recompresses the gas to return to the high-temperature region at $T_i(x)$. Subtracting the work done in the compression minus the work gained in the expansion and adding the work done by the gas in the two isothermal strokes gives the net energy W released for one cycle. The energy W is proportional to the differences $\Delta T/T(\Delta T/T - 2/3\Delta n/n)$. For the general value of the adiabatic gas constant the factor 2/3 is replaced by $\Gamma - 1$. Thus for the slab model of the ITG threshold, where there is only the parallel compressible motion, there is a one-degree-of-freedom compression and the threshold is raised to the value of $\Gamma - 1 = 2$. In toroidal systems there are three-degrees-of-freedom and the threshold starts at 2/3. The corresponding changes in the wave-particle resonances are shown in Fig. 12.2.

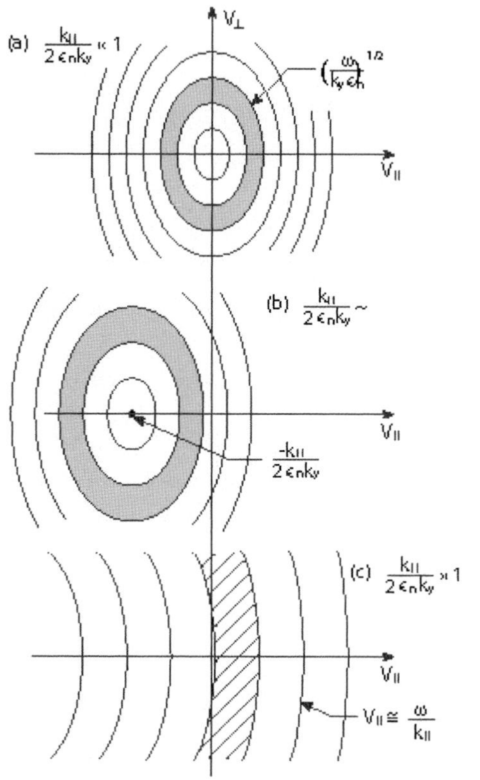

Fig. 12.2 The band of resonant ion velocities for a given drift-wave fluctuation $(\omega, k_y, k_\parallel)$. (a) the ideal delta function resonance curve $\omega_k = \omega_D$, (b) the shifted contours of resonant $\omega_k - k_\parallel v_\parallel = \omega_D$, and (c) the slab Landau resonance $\omega_k = k_\parallel v_\parallel$.

These threshold formulas are known to agree with the results from Nyquist diagrams for the stability condition of the electrostatic drift waves when other complications such radiative damping are absent.

Here some intermediate steps in the Carnot cycle construction are given for clarity. The different temperatures $T_i(x + \Delta x)$ and $T_i(x)$ are the driving force of the convection. In the ideal isentropic limit the work W received from the release of energy from the hot $T_2(x)$ plasma reservoir to the lower-temperature $T_1(x+\Delta x)$. Using conservation of energy (first law of thermodynamics) the $W = Q_2 - Q_1$ and $Q_2 = T_2 \Delta S_2$ and $Q_1 = -T_1 \Delta S_1$, with $\Delta S_1 + \Delta S_2 = 0$ gives the maximum work W

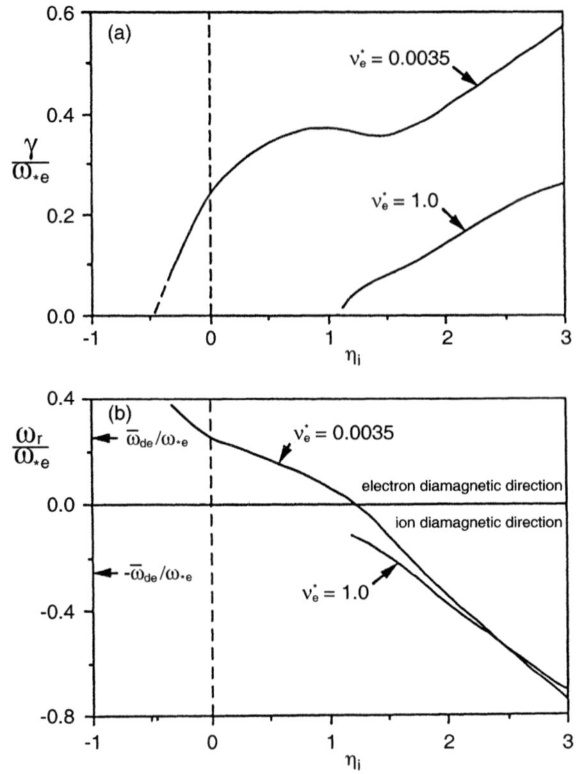

Fig. 12.3 The normalized (a) growth rate and (b) frequency of the complex kinetic toroidal eigenvalue problem. At high collisionality only the ITG mode appears for $\eta_i > 1.2$ while at low collisionality the trapped electron mode appears for $\eta_i < 1.5$ [Rewoldt, *et al.* (1998)].

that can be extracted as per

$$W = N\left(T_2 - T_1\right)\underbrace{\Delta \ell n\left(T^{3/2}/N\right)}_{\Delta S_2} = N\Delta T\left(\frac{3}{2}\frac{\Delta T}{T} - \frac{\Delta N}{N}\right) \tag{12.3}$$

from the gradients of $T(x)$ and $N(x)$.

Thus, in the absence of Landau damping and other non-ideal loss processes such as wave radiation from the convection region in the calculation, the maximum energy per cycle that the drift wave can extract from the distribution satisfies the condition in Eq. (12.3). We will return to the generalization of Eq. (12.3) to non-Maxwellian $f_0(H, x)$ distributions later.

The complete dispersion relation from solving the eigenvalue problem in the toroidal geometry was developed into computer codes. An early example for a TFTR plasma is shown in Fig. 12.3 from Rewoldt, *et al.* (1998). The fastest growing eigenmode from this numerical solution with the ITG and trapped electron (TEM) physics for typical discharge parameters in TFTR is shown in Fig. 12.3 from Rewoldt, *et al.* (1998).

One sees in Fig. 12.3(a) that for the collisionality of $\nu_{*e} = 1$ the mode onsets with a well-defined ion temperature gradient of $d\ln T_i/d\ln n_i = 1$ and rotates in the plasma rest frame in the ion diamagnetic direction. For the higher temperatures where $\nu^* = 3.5 \times 10^{-3}$ there is a substantial fraction of trapped electrons Fig. 12.3(b) shows that the mode is unstable even with no ion temperature gradient $d\ln T_i/d\ln n_i = 0$ from the eigenvalue code. This instability is the trapped electron mode (TEM) and has growth rate with $\gamma_k/\omega_{*e} = 0.25$ with even for no ion temperature gradient.

In the next section we show how the electrons respond and destabilize the drift wave in this weakly collisional regime where electrons with large pitch angles are trapped by the mirror force in the low-B field region on the outside of the torus. These types of modes exist in all toroidal confinement devices and produce drift wave instabilities that are universal to toroidal confinement geometries. The growth rate in Fig. 12.2(a) for $d\ln T_i/d\ln n_i = 0$ is from the mirror trapped electrons resonantly interacting with the drift wave.

The change in the unstable region for $k_\perp L_n$ versus $k_y\rho_i$ in a stability diagram with varying values of $\eta_\perp$ and $\eta_\parallel$ is described in Hojo (1995). At low values of $\eta_\perp$, the critical condition is a value of $\eta_\parallel$ equal to 3 as given by the thermodynamics of the one-degree of freedom gas. The change in the unstable region for $k_\perp L_n$ versus $k_y\rho_i$ in a stability diagram with varying values of $\eta_\perp$ and $\eta_\parallel$ is worked out in Hojo (1995). The result shows that at low $\eta_\parallel$ the critical condition is value of $\eta_\perp > 1$ to 2 consistent with the thermodynamics for 2 to 3 degrees of freedom in ion gas.

For the critical gradients over all k-modes Hojo (1995) explains how plasma with high T_i/T_e can remain stable to the ITG modes for $\eta_\perp$ moderately large. The analysis is done specifically for the high T_i/T_e ratios relevant to the mirror trapped plasmas. In Hojo's example the GAMMA 10 mirror machine contains a 5 KeV perpendicular ion temperature plasma in a relatively small radius ($a < 0.5\,\mathrm{m}$) cylindrical plasma. So one might expect a strong ITG instability. This analysis and related analysis by Kim (1992) shows why these systems are stable to ITG modes [Chen and Migliuolo (1992)]. A similar stability to ITG at high T_i/T_e ratios was reported in the TFTR supershots [Horton, *et al.* (1992)], although the analysis for the stability of the waves is rather different for the tokamak.

The kinetic stability analysis of the temperature gradient driven modes is determined by kinetic plasma response functions $P(\omega, k)$, $Q(\omega, k)$, $R(\omega, k)$ which are integrals of the gradients of the phase space density function $f(\psi, p_\perp, p_\parallel)$ weighted by the Bessel function $J_0^2, v_\perp J_1 J_0$ and $v_\perp^2 J_1^2$ where $J_n = J_n(k_\perp v \sin\alpha/\Omega)$ with α the pitch angle and $v = (2H_0/m)^{1/2}$ the particle speed. We use the symbol H_0 for the kinetic energy of the ion of mass m and charge e. To start with we will take $\sin\alpha$ as a parameter by using $\lambda = k_\perp v \sin\alpha/\Omega$ as the dimensionless variable for the mean or characteristic speed v_0 in the $f(H_0, x)$ ion distributions.

The simplest and most studied response function $P(\omega/k_y, \lambda)$ describes the electrostatic response of the plasma. For electrostatic waves the ion temperature gradient modes are given by dispersion relation

$$D_{\mathrm{ES}}(k,\omega) = 1 + \frac{T_e}{T_i}[1 - P(k,\omega)] = 0. \tag{12.4}$$

The function $P(\omega/k_y, \lambda)$ may be thought of as the generalization of the plasma dispersion function $Z(\omega/k_\| v_{th})$ when the grad-B and curvature drifts dominate the parallel phase velocity resonance.

The $P(\omega/k_y, \lambda)$ function is defined by

$$P(\omega/k_y, \lambda) = -\int \frac{dH_0 H_0^{1/2}\left(\omega \dfrac{\partial f}{\partial H_0} + \dfrac{k_y}{eB_z}\dfrac{\partial f}{\partial x}\right) J_0^2}{\omega - k_y \overline{v}_D H_0 + i0^+}. \tag{12.5}$$

We may define an effective temperature T (or $k_B T$) from the mean ion energy in the distribution function $f(H_0, x)$ such that T becomes the thermodynamic temperature when $f(H_0, x) \Rightarrow f_{\mathrm{Max-Boltzmann}}$.

For a Maxwellian $f_{\mathrm{Max-Boltzmann}} = N\exp(-H_0/T)/(2\pi T/m)^{3/2}$ the P-function is thoroughly studied in the literature [Similon, *et al.* (1984)]. Mirror machines and space plasmas have more general $f(p_\perp, p_\| \psi)$ distribution functions.

The $\overline{v}_D$ in equation Eq. (12.5) is the sum of the grad-B and curvature drift frequencies bounce-averaged and taken at some representative pitch angle (typically the variation with the pitch angle α is weak). The transition from the toroidal response function P to the slab model response function Z is given in Kim and Horton (1991).

The gradient $\partial/\partial x$ is with respect to the flux function ψ with $B_z dx = d\psi$ in the magnetic flux coordinate (per unit of dy where ∇y is the bi-orthogonal direction to the magnetic field vector and the radial gradient direction).

We may first approximate the spatial gradient as

$$\frac{\partial f}{\partial x} \equiv \frac{1}{L_n} f(H_0) \tag{12.6}$$

defining the density gradient scale length L_n. This local approximation is adequate for the first stage of analysis as seen comparing with case of a Maxwellian where

$$\frac{\partial f_M}{\partial x} = \left[\frac{1}{M}\frac{\partial N}{\partial x} + \frac{1}{T}\frac{\partial T}{\partial x}\left(\frac{H_0}{T} - \frac{3}{2}\right)\right] f_M \equiv \frac{f_M}{L_n}\left[1 + \eta\left(\frac{H_0}{T} - \frac{3}{2}\right)\right] \tag{12.7}$$

the energy dependence of $\eta(H_0/T - 3/2)$ arises describing that higher energy ions $L_T \ll L_n(H_0/T - 3/2)$ have a shorter gradient scale length than that of the density itself. The instabilities from this effect are often called η_i-modes due to the historical definition of η as

$$\eta_i = \frac{d_x \ell n\, T_i}{d_x \ell n\, N}. \tag{12.8}$$

For the Maxwellian distribution the electrostatic instability starts at

$$\eta_i > \eta_{\mathrm{crit}} = \frac{2}{3}. \tag{12.9}$$

This critical ratio of the temperature gradient instability threshold condition is understood by constructing a Carnot cycle between the different temperatures $T_i(x + \Delta x)$ and $T_i(x)$ and asking for the ideal isentropic limit of the work W done from the release of energy from the hot reservoir as explained with Eq. (12.3).

The analysis also shows that the case of inverted temperature and density gradients is strongly unstable. This regime of negative $\eta_{\rm crit}$ occurs when the pressure gradient is negligible across a region with a temperature and density gradient. The same situation occurs in the electron temperature gradient driven modes where the negative η_e instability has been demonstrated in basic laboratory experiments. The situation is common in between two different plasmas across transition layers as the plasma pause and the chromosphere-corona boundary. Using conservation of energy (first law of thermodynamics) the $W = Q_2 - Q_1$ and $Q_2 = T_2 \Delta S_2$ and $Q_1 = T_1 \Delta S_1$, with $\Delta S_1 + \Delta S_2 = 0$ gives the maximum work W that can be extracted by the turbulence per convection cycle as given in Eq. (12.3). The formula leads to the maximum turbulence energy density formula

$$\frac{W^{\rm turb}(t)}{nT} = \frac{3}{2}\frac{\Delta T}{T}\left(\frac{\Delta T}{T} - \frac{2}{3}\frac{\Delta n}{n}\right) \tag{12.10}$$

that can be a useful real-time diagnostic for feedback control of the auxiliary heating power $P_{\rm aux}(t)$.

Thus, without Landau damping and other loss processes we have a bound, valid nonlinearly, on the maximum turbulent energy W that can be extracted from the distributions. We will return to the generalization of Eq. (12.10) for non-Maxwellian $f_0(H, x)$ distributions in Chapter 13.

12.2.1 *Nyquist analysis*

In the kinetic stability analysis, condition Eq. (12.3) results from the Nyquist analysis of the P-function. Namely, we must find the $\omega_{\rm crit}$ at which Im $P(\omega) = 0$ and then evaluate the Re P at $\omega = \omega_{\rm crit}$.

For real ω the resonant ion contribution is fixed for each value of ω/k_y through

$$\text{Im } P = \pi \int_0^\infty \delta\left(\omega - k_y \bar{v}_D H_0\right) J_0^2 \left(\omega \frac{\partial f_0}{\partial H} + \frac{k_y}{e}\frac{\partial f_0}{\partial Bx}\right) H_0^{1/2} dH_0. \tag{12.11}$$

Only phase velocities ω/k_y in the same direction as the pitch-angle averaged-drift velocity contribute to Im P through the resonant ions of energy $H_0 = \omega/k_y\bar{v}_D$. If we use the approximation in Eq. (12.6) then

$$\text{Im } P = \pi \left(\frac{\omega}{k_y v_D}\right)^{1/2} \frac{J_0^2}{|k_y v_D|}\left(\omega\frac{\partial f_0}{\partial H} + \frac{k_y f_0}{eBL_n}\right)\Bigg|_{H_0 = \omega/k_y \bar{v}_D}. \tag{12.12}$$

The critical character frequency label by point c in Fig. 12.4 is

$$\omega_* = -\frac{k_y}{eL_n}\frac{f_0(H)}{(\partial f_0/\partial H)} \tag{12.13}$$

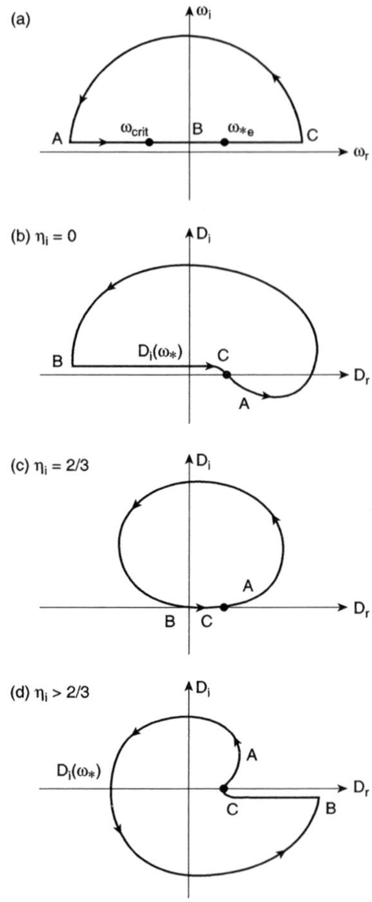

Fig. 12.4 The Nyquist diagram for the ion temperature gradient driven instability. Frame (a) is the contour enclosing the unstable upper-half plane that is mapped by the dispersion function in frames (b), (c) and (d) for three values of the stability parameter η_i. For the first case in (b) there is no temperature gradient and the mapped contour does not encircle the origin, so there is no unstable root. In frame (c) with $\eta_i > 2/3$ there is a marginally stable root. For frame (c) with $\eta_i > 2/3$ the D-function contour encircles the origin proving that there is an unstable root within the A-B-C contour in frame (a) of the complex omega plane.

as a function of H. The wave-ion drift resonance condition that

$$\omega = k_y \overline{v}_D H_0 = \omega_* \tag{12.14}$$

which is the critical frequency for the point (labeled as in Fig. 12.4) of crossing the imaginary axis in the complex P-plane. The change in the sign of the imaginary part Im $P(\omega)$ at this critical frequency reverses the flow of energy between the waves and the resonant ions.

Now that the critical frequency is determined, we evaluate the Re P, which now is the principal value integral with a well-defined value for $\omega = \omega_{\text{crit}}$ to complete the Nyquist diagram in Fig. 12.4.

In the simplest case of electrostatic ITG modes the electron density response is adiabatic with respect to the fluctuating potential $e\widetilde{\phi}$ giving

$$\widetilde{n}_e = \frac{Ne\widetilde{\phi}}{T_e}. \tag{12.15}$$

The ion kinetic density perturbation is

$$\widetilde{n}_i = \frac{-Ne\widetilde{\phi}}{T_i}[1 - P(k,\omega)] \tag{12.16}$$

where $\widetilde{\phi}$ is the fluctuation electrostatic potential ($\mathbf{E} = -\nabla\widetilde{\phi}$) and T_i is a characteristic ion energy for the distribution f_0. Thus, the quasi-neutrality condition yields the dispersion relation

$$D_{k\omega}^{\rm ES} = 1 + \frac{T_e}{T_i}[1 - P(k,\omega)] = 0 \tag{12.17}$$

for the ITG-ES waves that largely control turbulent ion thermal transport in toroidal magnetic confinement systems.

12.2.2 *Full 3 × 3 electromagnetic ion temperature gradient modes*

The electromagnetic wave fluctuations satisfy the matrix equation $\boldsymbol{A} \cdot \boldsymbol{X} = 0$ defined by

$$\begin{bmatrix} a & b & c \\ b & d & e \\ c & e & f \end{bmatrix} \begin{bmatrix} \phi \\ \psi \\ \dfrac{\delta B_\parallel}{B} \end{bmatrix} = 0 \tag{12.18}$$

where the six complex response functions are

$$\begin{aligned} a &= -1 + \frac{T_e}{T_i}(P-1) \\ b &= 1 - \frac{\omega_{*e}}{\omega} \\ c &= Q \end{aligned} \tag{12.19}$$

$$\begin{aligned} d &= \frac{k_\perp^2 \rho_s^2 \omega_A^2}{\omega^2} - \left(1 - \frac{\omega_{*e}}{\omega}\right) + \frac{\omega_{De}}{\omega}\left(1 - \frac{\omega_{*p_e}}{\omega}\right) \\ e &= -\left(1 - \frac{\omega_{*p_e}}{\omega}\right) \\ f &= \frac{2}{\beta_e} + \frac{T_i}{T_e}R. \end{aligned}$$

The polarization of the mode is given by the ratio of the components of the vector $X^T = (\phi, \psi, \delta B_\parallel/B)$.

The physics contained in Eq. (12.18) is the following: The first row of the matrix equation is the condition of quasi-neutrality $\delta\widetilde{n}_e = \delta\widetilde{n}_i$. The second row is from Ampére's law, $\nabla_\perp^2 A_\parallel = -\mu_0 \delta j_\parallel$, where $k_\perp^2 \widehat{\omega}_A^2 = -\partial_s v_A^2 k_\perp^2(s)\partial_s$ is the line-bending operator. The third row is from the "radial" ($\nabla\Psi$) component of Ampére's law.

Here the three ion kinetic response functions for ($|\omega| \gg k_\parallel v_i$) are given by P, Q, and R with analytic continuations from the upper-half of the complex ω plane

$$P = \left\langle \frac{\omega - \omega_{*i}(\epsilon)}{\omega - \omega_{Di}} J_0^2 \right\rangle, \tag{12.20}$$

$$Q = \left\langle \frac{\omega - \omega_{*i}(\epsilon)}{\omega - \omega_{Di}} \left(\frac{m_i}{b_i T_i}\right)^{1/2} v_\perp J_0 J_1 \right\rangle, \tag{12.21}$$

$$R = \left\langle \frac{\omega - \omega_{*i}(\epsilon)}{\omega - \omega_{Di}} \cdot \frac{m_i v_\perp^2}{b_i T_i} J_1^2 \right\rangle. \tag{12.22}$$

Here $\omega_{*i}(\epsilon) = \omega_{*i}[1 + \eta_i(\epsilon - 3/2)]$ with $\eta_i = d\ell n\, T_i/d\ell n\, n_i$, and likewise of ω_{*e} with $\eta_e = d\ell n\, T_e/d\ell n\, n_e$. The quantities $\omega_{*pj} = \omega_{*j}(1+\eta_j)$, where p stands for the pressure gradient and j denotes the species. Similon, *et al.* (1984) describe the calculation and properties of the analytic P, Q, R response functions.

12.2.3 *Limiting cases of the electromagnetic ITG dispersion relation*

Let us consider various limits of $D(\omega) = \det|A| = 0$ from Eq. (12.18). For a low-β plasma, $f \gg 1$ and the determinant D of Eq. (12.18) is

$$D = \left(ad - b^2\right) f \simeq 0. \tag{12.23}$$

For this system the MHD modes have $a \simeq -b$ and $d \cong -b$. Equation (12.23) for $f \neq 0$ gives the kinetically-modified MHD modes $\omega^2 - \omega\omega_{*pi} - k_\parallel^2 v_A^2 + \gamma_{\rm mhd}^2 = 0$ with small $E_\parallel$. There are electron drift modes with finite $E_\parallel$ given by

$$\omega^2\left(1 + k_\perp^2 \rho_s^2\right) - \omega\omega_{*e} - k_\parallel^2 v_A^2 k_\perp^2 \rho_s^2 = 0. \tag{12.24}$$

The $\delta B_\parallel \neq 0$ compressional Alfvén mode has $f(\omega) \simeq 0$ and is stable until the mirror mode instability condition $\beta(p_\perp/p_\parallel - 1) > C_m \approx 1$ is satisfied. The mirror mode instability is discussed in Chapter 5.2 and widely observed in space plasmas.

12.2.4 *Ion temperature gradient and trapped electron transport*

Now we consider the general case of the electron response $\widetilde{n}_e(k,\omega)$ to the potential fluctuations $\widetilde{\phi}_k(\omega)$. The full response is called the "non-adiabatic" electron response.

The stability and the transport produced by the ion temperature gradient (ITG) and the trapped electron modes (TEM) are now coupled. An analytical model [Zhu, *et al.* (1999, 2000)] as well as the numerical Weiland model [Guo and Weiland (1997, Jarmen, *et al.* (1987), Nordman, *et al.* (1990), Weiland (2000)] are used as tools to calculate the local stability of these complex drift modes. Other tools for the TEM-ITG analysis are an analytical model [Zhu, *et al.* (1999, 2000)] and the fluid-based Weiland model [Guo and Weiland (1997, Jarmen, *et al.* (1987), Nordman, *et al.* (1990), Hammett, *et al.* (1993), Weiland (2000)].

12.3 Analytical TEM-ITG Drift-Wave Model

Without the ion-wave resonances, the dispersion relation $D(\boldsymbol{k}, \omega, \boldsymbol{P})$ with temperature and density gradient driven drift-waves ($\omega \sim \omega_* = k_y T_e / eBL_n$) is given [Zhu, *et al.* (1999, 2000)] by

$$D(\boldsymbol{k}, \omega, \boldsymbol{P}) = 1 + \frac{T_e}{T_i} - \left(\frac{T_e}{T_i} + \frac{\omega_*}{\omega}\right)\Gamma_0 - \eta_i \frac{\omega_*}{\omega} b_i (\Gamma_1 - \Gamma_0)$$
$$- \int_0^\infty 2\pi v^2 dv F_e^M(v) \int_{-1}^{+1} d\mu h_e(v, \mu, \omega) = 0 \qquad (12.25)$$

where the perturbed electron phase-space distribution fraction is $\widetilde{f} = (e\phi/T_e)$ $F_e^M(1 - h_e)$ with $F_e^M(v)$ the Maxwell distribution for the electron velocity. Moreover, $\eta_{i,e} = d\log T_{i,e}/d\log n_{i,e}$ and the usual finite ion gyro-radius Bessel function $\Gamma_n(b_i) = e^{-b_i} I_n(b_i)$ where $b_i = k_y^2 \rho_i^2$. The non-adiabatic part of the electron distribution $h_e(v, \mu)$ satisfies the kinetic equation

$$\nu(v) \frac{\partial}{\partial \mu}\left(1 - \mu^2\right)\frac{\partial h_e}{\partial \mu} + i\left[\omega - \omega_{De}\frac{v^2}{v_e^2} - k_\parallel v \mu\right] h_e$$
$$= i\left[\omega - \omega_*\left[1 + \eta_e\left(\frac{v^2}{v_e^2} - \frac{3}{2}\right)\right]\right] \qquad (12.26)$$

where $\nu(v) = \nu_e(v_e/v)^3$. $\omega_{De} \simeq \omega_*(2L_n/R) = 2w_*\epsilon_n$, $\eta_e = d\log T_e/d\log n_e \neq 0$, $v_e = (2T_e/m_e)^{1/2}$, and $\mu = v_\parallel/v$.

The region of resonant wave-particle interactions is shown as the shaded elliptical region in Fig. 12.2 in frame (a). As the wave develops a positive parallel wavenumber $k_\parallel$ the resonance shifts to the left as shown in frame (b) and eventually becomes the usual Landau resonance as shown in frame (c) when the $k_\parallel\, v_\parallel \gg \omega_D$.

The model has the relevant electron temperature gradients and the density gradient as the primary driving forces for instability in this strongly heated plasma with no toroidal momentum input. The theoretical studies cited in Chapter 13 show that the ETG is the key instability for producing a strong electron thermal turbulent transport without large particle transport. The instability is insensitive to the details of magnetic shear when the q-profile is monotonically increasing and the shear parameter $s < 1$. The case of hollow or inverted q-profiles that occur in electron transport barrier are explained in Horton, *et al.* (2005) as discussed in Chapter 13.

The dispersion function $D(\boldsymbol{k}, \omega, \boldsymbol{P}_d)$ depends on the $d = 8$ dimensional system parameter vector $\boldsymbol{P}$, which in dimensionless form is

$$\boldsymbol{P}_q = \left(\eta_i, \eta_e, \epsilon_n = \frac{L_n}{R}, \epsilon = \frac{r}{R}, q, \widehat{s} = \frac{\mathrm{dln}\, q}{\mathrm{dln}\, r}, \nu_* = \frac{\nu_e}{\epsilon\omega_{be}}, \tau = \frac{T_e}{T_i}\right), \qquad (12.27)$$

where $\omega_{be}^T = v_e \epsilon^{1/2}/L_c$ is the effective bounce frequency, $\epsilon = r/R$, $L_c = qR$ is the connection length. The reciprocal of the collisionality ν_* is the number of electron bounces before an effective ($\Delta\theta = \epsilon^{1/2}$) detrapping collision. Thus, there is a large dimensionless set of the response functions, at least $2^8 = 256$ for high and low values of the eight parameters.

12.3.1 *Nonadiabatic electron response functions*

In the classic ITG instability the electron response is taken as adiabatic and produces the usual drift wave fluid dispersion relation in the ITG stable regime. In reality in the toroidal plasma, however, the drift wave in this regime becomes unstable if there is a resonance with the electron motion. The most relevant electron-wave resonance at these frequencies is the trapped electron mode. In this instability the wave frequency is resonant with the frequency of the bounce motion of the electrons that are trapped between the regions of $\max B$ on the in-board side, or at the smallest major radius $R_{\rm min}$ of the torus. For typical aspect ratio tokamaks there is of order 30-50% of the electron population executing these trapped particle orbits. These orbits are the ones that give the "banana" shape figure for the guiding center projected on to the poloidal cross-sectional plane. In the spherical tokamaks like NSTX and MAST the trapped electron traction can be the dominant set of electron orbits due the large mirror ration $R_M = B_{\rm max}/B_{\rm min}$ on the outer magnetic surfaces. The suprathermal and thus less collisional electrons, are the first to be trapped by R_M. As the temperature increases more and more of the electron population becomes collisionless and trapped. Thus the trapped electron instability has a strong dependence on the collisionality parameter $\nu_{\rm eff} = \nu_e/\omega_{be}$.

The perturbed, fluctuating, electron distribution functions δf_e is conveniently described with a decomposition in Legendre polynomials in the transport calculations of electrical conductivity from collisions. These basis functions diagonalize the second-order PDE of the dominant part of the collision operator.

To calculate the nonadiabatic part of the fluctuating electron distribution function we proceed as follows. We write

$$h_e(v, \mu, \boldsymbol{k}, \omega) = \Sigma_n h_{e,n}(v) P_n(\mu). \tag{12.28}$$

The zeroth order, $n = 0$, is isotropic with no the pitch-angle dependence. In general there are three velocity, or energy, ranges for the electrons for a given wave ω, k. We divide the electron energy ranges as follows depending on the collision frequency and the wave parameters:

(1) Low-electron energies: the parallel electron motion is collision-dominated for,

$$\frac{v}{v_e} \leq \left(\frac{\nu_e}{k_\parallel v_e}\right)^{1/4} \to h_{e_0}(v) = \frac{\omega - \omega_* \left[1 + \eta_e \left(v^2/v_e^2 - 3/2\right)\right]}{\omega + i\left(k_\parallel^2 v_e^2/6\nu_e\right)(v/v_e)^5}. \tag{12.29}$$

The factor of six multiplying the collision frequency ν_e in these response function arises from the eigenvalues $L(L+1)$ of the Lorentz pitch angle scattering operator in Eq. (12.26) where both the $L = 1$ and $L = 2$ eigenmodes arises in the fluctuating distribution function of the electrons.

(2) Intermediate electron energies: the electron Landau resonance is formed:

$$\left(\frac{\nu_e}{k_\| v_e}\right)^{1/4} < \frac{v}{v_e} \leq \left(\frac{\nu_e}{k_\| v_e \epsilon^{3/2}}\right)^{1/4}$$

$$h_{e_0}(v) = \frac{-i\pi}{2} \frac{\omega - \omega_* \left[1 + \eta_e\left(\frac{v^2}{v_e^2} - \frac{3}{2}\right)\right]}{|k_\||v_e\left(\frac{v}{v_e}\right)}. \tag{12.30}$$

(3) High-electron energies: the electrons divide into passing and trapped populations. The trapped electron response is bounce-averaged, such as $\overline{\phi} = (1/\tau)\oint ds\phi(s)/v_\|$.

$$\frac{v}{v_e} \geq \left(\frac{\nu_e}{k_\| v_e \epsilon^{3/2}}\right)^{1/4} \rightarrow h_{e_0}(v) = \left(\frac{\epsilon}{2}\right)^{1/2} \frac{\omega - \omega_*\left[1 + \eta_e(v^2/v_e^2 - 3/2)\right]}{\omega - \omega_{De} v^2/v_e^2 + i(\nu_e/\epsilon)(v_e/v)^3} \tag{12.31}$$

where the passing electron contribution averages to zero.

Substituting Eqs. (12.29)–(12.31) into (12.25) yields the dispersion relation

$$D_0(\boldsymbol{k}, \omega, \boldsymbol{P}) = 1 + \frac{T_e}{T_i} - \left(\frac{T_e}{T_i} + \frac{\omega_*}{\omega}\right)\Gamma_0 - \eta_i \frac{\omega_*}{\omega} b\left(\Gamma_1 - \Gamma_0\right) - \sum_m^3 \left(\Delta h_m^0 + \eta_e h_m^1\right) = 0 \tag{12.32}$$

where $\omega = \omega_0 + i\gamma$, $\Delta = 1 - \omega/\omega_*$ and the dimensionless energy integral Green's functions are

$$h_1^n = -\frac{2}{\pi^{1/2}} \int_0^{t_1} \frac{dt t^{1/2} e^{-t}(t - 3/2)^n}{1 - \Delta + i(t/t_0)^{5/2}} \tag{12.33}$$

$$h_2^n = i\pi^{1/2} \frac{\omega_*}{|k_\||v_e} \int_{t_1}^{t_2} dt e^{-t}\left(t - \frac{3}{2}\right)^n \tag{12.34}$$

$$h_3^n = -\left(\frac{\epsilon}{2}\right)^{1/2} \frac{2}{\pi^{1/2}} \int_{t_2}^{\infty} dt t^{1/2} \frac{e^{-t}(t - 3/2)^n}{1 - \Delta - \epsilon_n t + i(t_3/t)^{3/2}}, \tag{12.35}$$

with

$$t_0 = \left(\frac{6\nu_e\omega_*}{k_\|^2 v_e^2}\right)^{2/5}, \; t_1 = c_1\left(\frac{\nu_e}{k_\| v_e}\right)^{1/2}, \; t_2 = c_2\left(\frac{\nu_e}{k_\| v_e \epsilon^{3/2}}\right)^{1/2}, \; t_3 = \left(\frac{\nu_e}{\epsilon\omega_*}\right)^{2/3}. \tag{12.36}$$

h_1^n, h_2^n and h_3^n correspond to the collisional drift wave, the collisionless drifts wave and the trapped electron mode, respectively.

The electron energy dependence of the dissipative part of the fluctuating phase space density that gives the growth-damping contributions to the growth rate is shown in Fig. 12.5. The solid line gives the response when there is an electron temperature gradient of 2.3 times the density gradient and the dashed line when there is no electron temperature gradient. For the case with the temperature gradient ones sees the change in the direction of the energy flow from the electrons to the wave at an energy of $E/T_e \sim 2$ where the phase space gradient of the electron distribution

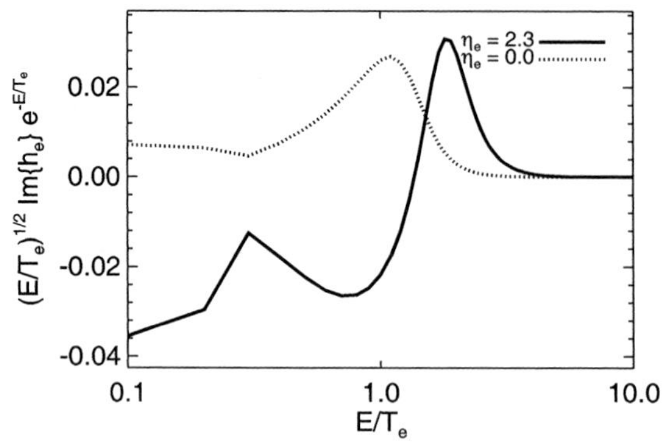

Fig. 12.5 Comparison of the nonadiabatic TEM electron response with and without the electron temperature gradient. The plasma parameters are from a TCV discharge discussed in detail in Chapter 13. The case $\eta_e = 2.3$ (solid) has both inward at low-electron energy and outward particle transport at high-electron energies. Whereas the reference case with $\eta_e = 0$ (dotted) has only outward particle transport. The discontinuous slope at $E/T_e \sim 0.5$ arises from the transitions between collisionless drift wave and trapped electron responses.

has a steeper unstable gradient owing to the temperature gradient. At subthermal energies the phase space density varies as $n_e/T_e^{3/2}$ and thus is inverted and stable for the $\eta_e = 2.3$ regime.

With the assumption of $\gamma/\omega_0 \ll 1$ and $h_e \ll 1$, the mode frequency ω_0 and growth rate γ are

$$\omega_0(k, \boldsymbol{P}) = \omega_* \frac{[\Gamma_0 + \eta_i b_i(\Gamma_1 - \Gamma_0)]}{1 + (T_e/T_i)(1 - \Gamma_0)} \tag{12.37}$$

with the parameter vector $\boldsymbol{P}$ in Fig. 12.27 and

$$\frac{\gamma\omega_*}{\omega_0^2}[\Gamma_0 + \eta_i b_i(\Gamma_1 - \Gamma_0)] = \Delta \sum_{m=1}^{3} \text{Im } h_m^0 + \eta_e \sum_{m=1}^{3} \text{Im } h_m^1 \tag{12.38}$$

where

$$\Delta = 1 - \frac{\omega_0}{\omega_*}. \tag{12.39}$$

Im h_m^n for even n are positive definite and determine the electron particle diffusivity D and the heat diffusivity χ_e. For odd n, Im h_m^n are indefinite in sign. Also, the contribution of Δh_m^0 and $\eta_e h_m^1$ to the growth rates determines which is the more critical parameter between the density gradient and the temperature gradient.

In Chapter 13 we see that the electron temperature gradient also drives drift wave unstable at smaller space-time scales. These higher k, ω turbulence is called the ETG mode and as with the ITG mode occurs both in cylindrical and toroidal plasmas under conditions with $\eta_e > \eta_{\text{crit}}$. The frequency and growth rate now scale with v_e/L_{T_e} where $v_e = \sqrt{(T_e/m_e)}$ and the electron gyroradius ρ_e or $\rho_{e,i}$ where the later gyroradius is the analog of the ρ_s of the ITG modes that has the mixed measure of electron temperature and the ion mass. Here the $\rho_{e,i} = (m_e T_i)^{1/2}/eB$

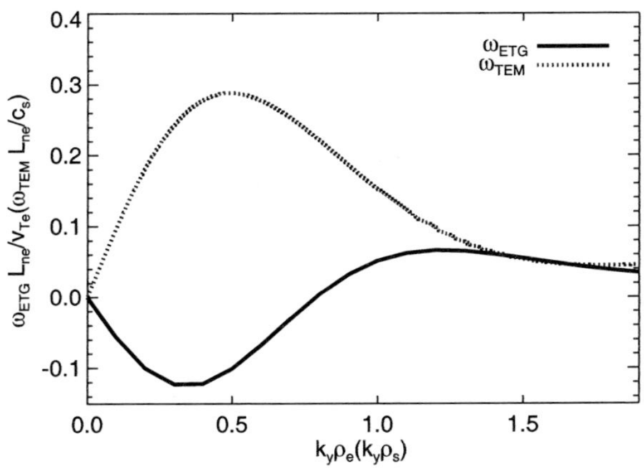

Fig. 12.6 The ETG (solid) and TEM (dotted) mode frequencies with increasing $k_y\rho_s(k_y\rho_e)$ at $r/a = 0.7$ and $t = 1.0$ s for TCV discharge 29892 in Table 13.1. The frequencies are respectively normalized by v_{T_e}/L_n and c_s/L_n. The high-frequency-small scale ETG mode, analyzed in Chapter 13, has many features similar to the TEM-ITG as shown here.

so that there is an interchange of the roles of electrons and ions in-going from the ITG to the ETG instability. A important difference is that owing to the small scales of the ETG the quasineutral approximation that is well satisfied for the ITG drift waves is marginal and often not satisfied for the ETG modes. The condition for the ETG to be quasineutral is that $k_\perp \lambda_{De} \ll 1$ which requires that $\omega_{pe} \gg \omega_{ce}$.

For comparison of the two temperature gradient driven instabilities we show in Fig. 12.6 the results of the ETG modes derived in the next chapter plotted along with the TEM results for a TCV discharge driven strongly with third harmonic electron cyclotron heating in the core plasma.

Carefully note the change in the scaling of the frequencies and the wavenumbers in Fig. 12.6 and reversal of the direction of rotation of the waves. The longer wave length modes of the TEM may cause a larger transport when present with sufficient amplitude however, the smaller scale ETG modes grow much faster and so the TEM modes live in a bath of the ETG modes. The ETG modes have an inverse cascade of their energy to wave lengths comparable to the ion gyroradius where they can feed the lower-frequency drift waves. Both forms of turbulence exist in plasmas with steep electron temperature gradients; however, the smaller scale modes are not sensitive to the details of the geometry, such as the fraction of trapped electrons, and are well above the usual electron collision frequencies. Thus, the universal nature of the anomalous electron transport across all magnetic confinement geometries is likely due to the small scale ETG modes. When the trapped electron mode at low collisionality with significant trapped electrons is present then the associated transport may mask that due to the ETG turbulence.

The role of zonal flows in controlling the level of transport from the trapped electron mode is explored in Ernst, *et al.* (2009). The work shows that the zonal flows can reduce the level of the electron thermal transport and the fluctuation level in general for the both the trapped electron modes and the modes driven by the ion temperature gradients. The method of investigation uses and compares the results of the particle-in-cell gyrokinetic codes and the continuum codes.

More generally, the works [Nevins, *et al.* (2005, 2006)] characterize the role of the interaction of the microturbulence and the sheared flows.

12.4 Internal Transport Barriers for ITG/TEM Models

The turbulent transport from the ion temperature gradient driven drift waves and the trapped electron modes transport the ion momentum flow in the plasma fluid in addition to the transport of thermal energy and particles. The first theories and computations of the flux of momentum from the turbulent $\mathbf{E} \times \mathbf{B}$ drift wave fluctuations were for cylindrical plasmas [Connor and Wilson (2000)]. In the neutral fluid literature the calculations of the momentum flux from turbulence is an long standing research area and the turbulent flux is called the Reynolds stress tensor given by $\rho_m \langle v_i v_j \rangle$ where ρ_m is the mass density and the components of the fluctuating velocity are v_i. The early calculations of Hamaguchi and Horton (1992) for the ITG turbulence included the calculation of the momentum stress tensor and showed that this flux can generate a mean sheared background flow.

A clear indication of the formation of a diffusion barrier is shown by integrating ensembles of test particles in a specified spectrum of drift waves existing on top of the background plasma with flow shear or reversed magnetic field. This method gives a direct picture of how the transport is reduced by the sheared fields. The reversed magnetic field breaks the Chirikov overlap condition so that the stochastic diffusion across the minimum $q_{\min}$ surface is not possible. There are trajectories that cross the $q_{\min}$ surface but these are better described by long-flight orbits than diffusion. Some works giving the details of the orbits in realistic fluctuation fields are Horton, *et al.* (1998), Evstatiev, *et al.* (2003), and Batista, *et al.* (2006).

Perpendicular shear flows are responsible for both symmetry breaking and suppression of turbulence, resulting in a shearing rate at which there is a maximum contribution to the momentum transport. The $\mathbf{E} \times \mathbf{B}$ momentum transport is shown to be quenched by increasing flow shear more strongly than the standard linear quench rule for turbulent heat diffusivity.

The analogous fluid turbulence calculations and experiment are found in Howard and Krishnamurti (1986). In their Bernard convection experiment in water, the low-thermal gradient driving the Bernard convection cells is analogous to the drift waves in the plasma interacting with one another to create a mean circulation in the box. A simple mathematical model with a few key k_x, k_y modes was derived by Howard and used to explain the bifurcation with increasing temperature gradient to a state of turbulence with mean sheared flows.

Here show the plasma physics analog of this fluid experiment in which an increasing ion temperature gradient drives turbulence which in turn drives a mean ion $\mathbf{E}\times\mathbf{B}$ flow that reacts back on the turbulence. Since the reaction of the sheared flow on the ITG turbulence is stabilizing this situation has been characterized as a "predator-prey" situation. Here the sheared flow feeds on the turbulence and at some point when sufficiently strong sheared flows are present the ITG turbulence as the prey is strongly reduced. This process can go through cycles as often seen in the simulations and in the dithering of plasmas between the two states of low-mode L and high-mode H of confinement [Diamond and Kim (1991)].

To understand and make a mathematical model of this problem we consider the dynamics of the ITG driven drift wave turbulent energy $W(t)$ in the toroidal magnetic geometry. The standard representation for the toroidal axisymmetric magnetic field in the torus is

$$\mathbf{B} = I(\psi)\nabla\phi + \nabla\phi \times \nabla\psi \tag{12.40}$$

where the toroidal angle is ϕ and the poloidal magnetic flux function is ψ and the toroidal field is $I(\psi)/R$ where $\nabla\phi$ is the unit vector in the toroidal direction divided by R the distance from the axis of symmetry to the spatial point x, y, z under consideration for the $\boldsymbol{x}$ field.

From the analysis of the ITG turbulence in this chapter we can write that the thermal ion flux q_i and the turbulent wave energy density $W(r,t)$ are given by

$$q_i = -n_i\chi_i(W)\left(\frac{dT_i}{dr}\right) \tag{12.41}$$

$$\gamma(W) = \gamma^{\ell} - \gamma^{n\ell}W \tag{12.42}$$

$$W = \sum_k \omega_k(\partial\epsilon_k/\partial\omega_k)\frac{|E_k|^2}{8\pi n_e T_e} \simeq \frac{1}{2}\sum_k \frac{e^2|\phi_k|^2}{T_e^2} \tag{12.43}$$

$$\frac{dW}{dt} = \left[2\gamma_0\left(\frac{R}{L_{T_i}} - \left(\frac{R}{L_{T_i}}\right)_{\rm crit}\right) - 2\gamma^{n\ell}W\right]W \tag{12.44}$$

$$\frac{3}{2}\frac{\partial}{\partial t}(n_iT_i) = \frac{1}{r}\frac{\partial}{\partial r}r\left(n_i\chi_0 W\frac{\partial T_i}{\partial r}\right) + P_E(r,t) \tag{12.45}$$

$$P_E' = \frac{\partial}{\partial r}\left(\frac{2P_E(r)}{3n_i(r)}\right). \tag{12.46}$$

The radial gradient of the power deposited per ion units of KeV/m per ion per second is given in Eq. (12.46).

The ion temperature profile is determined by the combination of the injected heating power $P_E(r,t)$, turbulent profile, and the turbulent flux $q_i(W, dT_i/dr)$ defined in Eqs. (12.41)–(12.45). Clearly the profile of the injected heating power $P_E(r,t)$ is important in determining the temperature gradient. In this simple model the spatial dependence of the heating power is reduced to the gradient defined in Eq. (12.46).

The transport problem is now reduced to a system of spatially local ODEs for the turbulence level dW/dt and the temperature gradient dT_i/dr. To this we add the equation for the momentum transport. Already at this simple stage there is interesting dynamics of the evolution of the turbulence and temperature gradient. There is a stable fixed point the system evolves to given by the formulas (12.47)–(12.48). The energy confinement time is then determined by the formula in Eq. (12.49).

$$W^*(P'_E) = \left(\frac{\gamma^0}{2\gamma^{n\ell}}\right)\left[\sqrt{\left(\frac{T'_i}{T_i}\right)_c^2 + \frac{6\Delta^2\gamma^{n\ell}P'_E}{\gamma^0\chi_0 T_i}} - \left(\frac{T'_i}{T_i}\right)_c\right] \tag{12.47}$$

$$\frac{T'_i(P'_E)}{T_i} = \frac{1}{2}\left[\sqrt{\left(\frac{T'_i}{T_i}\right)_c^2 + \frac{6\Delta^2\gamma^{n\ell}P'_E}{\gamma^0\chi_0 T_i}} - \left(\frac{T'_i}{T_i}\right)_c\right]. \tag{12.48}$$

The energy confinement time varies as

$$\tau_E = a\left(\frac{\gamma^{n\ell}}{\chi_0\gamma^0}\right)^{1/2}\left(W_*^{-1/2}\right). \tag{12.49}$$

The flow velocity is

$$\boldsymbol{u} = \frac{\mathbf{B}}{B_0}\left(u_\parallel + \frac{B_T}{B_p}u_\perp\right) - \frac{u_\perp B}{R_0 B_p}R^2\nabla\phi.$$

We break the flow velocity into poloidal and toroidal components given by

$$u_p = \left(\frac{B_t}{B}\right)u_\perp - \left(\frac{B_p}{B}\right)u_\parallel \tag{12.50}$$

$$u_t = \left(\frac{B_t}{B}\right)u_\parallel + \left(\frac{B_p}{B}\right)u_\perp. \tag{12.51}$$

The critical value of the flow shear divided by magnetic shear is

$$\left(\frac{L_s u'_\perp}{c_s}\right)_c = \gamma_{\text{crit}} = 2\sqrt{\frac{T_i}{T_e}}(\eta_i - \eta_{ic}). \tag{12.52}$$

For the JT-60U data we find the balance in Eq. (12.52) is

$$\left(\frac{L_s u'_\perp}{c_s}\right)_c = 2\sqrt{\frac{4\,\text{KeV}}{2\,\text{KeV}}}(2-1) = 2.8 \approx 3. \tag{12.53}$$

The local gradients of the momentum injection profiles are

$$P'_t = \frac{\partial}{\partial r}\left(\frac{P_t}{m_i n_i}\right) \quad \text{and} \quad P'_p = \frac{\partial}{\partial r}\left(\frac{P_p}{m_i n_i}\right) \tag{12.54}$$

and strongly influence the amount of flow shear. The nonlinear dynamics of the heating and flow shear are given by the four coupled ODEs:

$$\frac{dW}{dt} = 2\gamma_0\left[(\mu - \mu_c) + c_1 F_\parallel - c_2 F_\perp - \gamma^{n\ell}W\right]W \tag{12.55}$$

$$\frac{d\mu}{dt} = -\chi_x W_\mu + P'_E \tag{12.56}$$

$$\frac{dF_\perp}{dt} = -\nu^{\text{nc}}F_\perp + \bar{c}_2 W F_\perp + \sqrt{F_\perp}\,P'_\perp \tag{12.57}$$

$$\frac{dF_\parallel}{dt} = -\bar{c}_1 W F_\parallel + \sqrt{F_\parallel}\,P'_\parallel. \tag{12.58}$$

A typical model for the NBI power deposition is given by

$$P_{bi}(r) = \frac{P_T n\sigma}{(2\pi r)(2\pi r)\sin\theta}\left[e^{-(a-r)n\sigma/\sin\theta} + e^{-(r+a)n\sigma/\sin\theta}\right] \tag{12.59}$$

where θ is the angle of injection with tangential injection having $\theta_{\min} \simeq (a/R)^{1/2}$. The condition of depositing the power in the core plasma puts a narrow range on $a/\lambda_{\text{eff}} = an\sigma(E_0)/\sin\theta$ so that we can estimate $a/\lambda_{\text{eff}} \approx 2$ in calculations with Eq. (12.58). Using this estimate for $\lambda_{\text{eff}} \simeq a/2$ yields the coefficient in Eq. (12.59):

$$\frac{P_{\text{tot}}}{(2\pi R)(\pi r a)} \simeq \frac{20\times 10^6\,\text{Watts}}{2\times 10^7\,\text{cm}^3} \simeq 1\,\text{MW/m}^3.$$

The ion heating rate at $n = 3\times 10^{13}\,\text{cm}^{-3}$ is the

$$\frac{2P_{bi}}{3n_i} = 150\,\text{KeV/s} \tag{12.60}$$

For the TCV 29892 discharge we calculated the drift-waves and TEM modes at the radial position $r/a = 0.7$ at $t = 1.0\,\text{s}$, $(T_e, T_i) = (1.30, 0.76)\,\text{KeV}$, $n_e = 5.74\cdot 10^{19}\,\text{m}^{-3}$, $B_T = 1.09\,\text{T}$, $\eta_e = 2.3$ and $\widehat{s} = 0.65$. The relative plasma pressure is $\beta_e = 2\mu_0 p_e/B_T^2 = 0.025$ so that the $\omega \ll k_\parallel v_{\text{A}}$ and the modes are approximately electrostatic. Figure 12.7 shows the comparison of the growth rates between the

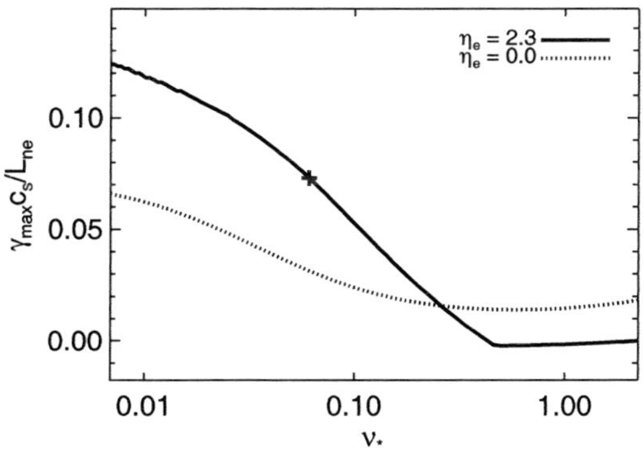

Fig. 12.7 The comparison of the maximum TEM growth rate versus the collisionality ν_* between the $\eta_e = 2.3$ (solid data) and $\eta_e = 0$ (dotted). The variation of collisionality ν_* is obtained by varying $T_e = 200 \sim 3000\,\text{eV}$ and with the fixed temperature ratio T_e/T_i.

case $\eta_e = 2.3$ (solid) and the reference case $\eta_e = 0$ (dotted) where in the absence of an electron temperature gradient where the density gradient drives turbulence. The comparison shows that the electron temperature gradient η_e leads to the distinct transition around $\nu_* = 0.45$ between a low growth rate collisional drift wave mode and the high growth rate trapped electron mode. At ν_* lower than 0.45, where the data point (+), $\nu_* = 0.06$, corresponds, the electron temperature gradient destabilizes the trapped electron mode by a factor 2 over the density driven TEM mode.

As far as the magnetic shear is concerned, the shear dynamics principally plays a role in determining the parallel wavelength $k_{\parallel} L_n = k_y \rho_s \hat{s}^{1/2}$. Therefore collisional and collisionless drift waves are affected by the magnetic shear through collisional dissipation and Landau damping. So, the magnetic shear dynamics has little influence over the trapped electron mode. The electron response functions for $\eta_e = 2.3$ and 0 with $\nu_* = 0.06$ of the experiment are compared in Figs. 12.8 and 12.9. The wave/guiding-center drift resonances occur at $E/T_e \sim \omega_0/\omega_{De}$. With the temperature driven case $\eta_e = 2.3$, the resonance shifts from $E/T_e = 1.1$ to 1.9 and it is clearly seen that the temperature gradient term $\eta_e(E/T_e - 3/2)$ stabilizes the lower-energy contribution ($E/T_e < 3/2$) to the TEM modes and destabilizes the higher-energy contribution ($E/T_e > 3/2$) to the TEM modes.

For all time slices and for all the radial profiles, the mode frequency, growth rate and electron response functions are calculated. The mode frequencies and the maximum growth rates for each radial point are shown at each time slice in Fig. 12.8.

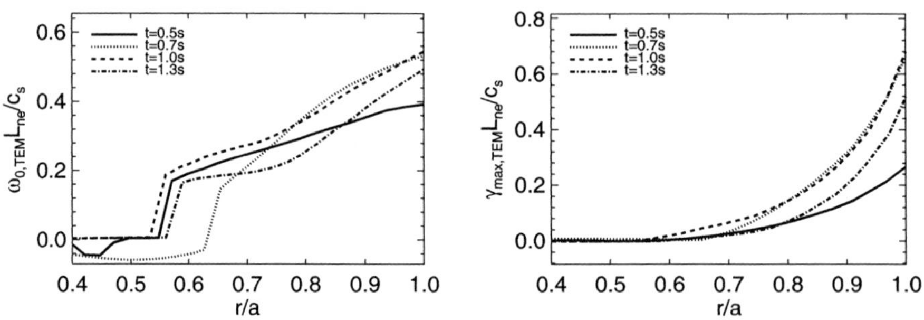

(a) TEM frequencies (b) Maximum TEM growth rates

Fig. 12.8 The TEM mode frequencies and the maximum growth rates of TCV 29892 at $t = 0.5$ (solid), 0.7 (dotted), 1.0 (dashed) and 1.3 s (dashed-dot). The maximum growth rate in panel (b) is obtained among the different k_y values.

At $r/a = 0.45 - 0.55$, there is a transition between drift waves and trapped electron modes. The drift waves propagate in the ion drift wave direction with low frequencies $\omega_0 \ll 0.1c_s/L_{ne}$ contrary to the TEM modes with higher frequencies $\omega_0 \sim 0.1c_s/L_{ne}$ at $r/a \gtrsim 0.5$, which propagate in the electron drift wave direction.

For each time slice, the drift wave modes are destabilized. Beyond the transition radial position $r/a = 0.5$, the trapped electron mode is η_e-dominated and the temperature gradient stabilizes the drift wave mode. Towards the edge, the density and temperature gradients contribute to the instability with comparable magnitudes.

The imaginary parts of the electron response function $h_e(E)$ are shown in Fig. 12.9. A positive value of the product of the response function and the sign

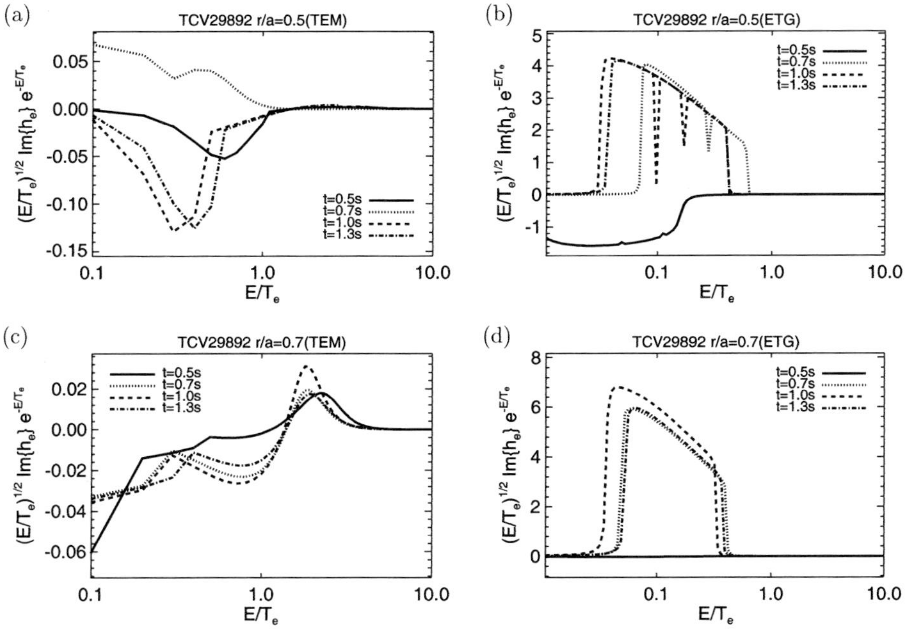

Fig. 12.9 The imaginary part of electron response functions at $r/a = 0.5$ (top panel) and $r/a = 0.7$ (bottom panel). The left and right panels are for TEM, Eqs. (12.29)–(12.30), and ETG, Eq. (12.31) respectively.

of the mode frequency ω_0 implies the destabilizing contribution of the trapped electrons. The sign of the functions in Fig. 12.9 indicates destabilizing contribution by positive values of the response function.

At $r/a = 0.5$ as in Fig. 12.9(a)–(b), for $t = 0.5$, 1.0, 1.3 s, the collisionless and collisional drift waves are stabilized by the electron temperature gradient, while for $t = 0.7$ s, the TEM/ITG mode propagates in the ion diamagnetic drift direction, $\omega_0 < 0$, due to the large ion temperature gradient $\eta_i = 3$ and the electron temperature gradient drives the TEM instability for the lower energy $E/T_e < 3/2$ and drift wave instability. The ETG modes have the broad energy range of the resonances due to electron Landau damping $\omega_0 - \omega_{De}(E/T_e) - k_\parallel v_\parallel = 0$ and strong resonances due to the lower collisional frequencies $\nu_e L_{n_e}/v_{T_e} \ll \nu_e L_{n_e}/c_s$ compared to the TEM modes.

At $r/a = 0.7$ as in Fig. 12.9(c)–(d), for all the times, the electron temperature drives the TEM/ITG instability at the higher energy $E/T_e > 3/2$ and stabilizes at the lower energy. The ETG modes in panel Fig. 12.9(d) are destabilized by the temperature gradient at the lower electron energy. The high-collisional frequency at $t = 0.5$ s weakens the resonance and the ETG instability. The stronger electron resonances at later times are consistent with the power balance transport [Erba, *et al.* (1999)].

12.5 The Weiland ITG/TEM Transport Model

The Weiland model [Jarmen, *et al.* (1987), Nordman, *et al.* (1990), Nordman, *et al.* (2005), Guo and Weiland (1997), Weiland (2000), Weiland and Holod (2005a), Weiland and Holod (2005b)] is often used to predict the radial global transport. In this section the model is used as a local stability tool to analyze the instabilities present in the plasma and their thresholds, and to calculate the local turbulent transport. The version of the Weiland model used here includes electromagnetic effects, collisions, impurity and magnetic shear effects but neglects varying correlation length [Weiland, *et al.* (2005a,b)] and toroidal and poloidal rotation effects [Weiland and Nordman (2006), Eriksson, *et al.* (2007)]. The two latter effects are included in a more refined version of the model.

In principle, the TEM ITG modes mainly drives electron and ion thermal transport. With a gyro-Bohm type mixing-length formula [Weiland and Nordman (1988)], the thermal diffusivities are

$$\chi^{\mathrm{W}}_{e,i} = \frac{R/L_{Te,i} - R/L_T^{\mathrm{crit}}(\widehat{w}_{\mathrm{re},i}, L_n)}{R/L_{Te,i}} \times \left\{ f_{e,i} \frac{m_i^{1/2}}{k_\theta \rho_s e^2 B^2 L_{ne}} T_e^{3/2}(eV) \sum_{g>0} \frac{\widehat{g}^3}{(\widehat{w}_{\mathrm{r}} \mp F_{e,i})^2 + \widehat{g}^2} \right\} \tag{12.61}$$

where

$$F_{e,i} = \frac{10}{3} \frac{L_{ne}}{R} \frac{T_{e,i}}{T_e} \tag{12.62}$$

and e and i denote electron and ion, respectively. Here $f_e = f_t$, is the trapped electron fraction and $f_i = 1$. The mode frequency, $\widehat{w}_{\mathrm{r}}$, and growth rate, $\widehat{g}$, are normalized to the diamagnetic drift frequency of the electrons ($w_* = K_\theta \rho_s c_s / L_{ne}$). The model assumes that the radial correlation length is of the same order as the poloidal one, i.e. $k_r^2 \approx k_\theta^2$. A heat pinch arises in Eq. (12.61) from the threshold function $p(\widehat{w}_{\mathrm{re},i}, L_n) = R/L_T^{\mathrm{crit}}$. The part of Eq. (12.61) within the curly brackets gives the mixing length diffusivity.

In Eq. (12.61), the sum over g is over all unstable modes. Hence, TEM modes can drive ion transport and ITG modes the electron transport depending on the relative sizes of the resonance factor, $(\widehat{w}_{\mathrm{r}} \mp F_{e,i})$, and the growth rate. In the case of ion transport, the resonance factor will decrease with F_i if both R/L_{ne} and T_e/T_i are much larger than unity. In such a case, a significant TEM drive of χ_i can be expected [Asp, *et al.* (2007)].

Worked Problem 13.5.1

Find the 2×2 matrix eigenvalue problem for polarization and stability of the compressional drift waves.

The perturbative motions induced by ϕ and ψ produce a compressional change in the magnetic field $\delta B_\parallel$ given by

$$\frac{\delta B_\parallel}{B} = -\frac{1}{f}(c\phi + e\psi) \tag{12.63}$$

where properly $1/f$ is the inverse of a complicated integral operator f involving bounce averages over the particle trajectories. Here f is taken as a local quantity. The compressional change in the magnetic field $\delta B_\parallel$ is dictated by Ampere's law from the third row of the matrix equation Eq. (12.18) with the plasma currents flowing across the magnetic field lines in the $\widehat{x} \propto \nabla\Psi$ direction.

Substituting Eq. (12.63) into the first and second rows of the full matrix equation Eq. (12.18) gives the reduced 2×2 symmetric matrix

$$\begin{bmatrix} a - \dfrac{c^2}{f} & b - \dfrac{ce}{f} \\ b - \dfrac{ce}{f} & d - \dfrac{e^2}{f} \end{bmatrix} \begin{bmatrix} \phi \\ \psi \end{bmatrix} = 0. \tag{12.64}$$

The dispersion relation given by the determinant is

$$D_k(\omega, \mu) = ad - b^2 - \frac{1}{f}\left(c^2 d - 2bce + e^2 a\right) = 0. \tag{12.65}$$

The compressional MHD limit is $a \simeq -b \simeq d$, which reduces Eq. (12.65) to

$$ad - b^2 - \frac{d}{f}(c+e)^2 = 0. \tag{12.66}$$

To find analytically the kinetic ballooning interchange drift mode and connect that mode with the variational formulas, we take $\eta_e = 0$ and note $a \simeq -b \simeq d \simeq e$ for the dominant terms owing to the small $E_\parallel$. Then the full determinant D factors as

$$ad - b^2 + \frac{\tau(1-\omega_*/\omega)[Q-(1-\omega_*/\omega)]^2}{\frac{2}{\beta_e} + \frac{T_i}{T_e}R} = 0 \tag{12.67}$$

as shown by Eq. (12.66). Here $\omega_{*e} = \omega_*$ for $\eta_e = 0$. The last term in Eq. (12.73) gives the kinetic compressional response. For the near-to-MHD regime, the response function ϕ in Eq. (12.13) reduces to

$$Q - 1 + \frac{\omega_*}{\omega} \simeq -\frac{\omega_{*pi} - \omega_*}{\omega} + i\Delta_Q = -\frac{\omega_{*p}}{\omega} + i\Delta_Q, \tag{12.68}$$

with ω_{*p} having the total pressure gradient and a wave-particle resonant part $i\Delta_Q$ from wave-particle resonance. A reasonable approximation for the compressional response function R in the region $\omega\omega_{Di} > 0$ is

$$R \cong c_0 \frac{[\omega - \omega_{*i}(1+2\eta_i)]}{\omega - \omega_{Di} + ic_1|\omega_{Di}|} \tag{12.69}$$

where c_0 and c_1 are positive fitting coefficients of order $c_0 \simeq 1$ and $c_1 \simeq 0.1$. Here $\Delta_Q = \operatorname{Im} Q(\omega)$ and $\Delta_R = \operatorname{Im} R(\omega)$ are real numbers giving the resonant ion contributions.

The resonant modes in the high-β region have

$$\omega = \omega_0 + i\gamma_k \simeq \omega_{*i}(1+\eta_i) + i\gamma_k \tag{12.70}$$

which is an ion diamagnetic drift wave. Here ω_0 and γ_k are real valued functions of k, P. A Taylor series expansion of the dominant terms in Eq. (12.67) gives

$$\begin{aligned} ad - b^2 &\simeq \left(\frac{\omega_*}{\omega} - 1\right)\left[\frac{\omega_A^2}{\omega_0^2} b - b\left(1 - \frac{\omega_{*pi}}{\omega_0}\right) - \frac{\omega_*\omega_D}{\omega_0^2}\right] \\ &+ \left(\frac{\omega_*}{\omega_0} - 1\right)\frac{i\gamma_k}{\omega_0}\left[\frac{-2\omega_A^2}{\omega_0^2} b - \frac{b\omega_{*pi}}{\omega_0} + \frac{2\omega_*\omega_D}{\omega_0^2}\right] \end{aligned} \tag{12.71}$$

for $\omega = \omega_0 + i\gamma_k$ with $|\gamma_k| \ll \omega_0$. Thus the growth rate γ_k is determined by

$$\frac{i\gamma_k}{\omega_0}\left[\frac{2\omega_A^2}{\omega_0^2} + \frac{\omega_{*pi}}{\omega_0}\right] b + \frac{T_e}{T_i}\frac{(\omega_{*p}/\omega_0 - i\Delta_Q)^2}{c_0[1 - \frac{\omega_{*i}(1+2\eta_i)}{\omega_0} - i\Delta_R]} = 0 \tag{12.72}$$

for $\Delta_Q \sim \Delta_R \ll 1$. Since $\omega_{*p}/\omega_0 \gtrsim 1$, the significant resonant contribution comes from $i\Delta_R$ in the denominator. Thus we obtain the growth rate formula

$$\frac{\gamma_k}{\omega_0}\left(\frac{2\omega_A^2}{\omega_{*p}^2} + 1\right) b = -\frac{T_e}{T_i}\frac{(\omega_*/\omega_0)^2 \Delta_R}{c_0\left[\left(\frac{\eta_i}{1+\eta_i}\right)^2 + \Delta_R^2\right]} \tag{12.73}$$

with $\gamma_k > 0$ for $\Delta_R \equiv \operatorname{Im}[R(\omega_0)] < 0$.

The results of applying the Weiland model to the TCV discharge 29892 are shown in Fig. 12.10. The trend of the diffusivities at $r/a = 0.7$ agrees with the trend of the experimental heat fluxes shown in Fig. 12.9 up to $t = 1.0$ s. At $t = 1.3$ s the MHD activity reduces the gradients in the outer part of the plasma, which results in a marked drop in the diffusivities at $r/a = 0.7$. This drop was also seen in the nonlinear ETG simulations for the same reason.

The diffusivities in Fig. 12.10 for $r/a = 0.5$ follows the same trend as for $r/a = 0.7$ with the exceptions that between $t = 1.0$ s and 1.3 s the diffusion only decreases by 20% and that the diffusivities at $t = 1.0$ s increases in the core..

During the ohmic phase at $t = 0.5$ s, the transport is much lower, as the ohmic heating of 0.49 MW is much smaller than the total heating with ECRH, $P_{\text{tot}} \sim 1.2$ MW. This ohmic phase is also the only phase in which the ITG instability is about twice as big as the TEM. As a consequence, the ITG modes drive most of the ion transport and the TEM most of the electron. In the following phases the TEM dominates both the ion and electron transport.

The TEM electron heat diffusivities, however, are higher than those from the power-balance calculations. In the next chapter on the ETG turbulence we show that the ETG turbulence gives closer agreement with the power-balance diffusivity.

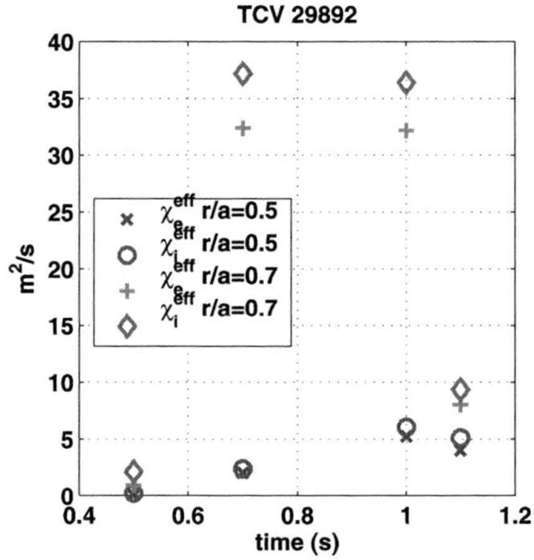

Fig. 12.10 Effective electron and ion diffusivities calculated by the Weiland model at $r/a = 0.5$ and 0.7 [Asp, *et al.* (2008)].

In Chapter 13 we introduce a metric called the Average Relative Variance to measure the degree to which a model and data agree.

An error analysis shows that the diffusivities can be reduced by, on average, $\sim 75\%$ for the ions and $\sim 65\%$ for the electrons by decreasing all gradients (R/L_{ne}, R/L_{Te} and R/L_{Ti}) by 30%, which is within the error bars of the ion gradients. This would give diffusivities of similar magnitude to the experiment in inner part of the plasma, whereas further out they would still be overestimated. The local version of the Weiland model hence gives an inadequate quantitative picture of the heat transport. However, the TEM dominance remains even when decreasing the gradients and hence the qualitative analysis remains valid. The local version can therefore give a good idea of which modes are present and how the transport behaves qualitatively.

When the magnetic shear is weak the hydrodynamics electron response is stronger and the simulations show a "bursty" time behavior of the heat flux. In this regime there are long periods with low levels of transport when the simulation is dominated by the nonlinearly excited velocity shear. These periods end when the shear flow drops sufficiently with a new bursts of high-level transport characterized by large numbers of modes interacting. In sheared-slab geometry, the only physics limiting the ($k_y = 0, k_z = 0$) mode is collisionality. However, in the toroidal plasma with a high fraction of trapped electrons the limit is more complex due to the trapped electrons which are hydrodynamic in their response interacting with the passing electrons that are near-to-adiabatic in their response.

Dorland and Hammett (1993) show a way to implement periodic radial boundary conditions in a sheared magnetic field to overcome the difficulty of edge effects where there is an absence of rational surfaces with Dirichlet boundary conditions. Periodic boundary conditions allow the heat flux leaving the box on the cold side to re-enter the simulation domain on the warm side. This recycling represents adequately the steady injection from heat sources on the high-temperature side of the box and a sink on the low-temperature side. The benefit of this method over simply zeroing out the ($k_y = 0, k_z = 0$) modes is that these modes are important in the nonlinear dynamics with respect to the generation and decay of the sheared flows and this mechanics is lost if these modes are completely removed from the simulation.

12.6 Zonal flows and ion temperature gradient instabilities in multiple-helicity magnetic fields

The effects of multiple-helicity magnetic fields on the ion temperature gradient (ITG) instability and on the zonal flow (ZF) evolution are studied with the linear gyrokinetic Vlasov code (GKV) [Watanabe and Sugama (2006)]. The model helical fields corresponding to the standard and inward-shifted axis configurations of the Large Helical Device [Motojima, *et al.* (2003)] are used to investigate how ITG mode properties and ZF evolution response to a given source are influenced by the field geometry. It is shown that, in the inward-shifted configuration, the ITG mode growth rate increases slightly while the ZF is sustained for a longer time. In addition, velocity-space structures of the ion perturbed distribution function are numerically obtained which illustrate the validity of the analytical prediction that the plasma inward shift retards the radial drift of the helically trapped particles leading to the enhancement of the ZF response. This supports the conjecture that anomalous transport can be reduced by the ZF generated in the configurations optimized to decrease the neoclassical transport.

Quantitative validation studies of flux-tube gyrokinetic Vlasov simulations on ion and electron heat transport are carried out for the JT-60U tokamak experiment. The ion temperature gradient (ITG) and/or trapped electron modes (TEM) driven turbulent transport and zonal-flow generations are investigated for an L-mode plasma in the local turbulence limit with a sufficiently small normalized ion thermal gyroradius and weak mean radial electric fields [Nakata, *et al.* (2016)].

Effects of equilibrium-scale radial electric fields (E_{r0}) on the zonal-flow response and the ion temperature gradient (ITG) instability in magnetically confined plasma with helical configurations are investigated by gyrokinetic simulations. Poloidally global simulations of the collisionless zonal-flow damping manifest enhancement of the residual amplitude by the uniform and constant (E_{r0}), and show agreement with the zonal-flow response function in the long wavelength limit analytically derived from the theory for helical plasmas with multiple-helicity confinement field components. A higher zonal-flow response is found with heavier ion mass for the same ion temperature and (E_{r0}), that is, the isotope effect, while the ITG mode frequency is Doppler shifted. Accordingly, the isotope dependence of the zonal-flow response through (E_{r0}) possibly leads to the ion mass (or the poloidal Mach number) dependence of the turbulent transport and fluctuations.

12.6.1 *Formation of ITB in gyrokinetic ITG turbulence*

Toroidal momentum injection produces an internal transport barrier (ITB) with a thermal flux driven by ion temperature gradient (ITG) turbulence. As described in Chapter 5, the momentum source changes the mean E_r through the radial force balance, leading to an ion thermal diffusivity that decreases to the neoclassical transport level. Co-current toroidal rotation produces the ITB formation in weak magnetic shear plasma, showing a qualitative agreement with the observations in the JET experiment. The underlying mechanism is a positive feedback loop between E_r shear and resultant toroidal momentum flux.

In the JET experiment, while strong temperature profile stiffness is observed around the nonlinear threshold of ion temperature gradient, the profile gradient can evolve and can be reduced by co-current toroidal rotation in a weak magnetic shear plasma [Mantica, *et al.* (2011)]. To investigate the complex transport mitigation mechanism the 5D global gyrokinetic code (GKNET) with heat and momentum sources [Obrejan, *et al.* (2015), Imadera and Kishimoto (2014)] was used. The simulations showed that a stiff temperature profile is established in the absence of the toroidal momentum source. The mean E_r driven by toroidal rotation plays an important role in triggering global transport by recovering the up-down symmetry of the ballooning mode structures. The analysis shows that the mean E_r can enhance the ion temperature profile stiffness.

The radial force balance relation in circular concentric tokamak configuration rewritten as

$$E_r + \frac{1}{e}\frac{dT_i}{dr} - \frac{rBv_{\|}}{qR} - \left(\frac{1}{n_i e}\right)\frac{dp_i}{dr} = 0$$

indicates that the toroidal rotation $rBv_\phi/qR = B_p v_\phi$ in outer region with small safety factor $|q|$ corresponding to weak or reversed magnetic shear case, can change the mean E_r.

For the JET parameters $a/R = 0.36, a/\rho_{ti} = 150$ and $\nu^* = 0.28$. The radial ion temperature profile in the three cases (i) without momentum input, (ii) with co- and (iii) counter-current input around $r = 0.6a$ were compared. Only the co-current momentum injection produced a strong local ion temperature build-up in NBI heat source region, where the ion thermal diffusivity decreases to the neoclassical transport thermal flux. Extensive simulations shows that co-input and weak magnetic shear regime produce the ITB formation consistent with the observations in the JET experiments [Mantica, *et al.* (2011)] and Citrin, *et al.* (2015).

The simulations show that co-current toroidal momentum injection strengthens the ITB formation through the modification of the mean E_r. The increased momentum pinch from the increased E_r gives rise to the ITB formation around $q_{\min}$ surface in reversed magnetic shear plasms.

Chapter 13

Electron Temperature Gradient Driven Turbulence

The problem of turbulent electron thermal losses has plagued all types of magnetic confinement experiments from the beginning of the fusion program. In the first reports of the Princeton Stellarator program led by Lyman Spitzer in the 1960s the electron temperature would not rise above a few hundred electron volts with intense heating.

There are two space-time scales for the electron temperature gradient driven turbulence so this chapter is longer than that on the ion temperature gradient driven turbulence. The small-scale turbulence is much faster than the ion temperature gradient driven turbulence and has much larger growth rates. The small-scale electron turbulence is also sensitive to the form of the electron phase-space momentum distribution function with respect to the parallel electron current, heat flux and the anisotropy of the temperatures parallel and perpendicular to the magnetic field. Thus, this chapter also discusses the change in the electron phase-space distribution function from the RF heating that determines the shape and gradients of the electron distribution in phase space. The fusion plasmas depend on high-power RF heating in several frequency ranges to keep the electrons at high temperatures and in the case of the tokamak and spherical torus to drive the plasma current to produce the confining magnetic field. Even with the stellarator and helical confinement systems the plasma is driven with high-power electron RF heating to maintain the plasma in the regime for nuclear fusion as explored in this lengthy chapter.

13.1 Electron Transport and the Critical Temperature Gradient

The confinement in these early experiments was characterized by the number of Bohm confinement times that could be achieved. The definition of the Bohm diffusivity is $T_e/16\,\mathrm{eB}$ where the coefficient of $1/16$ has no particular theoretical significance, but instead is historical from the early work of Bohm. Now, it is more conventional to define the Bohm D_B and related gyroBohm D_{gB} diffusivities by

$$D_B = \frac{k_B T_e}{eB} \qquad \text{and} \qquad D_{gB} = \frac{\rho_s k_B T_e}{aeB} \tag{13.1}$$

where the scale length a is to be associated with the scale length of the gradient driving the turbulence which for ETG would be the L_{T_e} scale length of the electron temperature gradient. In mks units for the charge and magnetic field and electron volts for the temperature, the Boltzmann constant is dropped. For example, $T_e = 1\,\mathrm{KeV}$ plasma in $B = 1\,T = 10^4\,\mathrm{G}$ magnetic field and minor radius $a = 1\,\mathrm{m}$ the formula for $D_B = 1000\,\mathrm{V}/1\ T = 10^3\,\mathrm{m}^2/\mathrm{s}$ and $\rho_s = 3\,\mathrm{mm}$ so the gyroBohm diffusivity is $D_{gB} = 3\,\mathrm{m}^2/\mathrm{s}$.

There is experimental evidence for both the Bohm and gyroBohm scaling in the magnetic confinement databases. Sometimes the transport is characterized as making a transition from one form to the other. When good confinement is obtained the numerical factor $\chi_e = c_B D_B$ in the Bohm diffusivity is given as a measure of the success of the confinement system. The toroidal quadrupoles and octupoles discussed in Chapter 6.4 were the first devices to obtain the low values of $c_B \lesssim 1/100$.

Scaling with the gyroBohm formula is easier to understand in terms of drift wave turbulence and brings in the scale length ρ_s of the turbulence driving the transport in a natural way. The formula for D_{gB} follows simply from using the drift wave wavelengths for the mixing lengths and the growth rate time scale for the decorrelation time scale. Thus, the numerical simulations discussed in Chapters 12–17 almost exclusively express their results in the units of D_{gB}. For Bohm turbulent diffusivity the mixing length increases from ρ_s to $(\rho_s L_T)^{1/2}$ and the correlations time remains of order L_T/c_s.

The electron thermal diffusivity, χ_e is reported in the first high electron temperature magnetic confinement experiments greatly exceeded the collisional transport value and this situation remains true today. Thus, it seems clear that the electron thermal transport is due to the drift wave turbulence over a wide range of conditions. Kadomtsev, in his 1992 monograph, devotes a chapter to the universality of the large, or anomalous value of the electron thermal diffusivity χ_e. Kadomtsev argues that intrinsic electromagnetic drift wave processes are responsible for this universal anomalous thermal transport.

Rebut and Lallia (1988) put forth a formula based on general arguments about electromagnetic fluctuations in magnetic confinement systems. The resulting formula was used extensively for a period to interpret the electron confinement data from JET.

The theoretical models for the anomalous electron transport based on drift waves are now almost exclusively used to interpret the electron transport data. There remain some disagreements on the particular type of drift wave fluctuations that dominate in producing the turbulent thermal diffusivity. In terms of broad categories, the two types of fluctuations that are found in theory and simulations are (i) the larger scale drift wave fluctuations on the scale of the ion inertial scale length ρ_s derived in the previous Chapter 12 and (ii) the smaller scale, faster growing electron gyroradius scale drift wave turbulence. There are cases where the

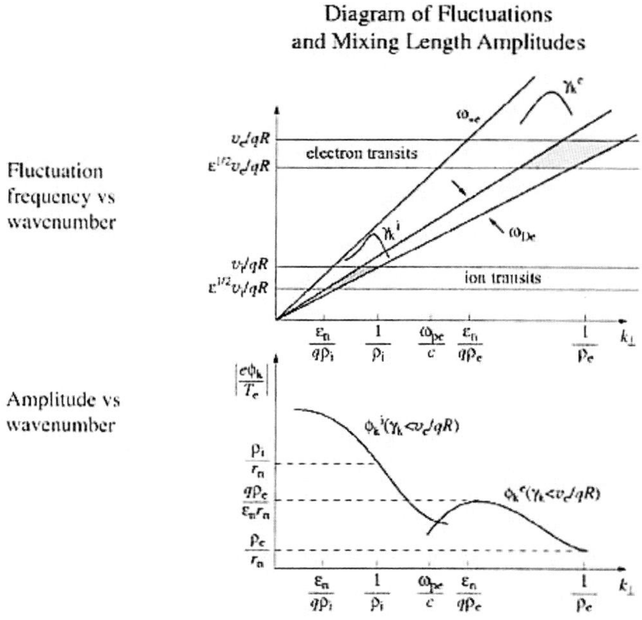

Fig. 13.1 A map for the distribution of the fluctuations and wave amplitudes ranging from the ion transit frequency through the ITG-TEM regime of Chapter 12 to the high-frequency, small-scale regime of the ETG turbulence. The upper part (a) gives the frequency versus wavenumber region of the turbulence and particle transit and bounce motions. The lower part (b) gives the amplitudes for the mixing length level of saturation, where the estimate is for the regime with $L_T \sim L_n < a \ll R$ for the typical well-confined tokamak plasma [adopted from Horton, *et al.* (1988)].

power balance studies compared with databases favor one or the other type of instabilities. Since the turbulence is on two different space-time scales separated by the $(m_i/m_e)^{1/2}$ both types of turbulence play a role in confinement experiments producing the measured anomalous electron thermal losses.

Electron energy confinement analysis of the Tore Supra, TCV, NSTX and MAST plasmas supports the conclusion that the ETG model is able to explain a wide range of anomalous electron transport data. Detailed thermal transport analysis supports the drift waves thermal loss models in ASDEX [Ryter, *et al.* (2001a,b)], Tore Supra [Hoang, *et al.* (2003)], and the Frascati Tokamak Upgrade (FTU) [Jacchia, *et al.* (2002)].

This chapter concentrates on the dynamics and transport from the small spatial scale-fast electron temperature gradient driven turbulence. The reader may observe that some of the ETG physics is similar to that of that of the ITG modes in Chapter 12 with the role of the electrons and ions interchanged.

The general scheme of the electron temperature gradient driven turbulence is shown in Fig. 13.1. The figure shows the small scale ETG is related to the larger scale ITG turbulence of Chapter 12. In the ETG turbulence the parallel wave phase velocity $\omega/k_\parallel$ is higher than and comparable to the electron thermal velocity in contrast to the lower phase velocity of the ITG-TEM turbulence. This makes the ETG turbulence less sensitive to the details of geometry than for the lower frequency ITG-TEM modes. The instability has a strong inverse cascade, as shown theoretically and in computer simulations, which transfers fluctuation energy to the ITG instability scale. Thus, wave energy from nonlinear interactions is transferred from the ETG to the ITG-TEM regime. ETG spectral simulations are typically in boxes of order $3\rho_i \times 3\rho_i$ so that the pile-up of low-k energy can be absorbed into the dynamics of the ITG modes at the box-size limit of the k-space. The formulas are for a tokamak geometry with parameters $\epsilon_n = L_n/R$ and magnetic shear $s \sim 1$. Here $\rho_i = v_i/\omega_{ci}$ is the thermal ion gyroradius and $\rho_e = v_e/\omega_{ce}$ is the electron thermal gyroradius. This range of scales and frequencies is typically two orders of magnitude and dictates the use of different diagnostic methods to measure the turbulence.

Owing to the faster parallel phase velocity the quiver velocity of the electrons along the magnetic field from the parallel electric field and pressure gradient is relatively high for a given wave amplitude. This builds up the wave parallel electric current giving a significant magnetic perturbation even at relative low electron plasma beta or thermal-to-magnetic pressure ratio $\beta_e = 2(\omega_{pe}v_e/\omega_{ce}c)^2 = 8\pi n_e T_e/B^2$. One introduces the parallel vector potential $A_\parallel$ to describe the electromagnetic parts of the ETG turbulence. The parallel vector potential is solved for from the parallel electron current by

$$\nabla^2 A_\parallel = -\mu_0 \delta j_\parallel. \tag{13.2}$$

A typical simulation for NSTX plasma is shown in Fig. 13.2 where contours of isopotentials are shown for the three fluctuating fields $\phi(x,y,t)$, $A_\parallel(x,y,t)$ and $\delta T_e(x,y,t)$ in the saturated state. The figure shows that there are different dominant scales for the three fields. The electron temperature fluctuations have the smallest spatial scale lengths of order ρ_e, the electrostatic potential has the intermediate scale length size of order $\sqrt{\rho_e L_{T_e}}$ to $q\rho_e$ and the vector potential field has the scale length of order c/ω_{pe}. The $\mathbf{E}\times\mathbf{B}$ flow velocities and the $\boldsymbol{\delta B}_\perp$ vectors are given by the local tangents to the isolines of the electric potential and the vector potential.

The directional derivatives along these vector fields are given by the Poisson bracket differential operators designated by $[\phi, f]$ and $[A_\parallel, f]$ respectively. As seen in Chapter 9 these convective nonlinearities create vortex structures in the fields, as is evident in Fig. 13.2 in the snapshot in time for this simulation for ETG turbulence.

The electromagnetic component of the turbulence introduces a magnetic flutter described by Callen (1977) and Horton (1984). The space scale of the magnetic flutter is the electron collisionless skin depth $\delta_e = c/\omega_{pe}$. The relationship of the skin depth scale to the electrostatic scale of the inverse cascaded electrostatic potential is important to determine the parameter scaling of the electron thermal diffusivity. In Fig. 13.1 the regime is shown with the skin depth larger than the electrostatic scale length $l_{es} = q\rho_e/\epsilon_n = qR\rho_e/L_n$. First one considers that $L_n \sim L_{T_e}$ typical of L-modes and I-modes. Then one considers H modes $L_n > L_{T_e}$ where the scale

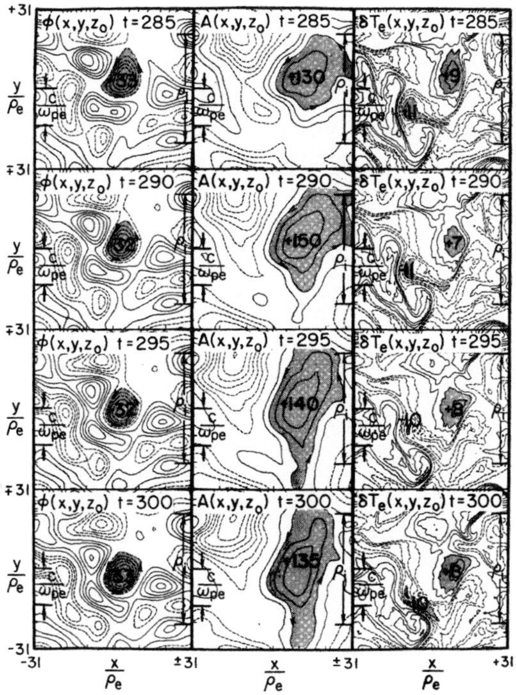

Fig. 13.2 ETG turbulence simulations used to interpret the electron thermal losses in NSTX [Kaye, *et al.* (2007)]. Contours of the electrostatic potential in the nonlinear saturated state showing the results of the inverse cascade and the formation of radially extended streamers.

length becomes L_n of order R which must be large compared with the temperature gradient scale length L_{T_e} for a confined hot plasma.

As the dimensionless electron pressure increases to exceed $\beta_{e,\mathrm{crit}}$ then the skin depth becomes smaller than the electrons scale length l_{es}. Now the mixing length is controlled by the electromagnetic skin depth and the theoretical turbulent diffusivity, χ_e, is reduced compared to the extrapolated high-T_e values given in the electrostatic turbulence. In the electromagnetic regime with

$$\beta_e > \beta_{e,\mathrm{crit}} = \frac{2m_e}{m_i} \tag{13.3}$$

one has $\delta_e > \rho_e$ and the mixing length, l_{em}, is now

$$l_{em} = \delta_e = \frac{c}{\omega_{pe}}. \tag{13.4}$$

The electromagnetic transport formulas for the electron thermal diffusivities with this transition are used in Horton, *et al.* (2002, 2003) and given for $\beta_e > \beta_{\mathrm{crit}}$

$$\chi_e^{em} = C_{em} q^{\nu} \frac{c^2}{\omega_{pe}^2} \frac{v_e}{(L_{T_e} R)^{1/2}} \left(\frac{R}{L_{T_e}} - \frac{R}{L_c} \right) \tag{13.5}$$

and for $\beta_c < \beta_{\rm crit}$ the thermal diffusivity returns to

$$\chi_e^{es} = C_e^{es} q^\nu \left(\frac{R}{L_{T_e}}\right)^{3/2} \left(\frac{\rho_e^2 v_e}{L_{T_e}}\right) \left(\frac{R}{L_{T_e}} - \frac{R}{L_c}\right). \tag{13.6}$$

The magnetic shear and q dependence of the growth rate are used in choosing the exponents ν in the thermal flux. The exponent ν varies between 1 and 2. The critical gradient $(\nabla T_e)_c$ is derived from the marginal stability analysis of the dispersion relation. The presence of a critical gradient in the thermal flux expressions gives the flux formula the form of a diffusivity plus a radial transport velocity $V_r(r,t)$ where the relationship with the critical gradient is

$$V_r = \chi_e \frac{(\nabla T_e)_c}{T_e} = -\frac{\chi_e}{L_{T_{e,c}}} \tag{13.7}$$

giving an inward thermal pinch velocity with $L_{T_{e,c}} V_r/\chi_e \sim 1$. The following power balance studies will yield $V_r = \chi_e^{pb}/L_{Te,c} \geq 10\,\text{m/s}$. The gyroBohm estimate for χ_e is $\chi_e/L_{Te} = c_s(\rho_s/L_{Te})^2$. The velocity V_r is inward giving rise to the off-diagonal heat pinch effect. The sheared slab regime has

$$V_r = -1.9|s|\chi_e/qR\,(1 + Z_{\rm eff}T_e/T_i) \tag{13.8}$$

from Eq. (13.7). In the absence of magnetic shear the compressional stabilization rather than electron Landau damping determines the transport velocity V_r with $(\nabla T_e)_c$. Peaked density profiles with small L_{ne} give an inward particle pinch velocity V_n. Pinch terms are equivalent to off-diagonal transport terms and are a standard feature of turbulent transport.

The formulas in Eqs. (13.5)–(13.8) are able to interpret the Tore Supra database since in the plasma profile the plasma β_e is above the $\beta_{\rm crit}$ in the core region and below the $\beta_{\rm crit}$ in the outer part of the plasma profile.

The transport formulas were used in the CRONOS transport code along with the WAVE code for the plasma heating profile from the Fast Wave Ion Cyclotron (FWIC) heating [Moroz and Colestock (1991)] to accurately interpret the Tore Supra plasma. The results showed a predictive performance as shown in Fig. 13.3 where the FWIC heating was first held at 3 MW and then ramped up to 6 MW during a long $\Delta t = 2\,$s pulse.

An important difference in the large-scale ITG turbulence and the small scale ETG turbulence is that the electron parallel fluctuating currents $j_\parallel(k,\omega)$ are relatively larger in terms of their parallel acceleration of the electron fluid. This means that the electron inertial is important in determining the phase of the accelerated electrons. Thus, the scale length $\delta_e = c/\omega_{pe}$ is an important dispersion scale length in addition to the ion inertial scale length $\rho_s = c_s/\omega_{ci}$ dominant in Chapter 12. As the plasma pressure increases these dispersion scale lengths become equal when the dimensionless electron pressure $\beta_e = \beta_{e,\rm crit} = 2m_e/m_i$ given in Eq. (13.3). At this $\beta_{e,\rm crit}$ pressure value the electron thermal pressure, the Alfvén wave velocity is equal to the electron thermal velocity, so it is natural that the waves and fluctuations become electromagnetic. When the cross field correlation length for the electron

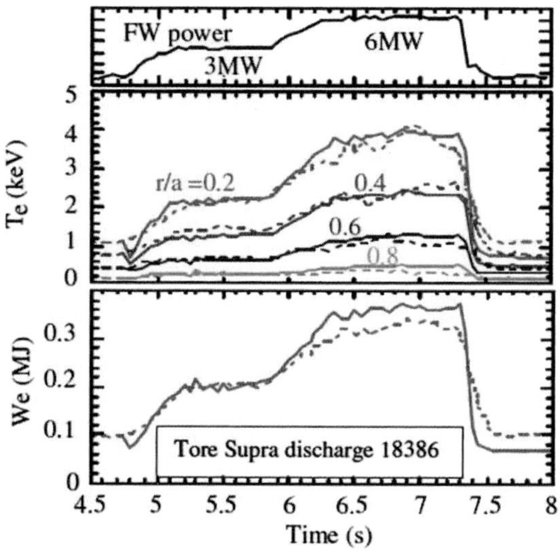

Fig. 13.3 Predictive simulation of an existing Tore Supra FWEH discharge [#18368, $I_p = 0.6\,\text{MA}$, $B = 2.2\,\text{T}$, $n_e(0) = 4 \times 10^{19}\,\text{m}^{-3}$], top panel with the fast wave heating power rising from 3 to 6 MW. Middle panel shows the time evolution of electron temperature (T_e) at various radii. The bottom panel shows the rise of the total electron energy content at (W_e).

transport is taken as the electron skin depth $\delta_e = c/\omega_{pe}$ then some of the features of the empirical scaling laws introduced by Rebut, *et al.* (1991) and Kadomtsev (1992) are present. The first principles understanding of this anomalous electron transport, however, comes from the electron temperature gradient driven drift waves and their nonlinear dynamics. Experiments in the Tore Supra device over a wide range of electron heating powers (ohmic heating to 8 MW of RF Fast Wave heating power) confirmed the ability of the ETG transport models to systematically explain the anomalous electron transport rates.

The dynamics of the large-scale structures in the toroidal ETG turbulence has been explained with theory and simulations. Simulations [Li and Kishimoto (2004)] have shown how the radially extended streamers emerge from the nonlinear interactions in the ETG turbulence.

The simulations using ETG formulas reproduced Tore Supra FWEH experiments and CLM experiments [Wei, *et al.* (2010)]. The most direct evidence of fluctuations in the tokamak follows from the electromagnetic scattering of microwaves off the ETG driven electron density fluctuations. The techniques are those described in Chapter 2 extended to smaller space scales by Mazzucato (2009, 2010) on NSTX. Earlier studies of the power balance on a large database of RF heated discharges in Tore Supra by Horton, *et al.* (2000, 2003), and Hoang, *et al.* (2001, 2003) showed that theory reproduces the $T_e(r,t)$ profiles measured with Thomson laser scattering for plasma heating powers from Ohmic heating to 8 MW of radio frequency heating power where the core temperature increased from about 1 KeV to 3 KeV. Both the spatial profile and the time dependence with a step-up in heating power are well reproduced. Similar good agreement with the profiles of $T_e(r,t)$ were found in the NSTX experiment with high-resolution Thomson scattering diagnostics in Kaye, *et al.* (2007). In the NSTX and MAST experiments the ion thermal transport flux remains close to the neoclassical collisional level while the electron transport is larger and requires the ETG level turbulent transport for predicting the measured electron temperature profiles from the deposited power. A typical example of the comparison of the ETG thermal flux calculated from simulations compared with the power balance flux for a fast wave heated NSTX discharge [Kaye, *et al.* (2007), Stutman, *et al.* (2009)] is shown in Fig. 13.4.

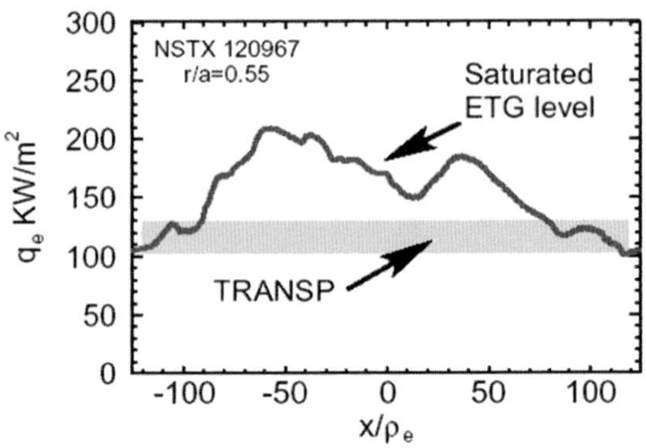

Fig. 13.4 Comparison of the power balance electron thermal flux computed for the NSTX parameters from the TRANSP data analysis code for discharge 120967 compared with the turbulent thermal flux computed from a local pseudo-spectral gyrofluid simulation. The system of partial differential equations used in the turbulence simulation are given Eqs. (13.21)–(13.25).

Measurements with coherent scattering of 280 GHz microwaves in NSTX with a five-channel heterodyne receiver showed that the presence of the ETG fluctuations at $k_\perp \rho_e \sim 0.2$–0.4 with amplitudes strongly correlated with the strength of the electron temperature gradient. The plasma beta is in the range of 3% to 6% so that the plasma density fluctuation producing the scattering has $k_\perp \delta_e \sim 2$. The electron temperature gradient varied from $R/L_{T_e} = 4$ to 8 in the experiments which produced an order of magnitude variation in the spectral density of the scattered microwave power.

In the KSTAR tokamak [Lee and KSTAR Team (2015)], experimental context for these calculations is not disruptions but stable discharges that last many tens of seconds while exhibiting a long-lived 2/1 mode under conditions similar to "advanced/hybrid scenarios" [Sips (2005)], with high-edge safety factor q_{sx} and a low-inductive current fraction. Investigations into the nature of this coherent mode discovered that in equilibria with peaked pressure profiles, either a benign 2/1 mode that saturates at a small amplitude or an explosively growing 2/1 occurs depending on the details of the pressure profile and $q_{\rm sx}$. Although the plasma-β is moderate

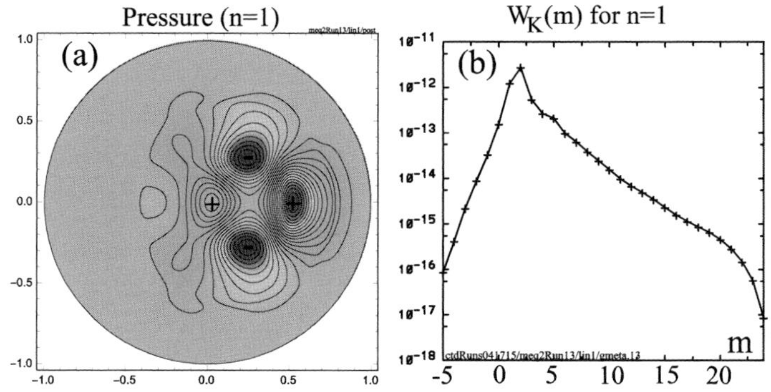

Fig. 13.5 Simulations of explosive growth of the ballooning mode in KSTAR. (a) Perturbed pressure contours for the $n = 1$ mode. Note the ballooning nature of the eigenfunction. (b) Poloidal spectrum of the kinetic energy for the $n = 1$ mode [Aydemir, *et al.* (2016)].

for these KSTAR experiments ($\beta_N \simeq 1.4$), the $n = 1$ mode is ideally unstable for both q-profiles.

The $n = 1$ eigenfunction for the pressure field is shown in Fig. 13.5(a) for the $q(r)$ profile. Figure 13.5(b) shows the poloidal mode number m spectrum. As seen in the figure, the poloidal structure of the mode is dominated by an $m = 2$ component, although smaller $m = 3$ effects are also visible. The high- and low-field side asymmetry of the perturbed pressure is indicative of a kink-ballooning mode, a low-n version of the infernal modes [Manickam, *et al.* (1987)]. The ballooning character of the mode is apparent in its poloidal energy spectrum in Fig. 13.5(b), showing a coupling that extends much beyond the immediate sidebands at $m = 2 \pm 1$.

A quadrupole pressure perturbation is crucial in the subsequent nonlinear evolution of the mode. At the toroidal location with this particular phase of the perturbation ($\theta = 0$ poloidal plane in this calculation), the flux and pressure surfaces that are nearly circular initially become elongated in the direction of the major radius ($+\boldsymbol{R}$) as a result of this perturbation. Continuing elliptical deformation of the plasma eventually leads to the formation of a ballooning finger on the low-field side, as seen in Fig. 13.6.

Before the nonlinear evolution of the $n = 1$ mode reaches the stage in Fig. 13.6, with a well-defined $m = 2$ ballooning finger, its growth becomes explosive. The finger proceeds to get narrower and more elongated, reaching in ~ 20 Alfvén times the final stage seen in the third frame in Fig. 13.6, where the hot plasma from the core has splashed against the wall. The end result, of course, is a disruption.

The mode evolves into an explosively growing ballooning finger and the disruption that accompanies the finger is driven by a 2/1 mode. Projection of the ballooning finger in Fig. 13.6 along the field lines to different poloidal planes globally exhibits an approximate 2/1 symmetry. For a linear high-n ballooning mode the pressure perturbation will satisfy $p \propto f(\zeta - q(r)\theta)g(\theta)h(r)$, where $g(\theta)$ is the ballooning eigenfunction that localizes the mode on the low-field, bad-curvature side

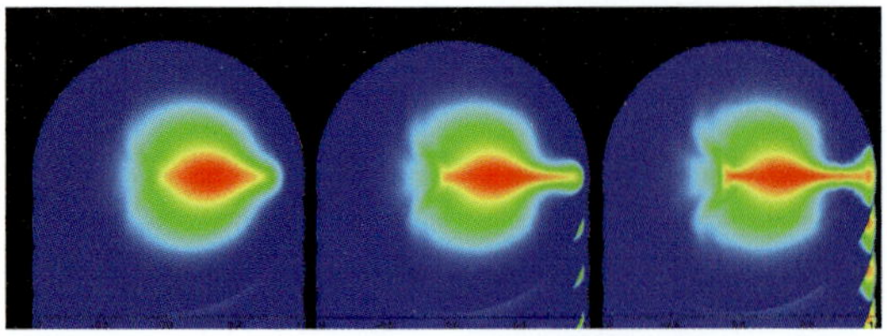

Fig. 13.6 Formation and nonlinear evolution of the "explosive finger" in the poloidal plane from a small $m/\ell = 2/1$ mode. The three figures are separated by approximately 20 Alfvén times in units of poloidal Alfvén time [Aydemir, *et al.* (2016)].

of the torus, and $h(r)$ is the radial envelope function. We see that the nonlinear ballooning structure rotates with $q \simeq q_s = 2/1$, except near the tip of the finger where the effective q is higher.

The explosive nature of the instability is also seen in plots of the total kinetic energy (without the $n = 0$ contribution). The growth rate starts increasing early in the nonlinear evolution beginning around $t \simeq 1420$ (normalized to the poloidal Alfvén time). In this explosive stage, $W_K(t)$ can be fitted with a curve of the form $W_K(t) = W_K(t_i)[(t_f - t_i)/(t_f - t)]$, where $t_i = 1417$, $t_f = 1502$, and the exponent $\nu = 3.37$. Note that the finite-time singularity model is a good fit to the numerical results except at the very end when the plasma starts coming in contact with the wall. The three panels in Fig. 13.6 shows the flux surfaces through the nonlinear stages. The core plasma remains intact, indicating the small amplitude of the $m = 2$ perturbation at intermediate times. The evolution of the explosive finger seen in Fig. 13.6 occurs in approximately 20 Alfvén times at the very end of the plots. Thus, this disruption would have only a brief precursor, much too small to be of practical experimental use. On the same time scale, the evolution of the ideal $n = 1$ mode leads to stochastic field lines from the disruption that started with the q-profile. Thus, parallel thermal transport also plays a significant role during thermal quench, although the loss of energy through this channel will not be explosive but occurs on a slower time scale. During the late stages of the disruption, only the plasma in the explosive finger seems to be well-confined, as the flux surfaces remain intact inside the finger.

This explosive regime was not the initial goal of the KSTAR studies, which was concentrated on the "long-lived" 2/1 mode observations in KSTAR [Lee and KSTAR Team (2015)]. Since it seemed clear that robustly unstable initial conditions would not yield a mode saturated at a small amplitude, as seen in the experiments, the research concentrated on equilibrium profiles near marginal stability where the $n = 1$ was only weakly unstable. Thus, the appearance of an eventual explosive regime when the initial linear growth rate $\gamma\tau_A \sim 10^{-3}$ or smaller was a surprising discovery. Further study showed that other the experiments have shown the same behavior with higher initial growth rates; apparently the high- resolution calculations for the low growth rates with $\gamma\tau_A \sim (10^{-2})$ were needed. The numerical results are in qualitative agreement with Cowley's theory [Cowley and Artun (1997);

Cowley, *et al.* (2003)] on the explosive growth of ballooning fingers (compare Fig. 2 of Cowley, *et al.* (2003) with Fig. 13.6). Subsequent analysis in full toroidal geometry by Wilson and Cowley (2004) predicted that the exponent $d\ell n\, W/d\ell n\, t = \nu_d$ for the finite-time singularity in the displacement ξ depends on the Mercier indices $\lambda_{\mathrm{L,S}} = 1/2 \pm \sqrt{1/4 - D_M}$, with the exponent given by $\nu_d = \lambda_S - \lambda_L$. Assuming $D_M \simeq 0$ in the circular geometry yields $\nu_d = 1$ and $\xi \sim 1/(t_f - t)$. Then using $W_K \sim |\partial\xi/\partial t|^2$ leads to the theoretical prediction $W_K \sim 1/(t_f - t)^4$ compared with $\nu = 3.37$ in the simulation. The relatively small discrepancy may be due to finite viscous dissipation and the perfectly conducting boundary conditions used in the simulations.

In summary, the KSTAR experiments are explained by the inherently explosive growth of an ideal $n = 1$ mode-driven ballooning finger, with a predominantly $m/n = 2/1$ component, that was found in the nonlinear toroidal simulations. Transition from exponential to explosive growth is driven nonlinearly by the characteristic quadrupole geometry of the $m = 2$ pressure perturbation leading to the formation of an outward ballooning finger. The plasma finger becomes narrower and it moves outward through flux surfaces. This rapid convective convection of mass and energy transport from the core to the exterior, in some tens of Alfvén times, provides an explanation for fast high-β disruptions that occur with little precursors.

Electron cyclotron emission (ECE) receivers are used to image the plasma structures in tokamaks. KSTAR has a pair of electron cyclotron emission imaging (ECEI) systems separated by 1/16th of a turn around the torus [Yun, *et al.* (2014). This allows visualization of the dynamics of the MHD instabilities from the plasma core to the edge. This flexible diagnostic capability was achieved by improvements in the large aperture quasi-optical microwave components and broad-band polarization rotators for imaging of the fundamental frequency ECE ordinary polarization waves and the 2nd harmonic extraordinary mode ECE waves.

Experiments on non-inductive plasma current start-up by electron cyclotron (ECH) heating and current drive (ECH/ECCD) were carried out on KSTAR by injecting the fundamental O-mode wave from the low-field side obliquely to the toroidal magnetic field. A plasma current up to 14.5 kA is generated by 180 kW of 84 GHz microwave power and the magnetic measurement shows the formation of a large last-closed flux surface with a diameter of 0.4 m. The soft x-ray emission profile and fast CCD images showed the existence of closed flux surfaces. The current of the cross-field-passing electrons (CFPEs) was calculated showing that the CFPE plasma current can produce the initial closed magnetic flux surfaces. The observed large increase of EC emission supports the generation of energetic electrons carrying the plasma current. After the formation of the closed flux surfaces, the pressure-driven current and CFPE currents do not flow on the same flux surfaces. Electron cyclotron (EC)-driven current ramps up merging with the plasma toroidal current. The data shows that the EC-driven current is about one third of the total plasma current flowing on the closed flux surface [Tanaka, *et al.* (2016)] of the plasma.

13.2 Electron Temperature Gradient Transport

In the high-$\boldsymbol{k}$ regime of ETG turbulence, the thermodynamics of the electron gas explains the direct drive of the turbulence from the electron temperature gradient $\boldsymbol{\nabla} T_e$. Thermodynamics shows that there is a critical temperature gradient $\boldsymbol{\nabla} T_e \equiv -T_e/L_{Te}^{\rm crit}$ for the ETG turbulence to extract energy from the temperature difference, $\Delta T = T_1 - T_2$, over the correlation length l_c.

The Carnot cycle shown in Fig. 13.6 gives the upper bound on the released turbulence energy $W \leq W_c = \Delta T \Delta S_e$, where ΔS_e in Eq. (13.9) is the change in the electron entropy density along the isothermal side T_1 in the core and T_2 in the lower temperature zone [Horton, *et al.* (2004)] of the nonlinear convective heat engine. The simulations of the associated partial differential equations for the electron gas directly confirm the onset of the turbulence above the critical gradient and the presence of elongated convection cells. As the gradient rises above the critical value, stronger nonlinear features appear in the fluctuations. In addition to thermal energy the fluctuations produce a transport of parallel electron momentum and magnetic flux that also develop into strong coherent structures for the parallel electron current and the magnetic flux function. The transport of magnetic flux results in the turbulence developing into a form of the small-scale tearing mode turbulence. Large-scale tearing modes are described in Chapter 14 are driven by the gradients of the background current density profile driving mechanism. Even in the absence of the dj/dr driven tearing modes the nonlinear ETG modes at finite electron pressures produce a small-scale, micro-tearing magnetic turbulence. This ∇T_e driven magnetic turbulence twists and shears the magnetic field lines which then reconnect releasing magnetic energy. This dynamics transforms a fraction of poloidal magnetic energy into the parallel electron acceleration and then into electron thermal energy.

Figure 13.6 shows the extended radial structures that convect the plasma in such a way as to take thermal energy from the high-temperature region T_1 to the lower temperature region T_2 across the convection cell. One can place an upper bound on the energy available to produce the turbulence and convective cells from the Carnot cycle theorem of thermodynamics.

On the adiabatic sides of the Carnot cycle, a gas constant, $\Gamma = (d+2)/d$, describes the change of temperature of electron gas, where $d = 1, 2, 3$ for the number of degrees of freedom active in the ETG dynamics. Kinetic theory guides the choice of d with the slab and toroidal modes having $d = 1$ and $d = 3$, respectively. The drift-wave vortex or streamer gives the convective $\mathbf{E} \times \mathbf{B}$ motion between (n_1, T_1) and (n_2, T_2) producing the Carnot cycle for the convection period. From the Carnot-cycle calculation, the change of entropy density is given by,

$$\Delta S_e = n_e k_B \left(\frac{3}{2} \frac{\Delta T_e}{T_e} - \frac{\Delta n_e}{n_e} \right) \tag{13.9}$$

for $d = 3$ where k_B is the Boltzmann constant. Using the correlation length ℓ_c and the gradients $T_1 - T_2 = -\ell_c \mathrm{d}T_e/\mathrm{d}r$ and $n_1 - n_2 = -\ell_c \mathrm{d}n_e/\mathrm{d}r$ gives the maximum released electron thermal energy density W_e released in the cycle as

$$W_e = \frac{3}{2} n_e T_e \ell_c^2 \frac{\mathrm{d}T_e}{\mathrm{d}r} \left[\frac{\mathrm{d}\ln T_e}{\mathrm{d}r} - \frac{2}{3} \frac{\mathrm{d}\ln n_e}{\mathrm{d}r} \right]. \tag{13.10}$$

In this ideal limit without dissipation in the system, the critical gradient expressed in terms of the major radius, R, of the torus is

$$\frac{R}{L_{Te}^{\mathrm{crit}}} = \frac{2}{d} \frac{R}{L_{ne}} \tag{13.11}$$

as follows from $W_e > 0$. Here d is the number of degrees of freedom in the electron dynamics. This critical gradient value R/L_{Te}^{crit} is expressed in terms of the temperature-to-density gradient ratio $\eta_e = L_n/L_T$ is known from the Nyquist stability analysis for both ITG and ETG dynamics. Adding wave dissipation induced by magnetic shear, $R/L_s = s/q$, increases the critical temperature gradient [Horton, *et al.* (2004)] in proportion to $|s|/q$. The corresponding linear contribution is given in Hahm and Tang (1989) eigenmode calculation for the offset in the critical gradient from magnetic shear.

13.2.1 *Two-space scales for electron transport*

The role of the two space scales in the electron thermal heat flux, $q_e(r, t) = -n_e \chi_e dT_e/dr$, is made clear from the kinetic theory formula for q_e. Kinetic theory gives the total energy flux as

$$Q_e = \frac{3}{2} T_e \Gamma_e + q_e \tag{13.12}$$

where Γ_e is the particle flux and q_e thermal conduction flux.

For a background plasma with local Maxwellian electrons parameterized by n_e, T_e and the energy density $\mathrm{g}(\epsilon)\mathrm{d}\epsilon$ for $\epsilon = m_e v^2/2T_e$, the cross field electron thermal flux is given by

$$q_e = n_e T_e \left[\sum_{q,\Omega} \frac{q_y |\overline{\Phi}_q|^2}{BT_e} f_t \int_0^\infty \mathrm{d}\epsilon \left(\epsilon - \frac{3}{2}\right) \mathrm{g}(\epsilon) \,\mathrm{Im}\left\{h(\epsilon, q, \omega)\right\} \right. \tag{13.13}$$

$$\left. + \sum_{k,\omega} \frac{k_y |\Phi_k|^2}{BT_e} \int_0^\infty \mathrm{d}\epsilon \left(\epsilon - \frac{3}{2}\right) \mathrm{g}(\epsilon) \,\mathrm{Im}\{h(\epsilon, k, \omega)\} \right] \tag{13.14}$$

where $\int_0^\infty \mathrm{d}\epsilon \mathrm{g}(\epsilon) = 1$, $\overline{\Phi}_q$ is the bounce-averaged potential fluctuation, f_t is the fraction of trapped electrons, $h(\epsilon, k, \omega)$ is the non-adiabatic electron phase-space-density response function and $\mathrm{Im}\{\cdot\}$ is the imaginary part. The notation is used that k and ω are the small-scale, high-frequency part of the fluctuation spectrum while Ω, q are for the large-scale, low-frequency part of the fluctuation spectrum.

In the thermal-flux formula, the low frequency large-scale contribution arises only from the trapped electrons, since the bounce-averaged potential $\overline{\Phi}_q(r) = 0$ for passing electrons (except for r exactly on a rational surface). The small-scale response Φ_k, involves resonance with both passing and trapped electrons through the response function,

$$h_k(\epsilon,\omega) = \frac{\{\omega - \omega_{*e}\left[1+\eta_e\left(\epsilon - \frac{3}{2}\right)\right]\} J_o^2\left(\frac{k_\perp v_\perp}{\omega_{ce}}\right) f_M}{\omega - \epsilon\omega_{De} - k_\parallel v_\parallel + i\delta^+} \tag{13.15}$$

where the wave frequency, $\omega \gg \Omega_q$ and is given approximately by

$$\omega_{\boldsymbol{k}} = -\frac{\omega_{*e}\left[I_o(b_e)\mathrm{e}^{-b_e} + \eta_e b_e(\mathrm{I}_1(b_e) - I_o(b_e))\mathrm{e}^{-b_e}\right]}{Z_{\text{eff}}T_e/T_i + 1 - I_o(b_e)\mathrm{e}^{-b_e} + k^2\lambda_{\text{De}}^2} \tag{13.16}$$

where $b_e = (k_y\rho_e)^2$. The formula Eq. (13.15) gives the nonadiabatic contribution to the fluctuating electron distribution function through $\delta f_{k,\omega} = [1-h_k(\epsilon,\omega)][e\phi_k/T_e]$. Equation (13.16) extends up to frequencies that are a fraction of $v_e/L_{Te} \leq 10^7$/s. Thus, thermal and supra-thermal electrons with large and small pitch angles carry the anomalous electron heat flux in the ETG turbulence.

For example, Eq. (13.16) yields the frequency in the case of TCV 29892 at $r/a = 0.7$. For the η_e values in this discharge, the ETG mode changes direction of rotation at $k_y\rho_e \simeq 0.3$–0.8 and this is where the linear ETG growth rate reaches a maximum. The transport flux is dominated by longer wavelength turbulence generated by the mode coupling nonlinearities as shown in the next subsection.

13.2.2 *Nonadiabatic ion response*

The effect of nonadiabatic ions occurs in the transitional wavenumber space scale region where $k_\perp\rho_i \sim 1$–5. The wave response is associated with the part of the frequency spectrum shown in Fig. 13.5 where the wave propagates in the ion diamagnetic direction. In the low-k part of the ETG spectrum the wave resonates with the guiding center drift motion of the ions with a resonant energy ϵ determined by

$$\omega_{k_y} = \omega_{Di}\epsilon + k_\parallel v_i \epsilon^{1/2}$$

where $\epsilon = m_i v^2/(2T_i)$ and α is the pitch angle. At $B = B_{\text{min}}$ the pitch angle is defined by $\cos\alpha = v_\parallel/v$. The guiding center drift dominates the resonance for the conditions present in TCV.

In the low-$k_y\rho_e$ part of the spectrum the waves are resonant with ion guiding center drifts. The resonant ions have energy $\epsilon_r = E_{\text{res}}/T_i = \omega_{k_y}/\omega_{Di} \lesssim 1$ in the low-k_y part of the spectrum and no resonance when the direction of rotation changes to that of the electrons. These resonant ions add the dissipative response to the ion density fluctuation through the nonadiabatic ion response,

$$\delta n_i = -\frac{Ze\phi}{T_i} n_{i0}\left\langle 1 - \frac{\omega_k - \omega_{*i}(\epsilon)}{\omega_k - \omega_{Di}\epsilon + i0^+} J_0^2 \right\rangle. \tag{13.17}$$

From Eq. (13.17) one calculates by making an expansion for $k_\perp\rho_i > 1$ in γ_k/ω_k, that the contribution of the resonant ions to the growth rate is

$$\gamma_i(k) = \frac{\pi^{1/2}\omega_k(\omega_{*i} - \omega_k)}{|\omega_{Di}|(1+b_i)^{1/2}}. \tag{13.18}$$

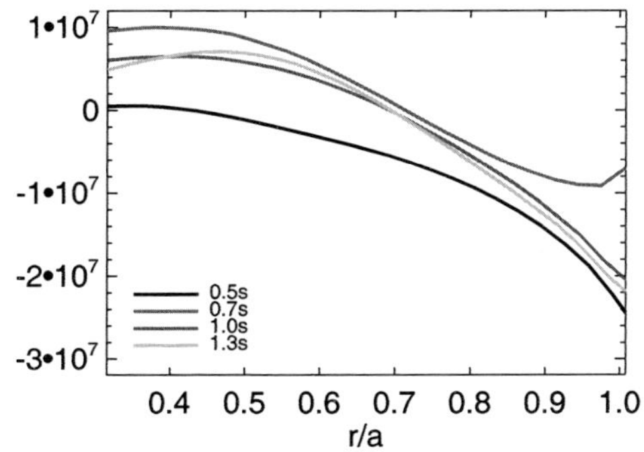

Fig. 13.7 The ETG mode frequencies along the radial positions r/a for $k_y\rho_e = 1.0$ of TCV 29892. In the core region the frequency increases monotonically in time with $t = 0.5, 0.7, 1.0$ from Eq. (13.16) until $t = 1.3$ where a magnetic island has formed.

Equation (13.18) applies only for rotation of the mode in the ion diamagnetic direction as shown in Fig. 13.7 at low-$k_y\rho_e$. Since the upper bound on this growth rate is given by v_e/L_{T_e} the contribution is small compared with the maximum ETG growth rate from the shorter wavelength part of the spectrum where the mode rotates in the electron direction. In many works on ETG turbulence the ions are taken as adiabatic. At the same time this ion contribution gives a small particle flux due to the difference in the phase between the density fluctuation and the potential fluctuation from Eq. (13.17).

13.3 Electron Thermal Transport in TCV

The Tokamak Configuration Variable (TCV) [Tonetti, *et al.* (1991)] is a plasma with high-power density (up to $8\,\mathrm{MW/m^3}$) deposited in the core of the plasma by electron cyclotron resonance heating (ECRH) at significant plasma densities ($\leq 7 \times 10^{19}\,\mathrm{m^{-3}}$) with a strong turbulent electron thermal transport χ_e.

The discharge chosen for analysis is distinguished by having four distinct high-confinement mode (H-mode) phases. An ohmic H-mode with type III edge localized modes (ELMs), which turns into a type I ELMy H-mode when the ECRH is switched on. The ELMs then vanish, which gives rise to a long quasi-stationary ELM-free H-mode. This ELM-free phase can be divided into two parts, one without magnetohydrodynamics (MHD oscillations) followed by a period with the oscillations. The MHD mode in the final phase case causes the plasma energy confinement to drop by $\sim$ 15%. For all four phases both large-scale trapped electron (TEM) and ion temperature gradient (ITG) modes and small-scale electron temperature gradient (ETG) modes are analyzed and found to co-exist. One finds, however, that the transport TEM formulas have difficulty in predicting both the magnitude and the radial profile of the electron thermal flux. Collisionality governs the drive of the TEM, which for the discharge in question implies the instability can be driven by either (1) the temperature or (2) density gradient. The TEM response function is calculated to be relatively small and to have sharp resonances in its energy dependence defined by the condition $\omega_k = \overline{\omega}_{De}$. The ETG turbulence, on the other hand, is driven solely by the electron temperature gradient. Both trapped and passing electrons add to the ETG instability and turbulent thermal flux.

Average Relative Variance

For a quantitative comparison of the results of the TEM and ETG modes a dimensionless error measure, the so-called average relative variance (ARV) is introduced. According to this ARV metric method the ETG model explains 70% of the variation in the electron heat diffusivity whereas the predictive capabilities of the TEM-ITG models are poor with an ARV of order unity. These results for TCV anomalous transport support the conclusion that the ETG model is able to explain a wide range of anomalous electron transport data, in addition to similar evidence from ASDEX [Ryter (2001a,b)], Tore Supra [Hoang, *et al.* (2003)], and the Frascati Tokamak Upgrade (FTU) [Jacchia, *et al.* (2002)].

The Tokamak Configuration Variable (TCV) [Tonetti, *et al.* (1991)] is a compact tokamak with high-power electron cyclotron resonance heating (ECRH) that is ideally suited for studies of turbulent electron thermal transport. In the configuration of TCV analyzed here, the plasma energy confinement measure gives $H_{98(2,y)} \geq 1$. The interesting feature of this particular discharge, TCV 29892, is that it contains no less than four different high-confinement mode (H-mode) phases whereof two are quasi-stationary without edge localized modes (ELMs) [Porte, *et al.* (2006)]. The differences between this quasi-stationary regime and other regimes obtained at DIII-D [Lohr, *et al.* (1988)] and Alcator C-Mod [Fielding, *et al.* (1996)], are that on TCV the confinement is achieved with pure electron heating at low-q_{95} (< 3), with no auxiliary ion heating, no active fueling and no cryo-pumping. This shot is an ITER-relevant plasma, as the plasma has a hybrid scenario-like flat q-profile with low magnetic shear.

RF current drive experiments on the Frascati Tokamak Upgrade (FTU) have shown that with the addition of electron cyclotron heating (ECRH) to the lower-hybrid current drive (LHCD) system the plasma with a runaway population can be controlled and the runaways eliminated by using the resonant electron heating to rotate the runaway electrons in the $p_{\|}, p_{\perp}$ phase space out of the RA region into region of perpendicular momentum heating. The fast-electron bremsstrahlung x-ray emission is measured from 20 to 200 KeV through a set of 17 horizontal lines of sight. The fast-electron bremsstrahlung x-ray emission shows electrons energies up to 16 MeV in the runaway regime. Then with the proper combination of the LH and ECRH power and beams, high-energy electron bremsstrahlung x-ray emission is essentially wiped out and the current is maintained at constant level without runaways by the lower-hybrid waves for periods of several seconds. Examples are given where a smooth transition from the Ohmic electric field-driven plasma current is transitioned to a lower-hybrid-driven current plasma and the bremsstrahlung x-ray emission drops to a relatively low level [Solis, *et al.* (2005)].

The TCV shot 29892 begins with a typical ohmic H-mode with type III ELMs. When the ECRH is turned on at $t = 0.7\,\mathrm{s}$, the plasma rapidly switches to a type I ELMy H-mode. When the ELMs disappear, a quasi-stationary ELM-free H-mode emerges. After a single ELM event at $t = 0.9\,\mathrm{s}$, the quasi-stationary ELM-free H-mode is recovered but the residual magnetohydrodynamic (MHD) activity keeps the energy confinement lower than in the previous phase. The suppression of the ELMs in this case coincides with the onset of the ECRH power modulation. The transition has also been observed to arise spontaneously [Porte, *et al.* (2006)]. The transition to an ELM-free H-mode is an interesting topic in itself.

13.3.1 *I-mode for improved confinement regimes*

An improved confinement mode with steep gradients in the scrape-off layer but without the ELMs for the H-mode were discovered in ASDEX [Wagner, *et al.* (1982)] and reproduced in more detail with extensive fluctuations measurements in Alcator C-Mod in Whyte, *et al.* (2010). In the I-mode the turbulent fluctuations and transport, in the edge region just inside the magnetic separatrix, decreases in the intermediate frequency 50–200 kHz turbulence level without the appearance of the ELMs. The inward radial electric field $E_r(r,t)$ in the I-mode is lower than for the H-mode but larger than in the L-mode plasmas.

As shown in Hubbard, *et al.* (2011), the electron temperature profile remains high and similar to that in the H-mode. This improved confinement mode offers an ELM-free or Type-III ELM improved confinement mode and is being explored as a possible tokamak fusion power system operating regime.

This discharge exhibits a higher density ($\sim 6 \cdot 10^{19}\mathrm{m}^{-3}$) than what is commonly found with ECRH heated tokamaks. The use of the third harmonic (X3) heating system instead of the second harmonic system allows for strong electron heating at plasma density up to approximately $10^{20}\mathrm{m}^{-3}$. This in turn also permits considerable ion heating due to the faster thermal equilibration between ions and electrons increasing with higher density. Unfortunately there is no charge exchange recombination spectroscopy (CXRS) measurements during this shot, that would provide profiles of carbon ion temperature, density and rotation. Such data is available for a sister shot.

The core electron power deposition in this discharge varies through the four phases with an average of 1–2 $\mathrm{MW/m^3}$ and has a maximum of 8 $\mathrm{MW/m^3}$ on-axis. Thus, the power deposition is strongly peaked on-axis. With a core electron density, $n_e \sim 6 \cdot 10^{19}\mathrm{m}^{-3}$ this gives a heating rate of 800 KeV/s per electron. The electron temperature is measured by Thomson scattering with 25 cords every 25 ms and the ECRH power deposition profile is calculated with the TORAY-GA ray-tracing code [Matsuda (1989)].

The electron heating is focused in the core plasma which gives rise to an accurately defined flow of thermal energy through the electron channel from the core plasma ($1 < T_e < 3\,\mathrm{KeV}$) to the edge plasma. This large thermal flow yields a precise estimate of the radial electron heat flux $q_e(r)$ and the electron heat diffusivity χ_e as there is accurate electron temperature profile data (5% error). The profiles are stationary for more than five energy confinement times, τ_E, and therefore the power balance χ_e is accurately determined. The greatest uncertainty in q_e and χ_e arrives from the ion temperature, T_i, and the uncertainties of the measurements of the plasma gradients. The error analysis performed shows that allowing T_i to vary between $0.33\,T_e$ to T_e only yields a 10% variation in the electron flux q_e in the ECRH heated phases. For the ohmic phase the error may be considerably larger.

The TCV deuterium plasma analyzed has electron plasma beta $\beta_e > 2\%$ at the core such that $\beta_e m_i/m_e \simeq 73$, where m_i and m_e are the masses of ions and electrons. The TCV 29892 plasma has a toroidal magnetic field $B \simeq 1.27\,\mathrm{T}$, plasma current $I_p \simeq 400\,\mathrm{kA}$ and Alfvén velocity $v_A \ll v_e = (T_e/m_e)^{1/2}$. This implies that we have an electromagnetic regime where the Alfvén wave falls in the low parallel velocity $v_\| < v_e$ part of the electron energy distribution where electron Landau damping is significant. Hence, wave damping is important and the coupling to drift waves is strong. Due to the steep electron temperature gradient and the electron plasma pressure we expect the electromagnetic electron temperature gradient (ETG) drift-wave turbulence to be an important transport mechanism. The plasma also enters the trapped electron (TEM) and ion temperature gradient (ITG) modes in the electron energy spectrum.

The tools for the TEM-ITG analysis are an analytical model [Zhu, *et al.* (1999, 2000)] and the fluid-based Weiland model [Guo and Weiland (1997, Jarmen, *et al.* (1987), Nordman, *et al.* (1990), Weiland (2000)]. For the ETG analysis we use a code based on the three-coupled gyrofluid differential equations, the linear modes of which describes well the ETG instability [Horton, *et al.* (2005), Kim, *et al.* (2006)]. This ETG code has proven itself by accurately modeling a large database of electron thermal transport from ASDEX [Ryter, *et al.* (2001a,b)], Tore Supra [Hoang, *et al.* (2003)] and the Frascati Tokamak Upgrade (FTU) [Jacchia, *et al.* (2002)].

13.3.2 *Average relative variance*

The metric called the average relative variance (ARV) [Weigend, *et al.* (1990), Detman and Vassiliadis (1997)] is introduced to obtain a quantitative measure of how well the theoretical models explain the observed heat flux,. The ARV metric measures the dimensionless error measure for inter-model comparisons. The ARV increases quadratically with the magnitude of the model χ_e so it increases with a mismatch in the model's relative error with respect to the database χ_e. The model ETG thermal flux is also in good agreement with the power balance as shown in a figure for the thermal diffusivities in Section 13.2. The ARV parameters for the analytical TEM model and the local Weiland model show that these models fail to predict quantitatively the thermal transport. The ETG model, on the other hand, can, according to this metric, explain 70% of the variation of electron heat diffusivity.

13.4 ECRH Driven Plasmas

A range of H-mode plasma regimes is produced with high-power electron cyclotron resonance heating and strong shaping of the magnetic flux surfaces. DIII-D, EAST, JET, FTU, and TCV have produced H-mode plasmas with ECRH. The different types of plasma confinement regimes are illustrated in a single long pulse discharge in basic tokamak experiment with extensive diagnostics. The TCV discharge 29892 analyzed in this chapter is described by Porte, *et al.* (2006) in the conference proceedings of the 21st IAEA Fusion Energy Conference [Porte, *et al.* (2006)]. This section gives the basic features of a quasi-steady-state plasma discharge that illustrates the underlying heat transport and energy confinement in a variety of plasma confinement states.

The TCV discharge 29892 analyzed in this chapter is described by Porte, *et al.* (2006) in the conference proceedings of the 21st IAEA Fusion Energy Conference [Porte, *et al.* (2006)]. This section gives the basic features of the experiment needed to understand the underlying heat transport and energy confinement. TCV is a tokamak with a high-power density electron cyclotron resonance heating (ECRH) system described in Table 13.1. In shot 29892, the third-harmonic X-mode (X3) heating system was used to enable significant electron heating at the plasma density $n_e \sim 6 \cdot 10^{19}$ m^{-3}.

The RF electron heating gave rise to a significant ion heating $P(e \to i)$ through thermal equilibration, $P_{e \to i} = 0.2 \pm 0.1$ MW. Ion dynamics may directly or indirectly influence the electron transport. Ion heating from the electrons is also of fundamental importance for ITER and in nuclear fusion producing machines in which the ion heating is from the alpha particles through the electrons. Electron-to-ion heat transfer is vital to sustain the fusion reactions. This TCV discharge features an ITER hybrid scenario q-profile with low magnetic shear.

The X3 (third harmonic extraordinary mode polarized electron cyclotron wave) heating is applied at $t = 0.6$ s to a target plasma consisting of an ohmic ELMy H-mode. The X3 ECH power is injected vertically from the top of the machine with the resonating surfaces close to the plasma axis which makes the heating peak in the inner core of the plasma as shown in Fig. 13.8. Moreover, the high temperature ($T_e \approx 1.4$ KeV) caused by the high-energy confinement time, ($\tau_E \approx 37$ ms), of the target plasma helps to increase the X3 absorption. The X3 heating does not produce a significant population of supra-thermal electrons that would have influenced the transport analysis. The ion temperature profile used for the calculation of the thermal exchange power density in Fig. 13.8 was taken from a sister shot where CXRS data was available. The $T_i(r,t)$ profiles were cross-checked with

Table 13.1 TCV machine and discharge 29892 parameters. Radial values are given at $r/a = 0.7$ in the Ohmic H-mode and the ELM-free H-mode.

Fixed Parameters		
Plasma major radius, R (m)	0.9	
Plasma minor radius, a (m)	0.22	
Plasma elongation	1.75	
Plasma triangularity	0.51	
Plasma volume (m^3)	1.5	
Plasma current, I_p (kA)	400	
Toroidal field at the magnetic axis (T)	1.3	
Time dependent parameters	$t = 0.5$ s	$t = 1.0$ s
Total heating, $P_{\rm tot}$ (MW)	0.49	1.13
Global energy confinement time, τ_E (ms)	37	32
H-factor $H_{98(2,y)}$	0.72	1.1
Safety factor, q	1.29	1.30
Safety factor, q_{95}	2.18	2.16
Magnetic shear, s	0.80	0.65
Density gradient length, L_{ne} (m)	0.15	0.21
Temperature gradient length, L_{Te} m)	0.12	0.09
Electron temperature, T_e (KeV)	0.57	1.3
Electron diamagnetic drift velocity, v_{de} (km/s)	2.0	4.5
Electron magnetic drift velocity, v_{De} (km/s)	0.5	1.3
Ion Larmor radius at T_e, ρ_s (mm)	2.7	4.1
Power-balance diffusivity, $\chi^{\rm PB}$ (m^2/s)	0.28	0.83
Gyro-Bohm diffusivity, $D_{gB} = \rho_s v_{de}$ (m^2/s)	5.4	18.5
Drift wave frequency, $\omega = k_y v_{de}$ (kHz) at $k_y \rho_s = 0.3$	341	353
Electron collisionality, ν_*	0.24	0.06
ω_{pe}/ω_{ce}	1.8	1.9

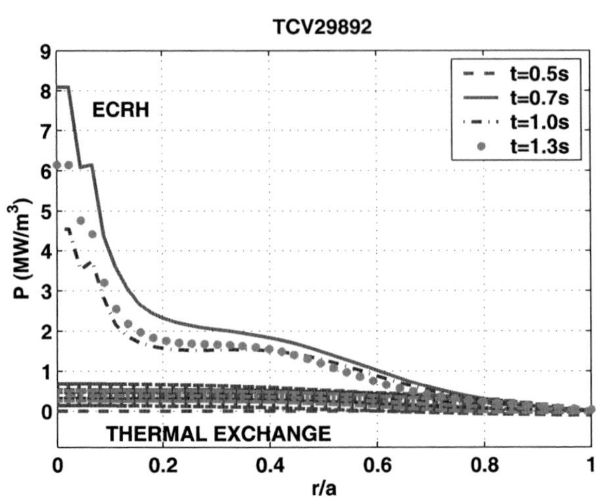

Fig. 13.8 Deposition of ECRH in TCV 29892 calculated by TORAY-GA [Matsuda (1989)] and the thermal equilibration power to the ions. The latter shows the range of values expected assuming a 5% error in n_e and T_e and within a 20% error in T_i.

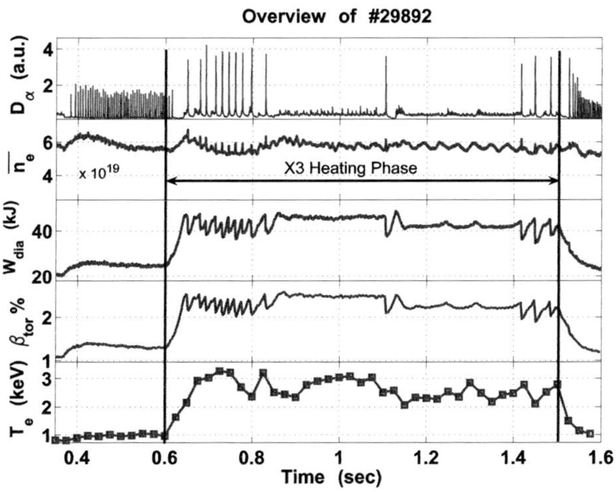

Fig. 13.9 Temporal evolution of TCV29892. From top to bottom are shown D_a emission, line-averaged density, total stored energy measured by the diamagnetic loop, toroidal β and on-axis electron temperature. At $t < 0.6$ s, standard TCV H-mode with Type I ELMs. Full ECRH power for $0.6 \leq t < 0.8$ s and Type I ELMy H-mode. Modulated ECRH power for $t \geq 0.8$ s and quasi-stationary ELM-free H-mode. Single-ELM event at $t = 1.1$ s triggers MHD activity which reduces the confinement as shown by the decrease in W_{dia} at the D_α-spike at $t = 1.1$ s.

Neutral Particle Analyzer (NPA) [Karpushov, *et al.* (2006)] measurements of the energy spectra of the deuterium ions escaping the plasma.

The electron heat flux is proportional to the difference between the integrated electron power deposition P_e and the integrated electron-ion thermal equilibration $e \to i$, energy transfer power from electrons to ions through collisions. During the ohmic phase at $t = 0.5$ s, the total thermal equilibration is comparable to the total power deposition $P_e = P_{\text{oh}} \sim 0.4$ MW. For the ohmic regime, a 10% error in the ion temperature produces a 20% error in the electron heat flux. In the ECRH X3-heated phases of the discharge, a 10% error in the ion temperature leads to approximately a 10% variation in the electron heat flux.

The time history of the discharge TCV2989 is shown in Fig. 13.9 with five panels for plasma-neutral ionization $D_\alpha(t)$, mean density $\overline{n}_e(t)$, stored plasma energy $W_{\text{dia}}(t)$, plasma beta and electron temperature.

The D_α line is from electron transitions from the $n = 3$ to the $n = 2$ quantum (Bohr) levels of the deuterium atom. The signal is used to monitor the interaction of the hot plasma with neutral gas component that surrounds the plasma in edge and scrape-off layer. When there is a burst of brightness in the alpha line radiation there is a excursion or a plume of the hot plasma into the outer layer of the confinement chamber. Thus, the edge localized modes called ELMs give rise to bursts of the red $\lambda = 656$ nm radiation visible light from the plasma. For hydrogen working gas the line radiation is from the neutral hydrogen atom and for the deuterium working gas the line is from the neutral deuterium atoms. Owing to the higher nuclear mass

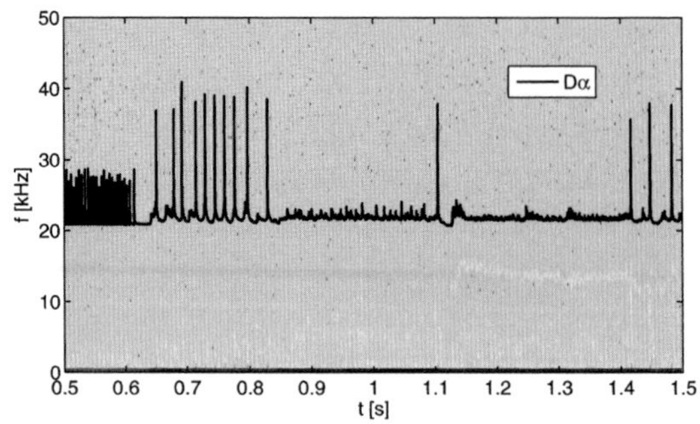

Fig. 13.10 Spectral temporal evolution of TCV29892 also including the D_a emission. The $m/n = 3/2$ mode with a frequency of 15 kHz emerges after the single ELM event at 1.1 s.

of deuterium the D_α line has a slight shorter wavelength of $\lambda = 656\,\text{nm}$ than the hydrogen H_α line.

The transport analysis is carried out in each of the four H-mode phases shown in Figs. 13.9 and 13.10. The first phase ($< 0.6\,\text{s}$) is a standard ohmic H-mode with Type III ELMs. At $t = 0.6\,\text{s}$ the ECRH X3 heating is switched on at full power and the plasma enters into a type I ELMy H-mode. The X3 heating is at full power ($\sim 1.35\,\text{MW}$ coupled to the plasma) until 0.8 s when the power from one gyrotron is fully modulated at a frequency of 127 Hz with a duty cycle of 50%. This modulation results in an average heating power reduction of approximately 17%. Although, in this case, there was a strong correlation between X3 power modulation and the onset of the ELM-free H-mode, power modulation is not necessary [Porte, *et al.* (2006)]. The ELM-free H-mode regime has been achieved without modulation. In contrast to other machines like DIII-D [Lohr, *et al.* (1988)] and Alcator C-Mod [Fielding, *et al.* (1996)] which have also produced ELM-free H-modes, the TCV tokamak accomplishes this with no direct ion heating, no active fueling and no cryo-pumping.

The ELM-free nature of the discharge is seen from the steady D_a signal in the top frame of Fig. 13.9 and Fig. 13.10 while the high-energy confinement is shown in the measurement of total energy content obtained from a diamagnetic loop, W_{dia} in the third frame of Fig. 13.9. The constant W_{dia} and line averaged density (second frame) show that this phase is quasi-stationary. The phase ends with a single ELM event at 1.1 s at which time a $m/n = 3/2$ magnetic island is triggered by a core $m = 1$ sawtooth oscillation as seen in Fig. 13.10. This MHD mode has a frequency of 15 kHz and a magnetic island width of about 5 cm [Turri, *et al.* (2008)]. A magnetic island of this width located at a normalized minor radius around $r/a = 0.6$ gives rise to a decrease of total stored energy of about 25% according to the model of Chang and Callen (1990, 1992). The observed energy drop in Fig. 13.9 third frame, is closer to 15%. There is therefore no doubt that it is the destabilization of the $m/n = 3/2$ MHD mode that causes the drop in confinement in this fourth and last H-mode phase of the discharge.

Note that although affected by this 3/2 MHD mode, this fourth phase is a quasi-stationary ELM-free H-mode. Moreover, $m/n = 1/1$ modes associated with sawteeth activity and intermittent $m/n = 2/1$ are present throughout the discharge [Turri, *et al.* (2008)].

Simulations of ELMs are performed with MHD codes like JOREK [Huysmans and Czarny (2007)]. Using models of the H-mode plasma profiles with sharp gradients in the pressure profile and the toroidal current profiles produce bursts of complex interchange modes called the peeling-ballooning mode that releases up to 10% of stored plasma energy in a short burst. These events are monitored by the light emission from the ionization-recombination lines of the neutral deuterium atoms in the edge plasma. Examples of the ELMs are shown in Fig. 13.10.

For axisymmetric tokamak geometry without external toroidal momentum input the radial electric field E_r is arbitrary within the ordering of $|v_{E_r}| \sim \rho_i v_i/a$. Only, small deviations from axisymmetry such as the toroidal field ripple lead to a particular E_r. The poloidal rotation velocity is of the order

$$v_\theta = C_{E_r} \frac{T_e}{eBL_{ne}} \tag{13.19}$$

with the numerical coefficient C_{E_r} bounded by 2. The standard observed tokamak core rotation is in the electron diamagnetic direction from $E_r < 0$. The heavy ion beam probe on TEXT [Hallock, *et al.* (1994)] measured this negative $E_r < 0$ directly. This magnitude of E_r is consistent with a collisional loss of ions in the edge or scrape off layer plasma. In outer edge region the sign of E_r flips to hold the electrons in resulting in a narrow shear poloidal flow shear layer at the edge of the plasma. Bengtson, *et al.* (1992) verified this picture with probes and the HIBP in the Ohmic heated L-mode TEXT tokamak.

In the single null divertor configurations the inward E_r shear is enhanced by having the grad-B and curvature drifts directed toward the separatrix divertor chamber. This breaks up-down symmetry and results in a lower threshold power for the transition from the L-mode to the H-mode presumably due to the stronger radial shear in the $E_r(r)$ scrape-off layer field.

This upper limit to E_r is used to analytically and numerically estimate the effect on the growth rates for ITG and ETG turbulence [Hamaguchi and Horton (1992), Sugama and Horton (1995)]. The effect is rather complicated since at first order there is a radial shift of the wave functions proportional to the E_r shear divided by the magnetic shear. The shear-flow effect is significant on weakly growing ITG modes and generally weak on the ETG modes. From the electron temperature T_e profile and the system parameters in Table 13.1, the range of the shearing frequency is $\mathrm{d}v_{E_r}/\mathrm{d}r = 5 \sim 8 \times 10^4\ \mathrm{s}^{-1}$. This shearing rate may stabilize some of the slower growing TEM modes given in Section 12.2.4 but is too low to stabilize the turbulence considered here. For ETG this level of sheared rotation is negligible due to the fact that the electrons are interacting with much higher frequency fluctuations.

13.5 Electron Temperature Gradient Turbulence Modeling

The electron temperature gradient drift wave frequency $\omega_{0,\mathrm{ETG}}$ is given in Eq. (13.16) arising from the $\mathbf{E} \times \mathbf{B}$ convection of the electron density in the presence of adiabatic ions $\delta n_i = -eZ_i n_i \phi/T_i$. The frequency of the mode is shown in Fig. 13.7 for $k_y \rho_e = 1.0$. The smaller electron temperature gradient toward the edge gives the change of the wave direction from electron diamagnetic to ion diamagnetic drift direction.

A gyrokinetic simulation for the ETG was carried out with by Jenko and Dorland (2002) using the codes they developed called GENE and GS2. The results were that the ETG thermal flux for usual tokamak parameters is large and universal in nature. The comparison with the slab ETG simulation is given and the toroidal curvature-driven flux is about 30 to 40 times larger than the slab ETG thermal flux for their particular set of parameters.

The slab ETG simulation with the GTC code are given by Fu, *et al.* (2012) and show a validation of the model with the measured $T_e(r)$ and $n_e(r)$ profiles and fluctuations in the steady-state hydrogen cylindrical Columbia Linear Machine (CLM) described in Wei, *et al.* (2010). The simulations required large computing power owing to the small scale of the ETG simulation in the initial growth phase of the ETC. The experiments require special micro-electronic detectors for the mega-Hertz frequencies and 100 micron wavelengths. The GTV simulations show that there is a strong inverse cascade from the high-mode numbers to the lower modes with amplitudes peaking at $m = 11$, 12, 13 modes. The agreement with the laboratory data is impressive in Fu, *et al.* (2012). This agreement between the data, the theory, and the gyrokinetic simulations validates the ETG model.

In this chapter we analyze the turbulence using gyrofluid equations for the electron dynamics with an electromagnetic field $A_\parallel$ to generate the associated magnetic turbulence. The gyrokinetic simulations use several simplifications including adiabatic ion response and electrostatic fields with a limited number of parameter variations.

Gyrofluid models of the ETG mode were developed to extend the convection fluid model for the study of drift wave turbulence [Horton, *et al.* (1988), Li and Kishimoto (2004, 2005), Holland and Diamond (2002)]. Here we describe the finite electron Larmor radius (gyrofluid) fluid model [Horton, *et al.* (1988, 1990), Horton and Tajima (1988)]. The model was used for the analysis of the Tore Supra discharges [Horton, *et al.* (2004, 2005)] and upgraded [Kim, *et al.* (2006)] with electron Landau damping effect [Hammett and Perkins (1990)] for NSTX. The model is constructed in a local toroidal geometry where the magnetic curvature and grad-$\mathbf{B}$ drifts are important. The magnetic shear is low and positive in these discharges which allows the approximation of a constant $k_\parallel = |s|/qR$ in the analysis. This constant $k_\parallel$ approximation is acceptable when the growth rate is not near marginal stability. The higher-order finite Larmor radius (FLR) dynamics is modeled by cross-field viscosities that absorb the energy cascaded to high-k fluctuation components. The kinetic Landau damping physics is represented by the Hammett and Perkins closure model [Hammett and Perkins (1990)], $q_{\parallel} = -\sqrt{8/\pi}\, k_{\parallel} v_e \delta T_e(k)/|k_{\parallel}|$, for the fast parallel electron thermal dynamics.

The ETG turbulence equations of the toroidal system are given by

$$\widehat{\boldsymbol{T}}\Phi = \boldsymbol{L}\Phi + \widehat{\boldsymbol{N}}(\Phi, \Phi), \tag{13.20}$$

where $\widehat{\boldsymbol{T}}$ is the first-order time derivative operator, the $\widehat{\boldsymbol{L}}$ operators are the linear spatial operator and $\widehat{\boldsymbol{N}}$ the nonlinear spatial operator, respectively.

The radial and poloidal position (x, y), the toroidal position z and time t are normalized by the electron gyroradius ρ_e, electron density gradient length L_n and L_n/v_{Te}. The 3-component field vector $\boldsymbol{\Phi} = (\phi, A_{\parallel}, \delta T_e)$ consists of the electrostatic fluctuation ϕ, the parallel magnetic potential fluctuation $A_{\parallel}$ and the electron temperature fluctuation δT_e, all of which are normalized by T_e/e, $\beta_e T_e/eV_{Te}$ and T_e with the additional factor ρ_e/L_n.

The time derivative operators in Eq. (13.20) are

$$\begin{aligned}
\widehat{\boldsymbol{T}}_{11} &= \left(1 - \nabla_{\perp}^2\right)\frac{\partial}{\partial t}, \\
\widehat{\boldsymbol{T}}_{22} &= \left(\frac{\beta_e}{2} - \nabla_{\perp}^2\right)\frac{\partial}{\partial t}, \\
\widehat{\boldsymbol{T}}_{33} &= \frac{\partial}{\partial t} \quad \text{and} \quad \widehat{\boldsymbol{T}}_{ij} = 0 \quad \text{where} \quad i \neq j.
\end{aligned} \tag{13.21}$$

The linear spatial derivative operators are given by

$$\widehat{\boldsymbol{L}}_{11} = \left(1 - 2\epsilon_n + (1+\eta_e)\nabla_{\perp}^2\right)\frac{\partial}{\partial y} - \mu\nabla^4, \widehat{\boldsymbol{L}}_{12} = \nabla_{\perp}^2\frac{\partial}{\partial z}, \ \widehat{\boldsymbol{L}}_{13} = 2\epsilon_n\frac{\partial}{\partial y} \tag{13.22}$$

$$\widehat{\boldsymbol{L}}_{21} = -\frac{\partial}{\partial z}, \ \widehat{\boldsymbol{L}}_{22} = -\frac{\beta_e}{2}(1+\eta_e)\frac{\partial}{\partial y} + \frac{\eta}{\mu_0}\nabla_{\perp}^2, \ \widehat{\boldsymbol{L}}_{23} = \frac{\partial}{\partial z} \tag{13.23}$$

$$\widehat{\boldsymbol{L}}_{31} = -\left(\eta_e - 4\epsilon_n(\Gamma - 1)\right)\frac{\partial}{\partial y}, \ \widehat{\boldsymbol{L}}_{32} = -(\Gamma - 1)\nabla_{\perp}^2\frac{\partial}{\partial z} \tag{13.24}$$

$$\widehat{\boldsymbol{L}}_{33} = -2\epsilon_n(\Gamma - 1)\frac{\partial}{\partial y} + \chi_{\perp}\nabla_{\perp}^2 + \frac{\partial^2}{\partial z^2} + (\Gamma - 1)\sqrt{\frac{8}{\pi}}\,|k_{\parallel}| \tag{13.25}$$

where electron plasma beta $\beta_e = 2\mu_0 p_e/B^2$, the ratio of electron and ion temperature $\tau = T_e/T_i$ and the ratio of electron density gradient length L_n and magnetic field gradient R, $\epsilon_n = L_n/R$. Background viscosity μ, resistive diffusivity η/μ_0 and heat diffusivity $\chi_{\perp}$ are included.

The linear operator matrix $\widehat{\boldsymbol{L}}_{ij}$ is a function of five-dimensional parameter vector $\{\mu^5\} = \{\eta_e, \epsilon_n, T_e/T_i, \beta_e, \Gamma\}$ and weakly on the three dissipation coefficients $\{\mu, \eta/\mu_0, \chi_{\perp}\}$.

The nonlinear derivative terms in the dynamics in Eq. (13.20) are given by

$$\widehat{\boldsymbol{N}}_1 = \widehat{\boldsymbol{N}}_{111} + \widehat{\boldsymbol{N}}_{122} = -\left[\phi, \nabla_{\perp}^2\phi\right] + \frac{\beta_e}{2}\left[A_{\parallel}, \nabla_{\perp}^2 A_{\parallel}\right]$$

$$\widehat{\boldsymbol{N}}_2 = \widehat{\boldsymbol{N}}_{212} + \widehat{\boldsymbol{N}}_{221} + \widehat{\boldsymbol{N}}_{223} = -\left[\phi, \nabla_\perp^2 A_\parallel\right] - \frac{\beta_e}{2}\left[A_\parallel, \phi - \delta T_e\right]$$

$$\widehat{\boldsymbol{N}}_3 = \widehat{\boldsymbol{N}}_{313} + \widehat{\boldsymbol{N}}_{322} = \left[\phi, \delta T_e\right] - (\Gamma - 1)\frac{\beta_e}{2}\left[A_\parallel, \nabla_\perp^2 A_\parallel\right]$$

where the Poisson bracket $[f, g] = (\partial f/\partial x)(\partial g/\partial y) - (\partial f/\partial y)(\partial g/\partial x)$.

The energy transfer equations without viscosity, resistivity and diffusion derived from the dynamics are

$$\begin{aligned}
\frac{\partial W_E}{\partial t} &= \left\langle \phi \widehat{\boldsymbol{L}}_{11}\phi \right\rangle + \left\langle \phi \widehat{\boldsymbol{L}}_{12} A_\parallel \right\rangle + \left\langle \phi \widehat{\boldsymbol{L}}_{13}\delta T_e \right\rangle - \left\langle \phi \widehat{\boldsymbol{N}}_{122} \right\rangle \\
&= \left\langle \phi \nabla_\perp^2 \frac{\partial A_\parallel}{\partial z} \right\rangle + 2\epsilon_n \left\langle \phi \frac{\partial \delta T_e}{\partial y} \right\rangle - \frac{\beta_e}{2}\left\langle \phi\left[A_\parallel, \nabla_\perp^2 A_\parallel\right]\right\rangle \\
\frac{\partial W_B}{\partial t} &= -\left\langle \nabla_\perp^2 A_\parallel \widehat{\boldsymbol{L}}_{21}\phi \right\rangle - \left\langle \nabla_\perp^2 A_\parallel \widehat{\boldsymbol{L}}_{22} A_\parallel \right\rangle - \left\langle \nabla_\perp^2 A_\parallel \widehat{\boldsymbol{L}}_{23}\delta T_e \right\rangle \\
&\quad + \left\langle \nabla_\perp^2 A_\parallel \left(\widehat{\boldsymbol{N}}_{221} + \widehat{\boldsymbol{N}}_{223}\right)\right\rangle \\
&= -\left\langle \phi \nabla_\perp^2 \frac{\partial A_\parallel}{\partial z} \right\rangle + \left\langle \delta T_e \nabla_\perp^2 \frac{\partial A_\parallel}{\partial z} \right\rangle + \frac{\beta_e}{2}\left\langle \phi\left[A_\parallel, \nabla^2 A_\parallel\right]\right\rangle \\
&\quad - \frac{\beta_e}{2}\left\langle \delta T_e\left[A_\parallel, \nabla^2 A_\parallel\right]\right\rangle \\
\frac{\partial W_{\delta T_e}}{\partial t} &= \frac{1}{\Gamma - 1}\left(\left\langle \delta T_e \widehat{\boldsymbol{L}}_{31}\phi \right\rangle + \left\langle \delta T_e \widehat{\boldsymbol{L}}_{32} A_\parallel \right\rangle + \left\langle \delta T_e \widehat{\boldsymbol{L}}_{33}\delta T_e \right\rangle - \left\langle \delta T_e \widehat{\boldsymbol{N}}_{322} \right\rangle\right) \\
&= \left(\frac{\eta_e}{\Gamma - 1} - 4\epsilon_n\right)\left\langle \phi \frac{\partial \delta T_e}{\partial y} \right\rangle - \left\langle \delta T_e \nabla_\perp^2 \frac{\partial A_\parallel}{\partial z} \right\rangle + \frac{\beta_e}{2}\left\langle \delta T_e\left[A_\parallel, \nabla_\perp^2 A_\}\right]\right\rangle
\end{aligned}$$

where the three components of the total turbulent energy are

$$\begin{aligned}
W_E &= \frac{1}{2}\left\langle \tau|\phi|^2 + |\boldsymbol{\nabla}_\perp \phi|^2 \right\rangle \\
W_B &= \frac{1}{2}\left\langle \frac{\beta_e}{2}|\boldsymbol{\nabla}_\perp A_\parallel|^2 + |\nabla_\perp^2 A_\parallel|^2 \right\rangle \\
W_{\delta T_e} &= \frac{1}{2(\Gamma - 1)}\left\langle |\delta T_e|^2 \right\rangle
\end{aligned}$$

and the integration $\langle\cdot\rangle$ is done over $dxdydz$. The total energy evolves as

$$\frac{\partial}{\partial t}\left(W_E + W_B + W_{\delta T_e}\right) = -\left(\frac{\eta_e}{\Gamma - 1} - 4\epsilon_n\right)\left\langle \delta T_e \frac{\partial \phi}{\partial y} \right\rangle - \left\langle \sqrt{\frac{8}{\pi}}|k_\parallel||\delta T_e|^2 \right\rangle \quad (13.26)$$

which in terms of the turbulent thermal flux

$$q_e = \langle v_{Ex}\delta T_e \rangle = -\left\langle \frac{\partial \phi}{\partial y}\delta T_e \right\rangle$$

which is the driving term in the energy balance equation given by

Energy Balance Theorem

$$\frac{d}{dt}\left(W_E + W_B + W_{\delta T_e}\right) = q_e\left(\frac{\eta_e}{\Gamma - 1} - 4\epsilon_n\right) - \left(\frac{8}{\pi}\right)^{1/2}\left\langle |k_\parallel|\,|\delta T_e(k)|^2\right\rangle. \quad (13.27)$$

Thus, the heat flux times electron temperature gradient beyond the compressional threshold drives up the turbulent fluctuations. The $\mathbf{E}\times\mathbf{B}$-induced turbulent energy flux $q_e = -\langle \delta T_e \partial\phi/\partial y\rangle$ times the threshold function determines the total energy evolution according to the power balance theorem in Eq. (13.21).

The linear stability analysis of the gyrofluid model shows the large increase of the growth rates from $t = 0.5\,\mathrm{s}$ to $0.7\,\mathrm{s}$ (Fig. 13.11) for the TCV discharge.

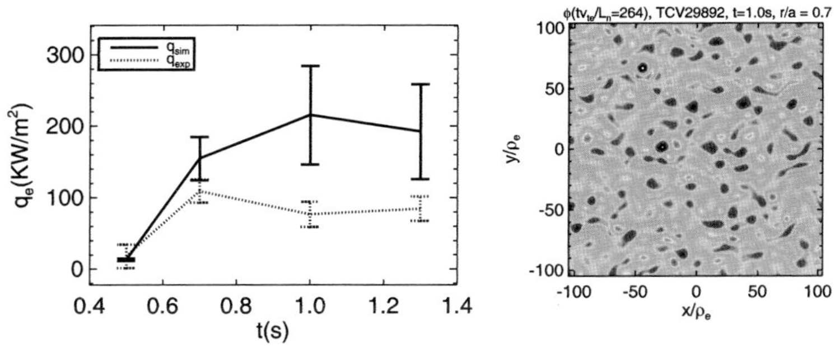

(a) Electron heat fluxes. (b) Electrostatic potential.

Fig. 13.11 Electron heat fluxes (a) from nonlinear ETG fluid simulations. The solid and dashed line represent the simulations and the experiment. The contours of electrostatic potential $\phi(x, y)$ (b) from the nonlinear simulation at TCV29892 $r/a = 0.7$ and $t = 1.0\,\mathrm{s}$ where $\rho_e = 75\mu\,\mathrm{m}$ and the box corresponds to $1.5\,\mathrm{cm} \times 1.5\,\mathrm{cm}$.

In the inner region $r/a < 0.3$ the mode is almost linearly stable. The strong ETG instabilities at the later times are consistent with the behavior of the heat diffusivities derived from the power balance analysis with accurately measured electron temperature profiles.

Gyrokinetic simulations of the ETG turbulence show similar results as the gyrofluid simulations described here, but require extensive computer simulation resources [Nakata, *et al.* (2010)]. An example giving similar comparisons of the power balance χ_e with the simulations from a gyrokinetic code are given in Joiner, *et al.* (2006).

13.6 Validation Analysis of the Electron Transport Modeling

Since there has historically been a multitude of models for the anomalous electron thermal transport problem a validation method is needed to rank the success of the various models. Here we briefly describe the use of one metric commonly used in weather systems to compare the TEM-ITG model and the ETG model for the discharge analyzed in detail in this chapter. There will be a need for a community wide method of comparing models in the future as the number of models and the associated number of simulations codes is continually expanding.

Validation Analysis of the ETG Model

When the nonlinear simulations are too heavy numerically to give the heat transport over the whole minor radius, one may use the theoretical scaling law to estimate the ETG-driven transport in the discharge. The gyroBohm heat diffusivity with length scale $q\rho_e$ and time scale L_{T_e}/v_{T_e} [Horton, *et al.* (2004)] is

$$\chi_e = \alpha_{\text{ETG}}\, q^2 \rho_e^2 v_{T_e} \left(\frac{1}{L_{T_e}} - \frac{1}{L_{T_e}^{\text{crit}}} \right) \tag{13.28}$$

where q is the safety factor and α_{ETG} is a scaling coefficient obtained from comparison with experimental data or with more rigorous theoretical calculations.

In the region $r/a > 0.4$, we calculated χ_e in Eq. (13.28) using the inverse of the critical temperature gradient length,

$$\frac{1}{L_{T_e}^{\text{crit}}} = \left(1 + Z_{\text{eff}} \frac{T_e}{T_i}\right) \left(\frac{1.33 + 1.9|s|}{q}\right) /R \tag{13.29}$$

The critical temperature length turns out to be five to ten times larger than the electron temperature gradient length, L_{T_e} at $r/a > 0.4$ derived from the measure electron temperature profiles.

The average relative variance (ARV) [Weigend, *et al.* (1990), Detman and Vassiliadis (1997)] is the degree to which the theoretical models for χ_e explain the power balance data χ^{pb} and is given in terms of the data and the model by

$$\text{ARV}_j = \frac{\sum_i \left(\chi_{e,i} - \chi_i^{\text{pb}}\right)^2}{\sum_i \left(\chi_i^{\text{pb}} - \left\langle \chi_i^{\text{pb}} \right\rangle\right)^2} \tag{13.30}$$

with the spatial and temporal indices i and j, respectively. For TCV 29892, the power balance diffusivities χ_i^{pb} are accurately known from the plasma data and the theoretical heat diffusivities $\chi_{e,i}$. Here $\left\langle \chi_i^{\text{pb}} \right\rangle$ is the average power balance diffusivity over all radial points for each temporal index j. For each temporal slice with index j, there is a total of 41 points whereof there are 30 radial points for $r/a > 0.4$.

The ARV is a widely-used metric measure for how well a model's prediction follows the behavior of experimental data. When the averages of $\chi_{e,i}$ and χ_i^{pb} are comparable, ARV < 1 indicates that a model behaves in the same way as the experiment as shown in Fig. 13.12 where χ_e^{pb} increases as χ_e^{ETG}. A smaller ARV means better agreement.

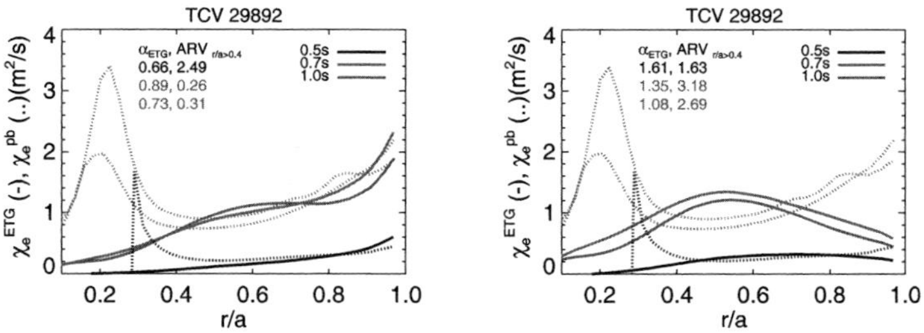

(a) (b)

Fig. 13.12 The comparison between the electron gyroBohm heat diffusivity, χ_e in Eq. (13.28), with ETG time and spatial scales (the lines with $*$) and the power balance heat diffusivity χ_e^{pb} (no mark) for each time slice $t = 0.5$ (solid), 0.7 (dotted) and 1.0 (dashed) are shown.

The model with $\alpha_{\rm ETG} \sim 0.8$ shows the best agreement beyond $r/a = 0.4$ in Fig. 13.12. The gyroBohm ETG heat diffusivity formula with a free parameter $\alpha_{\rm ETG}$, Eq. (13.28) explains not only the electron heat diffusivity increase with the H-mode transition but also gives ARV ~ 0.3 for $t = 0.7\,$s and $1.0\,$s. For $t = 0.5\,$s, the flat experimental heat flux profile gives a small denominator in Eq. (13.30) producing the large ARV. With $\alpha_{\rm ETG} \sim 1$ and the assumption that the dominant time scale is $\omega \sim v_{T_e}/L_{T_e}$, we can estimate the correlation length $l_c \sim 2\pi q \rho_e \sim 10\rho_e$, consistent with the nonlinear simulation result.

13.7 LHCD Driven Discharges and Anisotropic Electron Phase-Space Distribution Functions

In tokamak fusion plasmas, coherent fluctuations in the form of blobs or filaments are routinely observed in the scrape-off layer. Radio frequency (RF) electromagnetic waves, excited by antenna structures placed near the wall of a tokamak, have to propagate through the scrape-off layer before reaching the core of the plasma. While the effect of fluctuations on the properties of RF waves has not been quantified experimentally, it is of interest to carry out a theoretical study to determine if fluctuations can affect the propagation characteristics of RF waves. Lower-hybrid current drive demonstrated on PLT tokamak [Bernabei, *et al.* (1982)] is the most robust and efficient method of driving the tokamak current with external radio frequency waves in steady-state tokamak operation [Litaudon (2010)]. The method has been tested in Tore Supra for discharges at megawatts of RF power over periods of minutes [Hoang, *et al.* (2003)].

Lower-hybrid current drive suggested by Fisch (1978) and demonstrated on PLT tokamak [Bernabei, *et al.* (1982)] is a robust and efficient method of driving the tokamak current with external radio frequency waves in steady-state tokamak operation [Litaudon (2010)]. The method has been tested in Tore Supra for discharges at megawatts of RF power over periods of minutes [Hoang, *et al.* (2003)].

The Tore Supra power deposition code is ALCYON and the transport code is CRONOS, given in Artaud, *et al.* (2010). Simulations have been performed with CRONOS using ALCYON for the FW power deposition calculation.

Intense electron heating from electron cyclotron resonant waves (ECRH) and from lower hybrid waves (LHRH and LHCD) distort the electron phase space distribution functions producing significant changes in the unstable parameter space for the ETG turbulence. Heavy parallel velocity tails from LHCD are close to runaway [Knoepfel and Spong (1979)], some times called slide away, electron distribution functions have much lower rates of electron Landau damping and thus higher rates of radial gradient driven growth rates for ETG turbulence.

The electron cyclotron resonance heating in TCV showed through tomographic reconstruction of the ECE emission on the high-field side of the plasma a two-component energy distribution with the high-energy part having an effective temperature of 10 to 50 KeV. The high-energy part is essential in creating an increased absorption of the second 2X and third 3X harmonic heating with the extraordinary wave (X) polarization. The extraordinary wave has the wave electric field pointing perpendicular to the magnetic field, which then gives much stronger coupling to the plasma through mode conversions to electrostatic waves than the ordinary wave (O) with the wave electric field parallel to the toroidal magnetic field [Blanchard, *et al.* (2002), Laqua (2007)].

The electron distributions functions computed from LHCD wave codes produce skewed parallel velocity distributions as opposed to drifting Maxwellians. These RF driven skewed distributions have large parallel thermal fluxes $q_\parallel$ with steep radial gradients.

A velocity distribution with a large parallel heat flux $q_\parallel(r,t)$ has a velocity skewness that makes the distribution intrinsically different from that of a shifted, bi-Maxwellian. Thus, it is important to compute the kinetic Vlasov response functions for these skewed distribution functions. Methods for driving an electron drift velocity to maintain a steady-state tokamak plasma current I_p, will typically create a large parallel thermal flux $q_\parallel(r)$. The best known method of driving the plasma current is with microwaves focused by the antennas to carry the required unidirection momentum to be absorbed by the electrons. This involves the use of high-power levels of microwaves that become lower hybrid plasma waves. This method of current drive is designated as LHCD for lower hybrid current drive. Contours of the constant levels of a the electron distribution functions computed for the simulations of $P_{\mathrm{LHCD}} = 8\,\mathrm{MW}$ apply to the Tore Supra plasma using the CIEL2F3 simulation code. All the electric current $j_\parallel = -en_e u_\parallel$ and parallel thermal flux $q_\parallel$ is carried by the passing electrons. The plasma response function for the LHCD distribution is defined as $P_{\mathrm{LHCD}}(\omega, \mathbf{k}, f_e)$.

In Table 13.2 the drift wave stability parameters for the LHCD driven plasmas are given. The dimensionless parameter $\alpha_q = q_\parallel / n_e T_e v_e$ defined by the generalized drift wave driving term is given. For example, the LHCD distribution

Table 13.2 Parameters for LHCD driven plasmas.

Drift wave parameters in LHCD driven plasmas

$$\omega_{*e} = \frac{k_y T_e}{q_e B n_e}\frac{dn_e}{dr} = \frac{k_y T_e}{eBL_{n_e}} \text{ and } \omega_{*eT_e} = \eta_{e\|}\omega_{*e} = -\frac{k_y}{eB}\frac{dT_{e\|}}{dr} = \frac{k_y T_{e\|}}{eBL_{T_e}},$$

where $\eta_{e_\|} = \frac{\partial_r \ell n T_{e\|}}{\partial_r \ell n n_e} = \frac{L_{n_e}}{L_{T_e}}$ and $\eta_{e\perp} = \frac{\partial_r \ell n T_{e\perp}}{\partial \ell n n_e}$

The dangerous instabilities have: $\omega_k \leq |\omega_{*e}| \leq \frac{v_e}{L_{T_{e\|}}} \simeq (3-5)k_\| v_e$

and thus $k_\| L_{T_{e_\|}} \leq 1/3$ and $k_\perp \rho_e \lesssim 1$.

The nonuniform parallel thermal flux introduces the electron drift-gradient term

$\frac{k_y \phi}{B}\frac{dT_e}{dx} + \frac{k_\| k_y \phi}{n_e(\omega+i\varepsilon)B}\frac{dq_\|}{dx}$ where $\mathbf{E} = -\nabla\phi$.

Comparison of T_e and $q_{\|e}$ gradient terms $\frac{dT_e}{dx} : \frac{k_\|}{\omega+i\varepsilon}\frac{1}{n_{0e}}\frac{dq_\|}{dx}$

Thus, the comparison involves the scale length $\frac{1}{L_{q\|}} = \frac{1}{q_\|}\frac{dq_\|}{dx} \gg \frac{1}{L_{Te}}$

and the magnitude of the heat flux parameter $\alpha_q = \frac{q_\|}{n_e T_e v_e} \equiv \frac{u_{\|q}}{v_e}$

LHCD electron distributions have:

$u_{\|q} \gtrsim j_\|/n_e e$ where $j_\| = I_p/\pi\Delta^2$

is the driven plasma current density required for confinement.
The dimensionless α_q parameter measures the skewness of the $v_\|$-distribution.

of f_e and $q_\|$ have a large gradient of $q_\|(r)$, giving $L_{Te}/L_{q\|} \gtrsim 5$. The combined ray tracing/Fokker-Planck code DELPHINE of Peysson, *et al.* (2011) and Saoutic, *et al.* (1994) is used to compute the electron phase space distribution functions.

The $P^{\text{LHCD}}(\omega, k_\|, k_\perp)$ dispersion function reduces to the well-known $Z(\omega/k_\| v_T)I_0 e^{-b}$ response function when the electron distribution function is replaced by a non-relativistic bi-Maxwellian with $v_T = (T_\|/m_e)^{1/2}$ and $b = k_\perp^2 T_\perp/m\omega_c^2$. The complex relativistic electrostatic response function is given by

$$P(\omega, k_\|, k_\perp) = -\int_0^\infty \int_{-\infty}^{+\infty} \frac{\left(\frac{\omega}{v_\|}\frac{\partial f}{\partial p_\|} + \frac{k_y}{eB}\frac{\partial f}{\partial x}\right) J_0^2\left(\frac{k_\perp p_\perp}{eB}\right) dp_\| dp_\perp^2}{\omega - k_\| v_\| \omega_D + i\varepsilon} \tag{13.31}$$

with the Landau initial value rule that by $\varepsilon \to 0^+$ for real ω. For $\text{Im}(\omega) < 0$ the function $P(\omega)$ is the analytic continuation of the polarization function $P(\omega)$ defined in Eq. (13.31). The phase distribution function $f(p_\|, p_\perp, x)$ has relativistic electrons with $p/mc = \gamma > 1$.

13.7.1 *Comparison of T_e and $q_{\|e}$ gradients*

The contributions of the gradients of T_e and $q_\|$ to the RF wave dispersion matrix are given by

$$\frac{k_y}{\omega B}\frac{dT_e}{dx} \quad \text{and} \quad \frac{k_\|}{\omega + i\varepsilon}\frac{1}{n_{0e}}\frac{dq_\|}{dx}. \tag{13.32}$$

Thus, the comparison involves the two gradient scale lengths

$$\frac{1}{L_{q_\parallel}} = \frac{1}{q_\parallel}\frac{dq_\parallel}{dx} \gg \frac{1}{L_{T_e}} \tag{13.33}$$

and the magnitude of the heat flux parameter

$$\alpha_{q_\parallel} = \frac{q_\parallel}{n_e T_e v_e} \equiv \frac{u_{\parallel q}}{v_e} \tag{13.34}$$

of the parallel thermal flux. For the LHCD electron distributions we have that $u_{\parallel q} \gtrsim u_{\parallel d} \cong j_\parallel / n_e e$ where $j_\parallel = I_p/\pi\Delta^2$ is the driven plasma current density required to confine the tokamak with current I_p over the region $\Delta < a$.

The dimensionless α_q parameter measures the skewness of the $v_\parallel$-distribution. For example, the LHCD distribution f_e which typically has

$$\frac{L_{T_e}}{L_{q_\parallel}} \gtrsim 5 \tag{13.35}$$

from the gradient of parallel flux $q_\parallel(r)$. The combined ray tracing/Fokker-Planck code DELPHINE of Imbeaux and Peysson (2005) is used to compute the driven electron distributions $f(v_\parallel, v_\perp, x)$.

13.7.2 *Kinetic dispersion relation for LHCD plasma*

The electrostatic kinetic response function is given by

$$D_k = k^2 + \sum_i \frac{n_i e_i^2}{\varepsilon_0 T_i} + \frac{n_e e^2}{\varepsilon_0 T_{\parallel e}}\left[1 + \left(\frac{T_{\parallel e}}{T_{\perp e}} - 1\right)(1 - \Gamma_0(b_e)) - P_e^{\mathrm{LHCD}}\right] \tag{13.36}$$

where $b_e = k_\perp^2 T_{e\perp}/m_e\omega_{ce}^2$ and

$$P_e^{\mathrm{LHCD}}(\omega, k_\parallel, k_\perp)$$
$$= -T_{e\parallel}\int \frac{\frac{\partial f_e}{\partial E_\parallel} J_0^2 d^3p\left[\omega - \frac{k_y T_{e\parallel}}{eBLn_e}\left(1 + \eta_{e\parallel}\left(\frac{\epsilon_\parallel}{T_\parallel} - \frac{1}{2}\right)\right) + \eta_{e\perp}\left(\frac{\epsilon_\perp}{T_\perp} - 1\right)\right]}{\omega - k_\parallel v_\parallel - k_y \widehat{v}_D(\epsilon_\parallel + \epsilon_\perp/2)}. \tag{13.37}$$

In the limit of a two-temperature Maxwellian, this P function reduces to the standard toroidal $P_e^{\max}$ [Kim, *et al.* (1992)] in Eq. (13.5).

13.7.3 *Hydrodynamic-FLR limit of P^{LHCD}-function*

To gain insight into the form of the unstable modes driven by the LHCD distribution function we take the hydrodynamic or fluid limit of the integrals. Thus, we obtain the dispersion relation for the LHCD plasma

$$D_{f\ell}^{\mathrm{LHCD}} = k^2 + \sum_i \frac{n_i e_i^2}{\varepsilon_0 T_i} + \frac{ne^2}{\varepsilon_0 T_{\parallel e}}\left[1 + \left(\frac{T_\parallel}{T_\perp} - 1\right)k_\perp^2\rho_{\perp e}^2 - 1 + \frac{\omega_{*ne}}{\omega}\right]$$
$$- \left(1 - \frac{\omega_{*\pi e}}{\omega}\right)\frac{k_\parallel^2 T_\parallel}{m\omega^2} - \frac{k_y v_{De}}{\omega} \tag{13.38}$$

which has the form

$$= A + \frac{B}{\omega} + \frac{C}{\omega^2} + \frac{D}{\omega^3} \tag{13.39}$$

Slab Modes $k_y q \rho_{e\|} < 1$

The slab regime ETG modes are given by $k_\| v_{e\|} \sim (v_{e\|}/qR)(1+|s|) > k_y v_{De}$ with the growth rate given by $\gamma_{\text{slab}} = |k_\||v_{e\|}(L_{ne}/L_{Te} - \eta_{e,\text{crit}})^{1/2}$.

Toroidal ETG Modes $k_y q \rho_{e\|} > 1$

The anisotropic toroidal regime modes are given by

$$\gamma_{\text{toroidal}} = \frac{v_{e\|}}{R}\left(\frac{R}{L_{T_e}} - \frac{R}{L_{T_{e,\text{crit}}}}\right)^{1/2} \frac{|k_y|\rho_{e\|}}{\sqrt{1 + k_y^2 \rho_{e\|}^2 (1 + \omega_{ce}^2/\omega_{pe}^2)}}.$$

$$-\frac{k_\|^2 v_{e\|}^2}{\omega^2}\left(1 - \frac{\omega_{*e}(1+\eta_{e\|})}{\omega^2}\right) + \frac{2k_y^2 \epsilon_n}{\omega^2}\left[1 + \eta_{\|e} + \frac{T_e}{T_\|}(1 + \eta_{\perp e})\right]. \tag{13.40}$$

The maximum χ_e arises from longest wavelength with $\lambda_\perp \lesssim \rho_i$

$$\chi_e \le \rho_i^2 \frac{v_{\|e}}{R}\left(\frac{R}{L_{T_e}} - \frac{R}{L_{T_{r,\text{crit}}}}\right)^{1/2}.$$

There are two useful limiting forms of the LHCD-ETG waves from the fluid expansion of P. For small $k_\| L_{T_e}$ and $\overline{\omega} = \omega - k_\| u_\|$ we find from the fluid equations

$$\overline{\omega}(\overline{\omega} - \omega_{*i}) = \frac{k_\| k_y T_i}{eBZ_{\text{eff}}}\frac{du_\|}{dr} + \frac{k_\|^2 T_i}{m_e Z_{\text{eff}}}\left[1 - \frac{k_y T_e}{\overline{\omega} e B n_e}\frac{dn_e}{dr} - \frac{k_y T_e}{eB}\frac{1}{\overline{\omega} - k_\|^2 v_e^2/(\overline{\omega} + i\nu)}\right.$$
$$\left.\times \left[\left(\frac{1}{T_e}\frac{dT_e}{dr} - \frac{k_\| dq_\|/dr}{(\nu - i\omega)p_e}\right)\right]\right]. \tag{13.41}$$

The drift wave frequency ω_{*i} in Eq. (13.41) arises from the $\mathbf{E} \times \mathbf{B}$ convection of the electron density in the presence of the adiabatic ions $\widetilde{n}_i = -eZ_i n_i \phi/T_i$.

Due to the electron temperature gradient η_e, for frequency formula is approximately

$$\omega(k) = \frac{-\omega_{*e} S(b, \eta_{e\perp})}{\frac{Z_{\text{eff}} T_e}{T_i} + k_\perp^2 \rho_e^2 + k^2 \lambda_{De}^2}$$

with $S(b, \eta_e) = e^{-b} I_0 - \eta_{e\perp} b e^{-b}(I_0 - I_1) \simeq 1 - k_\perp^2 \rho_e^2 (1 + \eta_{e\perp})$.

For larger $k_\| L_{T_e} \gtrsim 1$ the waves are given by

$$\omega^2 = \frac{k_\|^2 T_i}{Z_{\text{eff}} m_e}\left[1 - \frac{k_y T_e}{eB\omega}\left[\frac{1}{T_e}\frac{dT_e}{dr} + \frac{k_\|}{\omega p_e}\frac{dq_\|}{dr}\right]\right]. \tag{13.42}$$

Often the temperature gradient is subdominant to the gradient of the parallel heat flux and the instability becomes

$$\omega_k = \omega_{*i} + i\left(\frac{k_\parallel^2 T_i}{2m_e}\right)^{1/3}\left|\frac{k_y T_{\parallel e}}{eBL_{T_e\parallel}}\frac{dq_\parallel}{dx}\right|^{1/3}.$$

These heat-flux gradient driven modes have

$$\omega_k = \left(\frac{k_\parallel^2 T_i}{Z_{\text{eff}} m_e}\right)^{1/4} |k_\parallel u_\parallel q\omega_{*i}|^{1/4} \propto k_\parallel^{3/4} k_y^{1/4} u_\parallel^{1/4} v_e^{3/4}. \tag{13.43}$$

The LHCD driven modes have a mean parallel wavenumber $\overline{k}_\parallel$ so the eigenmodes are shifted off the rational magnetic surfaces. Thus, the modes form magnetic islands in their nonlinear state when the plasma electron beta is appreciable and contribute to the radial transport of $j_\parallel$ and $q_\parallel$. Some of these are summarized in Table 13.2

In lower-hybrid current drive (LHCD) experiments on tokamaks, the parallel wavenumber of lower-hybrid waves is an important physics parameter that governs the wave propagation and absorption physics. However, this $n_\parallel^{\text{RF}}$ parameter has not been experimentally well-characterized in the high-density tokamaks, despite the advances in the wave physics modeling. The first measurements of the dominant parallel wavenumber of lower-hybrid waves in the scrape-off layer (SOL) were from the Alcator C-Mod tokamak [Greenwald, *et al.* (2014)] with an array of magnetic loop probes. The electric field strength measured with the probe in typical C-Mod plasmas is about one-fifth of that of the electric field at the mouth of the grill antenna. The amplitude and phase responses of the measured signals on the applied power spectrum are consistent with the expected wave energy propagation. At higher density, the $k_\parallel$ increases for the fixed antenna launched $k_\parallel$, and the wave amplitude decreases rapidly. This decrease correlates with the loss of LHCD efficiency at high density. Evidence of the RF power spectral broadening mechanisms is observed in the frequency spectra. However, no clear modifications in the dominant $k_\parallel$ are observed in the spectrally broadened wave components, as compared to the measured $k_\parallel$ at the applied frequency. The probes are in the SOL and with limited $k_\parallel$ resolution for the diagnostic. Future experiments are planned to investigate the roles of the observed spectral broadening mechanisms on the LH density limit problem in the strong single pass damping regime [Baek, *et al.* (2016)].

13.7.4 *Analytic quasilinear RF velocity diffusivity*

The intense LH-waves produce the parallel velocity diffusion coefficient $D_\parallel^{\text{LH}}$ that is modeled by a box function with

$$D_\parallel^{\text{LH}}(v_\parallel) = \widehat{D}\nu_e v_e^2 \qquad \text{for} \qquad v_1 \lesssim v_\parallel \lesssim v_2 \tag{13.44}$$

with $D_\parallel$ vanishing outside the interval $[v_1, v_2]$. For CIEL and other LHCD systems the upper parallel velocity $v_\parallel$ may be extended to $v_2 \lesssim c/2 = 1.5 \times 10^8$ m/s$\sim 13v_e$.

The dimensionless measure of the LH power is given by Stix parameter for RF power density

$$\widehat{D} = \frac{P^{\mathrm{RF}}}{n_e T_e \nu_e} \tag{13.45}$$

where ν_e is the Coulomb collision frequency for electron collisions that restore the distribution to the local Maxwellian.

In the region $[v_1, v_2]$ a useful analytic model is obtained by taking $\widehat{D} \gg 1$ so that the collisional effects are negligible and the absorbed RF power forces $\partial f/\partial v_\| = 0$ in this $v_\|$-interval. This model is referred to as the plateau LHCD model. In reality the electron Coulomb collisions will always be non-negligible due to the formation of boundary layers at the transition across the $v_\| = v_1$ and $v_\| = v_2$ layers.

The electron distribution function, in the absence of the mirror force and toroidal velocity drifts, is

$$\begin{aligned}\frac{\partial f}{\partial t} &= \frac{\partial}{\partial v_\|}\left[D_\|(v_\|)\frac{\partial f}{\partial v_\|}\right] + \frac{\nu_e}{v^2}\frac{\partial}{\partial v}\left[v^2 g(v)\left(fv + \frac{T_e}{m_e}\frac{\partial f}{\partial v}\right)\right] \\ &\quad + \nu_{ei} Z_{\mathrm{eff}}\left(\frac{v_e}{v}\right)^3 \frac{\partial}{\partial \mu}\left[(1-\mu^2)\frac{\partial f}{\partial \mu}\right]\end{aligned} \tag{13.46}$$

where ν_e is the electron collision frequency, $g(v)$ is the coefficient of drag (which we take as unity in the following discussion), ν_{ei} is the electron-ion (working gas) collision frequency and $Z_{\mathrm{eff}} = \Sigma_I Z_I^2 n_I/n_e$ gives the enhancement of the electron pitch-angle scattering rate from impurities I over that in the pure hydrogenic plasma for which $Z_{\mathrm{eff}} = 1$.

13.7.5 *High-power LHCD plateau F_e-distributions*

In the limit that $D_\|/\nu_e v_e^2 \gg 1$ in the interval (v_1, v_2), we may obtain the F_e phase space distribution thorough geometrical construction. The key point is that one may map the values of $f(v_\| = v_1, v_\perp) \to f(v_\| = v_2, v_\perp)$ since $\partial f/v_\| = 0$ in this interval. The coordinates of the locus of points $(v_2, v_\perp)$ parameterized by the pitch-angle α is $v_2 = v\cos\alpha$ and $v_\perp = v\sin\alpha = v_2\tan\alpha$. Thus, the value of f along the high velocity resonant plane $v_\| = v_2$ is given by $f(v_2, v_\perp) = f(v_2, v_2\tan\alpha)$ with $0 \le \alpha \le \pi/2$ parameterizing the position along the $v_\| = v_2$ surface.

Now the value of $f(v_2, v_\perp)$ is taken as the same as the value of $f(v_1, v_\perp)$ in this high-power limit due to $\partial f/\partial v_\| = 0$. So, on the surface $v_\| = v_1$ one has $v_\perp = v\sin\alpha$ and $v_\| = v\cos\alpha = v_1$. For parallel velocities below $v_\| = v_1$ the distribution is Maxwellian since this is the solution of $Cf = 0$ in the low-velocity region up to v_1 and for all $\pi/2 < \alpha \le \pi$ (co-current moving electrons). Thus, f is known through the region $[v_1, v_2]$ and on the surface $v_\| = v_2$. Pitch-angle scattering dominates for $v_\| > v_1$.

Typical values of the temperatures in the core plasma are $T_\| \sim 750\,\mathrm{KeV}$ to $1\,\mathrm{MeV}$, and $T_\perp \sim 150 - 300\,\mathrm{KeV}$ with lower energy Maxwellian with temperature of 50-100 KeV. The analytic model [Stevens, *et al.* (1985)] used in the literature is:

$$f_{3T} = \begin{bmatrix} C_n \exp\left(-\frac{p_\perp^2}{2T_\perp} - \frac{p_\parallel^2}{2T_{\parallel F}}\right) & \text{for} & p_\parallel > 0 \\ C_n \exp\left(-\frac{p_\perp^2}{2T_\perp} - \frac{p_\parallel^2}{2T_{\parallel B}}\right) & \text{for} & p_\parallel < 0 \end{bmatrix}$$

and $f_{3T} = 0$ for $E(p) = mc^2\gamma > E^* \sim 2\,\text{MeV}$. The temperatures are determined by the photon spectrum from the Bremsstrahlung spectroscopy.

The Nyquist stability analysis gives the marginal stability condition

$$-\frac{1}{T_{e\parallel}}\frac{dT_{e\parallel}}{dr} > \left[\frac{1}{L_{ne}} + \frac{1+2|s|/q}{R}\right] \times \left[1 + \frac{Z_{\text{eff}}T_{e\parallel}}{T_i} + \left(\frac{T_{\parallel e}}{T_{\perp e}} - 1\right)(1 - \Gamma_0(b_e)) + k_\perp^2\lambda_{De_\parallel}^2\right]. \quad (13.47)$$

In conclusion the LHCD can stimulate the (a) fast-growing toroidal modes $T_\parallel > T_{\perp i}$ and (b) fast-growing slab-like modes $k_\perp \rho_{e\parallel} \sim 1$. Secondly, one finds that (c) short wavelengths $b_e = k_\perp^2 \rho_{e\perp}^2 > 1$ are stabilized by $T_{\parallel e} > T$ and finally (d) that the effect of $Z_{\text{eff}} T_{e\parallel}/T_i$ is important in controlling the threshold.

Formulas for the anomalous thermal flux associated with the $\mathbf{E} \times \mathbf{B}$ motion and the magnetic $\delta B_\perp$ fluctuations for LHCD driven tokamaks remain an unsolved problem. The turbulence for the c/ω_{pe} scale electromagnetic turbulence is typical of the ∇T_e-driven short wavelength drift modes. The result is to be used in the formulas in Chapter 10 for the anomalous electron thermal energy transport due to $\langle \delta B_x^2 \rangle$.

The design for LHCD for ITER is given in Hoang, *et al.* (2009). Lower hybrid current drive produce record-breaking steady-state tokamak discharges in Tore Supra [Saoutic, *et al.* (1994)] and Alcator-C [Bonoli, *et al.* (1988)]. In divertor tokamaks, there is a turbulent plasma outside the last closed flux surface that produces refraction and scattering of the LHCD waves [Peysson, *et al.* (2011)].

13.7.6 *RF current drive in WEST*

The WEST (W – for tungsten – Environment in Steady-State Tokamak) machine is the current upgrade of Tore Supra from a limiter-based tokamak with carbon PFCs to an X-point divertor tokamak with full-tungsten armor shown in Fig. 13.13. The WEST project is testing for the first time the ITER technology in real plasma environment. The main divertor (i.e. the lower divertor) of the WEST project is an actively-cooled tungsten components [Lipa, *et al.* (2009)] that follow the design and the assembling technology used in the ITER divertor units [Missirlian, *et al.* (2014)].

The WEST ITER-like W divertor design was installed in 2016–17. The manufacturing of the series of WEST plasma-facing units (PFUs) was with the technology and designed planned for ITER and will contribute to the ITER divertor PFU

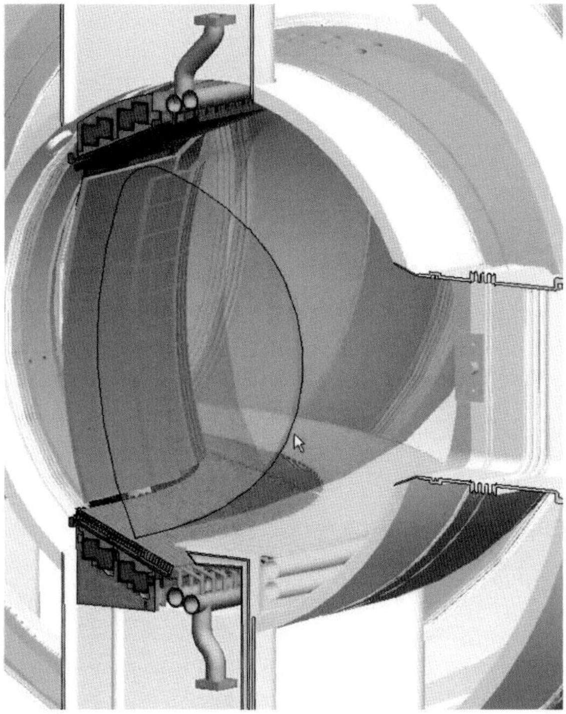

Fig. 13.13 Tore Supra in the WEST configuration, showing the main plasma-facing components and the divertor's lower target, using ITER technology.

qualification process promoting the early divertor plasma operation on ITER. The major points addressed through the WEST PFU manufacture were (i) the validation of non-destructive testing methods, (ii) experience in tungsten machining of and assembly of the monoblocks, and (iii) finding the machining margins required for meeting physics specifications [Boscary, *et al.* (2007)]. The WEST program provides valuable input to ITER operation issues. The device has actively-cooled divertor sectors. The initial phase of WEST operation will research the power handling of high-heat flux components to be studied for durations exceeding their thermal time constant. Critical power-handling issues will be investigated ahead of the end of the ITER divertor production where plasma heat fluxes ($10\,\mathrm{MW/m^2}$). The leading edges, tile-shaping studies, and operation with damaged W components are planned.

The installation of a full actively-cooled divertor in WEST will give access to long-pulse, high-fluence studies, including combined stationary heat loads with a large number of submelting threshold ELMs. The consequences of excursions above the W-recrystallization temperature and tests under extreme heat fluxes (up to $20\,\mathrm{MW/m^2}$) with high-fluence plasma wall interactions will be addressed on WEST well before the start of ITER divertor operation.

The RF simulations show how the antennas excite both the slow wave and the fast plasma waves in the fusion plasma. The start of the penetration of the rays for these two plasma eigenmodes is controlled by the complex grill antennas in Tore Supra (C3 and C4) that are now used in WEST. The WEST machine is a single-null divertor tokamak designed to study the tungsten (W) wall interactions with the 1–5 KeV electron temperature plasmas.

First the LH ray trajectories are calculated in the absence of the drift wave turbulence. The penetration to the core is shown to be controlled by combination of the launched azimuthal k_θ and axial k_z gigaHertz wave spectrum. The poloidal variation of the magnetic field vector and the radial density profile control the deposition of the RF power. One computes the RF wave propagation taking into account the drift wave fluctuations $(\boldsymbol{k}, \omega_{\boldsymbol{k}}, \gamma_k)$ producing the plasma density fluctuations that scatter the RF waves. The width of the poloidal and radial RF wave spectrum increases strongly in the presence of typical drift wave turbulence. The RF ray scattering from $\delta\widetilde{n}_e(\mathbf{x}, t)$ changes the RF wave polarization between the fast wave and slow plasma waves. The longer wavelength fast waves penetrate deep into higher density plasma while the slow waves are strongly absorbed in the plasma in the SOL and in the interior plasma close to the magnetic separatrix or pedestal plasma. The electron temperature gradient (ETG) drift wave spectrum is particularly effective in scattering the lower-hybrid waves due to its comparable wavelengths and high-parallel phase velocities. ETG turbulence (dT_e/dr) is also driven by the radial gradient of the electron current profile. The gradient $dj_\parallel/dr$ gives rise to an anomalous viscosity that spreads the lower-hybrid driven-plasma currents $j_\parallel(r, t)$. This feature is measured by the soft X-ray spectrum emitted by the 20–100 KeV suprathermal electrons. RF wave scattering is derived from a Fokker-Planck equation for the phase-space distribution of the ray trajectories with diffusivities from the drift wave fluctuations.

Core plasma current drive requires RF antennas with low azimuthal mode spectra peaked at $m = 0$ and low-$(|m|)$ azimuthal mode numbers. The broaden T_e and $j_\parallel$ profiles are reported in van Houtte, *et al.* (2004) which reports the record 6-min steady-state 1 GJ Tore Supra KeV plasmas. An overview of the steady-state LHCD experiments is given in Becoulet, *et al.* (2011), Hoang, *et al.* (2009) and van Houtte, *et al.* (2004) and for the EAST machine by Zhang, *et al.* (2014) (`http://english.ipp.cas.cn`).

The fast LH RF plasma waves $\omega(\boldsymbol{k}) = \omega(k_r, k_y, k_z, r, \theta, \phi)$ with $I_{\boldsymbol{k}}^{\mathrm{RF}}$ are a limit of the whistler waves observed universally in space plasmas. The whistler waves in the magnetosphere are excited by KeV-energy electrons with a temperature anisotropy $T_\perp > T_\parallel$. The waves are observed with various structures in space-time in the magnetosphere and efficiently accelerate electrons to MeV energies in the radiation belts [Summers, *et al.* (2002)]. In both the laboratory [Stenzel, *et al.* (2010)] and in space plasmas [Summers, *et al.* (2002)] the wave-particle inter-actions are analyzed with the method of rays for the trajectories of the dressed photons, or plasmons, and the electron scattering of the plasmons. Sophisticated antenna-wave control codes for the accelerations of electrons and ray tracing of the RF plasma waves are available for both space and laboratory plasmas. Examples are found in Summers, *et al.* (2002), Thorne (2010), Crabtree, *et al.* (2012), Peysson and Decker (2007), and Peysson, *et al.* (2011). Several earlier IFS graduates students [Crabtree, Mithaiwala, Mays, and Pratt] carried out PhD dissertations on the space physics problems of lower-hybrid and Whistler wave-particle interactions.

13.7.7 *Self-consistent calculation of LH wave damping and the steady-state electron distribution function*

The damping of RF waves on electrons produces a significant distortion of the electron distribution function $f(x, p_{\parallel}, p_{\perp}, t)$. Conversely, the damping of RF waves through resonant wave-particle interaction depends on the details of the distribution function. Thus, RF wave damping and the particle distribution function are calculated self-consistently.

Resonant interactions between electrons and RF waves are described by the quasilinear formalism which is valid under the following conditions: (i) the trapping time for the electron within the wave field must be much longer than the interaction time between the electron and the wave packet and (ii) the electrons follow their unperturbed orbits so that the phase information must be lost between two consecutive passes through the wave packet. The first condition, which sets a limit to the RF power, is equivalent to assuming a small perturbation to the electron orbits. This allows one to describe wave-particle interactions as a diffusive process. The second condition assumes that additional perturbations of the particle orbit due to collisions and turbulence are sufficient to randomize the wave-particle phase between consecutive electron passes through the wave packet. This condition may fail near low-n rational $q(\psi) = n/m$ flux-surfaces or when the interaction occurs within magnetic islands.

The electron distribution $f(x, p_{\parallel}, p_{\perp}, t)$ evolves from the balance between the distorting effect of wave-particle interaction and the restoring effect of electron collisions towards a Maxwellian distribution. The evolution is described by the bounce-averaged Fokker-Planck equation

$$\frac{\partial f}{\partial t} = C(f) + Q_{\mathrm{LH}}(f) \tag{13.48}$$

where $C(f)$ is the Fokker-Planck collision operator and $Q_{\mathrm{LH}}(f)$ is the quasilinear diffusion operator describing the interaction between electrons and LH waves. For an axisymmetric toroidal configuration and a low-collisionality regime, Eq. (13.48) is solved over the 3-D relativistic guiding-center phase-space where electron orbits may be parameterized by (ψ, p, ξ) where ψ is the poloidal flux, p is the electron momentum, and $\xi = p_{\parallel}/p$ is the cosine of the electron pitch angle evaluated at the position of minimum magnetic field on the flux-surface ψ.

Within the zero-orbit width approximation, which is generally valid for the interaction between electrons and LH waves, the quasilinear diffusion reduces to

$$Q_{\mathrm{LH}}(f) = \frac{\partial}{\partial p_{\parallel}} D_{\parallel} \frac{\partial f}{\partial p_{\parallel}} \tag{13.49}$$

within the ordering $\lambda \ll d \ll L$ where λ is the RF wavelength, d is the RF beam size, and L is the equilibrium nonuniformity scale length. The condition $\lambda \ll d$ for a well-collimated RF beam ensures that beam characteristics such as the polarization and the power flow are described by the ray propagation. The quasilinear diffusion coefficient $D_{\parallel}(\mathbf{x}, p_{\parallel}, p_{\perp})$ includes the contributions of all rays crossing a given magnetic flux-surface ψ, including multiple passes of any given ray. Decker, *et al.* (2014) model the ray diffusion in Eq. (13.49) as

$$D_{\parallel}(\psi, p_{\parallel}, p_{\perp}) = \sum_i \frac{\mathcal{P}_i}{\mathcal{P}_{i0}} \frac{D_{\parallel i}}{\sqrt{\pi}\,\Delta n_{\parallel i}} \exp\left[-\frac{(n_{\parallel i} - n_{\parallel \mathrm{res}})^2}{(\Delta n_{\parallel i}^2)}\right], \tag{13.50}$$

where $\mathcal{P}_i$ and $n_{\parallel i}$ are the power and parallel index of refraction of the ray fragment i on the flux-surface ψ, respectively. The resonance condition yields $n_{\parallel \mathrm{res}} = c/v_{\parallel}$, where $v_{\parallel}$ is the parallel velocity of the electron as it crosses the wave packet. The diffusion coefficient $D_{\parallel i}$ describes the intensity of the wave particle interaction. Within the well-collimated beam approximation $\lambda \ll d$, a narrow spectral width $\Delta n_{\parallel i} \sim c/\omega d \ll n_{\parallel i}$ is associated with the ray of $\mathcal{P}_i$. The spectral width $\Delta n_{\parallel i}$ enters the calculation of the quasilinear diffusion operator in Eq. (13.50), giving the width to the resonance region in momentum space [Decker, *et al.* (2010)]. An analysis of the range of validity of the ray diffusion and the transition to a wave diffraction description is developed with the Wigner-Moyal equation in Mendoca, *et al.* (2015).

13.7.8 *Steady-state RF heating and current drive plasmas in EAST*

The Experimental Advanced Superconducting Tokamak (EAST) is a superconducting torus with an advanced magnetic configuration that started operation in 2006. EAST research goal is to establish the scientific and technological basis for the next generation of tokamak reactors. The machine parameters are given in Table 13.3. Initial operation in 2006 Zhang, *et al.* (2014) used LHCD power of 2.0 MW at 2.45 GHz [Zhao, *et al.* (2010), Ding, *et al.* (2011)] and 1.5 MW ICRF system [Lin and Wan (2011), Zhang and Lin (2010)]. These LCH and ICRH heating systems were then upgraded to 4.0 MW and 6.0 MW, respectively.

The LHCD launcher and the protective limiters match the plasma boundary shape. The LHCD system delivers a continuous-wave power of 4.0 MW at the frequency of 2.4 GHz. Four 1.5 MWICRF transmitters in the frequency range of 25–70 MHz were operating in the 2012 experimental campaigns. As shown in Fig. 13.14 power the I-port antenna and the other two powered the B-port antenna. The Faraday shields of the B-port antennas are coated with B4C.

Table 13.3 Experimental Advanced Superconducting Tokamak (EAST) parameters

$B_t = 3.7\,\text{T}$, $I_p = 1\,\text{MA}$, $R = 1.7$–$1.9\,\text{m}$, $a = 0.45\,\text{m}$
shaping triangularity $\delta = 0.4$–0.7 elongation $\kappa = 0.2$ to 2.0
single null $= 2.0$ double null divertor configurations

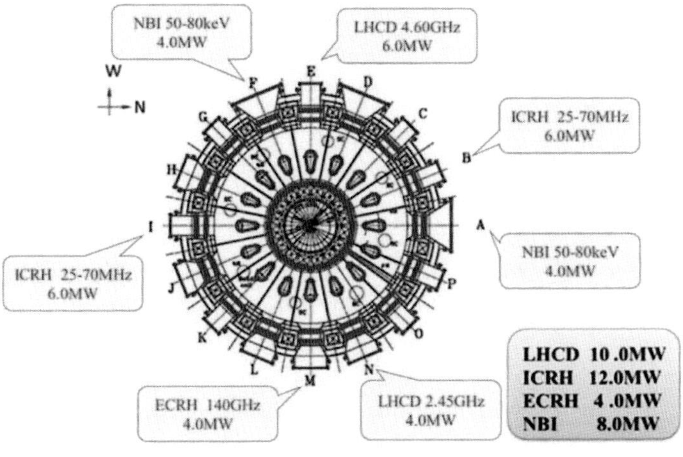

Fig. 13.14 Schematic showing the array of plasma heaters and the current drivers from LHCD, ICRH, ECRH and NBI for the Experimental Advanced Superconducting Tokamak (EAST) in Hefei, China [Zhang, *et al.* (2014)].

With the RF heating and current-drive, EAST achieved reproducible steady-state H-mode plasmas lasting longer than 30 s with lithium-coated walls in the 2012 experimental campaign.

The four plasma heaters and current drives are shown in Fig. 13.14 with their respective powers and locations. High electron temperature long-pulse plasmas were achieved in EAST. During the L-H transition where a steep density gradient profile formed inside the last closed flux surface in H-mode the LHCD coupling is observed to deteriorate. This effect lowers the density at the mouth of the antenna grill. EAST (Experimental Advanced Superconducting Tokamak) experiments have shown that sufficient gas puffing near the lower hybrid grill can reverse the local density depletion.

During 2010 and 2012 EAST experimental campaigns ICRF heating experiments were carried out at the fixed frequency of 27 MHz with power up to 6.0 MW achieving effective ions and electrons heating with the H minority heating (H-MH) mode. The H-MH mode produced H-mode plasmas using ICRF. In 2010, H-modes were generated and sustained by LHCD alone with lithium coating and gas puffing near the mouth of the LH launcher. In 2012, the combination of LHCD and ICRH power extended the H-mode duration up to over 30 s. H-modes with various types of edge localized modes (ELMs) were achieved with HIBP98(y, 2) ranging from

0.7 to over unity. An overview of LHCD and ICRF experiments during these two campaigns is presented in Zhang, *et al.* (2014).

High density is a challenge to the LHCD experiments in H-mode discharges. In high-density plasmas, lower-hybrid waves dissipate their power in the plasma periphery which limits the LHCD efficiency. High-density experiments on FTU [Cesario, *et al.* (2013)] have shown that lower-hybrid wave heating efficiency is recovered and the spectral broadening of the launched wave is reduced by increasing peripheral temperature [Zhang, *et al.* (2014)]. Theory predicts a reduced effect of parametric instability [Cesario, *et al.* (2004)] at higher plasma temperature. Different mechanisms can prevent lower-hybrid wave penetration at high density. For example, collisional absorption [Bonoli, and Englade (1986)], and parametric instability (PI) [Cesario, *et al.* (2004, 2011, 2013)] scattering from density fluctuations (SDF) [Peysson, *et al.* (2011)] reduce RF wave penetration. Broadening of the lower-hybrid power spectra [Cesario, *et al.* (2006)] in the parallel refractive index $n_{\|}(\mathbf{k}$ is produced by the drift wave turbulence [Horton, *et al.* (2016)] and parametric instability. Experiments performed in EAST show that current drive is improved with gas puffing and withlithium coating [Zhang, *et al.* (2014)]. These observations are consistent with the previous studies showing that lithium wall coating increases the current drive efficiency at high density, due to the lower-plasma recycling and higher-edge temperature obtained.

The wavenumber of the daughter lower-hybrid wave may reduce the current drive efficiency. The dependences of current drive effect on density are consistent with the onset of the frequency of ion-cyclotron (IC) sideband measured by a RF loop antenna located outside the machine, implying the sharp decay of HXR emission observed in high-density LHCD experiments is correlated with parametric decay instability (PDI) [Kong, *et al.* (2013)].

Nonlinear interactions involving the spectra of the launched lower hybrid waves in the edge plasma are important in determining the power absorption of the RF power in tokamak plasma [Castaldo, *et al.* (1997), Napoli, *et al.* (2013)].

Comparing the experimental data for different current and magnetic field shows that current drive improves with higher plasma current and higher toroidal magnetic field, in agreement with studies in Tore-Supra. This is because high plasma current increases the plasma temperature which improves the current drive efficiency. Comparing the hard X-ray spectrum (HXR) counts in L- and H- mode shows that at the same density the current drive efficiency in H-mode is larger than that in L-mode, indicating a profile and edge density dependence in current drive efficiency [Zhang, *et al.* (2014)].

Understanding the interaction of radio-frequency (RF) waves with the plasma in the near-field region of the high-power wave launcher is important for channeling of the RF waves through the plasma boundary and avoiding power losses in the edge plasma. A direct non-intrusive measurement of a near-antenna RF electric field in the range of lower-hybrid (LH) frequencies was achieved by measuring the

Balmer series deuterium spectral lines with a time-dependent Stark effect modeling [Klepper, *et al.* (2013)]. Through the analysis of numerous Tore Supra discharges quantitative agreement was shown for the measured and launched power when the launched power exceeded 0.5 MW. For low power the measurement becomes inaccurate. The passive spectroscopic technique used to measure the launched waves becomes dominated by the spectral noise at low RF power.

In the long-pulse tokamak Tore Supra experiments, the electric field strengths and direction derived from fitting the spectra to a model that included the time-dependent Stark effect and the tokamak-range magnetic field Zeeman-effect were shown to be in good agreement with full-wave modeling. The H-mode was achieved with LHCD H-mode with Type-III edge localized modes (ELMs) [Xu, *et al.* (2011)]. The H-factor of HIBP98(y,2) $= 0.8 \pm 0.2$ was obtained at a power level close to the threshold power (1 MW) predicted by the international tokamak scaling. Reduced recycling by lithium (Li) wall coating was achieved and is a key factor to accessing the H-mode at low power levels. The LHCD efficiency in such regimes is due to the improved wave penetration and reduced parametric instabilities [Cesario, *et al.* (2013)].

In the 2010 EAST campaign, H-mode plasmas by ICRF heating were only accessed together with LH power due to the low ICRF power absorption. In the 2012 campaign, H mode plasmas were achieved by ICRF power alone. These H-modes had Type-III edge localized modes (ELMs) [Zhang, *et al.* (2013)] and the confinement factor HIBP98(y,2) $= 0.7$.

With hydrogen minority heating 1.0 MW of ICRF power at 27 MHz injected from a two-strap antenna into a $B_t = 2.0\,\mathrm{T}$ and $I_p = 500\,\mathrm{kA}$ plasma, the electron density dropped by about 20% after the ICRF power was applied [Zhang, *et al.* (2012)] and the electron temperature increased by 1.0 KeV and the stored plasma energy increase by 30 kJ.

Worked Problem 13.1:

Problem:

Find the 2×2 matrix equations for compressional drift waves.
Use the kinetic matrix formulation in Chapter 12.2.2.

Solution:

The perturbative motions induced by the electric potential ϕ and the magnetic flux ψ produce a compressional change in the magnetic field $\delta B_\parallel$ given by

$$\frac{\delta B_\parallel}{B} = -\frac{1}{f}(c\phi + e\psi) \tag{13.51}$$

where $1/f$ is the inverse of a complicated integral operator f involving bounce averages over the particle trajectories. Here the diagonal compressional mode dispersion function f is taken as a local quantity defined in Eq. (12.19). The compressional change in the magnetic field $\delta B_\parallel$ is dictated by Amperé's law from the third row of the matrix equation Eq. (12.18) with the plasma currents flowing across the magnetic field lines in the local $\widehat{x} \propto \nabla\Psi$ direction.

Substituting Eq. (13.51) into the first and second rows of the full matrix equation (12.18) gives the reduced 2×2 symmetric matrix

$$\begin{bmatrix} a - \dfrac{c^2}{f} & b - \dfrac{ce}{f} \\ b - \dfrac{ce}{f} & d - \dfrac{e^2}{f} \end{bmatrix} \begin{bmatrix} \phi \\ \psi \end{bmatrix} = 0. \tag{13.52}$$

The dispersion relation given by the determinant is

$$D_k(\omega, \mu) = ad - b^2 - \frac{1}{f}\left(c^2 d - 2bce + e^2 a\right) = 0. \tag{13.53}$$

The compressional kinetic-MHD limit is $a \simeq -b \simeq d$, which reduces Eq. (13.53) to

$$ad - b^2 - \frac{d}{f}(c+e)^2 = 0. \tag{13.54}$$

To find analytically how the kinetic ballooning interchange drift mode connects to the compressional mode one takes $a \simeq -b \simeq d \simeq e$ for the dominant terms owing to the small $E_\parallel$. Then the full determinant D_k in Eq. (13.53) factors as

$$ad - b^2 + \frac{\tau(1 - \omega_*/\omega)[Q - (1 - \omega_*/\omega)]^2}{\frac{2}{\beta_e} + \frac{T_i}{T_e} R} = 0 \tag{13.55}$$

follows from Eq. (13.54). Here $\omega_{*e} = \omega_*$ for $\eta_e = 0$. The last term in Eq. (13.55) gives the kinetic compressional response. For the near-to-MHD regime, the response function $Q(k, \omega)$ in Eq. (13.55) reduces to

$$Q - 1 + \frac{\omega_*}{\omega} \simeq -\frac{\omega_{*pi} - \omega_*}{\omega} + i\Delta_Q = -\frac{\omega_{*p}}{\omega} + i\Delta_Q \tag{13.56}$$

with ω_{*p} having the total pressure gradient and a wave-particle resonant part $i\Delta_Q$ from wave-particle Landau resonance at $\omega = \omega_{De}$. An adequate approximation for the compressional response function R in the region $\omega\omega_{Di} > 0$ is

$$R \cong c_0 \frac{[\omega - \omega_{*i}(1 + 2\eta_i)]}{\omega - \omega_{Di} + ic_1|\omega_{Di}|} \tag{13.57}$$

where c_0 and c_1 are positive fitting coefficients of order $c_0 \simeq 1$ and $c_1 \simeq 0.1$. Let us define the resonant contributions by $\Delta_Q = \operatorname{Im} Q(\omega)$ and $\Delta_R = \operatorname{Im} R(\omega)$ as real numbers giving the resonant ion contributions.

The resonant modes in the high-β region have

$$\omega = \omega_0 + i\gamma_k \simeq \omega_{*i}(1 + \eta_i) + i\gamma_k \tag{13.58}$$

which is an ion diamagnetic drift wave. Here ω_0 and γ_k are real valued functions of k and the system parameters $\{\mu\}$. A Taylor series expansion of the dominant terms in Eq. (13.55) gives

$$ad - b^2 \simeq \left(\frac{\omega_*}{\omega_0} - 1\right)\left[\frac{\omega_A^2}{\omega_0^2}b - b\left(1 - \frac{\omega_{*pi}}{\omega_0}\right) - \frac{\omega_*\omega_D}{\omega_0^2}\right]$$

$$+ \left(\frac{\omega_*}{\omega_0} - 1\right)\frac{i\gamma_k}{\omega_0}\left[\frac{-2\omega_A^2}{\omega_0^2}b - \frac{b\omega_{*pi}}{\omega_0} + \frac{2\omega_*\omega_D}{\omega_0^2}\right] \tag{13.59}$$

for $\omega = \omega_0 + i\gamma_k$ with $|\gamma_k| \ll \omega_0$. Thus the growth rate γ_k is determined by

$$\frac{i\gamma_k}{\omega_0}\left[\frac{2\omega_A^2}{\omega_0^2} + \frac{\omega_{*pi}}{\omega_0}\right]b + \frac{T_e}{T_i}\frac{(\omega_{*p}/\omega_0 - i\Delta_Q)^2}{c_0[1 - \frac{\omega_{*i}(1+2\eta_i)}{\omega_0} - i\Delta_R]} = 0 \tag{13.60}$$

for $\Delta_Q \sim \Delta_R \ll 1$. Since $\omega_{*p}/\omega_0 \gtrsim 1$, the significant resonant contribution comes from $i\Delta_R$ in the denominator. Thus one obtains the growth rate formula

$$\frac{\gamma_k}{\omega_0}\left(\frac{2\omega_A^2}{\omega_{*p}^2} + 1\right)b = -\frac{T_e}{T_i}\frac{(\omega_*/\omega_0)^2\Delta_R}{c_0\left[\left(\frac{\eta_i}{1+\eta_i}\right)^2 + \Delta_R^2\right]} \tag{13.61}$$

with $\gamma_k > 0$ for $\Delta_R \equiv \mathrm{Im}\,[R(\omega_0)] < 0$.

Worked Problem 13.2:

Problem: Define or formulate the five-field model from the Braginskii equations derived in the five-field model developed in Zeiler and Drake (1997):

Solution:

Define the six dimensionless fluctuation fields

$$T_e = \frac{\widetilde{T}_e}{T_{e0}},\ p_i = \frac{\widetilde{p}_i}{p_{i0}},\ n = \frac{\widetilde{n}}{n_0},\ \phi = \frac{e\widetilde{\phi}}{T_{e0}},\ v_\parallel = \frac{\widetilde{v}_{\parallel i}}{c_s}$$

$$\psi = \frac{\omega_{ci}V_A\widetilde{\psi}}{c_s^2 B_0},\ J = V_A\frac{c_s^2}{\omega_{ci}^2}\nabla_\perp^2\psi, \tag{13.62}$$

and define

$$h = \phi - p_e - 0.71\,T_e, \qquad \tau = \frac{T_{i0}}{T_{e0}}, \quad \sigma = \frac{T_{e0}}{\eta_\parallel n_0 e^2}. \tag{13.63}$$

Then reduce the Braginskii equations following Zeiler and Drake (1997) to derive the six coupled pde system

$$\frac{c_s^2}{\omega_{ci}^2}\nabla_\perp\cdot\frac{d}{dt}\nabla_\perp(\phi + \tau p_i) - \nabla_\parallel J - 2\frac{c_s^2}{\omega_{ci}}\mathbf{b}\times\kappa\cdot\nabla(p_e + \tau p_i) = 0, \tag{13.64}$$

$$\frac{dn}{dt} + \frac{c_s^2}{\omega_{ci}}\frac{1}{L_n}\frac{\partial\phi}{\partial y} + 2\frac{c_s^2}{\omega_{ci}}\mathbf{b}\times\kappa\cdot\nabla(\phi - p_e) + c_s\nabla_\parallel v_\parallel - \nabla_\parallel J = 0, \tag{13.65}$$

$$\frac{3}{2}\frac{dT_e}{dt} - \frac{dn}{dt} + \frac{c_s^2}{\omega_{ci}^2}\left(\frac{3}{2}\frac{1}{L_{T_e}} - \frac{1}{L_n}\right)\frac{\partial\phi}{\partial y} - \frac{\kappa_\parallel}{n_0}\nabla_\parallel$$

$$\times\left(\nabla_\parallel T_e + \frac{c_s^2}{\omega_{ci}^2}V_A L_{T_e}\frac{\partial\psi}{\partial y}\right) - 0.71\nabla_\parallel J = 0, \tag{13.66}$$

$$\frac{3}{2}\frac{dp_i}{dt} + \frac{3}{2}\frac{c_s^2}{\omega_{ci}}\frac{1}{L_{p_i}}\frac{\partial\phi}{\partial y} + 5\frac{c_s^2}{\omega_{ci}}\mathbf{b}\times\kappa\cdot\nabla(\phi+\tau p_i) + \frac{5}{2}c_s\nabla_\parallel v_\parallel$$

$$-\frac{5}{2}\frac{c_s^2}{\omega_{ci}^2}\nabla_\parallel\cdot\frac{d}{dt}\nabla_\parallel(\phi+\tau p_i) = 0, \tag{13.67}$$

$$\frac{d}{dt}v_\parallel - c_s\left(\nabla_\parallel(p_e+\tau p_i) + \frac{c_s^2}{\omega_{ci}V_A}\left(\frac{1}{L_{p_e}} + \frac{\tau}{L_{p_i}}\right)\frac{\partial\psi}{\partial y}\right), \tag{13.68}$$

$$\frac{1}{V_A}\frac{\partial\psi}{\partial t} + \frac{c_s^2}{\omega_{ci}V_A}\left(\frac{1}{L_{p_e}} + \frac{0.71}{L_{T_e}}\right)\frac{\partial\psi}{\partial y} - \nabla_\parallel h = \frac{J}{\sigma}, \tag{13.69}$$

where the total derivatives and curvature terms are

$$\frac{d}{dt} = \frac{\partial}{\partial t} + \frac{c_s^2 R_0}{\omega_{ci}V_A}\nabla\xi\times\nabla\phi\cdot\nabla, \tag{13.70}$$

$$\nabla_\parallel = \nabla_{\parallel 0} + \frac{c_s^2 R_0}{\omega_{ci}V_A}\nabla\xi\times\nabla\psi\cdot\nabla, \tag{13.71}$$

$$\mathbf{b}\times\kappa = \nabla\mathbf{R}\times\nabla\zeta. \tag{13.72}$$

Derive the energy conservation laws for the six-field system as

$$\frac{d}{dt}\frac{1}{2}\int dV\left[\frac{c_x^2}{\omega_{ci}^2}[|\nabla_\perp(\phi+\tau p_i)|^2 + [\nabla_\perp\psi|^2] + v_\parallel^2 + n^2 + \frac{3}{2}T_e^2 + \frac{3}{5}\tau p_i^2\right]$$

$$= -\int dV\left[\frac{J^2}{\sigma} + \frac{\kappa_\parallel}{n_0}(\nabla_\parallel T_e)^2\right], \tag{13.73}$$

The coupling of the ETG turbulence to the ITG turbulence is studied in extended gyrokinetic simulations by Howard, *et al.* (2015) with what are named "cross-scale" simulations that required new gyrokinetic code developments described further in Chapter 17. The cross-scale simulations show that the ETG streamers produce a driving force in the scale-turbulent transport through the cross-scale coupling. The ETG streamers are well documented in their role of producing 50% of the electron transport in the ETG-driven thermal transport. The streamers also are important in determining the ion turbulent thermal transport. Both the fast ETG turbulence coupled to the slow ITG turbulence is required to reproduce the electron and ion thermal fluxes shown in Fig. 2 of Howard, *et al.* (2016).

The first detailed correlation between the measured ETG wave spectrum with a series of controlled, steady-state temperature gradients experiments was with large computer simulations was carried out in the IFS lead validation research with the high-resolution steady-state Columbia Linear Machine (CLM) plasma. The machine was equipped with small-scale space-time Langmuir probes in a 10 eV hydrogen plasma with continuously variable values of the ETG parameters. The data was analyzed both with IFS spectral codes and the gyrokinetic Vlasov (GKV) codes. The validation of the ETG theory and simulations where completed in this simple low-beta plasma. The theory and simulations showed why the turbulence was not observed earlier since the linear growth starts at small spatial scales and evolves with an inverse cascade to the steady-state spectrum measured in the CLM machine at scales near the ion inertial scale length. The Sci-DAC results were established in detail by IFS-UC Irvine-Columbia University team and was published as "Validation of the electron temperature gradient turbulence in the Columbia Linear Machine (CLM)" [Fu, *et al.* (2012)]. In tokamaks, the most comparable direct measurement of the different wavelength regions of the ETG turbulence appears to be those in DIII-D [Rhodes, *et al.* (2007)] with a complex three-regime fluctuation spectral diagnostic system described in Chapter 15. A definitive test of the hypothesis that the short scale $\nabla T_e(r,t)$-driven turbulence produces the high level of electron thermal transport in tokamaks is given in a relatively long ECH driven discharge in DIII-D. The discharge has multiple steps of electron heating starting with an ohmic phase followed by a series of steps-ups in the RF heating power [Rhodes, *et al.* (2007)]. The fluctuation spectrum was measured by three different techniques which separated the different spatial scales of the turbulence. The results of the data analysis validate the ETG scale model by showing that it is the small scale turbulence that increases with a series of 0.5 MW step-ups in the ECH heating while the longer scale turbulence remains relatively steady. The ion temperature is relatively undisturbed while the electron temperature profile measured accurately responds as predicted by the ETG turbulence model.

Chapter 14

Magnetic Reconnection Turbulence

Current-carrying plasmas as in tokamaks and the plasma current sheets in Earth's magnetosphere have a rich variety of magnetic reconnection instability. The linear and nonlinear dynamics of those current-carrying plasmas are developed in this chapter.

Tearing instabilities are spontaneous magnetic-field reconnection that occur in sheared magnetic configuration, such as in a current sheet or at a rational surface in a toroidal magnetic confinement. Early theory showed that the magnetic reconnection instability is driven by the current density gradient in a current sheet was caused by resistivity [Furth, *et al.* (1963)]. In space physics the collisionless magnetic reconnection with wave-particle resonance replaces the resistivity.

14.1 Introduction to Magnetic Reconnection (MRC)

The simplest model of magnetic reconnection instability is based on resistive magnetohydrodynamics (MHD), which consists of the vorticity equation and the Ohm's law. The reduced dynamical equations are

$$\frac{\partial U}{\partial t} + \mathbf{v} \cdot \nabla U = \mathbf{B} \cdot \nabla J \qquad \frac{\partial \psi}{\partial t} = -\mathbf{B} \cdot \nabla \phi + \eta J \tag{14.1}$$

where $U = \nabla^2 \phi$ and $J = \nabla^2 \psi$ are the vorticity and the current density normal to the two-dimensional plane (along to the guide magnetic field) in two-dimensional MHD (reduced MHD). The parameter η stands for plasma resistivity which violates frozen-in magnetic field line dynamics of MHD. Assuming the growth time of instability is much slower than Alfvén time, one finds that most of the system is governed by the equation $\mathbf{B} \cdot \nabla \mathbf{J} = 0$ except the region where the equation is singular $\mathbf{B} \cdot \nabla = 0$. It is clearly shown by the linearized form of the magnetic equation, $\mathbf{B_0} \cdot \nabla\nabla^2\psi_1 + \mathbf{B_1} \cdot \nabla\nabla^2\psi_0 = 0$ called the external equation [See Eq. (4.16) in Biskamp (2000)] becomes singular. The first term, that is the highest derivative term, vanishes at $\mathbf{B_0} \cdot \nabla = 0$ where there is a neutral sheet or rational surface in toroidal plasma. At the singular surface one needs to solve the resistive layer equation (14.1), that is obtained by assuming $\partial/\partial x \gg \partial/\partial y$ in a resistive layer δ_k where $k_\parallel v_A \to 0$. The asymptotic matching between solutions of the regular external equation and and the resistive layer equation gives the dispersion relation of the reconnection instability called the tearing instability.

When a current sheet is sufficiently narrow it becomes unstable to long-wavelength perturbations and magnetic islands appear. The theory shows that the stability parameter of tearing instability Δ' determines the instability threshold [Furth, *et al.* (1963)] and thus the formation of magnetic islands. The stability parameter of tearing instability

$$\Delta' = \frac{L}{2}\left[\frac{d\ln\psi_1}{dx}\right]_{0-}^{0+} \tag{14.2}$$

is obtained by solving the external equations in Eq. (14.1).

The tearing mode growth rate is a hybrid of the Alfvén wave time scale $\tau_A = L/v_A$ and slow resistive time scale $\tau_R = \mu_0 L^2/\eta$, where L is the space scale of the current sheet, η is the collisional diffusion of magnetic flux and $v_A = B_0/\sqrt{4\pi m_i n_0}$ is the Alfvén wave velocity. The associated dimensionless parameter $S = \mu_0 v_A L/\eta$ is the Lundquist number. A typical time scale for the tearing instability is $\tau_A^{2/5}\tau_R^{3/5}$. This strong dependence on the resistivity is similar to the time scale of reconnection described by the Sweet-Parker model, which predicts the scaling of $\tau_A^{1/2}\tau_R^{1/2}$.

There are three types of reconnection instabilities; tearing instability, internal kink instability, and double tearing instability. The plasma displacement $\xi_r = -\psi_1/(d\psi_0/dr)$ in cylindrical geometry (r,θ) is shown in Fig. 14.1. The displacement

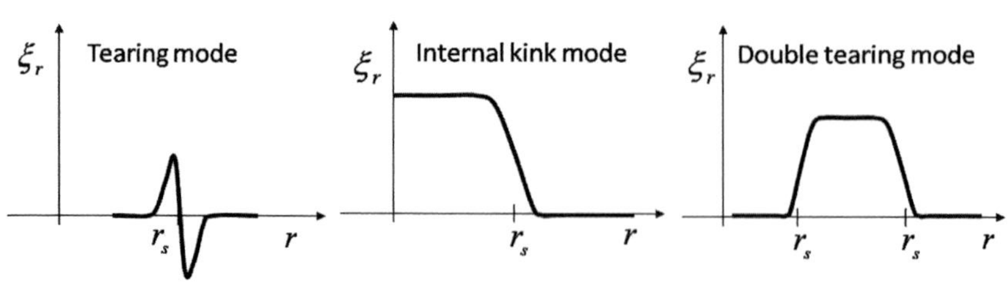

Fig. 14.1 Radial plasma displacement of three reconnection instabilities, tearing instability, internal kink instability, and double tearing instability. The r_s is the singular surface.

of tearing instability shows inflow to the reconnection point. The displacement of internal kink instability shows rigid body displacement inside r_s where the rational surface is located. The displacement of double tearing instability shows rigid body displacement between the two rational surfaces. The displacement of the latter two are similar at the rational surface and they cause faster reconnection than tearing instability with the growth time scaling as $\tau_A^{2/3}\tau_R^{1/3}$.

Fast magnetic reconnection, which is independent, or weakly depends, on the value of the resistivity, occurs when the current sheet thins to the ion inertial length δ_i or the ion Larmor radius ρ_i. Thin sheets require the Hall region around the reconnection point, $\mathbf{E} = -\mathbf{v}\times\mathbf{B} + \mathbf{J}\times\mathbf{B}/en$. In the Hall region, electrons are magnetized and move with magnetic field lines, $\mathbf{E} = -\mathbf{v_e}\times\mathbf{B}$, where $\mathbf{v_e} =$

$\mathbf{v_i} - \mathbf{J}/en$. Figure 14.2 shows that z-component of this equation $E_z = -\mathbf{v_e} \times \mathbf{B}|_z$ is valid within the ion skin depth d_i or the ion Larmor radius l_{mi} except the vicinity of reconnection point. On the other hand, ions are not tied to magnetic field lines in the Hall region, and they do not need to go through small scale electron layer, resulting in fast reconnection. The motion of electrons is decoupled from ions and this causes the Hall current in the region. Turbulence can arise around a thin current sheet, and its fluctuations can produce fast magnetic reconnection as shown in Fig. 14.2. The turbulence accelerates the magnetic reconnection and changes the threshold for magnetic island formation.

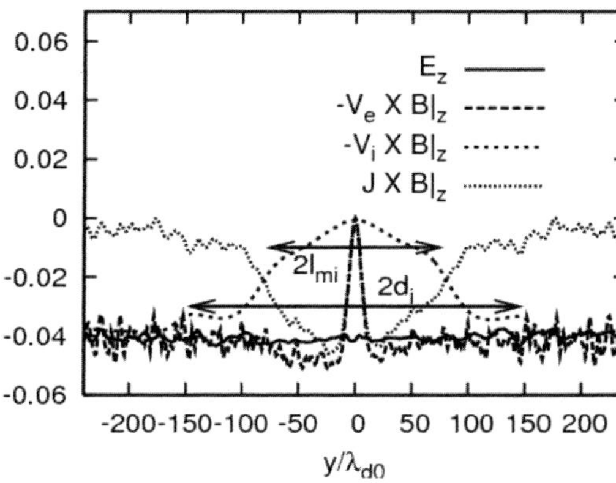

Fig. 14.2 The spacial profile of electric field E_z, $-\mathbf{v_e} \times \mathbf{B}_z$, $-\mathbf{v_i} \times \mathbf{B}_z$, and Hall term $-\mathbf{J} \times \mathbf{B}_z$, where d_i, l_{mi}, and λ_{d0} are the ion skin depth, local ion Larmor radius, and Debye length [Ishizawa and Horiuchi (2005)].

This chapter is divided into three parts, corresponding to the three instabilities. In each section, effects of turbulence and/or two-fluid effects, responsible for collisionless reconnection in the magnetic reconnecting instability are described. Section 15.2 describes nonlinear dynamics of internal kink instability with two-fluid dynamics. Section 15.3 describes effects of drift-wave turbulence on tearing instability. Section 15.4 describes two-fluid effects on tearing instability. Section 15.5 describes effects of drift-wave turbulence on double-tearing instability. Additional cases, such as tearing instabilities with the combined effects of resistivity, viscosity, and shear plasma flow, are not treated here, but are developed in detail in Chen and Morrison (1990a,b) and Ofman, *et al.* (1991).

Worked Problem 14.1:

(a) Derive the linearized equation that governs the exterior perturbed flux function $\delta\psi = \psi_{ky}(x)\cos(k_y)$ from $\mathbf{B} \cdot \nabla \mathbf{J} = 0$ for a general current sheet $J(x)$.

(b) Show that for a Harris sheet current profile $J(x) = J_0 \,\mathrm{sech}^2(x/L_s)$ the tearing mode stability parameter is $\Delta' = 2/L_s(1/(k_y L_s) - k_y L_s)$.

Solution:

(a) The linearized form of $\mathbf{B} \cdot \boldsymbol{\nabla} \boldsymbol{J} = 0$ is

$$B_y(x)\left(\partial_x^2 - k_y^2\right)\psi_{ky}(x) - J'(x)\psi_{ky}(x) = 0 \tag{14.3}$$

which has a singularity where the line bending stabilizing term (the first term) vanishes, $B_y(x) = \partial\psi_0/\partial x = 0$. The second term is the kink term, which drives tearing instability, where $J'(x) = d^3\psi_0/dx^3$ represents current density gradient. To resolve the $\psi_{ky}(x)$ functional at the singular the effect of magnetic flux diffusion $D_m = \eta/\mu_0(\text{mks}) = c^2\eta/4\pi(\text{cgs})$ must be included in a resistive plasma and the effect of electron inertia $(m_e/e)du_{e\|}/dt = -(m_e/ne^2)dJ_\|/dt)$ and parallel electron pressure $\nabla_\| p_e$ in the collisionless plasma.

(b) For the Harris sheet

$$B_y(x) = B_0 \tanh\left(\frac{x}{L_s}\right) \quad \text{and} \quad J(x) = J_0 \,\text{sech}^2\left(\frac{x}{L_s}\right)$$

the eigenvalue problem in Eq. (14.3) becomes

$$B_0 \tanh\left(\frac{x}{L_s}\right)\left(\frac{d^2}{dx^2} - k_y^2 + \frac{2}{L_s^2}\,\text{sech}^2\left(\frac{x}{L_s}\right)\right)\psi_{k_y}(x) = 0 \tag{14.4}$$

with boundary conditions, in the simplest case with no external forcing, of $\psi_{k_y}(x \to \pm\infty) = 0$ and finite $\psi_{k_y}(0)$, which is allowed by non-ideal MHD effects such as the resistivity or electron inertia. By substitution one finds that $\psi_{k_y}(x) = (\exp -k_y|x|)(1 + 1/k_y L_s \tanh(|x|/L_s))$. Substituting this solution into Eq. (14.2) gives the stability parameter. For a long wavelength perturbation $k_y/L_s < 1$, the parameter is positive $\Delta' > 0$, and thus tearing mode is unstable.

Worked Problem 14.2: Derive the growth rate from Δ'

Show that tearing mode is unstable when the stability parameter Δ' is positive and its growth rate depends on the resistivity as $\eta^{3/5}$ by following (a) and (b).

(a) Derive the linearized equation that governs the resistive layer (inner layer) equation for tearing modes by assuming that the equilibrium magnetic field is approximated to $B_y = B_0 x/L_s$, $\partial/\partial x \gg \partial/\partial y$, and the growth rate is small (constant-ψ approximation) used in Sec. 15.1.

(b) By using the asymptotic matching condition

$$\Delta' = \frac{1}{\psi_1}\int J_1 dx. \tag{14.5}$$

Solution:

(a) The linearized form of Eq. (14.1) is

$$\frac{\partial}{\partial t}\frac{\partial^2 \phi_1}{\partial x^2} = x\frac{\partial}{\partial y}\frac{\partial^2 \psi_1}{\partial x^2}, \qquad \frac{\partial \psi_1}{\partial t} = x\frac{\partial \phi_1}{\partial y} + \eta\frac{\partial^2 \psi_1}{\partial x^2}. \tag{14.6}$$

By assuming the solution has a form $f_1 = \widehat{f}_1(x)\exp(st + iky)$,

$$s^2\frac{d^2\widehat{\xi}_1}{dx^2} = xk^2\frac{d^2\widehat{\psi}_1}{dx^2}, \qquad s\widehat{\psi}_1 = -sx\widehat{\xi}_1 + \eta\frac{d^2\widehat{\psi}_1}{dx^2}, \tag{14.7}$$

where $\widehat{\xi} = -ik\widehat{\phi}$ is the radial plasma displacement. For small growth rate $s \ll 1$, we can expand the magnetic flux and the displacement as $\widehat{\psi}_1 = \widehat{\psi}_1^{(0)} + \widehat{\psi}_1^{(1)}$ and $\widehat{\xi}_1 = \widehat{\xi}_1^{(1)}$. In the 0-th order the first equation becomes $d^2\widehat{\psi}_1^{(0)}/dx^2 = 0$, so that $\widehat{\psi}_1^{(0)} = \Psi$ is constant because of tearing parity $\psi(-x) = \psi(x)$. In the first order the equations become

$$s^2\frac{d^2\widehat{\xi}_1^{(1)}}{dx^2} = xk^2\frac{d^2\widehat{\psi}_1^{(1)}}{dx^2}, \qquad s\Psi = -sx\widehat{\xi}_1^{(1)} + \eta\frac{d^2\widehat{\psi}_1^{(1)}}{dx^2}, \tag{14.8}$$

and they are combined into an equation, which is the resistive layer equation for tearing modes,

$$\left(\frac{d^2}{dz^2} - z^2\right)\chi = z, \tag{14.9}$$

where $\chi = \widehat{\xi}_1^{(1)}/(\alpha\Psi)$, $z = \alpha x$, and $\alpha = (\eta s/k^2)^{-1/4}$.

(b) Since $J = \frac{\partial^2\psi}{\partial x^2}$ in the layer, the matching condition is

$$\Delta' = \frac{1}{\Psi}\int_{-\infty}^{\infty}\frac{1}{x}\frac{d^2\widehat{\psi}_1^{(1)}}{dx^2}dx = C\frac{s^2}{k^2}\left(\frac{k^2}{\eta s}\right)^{3/4}, \quad C = \int_{-\infty}^{\infty}\frac{1}{z}\frac{d^2\chi}{dz^2}dz. \tag{14.10}$$

The growth rate is $s = \Delta'^{4/5}\eta^{3/5}k^{2/5}C^{-4/5}$, and thus tearing mode is unstable only when $\Delta' > 0$, and its growth rate depends on the resistivity as $\eta^{3/5}$ because the resistivity makes the unstable magnetic reconnection.

The regime where $\Delta' \leq 0$ is unstable to the development of magnetic islands in the nonlinear finite beta regime from the growth of magnetic interchange instabilities [La Haye and Sauter (1998), La Haye (2006), Ishizawa and Nakajima (2007, 2010), Fitzpatrick, *et al.* (2006), Waelbroeck, *et al.* (2009)].

14.1.1 *Nonlinear dynamics of the sawtooth events*

In high-temperature tokamak plasmas, when the safety factor $q(r) = rB_\phi/RB_\theta = 1$ in the core, there is a marginally stable $m = 1$ MHD mode or oscillation with a singular layer at the $m/n = 1/1$ rational surface. Inside this singular surface there is a solid-body displacement of the interior plasma that pushes the magnetic at the position of the maximum of the radial displacement $\xi(r,\theta,\phi)$ against the exterior flux. The condition for forced reconnection is created and one finds that adding a small amount of resistivity to the plasma allows the rapid growth of the $m = 1$ mode in the core of the plasma with minor distortion of the plasma outside the $m = 1$ plasma. The growth rate for this internal kink mode with the finite resistivity is fast occurring as $\gamma = (\eta/\mu_0 r_1^2)^{1/3}(v_A/r_1)^{(2/3)}$ where the radius r_1 is the position of the $q(r_1) = 1$ resonant surface.

It was expected that this internal tearing mode would continue to grow at this linear resistive Alfvén hybrid power of 1/3 of the resistive diffusion rate and the 2/3 power of the Alfvén time. However, Aydemir (1992) showed that reconnection accelerates in the nonlinear stage when the two-component fluid equations are used and the Ohm's law includes the parallel electron pressure gradient and the electron inertia. The discovery of this accelerated nonlinear $m = 1$ mode has been an important step in understanding the dynamics of the internal kink dynamics that creates the sawtooth oscillations in the tokamak.

Early analytical work on the sawtooth did not anticipate this accelerated growth and conjectured the mode would saturate as the $m \geq 2$ in what is a saturated magnetic island. Kadomtsev (1975) realized that the current sheet at the rational surface would develop with a rapid reconnection if resistive effects were included and derived a widely-used reconnection model. The method of calculating the growth of the magnetic islands was found by Rutherford (1973). Magnetic islands at the low order rational surfaces $q = 3/2, 2$ are commonly observed as shown in Chapter 13 using the ECH driven TCV plasmas for an example.

The different topology of the $m = 1$ MHD disturbance as a rigid displacement of the core plasma inside the $q = 1$ surface and a relatively undisturbed exterior plasma changes the dynamics and creates the topology for the fast reconnection.

Figure 14.3 shows the nonlinear growth computed from the two-component fluid model by Aydemir (1991, 1992). The phenomenon is robust and reproduced now in a number of simulations including high resolution Adaptive Mesh Simulations by Bhattacharjee, *et al.* (2005). Here we describe the equations used by Aydemir for his important discovery of an accelerated rate of growth, sometimes called explosive growth, of the reconnection dynamics. The nonlinear simulations are shown in Figs. 14.3 and 14.4 for the nonlinear growth and crash of the sawtooth oscillation.

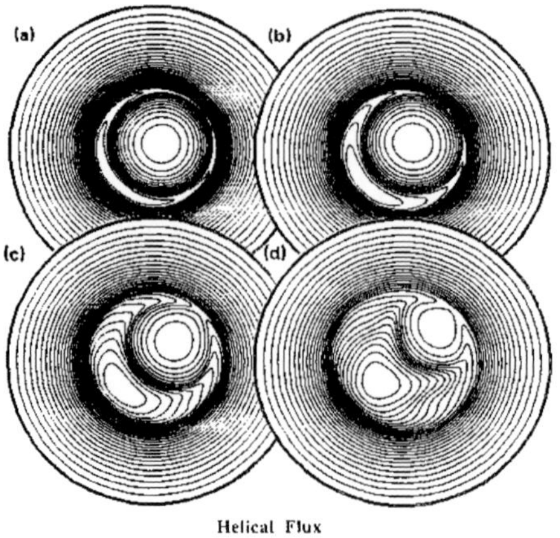

Fig. 14.3 The contours of the helical flux function at various points during the nonlinear evolution of the semi-collisional $m = 1$ mode, the growth rate of which is shown in Fig. 14.4.

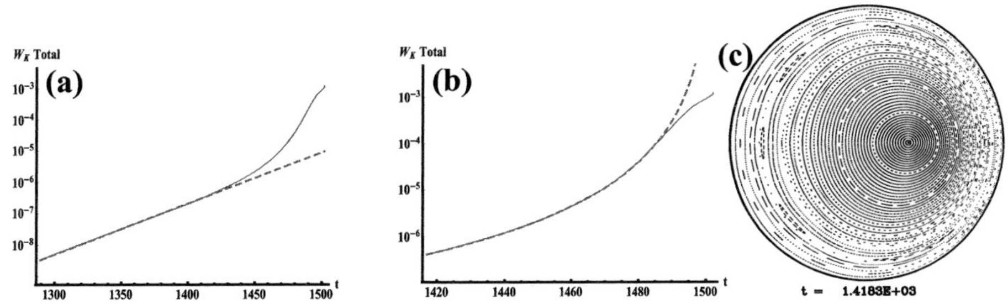

Fig. 14.4 (a) The accelerated growth of the total kinetic energy in $n \geq 1$ modes ($n_{\max} = 30$ in this calculation) in magnetic reconnection. For $t \geq 1420$, the growth rate max starts deviating from the purely exponential growth of the dashed line. (b) The dashed curve shows a numerical fit to the end phase of the disruption of the form $W_K(t) = W_K(t_i)[(t_f - t_i)/(t_f - t)]^\nu$, where $\nu = 3.37$. (c) Poincaré plot of the field lines at the beginning of the explosive growth phase [Aydemir, *et al.* (2016)].

The simplified model of the core of a tokamak that includes the effects of the long-mean-free-path electron dynamics and support of a parallel electric field by the electron pressure gradient as in drift waves follows from the reduce two component fluid equations in Chapter 4. Following Chapter 9 we introduce the parallel vector potential and the electrostatic potential to describe the coupled drift-Alfvén wave dynamics. The electron inertial term $m_e\, d\boldsymbol{v}/dt$ which introduces the collisionless skin depth $\delta_e = c/\omega_{pe}$ in the parallel component of the electron fluid acceleration equations since the dynamics becomes fast and develops on small space scales. The interchange driving terms from the toroidal curvature are weak contributors to the dynamics and are omitted in the following equations. The coupling to the parallel ion motion is retained but is not a key player in fast reconnection dynamics.

As in the nonlinear drift and Alfvén wave equations the divergence of the polarization current from the ions introduces the cross-field plasma vorticity $U = \nabla^2\phi$ and the Alfvén wave dynamics is described by the parallel component of the vector potential $A_\parallel = -\psi$ where the magnetic flux function ψ is introduced such that the parallel electric current is given by $J = \nabla^2\psi$. The electron pressure is $P_e = p$, and the parallel ion flow velocity v. The evolution of the four fields is given by

$$\frac{\partial U}{\partial t} + [\phi, U] + \nabla_\parallel J = 0 \tag{14.11}$$

$$\frac{\partial \psi}{\partial t} + \nabla_\parallel[\phi - \delta p] = \eta J + \delta_e^2\left\{\frac{\partial J}{\partial t} + [\phi, J]\right\} \tag{14.12}$$

$$\frac{\partial p}{\partial t} + [\phi, p] + \beta\nabla_\parallel[v + 2\delta J] = 0 \tag{14.13}$$

$$\frac{\partial v}{\partial t} + \frac{1}{2}\nabla_\parallel p + [\phi - \delta\tau p, v] = 0. \tag{14.14}$$

In this model there are four dynamical fields given by

$$\phi, \quad J = \nabla_\perp^2 \psi, \quad U = \nabla_\perp^2 \phi, \quad v \tag{14.15}$$

and the system parameters are

$$\beta = \frac{nkT_e}{B_T^2/2\mu_o} \quad \delta_i = \frac{c}{\omega_{pi} a}, \quad \delta_s = \frac{c/\omega_{pe}}{a} = \left(\frac{m_e}{m_i}\right)^{1/2} \delta_i. \tag{14.16}$$

The space-time variables are normalized as follows: $t \to t/\tau_A, \boldsymbol{r} \to \boldsymbol{r}/a, \eta = \tau_A/\tau_R$, where $\tau_A = a/v_A$, $\tau_R = \mu_o a^2/\eta$, and $v_A = B_{po}/\sqrt{\rho_o \mu_o}$. Here τ_A and τ_R are the poloidal Alfvén time and the resistive diffusion time defined in terms of the minor radius a, a characteristic poloidal field strength B_{po}, and plasma electrical resistivity η. See Hazeltine, *et al.* (1985, 1987) for more details.

The four brackets in Eqs. (14.11)–(14.14) are defined by $[\phi, U] = \widehat{\zeta} \cdot \nabla_\perp \phi \times \nabla_\perp U$, where $\widehat{\zeta}$ is a unit vector in the toroidal direction, and $\nabla_\perp$ is the 2D gradient in the plane perpendicular to the magnetic field. The parallel gradient operator is defined as $\nabla_\| J = \partial J/\partial \zeta + [J, \psi]$ for any scalar J. The parameter δ_i, in combination with the electron β, is related to the ion Larmor radius, as it can be shown that $\tau \delta_i^2 \beta = \rho_i^2$. Similarly, the ion Larmor radius based on electron temperature, ρ_s, is defined by $\rho_s^2 = \delta_i^2 \beta$. Both ρ_i and ρ_s are normalized to the minor radius. Finally, $\delta = \delta_s = c/\omega_{pe}$ is the collisionless skin depth.

14.1.2 *Electron inertia and hyper-resistivity*

When the finite-electron-inertia terms $m_e dJ/dt$ are present, it is useful to rewrite the parallel Ohm's law in terms with an auxiliary variable $\chi \equiv \psi - \delta_s^2 J$.

In terms of χ the electron acceleration equation along the reconnecting magnetic field becomes

$$\frac{\partial \chi}{\partial t} + [\phi, \chi] + \frac{\partial \phi}{\partial \zeta} = \delta \nabla_\| p + \eta J + \mu_e \nabla_\perp^2 \chi. \tag{14.17}$$

The last term on the right-hand side of Eq. (14.17) introduces an electron viscosity μ_e that controls the build-up of the smallest scale fluctuations in the simulations. The effective resistivity is $\eta_{\text{effec}} = \eta + \mu_e$ where μ_e is called the hyper-resistivity with $\eta_{\text{hyper}} = \mu_e \delta_s^2$. The effects of hyper-resistivity are analyzed in Aydemir (1990). The simulation shown here have $\mu_e \ll \eta$.

The equations are related to models of Drake and Lee's semi-collisional regime [Drake and Lee (1977)]. Observe that the intrinsic Coulomb resistivity is $\eta_0 = m_e \nu_{ei}/e^2 n_e$, where ν_{ei} is the electron-ion collision frequency. The definition of the resistive layer width is $x_\eta^2 = \eta_0/\omega$. A semi-collisional regime is defined by $\delta_s < x_\eta < \delta$. The collisionless regime is defined by $x_\eta < \delta_s$.

The nonlinear increase in the growth rate of the $m = 1$ mode shown in Fig. 14.3 is observed in both semi-collisional regime, where the electron inertia terms in the parallel Ohm's law can be ignored ($x_\eta \gg \delta_s$), and in the collisionless regime where the electron inertia dominates over the collisional effects ($x_\eta \ll \delta_s$).

The ion-sound terms, which come in through Eq. (14.14), are retained, although their effect was shown to be relatively unimportant Aydemir (1991)].

The four partial differential equations, (14.11)–(14.14), are solved in a cylindrical geometry, with periodically identified ends. The variables are Fourier-expanded in poloidal and toroidal directions; pseudo-spectral techniques are used [Orszag (1977)], keeping up to 256 modes. In the radial direction, a nonuniform grid with over 600 grid-points is used. In order to reduce the oscillatory behavior of fields near regions with sharp gradients, a third-order upwind-biased differencing is used for the convective derivatives.

Further differences between the semi-collisional/collisionless mode and its purely resistive counterpart can be seen also in the geometry of the island and the current sheet that develops around the separatrix. Figure 14.3 shows the helical flux contours at four times during the nonlinear evolution of the island.

The early nonlinear behavior is similar to that of a purely resistive island; the reconnection layer is poloidally extended and forms more of a "Y-point" [Waelbroeck (1989), Biskamp (1991), Zakharov and Rogers (1992)] than the classical X-point of an $m \geq 2$-island. However, as the island grows further, a well-defined X-point emerges, widening the outflow region of the reconnection layer.

The accelerated growth (faster reconnection) can be attributed to this change in the geometry of the layer, as it is easier to remove the reconnected flux with a "wider nozzle". The change in the geometry is clear as the extended current sheet forms the 'Y-layer' of early times breaks up and forms an "X-layer", clearly following the contours of the island separatrix. The current sheet that forms at the inner separatrix of the island shrinks and disappears as reconnection approaches completion; the sheet along the outer separatrix lengthens and eventually extends all the way around the circle, as predicted by Kadomtsev (1975).

This novel behavior exhibited by the semi-collisional/collisionless $m = 1$ mode can be traced to the coupling between the electron pressure gradient term in the Ohm's law, Eq. (14.12), and the parallel divergence of the electron velocity $(v+2\delta J)$ in Eq. (14.13). Linearly, these terms give rise to an adiabatic electron response in the long mean-free-path regime. The exact nonlinear mechanism that gives rise to an increase in the growth rate, however, is not clear at this point. Kleva (1992) has reported a similar accelerated growth during the reconnection of two flux bundles.

The nonlinear evolution of the $m = 1$ mode in high temperature plasmas produces a strong nonlinear acceleration of the linear magnetic reconnection rate. Both the electron inertia and electron pressure gradient through ρ_s support the parallel electric field $E_{\|}$ and are important aspects of the nonlinear dynamics of the sawteeth oscillations in tokamaks, as shown in detail in Bhattacharjee, *et al.* (2005).

In the magnetic flux topology of the reconnection dynamics at the $q = 1$ rational surface is one where the characteristic "Y" shape flux lines form as seen in many resistive MHD simulations. Concomitant with this change in the geometry of the reconnection layer, the growth rate of the reconnection process increases, resulting in a fast (explosive) reconnection of the remaining magnetic flux. This dynamics explains the observed growth of the sawtooth signals in the soft X-ray data and the subsequent rapid crash times.

Because of the importance of controlling the sawtooth oscillations in tokamaks, a real-time control system based on this relatively simple simulation dynamical model may provide a useful tool for feedback control of the core sawtooth activity in tokamaks and stellarators.

14.2 Explosive Reconnection in Reversed Magnetic Shear Profiles

For reversed magnetic shear in tokamaks with a minimum of the $q(r) \geq g_{\min}(r_0)$ there is an explosive tearing mode instability when $q_{\min}$ drops to a low-order rational. Experiments on several tokamaks have observed this behavior. On the KSTAR (Fig. 14.5) the phenomenon has now been measured and simulated in detail. As the $q_{\min}(r_0, t)$ approaches the low-order rational surface m/n – particularly the $q = 2$ surface a large increase in the nonlinear growth rate for the coupled neighboring $m/\ell = 2/\ell$ magnetic modes creates a wide spectrum of m-modes that produce a radially outward "finger" or "tongue" of hot, high pressure in a fast nonlinear event. Aydemir, *et al.* (2016) call the fast nonlinear release of the hot core plasma an "explosive finger", as shown in their nonlinear simulation in Fig. 14.4 and Fig. 14.5.

As shown in Fig. 14.6 the high-pressure vortex or finger rapidly formes and releases hot core plasma into the outside low-β plasma region. Figure 14.4 shows the sudden departure of the $n \geq 1$ modes from the earlier exponential growth of the vortex structure. In frame (c) of Fig. 14.4 the Poincaré plot of the field lines at the onset of the explosive growth is shown. The subsequent evolution of the hot plasma finger is shown in Fig. 14.6.

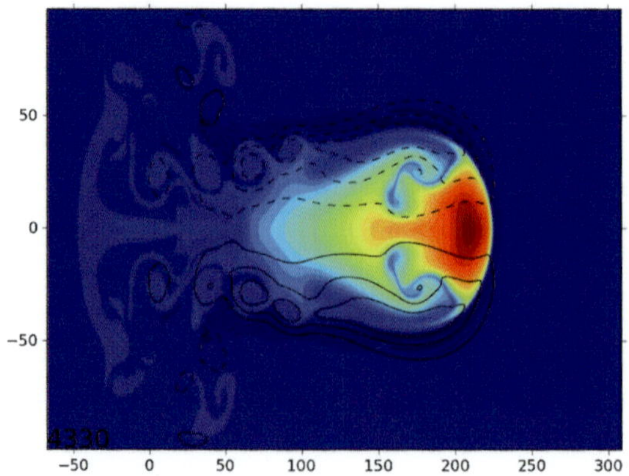

Fig. 14.5 Simulation of the nonlinear coherent structure from the longtime simulation of the plasma blob emitted from the $m = 2/l = 1$ helical mode a the $q = 2$ surface in the KSTAR tokamak [Aydemir, *et al.* (2016)].

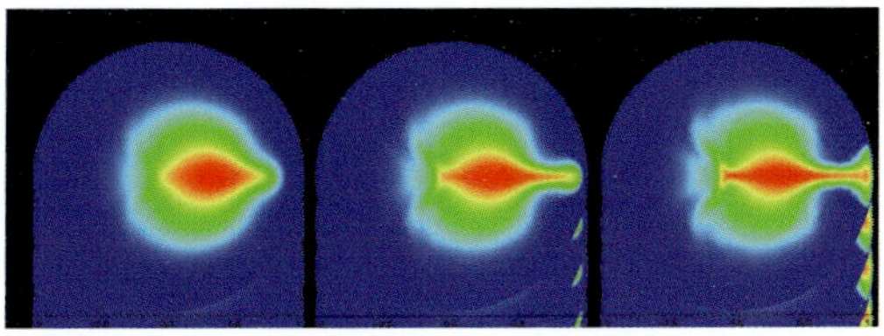

Fig. 14.6 Formation and nonlinear evolution of the "explosive finger" in the poloidal plane from a small $m/\ell = 2/1$ mode. The three figures are separated by approximately 20 Alfvén times in units of poloidal Alfvén time [Aydemir, *et al.* (2016)].

14.3 Effects of Drift Wave Turbulence on Magnetic Reconnection

Magnetic reconnection caused by turbulence in a current sheet is studied by means of numerical simulations of systems of nonlinear fluid equations [Ishizawa and Nakajima (2010)]. The nonlinear simulations show that drift wave turbulence evolves into magnetic reconnection and produces coherent magnetic islands in a current sheet, even when the sheet is so thick that there is no spontaneous magnetic reconnection, i.e. tearing modes are stable, $\Delta' < 0$. Thus, turbulence modifies the threshold of magnetic island formation predicted by the conventional linear theory. The large coherent magnetic islands are formed by the merging of small-scale magnetic islands produced in the electromagnetic drift wave turbulent fluctuations. This is consistent with the fact that the magnetic islands are caused by the energy transfer from small-scale electromagnetic turbulence in the Fourier space. In spite of the fact that the turbulence is driven by a short-wavelength instability due to temperature gradients, the length of the magnetic island is the same order as the system size. Even if one uses the MHD model, similar coherent magnetic islands are formed due to turbulence from the interchange instability which are generalized Rayleigh-Taylor instabilities [Ishizawa and Nakajima (2010), Muraglia, *et al.* (2011)].

Stronger turbulence causes wider magnetic islands, this suggests that the turbulence can trigger neoclassical tearing modes (NTMs). NTM is one of the primary MHD activities that limit the plasma pressure and are nonlinear instabilities. They require finite size magnetic islands, called seed islands, for overcoming a threshold for destabilization. The typical width of seed island is evaluated to be several times as large as the ion Larmor radius by using experimental data. Thus, such turbulence-driven magnetic island can be the seed island for NTM. Nonlinear simulations show that the magnetic islands propagate in the electron diamagnetic direction. The propagation of the island is also important for NTM destabilization because of polarization current effects. The appearance of magnetic islands as neoclassical tearing modes was reported in ASDEX Upgrade (`http://www.euro-fusion.org`). The magnetic signal showed that there was a substantial disturbance at the $q = 3/2$ rational surface that was identified as a NTM using a generalized Rutherford equation. The NTMs often form in ELMy H-mode plasmas as shown for the TCV discharge analyzed in Chapter 13. The control of NTM is a key current topic in fusion research.

A self-consistent fluid model for describing neoclassical tearing modes in global magnetohydrodynamic simulations is described and illustrates the dynamics of application of a simple toroidal configuration unstable to the (2,1) tearing mode. The island saturation is verified to increase with the bootstrap current fraction. New features shown are evidenced in the nonlinear in the unsteady saturated state of the island are the deformation to a droplet shape when the magnetic Prandtl number is not too high. Synthetic diagnostics demonstrate that diamagnetic and neoclassical effects have a measurable impact on the signature of magnetic islands [Maget, *et al.* (2016)].

The derivation of the generalized Rutherford equation for the growth and saturation of the magnetic islands is given in Fitzpatrick and Aydemir (1996) and Sauter, *et al.* (1997). There is a strong dependence on the beta-poloidal of the plasma as with the associated ELMs. The role of the drift wave turbulence on the NTM growth and propagation is given in Waelbroeck, *et al.* (2009). The stability of the $m = 1$ internal kink mode can be partially controlled by the resistivity of a close fitting metallic shell surrounding the plasma as shown by Fitzpatrick and Aydemir (1996).

14.3.1 *Two-component reconnection simulations with 5-fields*

Effects of turbulence driven by ion temperature gradient drift-wave instability on magnetic reconnection is studied by means of numerical simulations of a two-dimensional reduced set of ion and electron fluid equations [Ishizawa and Nakajima (2007)]. The model consists of the vorticity equation, the electron density equation, the parallel velocity equation, the generalized Ohm's law, and the ion temperature equation given by

$$\left(\frac{\partial}{\partial t} + \boldsymbol{v}\cdot\nabla\right)\nabla^2\phi = -\mathbf{B}\cdot\nabla J - \delta_i\beta\nabla\cdot(\boldsymbol{v}_{di}\cdot\nabla\nabla\phi) + \mu\nabla^4\phi \tag{14.18}$$

$$\left(\frac{\partial}{\partial t} + \boldsymbol{v}\cdot\nabla\right) n = -\mathbf{B}\cdot\nabla v_{e\|} + \mu\nabla^2 n \tag{14.19}$$

$$\left(\frac{\partial}{\partial t} + \boldsymbol{v}\cdot\nabla\right) v_\| = -\beta\mathbf{B}\cdot\nabla p + \mu\nabla^2 v_\| \tag{14.20}$$

$$\frac{\partial\psi}{\partial t} = -\mathbf{B}\cdot\nabla\phi + \delta_i\beta\mathbf{B}\cdot\nabla p_e + \eta J \tag{14.21}$$

$$\left(\frac{\partial}{\partial t} + \boldsymbol{v}\cdot\nabla\right) T_i = -(\Gamma-1)\left(\mathbf{B}\cdot\nabla v_\| + \kappa_L T_i\right) + \mu\nabla^2 T_i \tag{14.22}$$

where $\boldsymbol{v}\cdot\nabla f = [\phi, f]$, $\mathbf{B}\cdot\nabla f = -[\psi, f]$, $\boldsymbol{v}_{di}\cdot\nabla\nabla f = -[\nabla f, p_i]$, are written in terms of the Poisson bracket $[f,g] = \partial_x f\partial_y g - \partial_x g\partial_y f$, where $\nabla = (\partial_x, \partial_y)$, and $\mathbf{B} = B_0\boldsymbol{e}_z - \boldsymbol{e}_z\times\nabla\psi$, $\boldsymbol{v} = \boldsymbol{e}_z\times\nabla\phi$, $\boldsymbol{v}_{di} = \boldsymbol{e}_z\times\nabla p_i$ are the magnetic field, the $E\times B$

flow velocity, and the ion diamagnetic velocity respectively. In these equations ϕ, ψ, n, $J = -J_z = \nabla^2\psi$, $v_\parallel$, $v_{e\parallel} = v_\parallel + \delta_i J$, T_i, T_e, $p_i = nT_i$, $p_e = nT_e$, $p = p_i + p_e$, are the electrostatic potential, the magnetic flux function, the electron density, the negative of current density, the parallel ion velocity, the parallel electron velocity, the ion temperature, the electron temperature, the ion pressure, the electron pressure, and the total pressure, respectively.

Magnetic field configuration in (x, y)-plane is the Harris current sheet, $d\psi_{eq}(x)/dx = \tanh(x/L_s)$. The magnetic shear length L_s of the Harris sheet is associated with the gradient of current density and controls the strength of spontaneous magnetic reconnection called the tearing instability occurring for small L_s, i.e. for thin current sheets. The uniform current density model $d\psi_{eq}(x)/dx = x/L_s$, which represents an infinitely thick current sheet, is used as a limiting case. This constant current profile has the neutral sheet at $x = 0$ and is totally stable against tearing instability. The stability parameter of tearing instability $\Delta' = L/2[d \ln \widetilde{\psi}/dx]_{0-}^{0+}$ is obtained by solving the linearized form of $\mathbf{B} \cdot \nabla J = \mathbf{B} \cdot \nabla\nabla^2\psi = 0$. An equilibrium is stable (unstable) against tearing instability when Δ' is negative (positive).

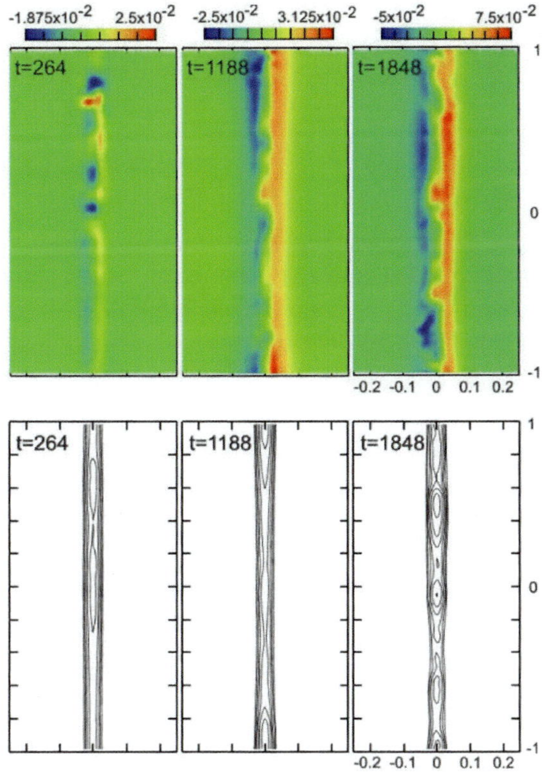

Fig. 14.7 (Color online) Color map of electrostatic potential and equicontours of magnetic flux that indicates long-wavelength magnetic islands even if there is no spontaneous magnetic reconnection, $\Delta' = -0.76$ [Ishizawa and Nakajima (2010)].

14.3.2 *Magnetic islands caused by turbulence*

Figure 14.7 shows map of electrostatic potential and equicontours of magnetic flux for the tearing mode stable Harris sheet with $L_s/L = 0.3125$, $\Delta' = -0.76$, and $\eta_i = 3.5$.

The stability parameter Δ' is calculated for the longest wavelength mode, $k = 1$, where the wavenumber is given by $k_y = \pi k/L$. The ion temperature gradient η_i drives the linear instability. At $t = 264$, narrow magnetic islands appear at the neutral sheet, and the electrostatic potential profile exhibits turbulent fluctuations. At $t = 1188$, the magnetic islands become large, and a zonal flow, which is uniform flow along the y-axis, appears. At the quasi-steady state, $t = 1848$, magnetic islands clearly appear on the neutral sheet $x = 0$. The length of the magnetic island is much longer than the ion Larmor radius and is the same order as the system size and the island width is several times the ion Larmor radius $\rho_i/L = 1/80$. The shape of the magnetic island is deformed and is different from that of a tearing instability because of the turbulent fluctuations in the magnetic field.

Figure 14.8 shows the magnetic island width as a function of the stability parameter of tearing instability Δ' for the case $\eta_i = 3.5$, where the width is time-averaged in the quasi-steady state. The magnetic island width is several times the ion Larmor radius, even if the current sheet is stable against tearing instability, $\Delta' < 0$. Hence, the ∇T_i-driven turbulence removes the island formation threshold $\Delta' = 0$ given by the conventional theory.

Figure 14.9 shows time trace of the reconnected flux which is the largest at the neutral sheet, $x = 0$, for each temperature gradient parameter η_i. When turbulence is strong, $\eta_i = 4.5$, the reconnected flux grows rapidly and attains a large value as the quasi-steady state is formed. On the other hand, when the instability is weak, $\eta_i = 2.5$, the reconnected flux grows slowly reaching a small value after which the growth saturates. Thus, stronger turbulence causes faster magnetic reconnection and wider magnetic islands.

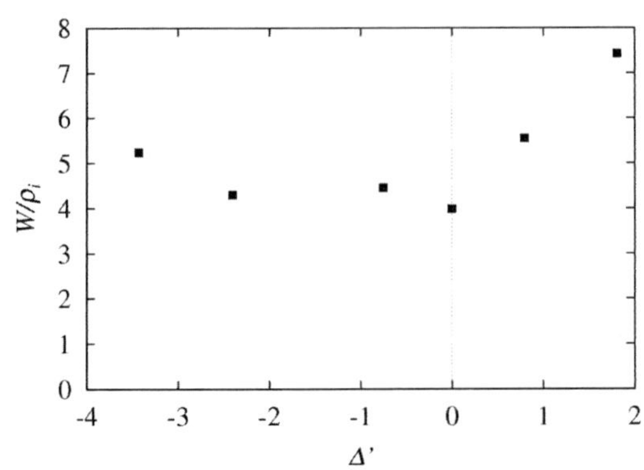

Fig. 14.8 Magnetic island width W as a function of the stability parameter Δ' of tearing instability [Ishizawa and Nakajima (2010)].

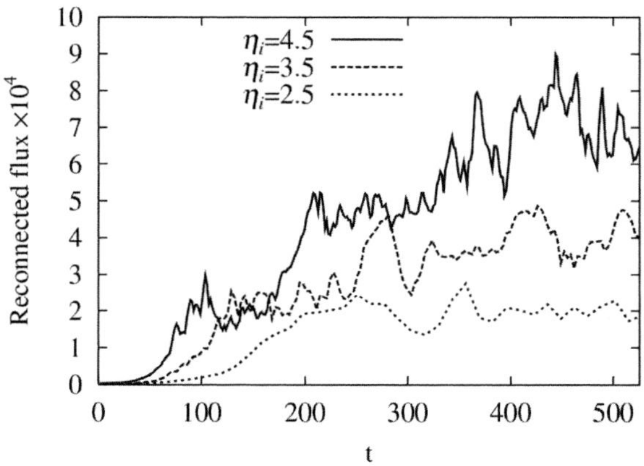

Fig. 14.9 Time trace of the reconnected magnetic flux for three values of the ion temperature gradient parameter η_i for the case $\Delta' = -3.4$ [Ishizawa and Nakajima (2010)].

14.3.3 *Formation process for magnetic islands from ITG turbulence*

The process of the magnetic island formation is examined. Figure 14.10 shows the time evolution of magnetic flux on the neutral sheet, $x = 0$, for the Harris sheet with $\Delta' = -2.4$, where red region represents positive flux, and blue negative flux, that represents the core parts of magnetic islands around O-points. Initially, the patterns show wave structures characterized by the wavelength of ion temperature gradient instability. After $t \approx 200$, large red and blue regions appear, and this means that long-wavelength magnetic islands are formed. Figure 14.11 shows the time evolution of the positions of X-points and O-points formed from δB in the initial Harris sheet with $\Delta' = -2.4$.

The red and black points represent X-points and O-points, respectively. Initially, there are many X- and O-points, and then some islands merge and disappear. In this process small-scale magnetic reconnection plays a role. After the large red and blue regions appear in Fig. 14.10, sometimes X-points and O-points appear and disappear, because the turbulent electromagnetic fluctuations cause random magnetic reconnection. Figure 14.10 and Fig. 14.11 also show the propagation of magnetic islands. Initially, the islands propagate in the negative y-direction, which is the ion diamagnetic direction, because the ion temperature gradient instability propagates in the ion diamagnetic direction. Then, after $t \cong 200$, the magnetic islands propagate in the positive y-direction, which is the electron diamagnetic direction.

The nonlinear dynamics of rotating low-m mode tearing modes with external resonant magnetic perturbations is developed by Fitzpatrick (1993, 1995), and Smolyakov, *et al.* (1995). The analysis shows that the magnetic island rotation slows down as they grow in amplitude. In the presence of a symmetry breaking external magnetic perturbation the magnetic islands become locked to the external

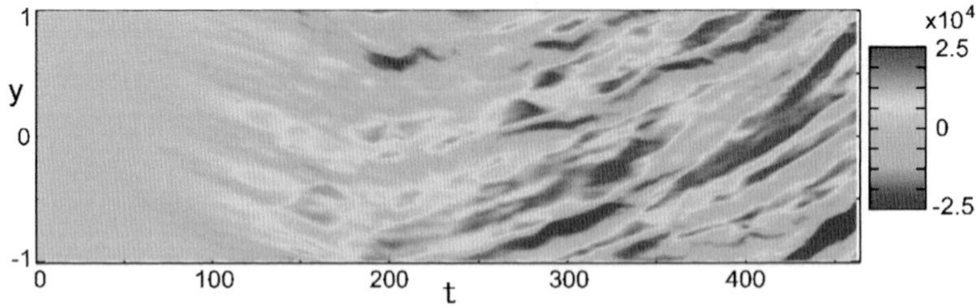

Fig. 14.10 Time evolution of magnetic flux on the neutral sheet, $x = 0$, for the case $\eta_i = 3.5$ and $\Delta' = -2.4$. Blue region represents the core parts of magnetic islands around O-points [Ishizawa and Nakajima (2010)].

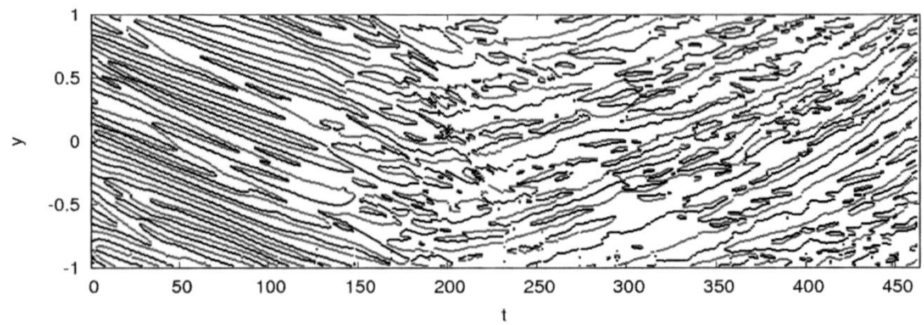

Fig. 14.11 Time traces of X-points and O-points, for the case $\eta_i = 3.5$ and $\Delta' = -2.4$. Black points are O-points and red points are X-points [Ishizawa and Nakajima (2010)].

perturbation. This locking effect can create a locked mode or a rotating magnetic island, depending on whether this a fixed machine associated asymmetry or an externally applied rotating magnetic perturbation (RMP) designed to mitigate the build-up of large ELMs (edge localized modes).

14.3.4 *Mixing of the eigenmode parities*

When Eqs. (14.18)–(14.22) are linearized by dividing field quantities into equilibrium part and perturbed part $F(x, y) = f_0(x) + f(x, y)$ with $f_0(x)$ having a symmetry, then the linearized equations also have symmetry. The symmetry yields two parities: one is tearing parity, $\psi(-x) = \psi(x)$, $\phi(-x) = -\phi(x)$, $v(-x) = v(x)$, $n(-x) = -n(x)$, $T_i(-x) = -T_i(x)$, the other is interchange parity, $\psi(-x) = -\psi(x)$, $\phi(-x) = \phi(x)$, $v(-x) = -v(x)$, $n(-x) = n(x)$, $T_i(-x) = T_i(x)$. The nonlinear process of how the drift wave turbulence which has interchange parity produces magnetic islands which has tearing parity is described and explained by Ishizawa and Nakajima (2007).

The mechanism is forced magnetic reconnection from the ∇T_i-driven fluctuations. Consider, for simplicity, resistive interchange mode in slab plasma governed by reduced MHD equations instead of ITG as shown in Fig. 14.12(a). The magnetic

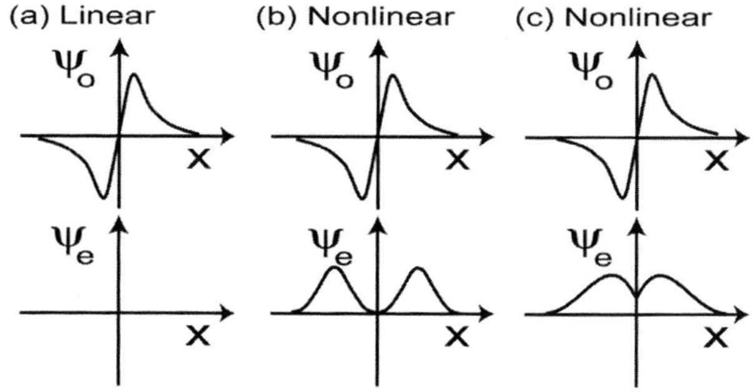

Fig. 14.12 Profile of magnetic flux showing the process turbulence which has interchange parity causes magnetic islands which has tearing parity, where ψ_o is odd part and ψ_e is even part; (a) linear evolution of interchange instability, (b) nonlinear evolution of interchange instability, (c) forced reconnection occurs and producing magnetic islands [Ishizawa and Nakajima (2007)].

flux ψ of interchange mode is odd function of x, ψ_o, in linear evolution, while the potential is even function of x, ϕ_e, and thus the interchange mode does not cause the magnetic flux of tearing mode that is finite at $x = 0$ and even function of x, ψ_e, as shown in Fig. 14.12(a), where $f_e(-x) = f_e(x)$ and $f_o(-x) = -f_o(x)$. Then, in the nonlinear evolution, the interchange mode produces the even magnetic flux ψ_e, and the odd potential ϕ_o, through the nonlinear terms of reduced MHD equations,

$$\frac{\partial \nabla_\perp^2 \phi_o}{\partial t} = -[\phi_e, \nabla_\perp^2 \phi_e] - [\phi_o, \nabla_\perp^2 \phi_o] + [\psi_e, \nabla_\perp^2 \psi_e] + [\psi_o, \nabla_\perp^2 \psi_o] \quad (14.23)$$

$$\frac{\partial \psi_e}{\partial t} = [\psi_o, \phi_e] + [\psi_e, \phi_o]. \quad (14.24)$$

This is described in Fig. 14.12(b). These ψ_e and ϕ_o from Eqs. (14.23) and (14.24) force magnetic flux to reconnect at $x = 0$, and then we have a finite amplitude of magnetic flux at $x = 0$ as shown in Fig. 14.12(c). The finite magnetic flux at $x = 0$ means that magnetic reconnection takes place.

Interchange-type modes exchange not only thermal and magnetic energies between flux tubes but also current. In a plasma with a current (or resistivity) gradient, such an interchange can create a current sheet at a mode resonance surface and result in the excitation of current-interchange tearing modes. To reveal this effect one just needs to include the current (or resistivity) gradient effect in magnetic diffusion equation. Inclusion of these gradients leads to the coupling of even (tearing) and odd (interchange) modes.

An adaptive eigenfunction independent solution-kinetic (AEGIS-K) shooting code has been developed. AEGIS-K is based on a fully kinetic (non-hybrid) and self-consistent (nonperturbative) description. It includes wave-particle resonances, shear Alfvén continuum damping, trapped particle effects, and parallel electric effects. AEGIS-K has been applied to study the kinetic and rotation stabilization of resistive wall modes. The analysis shows that the resistive wall modes in ITER configuration can be stabilized by subsonic plasma rotation.

14.3.5 *Electron viscosity or hyper-resistivity*

The electric field in the plasma is modified by the electron viscosity in addition to the resistivity. With electron viscosity, ohm's law becomes

$$\mathbf{E} = \eta \boldsymbol{j} - \frac{1}{c}\boldsymbol{V} \times \mathbf{B} - \frac{m_e}{n_e e^2}\nabla^2 \boldsymbol{j}. \tag{14.25}$$

The z-component of the curl of Eq. (14.25) gives the magnetic flux function dynamics

$$\frac{\partial \psi}{\partial t} = -\boldsymbol{V} \cdot \nabla \psi + \frac{c^2}{4\pi}\eta \nabla^2 \psi - \frac{m_e \mu_e c^2}{4\pi n_e e^2}\nabla^4 \psi \tag{14.26}$$

after using

$$\mathbf{B}_\perp = \nabla \psi \times \hat{\boldsymbol{z}} \tag{14.27}$$

and Faraday's law. The z-component of plasma vorticity equation remains as

$$\frac{\partial}{\partial t}(\nabla^2 \phi) = -(\boldsymbol{V} \cdot \boldsymbol{\nabla})\nabla^2 \phi + \frac{1}{4\pi\rho}\left[\nabla(\nabla^2 \psi) \times \nabla \psi\right] \cdot \hat{\boldsymbol{z}} \tag{14.28}$$

where ρ is the mass density of the plasma. The closed set of Eqs. (14.26) and (14.28) is the starting point for the stability analysis of nonuniform current profiles.

The addition of the viscosity is most important for the reversed $q(r)$ profiles so we limit the discussion here to that case where there are two neighboring rational surfaces for the same m, n mode straddling the minimum q-surface. There q-profiles have double or coupled tearing modes.

14.4 Double Tearing Mode in Plasmas with Anomalous Electron Viscosity

The linear behavior of the double tearing mode in plasmas with an anomalous electron viscosity μ_e is investigated within the framework of magnetohydrodynamic (MHD) theory. Consider Reynolds number $R = \tau_v/\tau_A$ where τ_v and τ_A are, respectively, the viscosity penetration time of the magnetic field and the Alfvén time for a plasma sheet of width a. The growth rate is found to scale as $R^{-1/5}$ if the two resonant surfaces, at $x = \pm x_s$, are close enough to satisfy $x_s/a \ll (k_y a)^{-11/15} R^{-1/15}$. For larger separation between the resonant surfaces, the growth rate transits to the standard $R^{-1/3}$ scaling. The transition occurs at $x_s/a \sim (k_y a)^{-11/15} R^{-1/15}$. The $R^{-1/5}$ dependence is shown to arise from the violation of the constant-ψ approximation. The nonlinear velocity perturbations associated with the unstable double tearing mode are shown to saturate at a level high enough to trigger the formation of transport barriers.

One of the characteristics for an advanced tokamak (AT) operation is a non-monotonic safety factor q-profile, providing a region with negative magnetic shear [Lao, *et al.* (1996), Koide, *et al.* (1994)]. Such a configuration is prone to the excitation of a double tearing mode (DTM), an instability peculiar to a plasma with multiple resonant surfaces ($\boldsymbol{k} \cdot \mathbf{B} = 0$). With magnetic islands on sufficiently close adjacent resonant surfaces interacting, the double tearing mode turns out to be a stronger instability than the standard (i.e., constant-ψ) tearing mode because of a large Δ'. Numerical simulations demonstrate that a macro-magnetohydrodynamic (macro-MHD) mode is excited as a result of multi-scale interaction in a quasi-steady equilibrium formed by a balance between micro-turbulence and the zonal flow dynamics.

The double tearing mode occurs in a reversed magnetic shear equilibrium which includes zonal flow and turbulence due to kinetic ballooning modes. In the steady equilibrium, a macro-scale fluctuation which has the same helicity as the double tearing mode is an intrinsic part of the turbulence. After a certain period of time, the macro-MHD mode begins to grow from the free energy of the equilibrium current density gradient. This complex macro-mode is destabilized by a positive feedback loop between zonal flow suppression and the magnetic island growth. Thus, once the macro-MHD mode emerges from the equilibrium, it may continue to grow steadily.

The double tearing mode, driven by plasma resistivity, has been studied by numerous authors [Pritchett, *et al.* (1980), Mahajan and Hazeltine (1982), Goodall and Wesson (1984), Kleva, *et al.* (1989), Ofman (1992), Persson and Dewar (1994), Yu (1996)]. The linear growth rate of the mode varies as $S^{-1/3}$ when the separation of the two resonant surfaces ($x = \pm x_s$) is sufficiently small that is satisfied.

$$\frac{x_s}{a} < (k_y a)^{-7/9} S^{-1/9}. \tag{14.29}$$

When the separation of the resonant surfaces is large so that the inequality in Eq. (14.29) is reversed, the islands do not interact strongly, and the growth rate of the double tearing mode scales as $S^{-3/5}$ as for the standard tearing mode [Pritchett, *et al.* (1980)]. Here, $S = \tau_r/\tau_A$ is the magnetic Reynolds number with $\tau_r = 4\pi a^2/c^2\eta$ and $\tau_A = a/v_A$ being the resistive diffusion time and the poloidal Alfvén time of a plasma column of scale width a, respectively; c is the speed of light, η is the plasma resistivity, v_A is the poloidal Alfvén velocity. The island growth of the double tearing mode in the early nonlinear stage is analyzed by Yu (1996).

In contrast to extensive studies on tearing modes driven by plasma resistivity, there are few investigations on tearing modes driven by plasma viscosity. Experiments indicate, however, that anomalous viscosity may be present in tokamak plasmas due to electromagnetic turbulence [Kadomtsev (1992), Scott, *et al.* (1989)]. Bootstrap drive of neoclassical tearing modes in the presence of anomalous viscosity is performed recently by Konovalov, *et al.* (2002). Tearing modes driven by anomalous electron viscosity due to braiding of the magnetic field lines have been considered as possible candidates responsible for disruptive instability [Rosenbluth and Rutherford (1981), Kaw, *et al.* (1979)].

The growth rates of $m \geq 2$ and $m = 1$ electron viscosity tearing modes scale as $\gamma \sim R^{-1/3}$ and $\gamma \sim R^{-1/5}$, respectively. Here, $R = \tau_v/\tau_A$ is the fluid dynamic Reynolds number, while $\tau_v = 4\pi a^4 n_e e^2/c^2\mu_e m_e = \omega_{pe}^2 a^4/c^2\mu_e$ is the viscosity diffusion time of plasma current over a sheet of width a, μ_e is the electron viscosity diffusion coefficient. In strong electromagnetic drift-wave turbulence the coefficient μ_e may be comparable to the electron thermal diffusivity which is a few m^2/s. The electron viscosity tearing mode was studied by Aydemir as a possible candidate for experimentally observed fast sawtooth crashes [Aydemir (1990)]. It was found that the electron viscosity driven modes had higher growth rates than the corresponding resistivity driven modes in present high-temperature fusion devices if the electron viscosity had a value comparable to that of the anomalous electron thermal diffusivity. The viscosity considered here contributes to the parallel electron flow equation increasing plasma current penetration across the magnetic flux surfaces.

Worked Problem 14.3: Linear Eigenmodes in MRC

Derive the two coupled linearized equations for the hyper-resistivity tearing modes.

Solution:

Assuming all perturbations in the form $f \sim f(x)\exp(ik_y y + \gamma t)$, we get the linearized version of Eqs. (14.26) and (14.28):

$$\gamma\psi_1 = V_x B_{0y}(x) + \frac{\eta c^2}{4\pi}\left(\frac{\partial^2\psi_1}{\partial x^2} - k_y^2\psi_1\right) - \frac{\mu_e c^2}{\omega_{pe}^2}\left(\frac{\partial^4\psi_1}{\partial x^4} + k_y^4\psi_1\right) \tag{14.30}$$

and

$$\rho\gamma\left(\frac{\partial^2\phi}{\partial x^2} - k_y^2\phi\right) = -\frac{i}{4\pi}k_y B_{0y}''(x)\psi_1 + \frac{ik_y}{4\pi}B_{0y}(x)\left(\frac{\partial^2\psi_1}{\partial x^2} - k_y^2\psi_1\right) \tag{14.31}$$

where η, ρ and μ_e have been taken as constants and the double prime denotes second derivative with respect to x.
Normalizing all lengths to a, time to τ_A, and the magnetic field to some standard measure B_0, Eqs. (14.30) and (14.31) convert to the following dimensionless from

$$\gamma\tau_A\psi_1 = \gamma\tau_A\xi B_{0y}(x) + \frac{1}{S}\left(\frac{\partial^2\psi_1}{\partial x^2} - \alpha^2\psi_1\right) - \frac{1}{R}\left(\frac{\partial^4\psi_1}{\partial x^4} + \alpha^4\psi_1\right) \tag{14.32}$$

and

$$(\gamma\tau_A)^2\left(\frac{\partial^2\xi}{\partial x^2} - \alpha^2\xi\right) = \alpha^2 B_{0y}''(x)\psi_1 - \alpha^2 B_{0y}(x)\left(\frac{\partial^2\psi_1}{\partial x^2} - \alpha^2\psi_1\right) \tag{14.33}$$

where $\alpha = k_y a, \xi = ik_y\phi/\gamma a, S = \tau_r/\tau_A, R = \tau_v/\tau_A$. It is easy to note from Eq. (14.26) that $\tau_v = a^4\omega_{pe}^2/\mu_e c^2$ is the viscosity diffusion time for a magnetic field over a plasma sheet of width a.

The current penetration time due to viscosity over the tearing layer of width Δ is estimated as

$$\tau_\Delta \simeq \frac{\Delta^4\omega_{pe}^2}{\mu_e c^2} = \Delta^4\tau_A R \tag{14.34}$$

from Worked Problem 14.3. The number of e-foldings from the growth rate in this current penetration time is

$$\gamma\tau_\Delta = \gamma\left(\frac{\gamma\tau_A}{\alpha^2 B_{0y}'^2 R}\right)^{2/3}\tau_A R = \widehat{\lambda}^{5/3}. \tag{14.35}$$

If the ideal MHD mode is marginally stable, then $\widehat{\lambda} \simeq 1$. The current penetration time is comparable with the tearing mode growing time and the constant-ψ approximation is not valid. On the other hand, if the ideal MHD mode is stable, then $\widehat{\lambda} \ll 1$. The tearing mode growth time is much longer than the current penetration time. In this case, ψ_1 may approximately be considered as a constant over the tearing layer.

14.5 Magnetic Reconnection

Mahajan and Yoshida (2000) have shown that in a collisionless two-fluid model, a combination of the Hall term and the fluid nonlinearity can lead to the formation of a self-organized singular layer that displays some of the observational features of the thin shear layer associated with the high-confinement (H-mode) tokamak discharges. One of the necessary conditions for the formation of such a layer is a poloidal velocity Mach number of order unity. A similar situation, i.e., the existence of a narrow layer with a strongly sheared flow, exists for tokamak discharges with internal transport barriers (ITB).

An important clue comes from the experimental observations that the ITBs are often formed around low mode number rational surfaces [Joffrin, *et al.* (2002)]. Coupled with the fact that the magnetic energy released in the reconnection process following the development of a tearing mode can drive large flows, this suggests that the tearing mode may provide a possible trigger for the formation of the ITBs.

Local radial zones of reduced transport called internal transport barriers are routinely produced in the large tokamaks. Internal transport barriers (ITBs) were first observed in the large tokamaks in the 1990s [Koide, *et al.* (1994), Mori, *et al.* (1994), Levinton, *et al.* (1995), Strait, *et al.* (1995), Gormezano and the Jet Team (1996), Söldner and the Jet Team (1997)]. The first discovery was reported in the Japan Atomic Energy Research Institute Tokamak-60 Upgrade (JT-60U) [Ninomiya and JT-60 Team (1992), Koide, *et al.* (1994), Mori, *et al.* (1994)] where a dramatic increase in the core ion temperature and the associated ion temperature gradient was measured without a large turbulent ion thermal diffusivity. The JT-60U news was both exciting and puzzling.

An ITB is a region where a steep temperature and/or density gradient forms inside a tokamak plasma, within the region $\rho \lesssim 0.7$, where ρ is the normalized minor radius. The ITB formed earlier in time than the edge barrier and during the evolution of the $q(r,t)$ profile. Typically there was a $q_{\min}(t)$ decreasing in time in the core plasma. Later in time the edge transport barrier would emerge.

In the few tokamaks with toroidal plasma velocity measurements, it was subsequently found that the region within an ITB was also the region where the toroidal rotation velocity gradient was steep. This high-performance regime occurred with a lower-power threshold than required for the high-confinement mode (H-mode) regime where the transport barrier is located at the plasma edge. Since the hydrogen ions H^+ and D^+ do not radiate in the KeV core plasma, theoretical transport formulas were first used to infer the hydrogen ion velocity profiles.

14.6 Internal and Edge Zones of Reduced Transport

A schematic of the multiscale physics for the formation of the internal and edge transport barriers along with the particle and thermal transport is shown in Fig. 14.13. In the remainder of this chapter we give some details of these multiscale transport simulations for the transport barrier dynamics for experiments in JET, ASDEX-U and DIII-D. The name "transport barrier", while well accepted in the fusion literature, is misleading since there is no actual barrier but rather there is a marked reduction in the dominant thermal transport across relatively narrow radial regions. These changes in the transport are associated with regions of steep gradients in the toroidal rotation velocity and the magnetic shear radial profiles. A local minimum in the magnetic shear enhances the effect of the rotational shear

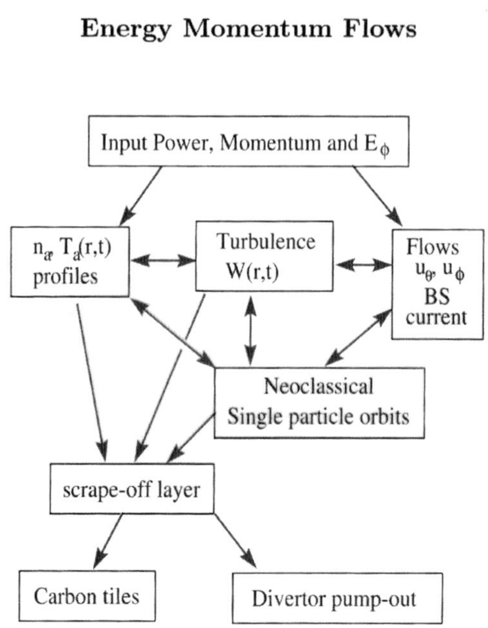

Fig. 14.13 A flow diagram of the complex weather-like dynamics of the driven toroidal plasma. The closed loops of couplings give rise to the bifurcations described in the simulations using the JETTO and MMM codes.

and is one method by which the small scale turbulent electron and ion transport can be strongly reduced.

Internal transport barriers were produced in two optimized magnetic shear (OS) discharges in the Joint European Torus (JET) [Rebut, *et al.* (1985)], shot #40542 and #40847 [Gormezano and the Jet Team (1996, Söldner and the Jet Team (1997), Parail, *et al.* (1999)]. In both shots, neutral beam injection (NBI) is the dominant auxiliary power source, although there is some RF heating as well. The q-profiles generated by current ramping in these two shots are flat and low in the core (near the magnetic axis) and steeply rising approaching the plasma edge, with the minimum value of q greater than unity throughout the plasma. In both discharges, the ITB forms first in the core while the edge remains in the low-confinement mode (L-mode). Then the ITB expands towards the edge. In one shot (#40542) an ELMy (edge localized mode) H-mode subsequently occurs and results in the co-existence of both an internal transport barrier and an edge transport barrier (ETB) until the ramping down of NBI power. In the other JET shot (#40847), the edge plasma first transits to an ELM-free H-mode and then enters an ELMy H-mode phase, during which the internal transport barrier decays. While it is an interesting feature for these two JET shots that both ITB and ETB co-exist, we focus on the ITB behavior in the simulations.

Balancing shearing frequency in the convection term and the viscous dissipation rate in the vorticity equation, we find $k_y V_y \sim k_\perp^4 \mu_e c^2/\omega_{pe}^2 = v_A(k_\perp^4 a^4/R\alpha$ where v_A is the poloidal Alfvén velocity and $k_\perp^2 = k_x^2 + k_y^2 \simeq k_x^2$. With $k_x \sim \Delta^{-1} \simeq (\lambda\sigma)^{-1/6} \simeq [\gamma\tau_A/(\alpha B'_{0y})^2 R]^{-1/6}$ and $\gamma\tau_A \sim \lambda \sim R^{-1/5}$. This leads to an estimate of the saturated poloidal shearing velocity compared with the thermal velocity is $V_y/v_T \sim R^{-1/5}$. For $R \sim 10^5$, this is the magnitude required [Mahajan and Yoshida (2000)] for strong sheared flows in the layer. In contrast the resistivity driven double tearing modes will generate similar levels of saturated poloidal flows for rather low $S \sim 10^3$ making them unlikely to be the sources for ITB creation. The viscosity driven double tearing mode, on the other hand is a viable candidate.

The underlying physical mechanisms that may lead to the anomalous electron viscosity needs further research. The braiding magnetic field lines [Aydemir (1990)] created by higher mode number MHD instabilities is one possible mechanism. Another possible source could be the unstable electromagnetic micro-modes characteristic to high β (= plasma pressure/magnetic pressure) plasmas [Dong, *et al.* (1999), Gao, *et al.* (2001)].

14.7 Turbulent Impulsive Magnetic Energy Release from Electron Scale Reconnection

Magnetic reconnection may occur as bursts of nonlinear plasma dynamics on the electron collisionless skin length scale $d_e = c/\omega_{pe}$ during which a large fraction of the magnetic energy is converted to plasma thermal energy and plasma flow energy. An example of such a bursty energy release event is given with a set of electron

Hall fluid equations. The energization mechanism is the cross-field compression of the electron gas between interacting magnetic islands. The electron energization appears to be consistent with the rapid electron energy flux changes measured by the Cluster spacecraft crossing thin current sheets at $-17\,R_E$ in the geotail. The analysis is relevant to spacecraft missions for measuring electron scale magnetic reconnection events in the geomagnetic tail.

In the domain of collisionless magnetic reconnection, there are two space/time scales of the dynamics. One is the ion tearing mode dynamics where the nonlinear structure and islands are comparable to ion gyroradius ρ_i and the ion inertial scale length $\rho_s = (m_i T_e)^{1/2}/eB$, and the second is the formation of small scale structures and islands on the scale of electron [Drake and Lee (1977)] skin length $d_e = c/\omega_{pe}$. The time scale for the electron event is corresponding shorter than that for the ion events, and they can have a larger power transfer from the magnetic energy releases $\Delta E_B < 0$ to the plasma thermal energy.

The short duration and large numbers of such events make them an ideal candidate to explain the fast, sometimes called explosive, energy release required by geomagnetic substorms. The time scale for the collisionless ion tearing modes is generally too long for the fast dynamics $\Delta t \ll 100L/v_A < 10\,\text{min}$ observed in geomagnetic substorms.

For geotail reconnection we use the Geocentric Solar Magnetospheric Coordinate System, called the GSM system, with x-axis pointing toward the sun, y-axis from dawn to dusk perpendicular to x-axis and the z-axis along the Earth's magnetic dipole axis. In general, the planet's magnetic dipole axis has a tilt angle θ with respect to the ecliptic plane of the Earth's rotation about the sun.

Here we derive a relatively simple nonlinear electron scale magnetic reconnection model based on two nonlinear partial differential equations for the magnetic flux $\psi(x,z,t)$ and the electrostatic potential $\phi(x,z,t)$. The reconnection dynamics evolves to a final state with the release of magnetic energy, $-\Delta E_B$, of order 50% available from the reversed field components, $\pm B_{x0}$ in a volume of $\Omega = L_x L_y L_z \sim 2000\,d_e^2 L_y$ in the geomagnetic tail with the current sheet $J_y(z)$.

During the reconfiguration period there is an interval of time, in the initial evolution of the magnetic island, during which the $\mathbf{E}\times\mathbf{B}$ flow energy dominates the parallel flow energy and the electron thermal energy. Subsequently the power transfer to the electron pressure $\delta P_e = \delta n T_e$ is dominant.

Indeed, in the early nonlinear phase, the magnetic island associated with the instability undergoes an explosive growth, which confirms the theoretical predictions of Ottaviani and Porcelli (1993) and Bhattacharjee, *et al.* (2005). In this first phase, most of the power transfer goes to the $\mathbf{E}\times\mathbf{B}$ flow (around 70% of the released magnetic energy $|\Delta E_B|$).

However, when islands on neighboring resonant surfaces start to interact, the system becomes turbulent. The $\mathbf{E}\times\mathbf{B}$ energy decreases initially in favor of the parallel electron flows energy, $K_\parallel = 1/2\, n_e m_e u_\parallel^2$, and finally a large fraction of the released energy goes into the electron thermal energy, $\Delta W_e = 3/2\,\Delta p_e \Omega$. As the energy unit scale in the theory we use the initial reversed field energy per unit length along the ambient current, $E_{B0} = B_{x0}^2 L_x L_z/(2\mu_0)$.

In the numerical example shown in Fig. 14.14, the energy release in the turbulent state, $-\Delta E_B$, is approximatively 70% of the initial reversed field energy E_{B0}.

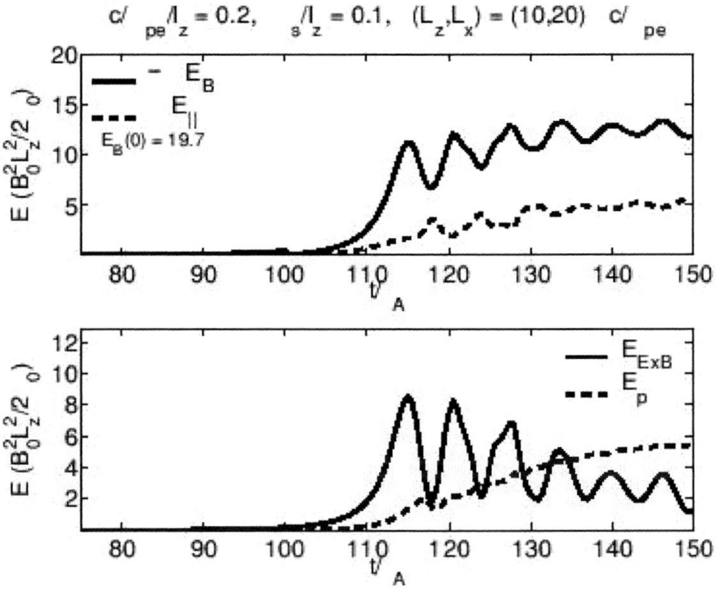

Fig. 14.14 The impulsive release of stored magnetic energy $-\Delta E_B$ from collisional reconnection into the parallel outflow energy of the electrons $\Delta E_{\|}$ in top figure in time of 30 Alfvén periods. Lower panel shows the transient increase and subsequent decrease in the $\mathbf{E} \times \mathbf{B}$ flow energy and the steady increase in the electron thermal energy E_p.

Of this released magnetic energy, approximately 44% goes to electron thermal energy, 47% to parallel electron flow kinetic energy, $K_{\|}$, and 9% to $\mathbf{E} \times \mathbf{B}$ flow kinetic energy after the reconnection or substorm event.

In the model, there are four nonlinear terms and two key kinds of power transfer terms, $J_{\|}\nabla_{\|}\phi$ and $v_{e\|}\nabla_{\|}p_e$. A third nonlinear transfer, the $\mathbf{E} \times \mathbf{B}$ convection of the plasma vorticity, $U = \nabla_{\perp}^2 \phi / B_0$, derived from $\vec{v}_E = \vec{E} \times \vec{B}/B_0^2$ with $\vec{E} = -\nabla\phi$ and $B_0\hat{\mathbf{y}}$ the out-of-plane guide field, arising from the ambient dawn-to-dusk magnetic field in the geomagnetic tail. The vorticity nonlinearity, $\vec{v}_E \cdot \nabla U$, is well known and recently measured by Perez, *et al.* (2006) in the large plasma device (LAPD). Its effect is to couple energy throughout the cross-field flow wavenumber spectrum, $|\phi_{\vec{k}}(t)|^2$, which appears in configuration space as vortex merging. Thus, this vorticity nonlinearity has only an indirect effect in the dynamics of the reconnection which accounts for its being dropped in the model of Ottaviani and Porcelli (1993).

14.7.1 *Electron scale magnetic reconnection model*

The developments of the electron-scale magnetic reconnection (MR) modes are given in Ottaviani and Porcelli (1993) and Bhattacharjee, *et al.* (2005) and Horton, *et al.* (2007). The impulsive reconnection model of Bhattacharjee has the field-line-breaking mechanism of electron inertial given by finite $d_e = c/\omega_{pe}$. Therefore, it is necessary to use the generalized ohm's Law

$$\vec{E} + \vec{v} \times \vec{B} = \frac{c^2 \mu_0}{\omega_{pe}^2} \frac{\mathrm{d}\vec{J}}{\mathrm{d}t} - \frac{\nabla p_e}{ne} + \frac{\vec{J} \times \vec{B}}{ne} + \eta \vec{J} \tag{14.36}$$

where $\vec{E}$ is the electric field, $\vec{B}$ is the magnetic field, $\vec{v}$ is the plasma flow velocity, c is the speed of light, $\vec{J}$ is the current density, p_e is the electron pressure, ω_{pe} is the electron plasma frequency, n is the electron density, e is the magnitude of the electron charge.

The simulation shown here in Fig. 14.14 takes the limit of zero plasma resistivity, η, and checks for energy conservation to five digits. High-order time integrator techniques allows for running simulations for finite times with zero resistivity.

The simplest electron tearing mode dynamics occurs when plasma flow is almost incompressible. This occurs when there is a guide field B_y. Although the pressure of the guide field is not thought to be important to the event, it is included in the treatment to make a simple justification of dropping the compressional dynamics of the electron pressure equation.

The small flow compression comes from the polarization drift of the ions and is given by $\partial U/\partial t$ where $U = \nabla_\perp^2 \phi / B_y$, with $\vec{E} = -\partial\psi/\partial t - \nabla\phi$ and $\vec{B} = \nabla\psi \times \widehat{\mathbf{y}} + B_y \widehat{\mathbf{y}}$. The constant B_y-field is called the guide field in space plasma physics. The vorticity, U, grows to feed plasma into the reconnection layer. The compressional Alfvén wave drops out of the dynamics in this regime owing to its higher frequency and different polarization.

The plasma is described in an electron-Hall fluid limit by the following two field equations

$$\frac{\partial F}{\partial t} + [\phi, F] = \widehat{\rho}_s^2 [U, \psi] \tag{14.37}$$

$$\frac{\partial U}{\partial t} + [\phi, U] = [J, \psi] \tag{14.38}$$

where $J = -\nabla_\perp^2 \psi$, $F = \psi + \widehat{d}_e^2 J$, $U = \nabla_\perp^2 \phi$, and the Poisson bracket is defined by $[\phi, F] = \widehat{\mathbf{y}} \cdot \nabla\phi \times \nabla F$. All quantities are dimensionless (this will be fixed with the proper dimensions). Time is normalized by Alfvén time scale $\tau_A = l_z / v_A$ where the Alfvén velocity $v_A = \sqrt{B^2/\mu_0 \rho}$ and $l_z = L_z / 2\pi$ is the scale over which the ambient current density varies. The dynamics depends on two dimensionless parameters $\widehat{d}_e$ and $\widehat{\rho}_s$ where the electron skin depth is $\widehat{d}_e = (c/\omega_{pe})/l_z$ and the ion sound gyroradius is $\widehat{\rho}_s = (c_s/\omega_{ci})/l_z$.

In the region of a uniform $B_x = \partial\psi/\partial z$ field, the linearized dynamics from Eqs. (14.37) and (14.38) gives $\omega F_k - k_x B_x \phi_k = \rho_s^2 k_x B_x U_k$ and $\omega U_k = -k_x B_x J_k$.

Therefore, we obtain the dispersion relation

$$\omega^2 = \frac{k_x^2 v_A^2 (1 + k_\perp^2 \widehat{\rho}_s^2)}{1 + k_\perp^2 \widehat{d}_e^2} \tag{14.39}$$

which is the well-known kinetic Alfvén wave dispersion relation. The properties of these waves in a uniform plasma have been studied experimentally by Vincena, *et al.* (2004) in the LAPD. Here we follow the formulation of Bhattacharjee, *et al.* (2005) and Ottaviani and Porcelli (1993), but note there is an important error in the magnetic flux function equation in Bhattacharjee, *et al.* (2005). That is corrected here in Eq. (14.37). The error in the $\rho_s^2[U, \psi]$-term would eliminate the energy conservation derived here. The same model in Eqs. (14.37) and (14.38) has been used in Cafaro, *et al.* (1998) and Grasso, *et al.* (2001) where the formation of fine structures and the saturation of a single magnetic island are investigated. In particular, Grasso, *et al.* (2001) present a numerical study of the energetics of an isolated magnetic island by using an Harris pinch equilibrium. A more complete Hamiltonian model, which necessarily conserves energy, is given in Tassi, *et al.* (2008).

Owing to the absence of boundaries in the central plasma sheet and the non-monotonic nature of the cross-tail current density $J_y(z)$, periodic box boundary conditions, as those in Bhattacharjee, *et al.* (2005), are appropriate in the local simulations.

The collisionless reconnection computation has been done with the spectral representations

$$\psi = \Sigma_{n,m} \psi_{m,n}(t) \exp(inz + im\epsilon x) \tag{14.40}$$

$$\phi = \Sigma_{n,m} \phi_{m,n}(t) \exp(inz + im\epsilon x) \tag{14.41}$$

where n, m are integers, $\epsilon = L_z/L_x$ = width of current sheet/length of current sheet is an aspect ratio of the system. The initial profile are the unstable current sheet $\psi_{0,1}$ with small perturbations $\psi_{1,4}$ and $\phi_{1,4}$, that is,

$$\psi(0) = \cos z + \delta\psi \cos(4z) \cos(\epsilon x)$$
$$\phi(0) = \delta\phi \sin(4z) \sin(\epsilon x).$$

For the example shown here, $\psi_{1,4} = \phi_{1,4} = 0.0001$ are 10 e-folding to reach the first maximum of $-E_B$ in Fig. 14.14.

Other initial conditions ($n = 2$) with a spectrum of small modes leads to the same conclusions.

The simulation box is $L_z \times L_x = 10\pi\, c/\omega_{pe} \times 20\pi\, c/\omega_{pe}$ and the unit of energy is $E_0 = B_0^2 L_z^2 / 2\mu_o$, which for 10 nT reversed field over a $l_z = 5\, c/\omega_{pe} = 50$ km is $2.5 \times 10^6 \mathrm{J}/R_E$.

Equations (14.37) and (14.38) are numerically solved in a double periodic box, with three resonant surfaces at $z = 0$ and $z = \pm\pi$. The periodic boundary conditions used in the numerical analysis are justified by the fact that one is interested in the dynamics of interacting microscale magnetic islands. The use of periodic boundary conditions is the standard model in condensed matter physics for eliminating the influence of surface effects in bulk matter calculations. Similarly, the physics of the boundary conditions at the lobe plasma and the magnetopause is eliminated by assuming periodicity at the edge of the numerical box when its size is large compared with c/ω_{pe}, but small compared to the dimensions of the current sheet.

Note that the theoretical models of Ottaviani and Porcelli (1993) and Bhattacharjee, *et al.* (2005) apply only to the early nonlinear phase of the numerical calculation, when the magnetic islands do not interact with each other. This is due to the fact that both models rely on the standard Rutherford nonlinear growth theory for single islands [Rutherford (1973)]. In particular, the existence of a dominant harmonic in the nonlinear magnetic flux is not a valid *Ansatz* for the interacting islands arising in the late phase of our simulation. For this reason, the numerical integration in Bhattacharjee, *et al.* (2005) is stopped when the width of the central magnetic island becomes comparable to the system size. The simulations presented here are continued farther in time, roughly twice as long as those in Bhattacharjee, *et al.* (2005).

The electron reconnection model conserves the total energy of electrons and the magnetic field. The energy component formulas and the conservation law are given by

$$\frac{\mathrm{d}E_{\mathrm{tot}}}{\mathrm{d}t} = \frac{\mathrm{d}}{\mathrm{d}t}\left(E_B + E_\parallel + E_{\mathbf{E}\times\mathbf{B}} + E_p\right) = \frac{1}{2}\frac{\mathrm{d}}{\mathrm{d}t}\int dxdz\Big[(\nabla\psi)^2 + d_e^2(\nabla^2\psi)^2 + (\nabla\phi)^2 + \left(\rho_s^2(\nabla^2\phi)\right)^2\Big] = 0 \tag{14.42}$$

where E_B is magnetic energy, $E_\parallel$ is electron parallel flow kinetic energy, $E_{\mathbf{E}\times\mathbf{B}}$ is the electron perpendicular flow energy and E_p is the electron thermal energy. There are four transfers between the different energy components. The magnetic energy E_B and the electron parallel flow energy $E_\parallel$ are transferred back and forth to the thermal energy E_p by thermal interaction $\langle v_{e\parallel}\nabla_\parallel p_e\rangle$. And the energies, E_B and $E_\parallel$, are transferred back and forth to the perpendicular flow energy $E_{\mathbf{E}\times\mathbf{B}}$ by the electric interaction $\langle j_\parallel \nabla_\parallel \phi\rangle$.

The time history of the energy relevant to the ϕ, U, ψ and J fields are shown in Fig. 14.15. During the saturation process, almost 70% of the magnetic energy is released and about 45-50% of the released energy is transferred to the electron thermal energy. In Fig. 14.15(a), we see the released magnetic energy ΔE_B of $14.1\,E_0$ and the increase of the parallel flow kinetic energy in the electrons is $6.2\,E_0$. In the bottom panel the perpendicular flow energy increases first to $8\,E_0$ and then drops to $1.3\,E_0$ while the electron thermal energy continually increases to $6.6\,E_0$. Energy is conserved by the 14.1 units of released magnetic energy going to 6.2+6.6+1.3 (E_0 unit) units of parallel flow, perpendicular flow and thermal plasma energies. In terms of fractional changes, this is approximately a 70% decrease of magnetic energy transformed into 9% $\mathbf{E}\times\mathbf{B}$ flow, 44% parallel flow and 47% electron thermal energy. The space and time scales are $10\,c/\omega_{pe}$ and $60L_z/v_A$. In the central plasma sheet, these scales are typically 100 km and 1 min.

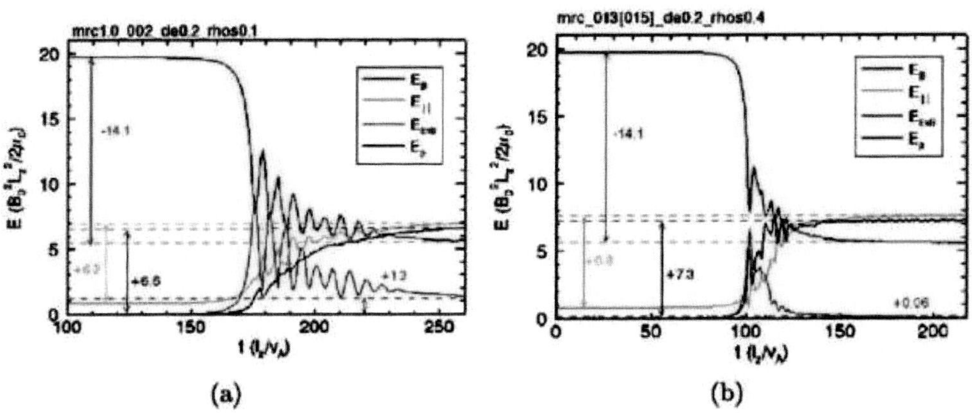

Fig. 14.15 Time evolution of magnetic energy E_B, parallel electron flow energy $E_{\|}$, perpendicular $\mathbf{E}\times\mathbf{B}$ energy and electron thermal energy E_p for small and large ρ_s values. (a) $\widehat{d}_e = 0.2$, $\widehat{\rho}_s = 0.1$ and (b) $\widehat{d}_e = 0.2$, $\widehat{\rho}_s = 0.4$ with initial mode $(k_x, k_z) = 2\pi(4/L_x, 1/L_x) = 2\pi(-.5, 4)/L_z$. Approximately 70% of initial magnetic energy is released in both cases (a) and (b) [Horton, *et al.* (1979)].

In Fig. 14.15(b), the case of $\widehat{d}_e = 0.2$, $\widehat{\rho}_s = 0.4$ shows the energy release 70% of the magnetic energy and the transfer of 40%, 0.4% and 52% to the parallel flow energy, the $\mathbf{E}\times\mathbf{B}$ energy and the thermal energy. The twice larger $\widehat{\rho}_s$ corresponds to a four times large $\nabla_{\|}p_e$-term in ohm's law which makes the transition faster and increases the final electron thermal energy from 6.6 to $7.3\,E_0$.

Figure 14.3 shows in more detail the 30 Alfvén time interval where the magnetic energy E_B drops and transfers to the thermal energy E_p and the parallel flow energy $E_{\|}$ in the collisionless reconnection event.

When Eqs. (14.37) and (14.38) are run as written with no dissipation, the magnetic energy drops sharply over a period of 100 to 150 Alfvén time and the high-k spectrum increases until subscale energy dissipation is required or the simulation is stopped in time. For a convergence test the simulations are then repeated with hyper-resistivity $\eta_4(\nabla^2)^4F$ and hyperviscosity $\mu_4(\nabla^2)^4U$ added to the right-hand side. Now wavenumber spectra are compared. The wavenumber spectra in the range $k_\perp = [1, 20]/L_z$ are essentially the same. For $k_\perp$ greater or equal to $30/L_z$, there is a rollover and a rapid drop off for the runs with high-k damping.

The background electron temperature gradient effect with two parallel heat closure models was investigated.

The first closure model is an electrostatic closure model. The parallel heat flux convected by $\mathbf{E}\times\mathbf{B}$ velocity without the consideration of the magnetic fluctuation is zero. This model uses a large but finite parallel heat diffusivity $\chi_{e\|}$. Thermal balance reduces to

$$n\vec{v}\cdot\nabla T_e + \nabla_{\|}q_{e\|} = 0 \tag{14.43}$$

with $q_{e\|} = -n\chi_{e\|}\nabla_\| T_e$ where Sugama, *et al.* (2001) give the formula for $\chi_{e\|}$. The electron temperature fluctuation is

$$\frac{\delta T_e}{T_e} = -i\frac{\omega_{*Te}}{\chi_{e\|}k_\|^2}\frac{e\phi}{T_e} \tag{14.44}$$

where $\omega_{*Te} = k_x T_e/eB_y L_{Te}$.

The second model is an electromagnetic closure model. The parallel thermal diffusivity is taken as infinitely large and the parallel heat flux $q_{e\|}$ is bounded from above. This means the magnetic fluctuation is governed by

$$\nabla_\| T_e = \frac{\partial \delta T_e}{\partial y} + \frac{\delta \vec{B}}{B_y}\cdot\nabla T_e = 0 \tag{14.45}$$

reducing to

$$\frac{\delta T_e}{T_e} = \frac{\omega_{*Te}}{ck_y}\frac{ce\psi}{T_e}. \tag{14.46}$$

At the small values of $\omega_{*Te}/\chi_{e\|}k_y^2$ and ω_{*Te}/ck_y, one does not observe the significant effect and we need to investigate further.

The electron reconnection scale produces fast, small scale releases of the reversed field magnetic energy. In laboratory experiments, the calculations are typically performed for a few, well separated magnetic islands, as observed and measured from the soft X-ray emissions in tokamaks. The pseudo-spectral simulations with high order time integration can resolve the strong nonlinear state down to the scale where microturbulence from lower-hybrid drift waves occurs. The current densities reported in the Cluster spacecraft geotail data have peak values with the electron drift velocities of order twice the ion thermal velocity.

The dynamics exhibits a long exponential growth, ending with a nonlinear, faster than exponential growth pulse. Finally, saturation is reached in a turbulent state. Typically, 70% of the initial magnetic energy, which is the difference of the magnetic energy between the initial and final magnetic field geometries, is redistributed into the sum of the parallel and perpendicular flow kinetic energies and the electron thermal pressure by the reconnection pulse.

The model developed in this section is missing the kinetic theory physics of the wave-particle interactions. Thus, the simulation alone is not able to capture the beam-like features of the ion and electron velocity distribution, reported by spacecraft particle detectors. Some of these features maybe captured by adding ensembles of test particles to be integrated with the evolving reconnection electromagnetic fields. An example of this technique can be found in Horton and Tajima (1991), where the strong ion energization is shown for a simple analytic model of the ion-tearing mode.

The physical nature of the ion tearing mode is less clear. A common formulation of ohm's law (ion Hall MHD) for the ion tearing mode is $\mathbf{E} = -\mathbf{v}\times\mathbf{B} + \mathbf{J}\times\mathbf{B}/ne$ which keeps $E_\| = \mathbf{E}\cdot\mathbf{b} = 0$ so that the magnetic flux is not reconnected [Biskamp (2000), p. 16]. In the case where the $\mathbf{B}\cdot\nabla\cdot\mathbf{P}_i$ is large, magnetic reconnection does occur. An example of this type of ion reconnection is analyzed in Horton and Tajima (1991). This dynamics is a chaotic ion viscous dissipation that frees the magnetic field from the assumption of the frozen-in theorem.

Some kinetic theory physics related to a finite ion gyroradius is added to the partial differential equations. The most important kinetic effect to be added to the fluid model is the electron temperature fluctuation, δT_e. This is driven by the $\widetilde{\mathbf{v}}_E \cdot \nabla T_e$-convection in the nonuniform background T_e field. The background temperature gradient introduces the drift-tearing mode physics, thought to be an important instability for electron thermal transport in high-β laboratory confinement experiments. The simplest version of this modification is to generalize the right-hand side of Eq. (14.37) to $\widehat{\rho}_s^2[U + \eta_e \partial_x \phi, \psi]$, where η_e measures the nonuniformity of the ambient electron temperature profile.

The electron reconnection events similar to that modeled here have been observed in the substorm of August 24, 2003 during a period where the Cluster spacecrafts crossed a thin current sheet $L_j \leq 2c/\omega_{pi} \approx 1000$ km and measured electron acceleration and tailward high speed flows. The event reported in the Cluster 2003 tail season data by Nakamura, *et al.* (2006). The current sheet is comparable to the ion collisionless skin depth in thickness as the four space craft with a spacing of 200 km flew through the geotail. The event was found by selecting data with high current densities. The current density for this event reached $40\,\mathrm{nA/m^2}$ which is about ten times the nominal value for the center of the current sheet. The space crafts were at approximately $(-17, -3.8, 3.3)\,\mathrm{R_E}$ in the geotail. The electron energy flux observed to increase from $0.5 - 0.8\,\mathrm{KeV}$ up to $5\,\mathrm{KeV}$ in crossing a turbulent reconnection zone. Electron acceleration measured during this reconnection event is suggestive of the electron reconnection dynamics analyzed in the simulation.

A second example [Wygant, *et al.* (2005)] of a microscale magnetic reconnection event, is described by a 15 minutes Cluster event with repeated crossings of a moving plasma sheet at distance $x = -16\,\mathrm{R_E}$ in the midnight sector. This event was measured during the geomagnetic storm of October 1, 2001. The electric and magnetic fields were measured every 50 ms and the particle distribution every 8 s. In the 15-minute interval, the signature of the c/ω_{pe}-scale reconnection dynamics are seen. The density is $0.1\,\mathrm{cm^{-3}}$, giving $c/\omega_{pe} = 20\,\mathrm{km}$. The measured magnetic field reverses from $B_x = \pm 15\,\mathrm{nT}$ over a current sheet of thickness $\sim 5\,c/\omega_{pe}$ in 4 s, from the motion of the spacecraft through the current sheet. The guide field, B_y, is approximately 5 nT. The plasma parameters upstream ($\beta \ll 1$), in the layer ($\beta \geq 100$), and downstream appear to be consistent with the magnetic reconnection model described in Sec. 14.7.1.

When both the electron temperature gradient and the current density gradient are retained in the nonlinear dynamics of small scale ETG and microtearing modes we find the following equations:

$$\left(1-\nabla_\perp^2\right)\frac{\partial\phi}{\partial t}=\left\{\left[1-2\epsilon_n+(1+\eta_e)\nabla_\perp^2\right]\frac{\partial}{\partial y}-\mu\nabla_\perp^4\right\}\phi \tag{14.47}$$

$$+\left(\nabla_\parallel\nabla_\perp^2-j_\parallel'\frac{\partial}{\partial y}\right)A+2\epsilon_n\frac{\partial\delta T_e}{\partial y}+\left[\phi,\nabla^2\phi\right] \tag{14.48}$$

$$\left(\nabla_\perp^2-\frac{\beta}{2}\right)\frac{\partial A}{\partial t}=\left(2\nabla_\parallel-j_\parallel'\frac{\partial}{\partial y}\right)\phi+\left[\frac{\beta}{2}(1+\eta_e)\frac{\partial}{\partial y}-\frac{\eta}{\mu_0}\nabla_\perp^2\right]A \tag{14.49}$$

$$-\nabla_\parallel\delta T_e-\left[\phi,\nabla_\perp^2A\right] \tag{14.50}$$

$$\frac{\partial\delta T_e}{\partial t}=-\left[\eta_e-4\epsilon_n(\Gamma-1)\right]\frac{\partial\phi}{\partial y}-(\Gamma-1)\left(\nabla_\parallel\nabla_\perp^2-j_\parallel'\frac{\partial}{\partial y}\right)A \tag{14.51}$$

$$-\left[2\epsilon_n(2\Gamma-1)\frac{\partial}{\partial y}-\chi_\perp\nabla_\perp^2-\chi_\parallel\nabla_\parallel^2\right]\delta T_e-\left[\phi,\delta T_e\right]. \tag{14.52}$$

Worked Problem 14.4: Six-Field Electromagnetic Model for Turbulence with Magnetic Reconnection

Analyze the six nonlinear partial differential equations derived in [Zeiler and Drake (1997)] that describe the combined dynamics of drift waves and magnetic reconnection. Derive the dispersion relations for the waves and discuss the relationship with the results in Chapters 3, 5 and 15.

Derive the two coupled linearized equations for the hyper-resistivity tearing modes.

Equations developed by Zeiler and Drake (1997) use the dimensionless variables given by

$$T_e=\frac{\widetilde{T}_e}{T_{e0}},\quad p_i=\frac{\widetilde{p}_i}{p_{i0}},\quad n=\frac{\widetilde{n}}{n_0},\quad \phi=\frac{e\widetilde{\phi}}{T_{e0}},\quad v_\parallel=\frac{\widetilde{v}_{\parallel i}}{c_s}$$

$$\psi=\frac{\omega_{ci}V_A\widetilde{\psi}}{c_s^2B_0},\quad J=V_A\frac{c_s^2}{\omega_{ci}^2}\nabla_\perp^2\psi, \tag{14.53}$$

$$h=\phi-p_e-0.71T_e,\quad \tau=\frac{T_{i0}}{T_{e0}},\quad \sigma=\frac{T_{e0}}{\eta_\parallel n_0e^2} \tag{14.54}$$

where $\tau=T_{i0}/T_{e0}$ and $\sigma=T_{e0}/\eta_\parallel n_0e^2$. The nonlinear equations in these six fields are given by

$$\frac{c_s^2}{\omega_{ci}^2}\nabla_\perp\cdot\frac{d}{dt}\nabla_\perp(\phi+\tau p_i)-\nabla_\parallel J-2\frac{c_s^2}{\omega_{ci}}\boldsymbol{b}\times\boldsymbol{\kappa}\cdot\boldsymbol{\nabla}(p_e+\tau p_i)=0, \tag{14.55}$$

$$\frac{dn}{dt}+\frac{c_s^2}{\omega_{ci}}\frac{1}{L_n}\frac{\partial\phi}{\partial y}+2\frac{c_s^2}{\omega_{ci}}\boldsymbol{b}\times\boldsymbol{\kappa}\cdot\boldsymbol{\nabla}(\phi-p_e)+c_s\nabla_\parallel v_\parallel-\nabla_\parallel J=0, \tag{14.56}$$

$$\frac{3}{2}\frac{dT_e}{dt} - \frac{dn}{dt} + \frac{c_s^2}{\omega_{ci}}\left(\frac{3}{2}\frac{1}{L_{T_e}} - \frac{1}{L_n}\right)\frac{\partial\phi}{\partial y} - \frac{\kappa_\parallel}{n_0}\nabla_\parallel$$

$$\times\left(\nabla_\parallel T_e + \frac{c_s^2}{\omega_{ci}}V_A L_{T_e}\frac{\partial\psi}{\partial y}\right) = 0.71\nabla_\parallel J = 0, \tag{14.57}$$

$$\frac{3}{2}\frac{dp_i}{dt} + \frac{3}{2}\frac{c_s^2}{\omega_{ci}}\frac{1}{L_{p_i}}\frac{\partial\phi}{\partial y} + 5\frac{c_s^2}{\omega_{ci}}\boldsymbol{b}\times\boldsymbol{\kappa}\cdot\boldsymbol{\nabla}(\phi+\tau p_i) + \frac{5}{2}c_s\nabla_\parallel v_\parallel$$

$$-\frac{5}{2}\frac{c_s^2}{\omega_{ci}^2}\nabla_\parallel\cdot\frac{d}{dt}\nabla_\parallel(\phi+\tau p_i) = 0, \tag{14.58}$$

$$\frac{d}{dt}v_\parallel - c_s\left(\nabla_\parallel(p_e+\tau p_i) + \frac{c_s^2}{\omega_{ci}V_A}\left(\frac{1}{L_{p_e}} + \frac{\tau}{L_{p_i}}\right)\frac{\partial\psi}{\partial y}\right), \tag{14.59}$$

$$\frac{1}{V_A}\frac{\partial\psi}{\partial t} + \frac{c_s^2}{\omega_{ci}V_A}\left(\frac{1}{L_{p_e}} + \frac{0.71}{L_{T_e}}\right)\frac{\partial\psi}{\partial y} - \nabla_\parallel h = \frac{J}{\sigma}, \tag{14.60}$$

where the derivatives are given by

$$\frac{d}{dt} = \frac{\partial}{\partial t} + \frac{c_s^2 R_0}{\omega_{ci}}\boldsymbol{\nabla}\zeta\times\boldsymbol{\nabla}\phi\cdot\boldsymbol{\nabla}, \tag{14.61}$$

$$\nabla_\parallel = \nabla_{\parallel 0} + \frac{c_s^2 R_0}{\omega_{ci}V_A}\boldsymbol{\nabla}\zeta\times\boldsymbol{\nabla}\psi\cdot\boldsymbol{\nabla}, \tag{14.62}$$

and the curvature vector $\boldsymbol{k}$ is given by

$$\boldsymbol{b}\times\boldsymbol{\kappa} = \boldsymbol{\nabla R}\times\boldsymbol{\nabla}\zeta. \tag{14.63}$$

The model is useful in plasmas with large vortex and wave structures with magnetic reconnection since the fields $\phi, n, p, v_\parallel, T_e$, and ψ are not split into background and fluctuating components as in Eqs. (15.95)–(15.100).

14.8 Rotating Magnetic Island and Locked Modes

As shown in the TCV data in Fig. 13.9 low-order magnetic islands arise spontaneous in the tokamak discharges. Their onset is often correlated with the sawteeth oscillations in the core electron temperature. In the absence of a symmetry breaking magnetic perturbation from the external coils these magnetic islands propagate in the toroidal direction at speeds associated with the electron diamagnetic drift velocity relative to the rest frame of the plasma. The islands, by definition, have an even magnetic perturbation δB_r at the magnetic surface that creates the closed helical flux surfaces of half-width $w = 2\sqrt{(L_s\psi(x_{m,n})/B_0)}$. Here the magnetic shear length $L_s = qR/s$ with s/q evaluated at the rational surface specified by $q = m/n$. Clearly, in weak magnetic shear confinement modes the magnetic island width is substantial. In the TCV example in Chapter 13 the profiles are in a normal ELM-free H-mode state and the magnetic island grows to be a to be several centimeters in width as inferred from the Thomson scattering diagnostics.

The theoretical framework for the formation of the magnetic islands and the locked modes is that of the tearing mode calculations and the existence of a symmetry breaking external magnetic field $\delta B_{\rm ext}$. Smolyakov, *et al.* (1995) develop two-component fluid equations for the rate of rotation for the island in the presence of the external magnetic field. The island rotation is determined by the projection of the momentum balance equation on to the toroidal direction with the parallel component of the momentum stress tensor evaluated as a function of the viscosity coefficients acting on the ion velocity components v_θ and v_ϕ. The calculations are complicated and not repeated here.

From the nonlinear equations the structures are projected on to the single helicity magnetic island

$$\Psi[r, \theta, \phi(t), t] = \psi(r, t)\cos[m\theta - \ell\phi(t)]. \tag{14.64}$$

A pair of nonlinear equations for the rate of change of the toroidal rotation velocity are derived that with the amplitude of the external perturbation being the control parameter. From these equations the critical value for the external magnetic perturbation for the locking of the magnetic island to the external perturbation is derived.

There are several mechanisms controlling the rotation of these magnetic island so there is observed a variety of behaviors [Fitzpatrick and Hender (1991), Fitzpatrick (1993), Nave and Wesson (1986)]. The comparison of theory and experiments on magnetic islands is given in Waelbroeck and Fitzpatrick (1997), Zohm, *et al.* (2001), and Gunter, *et al.* (1998).

The separatrix region of the magnetic island develops sharp plasma gradients and a strong radial electric field shown in the figures from analytic theory and the gyrokinetic simulations in Siccinio, *et al.* (2011). The 3/2-magnetic island shown in Chapter 13 for a TCV ECRH discharge is a ubiquitous feature of high-temperature discharges as reviewed by Zhom, *et al.* (2001) for ASDEX.

14.8.1 *Shielded 3D magnetic perturbations*

The 3D magnetic perturbations in this chapter can destabilize microinstabilities and enhance the associated anomalous transport. Bird and Hegna (2013) develop 3D equilibrium theory in shaped tokamak equilibria with small 3D deformations. In the vicinity of rational magnetic surfaces, the infinite-n ideal MHD ballooning stability boundary is strongly perturbed by the 3D modulations of the local magnetic shear associated with the presence of near-resonant Pfirsch-Schlüter currents. These currents are driven by 3D components of the magnetic field spectrum even when there is no resonant radial component. The infinite-n ideal ballooning stability boundary is often used as a proxy for the onset of virulent kinetic ballooning modes and associated stiff transport. These results show that the achievable pressure gradient may be lowered in the vicinity of low-order rational surfaces when 3D magnetic perturbations are applied. This mechanism may provide an explanation for the observed reduction in the peak pressure gradient at the top of the edge pedestal during experiments where edge localized modes have been completely suppressed by applied 3D magnetic fields [Bird and Hegna (2013)].

14.9 High-β Disruptions in Tokamaks from Explosive Instabilities

Intrinsically explosive growth of ballooning mode fingers has been demonstrated in nonlinear magnetohydrodynamic calculations of high-β disruptions in tokamaks. The explosive finger is formed by an unstable $n = 1$ mode, dominated by an $m/n = 2/1$ component. The quadrupole geometry of the 2/1 perturbed pressure field provides a generic mechanism for the formation of the initial ballooning finger and its subsequent transition from exponential to explosive growth, without relying on secondary processes. The explosive ejection of the hot plasma from the core and stochastization of the magnetic field occur on Alfvénic time scales, accounting for the extremely fast growth of the precursor oscillations and the rapidity of the thermal quench in some high-β disruptions. Some disruptions at high-β have very fast-growing precursors or no discernible precursors at all [Fredrickson, *et al.* (1995), Yoshino (2005)]. When they are present, the precursor oscillations, for example on the edge magnetic field signals, precede the thermal quench typically by less than a millisecond; time evolution of the signals suggests a mode growing explosively (faster than exponentially) well into the nonlinear regime.

The characteristic time $\tau \propto (\tau_A^2 \tau_A)^{1/3}$ is a geometric mean of the MHD (τ_A) and heating (τ_A) time scales. Under these conditions, faster than exponential growth of an ideal $n = 1$ mode was confirmed in numerical simulations of a particular DIII-D disruption [Chu, *et al.* (1996)] using the NIMROD code [Kruger, *et al.* (2005)]. However, this simulation used only the $n = 0, 1$ modes in the toroidal Fourier expansions; thus, is only a "quasilinear", not a full nonlinear calculation, which requires the presence of higher harmonics of the $n = 1$ mode. This particular path to explosive growth was shown [Cowley and Artun (1997)] to be implausible, requiring an unrealistically small initial perturbation amplitude to generate Alfvénic growth, due to the vast separation in time scale τ_A.

Some characteristics of the actual MHD modes involved in high-β disruptions were studied by Park, *et al.* (1995) and Kleva and Guzdar (1998, 1999). In the former a high-pressure front produced by a non-resonant $m/n = 1/1$ mode [Wesson (1986), Aydemir (1987)] was shown to be unstable to high-n ballooning mode, which nonlinearly lead to a disruption. The ideal $m/n = 2/1$ mode provides a mechanism for an explosive instability directly and thus may be responsible for those high-β disruptions with fast precursors. The numerical tool used is the CTD code [Aydemir (2015)] (and the references therein) that solves the nonlinear, resistive MHD equations in toroidal geometry. In CTD code the poloidal and toroidal directions are treated spectrally while a finite-difference mesh is used in the radial direction. Noncircular geometries are treated through a conformal transformation from the poloidal coordinates (r, θ) to the unit circle in conformal coordinates (ρ_c, θ_c) [Aydemir (1990)]. A resistive wall [Fitzpatrick and Aydemir (1996)] and diverted plasmas [Aydemir (1987)] can also be treated.

Chapter 15

Fusion Power in Multiscale Plasma Turbulence

In the past several decades the emphasis has shifted from achieving ignition to building a long-pulse fusion confinement machine that achieves a power amplification factor $Q_{\text{fus}} = P_{\text{DT}}/P_{\text{aux}} \geq 10$. Chapter 1 describes the three previous D-T tokamaks that produced significant levels of D-T fusion power P_{DT} from the injected power P_{aux} pulse. Chapter 1.8 with Table 1.4 describes the $800\,\text{m}^3$ ITER tokamak driven by 15 MA of plasma current under construction adjacent to the Cadarache Nuclear Laboratories (CEA) (`http://www-cadarache.cea.fr/index_gb.php`. The 300 volt.sec = 300 Wb central solenoidal that drives the plasma current is expected to enable the D-T plasma to "burn" producing $P_{\text{DT}}(t) \lesssim 500\,\text{MW}$ of fusion power for creating 100 Gigajoule of energy.

During the period that the international fusion science community debated the choice of next fusion power experiment, there was extensive design research for a short-pulse tokamak machine that would attempt to reach ignition within a ten to twenty second heating pulse. The design became advanced and is discussed in this chapter, along with other exploratory fusion machine concepts.

15.1 Confinement for Compact High-Field Tokamaks

Tokamak machines designed to reach ignition during a transient pulse driven by the transformer action of the primary voltage applied with coils in a central solenoid have a long history. Such tokamaks, however, are yet to be built on the scale required to test the ignition hypothesis. Tokamak designs in this class include the compact ignition torus (CIT), fusion ignition research experiment (FIRE) and IGNITOR. The compact high-field IGNITOR design is the most extensively developed design. In this section we show the performance projections for the IGNITOR machine made with the multiscale transport codes called JETTO and the MMM model. The parameters and characteristics of typical high-field tokamak designs are given. There are plans to build a version of this compact high-field tokamak machine in Troisk under an agreement between Italy and Russia an at MIT. The general description of IGNITOR is found at `http://www.frascati.enea.it/ignitor/`. The compact field tokamak Alcator C-Mod for $\Delta t = 1\,\text{s}$ reached record values of the plasma pressure $p \leq 2 \times 10^5\,\mathrm{J/m^3} \cong 2\mathrm{Pa}$ at $B \leq 8\,\mathrm{T}$ before closing in 2016 [Marmar, *et al.* (2015)].

The basic plan for IGNITOR/CIT tokamak is to use the natural ohmic heating to the maximum extent that technology will allow. The plasma current heating power is limited by the maximum toroidal magnetic field B_{tor}. The energy confinement time scaling law is most favorable for the intrinsic plasma heating produced by Coulomb collisions relaxing the transformer driven plasma current I_p driven by the toroidally-symmetric transformer voltage V_{loop}. When auxiliary power P_{aux} from neutral beams or radio frequency antennas is applied the energy confinement time degrades, as detailed first in the Kaye-Goldston scaling law in PLT. The confinement time is found to decrease with applied P_{aux} as $\tau_E \sim 1/(P_{\mathrm{aux}}^{1/2})$. The reason for this degradation, as explained in Chapters 12-14, is that as the auxiliary power P_{aux} increases, the level of the drift wave turbulent transport rapidly increases.

The issue with using the natural ohmic heating of ηj^2 to reach the temperature required for ignition, which is at least 10 KeV, is that the plasma resistivity η decreases with increasing temperature and thus the plasma temperature reaches a plateau for a given geometry and toroidal magnetic field strength. To reach the ignition temperature with ohmic heating alone, the plasma current density j must be increased to high levels. For a given major radius of the torus, the limit on the current density is from the strength of the toroidal magnetic field B_{tor} that can be applied. Thus, the pathway to an ohmic ignition experiment requires the use of the highest magnetic fields that can be maintained over relatively small plasma volumes. Volumes of a few cubic meters with fields of the level of $B_{\mathrm{tor}} \sim 10\,\mathrm{T}$ are sufficient. Alcator C-Mod reached $B_{\mathrm{tor}} \sim 8\,\mathrm{T}$ [Marmar, *et al.* (2015)].

Since superconducting wires make a spontaneous transition to normal conductivity for magnetic fields above a critical magnetic field strength, the high field ignition experiment have used normal conductivity copper wire for the toroidal field coil magnets. The ohmic dissipation in these copper coils then gives engineering limits to the maximum field strength in machines of spatial scale of a meter. New machine designs with high critical temperature superconducting coils are now under critical development [Marmar (2016), Sonborm, *et al.* (2015)].

The fact that one uses the transformer action to drive the toroidal electric field and associated loop voltage V_{loop}, limits the length of time $\Delta t_{\mathrm{pulse}}$ the ohmic heating is applied to the plasma. For example, the design values that were developed for IGNITOR and CIT would limit the ohmic heating pulses to the order $\Delta t_{\mathrm{pulse}} \sim 10\,\mathrm{s}$. The pulse limit is determined by the amount of magnetic flux swing $\delta\psi_{\mathrm{primary}}$ that can be produced in the primary external transformer coil.

15.1.1 *Central solenoid and plasma current densities*

The magnetic flux-swing in the primary is produced by currents in core winding around what is called the central solenoid (CS) in a tokamak. The compact ignition experiments have flux swings $\delta\psi_{\text{primary}} = \int E_{\text{loop}}dt = V_{\text{loop}} \sim 20\,\text{Volt.sec}$. For comparison the flux swing in the central solenoid of ITER is 300 volt.sec.

In addition the plasma current density is limited by the minor radius of the machine a and the total plasma current I_p by the maximum toroidal magnetic field and the aspect ratio through the condition in Chapter 6 for the onset of the kink instability. Recall the driving force of the kink is the radial gradient of the toroidal current density. For MHD kink stability the maximum current density is set by $j_{\max} = B_{\text{tor}}/qR\mu_0$. The safety factor q is limited to $q > 1$ for kink stability as confirmed in tokamaks and in another class of machines called the reversed field pinch (RFP). These toroidal RFP machines operate in the kink unstable regime with $q \leqq 0.5$ and have high levels of magnetic fluctuations [Escande, *et al.* (2000)].

From the above discussions on the required magnetic flux swing in volt-sec, the limit on the current density, and the Lawson condition in Chapter 1 one is lead to the design of a machine with a small major radius and the strongest practical toroidal magnetic field that can be maintained for the time periods of order a few seconds to reach ignition. The time interval available is limited by the volt-seconds of the central solenoid. When these limits are applied in transport codes with D-T plasmas, one achieves the ignition temperature and the Lawson condition as shown in the transport simulations in Fig. 15.1.

Achieving ignition depends on the details of the thermal diffusivities and the evolution of the plasma profiles. Thus sophisticated transport models and simulations are used in these ignition studies. The plasma density is limited both by the

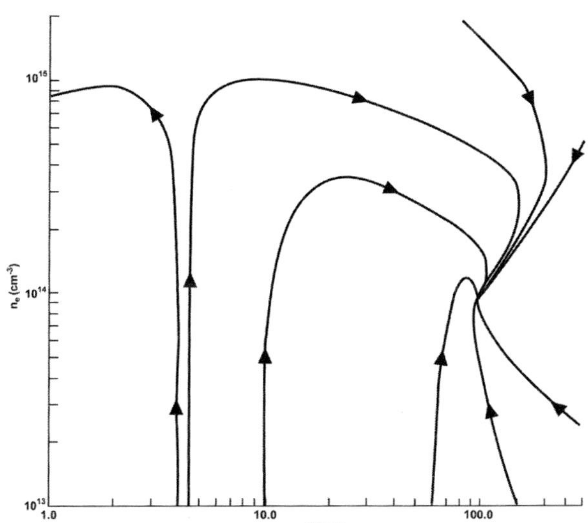

Fig. 15.1 Schematic for the path to ignition from integrating a pair of density and temperature equations with ohmic heating, gyro-Bohm losses and a suitable constant fueling rate. The low-order fusion model does not include magnetic field evolution and uses simple models for the rapid losses that set in when a plasma MHD beta limit is exceeded.

ballooning pressure gradient limit and by a radiative collapse density limit known as the Murakami limit or the Greenwald limit for the maximum confined plasma density. The fact that the simulations show ignition under suitable conditions has encouraged numerous designs for high-field compact ignition experiments.

Some compact high-field ignition designs include a heating boost from radio frequency RF power, usually at the ion cyclotron frequency, to control and push the deuterium-tritium plasma into thermonuclear ignition regime more rapidly at the end of the pulse. A design for such an ignition experiment was debated against the current ITER design in a high level panel meeting at the headquarters of the Commissaire Energie Atomic CEA in 1999 in Paris. The CEA panel chose the standard large tokamak design over the compact high-field design for two main reasons [Laval (2001)]. The dominant reason was that it is not clear how the intrinsically pulsed ignition experiment would be transformed into a steady-state electric power source. The second more pragmatic reason for the ITER choice was that the magnetic confinement database is very extensive for the standard large tokamak and rather limited for the compact high-field tokamak. The database for the compact high field approach comes from the Alcator experiments with $B_{\max} \leqq 6\,\mathrm{T}$ and the Frascati Tokamak Upgrade (FTU). The Alcator experiments hold the record for the highest values of the Lawson measure of a few times $n_e\tau_E \sim 10^{13}\,\mathrm{s/cm^3}$. The temperature in these record $n_e\tau_E$, discharges was about 1 KeV for both the ions and electrons in deuterium plasmas. The Frascati Tokamak Upgrade (FTU) has supporting experiments with pulsed high-field discharges maintaining the high densities by continuous high-speed pellet injections.

There was considerable excitement in the tokamak community over the high B_T Alcator results in the 1980s. The Alcator results lead to several proposals to the US Department of Energy from the tokamak community, including The University of Texas at Austin. The Princeton Plasma Physics Laboratory proposed the compact ignition torus (CIT) design and the earlier IGNITOR proposal from the Coppi and Brussard team was refined and upgraded.

The proposers reasoned that it is not clear what limits the Lawson parameter in the fusion tokamak experiments designed to achieve thermonuclear ignition in a deuterium-tritium plasma. The proposed experiments were well suited to reach ignition for a short time interval of a second or two. The results would bring a much clearer understanding to the ignition problem. The plan was to achieve the burning plasma for about ten energy confinement times. The energy confinement time would be about 100 ms under the extrapolation of the Alcator database and consistent with scaling from other tokamak data bases. Plasma ignition is defined as the interval where fusion alpha products heat the plasma sufficiently to compensate for the plasma energy losses. Computer simulations, described here in Fig. 15.1, show that a pulse of α particle power up to 10-20 MW is produced for a period over a few seconds. Crucial issues are the production of peaked density profiles on a time-scale of several energy confinement times, the control of the plasma current penetration for the optimization of the ohmic heating and the avoidance of sawtooth oscillations. In some simulations, the programmed application of a 10-20 MW ion cyclotron radio frequency heating and the operation of a high-speed pellet injectors were introduced to provide real-time control of the plasma dynamics.

The designs considered are the IGNITOR [Coppi, *et al.* (1991,1992)], the fusion ignition research experiment (FIRE) [Meade, *et al.* (2001)] and the later compact ignition toroidal experiment dubbed CIT [Greenwald, *et al.* (1995), Meade, *et al.* (1990)]. New designs are under development at MIT by Marmar, *et al.* (2015) using high-temperature superconductors.

15.1.2 *Simulations for IGNITOR*

Simulations with the TRANSP transport code with full alpha particle physics [Budny (2002)] calibrated with JET and TFTR data showed comparisons of the ignition dynamics in IGNITOR, FIRE and ITER-FEAT. The Budny work includes extensive tables of parameters for the five machines.

For the present purpose it is sufficient to consider the simulations for the IGNITOR tokamak machine with the parameters shown in Table 15.1. The relevant information that would be gained from such an experiment can be summarized as follows:

(i) Improved understanding of plasma turbulence and transport processes, by advanced exploration of high-plasma density, high-magnetic field regimes never accessed before. More specifically, a compact, high-field ignition experiment would provide insight into the ohmic confinement scaling laws and self-organized plasma profiles at multi-KeV electron temperatures. The experiments would provide new information on the conditions required for the spontaneous generation of plasma rotation through anomalous angular momentum transport. Plasma rotation plays an important role in determining the level of turbulent fluctuations controlling the radial transport processes in plasmas.

(ii) Alpha particle physics issues would be answered. The confinement of the fusion alpha particle product in fusion reactors may be limited by collective electromagnetic modes driven by the fusion alphas phase-space gradients. The large JET tokamak has a wide range of the high-energy deuteron and alpha particle driven instabilities [Nave, *et al.* (1995)]. In strong NBI heated tokamaks an outstanding alpha physics issue is when do the high-energy ions create *fishbone oscillations* and toroidal Alfvén eigenmodes (TAE) that degrade the quality of confinement [Coppi and Porcelli (1986)]. An outstanding physics issue is whether alpha particle heating causes the same degree of confinement degradation as found from auxiliary NBI plasma heating. Purely ohmic heated plasma has the highest energy confinement laws at lower levels of auxiliary heating power.

(iii) The experiment would enable the control of a fusion-burning plasma over physically significant time scales. With a few second flat top in the ten meg-ampere current range, the compact high-field experiments would provide a long enough

Table 15.1 Parameters of IGNITOR for a typical inductive operation scenarios. The first column is from Coppi, *et al.* (1999). The second column is the cross-over point (COP) where $P_\alpha \simeq P_{\rm OH}$ at $t = 4.8\,{\rm s}$ and the third column (BP) is at $t \simeq 6\,{\rm s}$ in the burning plasma.

Parameter	Coppi, *et al.* (1999)	COP	BP
R/a(m/m)	1.32/0.47	1.32/0.46	1.32/0.46
B_T(T)	13	13	13
I_p(MA)	11	12	12
κ_x/δ_x	1.83/0.4	1.83/0.4	1.83/0.4
q_a	3.5	3.5	3.5
β_p	0.26	0.11	0.16
$\beta_T\%$	1.2	1.6	2.3
β_N	0.67	0.71	0.99
$\langle n_e\rangle$ ($10^{20}/{\rm m}^3$)	5	8.9	9.2
$\overline{n}_e$	6.2	8.9	9.6
$n_e(0)$	9.5	8.3	9.7
$n_e/n_{\rm GW}$	0.4	0.5	0.5
$T_e(0)$(KeV)	11.5	12.0	18.9
$T_i(0)$(KeV)	10.5	10.8	15.5
$P_{\rm OH}$(MW)	11	9.8	7.1
$P_{\rm ICRH}$(MW)	0	0	0
$P_{\rm NBI}$(MW)	0	0	0
P_α(MW)	19	10	31
$P_{\rm BREMS}$(MW)	4	12	17
dW/dt(MW)	11	4.8	4.4
W_{th}(MJ)	12	15	20
$Q(Q^*)$	8.6(∞)	5.1 (10)	21.8 (57)
$Z_{\rm eff}$	1.2	1.1	1.1
$\tau_E(s)$	0.62	0.73	0.52
$n\tau_E$(10^{20} m^{-3} s)	6.0	6.5	4.8
$H_{\rm 97L}$	1.3	1.4	1.4

pulse length to thoroughly examine alpha particle physics and thermal transients in a DT burning plasma. The alpha particle slowing down time is short owing to the high-plasma density and relatively low-plasma temperature compared with that in conventional large-volume low-field tokamak plasmas.

(iv) The high-field experiment would give first indications about a possible development path to a tritium-poor, reduced neutron production of fusion power. Since the lifetime for the metal walls under the intense neutron fluxes in the DT reactor are unknown, exploring the alternatives without tritium are needed.

Ignition is intrinsically a time-evolving event. Thus a discharge of 10 seconds maybe sufficient to answer many questions if the fusion products thermalize in less than one second. In these demonstration experiments, the burning plasma state would last a few confinement times (2-5τ_E with $\tau_E \sim 0.5\,{\rm s}$) and many alpha particle slowing-down times (50-100$\tau_{\rm sd}$), which is adequate, from the physics point of view, to explore the burning fusion plasma state.

An example of a density versus temperature phase diagram for the transition to ignition from an ohmically-heated plasma with a sufficient continuous fueling rate is shown in Fig. 15.1. The dynamical trajectories of different paths of this ohmically-heated burning plasma follow from integrating a set of ordinary differential equations for the density and temperature with a source and Bohm loss rate and the energy equation with ohmic heating and Bremsstrahlung radiation. The system has a bifurcation at the ignition temperature of approximately 4 KeV and the dynamics leads to the stable fixed point between 10 and 100 KeV. The simple example shown does not include the evolving magnetic field so that to achieve MHD stability along higher nT-curves would take an unrealistically large magnetic field of over 15 T owing to the plasma pressure being about a megapascal at the fixed point. For a pressure limited loss rate the curves move down into a lower operating temperature. Numerous other practical considerations would need to be included for the engineering scenario.

The concept of ignition is defined [Miyamoto (1989)] as the transition to a new plasma state in which the heating power due to the nuclear fusion alpha particles compensates for all forms of power losses (due to anomalous transport and radiation). It is worthwhile to discuss this concept quantitatively with simulation codes, especially in view of dynamics of the transition given in the Miyamoto definition. Now we show some examples of the passage to a burning plasma.

15.1.3 *Transition to a burning D-T plasma*

Consider the power balance equation,

$$\frac{dW}{dt} = P_{\text{ohm}} + P_\alpha + P_{\text{aux}} - P_{\text{loss}} \tag{15.1}$$

where W is the plasma energy content, P_{ohm} is the ohmic power, P_α is the alpha particle heating power, P_{aux} is the auxiliary heating power, and P_{loss} is the loss power, including radiation losses. In the ignited state, $P_\alpha = P_{\text{loss}}$ and the auxiliary power, P_{aux}, may be switched off. Thus, ignition is an overheated state with $dW/dt = P_{\text{ohm}} > 0$. One may introduce a parameter

$$Q^* = \frac{P_{\text{fus}}}{(P_{\text{loss}} - P_\alpha)} \tag{15.2}$$

where the fusion power is given as $P_{\text{fus}} = (17.6\,\text{MeV}/3.5\,\text{MeV})P_\alpha \simeq 5P_\alpha$ for a D-T reacting plasma. With this definition, $Q^* = \infty$ at ignition. Alternatively, using the power balance relation in Eq. (15.1), one may rewrite Eq. (15.2) as

$$Q^* = \frac{P_{\text{fus}}}{\left(P_{in} - \frac{dW}{dt}\right)} \tag{15.3}$$

where $P_{in} = P_{\text{ohm}} + P_{\text{aux}}$.

A key objective of high-field experiment is to provide multi-megawatts of ohmic power without the symmetry-breaking from an auxiliary power P_{aux}.

The fusion power gain parameter Q^* is the measure of fusion power to the input power, taking into account the transient loading/unloading of the plasma energy $W(t)$. One can see that Q^* becomes equal to the widely-used thermonuclear gain parameter, $Q = P_{\rm fus}/P_{in}$, when $dW/dt = 0$. Thus Q^* is the natural extension of Q under non-steady-state operation.

During transient regimes, the difference between Q and Q^* becomes significant and should be kept in mind. For instance, for the simulation example in the first column labeled Coppi of Table 15.1 of an ignited discharge in IGNITOR, one finds $Q^* = \infty$, while $Q = 8.6$.

Table 15.1 gives the parameters of simulation from the IGNITOR Team in the first column. Values at two times called the cross-over point (second column labeled COP) and a burning plasma (third column labeled BP) from the first simulation given in Hu, *et al.* (2003).

A record value of $n_0\tau_E \sim 1 \times 10^{20}\,\mathrm{m}^{-3}\,\mathrm{s}$ was reached in ALCATOR C-Mod experiments with high peaked density, $n_0 \approx 2\times 10^{21}\mathrm{m}^{-3}$ and a confinement time of about $\tau_E = 0.05\,\mathrm{s}$. Similar record values of the Lawson parameter were subsequently achieved in high density ohmic discharges in TFTR [Meade, *et al.* (1990)] and in FTU [Frigione (2000)].

Experience from ALCATOR C-Mod [Greenwald, *et al.* (1995)] and from FTU [Frigione (2000)] at Frascati indicates that the worst-case discharges in these machines have an energy confinement time that follows the ITER89P L-mode scaling law. A more favorable confinement law called the neo-Alcator and improved ohmic confinement (IOC) scaling describes discharges at slightly lower densities. Regimes of improved confinement at high plasma density have been observed. Ohmic H-modes have been observed in limiter as well as divertor configurations. Enhanced confinement in L-mode operation, such as the improved ohmic confinement (IOC) regime observed in ASDEX-U [Bessenrodt-Weberpals, *et al.* (1991)], has now been observed in Alcator C [Greenwald, *et al.* (1984)] at high plasma densities with peaked density profiles. Frigione (2000) reports pellet-assisted ohmic discharges with very good confinement obtained in FTU, with a line-average density $\overline{n}_e \approx 4 \times 10^{20}\,\mathrm{m}^{-3}$ and $\tau_E \approx 0.1\,\mathrm{s}$.

As compared with the record Alcator C experiments, a value of τ_E higher by about a factor of about ten is needed to reach the target value $n_0\tau_E \geq 4\times 10^{20}\,\mathrm{m}^{-3}\,\mathrm{s}$. This factor reduces to five if the comparison is made against the FTU results.

High-field, high current operation in existing experiments are limited. As a result, the available database is limited compared with the low-field, high $P_{\rm aux}/P_{\rm ohm}$ experimental database. A new high-field, high-density database is needed. The new database would widen the margin of confidence for extrapolations to high-field burning plasma experiments.

A simple but universal method of taking into account the expectations of new regimes is to use scaling laws based on the invariance of the equations of the system. This scaling law method is widely used in engineering and complex systems, including weather systems. Lackner (1990) used that scaling law to predict the expected performance of two tokamak upgrades — namely that of ASDEX-U and Alcator to C-MOD and for JET. The data used for the prediction was from the

ASDEX, Textor, FT and Alcator C. The scaling law predictions for the future device were largely correct. The same method was used to choose the parameters of ITER in 1990s with some modifications made as the machine size was reduced from 8 m to 6 m major radius in the early 2000s.

An ion cyclotron range of frequencies (ICRF) system with power up to 6.0 MW and a lower-hybrid current drive (LHCD) system up to 4 MW have been applied for heating and current drive experiments in the new Experimental Advanced Superconducting Tokamak (EAST) in Hefei, China. Significant progress has been made with ICRF heating and LHCD for realizing high-performance H-mode plasma operation in EAST [Zhang, *et al.* (2014)].

15.2 Single X-point Divertor Tokamaks

The divertor model in Fig. 15.2 has been used to investigate the steady-state plasma driven by 10-20 MW lower-hybrid waves launched from the complex antenna used in Tore Supra to produce steady-state fusion-grade plasmas [Horton, *et al.* (2015)].

The new WEST chamber has a high-tech water-cooled tungsten target plate just below the X-point in Fig. 15.2 that is designed to take away the hot plasma transported across the magnetic separatrix into the scrape-off layers. This makes the machine a close model to the current design of ITER.

Plasma simulations of the RF waves launched from the complex antennas that transform in the plasma to lower-hybrid slow waves and their coupling to whistler fast waves is given in Horton, *et al.* (2016).

The WEST experiment is designed to learn how to minimize technology and operational risks for a long-pulse fusion reactor with actively cooled tungsten divertors as in ITER machine [Grosman, *et al.* (2013)].

We explore a method for constructing two-dimensional area-preserving, integrable maps associated with Hamiltonian systems, with a given set of fixed points and given invariant curves. The method is used to find an integrable Poincaré map for the field lines in a large aspect ratio tokamak with a poloidal single-null divertor. The divertor field is a superposition of a magnetohydrodynamic equilibrium with an arbitrarily chosen safety factor profile, with a wire carrying an electric current to create an X-point. This integrable map is perturbed by an impulsive perturbation that describes non-axisymmetric magnetic resonances at the plasma edge. The non-integrable perturbed map is applied to study the structure of the open field lines in the scrape-off layer, reproducing the main transport features obtained by integrating numerically the magnetic field line equations, such as the connection lengths and magnetic footprints on the divertor plate [Kroetz, *et al.* (2010)].

The drift wave turbulence in this complicated geometry is simulated using the conversion of the terms $k_y(dn/dx)$, $k_\| c_s$, $k_\| v_A$ with the Poisson brackets $[\psi_p, \varphi_{\mathbf{k}}]$ and $v_A[\partial_z \psi_p, \delta\psi_{\mathbf{k}}]$ for cross-field and parallel gradients where ψ_p is the poloidal flux function. In the spectral codes each time step in the wave fields are solved and saved in both x, y, z and k_x, k_y, k_z space. The fast FFT software on the paral-

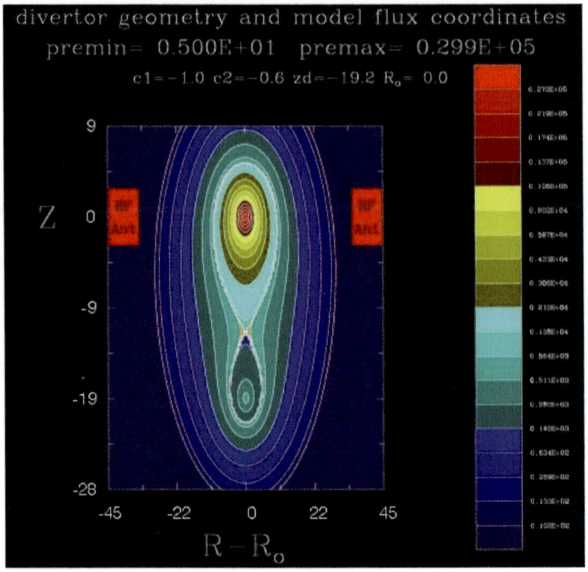

Fig. 15.2 Theoretical model for the magnetic flux surfaces for the divertor plasmas like WEST and EAST with color coding for the values of the plasma pressure. The hot plasma is at $R = R_0$ and $Z = 0$ is shown as red and the pressure decreases with a steep gradient outside the magnetic separatrix. The lower plasma between $Z = -10$ to -20 is in the "exhaust pipe" called the divertor chamber in the literature. This chamber is below the machine floor and not visible in the figures of WEST and other divertor tokamaks. Fast pumping of that exhaust plasma is needed to keep the upper fusion plasma with low-Z_{eff} in a steady state.

lel super computers provides the programmer with both the $\boldsymbol{x}$-space and $\boldsymbol{k}$-space representation of N-coupled plasma fields, e.g. $[n_{\mathbf{k}}, \varphi_{\mathbf{k}}, v_{\|\mathbf{k}}, T_{i\mathbf{k}}, T_{e\mathbf{k}}, j_{\|,\mathbf{k}}]$ at each time step. The development of the spectral simulations is described in Chapters 4 and 5.

For plasma turbulence and RF waves the understanding of the dynamics requires thinking about both $\boldsymbol{x}$-space and $\boldsymbol{k}$-space field representations as well appreciated from quantum theory and the Heisenberg uncertainty principle [Lindley (2008), Budker, *et al.* (2008)].

Fusion experiments from the 1990s to 2000 made it evident that the self-consistent coupling of the turbulence with the plasma profiles plays the crucial role in determining the effective plasma transport and power losses. Thus the plasma fusion Q_{fus} is determined by the drift wave turbulent transport in the magnetic field geometry. The strength of this turbulence depends on the growth rates derived from the gradients of the density and electron and the ion temperature profiles along with profiles of the safety factor $q(\psi, t)$. Thus the tokamak dynamics is like a "weather system" with feedback loops within this complex driven dynamical system. There are bifurcations that arise analogous to those well known in turbulent neutral fluids and atmospheres. As we have already listed in Chapters 12–14, there are a variety of plasma confinements states. In some cases similar system-control parameters result in discharges that take up rather different confinement states. A well-known example occurred for the matched discharges in TFTR, in which one state ultimately deviates to a second state through the bifurcation to a new state known as enhanced reversed shear (ERS) confinement [Mazzucato, *et al.* (1996)]. These issues of the bifurcations, or sharp transitions, between different plasma confinement states are of fundamental importance to plasma science and require external control to achieve a controlled fusion power generator.

The confinement issue for new fusion devices will be addressed in the future with a wider variety of methodologies including those of real-time control from engineering system control theory. The different perspectives and methodologies are required to achieve the plasma confinement required for a practical energy source from nuclear fusion. In the following sections of this chapter, we examine the projections based on multiscale physics in the transport simulation codes. There are now a wide range of turbulent transport simulation codes used to predict plasma behavior.

15.3 Tokamak Confinement Time Based on the L-mode Database

The L-mode database with approximately 3000 discharge entries is summarized by the ITER-97 L-mode formula [Kaye, *et al.* (2007)]:

$$\tau_E^{\mathrm{L97}} = 0.023\kappa^{0.64} R^{1.83} A^{0.06} B_T^{0.03} \overline{n}_e^{0.4} m_{\mathrm{eff}}^{0.2} I_p^{0.96} P_{\mathrm{loss}}^{-0.73}, \tag{15.4}$$

where $P_{\mathrm{loss}} = P_{in} + P_\alpha - dW/dt$ is the loss power, which includes radiation losses, and $P_{in} = P_{\mathrm{OH}} + P_{\mathrm{aux}}$. At ignition, $dW/dt = P_{in}$ and $P_{\mathrm{loss}} = P_\alpha$. For the reference case of Table 15.1, $P_{\mathrm{loss}} = 19\,\mathrm{MW}$ at ignition. Taking the remainder of the parameters from Table 15.1 gives $\tau_E^{97\mathrm{L}} = 0.47\,\mathrm{s}$; thus to obtain the required confinement time of $\tau_E \approx 0.62\,\mathrm{s}$, an enhancement H-factor of 1.3 would be required to reach ignition. Similarly if the older L-mode scaling expression ITER89-P scaling is used, then $\tau_{\mathrm{E89L}} = 0.4\,\mathrm{s}$ and an H factor of 1.55 is required for ignition.

The L97 scaling law has been compared with confinement in the superconducting steady-state tokamak Tore Supra [Horton, *et al.* (2000)]. The Tore Supra database has 50 discharges with Fast Wave ICRH that deposits wave power into the electrons, $P_{\mathrm{ICRH}} = P_0 \exp(-r/L_p)$, in the core plasma with $L_p \approx a/5$. Thus, the fast wave ICRH heating experiments are a guide to the simulation of the alpha power heating to the electrons. Tore Supra operated routinely in L-mode and with various levels of enhancement over the ITER-97 L-mode scaling law as a function of density profile peaking. The best Tore Supra discharges had confinement time enhancement factors of $H = 1.4$ to 1.7 with respect to the ITER-97-L mode formula.

A key scientific issue that the future high-field experiments will address is whether the ohmic power at multi-megawatt levels plays the same role as P_{aux} power in the confinement scaling laws as assumed in the above formulas for the confinement time. The same type of question arises for the alpha heating power P_{alpha} since this would become the dominant form of heating in burning plasmas.

15.4 Improved Confinement Regimes with Peaked Density Profiles: LOC, SOC and IOC Regimes

Since the high-field tokamak may reach ignition with ohmic heating alone, we start with consideration of ohmic confinement modes. The linear ohmic confinement (LOC) mode is a regime of ohmic confinement with a linear increase in the energy confinement time with increasing plasma density. This is also called the neo-ALCATOR scaling regime [Coppi and Mazzucato (1979)]. With increasing plasma density above a critical density, the LOC regime makes a transition either into a saturated ohmic confinement (SOC) mode confinement scaling or the L-mode confinement scaling as shown in Fig. 15.3. The critical density at which the transition between LOC and SOC regimes occurs is called the Shimomura density [Shimomura (1985)], whose expression is

$$n_S \approx \left(\frac{A_i}{2}\right)^{1/2} \frac{B_T}{Rq_\psi} \tag{15.5}$$

where A_i is the mean atomic mass number, B_T is expressed in Tesla, R in meters and n_S in $10^{20}\,\mathrm{m}^{-3}$. The density limit Eq. (15.5) can be re-expressed in terms of plasma current density limit with $\mu_0 j_p = B_p/a$ and $Rq_\psi = B_T a/B_p$ to give the Greenwald density limit formula [Greenwald, *et al.* (1984)].

Experiments on ASDEX and on Alcator C-Mod reached a new regime called the improved ohmic confinement (IOC) that has an internal transport barrier and extends the good confinement to the high-confinement regime of H-modes at higher plasma densities [Fiore, *et al.* (2012)]. One method of achieving this regime is to

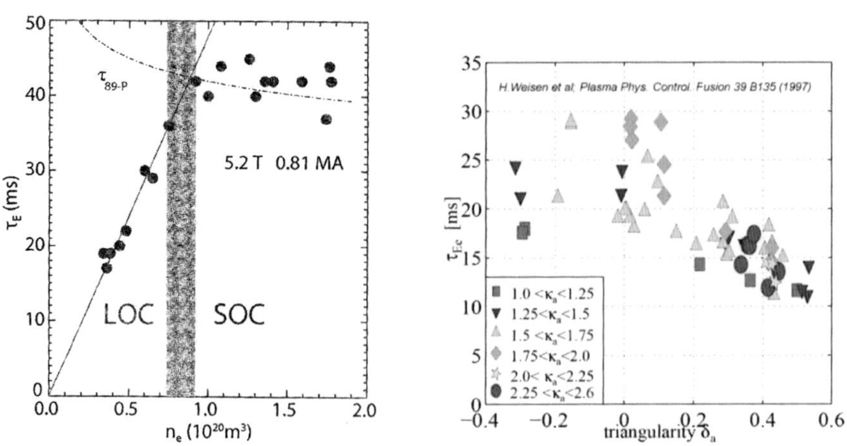

Fig. 15.3 An example of the ohmic confinement transition in Alcator C-Mod discharges ($B_t = 5.2\,\mathrm{T}, I_p = 0.81\,\mathrm{MA}$). The vertical axis is the energy confinement time [ms] and the horizontal axis is the line averaged electron density [$10^{20}\,\mathrm{m}^{-3}$]. (a) The energy confinement time saturates as density increases above shaded transition region. The solid line is fit to the date in the LOC regime, and the dash-dot line is the ITER 89P L-mode scaling [Rice, *et al.* (2012)], (b) Electron energy confinement time τ_E against triangularity δ_s for $\overline{n}_e \approx 6.3\times 10^{19}\mathrm{m}^{-3}$ [Weissen *et al.* (1997)].

control the L-H power threshold. Theoretically, this is understood as keeping the ratio of the temperature gradient to the density gradient parameterized in Chapters 3 to 13 by the ratios called "eta-i" and "eta-e" defined by $\eta_i = (d \ln T_i/dr)(d \ln n_i/dr)$ and $\eta_e = (d \ln T_e/dr)(d \ln n_e/dr)$ sufficiently close to unity.

These "eta" parameters η_i and η_e measure the peaking of the temperature profiles with respect to that of the density profile. For an adiabatic pressure profile $n^{5/3}$ their values are $\eta = 2/3$. Large values of these dimensionless parameters occur from flat density profiles and lead to strong turbulent mixing from the ITG and ETG instabilities described in Chapters 12 and 13 [Horton (1999), Horton, *et al.* (2000)].

Here, we summarize the improved confinement regimes that can be reached with peaked density profiles. These regimes are listed as follows:

(i) The improved ohmic confinement (IOC) mode initially discovered in ASDEX [Bessenrodt-Weberpals, *et al.* (1991)].

(ii) The radiative improved (RI) mode, discovered in TEXTOR [Weynants, *et al.* (1999)].

(iii) The P-mode, which is a mode of improved ohmic confinement both in ohmic and in auxiliary heated discharges, first obtained with the help of pellet injection in ALCATOR-C [Greenwald, *et al.* (1984)].

(iv) The supershot, or S-mode, experimentally discovered in TFTR with central neutral beam injection (NBI) with a strongly peaked density distributions [Meade, *et al.* (1990)].

(v) Fast Wave ICRH on Tore Supra [Hoang, *et al.* (2000), Horton, *et al.* (2000)].

In the experimental regimes listed above, energy confinement enhancement factors up to 2-3 over that for L-mode were obtained. Clearly, the relevant question is whether the compact high-field experiments can access these regimes.

The key feature common to these improved confinement regimes is the realization of peaked density profiles. Different injection methods and considerable tokamak operator skills have led to these enhanced confinement regimes. In ASDEX, peaked densities and the IOC mode were obtained after appropriate wall conditioning and decreased gas puffing. In TEXTOR, the transition from the L- to the RI-mode was obtained with impurity seeding. In this way, a strongly radiating plasma layer was established at the edge, with a corresponding decrease of the edge temperature and a steepening of the density gradient deeper inside the plasma. In ALCATOR C, pellet deposition in the plasma core resulted in peaked density profiles and the establishment of the P-mode. In TFTR, peaked densities were obtained in supershots with central NBI deposition.

A spontaneous density profile peaking called the I-mode was produced in Alcator C-Mod H-mode plasmas when the additional ICRH resonant layer was placed in the the high-field side of the magnetic axis [Hubbard, *et al.* (2011)]. Density peaking has also been observed in the same device during ohmic H-modes when the toroidal field was ramped down. These phenomena are not fully understood and are under experimental investigation, with the objectives of extending the duration of the improved energy confinement performance phase and simultaneously increasing the temperature and density peaking.

An example of the transition from the L-mode to the I-mode in ASDEX-Upgrade. In the I-mode discharge AUG#29744 one sees the lower and intermittent level of the turbulence just inside the separatrix in the I-mode compared to that in the L-mode.

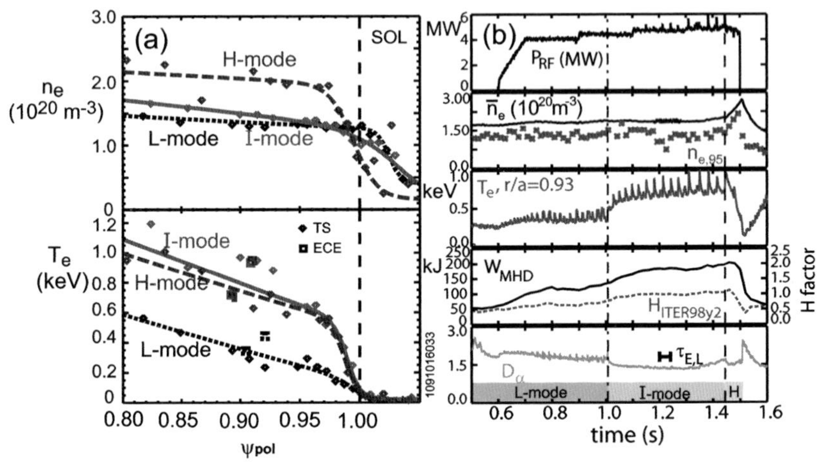

Fig. 15.4 (a) Comparison of the L-mode and I-mode radial profiles of density and temperature in C-Mod. Profiles of electron density (top) and temperature (bottom) for different phases of a 1.27 MA, 5.6 T C-Mod discharge (1091016033). The data was fit using modified tanh functions as described in Hughes, *et al.* (2007). The I-mode phase (1.41 s) has a steep T_e gradient, but a L-mode type of peaked density profile (0.78 s). The ELM-free H-mode (1.47 s) T_e is similar to that in the I-mode but the n_e profile in the I-mode is much steeper showing better density confinement. (b) Time histories of input ICRF power (top) and plasma parameters for the same discharge. In fourth panel, is stored energy, and is confinement normalized to IPB98y2 scaling, reaching 1 in I-mode [ITER Physics Expert (1999)].

Experiments on FTU in Frascati [Frigione (2000), Apicella, *et al.* (1997)] with high fields and pellet injections were successful in extending high-confinement times. Estimated confinement times of about 90–100 msec were reached with a magnetic field $B \approx 7\,\mathrm{T}$, a line-average density $\overline{n}_e \approx 4 \times 10^{20}\,\mathrm{m}^{-3}$ and a current $I_p \approx 0.8\,\mathrm{MA}$. The improvement in confinement lasts for more than several energy confinement times and is recovered after a second fueling pellet was injected.

From a theoretical viewpoint, peaked density profiles are known to have a beneficial effect on plasma confinement through the quenching of the ion temperature gradient (ITG) driven turbulence. Peaked density profiles reduce the two-dimensionless temperature gradient profile parameters, η_i and η_e, that are the driving terms for the instability of ion and electron temperature gradient turbulence as described in Chapters 12 and 13. For flatter density profiles, the ITG stability

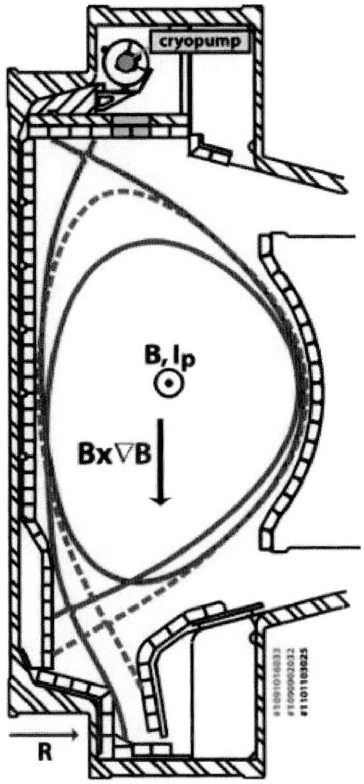

Fig. 15.5 (a) Typical shape of C-Mod last closed surface for I-mode experiments in unfavorable drift (upper x-point) and favorable drift (lower x-point, solid). Plasmas in more usual C-mod configurations (e.g., dashed curve) have not exhibited transitions to I-mode.

condition is that L_{T_i}/R exceed a critical value, which is typically not compatible with the overall required temperature difference between the edge and core plasmas. In the 1980s, several theoretical investigations of ITG modes supported the conclusion that the improved confinement in Alcator C pellet fueling experiments [Greenwald, *et al.* (1984)] was due to the simultaneous drop of the η_i parameter as described in Chapter 12. Numerous other machines have shown discharges with improved energy confinement from density peaking. For example, Tokar, *et al.* (1999) predict a suppression of the ion thermal flux due to ITG turbulence going from L-mode to the RI-mode in TEXTOR [Samm, *et al.* (1993), Winter, *et al.* (1993)].

The electron temperature gradient driven turbulence has a similar structure to the ITG but operates on the scale of the electron dynamics rather than that of the ions. There are two theoretical forms of the anomalous electron thermal diffusivity that are suppressed by high density: (i) the dissipative trapped electron turbulent diffusivity and (ii) the shorter wavelength electromagnetic diffusivity with mixing length proportional to the collisionless skin depth [Horton (1999), Horton, *et al.* (2000)].

For the high-field tokamak parameters given in Table 15.1, ignition would be reached with plasma densities well below the Greenwald density limit. This is a distinct advantage over conventional lower-field tokamak designs. Density peaking appears crucial to gain access to improved energy confinement regimes. Strong density peaking depends on the existence of an inward particle pinch. There are two theoretical models used to explain the particle pinch in tokamaks. The classical mechanism, known as the Ware pinch: an off-diagonal transport coefficient that is conjugate through the Onsager symmetries [Helander and Sigmar (2002)] to the experimentally verified bootstrap current transport coefficient. The second density pinch arises from ∇T-driven turbulence. An inward convection is required for transport modeling to be consistent with the measured density profiles. The density pinch effect is produced through drift wave turbulence driven by the temperature gradients, where again symmetries dictate an off-diagonal transport matrix. The turbulent inward particle transport is developed in numerous articles [Coppi and Spight (1978), Futatani, *et al.* (2010)].

A cold radiating mantle from the injection of neutral Argon gas, for example, is advantageous for the transition to peaked density profiles. Indeed, the "R" in the acronym RI-mode refers to this radiating mantle mode discovered in TEXTOR. The RI-mode was found when a cold plasma mantle concept was being established as a means to solve the reactor thermal exhaust problem [Messiaen, *et al.* (1996)]. Modeling the density profile with the RITM code showed that an essential ingredient for peaking of the density profile is the action of the radiative mantle on the anomalous inward pinch velocity $v_{\rm in}$, which has the form $v_{\rm in} = (1/2T_e)(dT_e/dr)D$, where $D(r)$ is the radial particle diffusivity. This formula for $v_{\rm in}$ may be justified by arguments of profile consistency and more generally justified as a fundamental off-diagonal contribution of the transport matrix for fluctuation-driven particle transport [Shaing (1988)]. This pinch velocity was also found to be essential in explaining the SOC-IOC transition in ASDEX [Becker (1990)]. Direct observation of an inward particle pinch is given by spectroscopic measures during silicon injection in TEXT and in Alcator C-Mod. The density data is interpreted with a simple diffusion-pinch velocity model from the Einstein relation for drift wave turbulence in Horton and Rowan (1994).

15.5 Dynamics of Reversed Magnetic Shear Confinement in Fast Current Ramp Discharges

Transient effects can be exploited in tokamaks to produce states with reversed magnetic shear. In ohmically-heated plasmas, a hollow current density $j_\phi(r,t)$ naturally produces a reversed magnetic shear profile $q(r,t)$ during the penetration of the driving toroidal electric field from the central solenoid. This regime is called the current-ramp where the plasma current is rising with $dI_p(t)/dt > 0$. In this regime the plasma current increases from induction loop voltage $V_l = 2\pi \mathrm{RE}_\phi$ from the rising magnetic flux in the central solenoid. Skin-layers of current driven by rising flux $d\Phi_{cs}/dt = 2\pi\,\mathrm{RE}_\phi$ in the central solenoid (CS) are continuously added to the outer surface of the plasma column.

The plasma current penetration time based on neoclassical plasma resistivity is slow and can be controlled with programmable currents in the central solenoid. As a consequence, non-monotonic q-profiles form during the current ramp remain for a significant fraction of the current flat top. Since ignition is expected to be achieved near the end of the current ramp, high-field ohmic tokamaks are well suited for a reversed shear (RS) mode of operation [Mazzucato, *et al.* (1996)]. The PEP mode [Hugon, *et al.* (1992)], which is the JET variant of the RS mode, was obtained with central ICRH and pellet injection. For these modes of operation, energy confinement enhancement factors of H $\approx$ 2-3 were achieved.

The physics of confinement enhancement for a non-monotonic q-profile clearly has basic differences with that in a monotonic q-profile. The reversed magnetic shear is one factor in achieving reduced core turbulence. Sheared plasma rotation and/or peaked density profiles also play important roles in controlling the core turbulent transport. The detailed shape of the current ramp maybe programmed with real-time control systems [Alvarado (2006)] with the assistance of pellet injection and core ICRH heating so as to enhance the reversed shear (RS) confinement regimes.

15.5.1 *High-confinement H-modes*

High-confinement H-modes of operation quickly occur in the presence of a magnetic X-point in the poloidal magnetic field structure and heating power levels above a threshold value called P_{LH}. The threshold power P_{LH} for the L-mode to H-mode transition is low for single null divertor (SND) geometry when the ion ∇B-drift towards the X-point. Below the X-point there is the divertor chamber designed to remove the escaping cool plasma.

The L-H power threshold formula used by the ITER-FEAT physics expert group was

$$P_{\mathrm{LH}}^{\mathrm{IPB99(5)}} = 3.24\overline{n}_e^{0.62} B^{0.75} R^{0.98} a^{0.81} m_{\mathrm{eff}}^{-1}. \tag{15.6}$$

When the H-mode is achieved with heating power above that in Eq. (15.6), the ITER ELMy H-mode scaling law applies and predicts a confinement time according to the formula

$$\tau_E^{\mathrm{IPB98}} = 0.0562\kappa^{0.78} R^{1.97} A^{-0.58} I_p^{0.93} B_T^{0.15} \overline{n}_e^{0.41} m_{\mathrm{eff}}^{0.19} P_{\mathrm{loss}}^{-0.69}. \tag{15.7}$$

For the IGNITOR parameters in Table 15.1, these formulae give a power threshold

$$P_{\mathrm{LH}}^{\mathrm{IPB98}}(\mathrm{IGNITOR}) \approx 20\ \mathrm{MW}. \tag{15.8}$$

This threshold power level would be exceeded by a combination of ohmic and ICRH heating with the resulting confinement time at ignition given by

$$\tau_E^{\mathrm{IPB98}} = 0.98\,\mathrm{s}\ \approx 1\,\mathrm{second} \tag{15.9}$$

from Eq. (15.7). The parameters for the examples in Eqs. (15.8) and (15.9) are from Table 15.1. With this longer confinement time $\tau_E = 1.0\,\mathrm{s}$ the plasma would reach ignition in one to two seconds.

H-mode quality plasmas have been obtained with the magnetic X-point just outside the vacuum vessel. Indeed, H-modes have been observed in limiter configurations [Weisen, *et al.* (1996)]. There is evidence from Alcator C-Mod of an enhanced L-mode with the magnetic X-point on the vacuum vessel wall [Greenwald, *et al.* (1995)]. In this experiment, the plasma was prevented from entering the H-mode by reversing the toroidal magnetic field so as to set the ∇B drift away from the single null divertor X-point. For power thresholds well in excess of those that would be required for an L-H transition (for ∇B drift in the favorable direction), confinement enhancement factors of 1.5 to 2.0 relative to L-mode have been obtained [Connor and Wilson (2000)]. Thus, improved confinement assisted by real-time X-point control operation systems is a promising future direction for exploring burning plasmas.

15.5.2 *Turbulent heat diffusivity in transport codes*

Let us examine the question of thermal confinement from the point of view of the heat diffusivity values needed for ignition, relative to models that are known to work well in standard tokamak experiments.

JET plasmas were described by the JETTO model that used either the drift wave models or the empirical Taroni-Bohm [Taroni, *et al.* (1994)] model χ_e^{TB} for the transport formula given as follows:

Taroni-Bohm Thermal Transport Coefficient

$$\chi_e^{\mathrm{TB}} = 0.33q^2 \left(\frac{-a\,\nabla p_e}{p_e} \right) \frac{T_e}{B}. \tag{15.10}$$

ITG formulas in the multi-mode model (MMM) give the major contribution to the anomalous thermal diffusivity in numerous toroidal plasmas. The FLR-fluid or gyro-fluid model [Nordman and Weiland (1992, 1993)] describes the transport by ITG turbulence.

The ETG turbulent electron transport has two regimes as a function of the plasma β_e. In the lower pressure region the χ_e is dominated by electrostatic (es) turbulence giving

$$\chi_e^{\mathrm{es}} = C_e^{\mathrm{es}} q^2 \left(\frac{R}{L_{T_e}} \right)^{3/2} \left(\frac{\rho_e^2 v_e}{T_e} \right) \left(|\nabla T_e - (\nabla T_e)_c| \right), \tag{15.11}$$

where

$$(\nabla T_e)_c \simeq 1.88 \left(\frac{|s| T_e}{qR} \right) \left(1 + Z_{\mathrm{eff}} \frac{T_e}{T_i} \right) \tag{15.12}$$

and $C_e^{\mathrm{es}} \simeq .05$. In the higher-pressure region there is a $\delta\mathbf{B}$ component.

Inside the separatrix the electromagnetic (EM) turbulent electron diffusivity based on electromagnetic turbulence driven by ETG and collisionless electron skin depth [Horton, *et al.* (2000), Zhu, *et al.* (2000)] dominates giving the (em) thermal diffusivity

$$\chi_e^{\rm em} = C_e^{\rm em} \frac{qc^2}{\omega_{pe}^2 \left(L_{T_e} R\right)^{1/2}} \left(|\nabla T_e - (\nabla T_e)_c|\right), \tag{15.13}$$

where $C_e^{\rm em} \simeq 2C_e^{\rm es}$. Transitional formulas from the ES to the EM regime are given in Hoang, *et al.* (2000b). The transitions give $\chi_e = \min(\chi_e^{es}, \chi_e^{em})$.

Formulae (15.11)-(15.13), with $C_e^{\rm es} = 0.05$ and $C_e^{\rm em} = 0.1$, were validated with the confinement database and showed good predictive/interpretative performance for the electron power balance in Tore Supra [Hoang, *et al.* (2000), Horton, *et al.* (2000)] and NSTX [Kaye, *et al.* (1997)]. The models are developed in Chapter 13.

15.6 Toroidal Simulations with Transport Codes

Transport codes for the plasma inside the magnetic separatrix are typically unable to match the data in the higher diffusivity edge plasma turbulent thermal loss rates. Thus, the transport codes use the edge or pedestal temperatures for the outer boundary condition. Pedestal temperatures are often used at 1 to 2 KeV. More optimistic values were used in the FIRE simulations with the pedestal $T_e = 3.0$ to 6.0 KeV. The fusion power performance dependence on the edge temperatures for ITER, AIRES-RS and FIRE is explored in Samm, *et al.* (1993). The MMM transport code has collisional turbulence and edge physics in the SOL plasma.

The simplest turbulent electron heat diffusivity model is a non-threshold parameterization defined as the Taroni-Bohm formula

$$\chi_e^{\rm TB} = \alpha_B \frac{q^2(r) a}{B_T} \left(-\frac{dT_e}{dr}\right) \tag{15.14}$$

where the constant α_B is calibrated with certain JET discharges and validated on other JET discharges. The model has good predictive power for JET, TS and JT-60U. The theoretical basis from particle-in-cell simulations is given in Kishimoto, *et al.* (1996) and Furnish, *et al.* (1999) for the q^2 modified Bohm scaling. The power balance of $P_L = q_r^{\rm TB} S = P_{\rm aux}$ with the heat flux q_r and surface area $S = 4\pi^2 Rr$ leads to a global τ_E scaling exponents close to the ITER-97L formula [Kaye, *et al.* (1997)]. Similar formulas with lower q-exponents than in Eq. (15.14) are found in the literature.

The multiple mode transport model (MMM) gives one of the interpretations for the ITER profile database. The transport MMM model contains descriptions for the entire standard forms of plasma turbulence including resistive-ballooning, TEM, CTEM, and ITG. To include E_r-shear and reversed q-profile formulas, the code was modified [Zhu, *et al.* (2000)]. The MMM95-version was calibrated with

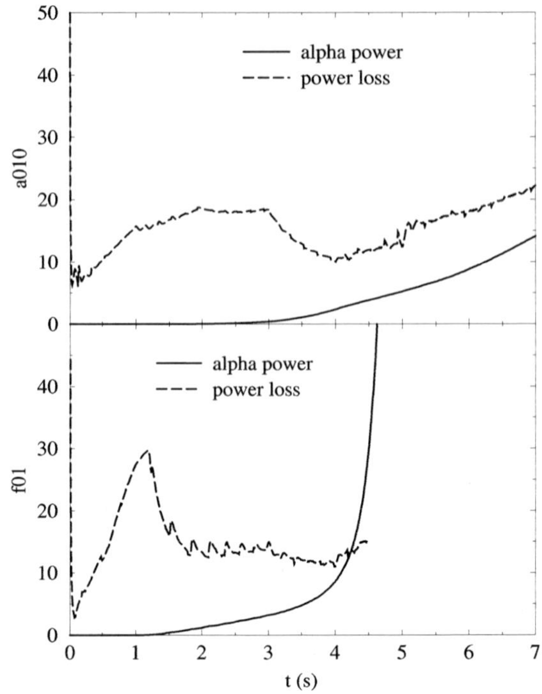

Fig. 15.6 Comparison of slow and fast current ramps for the high-field Ohmic tokamak. The time evolutions of the alpha power $P_\alpha(t)$ heating the plasma and the power losses from turbulent thermal diffusivity for the compact high-field tokamak with parameters in Table 15.1. The top panel shows the experiment with a slow current ramp and the lower panel the experiment with a fast current ramp where the cross-over point $P_a(t*) > P_{\text{loss}}(t_*)$ occurs at $t_* \simeq 4.25\,\text{s}$ [Hu, *et al.* (2003)].

circular TFTR discharges and tends to be strongly L-mode in character and thus is overly pessimistic for JET, JT-60U, and DIII-D. The two models (JETTO and MMM) tend to bracket the fusion plasma confinement data [Nordman and Weiland (1993)].

The highest performance discharges obtained with these mixed Bohm/gyroBohm transport models interpret well the JET confinement data. Thus, these transport codes have been applied to numerous experiments. Here we shown in Fig. 15.6 the time traces from JETTO simulations for several key parameters in an ohmic ignition discharge simulation for the IGNITOR parameters in Table 15.1. The time interval where the alpha heating power $P_\alpha(t)$ raises above $P_{\text{loss}} = W/\tau_E$ is $t_{ig} = 4\text{-}5\,\text{s}$ shown in Fig. 15.6 which compares slow (upper panel) and fast current ramp (lower panel) discharges in the upper and lower figures. Only the fast current ramp simulation reaches ignition.

Figure 15.6 shows how the fast current ramp allows the plasma to reach the cross-over time defined as the time when the alpha particle heating dominates the ohmic heating rate earlier than in the slow current ramp discharge. For the fast

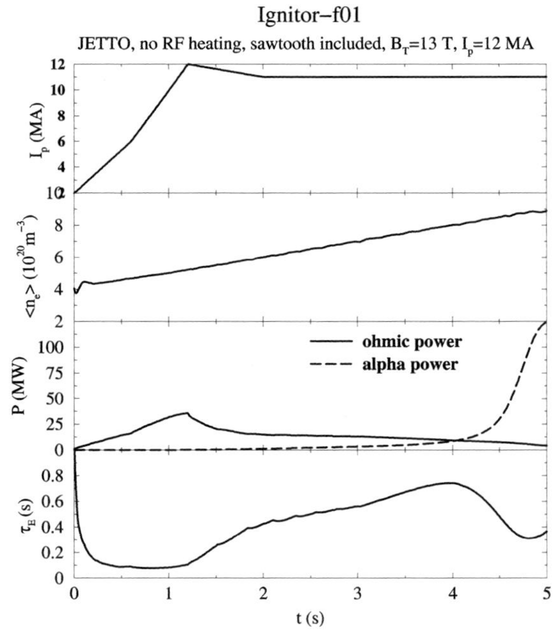

Fig. 15.7 Time history of a) plasma current I_{p}, b) volume-average density $\langle n_{\mathrm{e}} \rangle$, c) Ohmic power P_{oh} (solid line), α power P_α (dashed line), d) confinement time τ_{E} in the fast current ramp simulation run. The fast ramp maintains the peaked density profile giving ignition after four seconds in the high-field compact tokamak.

current ramp the alpha particle heating dominates shortly after the time $t = 4$ s. This 4 s point is defined as the cross-over time where one may refer to the plasma as self-heated from the nuclear fusion reactions.

The improvement in the fast-current ramp regime is due to the inward movement of the $q_{\min}$ surface to mid-radius from the plasma wall. This $q_{\min}$ surface is where the sawtoothing from the internal kink-tearing modes sets in which is now displaced away from the core peaked density and temperature region where the nuclear reactivity is peaked. This spatial displacement in early time allows the transition to ignition to occur well within the ten second pulse of the high-field pulsed tokamak discharge. The sawteeth periods and amplitudes at the $q(r,t) = 1$ surface seen in Fig. 15.6 are calculated following the modeling from Porcelli, *et al.* (1996) and Hu, *et al.* (2003).

Figure 15.7 shows an ignition run similar to the fast current ramp in Fig. 15.6 giving the plasma current $I_p(t)$ the plasma density $\langle n_e \rangle$ and the energy confinement time $\tau_E(t)$.

Transport modeling based on inclusion of turbulence spreading and residual stresses shows internal rotation reversals and polarity reversal of cold pulses, with a clear indication of nonlocal radial transport due to fast radial spreading of the turbulence intensity field [Hariri, *et al.* (2016)].

15.7 Lower-Hybrid and Electron Cyclotron RF Drivers

The propagation, scattering and absorption of the lower-hybrid and electron cyclotron RF waves are used to control fusion plasmas. The analysis in the previous Section 15.6 shows that maintaining a steady-state fusion plasma with ohmic heating is not feasible. Current drive is necessary to allow recharging of the central solenoid transformer and for maintaining quasi-steady operation of the tokamak. Lower-hybrid antennas have been successful in producing long-period tokamak plasmas with vanishing transformer electric fields.

Drift wave turbulence driven by steep ion and electron temperature gradients in H-mode divertor tokamaks produce strong density fluctuations that scatter the RF waves used for heating and for driving the toroidal current. Both the 3-5 GHz lower-hybrid (LH) and the 170 GHz electron cyclotron (EC) waves experience scattering and diffraction propagating through the statistically complex density of the plasma. Ray equations are used to calculate the spread of the rays and the associated change in the parallel phase velocity of the RF waves and the change of their polarization while propagating through the fusion plasma. A Fokker-Planck equation for the phase space of the RF plasmons, or rays, describes the spread of the RF wave power in the complex geometry of a divertor tokamak using ray tracing codes. Evolution of the electron phase-space distribution function from resonant electron-wave interactions is used for several fusion scenarios. The resulting X-ray spectrum from the fast electrons is broad, giving better agreement with the measured X-ray spectra than that calculated in the absence of the turbulent scattering of the RF waves [Horton, *et al.* (2013), Decker, *et al.* (2011, 2014)].

The simulations show that (1) control of the sawtoothing and (2) the density profile are critical elements for achieving the goals of ignition or long-pulse steady-state fusion Q_{fus}. While these two problems have been identified earlier new simulations make quantitative the degree of the problem. By comparing predictions of an empirical-based transport model with a theory-based model, the problem of extrapolating from the limited plasma database in terms of long-pulse fusion power operation or a prediction for ignition is made clear.

Since the high-field compact tokamaks are designed to reach ignition at relatively low plasma temperatures, the projected alpha particle pressure is relatively low. In particular the alpha pressure is lower than the threshold for the excitation of toroidal Alfvén eigenmodes (TAE). The design goal is to operate at such low poloidal beta β_p that neoclassical tearing modes, called NTMs, are expected to be stable or very weak. The NTM growth rate would be slow compared to the energy confinement time. To the extent that a burning plasma regime is achieved and sustained for a time of the order of the skin diffusion time, neoclassical tearing modes might be observed. Thus, neoclassical tearing modes are expected to be a minor issue for the high-field compact tokamak.

A fast current ramp example creates well-defined off-axis minimum in the $q(r)$ profile that improves energy confinement by the reversed shear (RS) regime shown in Fig. 15.6 (lower panel) and Fig. 15.7. Recall that for $q(r) = q_{\min} + 0.5(r - r_m)^2 d^2q/dr^2$ there is a wide gap in the density of rational surfaces so that the drift modes no longer have a significant radial overlapping [Beklemishev and Horton (1992)]. In Fig. 15.6 the slow current ramp-up in curve (a010), the density peaking or gradient continually forms as the plasma heats and grows in density and temperature. In the fast current ramp simulation the density gradient measured by the peaking ratio increases to 1.6 at $t = 4\,\mathrm{s}$ and then drops. The peaked density profile reduces the parameters driving the ion temperature gradient and the electron temperature gradient turbulence of Chapters 12 and 13. For the fast current ramp-up in Fig. 15.6 ($f\emptyset1$) the transport simulation predicts ignition.

Transport modeling based on inclusion of turbulence spreading and residual stresses shows internal rotation reversals and polarity reversal of cold pulses, with a clear indication of nonlocal transport effects due to the fast radial spreading in the turbulence intensity field [Hariri, *et al.* (2016)].

15.8 RF Control of Fusion Plasmas

Real-time control of plasma pressure profiles and the toroidal plasma current profiles requires high-power RF waves from antennas on the vessel walls driven by Klystrons and gyrotrons that feed power to the antennas from adjoining rooms. The multi-KeV W7-X plasmas produced by high-powered electron cyclotron gyrotrons alone provides a recent example.

15.8.1 *Plasma confinement in TORPEX*

TORPEX experiment shown in Fig. 15.8 is an example of an electron cyclotron RF driven plasma. The vacuum vessel has the major R and minor a radius of 1 m and 0.2 m. Flow-meters allow adjusting the injection rate of different gases. Toroidal coils are installed to generate a toroidal field up to 0.1 Tesla. Horizontal coils shown in Fig. 15.8 produce a vertical field component up to around 5 mT as well as for the induction of a loop voltage. A 2.45 GHz microwave source of is installed for the production and sustainment of the plasma. The RF power can be adjusted between 0.2 and 20 kW and modulated for sinusoidal, square and triangle waveforms. These control parameters allow the production of reproducible H, He and Ar plasmas with density, electron temperature, and plasma potential in the range $n_e \approx 10^{16} - 10^{17}\mathrm{m}^{-3}$, $T_e \approx 5 - 10$ eV and loop voltage $V_p \approx$ 10-20 V.

In TORPEX, the plasma is produced and sustained by injection of the microwaves perpendicular to the magnetic field from the low-field side (LFS). Before the plasma is fully formed, the naturally existing free electrons are accelerated, most efficiently at the electron cyclotron (EC) resonance layer where the electron cyclotron frequency ω_{ce} equals to the microwave frequency ω_{rf}. Impact ionization occurs and the plasma is formed. To sustain the plasma, the wave power is injected in the O-mode polarization to avoid the cutoff of the X-mode [Rypdal, *et al.* (1997)].

Fig. 15.8 The TORPEX [Podestá *et al.* (2005)] experiment with $R = 1\,\mathrm{m}$, $a = 0.2\,\mathrm{m}$ and $B_\phi = 1\,\mathrm{kg} = 0.1\,\mathrm{T}$.

There are two RF plasma wave polarizations of the electric field with respect to the confinement magnetic field. The simpler O-mode has the $\mathbf{E}_{\mathrm{RF}}$ parallel to the $\mathbf{B}$ while the X-mode has $\mathbf{E}_{\mathrm{RF}}$ perpendicular to the ambient $\mathbf{B}$-field.

After reflections on the high-field side (HFS) wall, both O-mode and X-mode waves propagate back into the plasma [Podestá, *et al.* (2005)] where diagnostics show that the RF power is absorbed at the upper hybrid (UH) layer. The X-mode $\mathbf{E}_{\mathrm{RF}}$ encounters the upper hybrid (UH) plasma resonance layer on the 3D spatial surface defined by

$$\frac{e^2 n_e(\boldsymbol{x})}{\epsilon_0 m_e} + \frac{e^2 B^2(\boldsymbol{x})}{m_e^2} = \omega_{\mathrm{rf}}^2. \tag{15.15}$$

The UH resonance layer lies more to the low-field surface (LFS) than the resonant EC layer ($\omega_{ce}(\mathbf{x}) = \omega_{\mathrm{RF}}$) where some of the RF power is deposited. When the injected microwave power is increased, the plasma density increases and the upper hybrid (UH) layer moves towards the lowest $|\mathbf{B}|$-field surfaces on the outside of the toroidal axis [Müller (2006)].

15.8.2 *Steady-state operation*

Gormezano, *et al.* (2007) investigate modes of operation that are candidates for steady-state conditions in a tokamak fusion reactor. The plasma parameters, domains of operation, scenarios and integration issues of the advanced scenarios are described by the Gormezano team. The actuators required for heating and current drive control in the advanced tokamak discharges are modeled for ITER. The specific control issues for steady-state operation are described using the current experimental results by a variety of machines. The parameters for the ITER steady-state and hybrid scenarios for heating and current-drive systems are modeled with actuators for controlling the toroidal plasma current profiles $j_\phi(\mathbf{r}, t)$ with the RF and NBI drivers.

ITER hybrid and target steady-state fusion burn scenarios are simulated using the PTRANSP integrated modeling code together with input from the TSC code. In the hybrid scenarios, the majority of the current is driven inductively; whereas, for the target steady-state scenarios, approximately 20% of the toroidal current at $t = 1000$ s is driven inductively with the remaining 80% of the current driven by the bootstrap, the neutral beams and the radio frequency antennas. Predictive simulations are carried out using either the new Multi-Mode or the GLF23 anomalous transport models. Toroidal momentum transport models are used to compute the toroidal angular rotation frequency profile which, in turn, is used to compute the self-consistent flow shear suppression of the turbulent transport. The simulations of the hybrid scenario indicate that the fusion power production at 1000 s will be approximately 500 MW corresponding to a fusion $Q = 9.4$.

The fusion power predicted in the simulations of the target steady-state scenarios was shown to depend on the time dependence of the input heating and the associated current drive. The fusion power obtained in the target steady-state scenarios, depends on the transport model and input injected power and ranged from 168 MW up to 226 MW corresponding to a fusion $Q_{\rm DT}$ ranging from 2.0 to 7.0.

15.8.3 *Ion cyclotron RF plasma control*

Ion cyclotron heating is an essential methods for controlling fusion plasmas. Complex computer codes are used to predict and control ICRH plasma heating. Both at low and higher cyclotron harmonics, accounting for finite Larmor radius effects is crucial during ion cyclotron resonance frequency heating. A number of codes are available to simulate ICRF mode conversion experiments including TORIC (full wave, finite Larmor radius code) [Brambilla (1998, 1999), Wright, *et al.* (2004)], ALCYON coupled with RAYS [Petrov, *et al.* (2000)], and AORSA2D (full wave, all orders code) [Jaeger, *et al.* (2002)].

TOMCAT-U is a tokamak plasma RF wave equation solver [Van Eester and Lerche (2013)] for arbitrary cyclotron harmonics and arbitrary wavelengths. Rather than using the particle position, the guiding-center position is used as the independent variable in the expressions for the dielectric response in the simulations. This choice of the independent variable yields intuitively clear expressions which link to the corresponding expressions in the RF diffusion operator appearing in the Fokker-Planck equation for the ion phase space distribution functions. The method guarantees with Maxwellian distributions a positive definite power transfer from waves to particles for the wave modes in plasmas. Rather than using a truncated Taylor series expansion of the dielectric response, an integro-differential operator retaining all finite Larmor radius effects is used in TOMCAT-U.

The ICRH simulations show that there is plasma heating at general cyclotron harmonics and arbitrary wavelengths both for thermal and high energy ion populations. The Fokker-Planck calculations with the guiding center distribution functions give expressions that ensure positive definite absorption in local equilibrium plasmas and allows verification with well-known results in the literature.

Both at low- and high-cyclotron harmonics, accounting for finite Larmor radius effects accurately is crucial in ion cyclotron resonance frequency heating scenarios that create the high-energy ion tails. TOMCAT-U [Van Eester and Koch (1998)] evaluates high harmonics and arbitrary wavelengths of the ion cyclotron plasma waves while keeping leading order terms in equilibrium plasma variation. The ion guiding-center position is used as the independent variable in the expressions for the dielectric response. This choice of the independent variable yields expressions involving the Kennel-Engelmann operator which can directly be linked to the corresponding expressions in the RF diffusion operator in the Fokker-Planck equation. This method guarantees a positive definite power transfer from waves to particles in a plasma with Maxwellian distributions. The integro-differential approach retains all finite Larmor radius effects. Examples illustrating the potential of the wave code are given in Van Eester and Lerche (2013) and Van Eester, *et al.* (2016).

15.8.4 *Single null and double null divertors*

The onset threshold for the H-mode and then subsequently for the ELMs is controlled by the shaping of the plasma and the positions of the single null or double null divertors. The lowest auxiliary power for the onset of the H-mode is found for single null divertor with the magnetic configuration with the vertical ∇B of the ions towards the X-point. This configuration produces the large radially-inward electric field that is required for the onset of the H-mode at low-auxiliary heating power. Buttery, *et al.* (2007) explore this physics for the baseline ITER experiments at high-plasma currents and strongly-shaped plasma cross-sections with experiments on JET.

JET experiments numbered 66105 and higher with the 2014 divertor with a more triangular shape than in earlier JET experiments showed access to low-collisionality H-modes with smaller ELMS. The ITER-like divertor plasmas driven with high NBI and higher ion cyclotron heating produced record plasma confinement in lower collisionality regimes approaching those that would be expected in ITER. Real-time pellet injection control was installed to test this method of controlling or preventing ELMs above a certain energy threshold. The improved ITER-like performance with increasing the triangularity parameter from $\delta = 0.14$ to 0.5 was demonstrated in JET. At the highest triangularity there is a transition from the Type I ELMs — which are the most damaging to the walls — to the smaller, more frequent Type III ELMs. Pellet pacing is a key method to maintain the smaller, more frequent Type III ELMs [Loarte, *et al.* (2014)].

15.9 Mirror Confinement Fusion Systems

Research on controlled nuclear fusion has been concentrated on plasma confinement using toroidal magnetic fields. Toroidal systems are complex. Simpler magnetic confinement systems may provide valuable platforms for understanding fusion plasmas. The linear mirror machine has delivered high performance with the potential of giving a direct conversion of nuclear energy into electric power. The GAMMA-10 linear mirror confinement system at Tsukuba University demonstrated the principle of the direct conversion of plasma energy into electric power on a small scale using the exhaust plasma entering the exterior end divertor chambers.

In tokamaks the plasma confinement is due to the magnetic field from the plasma current in I_p rather than that from the solenoidal coils. There is room for creative new solutions in the magnetic confinement for plasmas in linear systems using magnetic field from solenoid coils. Here we describe the approach to fusion power using a linear machine with high magnetic mirror fields and the direct conversion of the escaping high-temperature plasma into electric power.

The gas dynamic trap (GDT) shown in Fig. 15.9 has large expansion tanks collecting plasma escaping past the B_{max} field. The GDT team has reported record high electron temperatures of 430 eV inferred by Thomson scattering [Solomahin, *et al.* (2014)]. This GDT team is working on using the high electron temperature plasmas to design a neutron source for radiation damage experiments.

The simplicity of the axisymmetric GDT geometry in a linear cylindrical configuration is highly desirable. The mirror plasma is a high plasma energy density system with the plasma and magnetic energy densities comparable. In the Earth's magnetosphere, for example, stable high energy density plasmas are known to be confined by the planetary magnetic field through the mirror confinement principle. There are long-lived MeV electrons and 500 KeV protons confined in the inner magnetosphere. The technical reason for the enhanced stability of these devices compared with similar sized toroidal plasma is the absence of the destabilizing

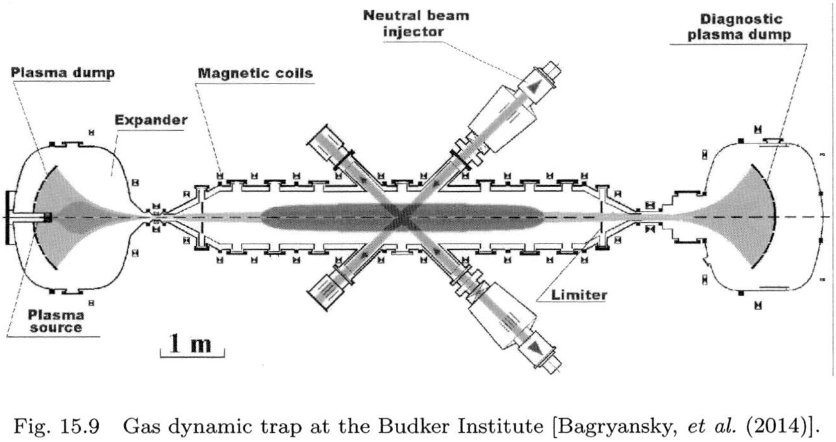

Fig. 15.9 Gas dynamic trap at the Budker Institute [Bagryansky, *et al.* (2014)].

magnetic curvature that covers the outer half volume of the torus. In the magnetic mirror there are locally unstable regions, but they are adjacent to — and strongly connected to — the stable magnetic curvature regions. The result is that the net plasma stability is strong up to higher plasma energy densities in both the laboratory and in the geophysical plasma system [Kesner, *et al.* (1987), Katanuma, *et al.* (2000)].

In the high-pressure (high-beta) mirror system the fluctuations are weaker or comparable to the collisional losses as shown by Pratt and Horton (2006) and Berk and Pratt (2011). The mirror confinement improves with higher heating power in the experiments reported from Tsukuba and the Budker GDT group. In contrast to the confinement time for toroidal systems derived from a large tokamak-database. The energy confinement time τ_E in the GDT and the Tri Alpha machines increases with increasing T_e.

The technologies (diagnostics, real-time controls, material radiation limits) are simpler in the open mirror linear magnetic confinement system. The linear confinement system has lower technological challenges. For example, the Earth's magnetosphere is a mirror confinement system that traps relativistic electrons and KeV ions for months. Japan and Russia have developed large mirror research programs with the construction of the GAMMA 10 (G-10) device at Tsukuba University in Japan and the gas dynamic trap (GDT) in Fig. 15.9 at the Budker Institute in Novosibirsk, Russia. Important advances were made from these two magnetic fusion plasma confinement devices.

The Tri Alpha plasma is mirror-confined plasma with magnetic field reversal confining the large ions shown in Fig. 15.10. The plasma current I_θ is in the theta direction in r, θ, z cylindrical coordinates with z along the length of the machine shown in Fig. 15.10. The C-2VY machine has eight NBI beams for a total of 15 MW injection and expects to achieve $T_e \lesssim 1$ KeV in 2017. The plasma length starts at 3 m and shrinks slowly to 1 m over 10 ms. The plasma pressure is approximately

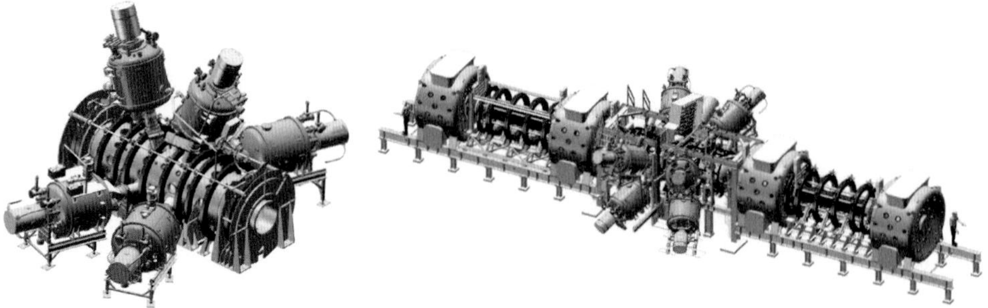

Fig. 15.10 The Tri Alpha C-2U mirror machine on the left with 6 neutral beam injectors at 15 kev injecting 10 MW for 8ms that produced 10ms fusion grade hot deuterium plasmas. On the right is schematic of the new C-2W machine currently under construction with 8 neutral beam injectors with a longer pulse length that is projected to produce significant D-D fusion power [Binderbauer, *et al.* (2016)].

10^3 Pa at 2 ms the separatrix radius $r_{\mathrm{sx}}^2 \simeq 3(c/\omega_{pi})L_z$ where L_z is the length of the hot plasma $r_{\mathrm{sx}} = 3\,L/\omega_{pi} \star \ell_s/r_{\mathrm{sx}}$.

The mirror physics confinement results have important implication for the plasma confinement time scaling laws and for the scaling laws for neutron sources. In particular, the mirror plasma experiments show energy confinement times increasing with electron temperature. The electron temperature has reached $T_e \lesssim 1$ KeV [Bagryansky, *et al.* (2014)] in the gas dynamic trap (GDT). These plasma confinement results are needed to make a decision on how to design a practical fusion reactor or a hybrid fusion-fission power system with linear machines. A new mirror plasma confinement experiment would give important results both for both 14 MeV neutron sources for plasma-wall interaction limits and for the hybrid fusion-fission power reactor design.

The linear mirror confinement machine is relatively simple and easily built compared with toroidal systems. The linear mirror machine is an excellent platform for advanced research in plasma science and the related topics of plasma sensors, actuators, and real-time controls. The mirror program would achieve the goals advocated by a leading mirror machine scientist, Dimitry Ryutov (2005), providing a high-powered neutron source needed to design the future development of the fusion power system.

The linear mirror geometry that could make a significant contribution for a better understanding and predictive tools for plasma science and fusion energy. The mirror system has low transverse (radial) ion transport — owing to the azimuthal symmetry which gives simpler and better confined particle orbits by the KAM theorems of classical mechanics. While the symmetric mirror geometry has less stability to certain plasma eigenmodes, the modes are not global and are amenable to several methods for stabilization. Methods for stabilizing the modes in an axisymmetric linear geometry are described by Ryutov, *et al.* (2011) and Post (2002) and validated in the GDT. Scientists in the Mirror Forum Group argue that the mirror confined plasma would lead to a simpler, high-powered neutron source in the near future at modest cost.

Modular, real-time control systems based on field-programmable-gate-arrays (FPGA) give an unprecedented set of fast, new control tools for adjusting the heating and fueling systems to maintain a stable, high temperature plasma. These FPGA systems are the fast microsecond to millisecond real-time input-output (RIO) systems used to control rockets, drones and many types of autonomous systems. The real-time input-output (RIO) systems are widely employed in industry and science (`http://www.ni.com`). Integrated, high-speed closed-loop control systems are available and broadly deployed in aviation and military/defense machinery. The plasma systems have real-time control of the mirror-confined plasma [Giannone, *et al.* (2009), Seo and Lee (2006)]. In addition to FPGAs, other hardware acceleration devices including graphics processing units (GPUs) and multi-core microprocessors would be incorporated in the diagnostic suite for feedback control of the mirror fusion plasma. The real-time control relies on the fast diagnostics from Thomson scattering, the radiation spectra from charge exchange recombination spectroscopy [Bespamyatnov, *et al.* (2010)], X-ray emission and neutron spectroscopy.

A multi-beam Thomson scattering system [LeBlanc, *et al.* (2003)] gives accurate, fast measurements of the electron temperature profiles. While high-ion temperatures achieved in the multi-KeV 5-10 KeV mirror plasmas, their interpretation has been clouded by the lack of this "gold standard" method for determining the electron temperature profiles. For example, earlier the GAMMA-10 device reported the core electron temperature of 800 eV [Cho, *et al.* (2006, 2008)]. The temperature was not verified and the x-ray radiation may have been due to the electron spectrum from a non-Maxwellian tail driven by the ECRH electron heating.

Future plans are to use 10 to 20 diagnostic laser beams through the radius of the cylindrical plasma of the mirror machine with longer laser pulses than used in the NSTX for collecting more scattered photons per pulse. Signal processing with FPGAs and GPUs [Wetzel, *et al.* (2005), Peng (2006)] would be performed in real-time using state-of-the-art instrumentation and computations made possible by the PIXI data acquisition system and the CompactRIO systems. Graphics processing units (GPUs) will be used in the high speed imagining of the plasma. These tools would allow real time solutions of MHD background plasma with the concomitant drift-wave turbulence that drives the anomalous radial transport. Real-time commands would be issued to the external neutral beam injectors and the RF heating systems to bring the plasma back to the target state.

Completion of a mirror machine with multi-beam real-time Thomson scattering and CXRS systems would give broad implications for both the main-line toroidal approach to fusion power and the alternative confinement systems. The diagnostic systems would validate the mirror fusion plasmas. Improved diagnostics and control systems are being implemented in the Large Helical Device system in Japan and the Wendelstein 7-X stellarator in Germany. The new diagnostic database will make significant contributions for better predicting the lifetime and performance of fusion power systems.

Tokamaks have achieved the highest product of density times energy confinement time at high plasma temperatures. The progress in tokamaks that was rapid in the 1980-1990s, but slowed appreciably in the 21st Century and has not lead to the demonstration of ignition nor to the production of a net fusion power amplification (fusion $Q_{\rm DT} > 1$). JET achieved the maximum ratio of fusion power to injected power for a few seconds in 1994 with this ratio $Q = 17\,\text{MW}/28\,\text{MW} = 2/3$.

The field reversed configuration is a prolate compact toroidal plasma with poloidal magnetic fields [Tuszewski (1988), Steinhauer (2011)]. The two Tri Alpha machines shown in Fig. 15.10 are linear with linear mirror machine coils with a system of six neutral beam drives create a large plasma j_θ-current that reverses the direction along the axis of the confinement cylinder.

The field reversed configuration provides high plasma pressure relative to magnetic pressure and in simple geometry with a natural divertor. As such, the FRC could lead to economic fusion reactors with high fusion power density, and capable of burning aneutronic fuels. However, as in other high-beta plasmas, there have been serious stability and confinement problems. Some of these problems are being solved by introducing and maintaining a significant large ion gyro-orbit population in the system. This is the conjecture first introduced by Norman Rostoker decades ago and adopted as the central design tenet in Tri Alpha Energy's (TAE) first of a kind C-2 and C-2U plasma machines [Binderbauer, *et al.* (2015)].

Recent achievements of the Tri Alpha FRC experiments and the future vision are given in the Proceedings of Norman Rostoker 2015 Memorial Symposium, 2015 AIP Conference `http://dx.doi.org/10.1063/1.4944010`.

Control of radial particle and thermal transport is instrumental for achieving and sustaining well-confined high-β plasma in a field-reversed configuration (FRC). Radial profiles of low-frequency ion gyro-scale density fluctuations ($0.5 \leq k\rho_s \leq 40$), consistent with drift- or drift-interchange modes, have been measured in the scrape-off layer (SOL) and core of the C-2 Field-Reversed Configuration (FRC), together with the toroidal $\mathbf{E} \times \mathbf{B}$ velocity. The axial electrostatic SOL biasing controls and reduces gyro-scale density fluctuations, resulting in the low-core fluctuation levels. When the radial $\mathbf{E} \times \mathbf{B}$ flow shearing rate decreases below the turbulence decorrelation rate, fluctuation levels increase substantially, concomitantly with onset of the $n = 2$ tilting instability with a rapid loss of diamagnetism. Low turbulence levels, improved energy/particle confinement and substantially increased plasma pressure is maintained when the $\mathbf{E} \times \mathbf{B}$ shear near the separatrix is controlled via axial plasma biasing using an annular washer gun [Schmitz, *et al.* (2016)].

15.10 JET Record Fusion Power Experiments and Future Directions

The Joint European Tokamak (JET) was designed in the 1970s and has produced the most fusion power from deuterium-tritium plasmas. The JET experiment, approved in 1975, was a major step up from the second-generation tokamaks that had plasma currents $I_p \simeq 300\,\text{kA}$ and volumes of order a few cubic meters. JET, called third-generation tokamak, was designed and first operated with plasma currents $I_p = 4\,\text{MA}$. The plasma chamber was designed to have a small aspect ratio $R/a \leq 3$ and height $b/a = 1.7$ allowing the large plasma current I_p for the given magnetic field $B_T \leq 4\,\text{T}$ and machine size $R = 3\,\text{m}$, $a = 1.15\,\text{m}$. The plasma volume is of order $100\,\text{m}^3$, and the plasma is confined by the magnetic field from the plasma current I_p.

Eventually, with an upgraded plasma heating system with 22 MW of neutral beam injection (NBI), 22 MW of ion cyclotron wave heating (ICRH), and 7.3 MW of lower-hybrid current device (LHCD) the plasma current I_p reached 7 MA which is approximately one half the designed current value (15 MA) of the ITER device currently under construction. Table 15.2 compares the main machine and plasma parameters of the JET machine to the ITER machine under construction.

Table 15.2 JET and ITER main parameters comparison [Shimada, *et al.* (2007)] of ITER Physics Basis).

	Maximum Values Achieved on JET Separately	ITER Design Values
R and a	3 m 1.25 m	6.2 m 2 m
Elongation	1.8	1.7
Plasma volume	100 m^3	840 m^3
Magnetic field on axis	4 T	5.3 T
Plasma current in D-shaped plasma	7 MA	16 MA
Plateau current	60 s (1 MA)	15 MA
Modes of operation	L, H and ELMy	ELMy
Plasma contact	beryllium/tungsten limiters-pumped divertor	pumped divertor
Neutral injection to the plasma	22 MW	33-50 MW
Coupled ICRH	22 MW	20-40 MW
ECRH	0	20-40 MW
Current drive	3 MA (LH)	
	50 V.s	300 V.s
Central density	2×10^{20} m^{-3}	10^{20} m^{-3}
Electron temperature	20 KeV	21 KeV
Ion temperature	40 KeV	18 KeV
Q value in DT plasma	0.6 (0.9 net)	10
Fusion power	16 MW	500 MW
Fusion energy	22 MJ in 4 s	120 GJ in 200 s

The JET machine was designed with a poloidal divertor and the divertor configuration has gone through numerous changes since 1990. All modes of plasma confinement states, the L-mode, the H-mode, and the ELMY-H-mode have been achieved in JET.

Prior to 1990, the metallic walls of the vacuum vessel were coated with carbon to withstand the plasma-surface interaction. At the highest heating power, the plasma became highly diluted with fully ionized carbon ions. Thus, in a rather inspired leap of judgment, the Director, Henri Rebut, changed the coating material to the lower-Z material of beryllium. The result was a hotter plasma with a lower dilution from the wall material and record values of the metrics used to measure the machine performance [Dietz, *et al.* (1990)].

The ITER project is the International Thermonuclear Experimental Reactor with the European Union and Japan playing the key roles and five partner countries. The 20 billion dollar project is scheduled to start plasma confinement experiments in 2020-2025. The foundation and numerous buildings were finished in 2012.

The temperature of the plasma at the high pressure pedestal formed just inside the magnetic separatrix was at a lower temperature and has different profile with the ILW chamber in JET. The plasma pedestal temperatures were lower than in previous JET experiments. Gyrokinetic simulations by Hatch, *et al.* (2016, 2017) showed a turbulence that lowered the temperature plasmas. The simulations showed microtearing turbulence in the MHD stable plasma. The magnetic turbulence and convection inside the magnetic separatrix convected the partially high-Z tungsten ions inward. There was then a slow build- up of highly-ionized tungsten ions in the core plasma. At higher temperatures there was an increase in ITG transport. Consistent with observations, the transport mechanism is strongly reduced by the change to a low-Z impurity (e.g. carbon or nitrogen at the level of $Z_{\rm eff} < 2$) impurity giving the higher pedestal temperatures.

Scaling studies with ρ^* gave simulations that recovered the historical record temperatures in JET. The simulations showed the observed degradation of confinement caused by gas puffing owing to the change in the density pedestal structure. Nonlinear simulation studies mapped out important regions of parameter space with optimal physical regimes for the recovery of high-pedestal temperatures on JET. The electromagnetic electron temperature gradient driven transport and microtearing modes are consistent with experimental power balance data across the pedestal. Kinetic ballooning modes are found to onset in a second-stability regime.

In the 21st Century fusion research has become much stronger with new toroidal fusion machines – EAST in China and KSTAR in South Korea – producing fusion grade plasmas and JT60-SuperUpgrade coming in Japan is currently under construction. In addition, fusion research is now growing in the private sector of research, as described in the Tri Alpha Energy series of machines reported in the previous section and several new private sector projects in the USA and abroad. In conclusion the development of controlled thermonuclear fusion power machines is steadily growing and will become a new source of power for future generations.

Bibliography

Allan, S. Y. (2016). Ion temperature measurements of L-mode filaments in MAST by retarding field energy analyser, *Plasma Phys. Control. Fusion* **58**, p. 045014, http://dx.doi.org/10.1088/0741-3335/58/4/045014.

Alvarado, I. (2006). Active learning of control theory using virtual instrumentation, 9th International Conference on Engineering Education-ICEE, July 23-28. http://www.icee.usm.edu/ICEE/conferences/icee2006/papers/3291.pdf.

Andrews, P. L. and Perkins, F. W. (1983). Spectral broadening of lower-hybrid waves by time-dependent density fluctuations, *Phys. Fluids* **26**, p. 2546, http://dx.doi.org/10.1063/1.864444.

Angioni, C., Peeters, A. G., Garbet, X., Manini, A., Ryter, F. and ASDEX Upgrade Team. (2004). Density response to central electron heating: theoretical investigations and experimental observations in ASDEX Upgrade, *Nucl. Fusion* **44**, p. 827, http://dx.doi.org/10.1088/0029-5515/44/8/003.

Angioni, C., Casson, F. J., Veth, C. and Peeters, A. G. (2012). Analytic formulae for centrifugal effects on turbulent transport of trace impurities in tokamak plasmas, *Phys. Plasmas* **19**, p. 122311, http://dx.doi.org/10.1063/1.4773051.

Angioni, C., Mantica, P. Pütterich, T., Valisa, M., Baruzzo, M., Belli, E. A., Belo, P., Casson, F. J., Challis, C., Drewelow, P., Giroud, C., Hawkes, N., Hender, T. C., Hobirk, J. , Koskela, T., Taroni, L. Lauro, Maggi, C. F., Mlynar, J., Odstrcil, T., Reinke, M. L., Romanelli, M. and JET EFDA Contributors (2014). Tungsten transport in JET H-mode plasmas in hybrid scenario, experimental observations and modeling, *Nucl. Fusion* **54**, p. 083028, http://dx.doi.org/10.1088/0029-5515/54/8/083028.

Angioni, C., Casson, F. J., Mantica, P. Pütterich, T., Valisa, M., Belli, E. A., Belo, P., Bilato, R., Giroud, C., Helander, P. and JET Contributors (2014). The impact of poloidal asymmetries on tungsten transport in the core of JET H-mode plasmas, *Phys. Plasmas* **22**, p. 055902, http://dx.doi.org/10.1063/1.4919036.

Anikeev, A. V., Bagrayansky, P. A., Deicheuli, P. P., Ivanov, A. A., Karpushov, A. N., Maximov, V. V., Pod'minogin, A., Stupishin, N. V. and Tsidulko, Yu. A. (1997). Observations of magnetohydrodynamic stability limit in a cusp-anchored gas-dynamic trap, *Phys. Plasmas* **4**, pp. 347-354, http://dx.doi.org/10.1063/1.872493.

Antar, G. Y., Krasheninnikov, S. I., Devynck, P., Doerner, R. P., Hollmann, E. M., Boedo, J. A., Luckhardt, S. C. and Conn, R. W. (2001). Experimental evidence of intermittent convection in the edge of magnetic confinement devices, *Phys. Rev. Lett.* **87**, p. 065001, http://dx.doi.org/10.1103/PhysRevLett.87.065001.

Antar, G. Y., Devynck, P., Garbet, X. and Luckhardt, S. C. (2001). Turbulence intermittency and burst properties in tokamak scrape-off laye, *Phys. Plasmas* **8**, p. 1612, http://dx.doi.org/10.1063/1.1363663.

Antar, G. Y. (2003). On the origin of "intermittency" in the scrape-off layer of linear magnetic confinement devices, *Phys. Plasmas* **10**, p. 3629, http://dx.doi.org/10.1063/1.1599855.

Antar, G. Y., Counsell, G., Ahn, J. W., Yang, Y., Price, M., Tabasso, A. and Kirk, A. (2006). The poloidal distribution of Type-III edge localized modes in the Mega Ampere Spherical Tokamak (MAST), *Phys. Plasmas* **13**, p. 052508, http://dx.doi.org/10.1063/1.2198210.

Antar, G. Y., Assas, S., Bobkov, V., Noterdaeme, J.-M., Wolfrum, E., Herrmann, A., Rohde, V. and ASDEX Upgrade Team. (2010). Convective transport suppression in the scrape-off layer using Ion cyclotron resonance heating on the ASDEX Upgrade Tokamak, *Phys. Rev. Lett.* **105**, p. 165001, http://dx.doi.org/10.1103/PhysRevLett.105.165001.

Apicella, M. L., Apruzzese, G., Borra, M., Bracco, G., Ciotti, M., Condrea, I., Crisanti, F., De Angelis, R., Ferro, C., Gabellieri, L., *et al.* (1997). Experiments in FTU with different limiter materials, *Nucl. Fusion* **37**, p. 381, http://dx.doi.org/10.1088/0029-5515/37/3/I12.

Arnas, C., Pardanaud, O. Martin, C., Roubin, P., De Temmerman, G. and Counsell, G. (2010). Analyses of dust samples collected in the MAST tokamak, *J. Nucl. Mater.* **401**, p. , http://dx.doi.org/10.1016/j.jnucmat.2010.04.010.

Artaud, J. F., Basiuk, V., Imbeaux, F., Schneider, M., Garcia, J., Giruzzi, G., Huynh, P., Aniel, T., Albajar, F., Ané, J. M., Bécoulet, A., Courdelle, C., Casati, A., Colas, L., Decker, J., Dumont, R., Eriksson, L. G., Garbet, X., Guirlet, R., Hertout, P., Hoang, G. T., Houlberg, W., Huysmans,G., Joffrin, E., Kim, S. H., Köchi, F., Lister, J., Litaudon, X., Maget, P., Masset, R., Pégourié, B., Peysson, Y., Thomas, P., Tsitrone, E. and Turco, F. (2010). The CRONOS suite of codes for integrated tokamak modeling, *Nucl. Fusion* **50**, p. 043001, http://dx.doi.org/10.1088/0029-5515/50/4/043001.

ASDEX Team. (1989). The H-Mode of ASDEX, *Nucl. Fusion* **29**, p. 1959, http://dx.doi.org/10.1088/0029-5515/29/11/010.

Aschwanden, M. J. and Nightingale, R. W. (2005). Elementary loop structures in the solar corona analyzed from TRACE Triple-Filter Images, *Ap. J.* **633**, p. 499, http://dx.doi.org/10.1086/452630.

Aschwanden, M. J. (2006). *Physics of the Solar Corona: An Introduction with Problems and Solutions* (Springer-Verlag) (March 21), ISBN-10:3540307656.

Asp, E., Weiland, J., Garbet, X., Parail, V., Strand, P. and the JET EFDA contributors. (2007). Critical gradient response of the Weiland model, *Plasma Phys. Control. Fusion* **49**, p. 1221, http://dx.doi.org/10.1088/0741-3335/49/8/009.

Asp, E., Kim, J.-H., Horton, W., Porte, L., Alberti, S., Karpushov, A., Martin, Y., Sauter, O., Turri, G. and the TCV TEAM. (2008). Electron thermal transport analysis in tokamak configuration variable, *Phys. Plasmas* **15**, http://dx.doi.org/10.1063/1.2965828

Aydemir, A. Y. (1987). Mechanism for rapid sawtooth crashes in tokamaks *Phys. Rev. Lett.* **59**, p. 649, https://doi.org/10.1103/PhysRevLett.59.649.

Aydemir, A. Y. (1990). Magnetohydrodynamic modes driven by anomalous electron viscosity and their role in fast sawtooth crashes, *Phys. Fluids B* **2**, pp. 2135-2142, http://dx.doi.org/10.1063/1.859433.

Aydemir, A. Y. (1991). Linear studies of $m = 1$ modes in high-temperature plasmas with a four-field model, *Phys. Fluids B* **3**, pp. 3025-3032, http://dx.doi.org/10.1063/1.859780.

Aydemir, A. Y. (1992). Nonlinear studies of $m = 1$ modes in high-temperature plasmas, *Phys. Fluids B* **4**, p. 3469, http://dx.doi.org/10.1063/1.860355.

Aydemir, A. Y. (2005). Convective transport in the scrape-off layer of tokamaks, *Phys. Plasmas* **12**, p. 062503, http://dx.doi.org/10.1063/1.1927539.

Aydemir, A. Y., Kim, J. Y., Park, B. H. and Seol, J. (2015). On resistive magnetohydrodynamic studies of sawtooth oscillations in tokamaks, *Phys. Plasmas* **22**, p. 032304, http://dx.doi.org/10.1063/1.4914090.

Aydemir, A. Y., Lee, H. H., Lee, S. G., Seol, J., Park, B. H. and In, Y. K. (2016). Role of explosive instabilities in high-β disruptions in tokamaks, *Nucl. Fusion* **56**, p. 054001, http://dx.doi.org/10.1088/0029-5515/56/5/054001.

Aymar, R. (1989). Equipe Tore Supra (Plasma Phys. Control. Nucl. Fusion Research, 1988).

Aymar, R., Chuyanov, V. A., Huguet, M., Shimomura, Y., ITER Joint Central Team and ITER Home Teams. (2001). Overview of ITER-FEAT – The future international burning plasma experiment, *Nucl. Fusion* **41**, p. 1301, http://dx.doi.org/10.1088/0029-5515/41/10/301.

Baek, S. G., Wallace, G. M., Shinya, T., Parker, R. R., Shiraiwa, S., Bonoli, P. T., Brunner, D., Faust, I., LaBombard, B. L., Takase, Y., and Wukitch, S. (2016). Measurements of the parallel wavenumber of lower-hybrid waves in the scrape-off layer of a high-density tokamak, *Phys. Plasmas* **23**, p. 050701, http://dx.doi.org/10.1063/1.4948555.

Bagryansky, P. A., Kovalenko, Yu. V., Savkin, V. Ya., Solomakhin, A. L. and Yakovlev, D. V. (2014). First results of an auxiliary electron cyclotron resonance heating experiment in the GDT magnetic mirror, *Nucl. Fusion* **54**, p. 082001, http://iopscience.iop.org/0029-5515/54/8/082001.

Bagryansky, P. A., Shalashov, A. G., Gospodchikov, E. D., Lizunov, A. A., Maximov, V. V., Prikhodko, V. V., Soldatkina, E. I., Solomakhin, A. L. and Yakovlev, D. V. (2015). Three-fold increase of the bulk electron temperature of plasma discharges in a magnetic mirror device, *Phys. Rev. Lett.* **114**, p. 205001, http://dx.doi.org/10.1103/PhysRevLett.114.205001.

Balescu, R. (1988). *Transport Processes in Plasmas* (North-Holland-Amsterdam), Vols. 1-2.

Balescu, R. (2005). Aspects of anomalous transport in plasmas, *Series in Plasma Physics* (Institute of Physics Publishing, Dirac House, Temple Back, Briston BS1 6BE, UK), ISBN:0 7503 1030 8.

Balmforth, N. J. and Morrison, P. J. (1999). A necessary and sufficient instability condition for inviscid shear flow, *Studies Appl. Math.* **102**, pp. 309-344, http://dx.doi.org/10.1111/1467-9590.00113.

Balmforth, N. J. and Morrison, P. J. (2002). Hamiltonian Description of Shear Flow, *Large-Scale Atmosphere-Ocean Dynamics II*, eds. J. Norbury and I. Roulstone (Cambridge, Cambridge) pp. 117-142.

Banerjee, D., O'Shea, E. and Doyle, J. G. (2000). Giant macro-spicule as observed by CDS on SOHO, *Astron. Astrophys.* **355**, pp. 1152-1159, `http://adsabs.harvard.edu/abs/2000A\%26A...355.1152B`.

Barnes, D. C. and Brackbill, J. U. (1977). Computation of magnetohydrodynamic flow in a magnetically confined plasma, *Nucl. Sci. Eng. (US)* **64**, pp. 18-32.

Basu, R., Naulin, V. and Rasmussen, J. Juul. (2003). Particle diffusion in anisotropic turbulence, *Comm. Nonl. Sci. Numer. Simul.* **8**, pp. 477-492, `http://dx.doi.org/10.1016/S1007-5704(03)00044-3`.

Batchelor, G. K. (1969). Computation of the energy spectrum in homogeneous two-dimensional turbulence, *Phys. Fluids* **12**, p. II-233, `http://dx.doi.org/10.1063/1.1692443`.

Bateman, G., Kritz, A. H., J. E. Kinsey, Redd, A. J. and Weiland, J. (1998). Predicting temperature and density profiles in tokamaks, *Phys. Plasmas* **5**, p. 1793, `http://dx.doi.org/10.1063/1.872848`.

Batista, A. M., Calderas, I. L., Lopes, S. R., Viana, R. L., Horton, W. and Morrison, P. J. (2006). Nonlinear three-mode interaction and drift-wave turbulence in a tokamak edge plasma, *Phys. Plasmas* **13**, p. 042510, `http://dx.doi.org/10.1063/1.2184291`.

Becker, G. (1990). Analysis of energy and particle transport and density profile peaking in the improved ohmic confinement regime, *Nucl. Fusion* **30**, p. 2285, `http://dx.doi.org/10.1088/0029-5515/30/11/006`.

Becoulet, A., Hoang, G. T., Artaud, J. F., Bae, Y. S., Belo, J., Berger-By, G., Bernard, J. M., Cara, Ph., Cardinali, A., Castaldo, C., Ceccuzzi, S., Cesario, R., Cho, M. H., Decker, J., Delpech, L., Do, H. J., Ekedahl, A., Garcia, J., Garibaldi, P., Goniche, M., Guilhem, D., Hamlyn-Harris, C., Hillairet, J., Huang, Q. Y., Imbeaux, F., Jia, H., Kazarian, F., Kim, S. H., Lausenaz, Y., Litaudon, X., Maggiora, R., Magne, R., Marfisi, L., Meschino, S., Milanesio, D., Mirizzi, F., Mollard, P., Namkung, W., Pajewski, L., Panaccione, L., Park, S., Park, H., Peysson, Y., Saille, A., Samaille, F., Schettini, G., Schneider, M., Sharma, P. K., Tuccillo, A., Tudisco, O. (2011). Steady state long pulse tokamak operation using lower hybrid current drive, *Fusion Eng. Des.* **86**, pp. 490-496, `http://dx.doi.org/10.1016/j.fusengdes.2011.01.098`.

Beer, M. A. and Hammett, G. W. (1996). Toroidal gyrofluid equations for simulations of tokamak turbulence, *Phys. Plasmas* **3**, 11, pp. 4046-4064, `http://dx.doi.org/10.1063/1.871538`.

Beer, M. A. and Hammett, G. W. (1999). *Proceedings of the Joint Varenna-Lausanne Intl. Workshop on Theory of Fusion Plasmas, Varenna, 1998*, eds. J. W. Connor, E. S. Indoni and J. Vaclavik (Bologna, Societa Italiana de Fisca).

Beklemishev, A. and Horton, W. (1992). Anomalous ion thermal diffusion from η_i modes, *Phys. Fluids B* **4**, p. 2176, `http://dx.doi.org/10.1063/1.860022`.

Beklemishev, A., Anikeev, A., Astrelin, V., Bagryansky, P., Burdakov, A., Davydenko, V., Gavrilenko, D., Ivanov, A., Ivanov, I., Ivantsivsky, M., Kandaurov, I., Polosatkin, S., Postupaev, V., Sinitsky, S., Shoshin, A., Timofeev, I. and Tsidulko, Yu. (2013). Novosibirsk project of gas-dynamic multiple-mirror trap, *Fusion Sci. Tech.* **63**, pp. 46-51.

Bell, M. G. and the TFTR Team. (1989). *Plasma Phys. Control. Nucl. Fusion Res., 1988* (International Atomic Energy Agency, Vienna), Vol. 1, pp. 27-40.

Belli, E. A. and Candy, J. (2008). Kinetic calculation of neoclassical transport including self-consistent electron and impurity dynamics, *Plasma Phys. Control. Fusion* **50**, p. 095010, http://dx.doi.org/10.1088/0741-3335/50/9/095010.

Bengtson, R. D., Freeman, M. R., Wootton, A. J., Wang, P. W., Myra, J. R. and Catto, P. J. (1992). Runaway electrons as a diagnostic of magnetic fluctuations, *Rev.Sci. Instrum.* **63:10**, pp. 4595-4598, http://dx.doi.org/10.1063/1.1143680.

Benkadda, S., Elskens, Y. and Doveil, F., eds. (1995). *Transport, Chaos and Plasma Physics: Institut Mediterranean De Technologie, Marseilles, France July 10-21* Advanced Series in Nonlinear Dynamics, (World Scientific Pub. Co. Inc).

Plasma Interaction in Controlled Fusion Devices, Editor S. Benkadda, AIP Conference Proceedings 1237 (2010), ISBN:978-0-7354-0781-7.

Berk, H. L., Cheng, C. Z., Rosenbluth, M. N. and Van Dam, J. W. (1983). Finite Larmor radius stability theory of ELMO Bumpy Torus plasmas, *Phys. Fluids* **26**, p. 2642, http://dx.doi.org/10.1063/1.864456.

Berk, H. L, Breizman, B. N. and Petviashvili, N. V. (1997). Spontaneous hole-clump pair creation in weakly unstable plasmas, *Phys. Lett. A* **234**, 3, pp. 213-218, http://dx.doi.org/10.1016/S0375-9601(97)00523-9; Erratum *Phys. Lett. A* **238**, 1998, p. 408.

Berk, H. L. and Pratt, J. (2011). Trapped particle stability for the kinetic stabilizer, *Nucl. Fusion* **51**, 8, p. 083025, http://dx.doi.org/10.1088/0029-5515/51/8/083025.

Bernabei, S., Daughney, C., Efthimion, P., Hooke, W., Hosea, J., Jobes, F., Martin, A., Mazzucato, E., Meservey, E., Motley, R., Stevens, J., Von Goeler, S. and Wilson, R. (1982). Lower-hybrid current drive in the PLT tokamak, *Phys. Rev. Lett.* **49**, pp. 1255-1258, http://dx.doi.org/10.1103/PhysRevLett.49.1255.

Bespamyatnov, I. O., Rowan, W. L., Liao, K. T. and Granetz, R. S. (2010). An integrated charge exchange recombination spectroscopy/beam emission spectroscopy diagnostic for Alcator C-Mod tokamak, *Rev. Sci. Instrum* **81**, p. 105F29, http://dx.doi.org/10.1063/1.3475707.

Bessenrodt-Weberpals, M., McCormick, K., Söldner, F. X., Wagner, F., Bosch, H. S., Gehre, O., Müller, E. R., Murmann, H. D., Neuhauser, J., Poschenrieder, W., Steuer, K.-H., Tsois, N. and ASDEX Team. (1991). The multiple facets of ohmic confinement in ASDEX, *Nucl. Fusion* **31**, p. 155, http://dx.doi.org/10.1088/0029-5515/31/1/014.

Bhattacharjee, A., Germaschewski, K. and Ng, C. S. (2005). Current singularities: Drivers of impulsive reconnection, *Phys. Plasmas* **12**, p. 042305,http://dx.doi.org/10.1063/1.1872893.

Biglari, H., Diamond, P. H. and Terry, P. W. (1988). Instability in low-collisionality plasmas, *Phys. Rev. Lett.* **60**, pp. 200-203 http://link.aps.org/doi/10.1103/PhysRevLett.60.200.

Binderbauer, M. W., Tajima, T., Steinhauer, L. C., Garate, E., Tuszewski, M., Schmitz, L., Guo, H. Y., Smirnov, A., Gota, H., Barnes, D., Deng, B. H., Thompson, M. C., Trask, E., Yang, X., Putvinski, S., Rostoker, N., Andow, R., Aefsky, S., Bolte, N., Bui, D. Q., Ceccherini, F., Clary, R., Cheung, A. H., Conroy, K. D., Dettrick, S. A., Douglass, J. D., Feng, P., Galeotti, L., Giammanco, F., Granstedt, E., Gupta,

D., Gupta, S., Ivanov, A. A., Kinley, J. S., Knapp, K., Korepanov, S., Hollins, M., Magee, R., Mendoza, R., Mok, Y., Necas, A., Primavera, S., Onofri, M., Osin, D., Rath, N., Roche, T., Romero, J., Schroeder, H., Sevier, L., Sibley, A., Song, Y., Van Drie, A. D., Walters, J. K., Waggoner, W., Yushmanov, P., Zhai, K., and the TAE Team (2015). A high performance field-reversed configuration, *Phys. Plasmas* **22**, p. 056110, `http://dx.doi.org/10.1063/1.4920950`.

Binderbauer, M. W., Tajima, T., Tuszewski, M., Schmitz, L., Smirnov, A., Gota, H., Garate, E., Barnes, D., Deng, B. H., Trask, E., Yang, X., Putvinski, S., Andow, R., Bolte, N., Bui, D. Q., Ceccherini, F., Clary, R., Cheung, A. H., Conroy, K. D., Dettrick, S. A., Douglass, J. D., Feng, P., Galeotti, L., Giammanco, F., Granstedt, E., Gupta, D., Gupta, S., Ivanov, A. A., Kinley, J. S., Knapp, K., Korepanov, S., Hollins, M., Magee, R., Mendoza, R., Mok, Y., Necas, A., Primavera, S., Onofri, M., Osin, D., Rath, N., Roche, T., Romero, J., Schroeder, H., Sevier, L., Sibley, A., Song, Y., Steinhauer, L. C., Thompson, M. C., Van Drie, A. D., Walters, J. K., Waggoner, W., Yushmanov, P. and Zhai, K. (2016). Recent breakthroughs on C-2U: Norman's legacy, *AIP Conf. Proc.* **1721**, p. 030003, `http://dx.doi.org/10.1063/1.4944019`.

Bird, T. M. and Hegna, C. C. (2013). A model for microinstability destabilization and enhanced transport in the presence of shielded 3D magnetic perturbations, *Nucl. Fusion* **53**, p. 013004, `http://dx.doi.org/10.1088/0029-5515/53/1/013004`.

Biskamp, D. (1991). Algebraic nonlinear growth of the resistive kink instability, *Phys. Fluids B* **3**, pp. 3353-3356, `http://dx.doi.org/10.1063/1.859766`.

Biskamp, D. (1993), *Nonlinear Magnetohydrodynamics* (Cambridge University Press).

Biskamp, D. (2000). *Magnetic Reconnection in Plasmas* (Cambridge University Press), `http://books.google.com/books?id=YkzgcUuLC8oC&lpg=PR13&ots=Z3BtXVwmVE&dq=Biskamp`.

Blanchard, P., Alberti, S., Coda, S., Weisen, H., Nikkola, P. and Klimanov, I. (2002). High-field side measurements of non-thermal electron cyclotron emission on TCV plasmas with ECH and ECCD, *Plasma Phys. Control. Fusion* **44**, p. 2231, `http://dx.doi.org/10.1088/0741-3335/44/10/310`.

Bohm, D., Burhop, E. H. S. and Massey, H. S. W. (1949). The use of probes for plasma exploration in strong magnetic fields, *The Characteristics of Electrical Discharges in Magnetic Field* eds. A. Guthrie and R. K. Wakerling, (New York, McGraw-Hill).

Bonoli, P. T. and Englade, R. C. (1986). Simulation model for lower-hybrid current drive, *Phys. Fluids* **29**, p. 2937, `http://dx.doi.org/10.1063/1.865494`.

Bonoli, P. T., Porkolab, M., Takase, Y. and Knowlton, S. F. (1988). Numerical modeling of lower-hybrid RF heating and current drive experiments in the Alcator-C tokamak, *Nucl. Fusion* **28**, p. 991, `http://dx.doi.org/10.1088/0029-5515/28/6/004`.

Boozer, A. H., Baldwin, D. E., Horton, W., Glasser, A. H., Krommes, J. A., Neilson, G. H., Shaing, K-C., Sadowski, W. L. and Weitzner, H. (1990). Altnerate transport, *Phys. Fluids B* **2**, 12, pp. 2870-2879, `http://dx.doi.org/10.1063/1.859357`.

Boris, J. P. and Book, D. L. (1973). *J. Comp. Phys.* **11**, pp. 38-69, `http://dx.doi.org/10.1016/0021-9991(73)90147-2`.

Boscary, J., Boswirth, B., Greuner, H., Grigull, P., Missirlian, M. and Plankensteiner, A. (2007). Fabrication and testing of W7-X pre-series target elements, *Phys. Scripta T* **128**, pp. 195-199, `http://dx.doi.org/10.1088/0031-8949/2007/T128/037`.

Bosch, H.-S. and Hale, G. M. (1992). Improved formulas for fusion cross-sections and thermal reactivities, *Nucl. Fusion* **32**, p. 611, http://dx.doi.org/10.1088/0029-5515/32/4/I07.

Bourdelle, C. (2005). Turbulent particle transport in magnetized fusion plasma, *Plasma Phys. Control Fusion* **47**, pp. A317-A326, http://dx.doi.org/10.1088/0741-3335/47/5A/023.

Bozhenkov, S. A., Fuchert, G., Niemann, H., Beurskens, M., Feng, Y., Ford, O. P., Geiger, J., Hirsch, M., Höfel, U., Jakubowski, M. W., Knauer, J., Kornejew, P., Langenberg, A., Laqua, H. P., Maassberg, H., Marushchenko, N. B., Moseev, D., Pablant, N., Pasch, E., Rahbarnia, K., Strange, T., Svensson, J., Trimino Mora,H. and Valson, P. (2016), Power balance analysis of Wendelstein 7-X plasmas using profile diagnostics, *43rd EPS Conf. Plasma Physics* (European Physical Society, 2016).

Braginskii, S. I. (1965). *Reviews of Plasma Physics* **1**, p. 205, [Authorized translation from the Russian by Herbert Lashinsky, University of Maryland, USA.] Edited by M. A. Leontovich (Consultants Bureau, New York).

Brambilla, M. (1998). Electron Landau damping of ion Bernstein waves in tokamak plasmas, *Nucl. Fusion* **38**, p. 1805, http://dx.doi.org/10.1088/0029-5515/38/12/306.

Brambilla, M. (1999). Numerical simulation of ion cyclotron waves in tokamak plasmas, *Plasma Phys. Control. Fusion* **41**, p. 1, http://dx.doi.org/10.1088/0741-3335/41/1/002.

Bravenec, R. V., Gentle, K. W., Richards, B., Ross, D. W., Sing, D. C., Wootton, A. J., Brower, D. L., Luhmann, N. C., Peebles, W. A., Yu, C. X., Crowley, T. P., Heard, J. W., Hickok, R. L., Schoch, P. M. and Yang, X. Z. (1992). Core turbulence and transport studies on the Texas Experimental Tokamak, *Phys. Fluids B* **4**, p. 2127, http://dx.doi.org/10.1063/1.860019.

Breizman, B. N. and Sharapov, S. E. (1995). Energetic particle drive for toroidicity-induced Alfvén eigenmodes and kinetic toroidicity-induced Alfvén eigenmodes in a low-shear tokamak, *Plasma Phys. Control. Fusion* **37**, p. 1057, http://dx.doi.org/10.1088/0741-3335/37/10/001.

Brizard, A. J. and Hahm, T. S. (2007). Foundations of nonlinear gyrokinetic theory, *Rev. Mod. Phys.* **79**, pp. 421-468, http://dx.doi.org/10.1103/RevModPhys.79.421.

Brower, D. L., Peebles, W. A., Kim, S. K., Luhmann, N. C., Tang, W. M. and Phillips, P. E. (1987). Observation of a high-density ion mode in tokamak microturbulence, *Phys. Rev. Lett.* **59**, pp. 48-51, http://dx.doi.org/10.1103/PhysRevLett.59.48.

Budker, D., Kimball, D. F. and DeMille, D. P. (2008). *Atomic Physics: An Exploration through Problems and Solutions* 2nd Ed., 512 pp., ISBN: 9780199532421.

Budny, R. V. (2002). Fusion alpha parameters in tokamaks with high DT fusion rates, *Nucl. Fusion* **42**, pp. 1383-1393, PII: S0029-5515(02)55646-5, http://dx.doi.org/10.1088/0029-5515/42/12/305.

Buneman, O. (1958). *Phys. Rev. Lett.* **1**, 8; *Phys. Rev.* **115**, p. 50, http://dx.doi.org/10.1103/PhysRevLett.1.8, Erratum *Phys. Rev. Lett.* **1**, p. 119 (1958).

Burgess, D. (1995). Chap. 5, Collisionless Shocks, *Introduction to Space Physics* eds. M. G. Kivelson and C. T. Russell, (Cambridge, MA) p. 129.

Burrell, K. H. (1997). Effects of $\boldsymbol{E} \times \boldsymbol{B}$ velocity shear and magnetic shear on turbulence and transport in magnetic confinement devices *Phys. Plasmas* **4**, p. 1499, http://dx.doi.org/10.1063/1.872367.

Buttery, R. J., Laorte, A., Lomas, P. J., Mantica, P., McDonald, D. C., Saarelma, S., Sartori, R., Gaibene, G. and JET-EFDA contributors, Progress towards an integrated solution for the ITER baseline scenario based on high current and highly shaped plasma operation at JET, *Plasma Phys. Control. Fusion* **49**, p. A59 http://dx.doi.org/10.1088/0741-3335/49/5A/S05.

Cafaro, E., Grasso, D., Pegoraro, F., Porcelli, F. and Saluzzi, A. (1998). Invariants and geometric structures in nonlinear Hamiltonian magnetic reconnection, *Phys. Rev. Lett.* **80**, p. 4430, http://dx.doi.org/10.1103/PhysRevLett.80.4430.

Callen, J. D. (1977). Drift-wave turbulence effects on magnetic structure and plasma transport in tokamaks, *Phys. Rev. Lett.* **39**, pp. 1540-1543, http://dx.doi.org/10.1103/PhysRevLett.39.1540.

Candy, J. (2009). A unified method for operator evaluation in local Grad-Shafranov plasma equilibria, *Plasma Phys. Control. Fusion* **51**, p. 105009, http://dx.doi.org/10.1088/0741-3335/51/10/105009.

Carbone, V., Einaudi, G. and Veltri, P. (1987). Effects of turbulence development in solar surges, *Sol. Phys.* **111**, pp. 31-41, \http://dx.doi.org/10.1007/978-94-009-3999-8_4.

Carreras, B. A., Garcia, L. and Diamond, P. H. (1987). Theory of resistive pressure-gradient-driven turbulence, *Phys. Fluids* **30**, p. 1388, http://dx.doi.org/10.1063/1.866518.

Carter, T. A. and Maggs, J. E. (2009). Modifications of turbulence and turbulent transport associated with a bias-induced confinement transition in the large plasma device, *Phys. Plasmas* , pp. 012304-012304.10, http://dx.doi.org/10.1063/1.3059410.

Casson, F. J., Peeters, A. G., Angioni, C., Camenen, Y., Hornsby, W. A., Snodin, A. P. and Szepesi, G. (2010). Gyrokinetic simulations including the centrifugal force in a rotating tokamak plasma, *Phys. Plasmas* **17**, p. 102305, http://dx.doi.org/10.1063/1.3491110.

Casson, F. J., Angioni, C., Belli, E. A., Bilato, R., Mantica, P., Odstrcil, T., Pütterich, T., Valisa, M., Garzotti, L., Giroud, C. (2014). Theoretical description of heavy impurity transport and its application to the modeling of tungsten in JET and ASDEX upgrade, *Plasma Phys. Control. Fusion* **57**, p. 014031, http://dx.doi.org/10.1088/0741-3335/57/1/014031.

Casson, F. J., Angioni, C., Belli, E. A., Bilato, R., Mantica, P., Odstrcil, T., Pütterich, T., Valisa, M., Garzotti, L., Giroud, C., Hobirk, J., Maggi, C. F., Mylnar, J., Reinke, M. L., *et al.* (2015). Theoretical description of heavy impurity transport and its application to the modeling of tungsten in JET and ASDEX Upgrade, *Plasma Phys. Control. Fusion* **57**, p. 014031 http://dx.doi.org/10.1088/0741-3335/57/1/014031.

Castaldo, C., Lazzaro, E., Lontano, M. and Sergeev, A. M. (1997). Spectral broadening of lower-hybrid waves due to ponderomotive effects, *Phys. Lett. A* **230**, p. 336, http://dx.doi.org/10.1016/S0375-9601(97)00268-5.

Catto, P. J., Bernstein, I. B. and Tessarotto, M. (1987). Ion transport in toroidally rotating tokamak plasmas, *Phys. Fluids* **30**, p. 2784, http://dx.doi.org/10.1063/1.866045.

Cesario, R., Cardinali, A., Castaldo, C., Paoletti, F. and Mazon, D. (2004). Modeling of a lower-hybrid current drive by including spectral broadening induced by parametric instability in tokamak plasmas, *Phys. Rev. Lett.* **92**, p. 175002, http://dx.doi.org/10.1103/PhysRevLett.92.175002.

Cesario, R., Cardinali, A., Castaldo, C., Paoletti, F., Fundamenski, F., Hacquin, S., and the JET-EFDA workprogramme contributors. (2006). Spectral broadening of lower-hybrid waves produced by parametric instability in current drive experiments of tokamak plasmas, *Nucl. Fusion 46*, p. 462, `http://dx.doi.org/10.1088/0029-5515/46/4/007`.

Cesario, R., Amicucci, L., Castaldo, C., Kempenaars, M., Jachmich, S., Mailloux, J., Tudisco, O., Galli, A., Krivska, A., and JET-EFDA contributors. (2011). Plasma edge density and lower-hybrid current drive in JET (Joint European Torus), *Plasma Phys. Control. Fusion* **53**, p. 085011, `http://dx.doi.org/10.1088/0741-3335/53/8/085011`.

Cesario, R., Amicucci, L., Cardinali, A., Castaldo, C., Marinucci, M., Panaccione, L., Santini, F., Tudisco, O., Apicella, M. L., Calabro, G., Cianfarani, C., Frigione, D., Galli, A., Mazzitelli, G., Mazzotta, C., Pericoli, V., Schettini, G., Tuccillo, A. A. and the FTU Team. (2013). Corrigendum: Current drive at plasma densities required for thermonuclear reactors, *Nat. Commun.* **4**, Art. No. 2282, `http://dx.doi.org/10.1038/ncomms3282`.

Chandrasekhar, S. (1961). *Hydrodynamic and Hydromagnetic Stability* (Oxford University Press), pp. 481-498, ISBN:0-486-64071-X.

Chen, F. F. (1965a). Resistive over-stabilities and anomalous diffusion, *Phys. Fluids* **8**, p. 912, `http://dx.doi.org/10.1063/1.1761335`.

Chen, F. F. (1965b). Excitation of drift instabilities in thermionic plasmas, *J. Nucl. Energy* Pt. C, p. 399, `http://dx.doi.org/10.1088/0368-3281/7/4/305`.

Chen, F. F. and Mosher, D. (1967). Shear stabilization of a potassium plasma, *Phys. Rev. Lett.* **18**, p. 639, doi:10.1103/PhysRevLett.18.639.

Chen, L. and Hasegawa, A. (1974). Plasma heating by spatial resonance of Alfvén waves. *Phys. Fluids* **17**, p. 1399, `http://dx.doi.org/10.1063/1.1694904`.

Chen, L. and Cheng, C. Z. (1980). Drift-wave eigenmodes in toroidal plasmas, *Phys. Fluids* **23**, p. 2242, `http://dx.doi.org/10.1063/1.862907`.

Chen, X. L. and Morrison, P. J. (1990). Resistive tearing instability with equilibrium shear flow, *Phys. Fluids B* **2**, pp. 495-507, `http://dx.doi.org/10.1063/1.859339`.

Chen, X. L. and Morrison, P. J. (1990). The effect of viscosity on the resistive tearing mode with the presence of shear flow, *Phys. Fluids B* **2**, pp. 2575-2580, `http://dx.doi.org/10.1063/1.859382`.

Chen, X. L. and Morrison, P. J. (1991). A sufficient condition for the ideal instability of shear flow with parallel magnetic field, *Phys. Fluids B* **3**, pp. 863-865, `http://dx.doi.org/10.1063/1.859841`.

Chen, J., Sen, A. K. and Migliuolo, S. (1992). Production and observation of the ion-temperature-gradient instability, *Phys Fluids B* **4**, pp. 512-520, `http://dx.doi.org/10.1063/1.860249`.

Chew, G. F., Goldberger, M. L. and Low. F. E. (1956). *Proceedings of the Royal Society of London. Series A*, Mathematical and Physical Sciences, **236**, 1204, pp. 112-118, `http://www.jstor.org/stable/99870`.

Chirikov, B. V. (1983). Chaotic dynamics in Hamiltonian systems with divided phase space, *Lecture Notes in Physics*, ISBN:978-3-540-39594-2, `\http://dx.doi.org/10.1007/3-540-12276-1_2`.

Chmyrev, V. M., Bilichenko, S. V., Pokhotelov, O. A., Marchenko, V. A., Lazarev, V. I.,

Streltsov, A. V. and Stenflo, L. (1988). Alfvén vortices and related phenomena in the ionosphere and the magnetosphere, *Physica Scripta* **38**, p. 841, http://dx.doi.org/10.1088/0031-8949/38/6/016.

Cho, T., Kohagura, J., Hirata, M., Numakura, T., Higaki, H., Hojo, H., Ichimura, M., Ishii, K., Islam, K. M., Itakura, A., Katanuma, I., Nakashima, Y., Saito, T., Tatematsu, Y., Yoshikawa, M., Takemura, Y., Kojima, A., Kobayashi, T., Yamaguchi, Y., Miyata, Y., Yokoyama, N., Tomii, Y., Miyake, Y., Kiminami, S., Shimizu, K., Kubota, Y., Saimaru, H., Higashizono, Y., Mase, A., Yasaka, Y., Ogura, K., Sakamoto, K., Yoshida, M., Pastukhov, V. P., Imai, T., Miyoshi, S. and Gamma 10 Group. (2005). Progress in potential formation and findings in the associated radially sheared electric-field effects on suppressing intermittent turbulent vortex-like fluctuations and reducing transverse losses, *Nucl. Fusion* **45**, pp. 1650-1657, http://dx.doi.org/10.1088/0029-5515/45/12/021.

Cho, T., Kohagura, J., Numakura, T., Hirata, M., Higaki, H., Hojo, H., Ichimura, M., Watanabe, O., Yoshikawa, M., Kojima, A., Miyake, Y., Miyata, Y., Shimizu, K., Tomii, Y., Yoshida, M., Sakamoto, K., Imai, T., Pastukhov, V. P., Miyoshi, S. and G. Group. (2006). Observation and control of transverse energy-transport barrier due to the formation of an energetic-electron layer with sheared $\boldsymbol{E} \times \boldsymbol{B}$-Flow, *Phys. Rev. Lett.* **97**, p. 055001, http://dx.doi.org/10.1103/PhysRevLett.97.055001; Errata *Phys. Rev. Lett.* **104**, p. 169904 (2010); *Phys. Rev. Lett.* **104**, p. 259902 (2010).

Cho, T., Pastukhov, V. P., Horton, W., Numakura, T., Hirata, M., Kohagura, J., Chudin, N. V. and Pratt, J. (2008). Active control of internal transport barrier formation due to off-axis electron-cyclotron heating in GAMMA-10 experiments, *Phys. Plasmas* **15**, p. 056120, http://dx.doi.org/10.1063/1.2906262.

Cho, T. and Hirata, M. (2016). Extended possibility of an active control of co-axially nested shear plasma formation due to electron cyclotron heatings, *J. Plasma Phys.* **82**, p. 595820501, http://dx.doi.org/10.1017/S0022377816000817.

Choi, D.-I. and Horton, W. (1980). Weakly localized two-dimensional drift modes, *Phys. Fluids* **23**, pp. 356-365, http://dx.doi.org/10.1063/1.862980.

Chu, M. S., Greene, J. M., Lao, L. L., Miller, R. L., Bondeson, A., Sauter, O., Rice, B. W., Strait, E. J., Taylor, T. S. and Turnbull, A. D. (1996). Resistive interchange modes in negative central shear tokamaks with peaked pressure profiles. *Phys. Rev. Lett.* **77**, p. 2710.

Cipiti, B. and Kulcinski, G. (2005). Helium and deuterium implantation in tungsten at elevated temperatures, *J. Nucl. Mater.* **347** (3), 298-306,http://dx.doi.org/10.1016/j.jnucmat.2005.08.009.

Citrin, J., Jenko, F., Mantica, P., Told, D., Bourdelle, C., Dumont, R., Garcia, J., Haverkort, J. W., Hogeweij, G. M. D., Johnson, T., Pueschel, M. J. and JET-EFDA contributors. (2014). Ion temperature profile stiffness: non-linear gyrokinetic simulations and comparison with experiment, *Nucl. Fusion* **54**, p. 023008, http://dx.doi.org/10.1088/0029-5515/54/2/023008.

Citrin, J., Garcia, J., Görler, Jenko, F., Mantica, P., Told, D., Bourdelle, C., Hatch, D. R., Hogeweij, G. M. D., Johnson, T., Pueschel, M. J., Schneider, M. and See the appendix of Romanelli F, *et al.* 2012 Proc. 24th IAEA Fusion Energy Conf. (San Diego, CA) (2015). Electromagnetic stabilization of tokamak microturbulence in a high-β regime *Plasma Phys. Control. Fusion* **57**, p. 014032, http://dx.doi.org/10.1088/0741-3335/57/1/014032.

Colas, L., Zou, X. L., Paume, M., Chareau, J., Guiziou, L., Hoang, G. T., Michelot, Y. and Grésillon, D. (1998). Internal magnetic fluctuations and electron heat transport in the Tora Supra Tokamak: Observation by cross-polarization scattering, *Nuclear Fusion* **38**, pp. 903-918, `http://dx.doi.org/10.1088/0029-5515/38/6/308`.

Connor, J. W., Hastie, R. J. and Taylor, J. B. (1978). Shear, periodicity, and plasma ballooning modes, *Phys. Rev. Lett.* **40**, p. 396, `http://dx.doi.org/10.1103/PhysRevLett.40.396`.

Connor, J. W., Hastie, R. J. and Taylor, J. B. (1979). High mode number stability of an axisymmetric toroidal plasma, *Proc. R. Soc. Lond. A* **365**, p. 1, `http://dx.doi.org/10.1098/rspa.1979.0001`.

Connor, J. W., Hastie, R. J. and Taylor, J. B. (1991a). Resonant magnetohydrodynamic modes with toroidal coupling. Part I: Tearing modes, *Phys. Fluids B* **3**, p. 1532, `http://dx.doi.org/10.1063/1.859724`.

Connor, J. W., Hastie, R. J. and Taylor, J. B. (1991b). Resonant magnetohydrodynamic modes with toroidal coupling. Part II: Ballooning-twisting modes, *Phys. Fluids B* **3**, p. 1539, `http://dx.doi.org/10.1063/1.859993`.

Connor, J. W. and Wilson, H. R. (2000). A review of theories of the L-H transition, *Plasma Phys. Control. Fusion* **42**, p. R1, `http://dx.doi.org/10.1088/0741-3335/42/1/201`.

Connor, J. W. (1993). Pressure gradient turbulent transport and collisionless reconnection, *Plasma Phys. Control. Fusion* **35**, pp. 757-763, `http://dx.doi.org/10.1088/0741-3335/35/6/009`.

Cooper, W. A., Bateman, G., Nelson, D. B. and Kammash,T. (1980). Beam-induced tensor pressure tokamak equilibria, *Nucl. Fusion* **20**, p. 985, `http://dx.doi.org/10.1088/0029-5515/20/8/005`.

Coppi, B., Laval, G., Pellat, R. and Rosenbluth, M. N. (1968). Collisionless microinstabilities in configurations with periodic magnetic curvature, *Plasma Phys.* **10**, pp. 1-22, `http://dx.doi.org/10.1088/0032-1028/10/1/302`.

Coppi, B., Rosenbluth, M. N. and Yoshikawa, S. (1968). Localized (ballooning) modes in multipole configurations, *Phys. Rev. Lett.* **20**, pp. 190-192, `http://dx.doi.org/10.1103/PhysRevLett.20.190`.

Coppi, B. and Spight, C. (1978). Ion-mixing mode and model for density rise in confined plasmas, *Phys. Rev. Lett.* **41**, p. 551, `http://dx.doi.org/10.1103/PhysRevLett.41.551`.

Coppi, B. and Mazzucato, E. (1979). Transport of electron thermal energy in high temperature plasmas, *Phys. Letts. A* **71**, p. 337, `http://dx.doi.org/10.1016/0375-9601(79)90071-9`

Coppi, B. and Porcelli, F. (1986). Theoretical model of fishbone oscillations in magnetically-confined plasmas, *Phys. Rev. Lett.* **57**, p. 2272-2275, `http://dx.doi.org/10.1103/PhysRevLett.57.2272`.

Coppi, B. and Pegraro, F. (1991). Symmetries and global transport equations, *Phys. Fluids B* **3**, p. 2582, `http://dx.doi.org/10.1063/1.859970`.

Coppi, B. and Detragiache, P. (1992). Magnetic topology transitions in collisionless plasmasm *Phys. Lett. A* **168**, pp. 59-64, `http://dx.doi.org/10.1016/0375-9601(92)90329-K`.

Coppi, B., *et al.* (1999). MIT (RLE) Report PTP 99/06.

Coppi, B. and Zhou, T. (2012). Interpretation of the I-Regime and transport associated with relevant heavy particle modes, *Phys. Plasmas* **19**, p. 012302, http://dx.doi.org/10.1063/1.3671944.

Cordey, J. G., Balet, B., Bartlett, D. V., Budnya, R. V., Christiansen, J. P., Conway, G. D., Eriksson, L.-G., Fishpool, G. M., Gowers, C. W., de Haasb, J. C. M., Harbour, P. J., Horton, L. D., Howman, A. C., Jacquinot, J., Kerner, W., Lowry, C. G., Monk, R. D., Nielsen, P., Righi, E., Rimini, F. G., Saibene, G., Sartori, R., Schunke, B., Sips, A. C. C., Smith, R. J., Stamp, M. F., Startc, D. F. H., Thomsen, K., Tubbing, B. J. D. and von Hellermann, M. G. (1999). Plasma confinement in JET H-mode plasmas with -, D, DT and T isotopes, *Nuclear Fusion* **39**, p. 301, http://iopscience.iop.org/0029-5515/39/3/301.

Cowley, S. C. and Artun, M. (1997). Explosive instabilities and detonation in magnetohydrodynamics, *Phys. Reports* **283**, pp. 185-211, http://dx.doi.org/10.1016/S0370-1573(96)00060-9.

Cowley, S. C. and Sudan, R. N. (1997). Turbulence and intermittency in plasmas, *Phys. Reports* **282**, 1-4, p. 3-4, ISSN:0370-1573, http://dx.doi.org/10.1016/S0370-1573(97)88184-7.

Cowley, S. C., Wilson, H., Hurricane, O. and Fong, B (2003). Explosive instabilities: from solar flares to edge localized modes in tokamaks, *Plasma Phys. Control. Fusion* **45**, p. A31, http://dx.doi.org/10.1088/0741-3335/45/12A/003.

Crabtree, C., Horton, W., Wong, H. V. and Van Dam, J. W. (2003). Bounce-averaged stability of compressional modes in geotail flux tubes, *J. Geophys. Res.* **108**, p. 10.1029,http://dx.doi.org/10.1029/2002JA009555.

Crabtree, C., Rudakov, L., Ganguli, G., Mithaiwala, M., Galinsky, V. and Shevchenko, V. (2012). Weak turbulence in the magnetosphere: Formation of whistler wave cavity by nonlinear scattering, *Phys. Plasmas* **19**, 12 pp. http://dx.doi.org/10.1063/1.3692092.

Cremaschini and Tessarotto. (2011). Kinetic description of rotating Tokamak plasmas with anisotropic temperatures in the collisionless regime, *Phys. Plasmas* **18**, p. 112502, http://dx.doi.org/10.1063/1.3656978.

Crew, G. B. and Ramos, J. J. (1983). Asymptotic theory of the internal kink mode in current-carrying toroidal plasmas, *Phys. Fluids* **26**, pp. 2621-2634, http://dx.doi.org/10.1063/1.864454.

D'Angelo, N. and Motley, R. W. (1963). Low-frequency oscillations in a potassium plasma, *Phys. Fluids* **6**, p. 422, http://dx.doi.org/10.1063/1.1706749.

da Silva, D. C., Caldas, I. L. and Viana, R. L. (2001). Field-line diffusion and loss in a tokamak with an ergodic magnetic limiter, *Phys. Plasmas* **8**, p. 2855, http://dx.doi.org/10.1063/1.1371769.

Dahlburg, R. B., Zang, T. A., Montgomery, D., and Hussaini, M. Y. (1983). *Proc. Nat. Acad. Sci USA* **80**, p. 5798.

Dahlburg, R. B., Zang, T. A. and Montgomery, D. (1986). Unstable transition properties of the driven magnetohydrodynamic sheet pinch, *J. Fluid Mech.* **169**, p. 71-107, http://dx.doi.org/10.1017/S002211208600054X.

Dahlburg, R. B. and Karpen, J. T. (1994). Transition to turbulence in solar surges, *Astrophys. J.* **434**, pp. 766-772 http://dx.doi.org/10.1086/174779.

Dahlburg, R. B. and Karpen, J. T. (1985). A triple current sheet model for adjoining coronal helmet streamers *J. Geophys. Res.* **100**, p. 23489-23498, `http://dx.doi.org/10.1029/95JA02496`.

Dahlburg, R. B., Boncinelli, P. and Einaudi, G. (1997). The evolution of plane current-vortex sheets, *Phys. Plasmas* **4**, p. **4**, 1213, `http://dx.doi.org/10.1063/1.872301`.

Dahlburg, R. B., Boncinelli, P. and Einaudi, G. (1998). The evolution of a plane jet in a neutral sheet, *Phys. Plasmas* **5**, p. 79 `http://dx.doi.org/10.1063/1.872677`.

Dahlburg, R. B. and Einaudi, G. (2000). The compressible plane current-vortex sheet, *Phys. Plasmas* **7**, p. 1356, `http://dx.doi.org/10.1063/1.873952`.

Dahlburg, R. B., Keppens, R. and Einaudi, G. (2001). The compressible evolution of the super-Alfvénic magnetized wake, *Phys. Plasmas* **8**, p. 1697, `http://dx.doi.org/10.1063/1.1339232`.

Dahlburg, R. B., Einaudi, G., LaRosa, T. N. and Shore, S. N. (2002). Modeling the galactic center nonthermal filaments as magnetized wakes, *Astrophys. J.* **568**, p. 220, `http://dx.doi.org/10.1086/338842`.

Dahlburg, R. B., Horton, W., Rowan, W. L., Correa, C. and Perez, J. C. (2009). Evolution of the bounded magnetized jet and comparison with Helimak experiments, *Phys. Plasmas* **16**, p. 072109, `http://dx.doi.org/10.1063/1.3166598`.

Decker, J., Peysson, Y., Brizard, A. J. and Duthoit, F.-X. (2010). Orbit-averaged guiding-center Fokker-Planck operator for numerical applications, *Phys. Plasmas* **17**, p. 112513, `http://dx.doi.org/10.1063/1.3519514`.

Decker, J., Peysson, Y., Hillairet, J., Artaud, J.-F., Basiuk, V., Becoulet, A., Ekedahl, A., Goniche, M., Hoang, G. T., Imbeaux, F., Ram, A. K. and Schneider, M. (2011). Calculations of lower-hybrid current drive in ITER, *Nucl. Fusion* **51**, p. 073025 `http://dx.doi.org/10.1088/0029-5515/51/7/073025`.

Decker, J., Peysson, Y., Artaud, J.-F., Nilsson, E., Ekedahl, A., Goniche, M., Hillairet, J. and Mazon, D. (2014). Damping of lower-hybrid waves in large spectral gap configurations, *Phys. Plasmas* **21**, p. 092504, `http://dx.doi.org/10.1063/1.4894749`.

Del-Castillo-Negrete, D. (2000). Chaotic transport in zonal flows in analogous geophysical and plasma systems, *Phys. Plasmas* **7**, p. 1702, `http://dx.doi.org/10.1063/1.873988`.

Dendy, R. O., Hastie, R. J., McClements, K. G. and Martin, T. J. (1995). A model for ideal $m = 1$ internal kink stabilization by minority ion cyclotron resonant heating, *Phys. Plasmas* **2**, p. 1623, `http://dx.doi.org/10.1063/1.871457`.

Detman, T. and Vassiliadis, D. (1997). Review of techniques for magnetic storm forecasting, *Geophysical Monograph, American Geophysical Union* **98**, pp. 253-266, `http://dx.doi.org/10.1029/GM098p0253`.

Dewar, R. L. and Glasser, A. H. (1983). *Phys. Fluids* **26**, p. 3038.

Diamond, P. H. and Kim, Y-B. (1991). Theory of mean poloidal flow generation by turbulence, *Phys. Fluids B* **3**, p. 1626, `http://dx.doi.org/10.1063/1.859681`.

Diamond, P. H., Lebedev, V. B., Newman, D. E., Carreras, B. A., Hahm, T. S., Tang, W. M., Rewoldt, G. and Avinash, K. (1997). Dynamics of transition to enhanced confinement in reversed magnetic shear discharges, *Phys. Rev. Lett.* **78**, pp. 1472-1475, `http://dx.doi.org/10.1103/PhysRevLett.78.1472`.

Diamond, P. H., Itoh, S-I., K. Itoh and Hahm, T. S. (2005). Zonal flows in plasma: A review, *Plasma Phys. Control. Fusion* **47**, pp. R35-R161, `http://dx.doi.org/10.1088/0741-3335/47/5/R01`.

Dietz, K. J. and The JET Team. (1990). Effect of beryllium on plasma performance in JET, *Plasma Phys. Control. Fusion* **32**, pp. 837-852, http://dx.doi.org/10.1088/0741-3335/32/11/002.

Dimits, A. M., Drake, J. F., Guzdar, P. N. and Hassam, A. B. (1991). Ion-temperature-gradient-driven turbulence and transport in a sheared magnetic field, *Phys. Fluids B* **3**, pp. 620-626, http://dx.doi.org/10.1063/1.859861.

Dimits, A. M., Bateman, G., Beer, M. A., Cohen, B. I., Dorland, W., Hammett, G. W., Kim, C., Kinsey, J. E., Kotschenreuther, M., Kritz, A. H., Lao, L. L. Andrekas, J. M., Nevins, W. M., Parker, S. E., Redd, A. J., Shumaker, D. E., Sydora, R. and Weiland, J. (2000). Comparisons and physics basis of tokamak transport models and turbulence simulations, *Phys. Plasmas* **7**, pp. 969-983, http://dx.doi.org/10.1063/1.873896.

Dimov, G. I. (2005). The ambipolar trap, *Physics-Uspekhi* **48**, pp. 1129-1149, http://dx.doi.org/10.1070/PU2005v048n11ABE005804.

Ding, B. J., Qin, Y. L., Li, W. K., Li, M. H., Kong, E. H., Zhang, L., Ekedahl, A., Peysson, Y., Decker, J., Wang, M., Xu, H. D., Hu, H. C., Xu, G. S., Shan, J. F., Liu, F. K., Zhao, Y. P., Wan, B. N., Li, J. G. and EAST Group. (2011). *Phys. Plasma* **18**, p. 082510, http://dx.doi.org/10.1063/1.3624778.

Doerner, R., Baldwin, M. and Stangeby, P. (2011). An equilibrium model for tungsten fuzz in an eroding plasma environment, *Nucl. Fusion* **51** (4), p. 043001, http://dx.doi.org/10.1088/0029-5515/51/4/043001.

Dong, J.-Q., Horton, W. and Kim, J-Y. (1992).Toroidal kinetic η_i-mode study in high-temperature plasmas, *Phys. Plasmas* **4**, pp. 1867-1876, http://dx.doi.org/10.1063/1.860040.

Dong, J.-Q. and Horton, W. (1993). Kinetic quasi-toroidal ion temperature gradient instability in the presence of sheared flows, *Phys. Fluids B* **5**, p. 1581, http://dx.doi.org/10.1063/1.860898.

Dong, J.-Q., Chen, L. and Zonca, F. (1999). Study of kinetic shear Alfvén modes driven by ion temperature gradient in tokamak plasmas, *Nucl. Fusion* **39**, p. 1041, http://dx.doi.org/10.1088/0029-5515/39/8/308.

Dorland, W. and Hammett, G. W. (1993). Gyrofluid turbulence models with kinetic effects, *Phys. Fluids B* **5**, pp. 812-835, http://dx.doi.org/10.1063/1.860934.

Doyle, E. J., Houlberg, W. A., Kamada, Y., Mukhovatov, V., Osborne, Polevoi, A., Bateman, G., Connor, J. W. Cordey, J. G., Fujita, T., Garbet, X., Hahm, T. S., Horton, L. D., Hubbard, A. E., Imbeaux, F., Jenko, F., Kinsey, J. E., Kishimoto, Y., Li, J., Luce, T. C., Martin, Y., Ossipenko, M., Parail, V., Peeters, A., Rhodes, T. L., Rice, J. E., Roach, C. M., Rozhansky, V., Ryter, F., Saibene, G., Sartori, R., Sips, A. C. C., Snipes, J. A., Sugihara, M., Synakowski, E. J., Takenaga, H., Takizuka, T., Thomsen, K., Wade, M. R., Wilson, H. R., ITPA Transport Physics Topical Group, ITPA Confinement Database and Modeling Topical Group, and ITPA Pedestal and Edge Topical Group. (2007). Chapter 2: Plasma confinement and transport, *Nucl. Fusion* **47**, S18-S127, http://dx.doi.org/10.1088/0029-5515/47/6/S02.

Drake, J. F. and Lee, Y. C. (1977). Kinetic theory of tearing instabilities, *Phys. Fluids* **20**, pp. 1341-1353, http://dx.doi.org/10.1063/1.862017.

Drake, J. F., Finn, J. M., Guzdar. P., Shapiro, V., Shevchenko, V., Waelbroeck, F. L., Hassam, A. B., Liu, C. S. and Sagdeev, R. Z. (1992). Peeling of convection cells and the generation of sheared flow, *Phys. Fluids B* **4**, p. 488, http://dx.doi.org/10.1063/1.860246.

Drazin, P. G. and Reid, W. H. (1981). *Hydrodynamic Stability* (Cambridge University Press).

Duane, B. H. (1972). Fusion Cross-Section Theory, Rept. BNWL-1685 (Brookhaven National Laboratory).

Dubuit, N., Garbet, X., Parisot, T., Guirlet, R. and Bourdelle, C. (2007). Fluid simulations of turbulent impurity transport, *Phys. Plasmas* **14**, p. 042301, http://dx.doi.org/10.1063/1.2710461.

Dudson, B. D., Umansky, M. V., Xu, X. Q., Snyder, P. B. and Wilson, H. R., (2009), BOUT++: A framework for parallel plasma fluid simulations, *Comput. Phys. Comm.* **180**, pp. 1467-1480, http://dx.doi.org/10.1016/j.cpc.2009.03.008.

Einaudi, G. and Rubini, F. (1986). Resistive instabilities in a flowing plasma: I. Inviscid case, *Phys. Fluids* **29**, p. 2563, http://dx.doi.org/10.1063/1.865548.

Einaudi, G. and Rubini, F. (1989). Resistive instabilities in a flowing plasma. II. Effects of viscosity, *Phys. Fluids B* **1**, p. 2224, http://dx.doi.org/10.1063/1.859038.

Einaudi, G., Boncinelli, P., Dahlburg, R. B. and Karpen, J. T. (1999). *J. Geophys. Res.* **104**, p. 521-534, http://dx.doi.org/10.1029/98JA02394.

Einaudi, G., Chibbaro, S., Dahlburg, R. B. and Velli, M. (2001). Plasmoid formation and acceleration in the solar streamer belt, *Astrophys. J.* **547**, p. 1167, http://dx.doi.org/10.1086/318400.

Ekedahl, A., K Rantamäki, K., Goniche, M., Mailloux, J., Petrzilka, V., Alper, B., Baranov, Y., Basiuk, V., Beaumont, P., Corrigan, G., Delpech, L., Erents, K., Granucci, G., Hawkes, N., Hobirk, J., Imbeaux, F., Joffrin, E., Kirov, K., Loarer, T., McDonald, D., Nave, M. F. F., Nunes, I., Ongena, J., Parail, V., Piccolo, F., Rachlew, E., Silva, C., Sirinelli, A., Stamp, M., Zastrow, K.-D. and JET-EFDA contributors, Effect of gas injection during LH wave coupling at ITER-relevant plasmawall distances in JET, *Plasma Phys. Control. Fusion* **51**, p. 044001, http://dx.doi.org/10.1088/0741-3335/51/4/044001.

Elmore, S., Allan, S. Y., Kirk, A., Fishpool, G., Harrison, J., Tamain, P., Kocan, M., Gaffka, R., Stephen, R., Bradley, J. W. and the MAST Team. (2012). Upstream and divertor ion temperature measurements on MAST by retarding field energy analyser, *Plasma Phys. Control. Fusion* **54**, p. 065001, http://dx.doi.org/10.1088/0741-3335/54/6/065001.

Elskens, Y. and Escande, D. (2003). *Microscopic Dynamics of Plasmas and Chaos* (Institute of Physics Publishing, Bristol and Philadelphia) ISBN:0-750306122.

Enomoto, N., Muto, S., Tanabe, T., Davis, J. and Haasz, A. (2009). Grazing-incidence electron microscopy of surface blisters in single and polycrystalline tungsten formed by h+, d+ and he+ irradiation, *J. Nucl. Mater.* **385**, pp. 606-614, http://dx.doi.org/10.1016/j.jnucmat.2009.01.298.

Equipe Tore Supra. (1989). *Plasma Phys. Control. Nucl. Fusion Res., 1988* **1**, pp. 9-26, (International Atomic Energy Agency, Vienna).

Erba, M., Aniel, T., Basiuk, V., Becoulet, A. and Litaudon, X. (1998). Validation of a new mixed Bohm/gyro-Bohm model for electron and ion heat transport against the ITER, Tore Supra and START database discharges, *Nucl. Fusion* **38**, p. 1013, http://dx.doi.org/10.1088/0029-5515/38/7/305.

Erba, M., Horton, W. and Ottaviani, M. (1999). Predictive tests of ITG based models of tokamak heat transport on ITER database discharges, *Nucl. Fusion* **39**, p. 495, http://dx.doi.org/10.1088/0029-5515/39/4/307.

Eriksson, A., Nordman, H., Strand, P., Weiland, J., Tala, T., Asp, E., Corrigan, G., Giroud, C., de Greef, M., Jenkins, I. and Knoops, H. C. M., Mantica, P., Rantamäki, K. M., de Vries, P. C., Zastrow, K-D. and JET EFDA Contributors (2007). *Plasma Phys. Control. Fusion* **49**, p. 1931, http://dx.doi.org/10.1088/0741-3335/49/11/012.

Ernst, D. R., Lang, J., Nevins, W. M., Hoffman, M., Chen, Y., Dorland, W. and Parker, S. (2009). Role of zonal flows in trapped electron mode turbulence through nonlinear gyrokinetic particle and continuum simulation, *Phys. Plasmas* **16**, p. 055906, http://dx.doi.org/10.1063/1.3116282.

Escande, D. F., Martin, P., Ortolani, S., Buffa, A., Franz, P., Marrelli, L., Martines, E., Spizzo, G., Cappello, S., Murari, A., Pasqualotto, R. and Zanca, P. (2000). Quasi-single-helicity reversed-field-pinch plasmas, *Phys. Rev. Lett.* **85**, pp. 1662-1665, http://dx.doi.org/10.1103/PhysRevLett.85.1662.

Evstatiev, E. G., Horton, W. and Morrison, P. J. (2003). Multi-wave model for plasma-wave interaction, *Phys. Plasmas* **10**, p. 4090, http://dx.doi.org/10.1063/1.1609989.

Fasoli, A., Borba, D., Bosia, G., Campbell, D. J., Dobbing, J. A., Gormezano, C., Jacquinot, J., Lavanchy, P., Lister, J. B., Marmillod, P., Moret, J.-M., Santagiustina, A. and Sharapov, S. E. (1995). Direct measurement of the damping of toroidicity-induced Alfvén eigenmodes, *Phys. Rev. Lett.* **75**, 4, pp. 645-648, http://dx.doi.org/10.1103/PhysRevLett.75.645.

Fedun, V., Verth, G., Jess, D. B. and Erdelyi, R. (2011). Frequency filtering of torsional Alfvén waves by chromospheric magnetic field, *Astrophys. J. Lett.* **740**, p. L46, http://dx.doi.org/10.1088/2041-8205/740/2/L46.

Field, A. R., McCone, L. and Conway, N. J., *et al.* (2009). Comparison of measured poloidal rotation in MAST spherical tokamak plasmas with neo-classical predictions, *Plasma Phys. Control. Fusion* **51**, p. 105002, http://dx.doi.org/10.1088/0741-3335/51/10/105002.

Fielding, S. J., Ashall, J. D., Carolan, P. G., Colton, A., Gates, D., Hugill, J., Morriss, A. W., Valovic, M. and the COMPASS-D and ECRH Teams. (1996). The H-mode in COMPASS-D, *Plasma Phys. Control. Fusion* **38**, p. 1091, http://dx.doi.org/10.1088/0741-3335/38/8/002.

Fiore, C. L., Ernst, D. R., Podpaly, Y. A., Mikkelsen, D., Howard, N. T., Lee, Jungpyo, Reinke, M. L., Rice, J. E., Hughes, J.W., Ma. Y., Rowan, W. L. and Bespamyatnov, I. O. (2012). Production of internal transport barriers via self-generated mean flows in Alcator C-Mod, *Phys. Plasmas* **19**, p. 056113, http://dx.doi.org/10.1063/1.3694668.

Fitzpatrick, R. and Hender, T. C. (1991). The interaction of resonant magnetic perturbations with rotating plasmas, *Phys. Fluids B* **3**, p. 644, http://dx.doi.org/10.1063/1.859863.

Fitzpatrick, R. (1993). Interaction of tearing modes with external structures in cylindrical geometry (plasma), *Nucl. Fusion* **33**, p. 1049, http://dx.doi.org/10.1088/0029-5515/33/7/I08.

Fitzpatrick, R. (1995). Helical temperature perturbations associated with tearing modes in tokamak plasmas, *Phys Plasmas* **2**, p. 825, `http://dx.doi.org/10.1063/1.871434`.

Fitzpatrick, R. and Aydemir, A. Y. (1996). Stabilization of the resistive shell mode in tokamaks, *Nucl. Fusion* **36**, p. 11, `http://dx.doi.org/10.1088/0029-5515/36/1/I02`.

Fitzpatrick, R., Waelbroeck, F. L. and Militello, F. (2006). The influence of the ion polarization current on magnetic island stability in a tokamak plasma, *Phys. Plasmas* **13**, p. 122507, `http://dx.doi.org/10.1063/1.2402914`.

Fredrickson, E. D., McGuire, Chang, Z., Janos, A., Bell, M., Budny, R. V., Bush, C. E., Manickam, J., Mynick, J., Nazikian, R. and Taylor, G. (1995). β-limit disruptions in the tokamak fusion test reactor, *Phys. Plasmas* **2**, p. 4216, `http://dx.doi.org/10.1063/1.871047`.

Fried, B. D. and Conte, S. S. (1961). *The Plasma Dispersion Function* (New York: Academic Press).

Friedberg, R. (1987). *Ideal Magnetohydrodynamics* (Cambridge University Press).

Friedberg, R. (2007). *Plasma Physics and Fusion Energy* (Cambridge University Press) ISBN:13-978-0-521-85107-7.

Frieman, E. and Rotenberg, M. (1960). *Rev. Mod. Phys.* **32**, pp. 898-902, `http://dx.doi.org/10.1103/RevModPhys.32.898`.

Frigione, D. (2000). *FTU, High-Density Performance* (International Atomic Energy Agency, Sorrento, Italy).

Frisch, U. (1996). *Turbulence* (Cambridge University Press, January 1996), ISBN:0-52145-713-0.

Frisch, U. (1995). *Turbulence: The Legacy of A. N. Kolmogorov* (Cambridge University Press) ISBN:0521-45103-5.

Fu, G. Y. and Van Dam, J. W. (1989). Excitation of the toroidicity-induced shear Alfvén eigenmode by fusion alpha particles in an ignited tokamak, *Phys. Fluids B* **1**, pp. 1949-1952, `http://dx.doi.org/10.1063/1.859057`.

Fu, G. Y., Ku, L. P., Cooper, W. A., *et al.* (2000). Magnetohydrodynamics stability of compact stellarators, *Phys. Plasmas* **7**, pp. 1809-2000, `http://dx.doi.org/10.1063/1.874002`.

Fu, X., Horton, W., Sen, A. K., Sokolov, V., Lin, Z. and Xiao, Y. (2012). Validation of electron temperature gradient turbulence in the Columbia Linear Machine, *Phys. Plasmas* **19**, p. 032303, `http://dx.doi.org/10.1063/1.3686148`.

Fu, X. R., Horton, W., Bespamyatnov, I. O., Rowan, W. L., Benkadda, S, Fiore, C. L., Futatani, S. and Liao, K. T. (2013). Turbulent impurity transport modeling for Alcator C-Mod, *J. Plasma Phys.* **79**, pp. 837-846, `http://dx.doi.org/10.1017/S0022377813000548`.

Fujita, T., Kamada, Y., Ishida, S., Neyatani, Y., Oikawa, T., Ide, S., Takeji, S., Koide, Y., Isayama, A., Fukuda, T., Hatae, T., Ishii, Y., Ozeki, T., Shirai, H. and JT-60 Team. (1999). High performance experiments in JT-60U reversed shear discharges, *Nucl. Fusion* **39**, p. 1627, `http://iopscience.iop.org/0029-5515/39/11Y/302`.

Furth, H. P., Killeen, J. and Rosenbluth, M. N. (1963). Finite-resistivity instabilities of a sheet pinch, *Phys. Fluids* **6**, p. 459, `http://dx.doi.org/10.1063/1.1706761`.

Furnish, G., Horton, W., Kishimoto, Y., LeBrun, M. J. and Tajima, T. (1999). Global gyrokinetic simulation of tokamak transport, *Phys. Plasmas* **6**, p. 1227, `http://dx.doi.org/10.1063/1.873366`.

Fusion Science and Technology (2011), editor J. W. Van Dam, **59**, http://www.ans.org.

Fülöp, T. and Helander, P. (1999). Nonlinear neoclassical transport in a rotating impure plasma with large gradients, *Phys. Plasmas* **6**, p. 3066, http://dx.doi.org/10.1063/1.873593.

Futatani, S., Horton, W., Benkadda, S., Bespamyatnov, I. O. and Rowan, W. L. (2010). Fluid models of impurity transport via drift wave turbulence, *Phys. Plasmas* **17**, p. 072512, http://dx.doi.org/10.1063/1.3459062.

Galeev, A. A. (1978). Stationary magnetic field line reconnection in a tearing-unstable plasma slab, *Phys. Fluids* **21**, p. 1353, http://dx.doi.org/10.1063/1.862376.

Galeev, A. A. and Zeleny, L. M. (1979). *Pis'ma Az. Eksp. Teor. Fis.* **29**, p. 669; *Space Res. Institute Report Pr-501* (Academy of Sciences, Moscow).

Galeev, A. A. (1984). Spontaneous reconnection of magnetic field lines in a collisionless plasma, *Handbook of Plasma Physics* **2**, eds. M.N. Rosenbluth and R. Z. Sagdeev (North-Holland, New York) pp. 305-335.

Gao, Z., Dong, J.-Q., Liu, G. J. and Ying, C. T. (2001). Study of electromagnetic instabilities driven by ion temperature gradient and parallel sheared flows in high-β plasmas, *Phys. Plasmas* **8**, p. 4080, http://dx.doi.org/10.1063/1.1388176.

Garbet, X., Garzotti, L., Mantica, P., Nordnam, H., Valovic, M., Weisen, H. and Angioni, C. (2003). Turbulent particle transport in magnetized plasmas, *Phys. Rev. Lett.* **91** p. 035001, http://dx.doi.org/10.1103/PhysRevLett.91.035001.

Gary, J. and Helgason, R. (1970). A matrix method for ordinary differential eigenvalue problems, *J. Comp. Phys.* **5**, pp. 169-187, http://dx.doi.org/10.1016/0021-9991(70)90058-6.

Gasparotto, M., Baylard, C., Bosch, H. S., Hartmann, D., Klinger, T., Vilbrandt, R. and Wegener, L. (2014). Wendelstein 7-X – Status of the project and commissioning planning, *Fusion Eng. Des.* **89**, pp. 2121-2127, http://dx.doi.org/10.1016/j.fusengdes.2014.02.075.

Gates, D. A, *et al.* (2002). *Phys. Plasmas* **10**, p. 1659.

Gates, D. A., Menard, J., Maingi, R., Kaye, S., Sabbagh, S. A., Diem, S., Wilson, J. R., Bell, M. G., Bell, R. E., Ferron, J., Fredrickson, E. D., Kessel, C. E., LeBlanc, B. P., Levinton, F., Manickam, J., Mueller, D., Raman, R. Stevenson, T. Stutman, D., Taylor, G. Tritz, K., Yu, H. and the NSTX Research Team. (2007). Progress towards steady state at low aspect ratio on the National Spherical Torus Experiment (NSTX), *Nucl. Fusion*, **47**, http://dx.doi.org/10.1088/0029-5515/47/9/040.

Gekelman, W. (1999). Review of laboratory experiments on Alfvén waves and their relationship to space observations, *J. Geophys. Res.* **104**, A7, pp. 14,417-14,435, http://dx.doi.org/10.1029/98JA00161.

Gentle, K. W. (1981). The Texas Experimental Tokamak (Text) Facility User's Facility, *Fusion Sci. Tech.* **1**, 4, pp. 479-485.

Gentle, K. W. and He, H. (2008). Texas Helimak, *Plasma Sci. Tech.* **10**, p. 284, http://dx.doi.org/10.1088/1009-0630/10/3/03.

Ginannone, L., Cerna, M., Cole, R., Fitzek, M., Kallenbach, A., Lddecke, K., McCarthy, P. J., Scarabosio, A., Schneider, W., Sips, A. C. C., Treutterer, W., Vrancic, A., Wenzel, L., Yi, H., Behler K., Eich, T., Eixenberger, H., Fuchs, J. C., Haas, G., Lexa, G., Marquardt, M., Mlynek, A., Neu, G., Raupp, G., Reich, M., Sachtleben, J., Schuhbeck, K. H., Zehetbauer, T., Concezzi, S., Debelle, T., Marker, B., Munroe,

M., Petersen, N., Schmidt, D. and NI Big Physics Team (2009). Data acquisition and real-time signal processing of plasma diagnostics on ASDEX Upgrade using LabVIEW RT, *Fus. Eng. Des.* **85**, pp. 303-307, `http://dx.doi.org/10.1016/j.fusengdes.2010.03.030`.

Gibson, A. and the JET Team. (1998). Deuterium-tritium plasmas in the Joint European Torus (JET): Behavior and implications, *Phys. Plasmas* **5**, p. 1839, `http://dx.doi.org/10.1063/1.872854`.

Gill, A. E. (1982). *Atmosphere-Ocean Dynamics, International Geophysics Series* (Academic Press), **30**, ISBN:0-12-283520-4.

Gladd, N. T. and Horton, W. (1973). Critical shear and growth rates for drift waves in a nonuniform current-carrying plasma, *Phys. Fluids* **16**, pp. 879-888, `http://dx.doi.org/10.1063/1.1694440`.

Goedblood, J. P., Keppens, R. and Poedts, S. (2010). *Advanced Magnetohydrodynamics* (Cambridge, UK: Cambridge University Press, 2010), pp. 269-304, ISBN-13:9780521705240.

Goldston, R. J., Mazzucato, E., Slusher, R. E. and Surko, C. M. (1977). *Plasma Phys. Control. Nucl. Fusion* (International Atomic Energy Agency, Vienna) **I**, p. 371.

Goodall, D. H. J. and Wesson, J. A. (1984). Cine observations of MHD instabilities in a tokamak, *Plasma Phys. Control. Fusion* **26**, p. 789, `http://dx.doi.org/10.1088/0741-3335/26/6/002`.

Goodman, T. P., Ahmed, S. M., Alberti, S., Andr'ebe, Y., Angioni, C., Appert, K., Arnoux, G., Behn, R., Blanchard, P., Bosshard, P., Camenen, Y., Chavan, R., Coda, S., Condrea, I., Degeling, A., Duval, B. P., Etienne, P., Fasel, D., Fasoli, A., Favez, J.-Y., Furno, I., Henderson, M., Hofmann, F., Hogge, J.-P., Horacek, J., Isoz, P., Joye, B., Karpushov, A., Klimanov, I., Lavanchy, P., Lister, J. B., Llobet, X., Magnin, J.-C., Manini, A., Marlétaz, B., Marmillod, P., Martin, Y., Martynov, An., Mayor, J.-M., Mlynar, J., Moret, J.-M., Nelson-Melby, E., Nikkola, P., Paris, P. J., Perez, A., Peysson, Y., Pitts, R. A., Pochelon, A., Porte, L., Raju, D., Reimerdes, H., Sauter, O., Scarabosio, A., Scavino, E., Seo, S. H., Siravo, U., Sushkov, A., Tonetti, G., Tran, M. Q., Weisen, H., Wischmeier, M., Zabolotsky, A. and Zhuang, G. (2003). An overview of results from the TCV tokamak, *Nucl. Fusion* **43**, pp. 1619-1631, `http://dx.doi.org/10.1088/0029-5515/43/12/008`.

Gorelenkov, N. N., Berk, H. L., Fredrickson, E., Sharapov, S. E. and JET EFDA Contributors. (2007). Predictions and observations of low-shear beta-induced shear Alfvén-acoustic eigenmodes in toroidal plasmas, *Phys. Letts. A* **370**, 1, pp. 70-77, `http://doi:10.1016/j.physleta.2007.05.113`.

Gormezano, C. and the JET Team, *Proceedings of the Sixteenth International Atomic Energy Agency Fusion Energy Conference, 1997* (Montréal, Canada, 1996). 1, p. 487.

Gormezano, C., Sips, A. C. C., Luce, T. C., Ide, S., Becoulet, A., Litaudon, X., Isayama, A., Hobirk, J., Wade, M. R. Wade, Oikawa, T., Prater, R., Zvonkov, A., Lloyd, B., Suzuki, T., Barbato, E., Bonoli, P., Phillips, C. K., Vdovin, V., Joffrin, E., Casper, T., Ferron, J., Mazon, D., Moreau, D., Bundy, R., Kessel, C.,Fukuyama, A., Hayashi, N., Imbeaux, F., Murakami, M., Polevoi, A. R. and St John H. E. (2007). Steady-state operation, *Nucl. Fusion* **47**, p. S285, `http://dx.doi.org/10.1088/0029-5515/47/6/S06`.

Grasso, D., Califano, F., Pegoraro, F. and Porcelli, F. (2001). Phase mixing and island saturation in Hamiltonian reconnection, *Phys. Rev. Lett.* **86**, p. 5051, http://dx.doi.org/10.1103/PhysRevLett.86.5051.

Greenwald, M., Gwinn, D., Milora, S., Parker, J., Parker, R., Wolfe, S., Besen, M., Camacho, F., Fairfax, S., Fiore, C., Foord, M., Gandy, R., Gomez, C., Granetz, R., LaBombard, B., Lipschultz, B., Lloyd, B., Marmar, E., McCool, S., Pappas, D., Petrasso, R., Pribyl, P., Rice, J., Schuresko, D., Takase, Y., Terry, J. and Watterson, R. (1984). Energy confinement of high-density pellet-fueled plasmas in the Alcator C tokamak, *Phys. Rev. Lett.* **53**, pp. 352-355, http://dx.doi.org/10.1103/PhysRevLett.53.352.

Greenwald, M., Boivin, R. L., Bonoli, P., Christensen, C., Fiore, C., Garnier, D., Goetz, J., Golovato, S., Graf, M., Granetz, R., Horne, S., Hsu, T., Hubbard, A., Hutchinson, I., Irby, J., Kurz, C., LaBombard, B., Lipschultz, B., Luke, T., Marmar, E., McCracken, G., Niemczewski, A., O'Shea, P., Porkolab, M., Rice, J., Reardon, J., Schachter, J., Snipes, J., Stek, P., Takase, Y., Terry, J., Umansky, M., Watterson, R., Wolfe, S., Bombarda, F., May, M. and Welch, B. (1995). Transport experiments in Alcator-C-Mod, *Phys. Plasmas* **2**, p. 2308, http://dx.doi.org/10.1063/1.871254.

Greenwald, M., Angioni, C., Hughes, J. W., Terry, J. and Weise,n H. (2007). Density profile peaking in low-collisionality H-modes: comparison of Alcator C-Mod data to ASDEX Upgrade/JET scalings, *Nucl. Fusion* **47**, p. L26, http://dx.doi.org/10.1088/0029-5515/47/9/L03.

Greenwald, M., Bader, A., Baek, S., Bakhtiari, M., Barnard, H., Beck, W., Bergerson, W., Bespamyatnov, I. O., Bonoli, P., Brower, D., Brunner, D., Burke, W., Candy, J., Churchill, M., Cziegler, I., Diallo, A., Dominguez, A., Duval, B., Edlund, E., Ennever, P., ernst, D., Faust, I., Fiore, C., Fredian, T., Garcia, O., Gao, C., Goetz, J., Golfinopoulos, T., Granetz, R., Grulke, O., Hartwig, Z., Horne, S., Howard, N., Hubbard, A., Hughes, J., Hutchinson, I., Irby, J., Izzo, V., Kessel, C., LaBombard, B., Lau, C., Li, C., Lin, Y., Lipschultz, B., Loarte, A., Marmar, E., Mazurenko, A., McCracken, G., McDermott, R., Meneghini, O., Mikkelsen, D., Mossessian, D., Mumgaard, R., Myra, J., Nelson-Melby, E., Ochoukov, R., Olynyk, G., Parker, R., Pitcher, S., Podpaly, Y., Porkolab, M., Reinke, M., Rice, J., Rowan, W., Schmidt, A., Scott, S., Shiraiwa, S., Sierchio, J., Smick, N., Sniptes, J. A., Synder, P., Sorbom, B., Stillerman, J., Sung, C., Takase, Y., Tang, V., Terry, J., Terry, D., Theiler, C., Tronchin-James, A., Tsujii, N., Vieira, R., Walk, J. R., Wallace, G., White, A., Whyte, D., Wilson, J., Wolfe, S., Wright, G., Wright, J., Wukitch, S. and Zweben, S. (2014). 20 years of research on the Alcator C-Mod tokamak, *Phys. Plasmas* **21**, p. 110501, http://dx.doi.org/10.1063/1.4901920.

Grierson, B. A., Burrell, K. H., Solomon, W. M. and Pablant, N. A. (2010). Deuterium velocity and temperature measurements on the DIII-D tokamak, *Rev. Sci. Instrum.* **81**, p. 10D735, http://dx.doi.org/10.1063/1.3491222.

Grosman, A., Bucalossia, J., Doceula, L., Escourbiac, F., Lipa, M., Merola, M., Missirlian, M., Pitts, R. A., Samaille, F. and Tsitrone, E. (2013). The WEST programme: Minimizing technology and operational risks of a full actively cooled tungsten divertor on ITER, *Fusion Eng. Design* **88**, pp. 497-500, http://dx.doi.org/10.1016/j.fusengdes.2013.02.039.

Grove, D. J. and Meade, D. M. (1985). Initial studies of confinement, adiabatic compression, and neutral-beam heating in TFTR, *Nucl. Fusion* **25**, p. 1167, http://dx.doi.org/10.1088/0029-5515/25/9/029.

Gruber, O., Bosch, H.-S., Günter, S., Herrmann, A., Kallenbach, A., *et al.* (1999). Overview of ASDEX Upgrade results, *Nucl. Fusion* **39**, p. 1321, `http://dx.doi.org/10.1088/0029-5515/39/9Y/309`.

Gruzinov, A. V., Isichenko, M. B. and Kalda, Ya. L. (1990). Two-dimensional turbulent diffusion, *Zh. Eksp. Teor. Fiz.* **97**, pp. 476-488 [*Sov. Phys. JETP* **70**, p. 263] ISSN:0044-4510.

Günter, S., Gude, A., Maraschek, M., Sesnic, S., Zohm, H. and ASDEX Upgrade Team. (1998). β scaling for the onset of neoclassical tearing modes at ASDEX Upgrade, *Nucl. Fusion* **38**, p. 1431, `http://hdl.handle.net/11858/00-001M-0000-0027-87F1-3`.

Guo, S. and Weiland, J. (1997). Analysis of η_i mode by reactive and dissipative descriptions and the effects of magnetic q and negative shear on the transport, *Nucl. Fusion* **37**, pp. 1095-1107, `http://dx.doi.org/10.1088/0029-5515/37/8/105`.

Gurchenko, A. D., Gusakov, E. Z., Altukhov, A. B., Stepanov, A. Yu., Esipov, L. A., Kantor, M. Yu., Kouprienko, D. V., Dyachenko, V. V. and Lashkul, S. I. (2007). Observation of the ETG mode component of tokamak plasma turbulence by the UHR backscattering diagnostics, *Nucl. Fusion* **47**, pp. 245-250, `http://dx.doi.org/10.1088/0029-5515/47/4/001`.

Hallock, G. A., Hickok, R. L. and Hornady, R. S. (1994). The TMX heavy ion beam probe, Plasma Science, *IEEE Transactions on Plasma Science* **22**, pp. 341-349, `http://dx.doi.org/10.1109/27.310639`.

Hamada, Y., Nishizawa, A., Ido, K, Watari, T., Kojima, M., Kawasumi, Y., Narihara, K., Toi, K. and JIPPT-IIU Group. (2005). Zonal flows in the geodesic acoustic mode frequency range in the JIPP T-IIU tokamak plasmas, *Nucl. Fusion* **45**, pp. 81-85, `http://dx.doi.org/10.1088/0029-5515/45/2/002`.

Hamaguchi, S. and Horton, W. (1990). Fluctuation spectrum and transport from ion temperature gradient driven modes in sheared magnetic fields,*Phys. Fluids B* **2**, pp. 1833-1851, `http://dx.doi.org/10.1063/1.859455`.

Hamaguchi, S. and Horton, W. (1991). Effects of sheared flows on ion-temperature-gradient-driven turbulent transport, *Phys. Fluids B* **4**, p. 319, `http://dx.doi.org/10.1063/1.860280`.

Hamaguchi, S. and Horton, W. (1992). Modeling of drift wave turbulence with a finite ion temperature gradient, *Plasma Phys. Control. Fusion* **34**, pp. 203-233, `http://dx.doi.org/10.1088/0741-3335/34/2/006`.

Hammett, G. W. and Perkins, F. W. (1990). Fluid moment models for Landau damping with application to the ion-temperature-gradient instability, *Phys. Rev. Lett.* **64**, pp. 3019-3022, `http://dx.doi.org/10.1103/PhysRevLett.64.3019`.

Hammett, G. W., Beer, M. A., Dorland, W., Cowley, S. C. and Smith, S. A. (1993). Developments in the gyrofluid approach to Tokamak turbulence simulations, *Plasma Phys. Control. Fusion* **35** p. 973, `http://dx.doi.org/10.1088/0741-3335/35/8/006`.

Han, M. K., Wang, Z-X., Dong, J. Q. and Du, H. (). Multiple ion temperature gradient driven modes in transport barriers, , .

Hariri, F., Naulin, V., Rasmussen, J. Juul, Xu, G. S. and Yan, N. (2016). Cold pulse and rotation reversals with turbulence spreading and residual stress, *Phys. Plasmas* **23**, p. 052512, `http://dx.doi.org/10.1063/1.4951023`.

Harrison, J. R., Lisgo, S. W., Gibson, K. J., Tamain, P., Dowling, J. and the MAST TEAM. (2011). Characterisation of detached plasmas on the MAST tokamak, *J. Nucl. Mat.* **415**, pp. S379-S382, http://dx.doi.org/10.1016/j.jnucmat.2010.12.226.

Hasegawa, A. and Mima, K. (1977). Stationary spectrum of strong turbulence in magnetized nonuniform plasma , *Phys. Rev. Lett.* **39**, p. 205, http://dx.doi.org/10.1103/PhysRevLett.39.205.

Hasegawa, A. and Wakatani, M. (1983). Plasma Edge Turbulence, *Phys. Rev. Lett.* **50**, pp. 682-686, http://dx.doi.org/10.1103/PhysRevLett.50.682.

Hassam, A. B., Antonsen, T. M., Drake, J. F. and Liu, C. S. (1991). Spontaneous poloidal spin-up of tokamaks and the transition to the *H* mode, *Phys. Rev. Lett.* **66**, p. 309, https://doi.org/10.1103/PhysRevLett.66.309.

Hassam, A. B. and Antonsen, T. M. (1994). Poloidal spin-up of tokamak plasmas from poloidal asymmetry of particle and momentum sources, *Phys. Plasmas* **1**, p. 337, http://dx.doi.org/10.1063/1.870836.

Hastie, R. J., Hesketh, K. W. and Taylor, J. B. (1979). Shear damping of two-dimensional drift waves in a large-aspect-ratio tokamak, *Nucl. Fusion* **19**, p. 1223, http://dx.doi.org/10.1088/0029-5515/19/9/006.

Hatch, D. R., Kotschenreuther, Mahajan, S., Valanju, P., Jenko, F., Told, D., Görler, T. and Saarelma, S. (2016). Microtearing turbulence limiting the JET-ILW pedestal, *Nucl. Fusion* **56**, p. 104003, http://dx.doi.org/10.1088/0029-5515/56/10/104003.

Hatch, D. R., Kotschenreuther, Mahajan, S., Valanju, P. and Liu, X. (2017). A gyrokinetic perspective on the JET-ILW pedestal, *Nucl. Fusion* **57**, pp. 036020-03636, \http://dx.doi.org/10.1088/1741-4326/aa51e1

Hawryluk, R. J. (1998). Results from deuterium-tritium tokamak confinement experiments, *Rev. Mod. Phys.* **70**, pp. 537-587, http://dx.doi.org/10.1103/RevModPhys.70.537.

Hazeltine, R. D., Lee, E. P. and Rosenbluth M. N. (1971). Resistive plasma rotation and shock formation in toroidal geometry, *Phys. Fluids* **14**, p. 361, http://dx.doi.org/10.1063/1.1693436.

Hazeltine, R. D., Holm, D. D. and Morrison, P. J. (1985). Electromagnetic solitary waves in magnetized plasmas, *J. Plasma Phys.* **34**, pp. 103-114, http://dx.doi.org/10.1017/S0022377800002713.

Hazeltine, R. D., Kotschenreuther, M. and Morrison, P. J. (1985). A four-field model for tokamak plasma dynamics, *Phys. Fluids* **28**, pp. 2466-2477, http://dx.doi.org/10.1063/1.865255.

Hazeltine, R. D., Hsu, C. T. and Morrison, P. J. (1987). Hamiltonian four-field model for nonlinear tokamak dynamics, *Phys. Fluids* **30**, pp. 3204-3211, http://dx.doi.org/10.1063/1.866527.

Hazeltine, R. D. and Waelbroeck, F. L. (1988). *The Framework of Plasma Physics* (Perseus Books, Reading, Massachusetts).

Hazeltine, R. D. and Meiss, J. D. (1992). *Plasma Confinement* (Addison-Wesley, Redwood City).

Heidbrink, W. W., Bol, K., Buchenauer, D., Fonck, R., Gammel, G., Ida, K., Kaita, R., Kaye, S., Kugel, H., LeBlanc, B., Morris, W., Okabayashi, M., Powell, E., Sesnic, S. and Takahashi, H. (1986). Tangential neutral-beam-driven instabilities in the

Princeton Beta Experiment, *Phys. Rev. Lett.* **57**, p. 835, `http://dx.doi.org/10.1103/PhysRevLett.57.835`.

Helander, P. and Sigmar, D. J. (2002). *Collisional Transport in Magnetized Plasmas* (Cambridge Monographs on Plasma Physics), 312 pp, vol. 4 (Cambridge: Cambridge University).

Helander, P. and Sigmar, D. J. (2005). *Collisional transport in magnetized plasmas* (Lavoisier).

Hendel, H. W., Chu, T. K. and Politzer, P. A. (1968). Collisional drift waves–Identification, stabilization, and enhanced plasma transport, *Phys. Fluids* **11**, p. 2426, `http://dx.doi.org/10.1063/1.1691833`.

Hennequin, P., Sabot, R., Honoré, C., Hoang, G. T., Garbet, X., Truc, A., Fenzi, C. and Quéméneur, A. (2004). Scaling laws of density fluctuations at high-k on Tore Supra, *Plasma Phys. Control. Fusion* **46**, p. B121, `http://dx.doi.org/10.1088/0741-3335/46/12B/011`.

Herken, G. (2002). Brotherhood of the Bomb: The Tangled Lives and Loyalties of Robert Oppenheimer, Ernest Lawrence, and Edward Teller (Henry Holt and Co.).

Hinton, F. L. and Horton, Jr., C. W. (1971). Amplitude limitation of a collisional drift wave instability, *Phys. Fluids* **14**, pp. 116-123, `http://dx.doi.org/10.1063/1.1693260`.

Hinton, F. L. and Hazeltine, R. D. (1976). Theory of plasma transport in toroidal confinement systems, *Rev. Mod. Phys.* **48**, pp. 239-308, `http://dx.doi.org/10.1103/RevModPhys.48.239`.

Hinton, F. L. and Wong, S. K. (1985). Neoclassical ion transport in rotating axisymmetric plasmas, *Phys. Fluids* **28**, p. 3082, `http://dx.doi.org/10.1063/1.865350`.

Hirshman, S. P. and Sigmar, D. J. (1997). *Phys. Fluids* **20**, p. 518.

Hoang, G. T., Bourdelle, C., Garbet, X., Aniel, T., Giruzzi, G., Ottaviani, M., Horton, W., Zhu, P. and Budny, R. V. (2000). *Electron transport and improved confinement on Tore Supra* (IAEA, Sorrento, Italy).

Hoang, G. T., Bourdelle, C., Garbet, X., Giruzzi, G., Aniel, T. and Ottaviani, M., Horton, W., Zhu, P. and Budney, R. V. (2001). Experimental determination of critical threshold in electron transport on tore supra, Phys. *Rev. Lett.* **87**, p. 125001, `http://dx.doi.org/10.1103/PhysRevLett.87.125001`.

Hoang, G. T., Bourdelle, C., Garbet, X., Artaud, J. F., Basiuk, V., Bucalossi, F., Clairet, F., Fenzi-Bonizec, C., Gil, C., Segui, J. L., Travere, J. M., Tsitrone, E. and Vermare, L. (2003). Parametric dependence of turbulent particle transport in Tore Supra plasmas, *Phys. Rev. Lett.* **93**, p. 135003, `http://dx.doi.org/10.1103/PhysRevLett.93.135003`.

Hoang, G. T., Horton, W., Bourdelle, C., Hu, B., Garbet, X. and Ottaviani, M. (2003). Analysis of the critical electron temperature gradient in Tore Supra, *Phys. Plasmas* **10**, pp. 405-413, `http://dx.doi.org/10.1063/1.1534113`.

Hoang, G. T., Bécoulet, A., Jacquinot, J., Artaud, J. F., Bae, Y. S., Beaumont, B., Belo, J. H., Berger-By, G., Bizarro, João, P. S., Bonoli, P., Cho, M. H., Decker, J., Delpech, L., Ekedahl, A., Garcia, J., Giruzzi, G., Goniche, M., Gormezano, C., Guilhem, D., Hillairet, J., Imbeaux, F., *et al.* (2009). A lower-hybrid current drive system for ITER, *Nucl. Fusion* **49**, p. 075001, `http://dx.doi.org/10.1088/0029-5515/49/7/075001`.

Hojo, H. (1995). Effect of resonant electrons on ion temperature gradient mode in anisotropic plasma, *Jpn. J. Appl. Phys.* **34**, pp. 650-65021, http://jjap.ipap.jp/link?JJAP/34/6501/.

Holland, C. and Diamond, P. H. (2002). Electromagnetic secondary instabilities in electron temperature gradient turbulence, *Phys. Plasmas* **9**, p. 3857, http://dx.doi.org/10.1063/1.1496761.

Hong, B. G. and Horton, W. (1989). Anomalous electron thermal conduction from magnetic turbulence, (Brief Communications) *Phys. Fluids B* **1**, p. 2527-2530, http://dx.doi.org/10.1063/1.859188.

Hong, B. G., Horton, W. and Choi, D-I. (1989a). Drift-Alfvén kinetic stability theory in the ballooning mode approximation, *Phys. Fluids B* **1**, p. 1589, http://dx.doi.org/10.1063/1.858937.

Hong, B. G., Horton, W. and Choi, D-I. (1989b). Pressure gradient-driven modes in finite beta toroidal plasmas, *Plasma Phys. Control. Fusion* **31**, p. 1291, http://dx.doi.org/10.1088/0741-3335/31/8/006.

Hong, B.-G., Horton, W., Hamaguchi, S., Wakatani, M., Yagi, M. and Sugama, H. (1991). Transition from resistive-G to η_i driven turbulence in stellarator systems, *Phys. Fluids B* **3**, p. 1638, http://dx.doi.org/10.1063/1.859683.

Hooper, B., Hallock, G. A. and Foote, J. H. (1982). Low-frequency oscillations in the central cell of the TMX tandem mirror experiment, *Phys. Fluids* **26**, p. 314, http://dx.doi.org/10.1063/1.863975.

Horton, C. W., Jr. and Varma, R. K. (1972). Electrostatic stability theory of tokamaks from two-component fluid equations, *Phys. Fluids* **15**, pp. 620-631, http://dx.doi.org/10.1063/1.1693957.

Horton, W. (1976). Spectral distribution of drift-wave fluctuations in tokamaks, *Phys. Rev. Lett* **37**, p. 1269, http://dx.doi.org/10.1103/PhysRevLett.37.1269.

Horton, W., Okuda, H., Cheng, C. Z., Kuo, Y. Y., Lee, W. W., Matsuda, Y. and True, M. (1977). *Plasma Phys. Control. Nucl. Fusion* (International Atomic Energy Agency, Vienna) **II**, pp. 467-479.

Horton, Jr., W. and Choi, Duk-In. (1979). Renormalized turbulence theory for the ion-acoustic problem, *Phys. Reports* **49**, pp. 273-410, http://dx.doi.org/10.1016/0370-1573(79)90056-5.

Horton, W., Jr., Choi, Duk-In and Estes, R. D. (1979). Diamagnetic frequency profile effects on toroidal drift waves, *Phys. Fluids* **22**, pp. 519-521, http://dx.doi.org/10.1063/1.862617.

Horton, W., Estes, R. and Biskamp, D. (1980). Fluid simulation of ion pressure gradient driven drift modes, *Plasma Phys.* **22**, pp. 663-678, http://dx.doi.org/10.1088/0032-1028/22/7/004.

Horton, W., Choi, D.-I. and Tang, W. M. (1981). Toroidal drift modes driven by ion pressure gradients, *Phys. Fluids* **24**, pp. 1077-1085, http://dx.doi.org/10.1063/1.863486.

Horton, W., Choi, D. I. and Hong, B. G. (1983). Electromagnetic drift modes driven by pressure gradients in tokamaks, *Phys. Fluids* **26**, p. 1461, http://dx.doi.org/10.1063/1.864316.

Horton, W. (1984). *Handbook of Plasma Physics, II*, M. N. Rosenbluth and R. Z. Sagdeev ed., pp. 384-402, (North-Holland, Amsterdam).

Horton, W. and Reichl, L. E., eds. (1984). *Statistical Physics and Chaos in Fusion Plasmas* (John Wiley and Sons, New York).

Horton, W. (1985). Onset of stochasticity and the diffusion approximation in drift waves, *Plasma Phys. Control. Fusion* **27**, p. 937, `http://dx.doi.org/10.1088/0741-3335/27/9/001`.

Horton, W., Tajima, T. and Kamimura, T. (1987). Kelvin-Helmholtz instability and vortices in magnetized plasma, *Phys. Fluids* **30**, pp. 3485-3495, `http://dx.doi.org/10.1063/1.866429`.

Horton, W., Choi, D. I., Yushmanov, P. N. and Parail, V. V. (1987). Electron diffusion in tokamaks due to electromagnetic fluctuations, *Plasma Phys.* **29**, p. 901, `http://dx.doi.org/10.1088/0741-3335/29/7/008`.

Horton, W. and Tajima, T. (1988). Linear theory of driven reconnection, *J. Geophys. Res.* **93**, pp. 2741-2748, `http://dx.doi.org/10.1029/JA093iA04p02741`.

Horton, W., Bekki, N., Berk, H. L., Hong, B.-G., LeBrun, M. J., Mahajan, S. M., Tajima, T. and Zhang, Y. Z. (1988). *12th Intl. Conf. Plasma Phys. Control. Nucl. Fusion Res.* (IAEA, London) CN-50/D-4-3.

Horton, W., Hong, B.-G. and Tang, W. M. (1988). Toroidal electron temperature gradient driven drift modes, *Phys. Fluids* **31**, p. 2971-2983, `http://dx.doi.org/10.1063/1.866954`.

Horton, W. (1989). Drift wave vortices and anomalous transport, *Phys. Fluids B* **1**, p. 524, `http://dx.doi.org/10.1063/1.859168`.

Horton, W., Hong, B.-G., Tajima, T., and Bekki, N. (1990). *Comm. Plasma Phys. Control. Fusion* **13**, p. 207.

Horton, W., Su, X. N and Morrison, P. J. (1990). Drift wave vortices and turbulent transfer, *Fizika Plazmy* **16**, Aug. 1990, pp. 969-977. In Russian Translation (ISSN 0367-2921).

Horton, W., Lindberg, D., Kim, J. Y., Dong, J. Q., Hammett, G. W., Scott, S. D., Zarnstorff, M. C. and Hamaguchi, S (1992). Ion-temperature-gradient-driven transport in a density modification experiment on the Tokamak Fusion Test Reactor, *Phys. Fluids B* **4**, pp. 953-967, `http://dx.doi.org/10.1063/1.860112`.

Horton, W. and Hasegawa, A. (1994). Quasi-two-dimensional dynamics of plasmas and fluids, *Chaos* **4**, p. 227, `http://dx.doi.org/10.1063/1.166049`.

Horton, W. and Rowan, W. (1994). Impurity transport studies in the Texas Experimental Tokamak (TEXT), *Phys. Plasmas* **1**, p. 901, `http://dx.doi.org/10.1063/1.870749`.

Horton, W. and Ichikawa, Y. (1996). *Chaos and Structures in Nonlinear Plasmas* (World Scientific) ISBN:81-7764-234-0.

Horton, W., Hu, G. and Laval, G. (1996). Turbulent transport in mixed states of convective cells and sheared flows, `http://dx.doi.org/10.1063/1.871651`.

Horton, W., Park, H-B., Kwon, J-M., Strozzi, D., Morrison, P. J. and Choi, D-I. (1998). Drift wave test particle transport in reversed shear profile, *Phys. Plasmas* **5**, pp. 3910-3917, `http://dx.doi.org/10.1063/1.873110`.

Horton, W. (1999). Drift waves and transport, *Rev. Mod. Phys.* **71**, p. 735, `http://dx.doi.org/10.1103/RevModPhys.71.735`.

Horton, W., Zhu, P., Hoang, G. T., Aniel, T., Ottaviani, M. and Garbet, X. (2000). *Phys. Plasmas* **7**, pp. 1494-1511, `http://dx.doi.org/10.1063/1.873969`.

Horton, W., Wong, H. V., Van Dam, J. W. and Crabtree, C. (2002). Stability of high-pressure geotail tubes, *J. Geophys. Res.* **106**, pp. 18803-18822, `http://dx.doi.org/10.1029/2000JA000415`.

Horton, W., Hu, B., Dong, J.-Q. and Zhu, P. (2003). Turbulent electron thermal transport in tokamaks, *New J. Phys.* **5**, p. 14, http://dx.doi.org/10.1088/1367-2630/5/1/314.

Horton, W., Hoang, G. T., Bourdelle, C., Garbet, X., Ottaviani, M. and Colas, L. (2004). Electron transport and the critical temperature gradient, *Phys. Plasmas* **11**, p. 2600, http://dx.doi.org/10.1063/1.1690761.

Horton, W., Wong, H. V., Morrison, P. J., Wurm, A., Kim, J.-H., Perez, J. C., Pratt, J., Hoang, G. T., LeBlanc, B. P. and Ball, R. (2005). Temperature gradient driven electron transport in NSTX and Tore Supra, *Nucl. Fusion* **45**, pp. 976-985, http://dx.doi.org/10.1088/0029-5515/45/8/025.

Horton, W., Perez, J. C., Carter, T., Bengtson, R. (2005).Vorticity probes and the characterization of vortices in the Kelvin-Helmholtz instability in the large plasma device experiment, *Phys. Plasmas* **12**, p. 022303, \http://dx.doi.org/10.1063/1.1830489.

Horton, W., Kim, J.-H. and Militello, F. (2007). Turbulent impulsive magnetic energy release from electron scale reconnection, *Phys. Plasmas* **14**, p. 012902, http://dx.doi.org/10.1063/1.2424555.

Horton, W., Kim, J.-H., Chagelishvili, G. D., Bowman, J. C. and Lominadze, J. G. (2010). Angular redistribution of nonlinear perturbations: A universal feature of nonuniform flows, *Phys. Rev. E* **81**, p. 066304, http://dx.doi.org/10.1103/PhysRevE.81.066304.

Horton, W. and Benkadda, S. (2015). *ITER Physics*, 248 pp, ISBN:978-981-4678-66-7.

Horton, W. (2012). Turbulent Transport in Magnetized Plasmas (World Scientific) http://dx.doi.org/10.1142/8362.

Horton, W., Goniche, M., Peysson, Y., Decker, J., Ekedahl, A. and Litaudon, X. (2013). Penetration of lower-hybrid current drive waves in tokamaks, *Phys. Plasmas* **20**, p. 112508, http://dx.doi.org/10.1063/1.4831981.

Horton, W., Michoski, C., Peysson, Y. and Decker, J. (2015). RF wave propagation and scattering in turbulent tokamak plasmas, *AIP Conf. Proc.* **1689**, p. 080009, http://dx.doi.org/10.1063/1.4936532.

Horton, W., Arefiev, A., Peysson, Y., Goniche, M. and Ekedahl, A. (2016). RF wave propagation and scattering in turbulent tokamaks, *Radio Frequency (RF) Conference.*

Howard, L. N. and Krishnamurti, R. (1986). Large-scale flow in turbulent convection: a mathematical model, *J. Fluid Mech.* **170**, pp. 385-410, http://dx.doi.org/10.1017/S0022112086000940.

Howard, N. T., Holland, C., White, A. E., Greenwald, M. and Candy, J. (2015). Multi-scale gyrokinetic simulation of tokamak plasmas: enhanced heat loss due to cross-scale coupling of plasma turbulence, *Nucl. Fusion* **56**, p. 014004, http://dx.doi.org/10.1088/0029-5515/56/1/014004.

Hu, B. and Horton, W. (1997). Minimal model for transport barrier dynamics based on ion-temperature-gradient turbulence, *Phys. Plasmas* **4**, p. 3262, http://dx.doi.org/10.1063/1.872467.

Hu, B., Horton, W., Zhu, P. and Porcelli, F. (2003). Density profile control with current ramping in a transport simulation of IGNITOR, *Phys. Plasmas* **10**, p. 1015, http://dx.doi.org/10.1063/1.1555623.

Huang, J., Lisgo, S., Maddison, G. and the MAST Team. (2010). Analysis of fuel retention on MAST by global gas balance, *Plasma Phys. Control. Fusion* **52**, p. 075012, `http://dx.doi.org/10.1088/0741-3335/52/7/075012`.

Hubbard, A. E., Whyte, D. G., Churchill, R. M., Cziegler, I., Dominguez, A., Golfinopoulos, T., Hughes, J. W., Rice, J. E., Bespamyatnov, I., Greenwald, M. J., Howard, N., Lipschultz, B., Marmar, E. S., Reinke, M. L., Rowan, W. L., Terry, J. L. and Alcator C-Mod Group. (2011). Edge transport barrier and turbulence in the I-mode regime on Alcator C-Mod, *Phys. Plasmas* **18**, p. 056115, `http://dx.doi.org/10.1063/1.3582135`.

Hughes, J. W., Hubbard, A. E., Mossessian, D. A., LaBombard, B., Biewer, T. M., Granetz, R. S., Greenwald, M., Hutchinson, I. H., Irby, J. H., Lin, Y., Marmar, E. S., Porkolab, M., Rice, J. E., Snipes, J. A., Terry, J. L., Wolfe, S. and Zhurovich, K. (2007). *Fusion Sci. Technol.* **51**, pp. 317-341.

Hughes, D. W. and Tobias, S. M. (2001). On the instability of magnetohydrodynamic shear flows, *Proc. R. Soc. London Ser. A* **457**, p. 1365, `http://dx.doi.org/10.1098/rspa.2000.0725`.

Hugon, M., van Milligen, B. Ph., Smeulders, P., Appel, L. C., Bartlett, D. V., Boucher, D., Edwards, A. W., Eriksson, L.-G., Gowers, C. W., Hender, T. C., Guysmans, G., Jacquinot, J. J., Kupschus, P., Porte, L., Rebut, P. H., Start, D. F. H., Tibone, F., Tubbing, B. J. D., Watkins, M. L. and Zwingmann, W. (1992). Shear reversal and MHD activity during pellet enhanced performance pulses in JET, *Nucl. Fusion* **32**, p. 33, `http://dx.doi.org/10.1088/0029-5515/32/1/I04`.

Huld, T., Nielsen, A. H., Pecseli, H. L. and Rasmussen, J. Juul. (1991). Coherent structures in two-dimensional plasma turbulence, *Phys. Fluids B* **3**, p. 1609, `http://dx.doi.org/10.1063/1.859680`.

Huysmans, G. T. A. and Czarny, O. (2007). MHD stability in X-point geometry: Simulation of ELMs, *Nucl. Fusion* **47**, pp. 659-666, `http://dx.doi.org/10.1088/0029-5515/47/7/016`.

Ida, K., Funaba, H., Kado, S., Narihara, K., Tanaka, K., Takeiri, Y., Nakamura, Y., Ohyabu, N., Yamazaki, K., Yokoyama, M., Murakami, S., Ashikawa, N., deVries, P. C., Emoto, M., Goto, M., Idei, H., Ikeda, K., Inagaki, S., Inoue, N., Isobe, M., Itoh, K., Kaneko, O., Kawahata, K., Khlopenkov, K., Komori, A., Kubo, S., Kumazawa, R., Liang, Y., Masuzaki, S., Minami, T., Miyazawa, J., Morisaki, T., Morita, S., Mutoh, T., Muto, S., Nagayama, Y., Nakanishi, H., Nishimura, K., Noda, N., Notake, T., Kobuchi, T., Ohdachi, S., Ohkubo, K., Oka, Y., Osakabe, M., Ozaki, T., Pavlichenko, R. O., Peterson, B. J., Sagara, A., Saito, K., Sakakibara, S., Sakamoto, R., Sanuki, H., Sasao, H., Sasao, M., Sato, K., Sato, M., Seki, T., Shimozuma, T., Shoji, M., Suzuki, H., Sudo, S., Tamura, N., Toi, K., Tokuzawa, T., Torii, Y., Tsumori, K., Yamamoto, T., Yamada, H., Yamada, I., Yamaguchi, S., Yamamoto, S., Yoshimura, Watanabe, K. Y., Watari, T., Hamada, Y., Motojima, O. and Fujiwara, M. (2001). Reduction of ion thermal diffusivity associated with the transition of the radial electric field in neutral-beam-heated plasmas in the Large Helical Device, *Phys. Rev. Lett.* **86**, p. 5297, `http://dx.doi.org/10.1103/PhysRevLett.86.5297`.

Ilic, D. B., Rognlien, T. D., Self, S. A. and Crawford, F. W. (1973). Low-frequency flute instabilities of a hollow cathode arc discharge: Theory and experiment, *Phys. Fluids* **16**, p. 1042, `http://dx.doi.org/10.1063/1.1694466`.

Imadera, K. and Kishimoto, Y., 25th Fusion Energy Conference. TH/P5-8, Oct. 16, 2014.

Imbeaux, F. and Peysson, Y. (2005). Ray-tracing and Fokker-Planck modelling of the effect of plasma current on the propagation and absorption of lower-hybrid waves, *Plasma Phys. Control. Fusion* **47**, p. 2041, http://dx.doi.org/10.1088/0741-3335/47/11/012.

ITER Physics Expert Groups on Confinement and Transport and Confinement Modeling and Database. (1999). *Nucl. Fusion* **39**, pp. 217-2175-2249, https://inis.iaea.org/search/search.aspx?orig_q=journal:\%22ISSN\%200029-5515\%22.

Isichenko, M. B. and Horton, W. (1991). Scaling laws of stochastic $\boldsymbol{E}\times\boldsymbol{B}$ plasma transport, *Comments Plasma Phys. Control. Fusion* **14**, 5, pp. 249-262, ISSN:0374-2806.

Isichenko, M. B., Horton, W., Kim, D. E., Heo, E. G. and Choi, D.-I. (1992). Stochastic diffusion and Kolmogorov entropy in regular and random Hamiltonians, *Phys. Fluids B* **4**, p. 3937, http://dx.doi.org/10.1063/1.860300.

Ishizawa, A. and Horiuchi, R. (2005). Suppression of Hall-term effects by gyroviscous cancellation in steady collisionless magnetic reconnection, *Phys. Rev. Lett.* **95**, p. 0450003, http://dx.doi.org/10.1103/PhysRevLett.95.045003.

Ishizawa, A. and Nakajima, N. (2007). Excitation of macro-magnetohydrodynamic mode due to multi-scale interaction in a quasi-steady equilibrium formed by a balance between micro-turbulence and zonal flow, *Phys. Plasmas* **14**, p. 040702, http://dx.doi.org/10.1063/1.2716669.

Ishizawa, A. and Nakajima, N. (2010). Turbulence driven magnetic reconnection causing long-wavelength magnetic islands, *Phys. Plasmas* **17**, p. 072308, http://dx.doi.org/10.1063/1.3463435.

ITER-EDA. (1998). Final Design Report (published in *Nucl. Fusion* **39**, p. 2775, 1999) (International Atomic Energy Agency, Vienna), http://www-jt60.naka.jaea.go.jp/english/index-e.html.

ITER Physics Expert Groups on Confinement and Transport and Confinement Modeling and Database, ITER Physics Basis Editors, and ITER EDA. (1999). *Nucl. Fusion* **39**, p. 2175, http://epub.iaea.org/fusion.

Ito, K., Itoh, S.-I. and Fukuyama, A. (1992). Theory of anomalous transport in high-aspect-ratio toroidal helical plasmas, *Phys. Rev. Lett.* **69**, p. 1050, http://dx.doi.org/10.1103/PhysRevLett.69.1050.

Itoh, K. and Itoh, S.-I. (1996). The role of the electric field in confinement, *Plasma Phys. Control. Fusion* **38**, pp. 1-49, http://dx.doi.org/10.1088/0741-3335/38/1/001.

Ito, A., Ramos, J. J. and Nakajima, N. (2007). Ellipticity of axisymmetric equilibria with flow and pressure anisotropy in single-fluid and Hall magnetohydrodynamics, *Phys. Plasmas* **14**, p. 062502, http://dx.doi.org/10.1063/1.2741391.

Ito, A. and Nakajima, N. (2011), Equilibria of toroidal plasmas with toroidal and poloidal flow in high-beta reduced magnetohydrodynamic models, *Nucl. Fusion* **51**, p. 123006, http://dx.doi.org/10.1088/0029-5515/51/12/123006.

Jacchia, A., Luca, F. D., Cirant, S., Sozzi, C., Bracco, G., Bruschi, A., Buratti, P., Podda, S. and Tudisco, O. (2002). Gradient length driven electron heat transport study in modulated electron cyclotron heating FTU tokamak, *Nucl. Fusion* **42**, p. 1116, http://dx.doi.org/10.1088/0029-5515/42/9/310.

Jackson, J. D. (1999). *Classical Electrodynamics* (Wiley, New York) 3rd Ed., pp. 708-724, ISBN:047130932X.

Jaeger, E. F., Berry, L. A., D'Acevedo, E., Batchelor, D. B., Carter, M. D., White, K. F.

and Weitzner, H. (2002). Advances in full-wave modeling of radio frequency heated, multidimensional plasmas, *Phys. Plasmas* **9**, p. 1873, `http://dx.doi.org/10.1063/1.1455001`.

Janssen, Peter A. E. M. (1982). Single mode saturation of the rotational instability in a low-β, finite Larmor radius plasma, *Phys. Fluids* **25**, pp. 316-324, `http://dx.doi.org/10.1063/1.863737`.

Jarmen, A., Andersson, P. and Weiland, J. (1987). Fully toroidal ion temperature gradient driven drift modes, *Nucl. Fusion* **27**, p. 941, `http://dx.doi.org/10.1088/0029-5515/27/6/006`.

Jenko, F. and Dorland, W. (2001). Nonlinear electromagnetic gyrokinetic simulations of tokamak plasmas, *Plasma Phys. Control. Fusion* **43**, pp. A141-A150, `http://dx.doi.org/10.1088/0741-3335/43/12A/310`.

Jha, R., Kwa, P. K., Mattoo, S. K., Rao, C. V. S., Saxena, Y. C. and ADITYA Team. (1992). Intermittency in tokamak edge turbulence, *Phys. Rev. Lett.* **69**, p. 1375, `http://dx.doi.org/10.1103/PhysRevLett.69.1375`.

Joffrin, E., Challis, C. D., Conway, G. D., *et al.* (2002). Internal transport barrier triggering by rational flux surfaces in tokamaks, *International Atomic Energy Agency 19*th *Fusion Energy Conf.* (Lyon, France), submitted to *Nucl. Fusion*.

Joiner, N., Applegate, D., Cowley, S. C., Dorland, W. and Roach, C. M. (2006). Electron temperature gradient driven transport in a MAST H-mode plasma, *Plasma Phys. Control. Fusion* **48**, pp. 685-697, `http://dx.doi.org/10.1088/0741-3335/48/5/013`.

Jones, W. B. and Thron, W. J. (1980). *Continued Fractions: Analytic Theory and Applications*, Encyclopedia of Mathematics and its Applications (Cambridge University Press, 1980), ISBN:9780521101523.

Jose, J. V. and Saletan, E. J. (2002). Classical Mechanics (Cambridge UP, Cambridge), p. 46.

Kadomtsev, B. and Pogutse, O. (1970). Turbulence in Toroidal Systems, *Rev. Plasma Phys.* **5**, ed. A. Leontovich, pp. 249-400, `http://dx.doi.org/10.1007/978-1-4615-7793-5_2`.

Kadomtsev, B. B. and Pogutse, O. P. (1974). Nonlinear helical perturbations of a plasma in the tokamak, *Sov. Phys. JETP* **38**, pp. 283-290.

Kadomtsev, B. (1975). Disruptive instability in tokamaks, *Sov. J. Plasma Phys.* **1**, pp. 389-391.

Kadomtsev, B. B. and Pogutse, O. P. (1978). *Proc. 7th Int. Conf. on Plasma Phys. Control. Nucl. Fusion Research* (International Atomic Energy Agency, Vienna), **I**, p. 649.

Kadomtsev, B. B. and Pogutse, O. P. (1984). *Plasma Phys. Control. Fusion* 24th Conf., 2, Vienna, IAEA, p. 69.

Kadomtsev, B. B. (1992). *Tokamak Plasma: A Complex Physical System* (Institute of Physics Pub., Bristol, UK and Philadelphia) p. 153, ISBN:0750302348.

Kajita, S., Sakaguchi, W., Ohno, N., Yoshida, N. and Saeki, T. (2009). Formation process of tungsten nanostructure by the exposure to helium plasma under fusion relevant plasma conditions, *Nucl. Fusion* **49**, (9), p. 095005, `http://dx.doi.org/10.1088/0029-5515/49/9/095005`.

Karpushov, A. N., Duval, B. P., Schlatter, C., Afanasyev, V. I.,and Chernyshev, F. V. (2006). Neutral particle analyzer diagnostics on the TCV tokamak, *Rev. Sci. Instrum.* **77**, p. 033503, `http://dx.doi.org/10.1063/1.2185151`.

Katanuma, I., Tatematsu, Y., Ishii, K., Tamano, T. and Yatsu, K. (2000). Ideal ballooning modes in the GAMMA 10 tandem mirror, *J. Phys. Soc. Jpn.* **69**, pp. 3244-3252, `http://dx.doi.org/10.1143/JPSJ.69.3244`.

Kaw, P. K., Valeo, E. J. and Rutherford, P. H. (1979). Tearing modes in a plasma with magnetic braiding, *Phys. Rev. Lett.* **43**, pp. 1398-1401, `http://dx.doi.org/10.1103/PhysRevLett.43.1398`.

Kaye, S. M., Greenwald, M., Stroth, U., Kardaun, O., Kus, A., Schissel, D., DeBoo, J., Bracco, G., Thomsen, K., Cordey, J., Miura, Y., Matsuda, T., Tamai, H., Takizuda, T., Hirayama, T., Kikuchi, H., Naito, O., Chudnovskij, A., Ongena, J. and Hoang, G. T. (1997). *Nucl. Fusion* **37**, p. 1657, `http://dx.doi.org/10.1088/0029-5515/37/9/I10`.

Kaye, S. M., Levinton, F. M, Stutman, D., Tritz, K., Yuh, H., Bell, M. G., Bell, R. E., Domier, C. W., Gates, D., Horton, W., Kim, J.-Y. LeBlanc, B. P., Luhman, Jr., N. C., Maingi, R., Mazzucato, E., Menard, J. E., Mikkelsen, D., Mueller, D., Park, H., Rewoldt, G., Sabbagh, S. A., Smith, D. R. and Wang, W. (2007). Confinement and local transport in the National Spherical Torus Experiment (NSTX), *Nucl. Fusion* **47**, p. 499, `http://dx.doi.org/10.1088/0029-5515/47/7/001`.

Kaye, S. M., Greenwald, M., Stroth, U., Kardaun, O., Kus, A., Schissel, D., DeBoo, J., Bracco, G., Thomsen, K., Cordey, J., Miura, Y., Matsuda, T., Tamai, H., Takizuda, T., Hirayama, T., Kikuchi, H., Naito, O., Chudnovskij, A., Ongena, J. and Hoang, G. T. (1997). ITER L mode confinement database, *Nucl. Fusion* **37**, pp. 1303-1328, `http://dx.doi.org/10.1088/0029-5515/37/9/I10`.

Keilhacker, M., Gibson, A., Gormezano, C., Lomas, P.J., Thomas, P. R., Watkins, M. L., Andrew, P., Balet, B., Borba, D., Challis, C. D., Coffey, I., Cottrell, G. A., De Esch, H. P. L., Deliyanakis, N., Fasoli, A., Gowers, C. W., Guo, H. Y., Huysmans, G. T. A., Jones, T. T. C., Kerner, W., König, R. W. T., Loughlin, M. J., Maas, A., Marcus, F. B., Nave, M. F. F., Rimini, F. G., Sadler, G. J., Sharapov, S. E., Sips, G., Smeulders, P., Söldner, F. X., Taroni, A., Tubbing, B. J. D., von Hellermann, M. G., Ward, D. J. and JET Team. (1999). High-fusion performance from deuterium-tritium plasmas in JET, *Nucl. Fusion* **39** p. 209, `http://dx.doi.org/10.1088/0029-5515/39/2/306`.

Kelley, M. C. (2009). The Earth's Ionosphere, *Plasma Physics and Electrodynamics, 2nd Ed.* (Elsevier), ISBN:978-0-12-088425-4.

Kendl, A. and Scott, B. D. (2005). Shear flow reduction by the geodesic transfer mechanism in tokamak edge turbulence, *Phys. Plasmas* **12**, p. 064506, `http://dx.doi.org/10.1063/1.1938147`.

Kent, A. (1966). Instability of laminar flow of a perfect magnetofluid, *Phys. Fluids* **9**, p. 1286, `http://dx.doi.org/10.1063/1.1761842`.

Kesner, J., Horne, S. F. and Pastukhov, V. P. (1987). Cusp-stabilized mirror-based neutron source, *J. Fusion Energy* **6**, pp. 401-411, `http://dx.doi.org/10.1007/BF01052075`.

Kim, D. E., Choi, D.-I., Horton, W., Yushmanov, P. N. and Parail, V. V. (1990). *Phys. Fluids B* **2**, p. 547.

Kim, J.-Y. and Horton, W. (1991). Transition from toroidal to slab temperature gradient driven modes, *Phys. Fluids B* **3**, pp. 1167-1170, `http://dx.doi.org/10.1063/1.859808`.

Kim, C-B., Horton, W. and Hong, B-G. (1991). Fluctuation and thermal energy balance for drift-wave turbulence, *Plasma Phys. Control. Fusion* **33**, p. 469, `http://dx.doi.org/10.1088/0741-3335/33/5/006`.

Kim, J-Y., Horton, W., Choi, D-I., Migliuolo, S. and Coppi, B. (1992). Temperature anisotropy effect on the toroidal ion temperature gradient mode, *Phys. Fluids B*, p. 152, http://dx.doi.org/10.1063/1.860428.

Kim, J.-Y., Horton, W. and Dong, J. Q. (1993). Electromagnetic effect on the toroidal ion temperature gradient mode, *Phys. Fluids B* **5**, p. 4030, http://dx.doi.org/10.1063/1.860623.

Kim, J-Y., Kishimoto, Y., Wakatani, M. and Tajima, T. (1996). Poloidal shear flow effect on toroidal ion temperature gradient mode: A theory and simulation, *Phys. Plasmas* **3**, p. 3689, http://dx.doi.org/10.1063/1.871939.

Kim, J.-H., Perez, J. C., Horton, W., Chagelishvili, G. D., Changishvili, R. G., Lominadze, J. G. and Bowman, J. C. (2006). Self-sustaining vortex perturbations in smooth shear flows, *Phys. Plasmas* **13**, p. 062304, http://dx.doi.org/10.1063/1.2209229.

Kinsey, J. E., Bateman, G., Kritz, A. H. and Redd, A. (1996a). Comparison of two resistive ballooning mode models in transport simulations, *Phys. Plasmas* **3**, pp. 561-570, http://dx.doi.org/10.1063/1.871883.

Kinsey, J. E. and Bateman, G. (1996b). Theory-based transport modeling of the gyroradius experiments, *Phys. Plasmas* **3**, pp. 3344-3357http://dx.doi.org/10.1063/1.871604.

Kishimoto, Y., T. Tajima, Horton, W., M. J. LeBrun and J.-Y. Kim. (1996). Theory of self-organized critical transport in tokamak plasmas, *Phys. Plasmas* **3**, pp. 1289-1307, http://dx.doi.org/10.1063/1.871754.

Kishimoto, Y., Kim, J-Y., Horton, W., Tajima, T., LeBrun, M. J., Dettrick, S. A., Li, J. Q. and Shirai, S. (2000). Discontinuity model for internal transport barrier formation in reversed magnetic shear plasmas, *Nucl. Fusion* **40**, p. 667, http://dx.doi.org/10.1088/0029-5515/40/3Y/330.

Kivelson, M. G. and Russell, C. T. (1983). The interaction of flowing plasmas with planetary ionospheres: A Titan-Venus comparison, *J. Geophys. Res.*, **88**, pp. 49-57, http://dx.doi.org/10.1029/JA088iA01p00049.

Kivelson, M. G. and Russell, C. T. (1995). *Introduction to Space Physics* (Cambridge University Press) ISBN:0-521-45714-9.

Klepper, C. C., Isler, R. C., Hillairet, J., Martin, E. H., Colas, L., Ekedahl, A., Goniche, M., Harris, J. H., Hillis, D. L., Panayotis, S., Pegourié, B., Lotte, Ph., Colledani, G., Martin, V. and Tore Supra lower-Hybrid Systems Technical Team. (2013). Dynamic Stark spectroscopic measurements of microwave electric fields Inside the plasma near a high-power antenna, *Phys. Rev. Lett.* **110**, p. 215005, http://dx.doi.org/10.1103/PhysRevLett.110.215005.

Kleva, R. G. and Drake, J. F. (1984). Stochastic $\boldsymbol{E} \times \boldsymbol{B}$ particle transport, *Phys. Fluids* **27**, p. 1686, http://dx.doi.org/10.1063/1.864823.

Kleva, R. G., Drake, J. F. and Denton, R. E. (1989). Magnetohydrodynamic simulations of density-limit disruptions in tokamaks, *Comm. Plasma Phys. Control. Fusion* **13**, p. 63, ISSN:0374-2806.

Kleva, Robert G. and Guzdar, Parvez N. (1998). β limit disruptions in tokamaks, *Phys. Rev. Lett.* **80**, p. 3081, https://doi.org/10.1103/PhysRevLett.80.3081.

Kleva, R. G. (1992). Stabilization of sawteeth in tokamaks by sheared poloidal flows *Phys. Fluids B* **4**, p. 218, http://dx.doi.org/10.1063/1.860436.

Kleva, Robert G. and Guzdar, Parvez N. (1999). Collisionless non-ideal ballooning modes, *Phys. Plasmas* **6**, p. 116, http://dx.doi.org/10.1063/1.873266.

Knoepfel, H. and. Spong, D. A. (1979). Runaway electrons in toroidal discharges, *Nucl. Fusion* **19**, p. 785, http://dx.doi.org/10.1088/0029-5515/19/6/008.

Koh, S., Chang, C. S., Ku, S., Menard, J. E., Weitzner, H and Choe, W. (2012). Bootstrap current for the edge-pedestal plasma in a diverted tokamak geometry, emphPhys. Plasmas **19**, p. 072505, http://dx.doi.org/10.1063/1.4736953.

Koide, Y., Kikuchi, M., Mori, M., Tsuji, S., Ishida, S., Asakura, N., Kamada, Y., Nishitani, T., Kawano, Y., Hatae, T., Fujita, T., Fukuda, T., Sakasai, A., Kondoh, T., Yoshino, R. and Neyatani, Y. (1994). Internal transport barrier on $q = 3$ surface and poloidal plasma spin-up in JT-60U high-β_p discharges, *Phys. Rev. Lett.* **72**, pp. 3662-3665, http://dx.doi.org/10.1103/PhysRevLett.72.3662.

Koide, Y., Takizuka, T., Takeji, S., Ishida, S., Kikuchi, M., Kamada, Y., Ozeki, T., Neyatani, Y., Shirai, H., Mori, M. and Tsuji-Iio, S. (1996). Internal transport barrier with improved confinement in the JT-60U tokamak, *Plasma Phys. Control. Fusion* **38**, p. 1011, http://dx.doi.org/10.1088/0741-3335/38/7/006.

Koide, Y. and JT-60 Team. (1997). *Progress in confinement and stability with plasma shape and profile control for steady-state operation in the Japan Atomic Energy Research Institute Tokamak-60 Upgrade, Phys. Plasmas* **4**, p. 1623, http://dx.doi.org/10.1063/1.872364.

Kong, E. H., Ding, B. J., Zhang, L., Liu, L., Qin, C. M., Gong, X. Z., Xu, G. S., Zhang, X. J., Wu, Z. G., Wang, H. Q., Li, M. H., Wei, W.,. Li, Y. C, Xu, L., Wu, J. H., He, Z. X., Shan, J. F., Liu, F. K., Wang, M., Xu, H. D., Zhao, Y. P., Zhao, L. M., Feng, J. Q., Yang, Y., Jia, H., Hu, H. C., Wang, X. J., Wu, D. J. and the EAST Team. (2013). Study and optimization of lower-hybrid wave coupling in the experimental advanced superconducting (EAST) tokamak, *Plasma Phys. Control. Fusion* **55**, p. 065007, http://dx.doi.org/10.1088/0741-3335/55/6/065007.

Konovalov, S. V., Mikhailovskii, A. B., Shirokov, M. S. and Tsypin, V. S. (2002). Transport threshold model of neoclassical tearing modes in the presence of anomalous perpendicular viscosity, *Phys. Plasmas* **9**, p. 4596, http://dx.doi.org/10.1063/1.1509455.

Kotschenreuther, M., Dorland, W., Beer, M. A. and Hammett, G. W. (1995a). Quantitative predictions of tokamak energy confinement from first-principles simulations with kinetic effects, *Phys. Plasmas* **2**, p. 2381, http://dx.doi.org/10.1063/1.871261.

Kotschenreuther, M., Rewoldt, G. and Tang, W. M. (1995b). Comparison of initial value and eigenvalue codes for kinetic toroidal plasma instabilities, *Comp. Phys. Comm.* **88**, pp. 128-140, http://dx.doi.org/10.1016/0010-4655(95)00035-E.

Kraichnan, R. H. (1964). Direct-interaction approximation for shear and thermally driven turbulence, *Phys. Fluids* **1**, pp. 1048-1062, http://dx.doi.org/10.1063/1.1711320.

Kraichnan, R. H. (1967). Inertial ranges in two-dimensional turbulence, *J. Fluid Mech.* **10**, p. 1417, http://dx.doi.org/10.1063/1.1762301.

Kraichnan, R. H. (1971). Inertial-range transfer in two- and three-dimensional turbulence, *J. Fluid Mech.* **47**, pp. 525-535, http://dx.doi.org/10.1017/S0022112071001216.

Krall, N. A. and Trivelpiece, A. W. (1968). *Advances in Plasma Physics, I*, eds. A. Simon and W. B. Thompson (Interscience Publishers, New York), p. 153.

Krall, N. A. and Trivelpiece, A. W. (1973). *Principles of Plasma Physics* (McGraw-Hill Inc.), ISBN:0-070035346-8.

Krasheninnikov, S. I. (2001). On scrape-off layer plasma transport, *Phys. Lett. A* **283**, pp. 368-370, http://dx.doi.org/10.1016/S0375-9601(01)00252-3.

Kraus, M., Tassi, E. and Grasso, D. (2016). Variational integrators for reduced magnetohydrodynamics, *J. Comp. Phys.* **321**, pp. 435-458, http://dx.doi.org/10.1016/j.jcp.2016.05.047.

Krishnamurti, R. and Howard, L. N. (1981). Large-scale flow generation in turbulent convection, *Proc. Nat. Acad. Sci., USA* **78**, pp. 1981-1985, http://dx.doi.org/10.1098/rsta.2009.0250.

Kroetz, T., Roberto, M., Caldas, I. L., Viana, R. L., Morrison, P. J. and Abbamonte, P. (2010). Integrable maps with non-trivial topology: application to divertor configurations, *Nucl. Fusion* **50**, p. 034003, http://dx.doi.org/10.1088/0029-5515/50/3/034003.

Krommes, J. A. Oberman, C. and Kleva, R. G. (1983). Plasma transport in stochastic magnetic fields. Part 3. Kinetics of test particle diffusion, *J. Plasma Phys.* **30**, p. 11, http://dx.doi.org/10.1017/S0022377800000982.

Krommes, J. A. (1999). Recent results on analytical plasma turbulence theory: realizability, intermittency, submarginal turbulence, and self-organized criticality, *Plasma Phys. Control. Fusion* **41**, pp. A641-A652, http://dx.doi.org/10.1088/0741-3335/41/3A/058.

Krommes, J. A. (2002). Fundamental statistical description of plasma turbulence in magnetic fields, *Phys. Reps.* **360**, pp. 1-353, http://dx.doi.org/10.1016/S0370-1573(01)00066-7.

Kruger, S. E., Schnack, D. D. and Sovinec, C. R. (2005), *Phys. Plasmas* **12**, p. 056113.

Kruskal, M. D. and Oberman, C. R. (1958). On the stability of plasma in static equilibrium, *Phys. Fluids* **1**, p. 275, http://dx.doi.org/10.1063/1.1705885.

Kulsrud, R. M. (1983). *Handbook of Plasma Physics*, eds. M. N. Rosenbluth and R. Z. Sagdeev (New York: North Holland).

Kulsrud, R. M. (2005). *Plasma Physics for Astrophysics*, (Princeton University Press, 2005), ISBN:9780691120737.

L-H Mode Database Working Group. (1994). *Nucl. Fusion* **34**, pp. 131-167.

La Haye, R. J. and Sauter, O. (1998). Threshold for metastable tearing modes in DIII-D, *Nucl. Fusion* **38**, p. 987, http://dx.doi.org/10.1088/0029-5515/38/7/303.

La Haye, R. J. (2006). Neoclassical tearing modes and their control, *Phys. Plasmas* **13**, p. 055501, http://dx.doi.org/10.1063/1.2180747.

Lackner, K. (1990). Physical equivalence of tokamak discharges, *Comm. Plasma Phys. Control. Fusion* **13**, pp. 163-171, https://inis.iaea.org/search/search.aspx?orig_q=journal:\%22ISSN\%200374-2806\%22.

Lanczos, C. (1956). *Applied Analysis* (Prentice-Hall, NY).

Landreman, M., Antonsen, Jr., T. M. and Dorland, W. (2015). Universal instability for wavelengths below the ion Larmor scale, *Phys.Rev. Lett.* **114**, p. 095003, http://dx.doi.org/10.1103/PhysRevLett.114.095003.

Lao, L. L., St. John, H., Stambaugh, R. D., Kellman, A. G. and Pfeiffer, W. (1985). Reconstruction of current profile parameters and plasma shapes in tokamaks, *Nucl. Fusion* **25**, http://dx.doi.org/10.1088/0029-5515/25/11/007.

Lao, L. L., Burrell, K. H., Casper, T. S., Chan, V. S., Chu, M. S., DeBoo, J. C., Doyle, E. J., Durst, R. D., Forest, Greenfield, C. M., Groebner, R. J., Hinton, F. L., Kawano, Y., Lazarus, E. A., Lin-Liu, Y. R., Mauel, M. E., Meyer, W. H., Miller, R. L., Navratil, G. A., Osborne, T. H., Peng, Q., Rettig, C. L., Rewoldt, G., Rhodes,

T. L., Rice, B. W., Schissel, D. P., Stallard, B. W., Strait, E. J., Tang, W. M., Taylor, T. S., Turnbull, A. D. and Waltz, R. E. and the DIII-D Team. (1996). Rotational and magnetic shear stabilization of magnetohydrodynamic modes and turbulence in DIII-D high performance discharges, *Phys. Plasmas* **3**, p. 1951, `http://dx.doi.org/10.1063/1.871991`.

Laqua, H. P. (2007). Electron Bernstein wave heating and diagnostic, *Plasma Phys. Control. Fusion* **49**, pp. R1-R42, `http://dx.doi.org/10.1088/0741-3335/49/4/R01`.

Laqua, H. P., Braune, H., Hirsch, M., Hoefel, U., Marsen, S., Moseev, D., Stange, T., Thomsen, H., Wauters, T., Wolf, R. and W7-X Team. (2016). ECRH and ECCD experiments at the Wendelstein7-X Stellarator, APS-DPP Meeting, San Jose, CA, November 2016.

Larichev, V. D. and Reznik, G. M. (1976). Two-dimensional Rossby soliton: An exact solution, *Dokl. Akad. Nauk USSR* **231.5**, pp. 1077-109.

Laval, G. (2001). Thermonuclear Tokamak Panel Report, CEA, Paris, France.

Lazzaro, E., Buttery, R. J., Hender, T. C., Zanca, P., Fitzpatrick, R., Bigi, M., Bolzonella, T., Coelho, R., DeBenedetti, M., Nowak, S., Sauter, O. and Stamp, M. (2002). Error field locked modes thresholds in rotating plasmas, anomalous braking and spin-up, *Phys. Plasmas* **9**, p. 3906, `http://dx.doi.org/10.1063/1.1499495`.

Leblanc, B. P., Bell, R. E., Johnson, D. W., Hoffman, D. E., Long, D. C. and Palladino, R. W. (2003). Operation of the NSTX Thomson scattering system, *Rev. Sci. Instrum.* **74**, pp. 1659-1662, `http://dx.doi.org/10.1063/1.1532763`.

Lee, J., Yun, G. S., Lee, J. E., Kim, M., Choi, M. J., Lee, W., Park, H. K., Domier, C. W., Luhmann, Jr., N. C., Sabbagh, S. A., Park, Y. S., Lee, S. G. and Bak, J. G. (2014). Toroidal mode number estimation of the edge-localized modes using the KSTAR 3-D electron cyclotron emission imaging system, *Rev. Sci. Instrum.* **85**, p. 063505, `http://dx.doi.org/10.1063/1.4883180`.

Lee, S. G., Seol, J., Lee, H. H., Aydemir, A. Y., Terzolo, L., Lee, K. D., Bae, Y. S., Bak, J. G. Choe, G. H., Yung, G. S., Yoo, J. W. and the KSTAR Team. (2015). Long-lived pressure-driven coherent structures in KSTAR plasmas, *Phys. Plasmas* **23**, p. 052511, `http://dx.doi.org/10.1063/1.4949768`.

Legras, B., Santangelo, P. and Benzi, R. (1988). High-resolution numerical experiments for forced two-dimensional turbulence, *Europhys. Lett.* **5**, p. 37, `http://dx.doi.org/10.1209/0295-5075/5/1/007`.

Leith, C. E. (1968). Diffusion approximation for two-dimensional turbulence, *Phys. Fluids* **11**, pp. 671-672, `http://dx.doi.org/10.1063/1.1691968`.

Lesieur, M. (2008). Turbulence in Fluids, *Series: Fluid Mechanics and its Applications* (Berlin: Springer) ISBN:978-1-6434-0.

Leslie, D. C. (1973). *Theory of Turbulence* (Clarendon Press, Oxford).

Levinton, F. M., Zarnstorff, M. C., Batha, S. H., Bell, M., Bell, R. E., Budny, R. V., Bush, C., Chang, Z., Fredrickson, E., Janos, A., Manickam, J., Ramsey, A., Sabbagh, S. A., Schmidt, G. L., Synakowski, E. J. and Taylor, G. (1995). Improved confinement with reversed magnetic shear in TFTR, *Phys. Rev. Lett.* **75**, pp. 4417-4420, `http://dx.doi.org/10.1103/PhysRevLett.75.4417`.

Li, J. and Kishimoto, Y. (2004). Numerical study of zonal flow dynamics and electron transport in electron temperature gradient driven turbulence, *Phys. Plasmas* **11**, pp. 1493-1511, `http://dx.doi.org/10.1063/1.1669397`.

Liao, K. T., Rowan, W. L., Bespamyatnov, I. O., Wukitch, S. J., Lin, Y., Bader, A., Tsujii, N. and the Alcator C-Mod Team. (2011). Minority ion measurements during ICRF experiments in Alcator C-Mod, *AIP Conf. Proc.* **1406**, p. 361, `http://dx.doi.org/10.1063/1.3664993`.

Lichtenberg, A. J. and Lieberman, M. A. (1992). Regular and Chaotic Dynamics, *Appl. Math. Sci.*.

Lin, Z., Hahm, T. S., Lee, W. W., W. M. Tang and R. B. White. (1998). Turbulent transport reduction by zonal flows: massively parallel simulations, *Science* **281**, pp. 1835-1837, `http://dx.doi.org/10.1126/science.281.5384.1835`.

Li, J. and Wan, B. for EAST Team and International Collaborators (2011). *Nucl.Fusion* **51**, p. 094007.

Lindley, D. (2008). *Uncertainty* (Anchor Books) ISBN: 9780307389480.

Lipa, M., Schlosser, J. and Escourbiac, F. (2009). Development of actively cooled plasma-facing components for Tore Supra, *Fusion Sci. Technol.* **56**, pp. 1124-1149.

Lipschultz, B., Whyte, D. and LaBombard, B. (2005). Comparison of particle transport in the scrape-off layer plasmas of Alcator C-Mod and DIII-D, *Plasma Phys. Control. Fusion* **47**, p. 1559, `http://dx.doi.org/10.1088/0741-3335/47/10/001`.

Litaudon. X. (2010). Real-time control of advanced scenarios for steady-state tokamak operation, *Fusion Sci. Tech.* **59**, pp. 469-485.

Liu, J. and Horton, W. (1986a). Electromagnetic solitary vortices in rotating plasma, *Phys. Fluids* **29**, pp. 1828-1835, `http://dx.doi.org/10.1063/1.865611`.

Liu, J. and Horton, W. (1986b). The intrinsic electromagnetic solitary vortices in magnetized plasma, *J. Plasma Phys.* **36**, pp. 1-24, `http://dx.doi.org/10.1017/S0022377800011557`.

Liu, S. F., Zhang, C. L., Kong, W., Guo, S. C., Dong, J. Q., Liu, L. M., Liu, Q., and Liu, Z. Y. (2012). Gyrokinetic study of impurity mode in reversed field pinch, *EPL*, **97**, p. 55004, `http://dx.doi.org/10.1209/0295-5075/97/55004`.

Loarte, A., Huijsmans, G., Futatani, S., Baylor, L. R., Evans, T. E., Orlov, D. M., Schmitz, O., Becoulet, M., Cahyna, P., Gribov, Y., Kavin, A., Sashala Naik, A., Campbell, D. J., Casper, T., Daly, E., Frerichs, H., Kischner, A., Laengner, R., Lisgo, S., Pitts, R. A., Saibene, G. and Wingen, A. (2014). Progress on the application of ELM control schemes to ITER scenarios from the non-active phase to DT operation, *Nucl. Fusion* **54** p. 033007 (18pp), `http://dx.doi.org/10.1088/0029-5515/54/3/033007`.

Lochte-Holtgreven, W. (Ed.) (1968) American Vacuum Society Classics, *Plasma Diagnostics*, OSTI Identifier: 4798641.

Lohr, J., Stallard, B. W., Prater, R., Snider, R. T., Burrell, K. H., Groebner, R. J., Hill, D. N., Matsuda, K., Moeller, C. P., Petrie, T. W., St. John, H. and Taylor, T. S. (1988). Observation of H-mode confinement in the DIII-D tokamak with electron cyclotron heating, *Phys. Rev. Lett.* **60**, pp. 2630-2633, `http://dx.doi.org/10.1103/PhysRevLett.60.2630`.

Longmire, C. L. and Rosenbluth, M. N. (1956). Diffusion of charged particles across a magnetic field, *Phys. Rev.* **103**, pp. 507-510, `http://dx.doi.org/10.1103/PhysRev.103.507`.

Lysak, R. L. and Lotko, W. (1996). On the kinetic dispersion relation for shear Alfvén waves, *J. Geophys. Res.* **101**, pp. 5085-5094, `http://dx.doi.org/10.1029/95JA03712`.

Maget, P., Fevier, O., Garbet, X., Lütjens, H., Luciani, J-F. and Marx, A. (2016). Extended magnetohydrodynamic model for neoclassical tearing mode computations, *Nucl. Fusion* **56**, p. 086004, http://dx.doi.org/10.1088/0029-5515/56/8/086004.

Mahajan, S. M. and Hazeltine, R. D. (1982). Linear theory of electromagnetic instabilities in plasmas with a hollow q-profile, *Nucl. Fusion* **22**, p. 1191, http://dx.doi.org/10.1088/0029-5515/22/9/005.

Mahajan, S. M. and Yoshida, Z. (2000). A collisionless self-organizing model for the high-confinement (H-mode) boundary layer, *Phys. Plasmas* **7**, p. 635, http://dx.doi.org/10.1063/1.873850.

Maisonnier, C. (1989). *Artsimovich Memorial Lecture*, *Plasma Phys. Control. Nucl. Fusion Res., 1988* **1**, pp. 3-8 (International Atomic Energy Agency, Vienna).

Manickam, J., Pomprey, N. and Todd, A. M. M. (1987), Ideal MHD stability properties of pressure-driven modes in low-shear tokamaks, *Nucl. Fusion* **27**, p. 1461, http://dx.doi.org/10.1088/0029-5515/27/9/009.

Mantica, P., Angioni, C., Challis, C., Colyer, G., Frassinetti, L., Hawkes, N., Johnson, T., Tsalas, M., deVries, P. C., Weiland, J., Baiocchi B., Beurskens, M. N. A., Figueiredo, A. C. A., Giroud, C., Hobirk, J., Joffrin, E., Lerche, E., Naulin, V., Peeters, A. G., Salmi, A., Sozzi, C., Strintzi, D., Staebler, G., Tala, T., Van Eester, D. and Versloot, T. (2011). A key to improved ion core confinement in the JET tokamak: Ion stiffness mitigation due to combined plasma rotation and low magnetic shear, *Phys. Rev. Lett.* **107**, p. 135004, http://dx.doi.org/10.1103/PhysRevLett.107.135004.

Marcus, F. A., Caldas, I. L., Guimarẽs-Filho, Z. O., Morrison, P. J., Horton, W., Kuznetsov, Yu. K. and Nascimento, I. C. (2008). Reduction of chaotic particle transport driven by drift waves in sheared flows, *Phys. Plasmas* **15**, p. 112304-1-9, http://dx.doi.org/10.1063/1.3009532.

Marmar, E. S., Rice, J. E., Terry, J. L. and Seguin, F. H. (1982). Impurity injection experiments on the Alcator C tokamak, *Nucl. Fusion* **22**, p. 1567, http://dx.doi.org/10.1088/0029-5515/22/12/003.

Marmar, E. S., Baek, S. G., Barnard, H., Bonoli, P., Brunner, D., Candy, J., Canik, J., Churchill, R. M., Cziegler, I., Dekow, G., Delgado-Aparicio, L., Diallo, A., Edlund, E., Ennever, P., Faust, I., Fiore, C., Gao, C., Golfinopoulos. T., Greenwald. M., Hartwig, Z. S., Holland, C., Hubbard, A. E., Hughes, J. W., Hutchinson, I. H., Irby, J., LaBombard, B., Lin, Yijun, Lipschultz, B., Loarte, A., Mumgaard., R., Parker, R. R., Porkolab, M., Reinke, M. L., Rice, J. E., Scott, S., Shiraiwa, S., Snyder, P., Sorbom, B., Terry, D., Terry, J. L., Theiler, C., Vieira, R., Walk, J. R., Wallace, G. M., White, A., Whyte, D., Wolfe, S. M., Wright, G. M., Wright, J., Wukitch, S. J. and Xu, P. (2015), Alcator C-Mod: research in support of ITER and steps beyond, *Nucl. Fusion* **55**, p. 104020, http://dx.doi.org/10.1088/0029-5515/55/10/104020.

Marmar, E. S. (2016). Overview of Hi-Field Divertor Tokamak Results from Alcator C-Mod, *IAEA Proceedings*,presentation OV/2.

Matsuda, K. (1989). Ray tracing study of the electron cyclotron current drive in DIII-D using 60 GHz, *IEEE Transactions on Plasma Science* **17**, p. 6, http://dx.doi.org/10.1109/27.21664.

Mattor, N. and Parker, S. E. (1997). Nonlinear kinetic fluid equations, *Phys. Rev. Lett.* **79**, pp. 3419-3422; Erratum *Phys. Rev. Lett.* **80**, p. 2256 (1998), http://dx.doi.org/10.1103/PhysRevLett.79.3419.

Mattor, N. (1999). Collisionless fluid equations: Lagrangian, Eulerian, and local forms, *Phys. Plasmas* **6**, p. 1065, `http://dx.doi.org/10.1063/1.873381`.

Maurel, A. and Petitjeans, P. (1999). *Vortex Structure and Dynamics* (Workshop, Rouen, France, Springer).

Mays, M. L., Horton, W., Kozyra, J., Zurbuchen, T. H., Huang, C. and Spencer, E. (2007). Effect of interplanetary shocks on the AL and Dst indices, *Geophys. Res. Lett.* **34**, p. L11104, `http://dx.doi.org/10.1029/2007GL029844`.

Mays, M. L. (2009). The Study of Interplanetary Shocks, Geomagnetic Storms, and Substorms with the WINDMI Model, PhD dissertation thesis, Department of Physics, The University of Texas at Austin.

Mazzucato, E. (1976). Small-scale density fluctuations in the adiabatic toroidal compressor, *Phys. Rev.* **36**, p. 792, `http://dx.doi.org/10.1103/PhysRevLett.36.792`.

Mazzucato, E. (1978). Low-frequency microinstabilities in the PLT tokamak, *Phys. Fluids* **21**, p. 1063, `http://dx.doi.org/10.1063/1.862316`.

Mazzucato, E., Batha, S. H., Beer, M., Bell, M., Bell, R. E., Budny, R. V., Bush, C., Hahm, T. S., Hammett, G. W., Levinton, F. M., Nazikian, R., Park, H., Rewoldt, G., Schmidt, G. L., Synakowski, E. J., Tang, W. M., Taylor, G. and Zarnstorff, M. C. (1996). Turbulent fluctuations in TFTR configurations with reversed magnetic shear, *Phys. Rev. Lett.* **77**, p. 3145, `http://dx.doi.org/10.1103/PhysRevLett.77.3145`.

Mazzucato, E., Bell, R. E., Ethier, S., Hosea, J. C., Kaye, S. M., LeBlanc, B. P., Lee, W. W., Ryan, P. M., Smith, D. R., Wang, W. X., Wilson, J. R. and Yuh, H. (2009). Study of turbulent fluctuations driven by the electron temperature gradient in the National Spherical Torus Experiment, *Nucl. Fusion* **49**, p. 055001, `http://dx.doi.org/10.1088/0029-5515/49/5/055001`.

Mazzucato, E. (2010). Study of turbulent fluctuations driven by the electron temperature gradient in the National Spherical Torus Experiment, *Nucl. Fusion* **50**, p. 029801, `http://iopscience.iop.org/0029-5515/50/2/029801`.

McCarthy, D. R., Drake, J. F., Guzdar, P. N. and Hassam, A. B. (1993). Formation of the shear layer in toroidal edge plasma, *Phys. Fluids B* **5**, p. 1188, `http://dx.doi.org/10.1063/1.860909`.

McDermott, R. M., Angioni, C., Dux, R., Gude, A., Pütterich, T., Ryter, F., Tardini, G. and the ASDEX Upgrade Team. (2011). Effect of electron cyclotron resonance heating (ECRH) on toroidal rotation in ASDEX Upgrade H-mode discharges, *Plasma Phys. Control. Fusion* **53**, p. 035007, `http://dx.doi.org/10.1088/0741-3335/53/3/035007`.

McDonald, D. D., Cordey, J. G., Thomsen, K., Kardaun, O. J. W. F., Snipes, J. A., Greenwald, M., Sugiyama, L., Ryter, F., Kus, A. and Stober, J. (2007). Recent progress on the development and analysis of the ITPA global H-mode confinement database, *Nucl. Fusion* **47**, p. 147, `http://dx.doi.org/10.1088/0029-5515/47/3/001`.

McGuire, K., *et al.* (1983). Study of high-beta magnetohydrodynamic modes and fast-ion losses in PDX, *Phys Rev. Lett.* **50**, p. 891, `http://dx.doi.org/10.1103/PhysRevLett.50.891`; Erratum *Phys. Rev. Lett.* **51**, 1925 (1983).

McWilliams, J. C. (1984). Emergence of isolated coherent vortices in turbulent flow, *J. Fluid Mech.* **146**, pp. 21-43 (Cambridge University Press), `http://dx.doi.org/10.1017/S0022112084001750`.

McWilliams, J. C. (1990). The vortices of two-dimensional turbulence, *J. Fluid Mech* **219**, pp. 361-385 (Cambridge University Press) http://dx.doi.org/10.1017/S0022112090002981.

McWilliams, J. C. (1990). The vortices of geostrophic turbulence, *J. Fluid Mech.* **219**, pp. 387-404 (Cambridge University Press) http://dx.doi.org/10.1017/S0022112090002993.

Meade, D. M. and Yoshikawa, S. (1967). Observation of a ballooning-type instability in a plasma confined by a quadrupole magnetic field, *Phys. Fluids* **10**, pp. 2649-2652, http://dx.doi.org/10.1063/1.1762088.

Meade, D. M., *et al.* (1990). *13th Int. Conf. on Plasma Phys. Control. Fusion Research* (Washington, D.C.) vol. 1, p. 9.

Meade, D. M., *et al.* (2001). Fusion Ignition Research Experiment (FIRE), *Fusion Tech.* **39**, p. 336, http://adsabs.harvard.edu/abs/1999APS..DPP.KP166M.

Meiss, J. D. and Horton, W. (1983). Solitary drift waves in the presence of magnetic shear, *Phys. Fluids* **26**, p. 990, http://dx.doi.org/10.1063/1.864251

Mendoca, J. T. and Galvao, R., Transport equations for lower hybrid waves in a turbulent plasma, *J. Plasma Phys.* **81**, p. 905810206, http://dx.doi.org/10.1017/S0022377814001032.

Messiaen, A. M., Ongena, J., Samm, U., Tokar, M. Z., Unterberg, B., Boucher, D., Finken, K. H., Hintz, E., Koch, R., Mank, G., Telesca, G., Vandenplas, P. E., van Oost, G., van Wassenhove, G., Winter, J., Wolf, G. H., Dumortier, P., Durodie, F., Esser, H. G., Euringer, H., Fuchs, G., Hillis, D. L., Hoenen, F., Konen, L., Koslowski, H. R., Kramer-Flecken, A., Lochter, M., Pospieszczyk, A., Rapp, J., Schweer, B., Soltwisch, H., Uhleman, R., Van Nieuwenhove, R., Vervier, M., Waidmann, G. and Weynants, R. R. (1996). Transport and improved confinement in high power edge radiation cooling experiments on TEXTOR, *Nucl. Fusion* **36**, p. 39, http://dx.doi.org/10.1088/0029-5515/36/1/I03.

Meyer, J. and Mahn, C. (1981). Microinstabilities in the Wendelstein VII-A Stellarator observed by CO_2-laser-light scattering, *Phys. Rev. Lett.* **46**, p. 1206, http://dx.doi.org/10.1103/PhysRevLett.46.1206.

Meyerson, D., Michoski, C., Waelbroeck, F. and Horton, W. (2014). Effect of field-line chaos on plasma filament dynamics and turbulence in the scrape-off layer, *Phys. Plasmas* **21**, p. 072310, http://dx.doi.org/10.1063/1.4890349.

Mikhailovskii, A. B. (1967). *Review of Plasma Physics* **2**, p. 159, ed. M.A. Leontovich (Consultants Bureau, New York).

Mikhailovskii, A. B. (1974). *Theory of Plasma Instabilities* **1** (Consultants Bureau, New York).

Mikhailovskii, A. B., Aburdzhaniya, G. D., Onishchenko, O. G. and Churikov, A. P. (1984). Alfvén vortex solution in a homogeneous magnetized plasma, *Phys. Lett. A* **101**, pp. 263-264, http://dx.doi.org/10.1016/0375-9601(84)90867-3.

Mikhailovskii, A. B., Lakhin, V. P., Mikhailovskaya, L. A. and Onishchenko, O. G. (1985). *Sov. Phys. JETP* **59**, p. 1198.

Mikhailovskii, A. B. (1992). *Electromagnetic Instabilities in an Inhomogeneous Plasma*, Vol. 1, (CRC Press, 1992).

Mikhailovskii, A. B. (1992). *Electromagnetic Instabilities in an Inhomogeneous Plasma* (Institute of Physics Publishing) ISBN:0-7503-0182-1.

Mikhailovskii, A. B.. Smolyakov, A. I, Kovalishen, E. A., *et al.* (2006). Zonal flows generated by small-scale drift-Alfvén modes, *Phys. Plasmas* **13**, p. 042507, `http://dx.doi.org/10.1063/1.2192755`.

Militello, F., Garzotti, L., Harrison, J., Omotani, J. T., Scannell, R., Allan, S., Kirk, A., Lupelli, I., Thornton, A. J. and the MAST Team. (2016). Characterisation of the L-mode scrape off layer in MAST: decay lengths, *Nucl. Fusion* **56**, pp. 16006-16020, `http://dx.doi.org/10.1088/0029-5515/56/1/016006`.

Misguich, J. H., Balescu, R. and Nakach, R. (1988). *Plasma Phys. Control. Nucl. Fusion Res.* Proc. 12th Int. Conf., Nice, 12-19, 3, p. 521, (International Atomic Energy Agency, Vienna, 1989).

Missirlian, M., Bucalossi, J., Corre, Y., Ferlay, F., Firdaouss, M., Garin, P., Grosman, A., Guilhem, D., Gunn, J., Languille, P., Lipa, M., Richou, M. and Tsitrone, E. (2014). The WEST project: Current status of the ITER-like tungsten divertor, *Fusion Eng. Design* **89**, pp. 1048-1053, `http://dx.doi.org/10.1016/j.fusengdes.2014.01.050`.

Miura, H. and Nakajima, N. (2010). Influences of ballooning modes with moderate wavenumber on MHD equilibrium in LHD, *Nucl. Fusion* **50**, p. 054006, `http://dx.doi.org/10.1088/0029-5515/50/5/054006`.

Miyamoto, K. (1976). Chapter II, pp. 121-193, ISBN:0-262-13145-5.

Miyamoto, K. (1978). Recent stellarator research, *Nucl. Fusion* **18**, p. 243, `http://dx.doi.org/10.1088/0029-5515/18/2/008`.

Miyamoto, K. (1989). *Plasma Physics for Nuclear Fusion* (MIT Press, Cambridge, Mass) pp. 8-9.

Monin, A. S. and Yaglom, A. M. (1965). *Statistical Fluid Mechanics* (MIT Press).

Moreau, R. (1983) Two-Dimensional Turbulence, ed. R. Journal de Mechanique/Journal of Theoretical and Applied Mechanics, supplement, ISBN:2-04-011785-7.

Mori, M., Ishida, S., Ando, T., Annoh, K., Asakura, N., Azumi, M., van Blokland, A. A. E., Frieling, G. J. Fujii, T., Fujita, T., Fukuda, T., Funahashi, A., Hatae, T., Hoek, M., Honda, M., Hosogane, N., Isei, N., Itami, K., Kamada, Y., Kawano, Y., Kikuchi, M., Kimura, H., Kimura, T., Kishimoto, H., Kitsunezaki, A., Kodama, K., Koide, Y., Kondoh, T., Kubo, H., Kuriyama, M., Matsuoka, M., Matsuzaki, Y., Miya, N., Nagami, M., Nagashima, A., Naito, O., Nakamura, H., Nemoto, M., Neyatani, Y., Ninomiya, H., Nishitani, T., Ohga, T., Ohmori, S., Saidoh, M., Sakasai, A., Sato, M., Shimada, M., Shimizu, K., Shirai, H., Sugie, T., Takeuchi, H., Tani, K., Tobita, K., Tsuji, S., Ushigusa, K., Yamada, M., Yonekawa, I., Yoshida, H. and Yoshino, R. (1994). Achievement of high fusion triple product in the JT-60U high β_p H mode, *Nucl. Fusion* **34**, p. 1045, `http://dx.doi.org/10.1088/0029-5515/34/7/I09`.

Moroz, P. E. and Colestock, P. L. (1991). Electron absorption of fast waves in global wave calculations, *Plasma Phys. Control. Fusion* **33**, pp. 417-431, `http://dx.doi.org/10.1088/0741-3335/33/5/002`.

Morrison, P. J. (1998). Hamiltonian description of the ideal fluid, *Rev. Mod. Phys.* **70**, pp. 467-521, `http://dx.doi.org/10.1103/RevModPhys.70.467`.

Moser, J. (1973). On a nonlinear problem in differential geometry. *Dynamical Systems* (New York: Academic Press).

Motojima, O., Ohyabu, N., Komori, A., Kaneko, O., Yamada, H., Kawahata, K., Nakamura, Y., Ida, K., Akiyama, T., Ashikawa, N., Cooper, W. A., Ejiri, A., Emoto, M.,

Ezumi, N., Funaba, H., Fukuyama, A., Goncharov, P., Goto, M., Idei, H., Ikeda, K., Inagaki, S., Isobe, M., Kado, S., Kawazome, H., Khlopenkov, K., Kobuchi, T., Kondo, K., Kostrioukov,A., Kubo, S., Kumazawa, R., Liang, Y., Lyon, J. F., *et al.* , *Nucl. Fusion* **43**, p. 1674, http://dx.doi.org/10.1088/0029-5515/43/12/013.

Motojima, O., Yamada, H., Komori, A., Ohyabu, N., Mutoh, T., Kaneko, O., Kawahata, K., Mito, T., Ida, K., Imagawa, S., Nagayama, Y., Shimozuma, T., Watanabe, K. Y. Masuzaki, S., Miyazawa, J., Morisaki, T., Morita, S., Ohdachi, S., Ohno, N., Saito, K., Sakakibara, S., Takeiri, Y., Tamura, N., Toi, K., Tokitani, M., Yokoyama, M., Yoshinuma, M., Ikeda, K., Isayama, A., Ishii, K., Kubo, S., Murakami, S., Nagasaki, K., Seki, T., Takahata, K., Takenaga, H. and the LHD Experimental Group. (2007). Extended steady-state and high-beta regimes of net-current free heliotron plasmas in the Large Helical Device, *Nucl. Fusion* **47**, http://dx.doi.org/10.1088/0029-5515/47/10/S15.

Müller, S. H. (2006). Turbulence in basic toroidal plasmas (Thesis), http://dx.doi.org/10.5075/epfl-thesis-3672.

Muraglia, M., Agullo, O., Yagi, M., Benkadda, S. Beyer, P., Garbet, X., Itoh S.-I., Itoh, K. and Sen, A. (2009). Effect of the curvature and the β parameter on the nonlinear dynamics of a drift tearing magnetic island, *Nucl. Fusion* **49**, p. 055016, http://dx.doi.org/10.1088/0029-5515/49/5/055016.

Muraglia, M., Agullo, O., Benkadda, S., Yagi, M., Garbet, X. and Sen, A. (2011). Generation and amplification of magnetic islands by drift interchange turbulence, *Phys. Rev. Lett.* **107**, p. 095003, http://dx.doi.org/10.1103/PhysRevLett.107.095003.

Nachtsheim, P. R. and Reshotko, E. (1965). NASA Technical Note d-3144 (National Aeronautics and Space Administration, Washington D.C.).

Nakamura, R., Baumjohann, W., Asano, Y., Runov, A., Balogh, A., Owen, C. J., Fazakerley, A. N., Fujimoto, M., Klecker, B. and Rème, H. (2006). Dynamics of thin current sheets associated with magnetotail reconnection, *J. Geophys. Res.* **111**, p. A11206, http://dx.doi.org/10.1029/2006JA011706.

Nakata, T.-H. Watanabe, H. Sugama, H. and Horton, W. (2010). Formation of coherent vortex streets and transport reduction in electron temperature gradient driven turbulence *Phys. Plasmas* **17**, p. 042306, http://dx.doi.org/10.1063/1.3356048.

Nakata, M., Honda, M., Yoshida, M., Urano, H., Nunami, M., Maeyama, S., Watanabe, T-H. and Sugama, H. (2016). Validation studies of gyrokinetic ITG and TEM turbulence simulations in a JT-60U tokamak using multiple flux matching, *Nucl. Fusion* **56**, p. 086010, http://dx.doi.org/10.1088/0029-5515/56/8/086010.

Nakata, M., Nunami, M., Sugama, H. and Watanabe, T-H. (2016). Impact of hydrogen isotope species on microinstabilities in helical plasmas, *Plasma Phys. Control. Fusion* **58**, p. 074008, http://dx.doi.org/10.1088/0741-3335/58/7/074008.

Napoli, F., Castaldo, C., Cesario, R. and Schettini, G. (2013). Modeling of the nonlinear mode coupling of lower-hybrid waves in tokamak plasmas, *-Plasma Phys. Control. Fusion* **55**, p. 095004, http://dx.doi.org/10.1088/0741-3335/55/9/095004.

Nave, M. F. F., Ali-Arshad, S., Alper, B., Balet, B., De Blank, H. J., Borba, D., Challis, C. D., Von Hellermann, M. G., Hender, T. C., Huysmans, G. T. A., Kerner, W., Kramer, G. J., Porcelli, F., O'Rourke, J., Porte, L., Sadler, G. J., Smeulders, P., Sips, A. C. C., Stubberfield, P. M., Stork, D., Reichle, R., Wessom, J. A. and Zwingmann, W. (1995). MHD activity in JET hot ion H-mode discharges, *Nucl. Fusion* **35**, p. 409, http://dx.doi.org/10.1088/0029-5515/35/4/I04.

Nave, M. F. F. and Wesson, J. A. (1986). *Nucl. Fusion* **18**, p. 617.

Nazikian, R., Fu, G. Y., Batha, S. H., Bell, M. G., Bell, R. E., Budny, R. V., Bush, C. E., Chang, Z., Chen, Y., Cheng, C. Z., Darrow, D. S., Efthimion, P. C., Fredrickson, E. D., Gorelenkov, N. N., Leblanc, B., Levinton, F. M., Majeski, R., Mazzucato, E., Medley, S. S., Park, H. K., Petrov, M. P., Spong, D. A., Strachan, J. D., Synakowski, E. J., Taylor, G., Von Goeler, S., White, R. B., Wong, K. L. and Zweben, S. J. (1997). Alpha particle driven toroidal Alfvén eigenmode in the tokamak fusion test reactor, *Phys. Rev. Lett.* **78**, p. 2976, `http://dx.doi.org/10.1103/PhysRevLett.78.2976`.

Nazikian, R., Kramer, G. J., Cheng, C. Z. and Gorelenkov, N. N., Berk and SHarapov, S. E. (2003). *Phys Rev. Lett.* **91**, p. 125003, `http://dx.doi.org/10.1103/PhysRevLett.91.125003`.

Nelson, G. O., Spies and Hedrick, C. L. (1978). Necessary stability criterion for plasma equilibrium with tensor pressure, *Phys. Fluids* **21**, p. 1742, `http://dx.doi.org/10.1063/1.862089`.

Neu, R., Dux, R., Geier, A., Greuner, H., Krieger, K., Maier, H., Pugno, R., Rohde, V., Yoon, S. W. and ASDEX Upgrade Team. (2003). New results from the tungsten programme at ASDEX Upgrade, *J. Nucl. Mater.* **313**, p. 116, `http://dx.doi.org/10.1016/S0022-3115(02)01386-7`.

Nevins, W. M., Hammett, G. W., Dimits, A. M., Dorland, W. and Shumaker, D. E. (2005). Discrete particle noise in particle-in-cell simulations of plasma microturbulence, *Phys. Plasmas* **12**, p. 122305, `http://dx.doi.org/10.1063/1.2118729`.

Nevins, W. M., Candy, J., Cowley, S., Dannert, T., Dimits, A., Dorland, W., Estrada-Milla, C., Hammett, G. W., Jenko, F., Pueschel, M. J. and Shumaker, D. E. (2006). Characterizing electron temperature gradient turbulence via numerical simulation, *Phys. Plasmas* **13**, p. 122306, `http://dx.doi.org/10.1063/1.2402510`.

Newman, D., Carreras, B., Lopez-Bruna, D., Diamond, P. H. and Lebedev, V. (1998). Dynamics and control of internal transport barriers in reversed shear discharges, *Phys. Plasmas* **5**, p. 938, `http://dx.doi.org/10.1063/1.872662`.

Newton, S. and Helander, P. (2006). Neoclassical momentum transport in an impure rotating tokamak plasma, *Phys. Plasma* **13**, p. 102505, `http://dx.doi.org/10.1063/1.2160518`.

Ninomiya, H. and the JT-60 Team. (1992). JT-60U latest results and future prospects, *Phys. Fluids B* **4**, p. 2070, `http://dx.doi.org/10.1063/1.860480`.

Nordman, H., Weiland, J. and Jarmén, A. (1990). Simulation of toroidal drift mode turbulence driven by temperature gradients and electron trapping, *Nucl. Fusion* **30**, p. 983, `http://dx.doi.org/10.1088/0029-5515/30/6/001`.

Nordman, H. and Weiland, J. (1992). Concept of marginal stability and recent experimental results from the TFTR tokamak, *Nucl. Fusion* **32**, p. 1653, `http://dx.doi.org/10.1088/0029-5515/32/9/I14`.

Nordman, H. and Weiland, J. (1993). Comments on Ion-temperature-gradient-driven transport in a density modification experiment on the Tokamak Fusion Test Reactor, *Phys. Fluids B* **5**, p. 1032, `http://dx.doi.org/10.1063/1.860507`. .

Nordman, H., Strand, P., Eriksson, A. and Weiland, J. (2005). Anomalous particle transport in D-T plasmas, *Plasma Phys. Control. Fusion* **47**, p. L11, `http://dx.doi.org/10.1088/0741-3335/47/6/L01`.

Novakovskii, S. V., Mikhailovskii, A. B. and Onishchenko, D. G. (1988). On the theory of Kolmogorov spectra of drift wave turbulence, *Phys. Letts. A* **132**, p. 33, http://dx.doi.org/10.1016/0375-9601(88)90434-3.

Nunami, M., Watanabe, T-H., Sugama, H. and Tanaka, K. (2011). Linear gyrokinetic analyses of ITG modes and zonal flows in LHD with high-ion temperature, *Plasma Fusion Res.* **6**, p. 1403001, http://dx.doi.org/10.1585/pfr.6.1403001.

Nunami, M., Watanabe, T-H., Sugama, H. and Tanaka, K. (2012). Gyrokinetic turbulent transport simulation of a high ion temperature plasma in large helical device experiment, *Phys. Plasmas* **19**, p. 042504, http://dx.doi.org/10.1063/1.4704568.

Obrejan, K., Imadera, K., Li, J-Q. and Kishimoto, Y. (2015). Development of a global toroidal gyrokinetic vlasov code with new real space field solver, *Plasma Fusion Res.* **10**, p. 3403042, http://dx.doi.org/10.1585/pfr.10.3403042.

Ohkawa, T. and Yoshikawa, M. (1968). Observation of fluctuations in the plasma confined in the toroidal octopole, *Phys. Fluids* **11**, p. 2039, http://dx.doi.org/10.1063/1.1692240.

Ohkawa, T. (1978). A transport model for Alcator scaling in tokamaks, *Phys. Lett. A* **67**, p. 35, http://dx.doi.org/10.1016/0375-9601(78)90560-1.

Okabayahsi, M., Matsunaga, G., deGrassie, J. S., Heidbrink, W. W., In, Y., Liu, Y. Q., Reimerdes, H., Solomon, W. M., Strait, E. J., Takechi, M., Asakura, N., Budny, R. V., Jackson, G. L., Hanson, J. M., La Haye, R. J., Lanctot, M. J., Manickam, J., Shinohara, K. and Zhu, Y. B. (2011). Off-axis fishbone-like instability and excitation of resistive wall modes in JT-60U and DIII-D, **18**, p. 056112, http://dx.doi.org/10.1063/1.3575159 .

Ofman, L., Chen, X. L., Morrison, P. J. and Steinolfson, R. S. (1991). Resistive tearing mode instability with shear flow and viscosity, *Phys. Fluids B* **3**, pp. 1364-1373, http://dx.doi.org/10.1063/1.859701.

Ofman, L. (1992). Double tearing instability with shear flow, *Phys. Fluids B* **4**, p. 2751, http://dx.doi.org/10.1063/1.860148.

Onishchenko, O. G., Pokhotelov, O. A., Horton, W., Scullion, E. and Fedun, V. (2015). Large-scale Alfvén vortices, *Phys. Plasmas* **22**, p. 122901, http://dx.doi.org/10.1063/1.4936978.

Orszag, S. A. (1971). Accurate solution of the Orr-Sommerfeld stability equation, *J. Fluid Mech.* **50**, pp. 689-703, http://dx.doi.org/10.1017/S0022112071002842.

Gottlieb, D. and Orszag, S. A. (1977). *Numerical analysis of spectral methods: theory and applications* [Society for Industrial and Applied Mathematics (SIAM)], Philadelphia, PA), ISBN: 978-0-898710-23-6.

Ott, E. (1979). Lower-hybrid wave scattering by density fluctuations *Phys. Fluids* **22**, p. 1732, http://dx.doi.org/10.1063/1.862809.

Ottaviani, M. and Porcelli, F. (1993). Nonlinear collisionless magnetic reconnection, *Phys. Rev. Lett.* **71**, pp. 3802-3805, http://dx.doi.org/10.1103/PhysRevLett.71.3802.

Parail, V. V., Baranov, Yu.F., Challis, C. D., Cottrell, G. A., Fischer, B., Gormezano, C., Huysmans, G. T. A., Litaudon, X., Sips, A. C. C., Söldner, F. X., Springmann, E. M., Taroni, A. and Ward, D. J. (1999). Predictive modeling of JET optimized shear discharges, *Nucl. Fusion* **39**, p. 429, http://dx.doi.org/10.1088/0029-5515/39/3/310.

Paris, R. B. and Sy, W. N-C. (1988). Influence of equilibrium shear flow along the magnetic field on the resistive tearing instability, *Phys. Fluids* **26**, p. 2966, http://dx.doi.org/10.1063/1.864061.

Park, W., Fredrickson, E. D., Janos, A., Manickam, J. and Tang, W. M. (1995). High-β disruption in tokamaks, *Phys. Rev. Lett.* **75**, p. 1763, https://doi.org/10.1103/PhysRevLett.75.1763.

Park, H., Mazzucato, E., Munsat, T., Domier, C. W., Johnson, M., Luhmann, Jr., N. C., Wang, J., Xia, Z., Classen, I. G. J., Donné, A. J. H. and van de Pol, M. J. (2004). Simultaneous microwave imaging system for density and temperature fluctuation measurements on TEXTOR (invited), *Rev. Scien. Instrum.*, **75**, pp. 3787-3792, http://dx.doi.org/10.1063/1.1794864.

Parker, S. E., Mynick, H. E. Artun, M., Cummings, J. C., Decyk, V., Kepner, J. V., Lee, W. W. and Tang, W. M. (1996). Radially global gyrokinetic simulation studies of transport barriers, *Phys. Plasmas* **3**, p. 1959, http://dx.doi.org/10.1063/1.871992.

Parra, F. I. and Catto P. J. (2010). Transport of momentum in full f gyrokinetics, *Phys. Plasmas* **17**, p. 056106, http://dx.doi.org/10.1063/1.3327127.

Pearlstein, L. D. and Berk, H. L. (1969). Universal eigenmode in a strongly sheared magnetic field, *Phys. Rev. Lett.* **23**, pp. 220-222, http://dx.doi.org/10.1103/PhysRevLett.23.220.

Pedlosky, J. (1987). *Geophysical Fluid Dynamics*, 2nd Ed., *Science*.

Peeters, A. G., Angioni, C. and the ASDEX Upgrade Team. (2005). Linear gyrokinetic calculations of toroidal momentum transport in a tokamak due to the ion temperature gradient mode, *Phys. Plasmas* **12**, p. 072515, http://dx.doi.org/10.1063/1.1949608.

Peeters, A. G., Angioni, C. and Strintzi, D. (2007). Toroidal momentum pinch velocity due to the Coriolis drift effect on small- scale instabilities in a toroidal plasma, *Phys. Rev. Lett.* **98**, p. 265003, http://dx.doi.org/10.1103/PhysRevLett.98.265003.

Peng, F. (2006). Expanding PXI to solve new applications with PXI Express, Autotestcon, IEEE, Date: 18-21 Sept. 2006, pp. 543-546, https://doi.org/10.1109/AUTEST.2006.283723.

Perez, J., Horton, W., Gentle, K., Rowan, W., Lee, K. and Dahlburg, R. (2006). Drift wave instability in the Helimak experiment, *Phys. Plasmas* **13**, pp. 032101.1-032101.10, http://dx.doi.org/10.1063/1.2168401.

Persson, M. and Dewar, R. L. (1994). Nonlinear destabilization of linearly stable tearing modes with multiple rational surfaces. *Phys. Plasmas* **1**, p. 1256, http://dx.doi.org/10.1063/1.870723.

Petrov, Yu., Becoulet, A. and Monakhov, I. (2000). Coupled full-wave and ray-tracing numerical treatment of mode conversion in a tokamak plasma, *Phys. Plasmas* **7**, p. 911, http://dx.doi.org/10.1063/1.873888.

Petty, T. J., Baldwin, M. J., Hasan, M. I., Doerner, R. P. and Bradley, J. W. (2015). Tungsten 'fuzz' growth re-examined: the dependence on ion fluence in non-erosive and erosive helium plasma, *Nucl. Fusion* **55**, p. 093033, http://dx.doi.org/10.1088/0029-5515/55/9/093033.

Peysson, Y. and Decker, J. (2007). Advanced lower hybrid current drive modeling, *AIP Conf. Proc.* **933**, pp. 293-296, http://dx.doi.org/10.1063/1.2800496.

Peysson, Y., Decker, J., Morini, L. and Coda, S. (2011). RF current drive and plasma fluctuations, *Plasma Phys. Control. Fusion* **53**, p. 124028, http://dx.doi.org/10.1088/0741-3335/53/12/124028.

Podestá, M., Fasoli, A., Labit, B. McGrath, M. Müller, S. H. and Poli F. M. (2005). Plasma production by low-field side injection of electron cyclotron waves in a simple magnetized torus, *Plasma Phys. Control. Fusion* **47**, pp. 1989-2002, http://dx.doi.org/10.1088/0741-3335/47/11/008.

Pokhotelov, O. A., Sagdeev, R. Z., Balikhin, M. A. and Treumann, R. A. (2004). Mirror instability at finite ion-Larmor radius wavelengths. *J. Geophys. Res.* **109**, pp. A09213-A09222, http://dx.doi.org/doi:10.1029/2004JA010568.

Porcelli, F., Boucher, D. and Rosenbluth, M. N. (1996). Model for the sawtooth period and amplitude, *Plasma Phys. Control. Fusion* **38**, http://dx.doi.org/10.1088/0741-3335/38/12/010.

Porkolab, M. (2015). Nonlinear phenomena in RF wave propagation in magnetized plasma: A review, *AIP Conf. Proc.* **1689**, p. 020002, http://dx.doi.org/10.1063/1.4936464.

Post, R. F. (2002). The kinetic stabilizer: Further calculations and options, *Fusion Sci. Tech.*, http://www.refdoc.fr/?traduire=en\&FormRechercher=submit\&FormRechercher_Txt_Recherche_name_attr=identifiantsDoc:\%20(1536-1055)

Porte, L., Coda, S., Alberti, S., Arnoux, G., Blanchard, P., Bortolon, A., Fasoli, A., *et al.* (2006). *21st International Atomic Energy Agency Fusion Energy Conf.* (Vienna) EX/P6-20.

Pratt, J. and Horton, W. (2006). Global energy confinement scaling predictions for the kinetically stabilized tandem mirror, *Phys. Plasmas* **13**, p. 042513, http://dx.doi.org/10.1063/1.2188913.

Pritchett, P. L., Lee, Y. C. and Drake, J. F. (1980). Linear analysis of the double-tearing mode, *Phys. Fluids* **23**, p. 1368, http://dx.doi.org/10.1063/1.863151.

Rappazzo, A. F., Velli, M., Einaudi, G. and Dahlburg, R. B. (2005). Diamagnetic and expansion effects on the observable properties of the slow solar wind in a coronal streamer, *Astrophys. J.* **633**, p. 474, http://dx.doi.org/10.1086/431916.

Rebut, P.-H., Bickerton, R. J. and Keen, B. E. (1985). The Joint European Torus: installation, first results and prospects, *Nucl. Fusion* **25**, p. 1011, http://dx.doi.org/10.1088/0029-5515/25/9/003.

Rebut, P. H. Lallia, P. P. and Watkins, M. L. (1988). The critical temperature gradient model of plasma transport: Applications to JET and future tokamaks, *Plasma Phys. Control. Nucl. Fusion* **2**, pp. 191-200, http://www.iaea.org/inis/collection/NCLCollectionStore/_Public/21/008/21008733.pdf.

Rechester, A. B. and Rosenbluth, M. N. (1978), Electron heat transport in a tokamak with destroyed magnetic surfaces, *Phys. Rev. Lett.* **40**, pp. 38-41, http://dx.doi.org/10.1103/PhysRevLett.40.38.

Redd, A. J., Kritz, A. H., Bateman, G. and Horton, W. (1998). Predictive simulations of tokamak plasmas with a model for ion-temperature-gradient-driven turbulence, *Phys. Plasmas* **5**, pp. 1369-1379, http://dx.doi.org/10.1063/1.872797.

Rewoldt, G., Tang, W. M. and Chance, M. S. (1982). Electromagnetic kinetic toroidal eigenmodes for general magnetohydrodynamic equilibria, *Phys. Fluids* **25**, pp. 480-490, http://dx.doi.org/10.1063/1.863760.

Rewoldt, G. and Tang, W. M. (1983). Beam-ion and alpha-particle effects on microinstabilities in tokamaks, *Phys. Fluids* **26**, pp. 3619-3623, http://dx.doi.org/10.1063/1.864122.

Rewoldt, G., Beer, M. A., Chance, M. S., Hahm, T. S., Lin, Z. H. and Tang, W. M. (1998). Sheared rotation effects on kinetic stability in enhanced confinement tokamak plasmas, and nonlinear dynamics of fluctuations and flows in axisymmetric plasmas, *Phys. Plasmas* **5**, pp. 1815-1821, `http://dx.doi.org/10.1063/1.872851`.

Rewoldt, G., Ku, L.-P., Tang, W. M. and Cooper, W. A. (1999). Drift mode calculations in nonaxisymmetric geometry *Phys. Plasmas* **6**, p. 4705, `http://dx.doi.org/10.1063/1.873757`.

Rhodes, T. L., Peebles, W. A., DeBoo, J. C., Prater, R., Kinsey, J. E., Staebler, G. M., Candy, J., Austin, M. E., Bravenec, R. V., Burrell, K. H., deGrassie, J. S., Doyle, E. J., Gohil, P., Greenfield, C. M., Groebner, R. J., Lohr, J., Makowski, M. A., Nguyen, X. V., Petty, C. C., Solomon, W. M., St John, H. E., Van Zeeland, M. A., Wang, G. and Zeng, L. (2007). Broad wavenumber turbulence and transport during Ohmic and electron cyclotron heating in the DIII-D tokamak, *Plasma Phys. Control. Fusion* **49**, p. B183, `http://dx.doi.org/10.1088/0741-3335/49/12B/S17`.

Rice, J. E., Greenwald, M. J., Podpaly, Y. A., Reinke, M. L., Diamond, P. H., Hughes, W., Howard, N. T., Ma, Y., Cziegler, I., Duval, B. P., Ennever, P. C., Ernst, D., Fiore, C. L., Gao, C., Irby, J. H., Marmar, E. S., Porkolab, M., suji, N and Wolfe, S. M. (2012). Ohmic energy confinement saturation and core toroidal rotation reversal in Alcator C-Mod plasmas, *Phys. Plasmas* **19**, p. 056106, `http://dx.doi.org/10.1063/1.3695213` (copyright 2013, American Institute of Physics).

Rognlien, T. D. (1973). Low-frequency flute instabilities of a bounded plasma column, *J. Appl. Phys.* **44**, pp. 3505-3512, `http://dx.doi.org/10.1063/1.1662794`.

Romanelli, M. and Ottaviani, M. (1998). Effects of density asymmetries on heavy impurity transport in a rotating tokamak plasma, *Plasma Phys. Control. Fusion* **40**, p. 1767, `http://dx.doi.org/10.1088/0741-3335/40/10/007`.

Rose, H. A. (1982). Test-particle transport in stochastic magnetic fields: A fluid representation, *Phys. Rev. Lett.* **48**, pp. 260-263, `http://dx.doi.org/10.1103/PhysRevLett.48.260`.

Rosenbluth, M. N., Krall, N. A. and Rostoker, N. (1962). Finite Larmor radius stabilization of "weakly" unstable confined plasmas, *Nucl. Fusion Suppl.* **143**, OSTI ID:4808729.

Rosenbluth, M. N. and Simon, A. (1965). Finite Larmor radius equations with nonuniform electric fields and velocities, *Phys. Fluids* **8**, p. 1300, `http://dx.doi.org/10.1063/1.1761402`.

Rosenbluth, M. N. and Rutherford, P. H. (1981). *Fusion* ed. E. Teller, (Academic Press, New York) **1**, Part A, p. 31.

Rosenbluth, M. N. and Hinton, F. L. (1998). Poloidal flow driven by ion-temperature-gradient turbulence in tokamaks, *Phys. Rev. Lett.* **80**, pp. 724-727, `http://dx.doi.org/10.1103/PhysRevLett.80.724`.

Rowan, W. L., Bespamyatnov, I. O. and Granetz, R. S. (2008). Wide-view charge exchange recombination spectroscopy diagnostic for Alcator C-Mod, *Rev. Sci. Instrum.* **79**, p. 10F529, `http://dx.doi.org/10.1063/1.2979865`.

Rowan, W. L., Bespamyatnov, Igor O. and Fiore, C. L. (2008). Light impurity transport at an internal transport barrier in Alcator C-Mod, *Nucl. Fusion* **48**, p. 105005, `http://dx.doi.org/10.1088/0029-5515/48/10/105005`.

Rutherford, P. H. (1973). Nonlinear growth of the tearing mode, *Phys. Fluids* **16**, pp. 1903-1909, `http://dx.doi.org/10.1063/1.1694232`.

Rutherford, P. H. (1974). Impurity transport in the Pfirsch-Schlüter regime, *Phys. Fluids* **17**, pp. 1782-1784, http://dx.doi.org/10.1063/1.1694975.

Rypdal, K., Fredriksen, Å., Olsen, O. M. and Hellbolm, K. G. (1997). Microwave-plasma in a simple magnetized torus, *Phys. Plasmas* **4**, p. 1468, http://dx.doi.org/10.1063/1.872320.

Ryter, F., Imbeaux, F., Leuterer, F., Fahrbach, H.-U., Suttrop, W. and ASDEX Upgrade Team. (2001a). Experimental characterization of the electron heat transport in low-density ASDEX Upgrade Plasmas, *Phys. Rev. Lett.* **86**, pp. 5498-5501, http://dx.doi.org/10.1103/PhysRevLett.86.5498.

Ryter, F., Leuterer, F., Pereverzev, G., Fahrbach, H.-U., Stober, J., Suttrop, W. and ASDEX Upgrade Team. (2001b). Experimental evidence for gradient length-driven electron transport in tokamaks, *Phys. Rev. Lett.* **86**, pp. 2325-2328, http://dx.doi.org/10.1103/PhysRevLett.86.2325.

Ryutov, D. D. (2005). Axial electron heat loss from mirror devices revisited, *Fusion Sci. Tech.* **47**, pp. 148-154, http://www.llnl.gov/tid/lof/documents/pdf/310973.pdf.

Ryutov, D. D., Berk, H. L., Hooper, E. B. and Thomassen, K. I. (2011). Magnetohydrodynamically stable axisymmetric mirrors, *Phys. Plasmas* **18**, pp. 092301, http://dx.doi.org/10.1063/1.3624763.

Sabot, R., Hennequin, P. and Colas, L. (2009). Fluctuating measurements and their link with the transport in Tore Supra, *Fusion Sci. Tech.* **56**, pp. 1253-1272, https://inis.iaea.org/search/search.aspx?orig_q=journal:\%22ISSN\%201536-1055\%22.

Sagdeev, R. Z. and Galeev, A. A. (1969). *Nonlinear Plasma Theory* (Benjamin, New York).

Samm, U., Bertschinger, G., Bogen, P., Hey, J. D., Hintz, E., Konen, L., Lie, Y. T., Pospieszczyk, A., Rusbuldt, D., Schorn, R. P., Schweer, B., Tokar, M. and Unterberg, B. (1993). Radiative edges under control by impurity fluxes, *Plasma Phys. Control. Fusion* **35**, p. B167, http://dx.doi.org/10.1088/0741-3335/35/SB/013.

Saoutic, B., Beaumont, B., Becoulet, A., Bizarro, J. P., Fraboulet, D., Garbet, X., Goniche, M., Guiziou, L., Hoang, G. T., Hutter, T., Joffrin, E., Kuus, H., Litaudon, X., Mollard, P., Moreau, D., Nguyen, F., Pecquet, A. L., Peysson, Y., Rey, G., van Houtte, D. and Zabiego, M. (1994). High-power ICRF and LHCD experiments on Tore Supra, *AIP* **289**, pp. 24-31, http://dx.doi.org/10.1063/1.45005.

Sato, T. and Walker, R. J. (1982). Magnetotail dynamics excited by the streaming tearing mode, *J. Geophys. Res.* **87**, pp. 7453-7459, http://dx.doi.org/10.1029/JA087iA09p07453.

Sauter, O., La Haye, R. J., Chang, Z., Gates, D. A., Kamada, Y., Zohm, H., Bondeson, A., Boucher, D., Callen, J. D., Chu, M. S., Gianakon, T. A., Gruber, O., Harvey, R. W., Hegna, C. C., Lao, L. L., Monticello, D. A., Perkins, F., Pletzer, A., Reiman, A. H., Rosenbluth, M. N., Strait, E. J., Taylor, T. S., Turnbull, A. D., Waelbroeck, F., Wesley, J. C., Wilson, H. R. and Yoshino, R. (1997). β limits in long-pulse tokamak discharges, *Phys. Plasmas* **4**, pp. 1654-1664, http://dx.doi.org/10.1063/1.872270.

Sauter, O. Angioni, C. and Lin-Liu, Y. R. (1999). Neoclassical conductivity and bootstrap current formulas for general axisymmetric equilibria and arbitrary collisionality regime, *Phys. Plasmas* **6**, p. 2834, http://dx.doi.org/10.1063/1.873240.

Scannell, R., Walsh, M. J., Dunstan, M. R., Figueiredo, J., Naylor, G., O'Gorman, T., Shibaev, S., Gibson, K. J. and Wilson, H. (2010). A 130 point Nd:YAG Thomson scattering diagnostic on MAST, *Rev. Sci. Instrum.* **81**, p. 10D520, `http://dx.doi.org/10.1063/1.3460628`.

Schepp, T. J., Pegoraro, F. and Kuvshinov, B. N. (1994). Generalized two-fluid theory of nonlinear magnetic structures, *Phys. Plasmas* **1**, p. 2843, `http://dx.doi.org/10.1063/1.870523`.

Schmitz, L., Ruskov, E., Deng, B. H., Binderbauer, M., Tajima, T., Gota, H., Tuszewski, M. and the TAE Team Control of ion gyroscale fluctuations via electrostatic biasing and sheared $\boldsymbol{E} \times \boldsymbol{B}$ flow in the C-2 field reversed configuration, *AIP Conf. Proc.* **1721**, p. 1721, `http://dx.doi.org/10.1063/1.4944018`.

Schoch, P. M., Forster, J. C., Jennings, W. C. and Hickok, R. L. (1986). TEXT heavy ion beam probe system, *Rev. Sci. Instr.* **57**, 8, pp. 1825-1827, `http://dx.doi.org/10.1063/1.1139141`.

Scott, S. D., *et al.* (1989). *Plasma Phys. Control. Nucl. Fusion* (International Atomic Energy Agency, Vienna) **1**, p. 655.

Seehafer, N. and Schumacher, J. (1998). Resistivity profile and instability of the plane sheet pinch, *Phys. Plasmas* **5**, p. 2363, `http://dx.doi.org/10.1063/1.872910`.

Sen, A. K., Chen, J. and Mauel, M. (1991). Production and identification of the ion-temperature-gradient instability, *Phys. Rev. Lett.* **66**, pp. 429-432, `http://dx.doi.org/10.1103/PhysRevLett.66.429`.

Sen, A. K., Reva, V. and Avinash, K. (2001). A hybrid ion temperature gradient and Kelvin-Helmholtz instability, *Phys. Plasmas* **8**, p. 4772, `http://dx.doi.org/10.1063/1.1409345`.

Seo, S-H. and Lee, T. G. (2006). Integration of PXI and VXI digitizers into the Hanbit data acquisition system based on MDSplus, *Fusion Eng. Design* **81**, pp. 1785-1788, `http://dx.doi.org/10.1016/j.fusengdes.2006.04.036`.

Shaing, K. C. (1988). Neoclassical quasilinear transport theory of fluctuations in toroidal plasmas, *Phys. Fluids* **31**, p. 2249, `http://dx.doi.org/10.1063/1.866626`.

Shapiro, S. L. and Teukolsky, S. A. (1983). (Cambridge Press) ISBN:0-471-87316-0.

Shimada, M., Campbell, Mukhovatov, V., Fujiwara, M., Kirneva, N., Lackner, K., Nagami, M., Pustovitov, V. D., Uckan, N., Wesley, J., Asakura, N., Costley, A. E., Donné, A. J. H., Doyle, E. J., Fasoli, A., Gormezano, C., Gribov, Y., Gruber, O., Hender, T. C., Houlberg, W., Ide, S., Kamada, Y., Leonard, A., Lipschultz, B., Loarte, A., Miyamoto, K., Mukhovatov, V., Osborne, T. H., Polevoi, A. and Sips, A. C. C. (2007). Chapter 1: Overview and summary, *Nucl. Fusion* **47**, p. S1, `http://dx.doi.org/10.1088/0029-5515/47/6/S01`.

Shimomura, Y. (1985). JAERI Report JAERI-M 87-080.

Shoji, M., Kasahara, H., Tokitani, M., Seki, T., Saito, K., Kamio, S., Seki, R., Tanaka, Y., Pigarov, A., Smirnov, R., Kawamura, G., Tanaka, H., Masuzaki, S., Uesugi, Y., Mutoh, T. and The LHD Experiment Group. (2015). Studies of dust transport in long pulse plasma discharges in the large helical device, *Nucl. Fusion* **5** (5), p. 053014, `http://dx.doi.org/10.1088/0029-5515/55/5/053014`.

Shukla, P. K., Yu, M. Y. and Varma, R. K. (1985a). Formation of kinetic Alfvén vortices, *Phys. Lett. A* **109**, pp. 322-324, `http://dx.doi.org/10.1016/0375-9601(85)90183-5`.

Shukla, P. K., Yu, M. Y. and Varma, R. K. (1985b). Drift-Alfvén vortices, *Phys. Fluids* **28**, p. 1719, http://dx.doi.org/10.1063/1.864964.

Shukla, P. K. and Mamun, A. A. (2002). Introduction to dusty plasma physics, *Plasma Phys. Control. Fusion* **44**, p. 395, http://dx.doi.org/10.1088/0741-3335/44/3/701.

Siccinio, M., Poli, E., Casson, F. J., Hornsby, W. A. and Peeters, A. G. (2011). Gyrokinetic determination of the electrostatic potential of rotating magnetic islands in tokamaks, *Phys. Plasmas* **18**, p. 122506, http://dx.doi.org/10.1063/1.3671964.

Similon, P., Sedlak, J. E., Stotler, D., Berk, H. L., Horton, W. and Choi, D-I. (1984). Guiding-center dispersion function, *J. Comp. Phys.* **54**, pp. 260-277, http://dx.doi.org/10.1016/0021-9991(84)90118-9.

Sips, A. C. C., for the Steady-State Operation and the Transport Physics topical groups of the International Tokamak Physics Activity. (2005). Advanced scenarios for ITER operation, *Plasma Phys. Control. Fusion* **47**, p. A19, http://dx.doi.org/10.1088/0741-3335/47/5A/003.

Smolyakov, A. I., Hirose, A., Lazzaro, E., Re, G. B. and Callen, J. D. (1995). Rotating nonlinear magnetic islands in a tokamak plasma, *Phys. Plasmas* **2**, pp. 1581-1598, http://dx.doi.org/10.1063/1.871308.

Smolyakov, A. (1998). Gyroviscous forces in a collisionless plasma with temperature gradients, *Can. J. Phys.* **76**, pp. 321-331, http://dx.doi.org/10.1139/p98-012.

Snipes, J. A., Granetz, R. S., Greenwald, M., Kardaun, O. J. W. F., Kus, A., Ryter, F., Stroth, U., Kollermeyer, J., Fielding, S. J. and Vatovic, M. (1996). H-mode power threshold database for ITER, *Nucl. Fusion* **36**, p. 1217, http://dx.doi.org/10.1088/0029-5515/36/9/I11.

Snyder, P. B., Osborne, T. H., Burrell, K. H., Groebner, R. J., Leonard, A. W., Nazikian, R., Orlov, D. M., Schmitz, O., Wade, M. R. and Wilson, H. R. (2012). The EPED pedestal model and edge-localized mode-suppressed regimes: Studies of quiescent H-mode and development of a model for edge-localized mode suppression via resonant magnetic perturbations, *Phys. Plasmas* **19**, p. 056115, http://dx.doi.org/10.1063/1.3699623.

Söldner, F. and the JET Team. (1997). Shear optimization experiments with current profile control on JET, *Plasma Phys. Control. Fusion* **39**, pp. B353-B369, http://dx.doi.org/10.1088/0741-3335/39/12B/027.

Martín-Solís, J. R., Esposito, B., Sanchez, R., Granucci, G. and Panaccione, L. (2005). Comparison of runaway dynamics in LH and ECRH heated discharges in the Frascati Tokamak Upgrade, *Nucl. Fusion* **45**, p. 1524, http://dx.doi.org/10.1088/0029-5515/45/12/007.

Solomahin, A., Bagryansky, P., *et al.* (2014). XLI Zvenigorod Int. Conf. on Plasma Phys. Contr. Fus., MC1-16.

Solomon, W. M., Burrell, K. H., deGrassie, J. S., Boedo, J. A., Garofalo, A. M., Moyer, R. A., Muller, S. H., Petty, C. C. and Reimerdes, H. (2011). Characterization of intrinsic rotation drive on DIII-D, *Nucl. Fusion* **51**, p. 073010, http://dx.doi.org/10.1088/0029-5515/51/7/073010.

Sonborm, B. (2015). *Fusion Eng. Des.* **100**, p. 378.

Squire, H. B. (1933). On the stability for three-dimensional disturbances of viscous fluid flow between parallel walls, *Proc. R. Soc. A* **142**, pp. 621-628, http://www.jstor.org/stable/96239.

Stanisic, M. M. (1985). *The Mathematical Theory of Turbulence*, (Universitext, Springer-Verlag, Berlin, Heidelberg, New York, Tokyo) (paperback), 429 pp, ISBN:0-387-96107-0.

Steinhauer, Loren C. (2011). Review of field-reversed configurations, *Phys. Plasmas* **18**, p. 070501, `http://dx.doi.org/10.1063/1.3613680`.

Stenzel, R. L. (1978). Experiments on current-driven three-dimensional ion sound turbulence. Part I: Return-current limited electron beam injection, *Phys. Fluids* **21**, p. 99, `http://dx.doi.org/10.1063/1.862085`.

Stenzel, O., Wilbrandt, S., Yulin, S., Kaiser, N., Held, M., Tünnermann, A., Biskupek, J. and Kaiser, U. (2010). Plasma ion assisted deposition of hafnium dioxide using argon and xenon as process gases, *Opt. Mater. Exp.* **1** (2), pp. 278-292, `http://dx.doi.org/10.1364/OME.1.000278`.

Stern, M. E. (1963). Joint instability of hydromagnetic fields which are separately stable, *Phys. Fluids* **6**, p. 636, `http://dx.doi.org/10.1063/1.1706793`.

Stevens, J., Von Goeler, S., Bernabei, S., Bitter, M., Chu, T. K., Efthimion, P., Fisch, N., Hooke, W., Hosea, J., Jobes, F., Karney, C., Meservey, E., Motley, R.,and Taylor, G. (1985). Modeling of the electron distribution based on Bremsstrahlung emission during lower-hybrid current drive on PLT, *Nucl. Fusion* **25**, p. 1529, `http://dx.doi.org/10.1088/0029-5515/25/11/002`.

Stix, T. H. (1973). Magnetic braiding in a toroidal plasma, *Phys. Rev. Lett.* **30**, pp. 833-835, `http://dx.doi.org/10.1103/PhysRevLett.30.833`.

Strait, E. J., Lao, L. L., Mauel, M. E., Rice, B. W., Taylor, T. S., Burrell, K. H., Chu, M. S., Lazarus, E. A., Osborne, T. H., Thompson, S. J. and Turnbull A. D. (1995). Enhanced confinement and stability in DIII-D discharges with reversed magnetic shear, *Phys. Rev. Lett.* **75**, pp. 4421-4424, `http://dx.doi.org/10.1103/PhysRevLett.75.4421`.

Stringer, T. E. (1969). Diffusion in toroidal plasmas with radial electric field, *Phys. Rev. Lett.* **22**, p. 770, `:http://dx.doi.org/10.1103/PhysRevLett.22.770`.

Stuart, J. T. (1954). On the stability of viscous flow between parallel planes in the presence of a co-planar magnetic field, *Proc. R. Soc. London Ser. A* **221**, p. 189, `http:/dx.doi.org/10.1098/rspa.1954.0015`.

Stutman, D., Delgado-Aparicio, L., Gorelenkov, N., Finkentha, M., Mazzucato, E. and Tritz, K. (2009). Correlation between electron transport and shear Alfvén activity in the National Spherical Torus Experiment, *Phys. Rev. Lett.* **102**, p. 115002, `http://dx.doi.org/10.1103/PhysRevLett.102.115002`.

Su, X. N., Horton, W. and Morrison, P. J. (1991). Drift wave vortices in inhomogeneous plasmas, *Phys. Fluids B* **3**, pp. 921-930, `http://dx.doi.org/10.1063/1.859848`.

Su, X. N., Horton, W. and Morrison, P. J. (1992). Drift wave vortices in nonuniform plasmas with sheared magnetic fields, *Phys. Fluids B* **4**, pp. 1238-1246, `http://dx.doi.org/10.1063/1.860131`.

Sudo, S., Tamura, N., Suzuki, C., Muto, S., Funaba, H. and LHD Group (2012). Multiple-tracer TESPEL injection for studying impurity behavior in a magnetically confined plasma, *Nucl. Fusion* **52**, p. 063012, `http://dx.doi.org/10.1088/0029-5515/52/6/063012`.

Sudo, S., Tamura, N., Funaba, H., Muto, S., Suzuki, C. and Murakami, I. (2013). Impurity transport study with TESPEL injection and simulation, *Plasma Fusion Res.* **9**, p. 2402059, `http://doi.org/10.1585/pfr.8.2402059`.

Sugama, H. and Horton, W. (1994). Transport suppression by shear flow generation in multiplicity resistive-g turbulence, *Phys. Plasmas* **1**, p. 2220, http://dx.doi.org/10.1063/1.870622.

Sugama, H. and Horton, W. (1995). L-H confinement mode dynamics in three-dimensional state space, *Plasma Phys. Control. Fusion* **37**, p. 345, http://dx.doi.org/10.1088/0741-3335/37/3/012.

Sugama, H., Okamoto, M., Horton, W. and Wakatani, M. (1996). Transport processes and entropy production in toroidal plasmas with gyrokinetic electromagnetic turbulence, *Phys. Plasmas* **3**, 6, pp. 2379-2394, http://dx.doi.org/10.1063/1.871922.

Sugama, H. and Horton, W. (1997). Neoclassical electron and ion transport in toroidally rotating plasmas, *Phys. Plasmas* **4**, p. 2215, http://dx.doi.org/10.1063/1.872385.

Sugama, H. and Horton, W. (1998). Nonlinear electromagnetic gyrokinetic equation for plasmas with large mean flows, *Phys. Plasmas* **5**, p. 2560, http://dx.doi.org/10.1063/1.872941.

Sugama, H., Watanabe, T.-H. and Horton, W. (2001). Collisionless kinetic-fluid closure and its application to the three-mode ion temperature gradient driven system, *Phys. Fluids* **8**, pp. 2617-2628, http://dx.doi.org/10.1063/1.1367319.

Sugama, H., Watanabe, T. H. and Horton, W. (2003). Comparison between kinetic and fluid simulations of slab ion temperature gradient driven turbulence, *Phys Plasmas* **10**, p. 726, http://dx.doi.org/10.1063/1.1544664.

Sugama, H., Watanabe, T. H. and Horton, W. (2007). Collisionless kinetic-fluid model of zonal flows in toroidal plasmas, *Phys. Plasmas* **14**, p. 022502, http://dx.doi.org/10.1063/1.2435329.

Summers, D., Ma, C., Meredith, N. P., Thorne, R. B., Thorne, R. M., Heynderickx, D. and Anderson, R. R. (2002). Model of the energization of outer-zone electrons by whistler-mode chorus during the October 9, 1990 geomagnetic storm, *Geophys. Res. Lett.* **29**, p. 25, http://dx.doi.org/10.1029/2002GL016039.

Sun, Y., Liang, Y., Liu, Y. Q., Gu, S., Yang, X., Guo, W., Shi, T., Jia, M., Wang, L., Lyu, B., Zhou, C., Liu, A., Zang, Q., Liu, H., Chu, N., Wang, H. H., Zhang, T., Qian, J., Xu, L., He, K., Chen, D., Shen, B., Gong, X., Ji, X., Wang, S., Qi, M., Song, Y., Yuan, Q., Sheng, Z., Gao, G., Fu, P. and Wan, B. (2016). Nonlinear transition from mitigation to suppression of the edge-localized mode with resonant magnetic perturbations in the EAST Tokamak, *Phys. Rev. Lett.* **117**, p. 115001, http://dx.doi.org/10.1103/PhysRevLett.117.115001.

Sydora, R. D., Decyk, V. and Dawson, J. M. (1996). Fluctuation-induced heat transport results from a large global 3D toroidal particle simulation model, *Plasma Phys. Control. Fusion* **38**, p. A281, http://dx.doi.org/10.1088/0741-3335/38/12A/021.

Sykes, A., Akers, R. J., Appel, L. C., Arends, E. R., Carolan, P. G., Conway, N. J., Counsell, G. F., Cunningham, G., Dnestrovskij, A., Dnestrovskij, Yu.N., Field, A. R., Fielding, S. J., Gryaznevich, M. P., Korsholm, S., Laird, E., Martin, R., Nightingale, M. P. S., Roach, C. M., Tournianski, M. R., Walsh, M. J., Warrick, C. D., Wilson, H. R., You, S., MAST Team and NBI Team. (2011). First results from MAST, *Nucl. Fusion* **41**, p. 1423, http://dx.doi.org/10.1088/0029-5515/41/10/310.

Synakowski, E. J., Bengtson, R. D., Ouroua, A., Wootton, A. J. and Kim, J. Y. (1989). Spectroscopic measurements of light impurity peaking after pellet injection on TEXT, *Nucl. Fusion* **29**, p. 311, http://dx.doi.org/10.1088/0029-5515/29/2/013.

Tajima, T., Horton, W., Morrison, P. J., Schutkeker, J., Kamimura, T., Mima, K. and Abe, Y. (1991). Instabilities and vortex dynamics in shear flow of magnetized plasmas, *Phys. Fluids B* **3**, pp. 938-954, `http://dx.doi.org/10.1063/1.859850`.

Tanaka, K., Michael, C. A., Vyacheslavov, L. N., Sanin, A. L., Kawahata, K., Akiyama, T., Tokuzawa, T. and Okajima, S. (2008). Two-dimensional phase contrast imaging for local turbulence measurements in large helical device, *Rev. Sci. Instrum.* **79**, p. 10E702, `http://dx.doi.org/10.1063/1.2988821`.

Tanaka, K., Michael, C., Vyacheslavov, L., Funaba, H., Yokoyama, M., Ida, K., Yoshinuma, M., Nagaoka, K., Murakami, S., Wakasa, A., Ido, T., Shimizu, A., Nishiura, M., Takeiri, Y., Kaneko, O., Tsumori, K., Ikeda, K., Osakabe, M., Kawahata, K. and LHD Experiment Group. (2010). Turbulence Response in the High Ti Discharge of the LHD, *Plasma Fus. Res.* **5**, p. S2053, `http://dx.doi.org/10.1585/pfr.5.S2053`.

Tanaka, H., Ohno, N., Tsuji, Y., Kajita, S., Masuzaki, S., Kobayashi, M., Morisaki, T., Komori, A. and the LHD Experimental Group. (2012). Blob/hole generation in the Divertor leg of the large helical device, *Plasma Fusion Res.* **7**, p. 1402152, `http://dx.doi.org/10.1585/pfr.7.1402152`.

Tanaka, H., Uchida, M., Maekawa, T., Bae, Y.-S., Joung, M., Jeong, J. H. and KSTAR team. (2016). Non-inductive initiation of closed flux surfaces by ECH/ECCD on KSTAR using an oblique fundamental O-mode injection from the low-field side, *Nucl. Fusion* **56**, p. 046003, `http://dx.doi.org/10.1088/0029-5515/56/4/046003`.

Taroni, A., Erba, M., Tibone, F. and Springmann, E. (1994). Global and local energy confinement properties of simple transport coefficients of the Bohm type, *Plasma Phys. Control. Fusion* **36**, p. 1629, `http://dx.doi.org/10.1088/0741-3335/36/10/003`.

Tassi, E., Morrison, P. J., Waelbroeck, F. L. and Grasso, D. (2008). Hamiltonian formulation and analysis of a collisionless fluid reconnection model, *Plasma Phys. Control. Fusion* **50**, p. 085014, `http://dx.doi.org/10.1088/0741-3335/50/8/085014`.

Terry, P. W. and Horton, W. (1982). Stochasticity and the random phase approximation for three electron drift waves, *Phys. Fluids* **25**, pp. 491-501, `http://dx.doi.org/10.1063/1.863761`.

Terry, P. W. (2000). Suppression of turbulence and transport by sheared flow, *Rev. Mod. Phys.* **72**, pp. 109-165, `http://dx.doi.org/10.1103/RevModPhys.72.109`.

Terry, J. L., Basse, N. P., Cziegler, I., Greenwald, M., Grulke, O., LaBombard, B., Zweben, S. J., Edlund, E. M., Hughes, J. W., Lin, L., Lin, Y., Porkolab, M., Sampsell, M., Veto, B. and Wukitch, S. J. (2005). Transport phenomena in the edge of Alcator C-Mod plasmas, *Nucl. Fusion* **45**, p. 1321, `http://dx.doi.org/10.1088/0029-5515/45/11/013`.

Thorne, R. M. (2010). Radiation belt dynamics: The importance of wave-particle interactions, *Geophys. Res. Lett.* **37**(22), `http://dx.doi.org/10.1029/2010GL044990`.

Todo, Y., Berk, H. L. and Breizman, B. N. (2010). Nonlinear magnetohydrodynamic effects on Alfvén eigenmode evolution and zonal flow generation, *Nucl. Fusion* **50**, p. 084016, `http://dx.doi.org/10.1088/0029-5515/50/8/084016`.

Tokar, M. Z., Rapp, J., Bertschinger, G., Konen, L., Koslowski, H. R., Kramer-Flacken, A., Philipps, V., Samm, U. and Unterberg, B. (1997). Nature of high-Z impurity accumulation in tokamaks, *Nucl. Fusion* **37**, p. 1691, `http://dx.doi.org/10.1088/0029-5515/37/12/I03`.

Tonetti, G., Heym, A., Hofmann, F., *et al.* (1991). *Proceedings of the 16th Symposium on Fusion Technology* (North-Holland, London, U.K.) R. Hemsworth, ed., p. 587.

Troyon, F., Gruber, R., Saurenmann, H., Semenzato, S. and Succi, S. (1984). MHD-Limits to plasma confinement, *Plasma Phys. Control. Fusion* **26**, pp. 209-215, http://dx.doi.org/10.1088/0741-3335/26/1A/319.

Troyon, F. (1987). Theoretical studies of MHD stability, *Phil. Trans. R. Soc. Lond. A* **322**, pp. 163-171, http://dx.doi.org/10.1098/rsta.1987.0045.

Tsytovich, V. N. and Havnes, O. (1993). Charging processes, dispersion properties and anomalous transport in dusty plasma, *Comm. Plasma Phys. Control. Fusion* **15**, pp. 267-280, ISSN:0374-2806.

Turri, G., Sauter, O., Porte, L., Alberti, S., Asp, E., Goodman, T. P., Martin, Y. R., Udintsev, V. and Zucca, C. (2008). The role of MHD in the sustainment of electron internal transport barriers and H-mode in TCV, *J. Phys.* **123**, p. 1, http://dx.doi.org/10.1088/1742-6596/123/1/012038.

Tuszewski, M. (1988). Field reversed configurations, *Nucl. Fusion* **28**, p. 2033, http://dx.doi.org/10.1088/0029-5515/28/11/008.

Ueda, Y., Shimada, T. and Nishikawa, M. (2004). Impacts of carbon impurities in hydrogen plasmas on tungsten blistering, *Nucl. Fusion* **44** (1), p. 62, http://dx.doi.org/10.1088/0029-5515/44/1/007.

Urban, J., Decker, J., Peysson, Y., Preinhaelter, J., Shevchenko, V., Taylor, G., Vahala, L. and Vahala, G. (2011). A survey of electron Bernstein wave heating and current drive potential for spherical tokamaks, *Nucl. Fusion* **51**, p. 083050, http://dx.doi.org/10.1088/0029-5515/51/8/083050.

Uzawa, K., Li, J. and Kishimoto, Y. (2009). Global characteristics of zonal flows due to the effect of finite bandwidth in drift wave turbulence, *Phys. Plasmas* **16**, p. 042304, http://dx.doi.org/10.1063/1.3097490.

Valisa, M., Carraro, L., Predebon, I., Puiatti, M. E., Angioni, C., Coffey, I., Giroud, C., Taroni, L. Lauro, Alper, B., Baruzzo, M., da Silva, P. Belo, Buratti, P., Garzotti, L., Van Easter, D., Lerche, E., Mantica, P., Naulin, V., Tala, T., Tsalas, M. and JET-EFDA contributors. (2011). Metal impurity transport control in JET H-mode plasmas with central ion cyclotron radio frequency power injection, *Nucl. Fusion* **51**, p. 033002 http://dx.doi.org/10.1088/0029-5515/51/3/033002.

Van Eester, D. and Koch, R. (1998). A variational principle for studying fast-wave mode conversion, *Plasma Phys. Control. Fusion* **40**, p. 1949, http://dx.doi.org/10.1088/0741-3335/40/11/010.

Van Eester, D. and Lerche, E. (2013). A 1D model for describing ion cyclotron resonance heating at arbitrary cyclotron harmonics in tokamak plasmas, *Plasma Phys. Control. Fusion* **55**, p 055008, http://dx.doi.org/10.1088/0741-3335/55/5/055008.

Van Eester, D. Lerche, E., Kazakov, Y., Jacquet, P., Bobkov, V., Crombé, K., Czarnecka, A., Dumont, R., Eriksson, J., Giacomelli, L., Giroud, C., Goniche, M., Hellesen, C., Kiptily, V. Koskela, T., Nocente, M., Santala, M., Salewski, M., Schneider, M., Weisen, H. and JET contributors, Recent ion cyclotron resonance heating experiments in JET in preparation of a DT campaign, IAEA paper (Kyoto, Japan, 2016).

van Houtte, D., Martin, G., Bécoulet, A. Bucalossi, J., Giruzzi, G., Hoang, G. T., Loarer, Th. and Saoutic, B. (on behalf of the Tore Supra Team). (2004). Recent fully non-inductive operation results in Tore Supra with 6 min, 1 GJ plasma discharges, *Nucl. Fusion* **44**, p. L11, http://dx.doi.org/10.1088/0029-5515/44/5/L01.

Van Zeeland, M. A., Gorelenkov, N. N., Heidbrink, W. W., Kramer, G. J., Spong, D. A., Austin, M. E., Fisher, R. K., Garcia Munoz, M., Gorelenkova, M., Luhmann, N., Murakami, M., Nazikian, R., Pace, D. C., Park, J. M., Tobias, B. J. and White, R. B. (2012). Alfvén eigenmode stability and fast ion loss in DIII-D and ITER reversed magnetic shear plasmas, *Nucl. Fusion* **52**, p. 094023, `http://dx.doi.org/10.1088/0029-5515/52/9/094023`.

Velikhov, E. P. (1959). Stability of an ideally conducting liquid flowing between rotating cylinders in a magnetic field, *Sov. Phys. JETP* **36**, p. 848.

Vincena, S., Gekelman, W. and Maggs, J. (2004). Shear Alfvén wave perpendicular propagation from the kinetic to the inertial regime, *Phys. Rev. Lett.* **93**, p. 105003, `http://dx.doi.org/10.1103/PhysRevLett.93.105003`.

Volkov, T. F. (1966). Hydrodynamic description of a collisionless plasma, *Rev. Plasma Phys.* **4**, pp. 1-21.

Voronov, G. S. (1997). A practical fit formula for ionization rate coefficients of atoms and ions by electron impact:Z= 1-28, *Atomic Data Nucl. Data Tables* **65**, pp. 1-35, `http://dx.doi.org/10.1006/adnd.1997.0732`.

Waelbroeck, F. L. and Fitzpatrick, R. (1997). Rotation and locking of magnetic islands, *Phys. Rev. Lett.* **78**, p. 1703-1706, `http://dx.doi.org/10.1103/PhysRevLett.78.1703`.

Waelbroeck, F. L. (1989). Current sheets and nonlinear growth of the $m = 1$ kink-tearing mode, *Phys. Fluids B* **1**, pp. 2372-2380, `http://dx.doi.org/10.1063/1.859172`.

Waelbroeck, F. L., Militello, F., Fitzpatrick, R. and Horton, W. (2009). Effect of electrostatic turbulence on magnetic islands, *Plasma Phys. Control. Fusion*, **51**, p. 015015, `http://dx.doi.org/10.1088/0741-3335/51/1/015015`.

Wagner, F., Becker, G., Behringer, K., Campbell, D., Eberhagen, A., Engelhardt, W., Fussmann, G., Gehre, O., Gernhardt, J., Gierke, G. V., Haas, G., Huang, M., Karger, F., Keilhacker, M., Klüber, O., Kornherr, M., Lackner, K., Lisitano, G., Lister, G. G., Mayer, H. M., Meisel, D., Müller, E. R., Murmann, H., Niedermeyer, H., Poschenrieder, W., Rapp, H., Röhr, H., Schneider, F., Siller, G., Speth, E., Staebler, A., Steuer, K. H., Venus, G., Vollmer, O. and Yü, Z. (1982). Regime of improved confinement and high beta in neutral-beam-heated divertor discharges of the ASDEX Tokamak, *Phys. Rev. Lett.* **49**, pp. 1408-1412, `http://dx.doi.org/10.1103/PhysRevLett.49.1408`.

Wakatani, M. (1998). *Stellarator and Heliotron Devices* (Oxford University Press) ISBN:0-19-507831-4.

Waltz, R. E. (1983). Numerical study of drift wave turbulence with simple models for wave-wave nonlinear coupling, *Phys. Fluids* 26, p. 169, `http://dx.doi.org/10.1063/1.864004`.

Waltz, R. E., Kerbel, G. D. and Milovich, J. (1994). Toroidal gyro-Landau fluid model turbulence simulations in a nonlinear ballooning mode representation with radial modes, *Phys. Plasmas* **1**, pp. 2229-2244, `http://dx.doi.org/10.1063/1.870934`.

Wang, S., Lee, L. C., Wei, C. Q. and Akosofu, S.-I. (1988). *Sol. Phys.* **117**, p. 157, `http://dx.doi.org/10.1007/BF00148579`.

Wang, S. (2016). Anomalous pinch of turbulent plasmas driven by the magnetic-drift-induced Lorentz force through the Stokes-Einstein relation, *Phys. Plasmas* **23**, p. 072509, `http://dx.doi.org/10.1063/1.4959812`.

Watanabe, T.-H. and Sugama, H. (2006). *Nucl. Fusion* **46**, p. 24.

Wedemeyer-Bohm, S., Scullion, van der Voort, L. R., Bosnjak, A. and Antolin, P. (2013). Are giant tornadoes the legs of solar prominences? *Astrophys. J.* **774**, p. 123, http://dx.doi.org/10.1088/0004-637X/774/2/123.

Wei, X., Sokolov, V. and Sen, A. K. (2010). Experimental production and identification of electron temperature gradient mode, *Phys. Plasmas* **17**, p. 042108, http://dx.doi.org/10.1063/1.3381070.

Weigend, A. S., Huberman, B. A. and Rumelhart, D. E. (1990). Predicting the future: A connectionist approach,, *Int. J. Neural Systems* **1**, pp. 193-209, http://dx.doi.org/10.1142/S0129065790000102.

Weiland, J. and Nordman, H. (1988). *Theory of Fusion Plasmas: Proceedings of the Joint R.K. Wakerling* (Editrice Compositori for Societa Italiana di fisica, Bologna) former key: WeiNor88.

Weiland, J., Jarmen, A. B. and Nordman, H. (1989). Diffusive particle and heat pinch effects in toroidal plasmas, *Nucl. Fusion* **29**, http://dx.doi.org/10.1088/0029-5515/29/10/015.

Weiland, J. (1999). *Collective Modes in Inhomogeneous Plasma: Kinetic and Advanced Fluid Theory* (CRC Press), ISBN: 9780750305891, 200 pp.

Weiland, J. (2000). *Collective modes in inhomogeneous plasma* (IOP Publishing Ltd., New York).

Weiland, J., Asp, E., Garbet, X., Mantica, P., Parail, V., Thomas, P., Suttrop, W., Tala, T., and the EFDA-JET Contributors. (2005). *Plasma Phys. Control. Fusion* **47**, p. 441.

Weiland, J. and Holod, I. (2005). Drift wave transport scalings introduced by varying correlation length, *Phys. Plasmas* **12**, p. 012505, http://dx.doi.org/10.1063/1.1828083.

Weiland, J. and Nordman, H. (2006). *3rd EPS Conf. Plasma Phys.* **301**, p. 2.186 (Rome: ECA).

Weisen, H., Hofmann, F., Dutch, M. J., *et al.* (1996). Ohmic H-modes in the TCV tokamak, *Plasma Phys. Control. Fusion* **38**, p. 1137, http://dx.doi.org/10.1088/0741-3335/38/8/006. 38, p. 2163, http://dx.doi.org/10.1088/0741-3335/38/12/010.

Weisen, H., Alberti, S., Berry, S., Behn, R., Blanchard, P., Bosshard, P., Bühlmann, F., Chavan, R., Coda, S., Deschenaux, C., Dutch, M. J., Duval, B. P., Fasel, D., Favre, A., Franke, S., Furno, I., Goodman, T., Henderson, M., Hofmann, F., Hogge, J-Ph., Isoz, P.-F., Joye, B., Lister, J. B., Llobet, X., Magnin, J.-C., Mandrin, P., Marletaz, B., Marmillod, P., Martin, Y., Mayor, J.-M., Moret, J.-M., Nieswand, Ch., Paris, P., Perez, A., Pietrzyk, Z. A., Piffl, V., Pitts, R. A., Pochelon, A., Razumova, K., Reimerdes, H., Refke, A., Rommers, J., Roy, I., Sauter, O., Suttrop, W., van Toledo, W., Tonetti, G., Tran, M. Q., Royon, F., Vyas, P. and Ward, D. J. (1997). Effect of plasma shape on confinement and MHD behaviour in TCV, *Plasma Phys. Control. Fusion* **39**, p. B135, http://dx.doi.org/10.1088/0741-3335/39/12B/011.

Wesson, J. A. (2004). *Tokamaks (The International Series of Monographs on Physics)* (Oxford University Press) ISBN:0198509227.

Wesson, J. A. (1986). *Plasma Phys. Controlled Fusion* **28**, p. 243, http://dx.doi.org/10.1088/0741-3335/28/1A/022.

Wetzel, G.,*et al.* (2005). PXI Express: Extending backplanes to 6 Gbytes/s while maintaining backward compatibility, Autotestcon, IEEE, 26-29 Sept., pp. 231-234.

Weynants, R. R., Messiaen, A. M., Ongena, J., Unterberg, B., Bonheure, G., Dumortier, P., Jaspers, R., Koch, R., Koslowski, H. R. and Krämer-Flecken, A. (1999). Overview of radiative improved mode results on TEXTOR-94, *Nucl. Fusion* **39**, p. 1637, `http://dx.doi.org/10.1088/0029-5515/39/11Y/303`.

White, Roscoe B. (2014). The Theory of Toroidally Confined Plasmas: 3rd Edition Hardcover, ISBN-13:978-1783263639, ISBN-10:1783263636, 3rd Edition.

Whyte, D. G., Hubbard, A. E., Hughes, J. W., Lipschultz, B., Rice, J. E., Marmar, E. S., Greenwald, M., Cziegler, I., Dominguez, A., Golfinopoulos, T., Howard, N., Lin, L., McDermott, R. M., Porkolab, M., Reinke, M. L., Terry, J., Tsujii, N., Wolfe, S., Wukitch, S., Lin, Y. and the Alcator C-Mod Team. (2010). I-mode: an H-mode energy confinement regime with L-mode particle transport in Alcator C-Mod, **50**, p. 105005, `http://dx.doi.org/10.1088/0029-5515/50/10/105005`.

Wilson, H. R. and Cowley, S. C. (2004). Theory for explosive ideal magnetohydrodynamic instabilities in plasmas, *Phys. Rev. Lett.* **92**, pp. 175006-175006.4, `https://doi.org/10.1103/PhysRevLett.92.175006`.

Winter, J., Esser, H. G., Jackson, G. L.; Könen, L., Messiaen, A., Ongena, J., Philipps, V., Pospiesczcyk, A., Samm, U., Schweer, B., Unterberg, B. and Textor Team (1993). Improved plasma performance in TEXTOR with silicon coated surfaces, *Phys. Rev. Lett.* **71**, pp. 1549-1552, `http://dx.doi.org/10.1103/PhysRevLett.71.1549`.

Wolf, R. C. (2003). Internal transport barriers in tokamaks plasmas, *Plasma Phys. Control. Fusion* **45**, pp. R1-R91, `http://dx.doi.org/10.1088/0741-3335/45/1/201`.

Wong, S. K. (1987). Transport of impure plasma with arbitrary toroidal rotation, *Phys. Fluids* **30**, p. 818, `http://dx.doi.org/10.1063/1.866333`.

Wong, K. L., Fonck, R. J., Paul, S. F., Roberts, D. R., Fredrickson, E. D., Nazikian, R., Park, H. K., Bell, M., Bretz, N. L., Budny, R., Cohen, S., Hammett, G. W., Jobes, F. C., Meade, D. M., Medley, S. S., Mueller, D., Nagayama, Y., Owens, D. K. and Synakowski, E. J. (1991). Excitation of toroidal Alfvén eigenmodes in TFTR, *Phys. Rev. Lett.* **66**, pp. 1874-1877, `http://dx.doi.org/10.1103/PhysRevLett.66.1874`.

Wright, J. C., Bonoli, P. T., Brambilla, M., Meo, F., D'Azevedo, E., Batchelor, D. B., Jaeger, E. F., Berry, L. A., Phillips, C. K. and Pletzer, A. (2004). Full wave simulations of fast wave mode conversion and lower-hybrid wave propagation in tokamaks, *Phys. Plasmas* **11**, p. 2473, `http://dx.doi.org/10.1063/1.1652731`.

Wygant, J. R., Cattell, C. A., Lysak, R., Song, Y., Dombeck, J., McFadden, J., Mozer, F. S., Carlson, C. W., Parks, G., Lucek, E. A., Balogh, A., Andre, M., Reme, H., Hesse, M. and Mouikis, C. (2005). Cluster observations of an intense normal component of the electric field at a thin reconnecting current sheet in the tail and its role in the shock-like acceleration of the ion fluid into the separatrix region, *J. Geophys. Res.* **110**, p. A09206, `http://dx.doi.org/10.1029/2004JA010708`.

Xanthopoulos, P. and Jenko, F. (2007). Gyrokinetic analysis of linear microinstabilities for the stellarator Wendelstein 7-X, *Phys. Plasmas* **14**, p. 042501, `http://dx.doi.org/10.1063/1.2714328`.

Xu, G. S., Wan, B. N., Li, J. G., Gong, X. Z., Hu, J. S., Shan, J. F., Li, H., Mansfield, D. K., Humphreys, D. A. and Naulin, V. for EAST Team and International Collaborators (2011). Study on H-mode access at low density with lower-hybrid current drive and lithium-wall coatings on the EAST superconducting tokamak, *Nucl. Fusion* **51**, p. 072001, http://dx.doi.org/10.1088/0029-5515/51/7/072001.

Yagi, M., Wakatani, M., Sugama, H., Hong, B-G. and Horton, W. (1989). Ion-temperature-gradient-driven drift mode coupled to the resistive interchange mode in a heliotron/torsatron, *J. Phys. Soc. Jpn.* **12**, pp. 4265-4268, http://dx.doi.org/10.1143/JPSJ.58.4265.

Yamada, H., Harris, J. H., Dinkalge, A., Ascasibar, E., Sano, F., Okamura, S., Talmadge, J., Stroth, U., Kus, A., Murakami, S., Yokoyama, M., Beidler, C. D., Tribaldos, V., Watanabe, K. Y. and Suzuki, Y. (2005). Characterization of energy confinement in net-current free plasmas using the extended International Stellarator Database, *Nucl. Fusion* **45**, pp. 1684-1693, http://dx.doi.org/10.1088/0029-5515/45/12/024.

Yang, Y., Counsell, G. F. and The MAST Team. (2003). Observations with a mid-plane reciprocating probe in MAST, *J. Nucl. Mat.* **313-316**, pp. 734-737, http://dx.doi.org/10.1016/S0022-3115(02)01574-X.

Yoshino, R. (2005). Neural-net predictor for β limit disruptions in JT-60U, *Nucl. Fusion* **45**, p 1232, http://dx.doi.org/10.1088/0029-5515/45/11/003.

Yu, Q. (1996). Nonlinear growth of the double tearing mode, *Phys. Plasmas* **3**, p. 2898, http://dx.doi.org/10.1063/1.871649.

Yun, G. S., Lee, W., Choi, M. J., Lee, J., Kim, M., Leem, J., Nam, Y., Choe, Park, H. K., Park, H., Woo, D. S., Kim, K. W., Domier, C. W., Luhmann, Jr., N. C., Ito, N., Mase, A. and Lee, G. S. (2014). Quasi 3D ECE imaging system for study of MHD instabilities in KSTAR, *Rev. Sci. Instrum.* **85**, p. 11D820, http://dx.doi.org/10.1063/1.4890401.

Zakharov, L. and Rogers, B. (1992). Two-fluid magnetohydrodynamic description of the internal kink mode in tokamaks, *Phys. Fluids B* **4**, pp. 3285-3301, http://dx.doi.org/10.1063/1.860384.

Zeiler, A., Drake, J. F. and Rogers, B. (1997), Nonlinear reduced Braginskii equations with ion thermal dynamics in toroidal plasma, *Phys. Plasmas* **4**, p. 2134, http://dx.doi.org/10.1063/1.872368.

Zeta Discussion, *Proceedings of the IEE - Part A: Power Engineering* **106**, pp. 82-84, http://dx.doi.org/10.1049/pi-a.1959.0012.

Zhang, W., Decyk, V., Holod, I., Xiao, Y., Lin, Z. and Chen, Liu. (2010). Scalings of energetic particle transport by ion temperature gradient microturbulence, *Phys. Plasmas* **17**, p. 055902, http://dx.doi.org/10.1063/1.3379471.

Zhang, Y. Z. and Mahajan, S. M. (1988). Renormalized perturbation theory: Vlasov-Poisson system, weak turbulence limit, and gyrokinetics, *Phys. Fluids* **31**, p. 2894, http://dx.doi.org/10.1063/1.866998.

Zhang, X. J., Zhao, Y. P., Wan, B. N., Gong, X. Z., Mao, Y. Z., Yuan, S., Xue, D. Y., Wang, L., Qin, C. M., Ju, S. Q., Chen, Y., Qian, J. P., Hu, L., Li, J. G., Song, Y. T., Lin, Y., Wukitch, S., Noterdaeme, J. M., Kumazawa, R., Seki, T., Saito, J. and Kasahara, H. (2012). Initial results on plasma heating experiments in the ion cyclotron range of frequencies on EAST, *Nucl. Fusion* **52**, p. 032002, http://dx.doi.org/10.1088/0029-5515/52/3/032002.

Zhang, X. J., Zhao, Y. P., Wan, B. N., Gong, X. Z., Li, J. G., Lin, Y., Qin, C. M., Taylor, G., Xu, G. S., Sun, Y. W., Gao, B. X., Qian, J. P., Wang, F. D., Lu, B., Luo, C., Zhang, L., Hu, L. Q., Song, Y. T., Yu, C. X., Liu, W. D., Wukitch, S., Wilson, J.

R. and Hosea, J. C. (2013). First results from H-mode plasmas generated by ICRF heating in the EAST, *Nucl. Fusion* **53**, p. 023004, http://dx.doi.org/10.1088/0029-5515/53/2/023004.

Zhang, X. J., Wan, B. N., Zhao, Y. P., Ding, B. J., Xu, G. S., Gong, X. Z., Li, J. G., Lin, Y., Taylor, G., Noterdaeme, J. M., Braun, F., Wukitch, S., Magne, R., Litaudon, X., Kumazawa, R., Kasahara, H. and EAST Team. (2014). Lower-hybrid current drive and ion cyclotron range of frequencies heating experiments in H-mode plasmas in Experimental Advanced Superconducting Tokamak, *Phys. Plasmas* **21**, p. 061501, http://dx.doi.org/10.1063/1.4884356.

Zhao, L. M., Shan, J. F. and Liu, F. K., (2010). *Plasma Sci. Technol.* **12**, p. 118.

Zheng, L. J. and Furukawa, M. (2010). Current-interchange tearing modes: Conversion of interchange-type modes to tearing modes, *Phys. Plasmas* **17**, p. 052508, http://dx.doi.org/10.1063/1.3425857.

Zheng, L., Horton, W., Miura, H., Shi, T. H. and Wang, H. Q. (2016). Non-neutralized charge effects on tokamak edge magnetohydrodynamic stability, *Phys. Lett. A* **380**, pp. 2654-2657, http://dx.doi.org/10.1016/j.physleta.2016.06.027.

Zhu, P., Horton, W. and Sugama, H. (1999). The radial electric field in a tokamak with reversed magnetic shear, *Phys. Plasmas* **6**, p. 2503, http://dx.doi.org/10.1063/1.873522.

Zhu, P., Bateman, G., Kritz, A. H. and Horton, W. (2000). Predictive transport simulations of internal transport barriers using the Multi-Mode model, *AIP* **7**, pp. 2898-2908, http://ojps.aip.org/pop/popcr.jsp.

Zhu, Ping, Sovinec, C. R. and Hegna, C. C. (2015). The formation of blobs from a pure interchange process, *Phys. Plasmas* **22**, p. 022311, http://dx.doi.org/10.1063/1.4913476.

Zohm, H., Gantenbein, G., Gude, A., Gunter, S., Leuterer, F., Maraschek, M., Meskat, J., Suttrop, W. and Yu, Q. (2001). Neoclassical tearing modes and their stabilization by electron cyclotron current drive in ASDEX Upgrade, *Phys. Plasmas* **8**, pp. 2009-2016, http://dx.doi.org/10.1063/1.1344564.

Index